T0236455

Lecture Notes in Artificial Intelligence   8589

Subseries of Lecture Notes in Computer Science

Lecture Notes in Artificial Intelligence 8588

Subseries of Lecture Notes in Computer Science

LNAI Series Editors

Randy Goebel
University of Alberta, Edmonton, Canada
Yuzuru Tanaka
Hokkaido University, Sapporo, Japan
Wolfgang Wahlster
DFKI and Saarland University, Saarbrücken, Germany

LNAI Founding Series Editor

Joerg Siekmann
DFKI and Saarland University, Saarbrücken, Germany

De-Shuang Huang   Kang-Hyun Jo
Ling Wang (Eds.)

Intelligent Computing Methodologies

10th International Conference, ICIC 2014
Taiyuan, China, August 3-6, 2014
Proceedings

 Springer

Volume Editors

De-Shuang Huang
Tongji University
Machine Learning and Systems Biology Laboratory
School of Electronics and Information Engineering
4800 Caoan Road, Shanghai 201804, China
E-mail: dshuang@tongji.edu.cn

Kang-Hyun Jo
University of Ulsan
School of Electrical Engineering
680-749 No. 7-413, San 29, Muger Dong, Ulsan, South Korea
E-mail: jokanghyun@gmail.com

Ling Wang
Tsinghua University
Department of Automation
Bejing, China
E-mail: wangling@tsinghua.edu.cn

ISSN 0302-9743   e-ISSN 1611-3349
ISBN 978-3-319-09338-3   e-ISBN 978-3-319-09339-0
DOI 10.1007/978-3-319-09339-0
Springer Cham Heidelberg New York Dordrecht London

Library of Congress Control Number: 2014943591

LNCS Sublibrary: SL 7 – Artificial Intelligence

Typesetting: Camera-ready by author, data conversion by Scientific Publishing Services, Chennai, India

Printed on acid-free paper

Springer is part of Springer Science+Business Media (www.springer.com)

Preface

The International Conference on Intelligent Computing (ICIC) was started to provide an annual forum dedicated to the emerging and challenging topics in artificial intelligence, machine learning, pattern recognition, bioinformatics, and computational biology. It aims to bring together researchers and practitioners from both academia and industry to share ideas, problems, and solutions related to the multifaceted aspects of intelligent computing.

ICIC 2014, held in Taiyuan, China, during August 3–6, 2014, constituted the 10th International Conference on Intelligent Computing. It built upon the success of ICIC 2013, ICIC 2012, ICIC 2011, ICIC 2010, ICIC 2009, ICIC 2008, ICIC 2007, ICIC 2006, and ICIC 2005 that were held in Nanning, Huangshan, Zhengzhou, Changsha, China, Ulsan, Korea, Shanghai, Qingdao, Kunming, and Hefei, China, respectively.

This year, the conference concentrated mainly on the theories and methodologies as well as the emerging applications of intelligent computing. Its aim was to unify the picture of contemporary intelligent computing techniques as an integral concept that highlights the trends in advanced computational intelligence and bridges theoretical research with applications. Therefore, the theme for this conference was "**Advanced Intelligent Computing Technology and Applications**". Papers focused on this theme were solicited, addressing theories, methodologies, and applications in science and technology.

ICIC 2014 received 667 submissions from 21 countries and regions. All papers went through a rigorous peer-review procedure and each paper received at least three review reports. Based on the review reports, the Program Committee finally selected 235 high-quality papers for presentation at ICIC 2013, included in three volumes of proceedings published by Springer: one volume of *Lecture Notes in Computer Science* (LNCS), one volume of *Lecture Notes in Artificial Intelligence* (LNAI), and one volume of *Lecture Notes in Bioinformatics* (LNBI).

This volume of *Lecture Notes in Artificial Intelligence* (LNAI) includes 85 papers.

The organizers of ICIC 2014, including Tongji University and North University of China, Taiyuan Normal University, Taiyuan University of Science and Technology, made an enormous effort to ensure the success of the conference. We hereby would like to thank the members of the Program Committee and the referees for their collective effort in reviewing and soliciting the papers. We would like to thank Alfred Hofmann, executive editor from Springer, for his frank and helpful advice and guidance throughout and for his continuous support in publishing the proceedings. In particular, we would like to thank all the authors for contributing their papers. Without the high-quality submissions from the authors, the success of the conference would not have been possible. Finally, we are especially grateful to the IEEE Computational Intelligence Society, the International Neural Network Society, and the National Science Foundation of China for their sponsorship.

May 2014

De-Shuang Huang
Kang-Hyun Jo
Ling Wang

ICIC 2014 Organization

General Co-chairs

De-Shuang Huang, China
Vincenzo Piuri, Italy
Yan Han, China

Jiye Liang, China
Jianchao Zeng, China

Program Committee Co-chairs

Kang Li, UK
Juan Carlos Figueroa, Colombia

Organizing Committee Co-chairs

Kang-Hyun Jo, Korea
Valeriya Gribova, Russia

Bing Wang, China
Xing-Ming Zhao, China

Award Committee Chair

Vitoantonio Bevilacqua, Italy

Publication Chair

Phalguni Gupta, India

Workshop/Special Session Co-chairs

Jiang Qian, USA
Zhongming Zhao, USA

Special Issue Chair

M. Michael Gromiha, India

Tutorial Chair

Laurent Heutte, France

International Liaison

Prashan Premaratne, Australia

Publicity Co-chairs

Kyungsook Han, Korea
Ling Wang, China

Abir Hussain, UK
Zhi-Gang Zeng, China

Exhibition Chair

Chun-Hou Zheng, China

Program Committee Members

Khalid Aamir, Pakistan
Andrea F. Abate, USA
Sabri Arik, Korea
Vasily Aristarkhov, Australia
Costin Badica, Japan
Waqas Bangyal, Pakistan
Vitoantonio Bevilacqua, Italy
Shuhui Bi, China
Jair Cervantes, Mexico
Yuehui Chen, China
Qingfeng Chen, China
Wen-Sheng Chen, China
Xiyuan Chen, China
Guanling Chen, USA
Yoonsuck Choe, USA
Ho-Jin Choi, Korea, Republic of
Michal Choras, Colombia
Angelo Ciaramella, China
Youping Deng, Japan
Primiano Di Nauta, Italy
Salvatore Distefano, USA
Ji-Xiang Du, China
Jianbo Fan, China

Minrui Fei, China
Juan Carlos Figueroa-García, Colombia
shan Gao, China
Liang Gao, China
Dun-wei Gong, India
Valeriya Gribova, China
Michael Gromiha, China
Xingsheng Gu, China
Kayhan Gulez, USA
Ping Guo, China
Phalguni Gupta, India
Kyungsook Han, Korea
Fei Han, China
Laurent Heutte, France
Wei-Chiang Hong, Taiwan
Yuexian Hou, China
Jinglu Hu, China
Tingwen Huang, Qatar
Peter Hung, Taiwan
Abir Hussain, UK
Saiful Islam, India
Li Jia, China

Reviewers

Jakub Šmíd
Pankaj Acharya
Erum Afzal
Parul Agarwal
Tanvir Ahmad
Musheer Ahmad
Syed Ahmed
Sabooh Ajaz
Haya Alaskar
Felix Albu
Dhiya Al-Jumeily
Israel Alvarez Villalobos
Muhammad Amjad
Ning An
Mary Thangakani Anthony
Masood Ahmad Arbab
Soniya be
Sunghan Bae
Lukas Bajer
Waqas Bangyal
Gang Bao
Donato Barone
Silvio Barra
Alex Becheru
Ye Bei
Mauri Benedito Bordonau
Simon Bernard
Vitoantonio Bevilacqua
Ying Bi
Ayse Humeyra Bilge
Honghua Bin
Jun Bo
Nora Boumella
Fabio Bruno
Antonio Bucchiarone
Danilo Caceres
Yiqiao Cai
Qiao Cai
Guorong Cai
Francesco Camastra
Mario Cannataro
Kecai Cao
Yi Cao

Giuseppe Carbone
Raffaele Carli
Jair Cervantes
Aravindan Chandrabose
Yuchou Chang
Deisy Chelliah
Gang Chen
Songcan Chen
Jianhung Chen
David Chen
Hongkai Chen
Xin Chen
Fanshu Chen
Fuqiang Chen
Bo Chen
Xin Chen
Liang Chen
Wei Chen
Jinan Chen
Yu Chen
Junxia Cheng
Zhang Cheng
Feixiong Cheng
Cong Cheng
Han Cheng
Chi-Tai Cheng
Chengwang Xie
Seongpyo Cheon
Ferdinando Chiacchio
Cheng-Hsiung Chiang
Wei Hong Chin
Simran Choudhary
Angelo Ciaramella
Azis Ciayadi
Rudy Ciayadi
Danilo Comminiello
Carlos Cubaque
Yan Cui
Bob Cui
Cuco Curistiana
Yakang Dai
Dario d'Ambruoso
Yang Dan

Farhan Dawood
Francesca De Crescenzio
Kaushik Deb
Saverio Debernardis
Dario Jose Delgado-Quintero
Sara Dellantonio
Jing Deng
Weilin Deng
M.C. Deng
Suping Deng
Zhaohong Deng
Somnath Dey
Yunqiang Di
Hector Diez Rodriguez
Rong Ding
Liya Ding
Sheng Ding
Sheng Ding
Shihong Ding
Dayong Ding
Xiang Ding
Salvatore Distefano
Chelsea Dobbins
Xueshi Dong
Vladislavs Dovgalecs
Vlad Dovgalecs
Gaixin Du
Dajun Du
Kaifang Du
Haibin Duan
Durak-Ata Lutfiye
Malay Dutta
Tolga Ensari
Nicola Epicoco
Marco Falagario
Shaojing Fan
Ming Fan
Fenghua Fan
Shaojing Fan
Yaping Fang
Chen Fei
Liangbing Feng
Shiguang Feng
Guojin Feng
Alessio Ferone

Francesco Ferrise
Juan Carlos Figueroa
Michele Fiorentino
Carlos Franco
Gibran Fuentes Pineda
Hironobu Fujiyoshi
Kazuhiro Fukui
Wai-Keung Fung
Chun Che Fung
Chiara Galdi
Jian Gao
Yushu Gao
Liang Gao
Yang Gao
Garcia-Lamont Farid
Garcia-Marti Irene
Michele Gattullo
Jing Ge
Na Geng
Shaho Ghanei
Rozaida Ghazali
Rosalba Giugno
Fengshou Gu
Tower Gu
Jing Gu
Smile Gu
Guangyue Du
Weili Guo
Yumeng Guo
Fei Guo
Tiantai Guo
Yinan Guo
Yanhui Guo
Chenglin Guo
Lilin Guo
Sandesh Gupta
Puneet Gupta
Puneet Gupta
Shi-Yuan Han
Fei Han
Meng Han
Yu-Yan Han
Zhimin Han
Xin Hao
Manabu Hashimoto

Tao He

Selena He

Xing He

Feng He

Van-Dung Hoang

Tian Hongjun

Lei Hou

Jingyu Hou

Ke Hu

Changjun Hu

Zhaoyang Hu

Jin Huang

Qiang Huang

Lei Huang

Shin-Ying Huang

Ke Huang

Huali Huang

Jida Huang

Xixia Huang

Fuxin Huang

Darma Putra i Ketut Gede

Haci Ilhan

Sorin Ilie

Saiful Islam

Saeed Jafarzadeh

Alex James

Chuleerat Jaruskulchai

James Jayaputera

Umarani Jayaraman

Mun-Ho Jeong

Zhiwei Ji

Shouling Ji

Yafei Jia

Hongjun Jia

Xiao Jian

Min Jiang

Changan Jiang

Tongyang Jiang

Yizhang Jiang

He Jiang

Yunsheng Jiang

Shujuan Jiang

Ying Jiang

Yizhang Jiang

Changan Jiang

Xu Jie

Jiening Xia

Taeseok Jin

Jingsong Shi

Mingyuan Jiu

Kanghyun Jo

Jayasudha John Suseela

Ren Jun

Fang Jun

Li Jun

Zhang Junming

Yugandhar K.

Tomáš Křen

Yang Kai

Hee-Jun Kang

Qi Kang

Dong-Joong Kang

Shugang Kang

Bilal Karaaslan

Rohit Katiyar

Ondrej Kazik

Mohd Ayyub Khan

Muhammed Khan

Sang-Wook Kim

Hong-Hyun Kim

One-Cue Kim

Duangmalai Klongdee

Kunikazu Kobayashi

Yoshinori Kobayashi

Takashi Komuro

Toshiaki Kondo

Deguang Kong

Kitti Koonsanit

Rafal Kozik

Kuang Li

Junbiao Kuang

Baeguen Kwon

Hebert Lacey

Chunlu Lai

Chien-Yuan Lai

David Lamb

Wei Lan

Chaowang Lan

Qixun Lan

Yee Wei Law

Tien Dung Le
My-Ha Le
Yongduck Lee
Jooyoung Lee
Seokju Lee
Shao-Lun Lee
Xinyu Lei
Gang Li
Yan Li
Liangliang Li
Xiaoguang Li
Zheng Li
Huan Li
Deng Li
Ping Li
Qingfeng Li
Fuhai Li
Hui Li
Kai Li
Longzhen Li
Xingfeng Li
Jingfei Li
Jianxing Li
Keling Li
Juan Li
Jianqing Li
Yunqi Li
Bing Nan Li
Lvzhou Li
Qin Li
Xiaoguang Li
Xinwu Liang
Jing Liang
Li-Hua Zhang
Jongil Lim
Changlong Lin
Yong Lin
Jian Lin
Genie Lin
Ying Liu
Chenbin Liu
James Liu
Liangxu Liu
Yuhang Liu
Liang Liu

Rong Liu
Liang Liu
Yufeng Liu
Qing Liu
Zhe Liu
Zexian Liu
Li Liu
Shiyong Liu
Sen Liu
Qi Liu
Jin-Xing Liu
Xiaoming Liu
Ying Liu
Xiaoming Liu
Bo Liu
Yunxia Liu
Alfredo Liverani
Anthony Lo
Lopez-Chau Asdrúbal
Siow Yong Low
Zhen Lu
Xingjia Lu
Junfeng Luo
Juan Luo
Ricai Luo
Youxi Luo
Yanqing Ma
Wencai Ma
Lan Ma
Chuang Ma
Xiaoxiao Ma
Sakashi Maeda
Guoqin Mai
Mario Manzo
Antonio Maratea
Erik Marchi
Carlos Román Mariaca Gaspar
Naoki Masuyama
Gu Meilin
Geethan Mendiz
Qingfang Meng
Filippo Menolascina
Muharrem Mercimek
Giovanni Merlino
Hyeon-Gyu Min

Martin Renqiang Min

Minglei Tong

Saleh Mirheidari

Akio Miyazaki

Yuanbin Mo

Quanyi Mo

Andrei Mocanu

Raffaele Montella

Montoliu Raul

Tsuyoshi Morimoto

Mohamed Mousa Alzawi

Lijun Mu

Inamullah Muhammad

Izharuddin Muhammed

Tarik Veli Mumcu

Francesca Nardone

Fabio Narducci

Rodrigo Nava

Patricio Nebot

Ken Nguyen

Changhai Nie

Li Nie

Aditya Nigam

Evgeni Nurminski

Kok-Leong Ong

Kazunori Onoguchi

Zeynep Orman

Selin Ozcira

Cuiping Pan

Binbin Pan

Quan-Ke Pan

Dazhao Pan

Francesco Pappalardo

Jekang Park

Anoosha Paruchuri

Vibha Patel

Samir Patel

Lizhi Peng

Yiming Peng

Jialin Peng

Klara Peskova

Caroline Petitjean

Martin Pilat

Surya Prakash

Philip Pretorius

Ali Qamar

Xiangbo Qi

Shijun Qian

Pengjiang Qian

Bin Qian

Ying Qiu

Jian-Ding Qiu

Junfeng Qu

Junjun Qu

Muhammad Rahman

Sakthivel Ramasamy

Tao Ran

Martin Randles

Caleb Rascon

Muhammad Rashid

Haider Raza

David Reid

Fengli Ren

Stefano Ricciardi

Angelo Riccio

Alejo Roberto

Abdus Samad

Ruya Samli

Hongyan Sang

Michele Scarpiniti

Dongwook Seo

Shi Sha

Elena Shalfeeva

Li Shang

Linlin Shen

Yehu Shen

Haojie Shen

Jin Biao Shen

Ajitha Shenoy

Jiuh-Biing Sheu

Xiutao Shi

Jibin Shi

Fanhuai Shi

Yonghong Shi

Yinghuan Shi

Atsushi Shimada

Nobutaka Shimada

Ji Sun Shin

Ye Shuang

Raghuraj Singh

Dushyant Kumar Singh
Haozhen Situ
Martin Slapak
Sergey Smagin
Yongli Song
Yinglei Song
Meiyue Song
Bin Song
Rui Song
Guanghua Song
Gang Song
Sotanto Sotanto
Sreenivas Sremath Tirumala
Antonino Staiano
Jinya Su
Hung-Chi Su
Marco Suma
Xiaoyan Sun
Jiankun Sun
Sheng Sun
Yu Sun
Celi Sun
Yonghui Sun
Zengguo Sun
Jie Sun
Aboozar Taherkhani
Shinya Takahashi
Jinying Tan
Shijun Tang
Xiwei Tang
Buzhou Tang
Ming Tang
Jianliang Tang
Hissam Tawfik
Zhu Teng
Sin Teo
Girma Tewolde
Xiange Tian
Yun Tian
Tian Tian
Hao Tian
Gao Tianshun
Kamlesh Tiwari
Amod Tiwari
Kamlesh Tiwari

Andrysiak Tomasz
Torres-Sospedra Joaquín
Sergi Trilles
Yao-Hong Tsai
Naoyuki Tsuruta
Fahad Ullah
Pier Paolo Valentini
Andrey Vavilin
Giuseppe Vettigli
Petra Vidnerová
Villatoro-Tello Esaú
Ji Wan
Li Wan
Lin Wan
Quan Wang
Zixiang Wang
Yichen Wang
Xiangjun Wang
Yunji Wang
Hong Wang
Jinhe Wang
Xuan Wang
Xiaojuan Wang
Suyu Wang
Zhiyong Wang
Xiangyu Wang
Mingyi Wang
Yan Wang
Zongyue Wang
Huisen Wang
Yongcui Wang
Xiaoming Wang
Zi Wang
Jun Wang
Aihui Wang
Yi Wang
Ling Wang
Zhaoxi Wang
Shulin Wang
Yunfei Wang
Yongbo Wang
Zhengxiang Wang
Sheng-Yao Wang
Jingchuan Wang
Qixin Wang

Yong Wang

Fang-Fang Wang

Tian Wang

Zhenzhong Wang

Panwen Wang

Lei Wang

Qilong Wang

Dong Wang

Ping Wang

Huiwei Wang

Yiqi Wang

Zhixuan Wei

Zhihua Wei

Li Wei

Shengjun Wen

Shiping Wen

Di Wu

Yonghui Wu

Chao Wu

Xiaomei Wu

Weili Wu

Hongrun Wu

Weimin Wu

Guolong Wu

Siyu Xia

Qing Xia

Sen Xia

Jin Xiao

Min Xiao

Qin Xiao

Yongfei Xiao

Zhao Xiaoguang

Zhuge Xiaozhong

Minzhu Xie

Zhenping Xie

Jian Xie

Ting Xie

Chao Xing

Wei Xiong

Hao Xiong

Yi Xiong

Xiaoyin Xu

Dawen Xu

Jing Xu

Yuan Xu

Jin Xu

Xin Xu

Meng Xu

Li Xu

Feng Xu

Zhenyu Xuan

Hari Yalamanchili

Atsushi Yamashita

Mingyuan Yan

Yan Yan

Zhigang Yan

Bin Yan

Zhile Yang

Dan Yang

Yang Yang

Wankou Yang

Wenqiang Yang

Yuting Yang

Chia-Luen Yang

Chyuan-Huei Yang

Deng Yanni

Jin Yao

Xiangjuan Yao

Tao Ye

Xu Ye

Fengqi Yi

Wenchao Yi

Kai Yin

James j.q. Yu

Helen Yu

Jun Yu

Fang Yu

Xu Yuan

Lin Yuan

Jinghua Yuan

Lin Yuling

Faheem Zafari

Xue-Qiang Zeng

Haimao Zhan

Yong-Wei Zhang

Jing Zhang

Shuyi Zhang

Kevin Zhang

Ming Zhang

Zhenmin Zhang

Xuebing Zhang
Minlu Zhang
Jianping Zhang
Xiaoping Zhang
Yong Zhang
Qiang Zhang
Guohui Zhang
Chunhui Zhang
Yifeng Zhang
Wenxi Zhang
Xiujun Zhang
Long Zhang
Jian Zhang
Boyu Zhang
Hailei Zhang
Hongyun Zhang
Jianhua Zhang
Chunjiang Zhang
Peng Zhang
Jianhai Zhang
Hongbo Zhang
Lin Zhang
Xiaoqiang Zhang
Haiying Zhang
Jing Zhang
Wei Zhang
Qian Zhang
Hongli Zhang
Guohui Zhang
Liping Zhang
Hongbo Zhang
Sen Zhao
Yaou Zhao
Jane Zhao
Liang Zhao

Yunlong Zhao
Miaomiao Zhao
Xinhua Zhao
Xu Zhao
Guodong Zhao
Liang Zhao
Feng Zhao
Juan Zhao
Junfei Zhao
Changbo Zhao
Yue Zhao
Min Zheg
Dong Zhen
Guang Zheng
Xiaolong Zheng
Huanyu Zheng
Shenggen Zheng
Shan Zhong
Qi Zhou
Mian Zhou
Yinzhi Zhou
Jiayin Zhou
Songsheng Zhou
Bo Zhou
Qiang Zhou
Lei Zhu
Lin Zhu
Yongxu Zhu
Nanli Zhu
Xiaolei Zhu
Majid Ziaratban
Xiangfu Zou
Brock Zou
Tanish
Qiqi Duan

Table of Contents

Kernel Methods and Supporting Vector Machines

Machine Learning

Fuzzy Theory and Algorithms

Image Processing

Intelligent Computing in Computer Vision

Intelligent Computing in Communication Networks

Intelligent Image/Document Retrievals

Intelligent Data Analysis and Prediction

Intelligent Agent and Web Applications

Intelligent Fault Diagnosis

Knowledge Representation/Reasoning

Knowledge Discovery and Data Mining

Natural Language Processing and Computational Linguistics

Special Session on Complex Networks and Their Applications

Special Session on Time Series Forecasting and Analysis Using Artificial Neural Networks

Special Session on Computer Human Interaction Using Multiple Visual Cues and Intelligent Computing

Special Session on Biometric System and Security for Intelligent Computing

Affective Tutoring System for Android Mobiles

Ramón Zatarain-Cabada[1], M.L. Barrón-Estrada[1], José Luis Olivares Camacho[1],
and Carlos A. Reyes-García[2]

[1] Instituto Tecnológico de Culiacán, Juan de Dios Bátiz s/n, Col. Guadalupe,
Culiacán Sinaloa, 80220, México
[2] Instituto Nacional de Astrofísica, Óptica y Electrónica (INAOE)
Luis Enrique Erro No. 1, Sta. Ma. Tonanzintla, Puebla, 72840, México
{rzatarain,lbarron,jolivares}@itculiacan.edu.mx,
kargaxxi@inaoep.mx

Abstract. Detecting and responding to affective states may be more influential than intelligence for tutoring success. This paper presents a software system that recognizes emotions of users using Android Cell Phones. The system software consists of a feature extractor, a neural network, and an intelligent tutoring system. The tutoring system, the neural network, and the emotion recognizer were implemented for running in Android devices. We also incorporate a novel fuzzy system, which is part of the intelligent tutoring system that takes actions depending of pedagogical and emotional states. The recognition rate of the emotion classifier was 96 %.

Keywords: Intelligent Tutoring Systems, Affective Computing, Learning Technologies, Artificial Neural Networks, Education.

1 Introduction

The rapid advancement of technology in recent decades has allowed imitate science fiction since today there are computer systems that can recognize the emotional states of people and still react to them with related actions.

Affective computing is a term coined by Rosalind Picard in 1995 [1] to define a field of research that integrates different scientific disciplines, seeking to make computers have the ability to behave intelligently, interacting naturally with users through the ability to recognize, understand and express emotions. Knowing the emotional state of a person provides relevant information about their psychological state and gives a software system, the possibility to decide on how to respond to it.

Research in the area of affective computing aims to create systems to identify and respond to the emotions of a user (e.g. a client). Emotions are detected by special devices (pc camera, pc microphone, special mouse, neuro-headset, etc.) that can be placed in a computer or a person [2]. These devices are responsible for picking up signals (facial image, voice, mouse applied pressure, heart rate, stress level, etc.) of a user and of sending them to the computer to be processed and then get the resulting

D.-S. Huang et al. (Eds.): ICIC 2014, LNAI 8589, pp. 1–10, 2014.
© Springer International Publishing Switzerland 2014

emotional state in real time. In the area of education, an affective system seeks to change in an individual, a negative emotional state (e.g. confused) into a positive state (e.g. committed), in order to facilitate an appropriate emotional state for learning.

An Intelligent Tutoring System (ITS) is a computer program that incorporates AI techniques and pedagogy to provide tutors that know what they teach, who will teach and how to teach [3]. An ITS goal is to provide tutoring services that support learning [4]. Because the teaching-learning process is a complex area, ITS have support from other areas of knowledge such as education, psychology and computer science, which are combined to develop robust applications that are efficient tools for education. A traditional Intelligent Tutoring System (ITS) reacts from a significant set of pedagogical or cognitive strategies (i.e. a wrong answer or a help petition). The latest related works on emotion recognition in ITS incorporates different methods (hardware and software-based) to recognize student emotions [2, 5, 6, 7].

In this work, we present a system to incorporate emotion recognition to an ITS running into a Mobile Device. We have integrated two methods for selecting the learning style and emotional state of a student and to consider them in the ITS response. For recognizing the learning style and the affective state, we implemented two neural networks. During a training session, the first network (a SOM or Kohonen network) used for detecting the learning styles, receives a number of different input patterns (the student learning style obtained from an Inventory Learning Style Questionnaire (ILSQ), the learning style of three defined courses, and the student's grade in each course), discovers significant features in these patterns and learns how to classify the input patterns. The second network (a back-propagation neural network), which is used to detect affective or emotional states, is trained with a corpus of faces representing different emotional states. The affective and learning style recognizers are used into a fuzzy system, which is part of an ITS.

2   System General Structure and Architecture

Figure 1 shows the general structure of the system software. We have an application for Android devices, allowing the Intelligent Tutoring Systems accessed from mobile devices, taking pictures that allow the extraction of facial features and determine the student's emotion. The students will access the ITS through the web browser on a device with Android operating system, which can be a Smartphone or Tablet with Front Camera. The intelligent tutor requests the execution of the application for extracting facial features, when necessary. The ITS request is through the web browser, and is an invocation to the program installed on the mobile device, which takes a picture to obtain features of the eyes and mouth which are submitted to the ITS and then to a server with a neural network to determine the corresponding emotion, providing such information to the Tutor. Once the emotion is obtained, the ITS may take actions that allow students to improve their learning.

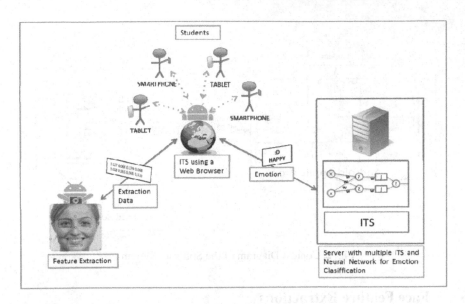

Fig. 1. General Structure of the System Software

Figure 2 shows the context diagram of the system. The context diagram allows us to define the communication that our system can have with other systems. In the context diagram shown in Figure 2 we have the Android application with the system that extracts facial features, which contains four components: Face, Right Eye, Left Eye and Mouth. The feature extraction application communicates with the Intelligent Tutoring System, when asked taking pictures with the front camera of the device. The **Face** component is responsible for finding the human face in the picture, using the Haar-like features cascades method implemented in OpenCV library (www.opencv.org). If the search is successful, the search for other objects (mouth, left and right eyes) is performed. Once the student's face is detected, we proceed to find the components of the face, which are the mouth, right eye and left eye. For optimal image processing, the search method ROI (Region of Interest) is used, which specifies the regions of interest in which you want to search, discarding the rest of the image. Once found the objects in the image, a set of transformations are performed, facilitating the search for object edges to calculate distances opening (mouth and eyes). This data (the distances) is provided to a neural network for classification of emotions.

The ITS has three components like a traditional ITS: the domain module which holds all knowledge of the area or field, which in our case is in the area of basic math for elementary level. The student module has the student knowledge (basic, intermediate or advanced level) of the student. A diagnostic test determines the level of the student. The tutor module defines the next exercise to be solved by the student. To do this, the ITS uses their pedagogical results (e.g. the time taken to solve a problem) and their affective or emotional state. For reaching this decision, the tutor uses an emotion recognizer and an affective module which in turn use an artificial neural network and a fuzzy system.

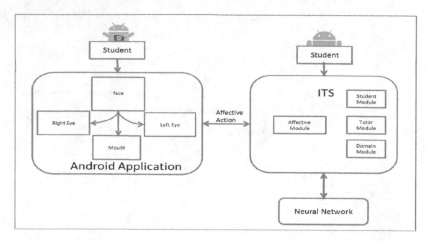

Fig. 2. Context Diagram of the Software System

3   Face Feature Extractions

For the application development, we have used some libraries that allow image manipulations and give support to create interfaces to communicate with different programming languages. The main library to manipulate images is OpenCV. This library has C++, C, Python and Java interfaces and supports Windows, Android and other operating systems. It is also popular because is open source and a standard in digital cameras. The application development also used JavaCV, a Java interface of OpenCV that facilitates the development of Android applications.

3.1   Algorithm for Facial Feature Extraction

I.   The Android device takes a picture with the front camera, in which the face of the user is located.

II.   The face of the user is found, and then we store the coordinates that describe the size and location of the face image.

III.   The search of the mouth starts using the ROI (Region of Interest) method, starting from the coordinates where the face is located. We show the ROI method, the area where we want to find the mouth, discarding the rest of the image.

IV.   A set of transformations needed to know the edges of the mouth are carried out, starting with a Gaussian operator for image refinement.

V.   Once the image is refined, it comes to know the level of brightness in the image by adjusting the pixels to apply the threshold operator.

VI.   The Threshold operator is applied to the image, so it contains only black and white pixels, allowing detection in a simpler way, to the edge of the objects to calculate opening distances.

VII.   Calculate opening distances using the Pythagorean Theorem.

VIII.   After obtaining the distances of openness, the ROI method is restored to the coordinates of the face.

IX.   Steps III to VIII are performed for the left and right eyes.

To calculate the distance (points) of the opening of the mouth, left eye and right eye, different transformations were performed in regions of interest where the objects are found in the image. These modifications allow the application to perform feature extraction, with optimal performance in image size, besides image noise cleaning and handling of certain pixels to identify objects in regions of interest.

The Gaussian average operator was considered to allow the method to clean the image obtained by the application. Initially we worked with Gaussian g where the coordinates x, y are controlled by the difference $\sigma 2$ according to equation 1 [8]:

$$g(x, y, \sigma) = \frac{1}{2\pi\sigma} e^{-\left(\frac{x^2 + y^2}{2\sigma^2} \right)} \tag{1}$$

The results obtained using the Gaussian average operator were smoother images, removing details of the photography, allowing a greater focus on long structures.

Another influential factor in the images is the brightness, which may hinder the transformation process because it does not allow the definition of some edges and structures. So we have applied the histogram equalization process where the image obtained pass through a non-linear process, which tends to emphasize the brightness in a particular way to make it more suitable for recognition. With this application the process makes changes producing an image with a flat histogram, where all levels are equally likely. Then, for a range of M levels the histogram draws the points. For the input (old) and output (new) image, the number of points per level is denoted as $O(l)$ and $N(l)$ (for $O \leq l \leq N$) respectively (Equation 2).

$$\sum_{l=0}^{M} O(l) = \sum_{l=0}^{M} N(l) \tag{2}$$

Since the output of the histogram is uniformly smooth, the cumulative histogram to the p level (for an arbitrarily chosen level p) should be a fraction of the total sum. Then the number of points in the output image is the ratio of the number of points in the range of levels of the output image (Equation 3).

$$N(l) = \frac{N^2}{N_{max} - N_{min}} \tag{3}$$

The last transformation is a *thresholding* that allows us to distinguish the starting point for calculating opening distances, for determining the border points of the mouth, right eye, and left eye. Specifying a certain level, the pixels are set only in two colors, white for high-level and black for low-level (figure 3). For representation of the probability of distribution of the intensity levels, the following equation (equation 4) is used.

$$P(l) = \frac{N(l)}{N^2} \tag{4}$$

Fig. 3. Application of Thresholding operator

3.2   The Neural Network for Recognizing Emotional States

The method used for the detection of visual emotions is based on Ekman's theory [9]. The recognition system was built in three stages: the first one was an implementation to extract features from face images in a corpus used to train the neural network. The second one consisted of the implementation of the neural network. The third stage integrated extraction and recognition into a fuzzy system which is part of the ITS (see figure 4). For training and testing the neural network we used the corpus RAFD (Radboud Faces Database) [10] which is a database with 8040 different facial expressions, which contains a set of 67 models including men and women. Once the emotion state is extracted from the student the state is sent to the fuzzy system.

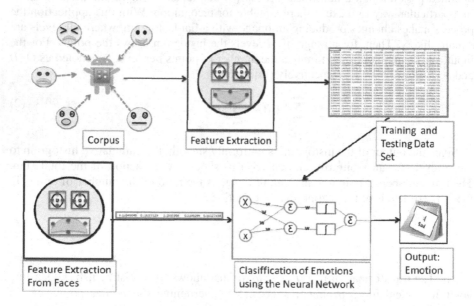

Fig. 4. Feature Extraction and Emotion Classification

4   The Fuzzy Expert System

The student module in the ITS provides information about student knowledge and learning aptitudes. The module identifies what the student's knowledge is through a diagnostic test. The student knowledge can be seen as a subset of all knowledge possessed by the expert in the domain (module) and this is stored in a student profile. In the ITS, a fuzzy expert system was implemented with a new knowledge tracing algorithm, which is used to track student's pedagogical states, applying a set of rules [11]. The benefit of using fuzzy rules is that they allow inferences even when the conditions are only partially satisfied. The fuzzy system uses the four input linguistic variables *error, help, time*, and Emotion (down side of Figure 5). These variables are loaded when the student solves an exercise. The output variable of the fuzzy system is the difficulty and type of the next exercise. In figure 6 we can see in the middle, the interface of the ITS with a division operation, a pedagogical agent (left side), which is used to help and motivate the students.

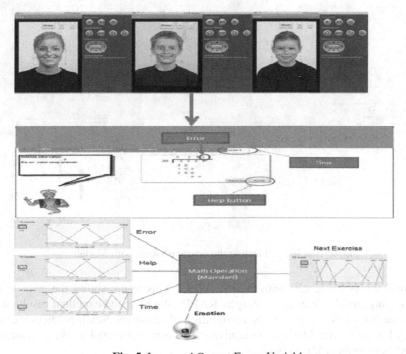

Fig. 5. Input and Output Fuzzy Variables

5   Result Interpretation and Conclusions

In our evaluation design method (Pretest –intervention- posttest) [12], we tested the affective tutoring system with different students and the accuracy of the classification of the neural network with two different tools: Weka (www.cs.waikato.ac.nz/ml/weka/) and Matlab (http://www.mathworks.com/).

In figure 6 we can see that with Matlab we created a two-layer feed-forward network with sigmoid hidden neurons and linear output neurons. The network was trained with the Levenberg-Marquardt back-propagation algorithm. Regression Values that measure the correlation between outputs and targets were values very close to 1 meaning an almost perfect lineal association between target and actual output values (independent and dependent variables) (figure 6). In other words, predicted values Y (actual output), from X (target values) according to the regression model coincide almost exactly with the values observed in Y, and very few prediction error will occur.

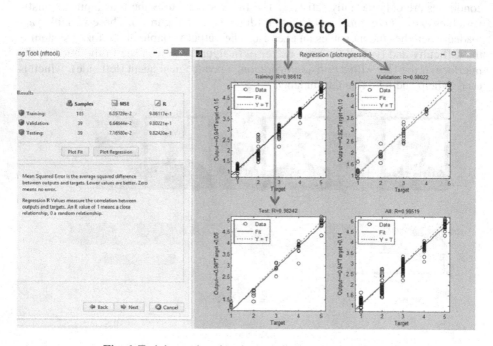

Fig. 6. Training and testing the neural network with Matlab

Figure 7 shows the results with the tool Weka. This figure shows the error levels when applying the classifier to the corpus RAFD. We obtained excellent results with a successful rate of 96.9466 % in the recognition of emotions. Small prediction errors shown in Figure 6 with Matlab, are equivalent to errors detected in Weka instances classified as incorrect (3.0534 %).

Based on the results obtained with Weka and because this tool is open source, we decided to integrate this classifier with the feature extractor and the intelligent tutoring system. Another reason to choose Weka for implementation of the tutoring system is that the code used to implement Weka is Java, which is the language to implement applications for Android.

As mentioned before, the feature extractor and the classifier were implemented in Java. The intelligent tutoring system and the fuzzy expert system were implemented with CCS3, HTML 5 and JSP. We now are starting to test the whole application. The next step in this work is to test the whole application with real students of different schools and create our own corpus of emotions from students with emotions oriented to the teaching-learning process.

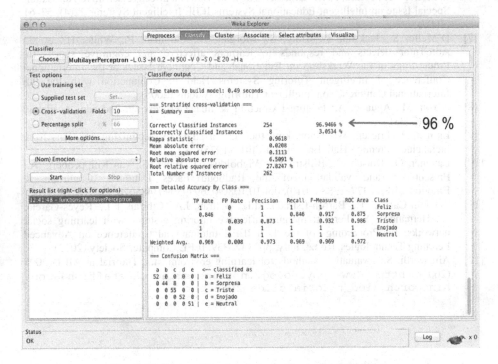

Fig. 7. Training and testing the neural network with Weka

Acknowledgment. The work described in this paper is fully supported by a grant from the DGEST (Dirección General de Educación Superior Tecnológica) in México under the program "Projects of scientific research and technological innovation".

References

[1] Picard, R.W.: Affective Computing. M.I.T Media Laboratory Perceptual Computing Section Technical Report No. 321 (1995)
[2] Arroyo, I., Woolf, B., Cooper, D., Burleson, W., Muldner, K., Christopherson, R.: Emotions sensors go to school. In: Proceedings of the 14th International Conference on Artificial Intelligence in Education, pp. 17–24. IOS Press, Amsterdam (2009)
[3] Stankov, S., Glavinic, V., Rosic, M.: Intelligent Tutoring Systems in E-Learning Environment: Design, Implementation and Evaluation. Information Science Reference (2011)
[4] Nkambou, R., Boudeau, J., Mizoguchi, R.: Introduction: What Are Intelligent Tutoring Systems, and Why This Book? Advances in Intelligent Tutoring Systems 308, 1–12 (2010)

[5] Conati, C., Maclare, H.: Evaluating a probabilistic model of student affect. In: Lester, J.C., Vicari, R.M., Paraguaçu, F. (eds.) ITS 2004. LNCS, vol. 3220, pp. 55–66. Springer, Heidelberg (2004)

[6] D'Mello, S.K., Picard, R.W., Graesser, A.C.: Towards an affective-sensitive AutoTutor. Special Issue on Intelligent Educational Systems IEEE Intelligent Systems 22(4), 53–61 (2007)

[7] D'Mello, S., Jackson, T., Craig, S., Morgan, B., Chipman, P., White, H., et al.: AutoTutor detects and responds to learners affective and cognitive states. In: Proceedings of the Workshop on Emotional and Cognitive Issues at the International Conference of Intelligent Tutoring Systems Held in Conjunction with the 9th International Conference on Intelligent Tutoring Systems (2008)

[8] Nixon, M., Aguado, A.: Feature Extraction & Image Processing, 2nd edn. Academic Press (2008)

[9] Ekman, P., Friesen, W.: Unmasking the face: a guide to recognizing emotions from facial clues. Prentice-Hall, Englewood Cliffs (1975)

[10] Langner, O., Dotsch, R., Bijlstra, G., Wigboldus, D., Hawk, S., van Knippenberg, A.: Presentation and validation of the Radboud Faces Database. Cognition & Emotion 24(8), 1377–1388 (2010), doi:10.1080/02699930903485076

[11] Zatarain-Cabada, R., Barrón-Estrada, M.L., Beltrán, J.A., Cibrian, F.L., Reyes-García, C., Hernández, Y.: Fermat: merging affective tutoring systems with learning social networks. In: Proceedings of the 12th IEEE International Conference on Advanced Learning Technologies, Rome, Italy, pp. 337–339. IEEE Computer Society (2012)

[12] Ainsworth, S.: Evaluation methods for learning environments. Tutorial at AIED 2005 (2005), http://www.psychology.nottingham.ac.uk/staff/Shaaron. Ainsworth/aied_tutorialslides2005.pdf

Solving Three-Objective Flow Shop Problem with Fast Hypervolume-Based Local Search Algorithm

Rong-Qiang Zeng[1,2] and Ming-Sheng Shang[2]

[1] School of Mathematics, Southwest Jiaotong University, Chengdu,
Sichuan 610031, P.R. China
zrq777@gmail.com
[2] School of Computer Science and Engineering, University of Electronic, Science and
Technology of China, Chengdu, Sichuan 610054, P.R. China
msshang@uestc.edu.cn

Abstract. In this paper, we present a fast hypervolume-based multi-objective local search algorithm, where the fitness assignment is realized by the approximating computation of hypervolume contribution. In the algorithm, we define an approximate hypervolume contribution indicator as the selection mechanism and apply this indicator to an iterated local search. We carry out a range of experiments on three-objective flow shop problem. Experimental results indicate that our algorithm is highly effective in comparison with the algorithms based on the binary indicators and the exact hypervolume contribution indicator.

Keywords: multi-objective optimization, approximate hypervolume contribution, local search, flow shop problem.

1 Introduction

The hypervolume indicator was originally proposed by Zitzler and Thiele [14] as a measure of comparing the performance of multi-objective evolutionary algorithms (MOEAs). It is also a popular metric for the fitness assignment in the design of multi-objective optimization algorithms. The main purpose of the present study is to integrate an hypervolume indicator into a population-based multi-objective iterated local search.

In single objective optimization, a total order relation can be easily used to rank the solutions. However, such a natural total order relation does not exist in multi-objective optimization. Indeed, there usually does not exist one optimal but a set of solutions called Pareto optimal solutions or efficient solutions. The aim is to generate a set of Pareto optimal solutions called Pareto optimal set, which keeps the best compromise solutions among a set of objective functions.

In this paper, we define an approximate hypervolume contribution indicator to assign a fitness value to each solution. Then, we integrate this indicator into an iterated local search as the selection mechanism so as to obtain an hypervolume-based multi-objective optimization algorithm. Based on the hypervolume contribution principle,

D.-S. Huang et al. (Eds.): ICIC 2014, LNAI 8589, pp. 11–25, 2014.

we propose a fast hypervolume-based multi-objective local search algorithm. In order to evaluate the effectiveness of the proposed algorithm, we apply it to three-objective flow shop problem, which is an extension of benchmark instances of bi-objective flow shop problem.

The rest of this paper is organized as follows: Section 2 is devoted to introducing the definitions of multi-objective optimization. In section 3, we briefly investigate the indicator-based optimization which is related to the present study. Then, we propose the fast hypervolume-based multi-objective local search algorithm in section 4. Afterwards, section 5 presents the computational results and the performance analysis of the algorithms. Finally, the conclusions and perspectives for the future work are discussed in the last section.

2   Multi-Objective Optimization

First, let us recall some useful notations and definitions of multi-objective optimization problems (MOPs), which are taken from [14]. Let X denote the search space of the optimization problem under consideration and Z the corresponding objective space. Without loss of generality, we assume that $Z = \Re^n$ and that all n objectives are to be minimized. Each $x \in X$ is assigned exactly one objective vector $z \in Z$ on the basis of a vector function $f : X \to Z$ with $z = f(x)$. The mapping f defines the evaluation of a solution $x \in X$, and often one is interested in those solutions that are Pareto optimal with respect to f. The relation $x_1 \succ x_2$ means that the solution x_1 is *preferable* to x_2. The dominance relation between two solutions x_1 and x_2 is usually defined as follows [14]:

Definition 1. *A decision vector x_1 is said to dominate another decision vector x_2 (written as $x_1 \succ x_2$), if $f_i(x_1) \leq f_i(x_2)$ for all $i \in \{1, \cdots, n\}$ and $f_j(x_1) < f_j(x_2)$ for at least one $j \in \{1, \cdots, n\}$.*
Definition 2. *$x \in X$ is said to be Pareto optimal solution if and only if a solution $x_i \in X$ which dominates x does not exist.*

Since in most cases, it is not possible to compute the Pareto optimal set in a reasonable time, we are interested in computing a set of non-dominated solutions which is as close to the Pareto optimal set as possible. Therefore, the whole goal is often to identify a good approximation of the Pareto optimal set.

3   Indicator-Based Optimization

The indicator-based optimization principle was initially proposed by Zitzler and Künzli [14], which extends the idea of flexible integration of preference information by Fonseca and Fleming [10]. They define the quality indicators which are used as a measure during the selection process. The quality indicator can be represented as a function that assigns each solution a fitness value reflecting its quality. Then, this principle transforms a multi-objective problem into a single objective one, since this indicator function induces a total order that can be used to rank the solutions of population.

3.1 Binary Indicator

Indeed, Zitzler and Künzli [14] proposed the Indicator-Based Evolutionary Algorithm (IBEA). In this algorithm, they define two binary indicators I_ε and I_{Hyp}, which are described as follows:

$$I_\varepsilon(x_1, x_2) = max_{i \in \{1, \cdots, n\}}(f_i(x_1) - f_i(x_2)) \tag{1}$$

I_ε (x_1, x_2) (where $x_1 \in X$ and $x_2 \in X$) represents the minimal translation (in the objective space) on which to execute x_1 so that it dominates x_2 (see Fig. 1). Let us note that the translation could take negative values.

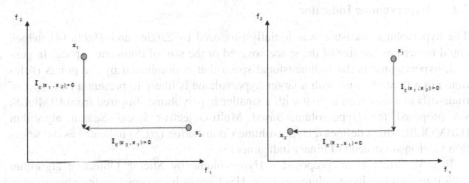

Fig. 1. Illustration of the I_ε indicator applied to two solutions x_1 and x_2 (left hand side: no dominance relation between x_1 and x_2; right hand side: $x_2 \succ x_1$) [2].

$$I_{Hyp}(x_1, x_2) = \begin{cases} H(x_2) - H(x_1), & if \ x_2 \succ x_1 \\ H(\{x_1, x_2\}) - H(x_1), & otherwise \end{cases} \tag{2}$$

H (x_1) denotes the volume of objective space dominated by x_1, I_{Hyp} (x_1, x_2) represents the volume of objective space that is dominated by x_2, but not by x_1 (see Fig. 2).

In fact, both indicators evaluate the quality of a solution x_1 with respect to another solution x_2. The experimental results showed that IBEA could significantly improve the quality of the generated Pareto approximation set with respect to the considered optimization goal.

In [2], Basseur et al. proposed a simple and generic Indicator-Based Multi-Objective Local Search algorithm (IBMOLS), where the selection mechanism is realized based on two binary indicators defined in [14]. This algorithm is easily adaptable, parameter independent and has a high convergence rate. The experiments show IBMOLS outperforms some classical metaheuristics.

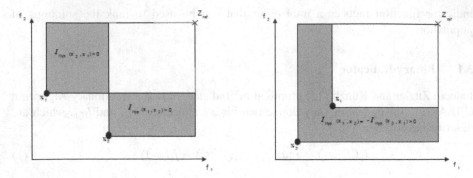

Fig. 2. Illustration of the I_{Hyp} indicator applied to two solutions x_1 and x_2 (left hand side: no dominance relation between x_1 and x_2; right hand side: $x_2 \succ x_1$) [2].

3.2   Hypervolume Indicator

The hypervolume measure was initially proposed by Zitzler and Thiele [4], which could be seen as the size of the space covered or the size of dominated space. In general, hypervolume is the n-dimensional space that is dominated by the points (solutions) in a front. A front with a larger hypervolume is likely to present a better set of trade-offs to a user than a front with a smaller hypervolume. Inspired from IBMOLS, we proposed the Hypervolume-Based Multi-Objective Local Search algorithm (HBMOLS), which defines a Hypervolume Contribution (*HCS*) indicator as the selection mechanism instead of binary indicators [3].

In [13], While et al. proposed a Hypervolume by Slicing Objective algorithm (HSO) to calculate hypervolume exactly. HSO works by proceeding the objectives in a front rather than the points. It divides the n D-hypervolume to be measured into separate $n-1$ D-slices through one of the objectives, then it calculates the hypervolume of each slice and sums these values to derive the total.

To improve the performance of HSO, Bradstreet, While and Barone [5] presented a fast incremental hypervolume algorithm (IHSO), which calculates the exclusive hypervolume of a point p relative to a set of points S. IHSO minimizes the number of slices that have to be proceeded and orders the objectives intelligently. Then, the exact hypervolume of each point in a set is calculated by repeated application of IHSO.

However, hypervolume computation is exponential in the number of objectives in the worst cases. In order to rank the solutions of population more effectively, Bader and Zitzler [1] put forward a new thought: the ranking of solutions induced by the hypervolume indicator is more important than the actual indicator values. In other words, it is not necessary to compute an exact HC value but to compute an approximate HC value in high dimensions. Based on this thought, they proposed a fast search algorithm (HypE), which uses Monte Carlo simulation to approximate the exact hypervolume values. This novel hypervolume-based multi-objective evolutionary algorithm well trade off between the accuracy of the estimates and the overall computing time budget.

4   Fast Hypervolume-Based Multi-Objective Local Search Algorithm

In this section, we present a Fast Hypervolume-Based Multi-Objective Local Search algorithm (FHBMOLS), which uses an Approximate Hypervolume Contribution (AHC) indicator as the selection mechanism during the search process. The outline of this algorithm is illustrated in Algorithm 1.

Algorithm 1 Fast Hypervolume-Based Multi-Objective Local Search

Input: N (Population size)
Output: A (Pareto approximation set)
Initialization: $P \leftarrow N$ randomly generated solutions
        $A \leftarrow$ Non-dominated solutions of P
While running time is not reached **do**
    1) Fitness Assignment: compute a fitness value for each $x \in P$, i.e., $Fit(x) \leftarrow AHC(x, P)$
    2) For each $x \in P$ do:
        a) $x^* \leftarrow$ one randomly chosen unexplored neighbors of x
        b) Progress $\leftarrow$ **Hypervolume Contribution Selection** (P, x^*)
        until all neighbors are explored or Progress $=$ True
End while
Return A

In Algorithm 1, the individuals of the initial population are generated randomly, i.e., each individual is initialized with a random permutation. Then, each solution in the population is assigned a fitness value by the AHC indicator (Section 4.2). Afterwards, the entire population is optimized by the Hypervolume Contribution Selection (HCS) illustrated in Algorithm 2.

Algorithm 2 Hypervolume Contribution Selection

Step:
    1) $P \leftarrow P \bigcup x^*$
    2) Compute x^* fitness: $AHC(x^*, P)$, then update all $z \in P$ fitness values:
        $Fit(z) \leftarrow AHC(z, P)$
    3) $w \leftarrow$ worst individual in P
    4) $P \leftarrow P \backslash \{w\}$, then update all $z \in P$ fitness values: $Fit(z) \leftarrow AHC(z, P \backslash \{w\})$
    5) if $w \neq x^*$, return True

Algorithm 2 shows the pseudo-code of the HCS procedure. In this procedure, a fitness value computed by the AHC indicator is assigned to the solution x^*, which is one of the unexplored neighbors of x. If x^* is dominated, the fitness values of solutions in the non-dominated set are not modified; if x^* is non-dominated, the fitness values of some solutions, which are the neighbors of x^* in the objective space, need to be updated. Then, the solution with the worst fitness value is removed from the population P. If this solution is dominated, nothing will be changed in the non-dominated set; otherwise, we should update the fitness values of the neighbors of this solution in the objective space.

Algorithm 3 Iterated FHBMOLS Algorithm

Input: N (Population size)
Output: PO (Pareto approximation set)
Step 1: $PO \leftarrow \Phi$ (Empty set)
Step 2:
While running time is not reached **do**
  1) $P \leftarrow$ Generate a new random initial population of size N using PO
  2) $A \leftarrow$ FHBMOLS (initialized with P)
  3) $PO \leftarrow$ Non-dominated solutions of $PO \bigcup A$
End while
Step 3: Return PO

Furthermore, the local search is executed in an iterative way shown in Algorithm 3, the Pareto approximation set PO is maintained and actualized with the solutions found by FHBMOLS. Specifically, the FHBMOLS algorithm is used to improve the quality of the initial population P and generate a Pareto approximation set A during the local search process. After each local search, the random mutation is used to create a new initial population for the next FHBMOLS execution. The whole process will repeat until a pre-defined running time is reached.

4.1 Hypervolume Computation

Hypervolume is described as the Lebesgue measure VOL of the union of hypercubes $[Z_{ref}, x_i]$ defined by a non-dominated point x_i and a reference point Z_{ref} [6, 7], which is illustrated in Equation 3 and Figure 3 respectively.

$$Hyp(P) = VOL([Z_{ref}, x_1] \cup \cdots \cup [Z_{ref}, x_n]) \tag{3}$$

The hypervolume contribution can be seen as the measure of the space that is only dominated by x. According to Equation 3, the hypervolume contribution $HypC$ of a solution x to a population P is defined as follows:

$$HypC(x, P) = Hyp(P) - Hyp(P\{x\}) \tag{4}$$

Based on the definition of the hypervolume contribution, we define the Hypervolume Contribution (HC) indicator, where we distinguish two cases:

$$HC(x, P) = \begin{cases} HypC(x, P), & \text{if } x \text{ is non-dominated} \\ -max_{y \in P, y > x}(VOL([y, x])), & otherwise \end{cases} \tag{5}$$

Actually, the HC indicator includes a special case when x is dominated by at least one solution of P. An example is illustrated in Figure 4. More precisely, we divide the entire population into two sets: one is the set composed of non-dominated solutions; the other one is the set composed of dominated solutions (at least dominated by one solution in the first set). According to the definition of HC indicator, the fitness computation for the dominated solutions depends on the solutions located in the non-dominated set.

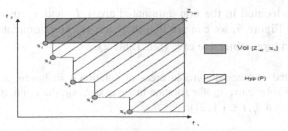

Fig. 3. The hypervolume $Hyp(P)$ is the area which is enclosed by a population P (including the solutions x_i, $i \in \{1, \cdots, 5\}$), according to a reference point Z_{ref}; The hypervolume of the solution x_1 is denoted as $[Z_{ref}, x_1]$, which is colored in grey.

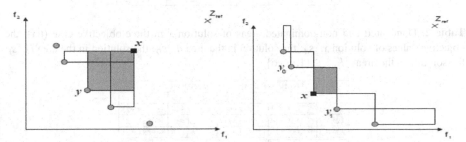

Fig. 4. Left hand side: $HC(x,P)$ of a dominated solution x to a population P: maximum dominance area (grey box) computed between x and y ($y > x$, $y \in P$). Right hand side: $HC(x,P)$: Hypervolume contribution of a solution x to a population P, the area dominated only by the solution x is colored in grey (y_0 and y_1: two non-dominated solutions next to x). Z_{ref}: the reference point.

4.2   Fast Computation of Hypervolume Contribution in Three-Objective Case

As mentioned in the previous section, hypervolume computation in N-objective case ($N \geq 3$) usually requires high computational efforts. Even in the three-objective case, it is very complicated to compute the hypervolume contribution for solutions located in the non-dominated set. As shown in Figure 5, the solutions which are incomparable

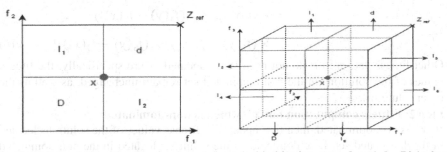

Fig. 5. Left hand side: the non-dominated areas in bi-objective case: I_1 and I_2. Right hand side: the non-dominated areas in three-objective case: I_j, $j \in \{1, \cdots, 6\}$. D: the dominating area where x is dominated; d: the dominated area where all the solution are dominated by x.

to x can only be located in the non-dominated areas I_1 and I_2 in bi-objective case (see Table 1). In Figure 5, we can see that there are six non-dominated areas of x (I_j, $j \in \{1, \cdots, 6\}$) in three-objective case (see Table 2).

Table 1. Dominated and non-dominated areas of solution x in bi-objective case ($f(x)$): the objective values of solution x, s_d: the solution in the area d, s_D: the solution in the area D, s_j: the solution in the area $I_j, j \in \{1, 2\}$).

$f(x)$	Dominated and non-dominated areas			
	d	D	I_1	I_2
$f_1(x)$	$\leq f_1(s_d)$	$\geq f_1(s_D)$	$> f_1(s_1)$	$< f_1(s_2)$
$f_2(x)$	$\leq f_2(s_d)$	$\geq f_2(s_D)$	$< f_2(s_1)$	$> f_2(s_2)$

Table 2. Dominated and non-dominated areas of solution x in three-objective case ($f(x)$): the objective values of solution x, s_d: the solution in the area d, s_D: the solution in the area D, s_j: the solution in the area $I_j, j \in \{1, \cdots, 6\}$.

$f(x)$	Dominated and non-dominated areas			
	d	D	I_1	I_2
$f_1(x)$	$\leq f_1(s_d)$	$\geq f_1(s_D)$	$> f_1(s_1)$	$\geq f_1(s_2)$
$f_2(x)$	$\leq f_2(s_d)$	$\geq f_2(s_D)$	$< f_2(s_1)$	$\geq f_2(s_2)$
$f_3(x)$	$\leq f_3(s_d)$	$\geq f_3(s_D)$	$< f_3(s_1)$	$< f_3(s_2)$
	I_3	I_4	I_5	I_6
$f_1(x)$	$< f_1(s_3)$	$\geq f_1(s_4)$	$< f_1(s_5)$	$< f_1(s_6)$
$f_2(x)$	$> f_2(s_3)$	$< f_2(s_4)$	$< f_2(s_5)$	$\geq f_2(s_6)$
$f_3(x)$	$< f_3(s_3)$	$\geq f_3(s_4)$	$> f_3(s_5)$	$\geq f_3(s_6)$

In our situation, we consider using the methods to compute an approximate value of the hypervolume contribution for each solution. For this purpose, we propose an Approximate Hypervolume Contribution indicator (AHC) and focus on the AHC computation in three-objective case. The main steps of AHC computation are presented in detail as follows:

- **Step 1: Fitness computation for x^* which is dominated:**
  If x^* is dominated, then according to the equation 5,

$$AHC(x^*, P) = -max_{y \in P, y > x^*}(VOL([y, x^*]))$$

$$= -max_{y \in P, y > x^*}|(f_1(y) - f_1(x^*))$$

$$\times (f_2(y) - f_2(x^*)) \times (f_3(y) - f_3(x^*))| \quad (6)$$

Other fitness values in P do not need to be updated. More specifically, the fitness values of the solutions in the non-dominated set remain unchanged, as well as the other solutions.

- **Step 2: Fitness computation for x^* which is non-dominated:**
  If x^* is non-dominated, then first update the fitness values of the solutions located in the dominated set. If x^* dominates some solutions located in the non-dominated set, compute their new negative fitness values according to x^*, update the solutions in the non-dominated set. Afterwards, compute the fitness values for x^* and the other solutions in the non-dominated set.

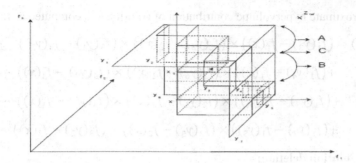

Fig. 6. The *AHC* computation in three-objective case, the two parts *A* and *B* are described in Figures 7 and 8 respectively

We compute the hypervolume contribution of x^* by using the solutions located in the non-dominated set. Among these solutions which are possibly distributed in the six non-dominated areas of *x*, part of them are useful for the *AHC* computation. Actually, these useful solutions will slice the hypervolume of *x* in the three-dimension space, in order to obtain an approximate value of the hypervolume contribution of x^*.

Now, we give an example to show how to compute the fitness value for x^*. All the neighbors of solution *x* are illustrated in Figure 6, which are denoted as y_1, y_2, y_3, y_4, y_5, y_6 and y_7. The point x_0 in Figure 6 is the projection of solution *x* on the plane *C*. The relation between solution *x* and its neighbors is summarized in Table 3.

Table 3. The neighbors of solution *x* in Figure 6 (solutions y_4 and y_5 are non-dominated)

Objective	The non-dominated neighbours of solution x						
	y_1	y_2	y_3	y_4	y_5	y_6	y_7
area	I_2	I_1	I_3	I_5	I_5	I_4	I_6
$f_1(x)$	$> f_1(y_1)$	$> f_1(y_2)$	$< f_1(y_3)$	$< f_1(y_4)$	$< f_1(y_5)$	$> f_1(y_6)$	$< f_1(y_7)$
$f_2(x)$	$> f_2(y_1)$	$< f_2(y_2)$	$> f_2(y_3)$	$< f_2(y_4)$	$< f_2(y_5)$	$< f_2(y_6)$	$> f_2(y_7)$
$f_3(x)$	$< f_3(y_1)$	$< f_3(y_2)$	$< f_3(y_3)$	$> f_3(y_4)$	$> f_3(y_5)$	$> f_3(y_6)$	$> f_3(y_7)$

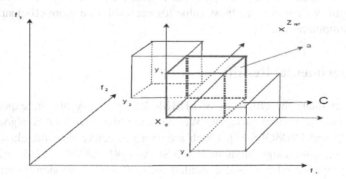

Fig. 7. The *AHC* computation in three-objective case: part *A*, the approximate hypervolume contribution of solution *x* in this part is computed by Equation *a*

The approximate hypervolume contribution of solution x is computed as follows:

$$AHC(x, P) = \left(f_1(y_3) - f_1(x)\right) \times \left(f_2(y_2) - f_2(x)\right) \times \left(f_3(y_1) - f_3(y_2)\right) \cdots \cdots a$$

$$+\left(f_1(y_4) - f_1(x)\right) \times \left(f_2(y_6) - f_2(x)\right) \times \left(f_3(y_2) - f_3(x)\right) \cdots \cdots b$$

$$+\left(f_1(y_5) - f_1(y_4)\right) \times \left(f_2(y_4) - f_2(x)\right) \times \left(f_3(y_2) - f_1(x)\right) \cdots \cdots c$$

$$+\left(f_1(y_7) - f_1(y_5)\right) \times \left(f_2(y_5) - f_2(x)\right) \times \left(f_3(y_2) - f_3(x)\right) \cdots \cdots d \quad (7)$$

- **Step 3: Solution deletion:**
  The solution w with the worst fitness value ($AHC(w) = min_{z \in P}(AHC(z, P))$) is deleted from the population P. If w is dominated, the fitness values of the remaining solutions do not need to be updated. If w is non-dominated, the fitness values of the solutions in the non-dominated set need to be re-computed as we do in Step 2.

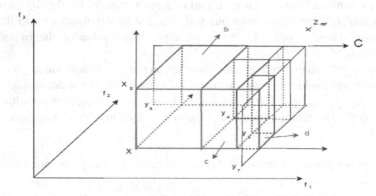

Fig. 8. The AHC computation in three-objective case: part B, the approximate hypervolume contribution of solution x in this part has three components, which are computed by Equation b, Equation c and Equation d respectively

The AHC computation plays an important role in the FHBMOLS algorithm. By the AHC indicator, we compute a fitness value for each solution more efficiently than the exact HC computation in [3].

5   Experimental Results

In order to evaluate the efficiency of FHBMOLS, we carry out the experiments on three-objective flow shop problem. We compare the FHBMOLS algorithm with IBMOLS [2] and HBMOLS [3], which are more effective than the classical multi-objective optimization algorithms such as NSGA-II [8] and SPEA2 [15]. All the algorithms are programmed in C and compiled using Dev-C++ version 2 compiler on a PC running Windows XP. Computational runs were performed on an Intel Core 2 5000B (2×2.61 GHz) machine with 2.00 GB RAM.

5.1 Bi-objective Flow Shop Problem

The Flow Shop Problem (FSP) is one of the most thoroughly studied machine sched-uling problems, which schedules a set of jobs on a set of machines according to a specific order. The bi-objective FSP can be presented as a set of N jobs { J_1, J_2,$\cdots$, J_n} to be scheduled on M machines {M_1, M_2, $\cdots$, M_m}. Machines are critical re-sources, i.e., one machine can not be assigned to two jobs simultaneously. Each job J_i is composed of m consecutive tasks t_{i1}, t_{i2}, $\cdots$, t_{im}, where t_{ij} represents the jth task of the job J_i requiring the machine m_j. To each task, t_{ij} is associated with a processing time p_{ij}. Each job J_i has a due date d_i. Then, we aim to mini-mize two objectives [4]: C_{max} (Makespan: Total completion time) and T (Total tardiness).

The task t_{ij} is scheduled at the time s_{ij}, two objectives can be computed as fol-lows [4]:

$$f_1 = C_{max} = max_{i \in [1, \cdots, n]} \{s_{im} + p_{im}\} \tag{8}$$

$$f_2 = T = \sum_1^n max\,(0, s_{im} + p_{im} - d_i) \tag{9}$$

Both of them have been proven to be NP-hard [11, 9]. In addition, all the FSP in-stances used in this paper are taken from Taillard benchmark instances and extended into bi-objective case [12][1].

5.2 Three-Objective Flow-Shop Problem Instance

Generally, there are no established benchmark instances for flow shop problem with more than two objectives. In our situation, we have to extend bi-objective benchmark instances to three-objective case. More specifically, we use two bi-objective instances with the same size to generate one three-objective instance. For example, C_{max} and T computed for the first instance represent the first two objectives, C_{max} computed for the second instance represents the third objective. Furthermore, we denote a three-objective instance as i_j_III with i jobs and j machines. For example, 50_15_III de-notes a three-objective instance generated by two bi-objective instances 50_15_01 and 50_15_02.

5.3 Parameter Settings

The proposed algorithm requires to set a few parameters, we mainly discuss two im-portant ones: running time and population size.

— **Running time:** The running time T is a key parameter in experiments. We define the time T for each instance by Equation 10, in which N_{Job}, N_{Mac} and N_{Obj} rep-resent the number of jobs, the number of machines and the number of objectives in an instance respectively (see Table 4).

---

[1] Benchmarks available at http://www.lifl.fr/~liefooga/benchmarks/index. html

$$T = \frac{N_{Job}^2 \times N_{Mac} \times N_{obj}}{100} \, sec \tag{10}$$

T is defined according to the "difficulty" of the instance. Indeed, N_{Job} defines the size of the search space, since it is of size $N_{Job}!$. The roughness of the landscape is strongly related with N_{Mac}. We use this formula to obtain a good balance between the problem difficulty and the time allowed.

— **Population size:** According to the results obtained in [2], the experiments realized previously on the IBMOLS algorithm showed that the best results are achieved with a small population size N. We set this size from 10 to 40 individuals by Equation 11, relative to the size of tested instance (see Table 4).

$$|N| = \begin{cases} 10 : 0 < |N_{Job} \times N_{Mac}| < 500 \\ 20 : 500 < |N_{Job} \times N_{Mac}| < 1000 \\ 30 : 1000 < |N_{Job} \times N_{Mac}| < 2000 \\ 40 : 2000 < |N_{Job} \times N_{Mac}| < 3000 \end{cases} \tag{11}$$

Table 4. Parameter values used for three-objective FSP instances (i_j_k represents the kth bi-objective instance with i jobs and j machines): population size (N) and running time (T)

Instance 1	Instance 2	Size	N	T
20_05_01_ta001	20_05_02_ta002	20 × 5	10	1'
20_10_01_ta011	20_10_02_ta012	20 × 10	10	2'
20_15_01	20_15_02	20 × 15	10	3'
20_20_01_ta021	20_20_02_ta022	20 × 20	10	4'
30_05_01	30_05_02	30 × 5	10	2'15"
30_10_01	30_10_02	30 × 10	10	4'30"
30_15_01	30_15_02	30 × 15	10	6'45"
30_20_01	30_20_02	30 × 20	20	9'
50_05_01_ta031	50_05_02_ta032	50 × 5	10	6'15"
50_10_01_ta041	50_10_02_ta042	50 × 10	20	12'30"
50_15_01	50_15_02	50 × 15	20	18'45"
50_20_01_ta051	50_20_02_ta052	50 × 20	30	25'
100_05_01_ta061	100_05_02_ta062	100 × 5	20	25'
100_10_01_ta071	100_10_02_ta072	100 × 10	30	50'
100_15_01	100_15_02	100 × 15	30	75'
100_20_01_ta081	100_20_02_ta082	100 × 20	40	100'

5.4   Performance Assessment Protocol

We evaluate the effectiveness of the FHBMOLS algorithm by using a test procedure that has been undertaken with the performance assessment package provided by Zitzler et al.[2].

The quality assessment protocol works as follows: we first create a set of 20 runs with different initial populations for each algorithm and each benchmark instance. Afterwards, we calculate the set PO^* in order to determine the quality of k different sets $A_0, \cdots, A_{k-1}$ of non-dominated solutions (The set PO^* is generated by removing the dominated solutions from the union of k different sets, more details can be

---

[2] http://www.tik.ee.ethz.ch/pisa/assessment.html

found in [16]). Furthermore, we define a reference point $z = [w_1, w_2]$, where w_1 and w_2 represent the worst values for each objective function in $A_0 \cup \cdots \cup A_{k-1}$. Then, the evaluation of a set A_i of solutions can be determined by finding the hypervolume difference between A_i and PO^* [16], which has to be as close to zero as possible.

For each algorithm, we compute 20 hypervolume differences corresponding to 20 runs, and perform the Mann-Whitney statistical test on the sets of hypervolume difference. In our experiments, we say that an algorithm $\mathcal{A}$ outperforms an algorithm $\mathcal{B}$ if the Mann-Whitney test provides a confidence level greater than 95%. The computational results are summarized in Table 5. In this table, each line contains at least a value **in grey** for each instance, which corresponds to the best average hypervolume difference obtained by the corresponding algorithm. The values both **in italic** and **bold** mean that the corresponding algorithms are **not** statistically outperformed by the algorithm which obtains the best result (with a confidence level greater than 95%).

5.5   Computational Results

The computational results are summarized in Table 5. In this table, we observe that FHBMOLS and HBMOLS both statistically outperform IBMOLS almost on all the instances (HBMOLS applied on the instance 100_20_III is an exception). Especially, HBMOLS obtains the best results on the first 7 instances except for the instance 20_05_III, while FHBMOLS obtains the best results on the remaining 9 instances from 30_20_III to 100_20_III.

Table 5. Hypervolume differences of FSP using *AHC* indicator in three-objective case

Instance	Algorithm			
	IBMOLS(I_ε)	IBMOLS(I_{Hyp})	HBMOLS(I_{HC})	FHBMOLS(I_{AHC})
20_05_III	0.008764	0.018905	*0.006219*	**0.005508**
20_10_III	0.037505	0.058672	*0.013197*	0.014477
20_15_III	0.048403	0.056849	*0.018632*	0.023745
20_20_III	0.046089	0.044354	*0.014347*	0.016805
30_05_III	0.052788	0.061270	*0.037125*	*0.038813*
30_10_III	0.122063	0.129725	*0.073722*	0.079506
30_15_III	0.121436	0.123352	*0.057563*	0.070556
30_20_III	0.118505	0.139560	0.073522	**0.070635**
50_05_III	0.135209	0.121354	0.040647	**0.037380**
50_10_III	0.127808	0.146287	0.107813	**0.094505**
50_15_III	0.132819	0.153093	0.098893	**0.092594**
50_20_III	0.134463	0.131359	0.125588	**0.099686**
100_05_III	0.125570	0.131555	0.058073	**0.051081**
100_10_III	0.143721	0.175792	0.134927	**0.106757**
100_15_III	0.167360	0.247699	0.161428	**0.125419**
100_20_III	0.144680	0.203293	0.153831	**0.119110**

Indeed, the population size of the experiments has an effect on the performance of FHBMOLS. When the population size is 10, there does not exist enough solutions used for the *AHC* computation, which usually leads to low effectiveness of FHBMOLS on these instances. On the other hand, the instance size also has an influence on the performance of FHBMOLS. For the large instances, it often takes more time to compute the exact *HC* values and less time in local search, which obviously affects the effectiveness of HBMOLS. In FHBMOLS, it is much faster to obtain an approximate *HC* value for each solution, which means FHBMOLS could keep a better balance between the *HC* computation and the local search. That explains why FHBMOLS statistically outperforms HBMOLS on the large instances. For this reason, even though the population size is 10, FHBMOLS still obtains the best result on the instance 50_05_III.

6   Conclusions and Perspectives

In this paper, we have defined an approximate hypervolume contribution indicator in order to compare and select a set of non-dominated solutions for multi-objective optimization algorithms. We have used the *AHC* indicator to settle the selection process and proposed a fast hypervolume-based multi-objective local search algorithm.

We have performed the experiments on three-objective flow shop problem, which is an extension of benchmark instances of bi-objective flow shop problem. Experimental results indicate the superiority of *AHC* indicator over two binary indicators I_ε and I_{Hyp}. In comparison with HBMOLS based on the *HC* indicator, the computational results obtained by FHBMOLS are also highly competitive, especially on the large instances.

The performance analysis of FHBMOLS sheds light on the ways to future research. One immediate possibility is to apply this algorithm to other multi-objective problems such as multi-objective quadratic assignment problem or multi-objective traveling salesman problem. Additionally, it would be very interesting to propose more effective methods to compute the approximate *HC* values for further improvements of the fitness assignment in higher dimensions.

Acknowledgment. The work in this paper was supported by the Fundamental Research Funds for the Central Universities (Grant No. A0920502051408-25) and National Natural Science Foundation of China (Grant No. 61073099). The authors would like to thank the anonymous referees for their valuable comments and suggestions.

References

1. Bader, J., Zitzler, E.: HypE: An algorithm for fast hypervolume-based many-objective optimization. IEEE Transactions on Evolutionary Computation 19(1), 45–76 (2011)
2. Basseur, M., Liefooghe, A., Le, K., Burke, E.: The efficiency of indicator-based local search for multi-objective combinatorial optimization problems. Journal of Heuristics 18(2), 263–296 (2012)
3. Basseur, M., Zeng, R.-Q., Hao, J.-K.: Hypervolume-based multi-objective local search. Neural Computing and Applications 21(8), 1917–1929 (2012)

4. Basseur, M., Seynhaeve, F., Talbi, E.-G.: Design of multi-objective evolutionary algorithms: application to the flow-shop scheduling problem. In: Proceedings of the IEEE Congress on Evolutionary Computation (CEC 2002), Honlulu, USA, vol. 2, pp. 1151–1156 (2002)
5. Bradstreet, L., While, L., Braone, L.: A fast incremental hypervolume algorithm. IEEE Transactions on Evolutionary Computation 12(6), 714–723 (2008)
6. Bringmann, K., Friedrich, T.: Don't be greedy when calculating hypervolume contributions. In: FOGA 1990: Proceedings of the Tenth ACM SIGEVO Workshop on Foundations of Genetic Algorithms, pp. 103–112. ACM, New York (2009)
7. Coello, C.A., Lamont, G.B., Van Veldhuizen, D.A.: Evolutionary Algorithms for Solving Multi-Objective Problems (Genetic and Evolutionary Computation). Springer-Verlag New York, Inc., Secaucus (2006)
8. Deb, K., Pratap, A., Agarwal, S., Meyarivan, T.: A fast elitist multi-objective genetic algorithm: NSGA-II. IEEE Transactions on Evolutionary Computation 6, 182–197 (2000)
9. Du, J., Leung, J.Y.-T.: Minimizing total tardiness on one machine is NP-hard. Mathematics of Operations Research 15, 483–495 (1990)
10. Fonseca, C.M., Fleming, P.J.: Multiobjective optimization and multiple constraint handling with evolutionary algorithms – part II: application example. Mathematics of Operations Research 28(1), 38–47 (1998)
11. Graham, R.L., Lawler, E.L., Lenstra, J.K., Rinnooy Kan, A.H.G.: Optimization and approximation in deterministic sequencing and scheduling: a survey. Annals of Discrete Mathematics 5, 287–326 (1979)
12. Taillard, E.: Benchmarks for basic scheduling problems. European Journal of Operational Research 64, 278–285 (1993)
13. While, R.L., Hingston, P., Barone, L., Huband, S.: A faster algorithm for calculating hypervolume. IEEE Transactions on Evolutionary Computation 10(1), 29–38 (2006)
14. Zitzler, E., Künzli, S.: Indicator-based selection in multiobjective search. In: Yao, X., et al. (eds.) PPSN VIII. LNCS, vol. 3242, pp. 832–842. Springer, Heidelberg (2004)
15. Zitzler, E., Laumanns, M., Thiele, L.: SPEA2: Improving the strength Pareto evolutionary algorithm for multiobjective optimization. TIK Report 103, Computer Engineering and Networks Laboratory (TIK), ETH Zurich, Zurich, Switzerland (2001)
16. Zitzler, E., Thiele, L., Laumanns, M., Fonseca, C.M., Grunert da Fonseca, V.: Performance assessment of multiobjective optimizers: an analysis and review. IEEE Transactions on Evolutionary Computation 7(2), 117–132 (2003)

A Learning Automata-Based Singular
Value Decomposition and Its Application
in Recommendation System

Yuchun Jing, Wen Jiang, Guiyang Su, Zhisheng Zhou, and Yifan Wang

Department of Electronic Engineering, Jiao Tong University, Shanghai, China
{jingyc,wenjiang,sugy,zhouzhisheng,wangyifan_1123}@sjtu.edu.cn

Abstract. Recommendation system nowadays plays an important role in e-commerce, by helping consumers to find their preference from tens of thousands of goods and at the same time bringing large profits to e-commerce companies. Till now many different recommendation algorithm have been proposed and achieved good effect. In the context Netflix Prize in 2006, Simon Funk proposed a matrix factorization-based recommendation algorithm named Funk-SVD, which caused a widespread concern about the use of SVD model in recommend algorithm. Traditional SVD-based recommendation algorithm employs gradient descent algorithm as its optimization strategy. In this paper, we proposed a CALA-based algorithm to perform Funk-SVD, taking into consideration that CALA, as a kind of reinforcement learning model, has a superior performance on continues parameter optimization, especially in a unknown environment. As far as we known, the whole concept of CALA-based SVD is novel and unreported in the literature. To analyze the new algorithm, we tested it on the data set of film rating and achieved an average RMSE of 0.85, which is comparable with the former algorithm.

Keywords: CALA, Learning Automata (LA), SVD, recommendation system.

1 Introduction

Today, the recommendation system plays a more and more important role in e-commerce. A well-performed recommendation algorithm can accurately predict a user's preference for a certain goods. so to bring in huge profits to the e-company. A user-based collaborative filtering (UserCF) algorithm [1], which was proposed early in 1992, aims to find the 'nearest' neighbor of a user, and predict this user's preference based on his/her neighbors' preference. Another item-based collaborative filtering (ItemCF) algorithm [2] propose by Amazon Co in 2003, has now become the most widely used recommendation algorithm. ItemCF focuses on the similarity between items instead of that between users. Association rule-based algorithm[3] is another item-based algorithm, which mainly focuses on digging out the potential association between two items. Since the context Netflix Prize in 2006, Latent Factor Model (LFM)[4]has become a hot spot owning to its high accuracy of recommendation. The key point of LFM is to fulfill the blank of the sparse rating matrix using Singular Value

D.-S. Huang et al. (Eds.): ICIC 2014, LNAI 8589, pp. 26–32, 2014.

Decomposition (SVD)[5], so to predict a user's rating for a particular goods. Traditional SVD-based recommendation algorithm employs gradient descent algorithm as its optimization strategy.

In this paper we propose a novel method to perform SVD, which is based on Learning Automata (LA)[6][7]. LA is a self-adaptive decision-making method that operates in an unknown random environment and progressively improve their performance via reinforcement learning. In each iteration, the automata randomly chooses one of the actions according to their probability density distribution, and apply the chosen action into the environment. LA uses environment feedback to update the action probability distribution and choose the next action for reinforce learning. This procedure continues until an optimal action is chosen.LA searches for the optimal parameter within the space of probability distribution rather than within the parameter space, so it works well as a global optimizer, especially in a random and noisy environment. Here we employ a continuous Action-set Learning Automate (CALA), which works in a continuous space. Experiment evidence show that the proposed CALA-SVD can achieve a good prediction result.

The paper is organized as follows: Section 2 provides a brief introduction of LA and CALA. Section 3 presents the CALA-based solution to perform SVD and its application in recommendation system. Section 4 presents the experimental results and analysis. Section 5 gives the conclusions.

2   Learning Automata

2.1   Introduction of Learning Automata (LA)

Learning Automata(LA) is a kind of artificial intelligence, which can be viewed in the framework of reinforcement learning. A Learning Automata interacts with a random environment and try to recognize the optimal action. There have been many researches on different type of learning models since 1974, such as parameterized learning automata (PLA), generalized learning automata (GLA), and continuous action-set learning automata (CALA)[7]. So far LA has been applied to solve many different practical problems.

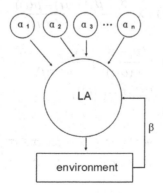

Fig. 1. The structure of LA

A typical finite-action LA is defined by a four parameters set $\{\alpha, \beta, p, T\}$ in which α is the action set $\{\alpha_1, \alpha_2, \alpha_3, ... \alpha_n\}$ (see Fig.1), p is the probability set of α, $\{p_1, p_2, p_3, ... p_n\}$, β denotes the environment feedback which is typically a variable between 0 and 1, T is the update scheme of the automata, which decides how the probability set $\{p_1, p_2, p_3, ... p_n\}$ will be updated after learning from the environment.

At time ' n ', LA stochastically choose one of its actions, α_i, according to their probability distribution, $p(n)$, apply the chosen α_i into the environment and get a feedback, $\beta(n)$. In a P model LA, $\beta(n) = 0$ means a reward from the environment while $\beta(n) = 1$ means getting a penalty. LA will then uses $\beta(n)$ to update the probability distribution $p(n+1)$, according to the update scheme T.

It has been proved that after finite iterations, the probability of one of these actions will be very close to 1, while the others will be down to 0. In this scenario we believe the LA has converged. If the selected 'optimal' action is actually optimal, then we say it is correctly converged. Different LA models have different converge accuracy and speed.

2.2   Continuous Action-set Learning Automata (CALA)

Finite action-set LA works in a discrete domain. If we need to optimize parameter in a continuous domain, then we must discretize the continuous space into several small intervals, such operation will reduce the accuracy. In fact there is a more satisfying LA solution to such models, that is continues action-set Learning Automata (CALA) [8][9].

The action probability distribution of CALA obeys a normal distribution $N(\mu(t), \sigma(t))$. At a certain time t, CALA stochastically chooses a point $x(t)$ on function $f(x)$ according to the probability distribution. Particularly, both $x(t)$ and $\mu(t)$ will be applied into the environment to get the feedback $\beta_x(t)$ and $\beta_\mu(t)$, then CALA will update $\mu(t+1)$ and $\sigma(t+1)$ according to a certain update scheme.

CALA use an update scheme as below:

$$\mu(t+1) = \mu(t) + \lambda \frac{(\beta_x - \beta_\mu)}{\varphi(\sigma(t))} \frac{(x(t) - \mu(t))}{\varphi(\sigma(t))} \tag{1}$$

$$\sigma(t+1) = \sigma(t) + \lambda \frac{(\beta_x - \beta_\mu)}{\varphi(\sigma(t))} [(\frac{x(t) - \mu(t)}{\varphi(\sigma(t))})^2 - 1]$$
$$+ \lambda\{C[\sigma_l - \sigma(k)]\} \tag{2}$$

where

$$\phi(\sigma) = \sigma_l \qquad for\ \sigma \le \sigma_l$$
$$= \sigma \qquad for\ \sigma > \sigma_l$$

To avoid non-optimal convergence, we should guarantee that $\sigma(t)$ will not converge to 0, so here a parameter $\sigma_l > 0$ (sufficiently small) is set as the lower bound of $\sigma(t)$.

It is proven [9] that with large probability, $\mu(t)$ will converge close to the maximum point of $f(x)$ and $\varphi(\sigma(t))$ will converge close to σ_l.

3 Proposed Algorithm

3.1 Latent Factor Model (LFM)

So far, many researchers have been done on classic recommendation algorithms, such as UserCF , ItemCF and Association Rule-based Algorithm. In the context Netflix Prize in 2006, Simon Funk proposed a new recommendation algorithm based on *Singular Value Decomposition* (SVD), which was later named Latent Factor Model Algorithm (LFM).Since then the application of SVD in recommendation algorithm have attracted a lot of attention. The key point of LFM is to discover the potential relevance between user preference and items by digging out the "latent factor", the model is as followers :

$$preference(u,i) = r_{ui} = p_u^T q_i = \sum_{k=1}^{F} p_{u,k} q_{i,k} \tag{3}$$

where F is the number of latent factor, $p_{u,k}$ denotes the u-th user's preference on the k-th latent factor, $q_{i,k}$ denotes the relevance between the i-th item and the k-th latent factor. Considering that in a real model, a rating system may has some inherent property which has nothing to do with users and items, a user may has some inherent property which has nothing to do with items, also an item may has some inherent property which has nothing to do with users, a modified Bias-SVD algorithm was proposed to Simulate the real situation :

$$preference(u,i) = \mu + b_u + b_i + p_u^T \cdot q_i = \mu + b_u + b_i + \sum_{k=1}^{F} p_{u,k} q_{i,k} \tag{4}$$

where μ is a mean value of all ratings in the training set, b_u is the average rating level of a particular user, and b_i is the average rating of a particular item. To compute parameter p and q, a optimization function in equation (5) is employed. Traditional algorithm

$$C = \sum_{(u,i) \in K} (r_{ui} - \hat{r}_{ui})^2 = \sum_{(u,i) \in K} (r_{ui} - \sum_{k=1}^{F} p_{u,k} q_{i,k})^2 + \lambda \|p_u\|^2 + \lambda \|p_i\|^2 \tag{5}$$

uses Gradient Descent Algorithm to gradually reduce the error, in this paper we will present another algorithm to perform this, which is referred to as CALA-SVD. The following experiment proved that the novel algorithm can achieve a good effect and CALA can perform well as a parameter optimizer in SVD.

3.2 Proposed CALA-SVD Algorithm

In the proposed algorithm, each value which needs training is regarded as an independent CALA, a mean value $\mu_{p,u,k}$ and a variance $\sigma_{p,u,k}$ is kept for each automata. The iteration step of CALA-SVD is showed as followers:

step1: Initialize each pair ($\mu_{p,u,k}$, $\sigma_{p,u,k}$) and ($\mu_{q,i,k}$, $\sigma_{q,i,k}$) randomly.

step2: For a certain user u and a certain item i, select $p_{u,k}$ and $q_{i,k}$ randomly according to their Gaussian probability distribution ($\mu_{p,u,k}$, $\sigma_{p,u,k}$) and ($\mu_{q,i,k}$, $\sigma_{q,i,k}$).

step3: Calculate the two estimates of $r_{u,i}$ respectively:

$$\hat{r}_{u,i} = \mu + b_u + b_i + \sum_{k=1}^{F} p_{u,k} q_{i,k} \tag{6}$$

$$\hat{r}_{u,i}(\mu) = \mu + b_u + b_i + \sum_{k=1}^{F} \mu_{p,u,k} \mu_{q,i,k} \tag{7}$$

step4: Use the environment feedback to update (μ, σ) pair by employing the CALA model:

$$\beta_r = (\hat{r}_{u,i} - r_{u,i})^2 \tag{8}$$

$$\beta_\mu = (\hat{r}_{u,i}(\mu) - r_{u,i})^2 \tag{9}$$

$$\mu_{p,u,k} = \mu_{p,u,k} + \lambda \frac{(\beta_r - \beta_\mu)}{\varphi(\sigma_{p,u,k})} \frac{(p_{u,k} - \mu_{p,u,k})}{\varphi(\sigma_{p,u,k})} \tag{10}$$

$$\mu_{q,i,k} = \mu_{q,i,k} + \lambda \frac{(\beta_r - \beta_\mu)}{\varphi(\sigma_{q,i,k})} \frac{(q_{i,k} - \mu_{q,i,k})}{\varphi(\sigma_{q,i,k})} \tag{11}$$

$$\sigma_{p,u,k} = \sigma_{p,u,k} + \lambda \frac{(\beta_r - \beta_\mu)}{\varphi(\sigma_{p,u,k})} [(\frac{(p_{u,k} - \mu_{p,u,k})}{\varphi(\sigma_{p,u,k})})^2 - 1] \tag{12}$$
$$+ \lambda \{C[\sigma_l - \sigma_{p,u,k}]\}$$

$$\sigma_{q,i,k} = \sigma_{q,i,k} + \lambda \frac{(\beta_r - \beta_\mu)}{\varphi(\sigma_{q,i,k})} [(\frac{(q_{i,k} - \mu_{q,i,k})}{\varphi(\sigma_{q,i,k})})^2 - 1] \tag{13}$$
$$+ \lambda \{C[\sigma_l - \sigma_{q,i,k}]\}$$

where parameters C, σ_l and λ have the same meaning as is introduced before in section 2.

step5: Go back to step 2 and iterates until β_r converges to a sufficiently small error.

Considering that each value to be trained reflects the relevance between a latent factor and an item or a user, it should never be a negative, so we give a threshold to the randomly generated $p_{u,k}$ and $q_{i,k}$, once they are out of range $[0,a]$, the algorithm will 'pull' them back by reinitiate them, here 'a' is a positive number.

Later experiment will show that each CALA can work independently and finally achieve a global convergence.

4   Experimental Results

To evaluate the proposed algorithm, we apply it to the dataset provided by MovieLens, whose movie rating data are all gathered from real users. Main parameters of this dataset are as follows:

1. User ID ranges between 1 and 6040.
2. Movie ID ranges between 1 and 3952.
3. Ratings are made on a 5-star scale (whole-star ratings only).
4. Each user has at least 20 ratings.
5. A total number of 1000209 rating pairs is included in the dataset.

Since we aim to get a 6040×3952 matrix as a training result, there is actually a 95.8% vacancy left to be filled in the original matrix.

As a preparation, we divide the dataset into two parts: the training set and the test set. The test set includes 500 users' rating which are randomly selected from the entire data set. For each selected user, we randomly select a half numbers of his/her ratings into the test set, while leaving the other half in the training set.

As same as Netflix Prize, here we employ RMSE to evaluate the average error of the test results, it is defined in equation (14):

$$RMSE = \frac{1}{N} \sum_{u=1}^{m} \sum_{i=1}^{n_u} abs(\hat{r}_{ui} - r_{ui}) \tag{14}$$

where m is the user numbers in test set, n_u is the rating numbers of the u-th user, abs(*) is the absolution function and $\hat{r}_{ui}$ denotes the reconstruction value of the i-th rating of the u-th user, while r_{ui} is the corresponding true value. $N = \sum_{u=1}^{m} n_u$ denotes the total number of ratings. Table 1 shows the experimental result:

Table 1. The experimental result

	RMSE
CALA-SVD	0.85
Bias-SVD	0.79

from Table 1 we can see that the novel CALA based algorithm can achieve a comparable RMSE with the traditional algorithm. Experiment result proved that it makes sense to apply CALA into SVD and recommendation system. Of course, further experimental is need to optimize the new algorithm.

5   Conclusion

This paper proposes a CALA-based algorithm to perform Singular value decomposition (SVD), and give a example of its application in recommendation system. To analyze the performance of the new algorithm, we tested it on the movie rating dataset and get a comparable result with the classical Bias-SVD algorithm, which employs gradient descent algorithm as optimization strategy. The experimental evidence has demonstrated that as a reinforcement learning model CALA can obtain a good convergence when applied in SVD. However, this paper does not aim to beat all the old methods proposed earlier, but to show that the CALA algorithm can effectively serve as an attractive method to successfully perform SVD in a Latent Factor Model and recommendation system. Of course there is still large space for further improvement.

Acknowledgement. This work is funded by National Science Foundation of China(61271316), 973 Program (2010CB731403, 2010CB731406, 2013CB329605) of China, Chinese National "Twelfth Five-Year" Plan for Science & Technology Support(2012BAH38 B04), Key Laboratory for Shanghai Integrated Information Security Management Technology Research, and Chinese National Engineering Laboratory for Information Content Analysis Technology.

References

[1] Breese, J.S., Heckerman, D., Kadie, C.: Empirical Analysis of Predictive Algorithms for Collaborative Filtering. In: Proceedings of the Fourteenth Conference on Uncertainty in Artificial Intelligence, pp. 43–52. Morgan Kaufmann Publishers Inc. (1998)

[2] Linden, G., Smith, B., York, J.: Amazon. com Recommendations: Item-to-Item Collaborative Filtering. IEEE Internet Computing 7(1), 76–80 (2003)

[3] Mobasher, B., Dai, H., Luo, T., et al.: Effective Personalization Based on Association Rule Discovery from Web Usage Data. In: Proceedings of the 3rd International Workshop on Web Information and Data Management, pp. 9–15. ACM (2001)

[4] Bell, R.M., Koren, Y.: Lessons from the Netflix prize challenge. ACM SIGKDD Explorations Newsletter 9(2), 75–79 (2007)

[5] Billsus, D., Pazzani, M.J.: Learning Collaborative Information Filters. In: ICML, vol. 98, pp. 46–54 (1998)

[6] Narendra, K.S., Thathachar, M.A.L.: Learning Automata:ASurvey. IEEE Trans. Syst. Man, Cybern. SMC-14, 323–334 (1974)

[7] Thathachar, M., Sastry, P.S.: Varieties of Learning Automata: An Overview. IEEE Transactions on Systems, Man, and Cybernetics, Part B: Cybernetics 32(6), 711–722 (2002)

[8] Santharam, G., Sastry, P.S., Thathachar, M.A.L.: Continuous Action-Set Learning Automata for Stochastic Optimization. J. Franklin Inst. 331, 607–628 (1994)

[9] Beigy, H., Meybodi, M.R.: A New Continuous Action-Set Learning Automaton for Function Optimization. Journal of the Franklin Institute 343(1), 27–47 (2006)

A Novel 2-Stage Combining Classifier Model with Stacking and Genetic Algorithm Based Feature Selection

Tien Thanh Nguyen[1], Alan Wee-Chung Liew[1], Xuan Cuong Pham[2], and Mai Phuong Nguyen[3]

[1] School of Information and Communication Technology, Griffith University, Australia
[2] School of Information and Communication Technology,
Hanoi University of Science and Technology, Vietnam
[3] College of Business, Massey University, New Zealand
tienthanh.nguyen2@griffithuni.edu.au

Abstract. This paper introduces a novel 2-stage classification system with stacking and genetic algorithm (GA) based feature selection. Specifically, Level1 data is first generated by stacking on the original data (called Level0 data) with base classifiers. Level1data is then classified by a second classifier (denoted by C) with feature selection using GA. The advantage of applying GA on Level1 data is that it has lower dimension and is more uniformity than Level0 data. We conduct experiments on both 18 UCI data files and CLEF2009 medical image database to demonstrate superior performance of our model in comparison with several popular combining algorithms.

Keywords: multi-classifier system, classifier fusion, combining classifiers, feature selection.

1 Introduction

Combining classifiers to improve classification performance has been shown to be an effective strategy [1-4]. In general, the more the differences in the training sets, base classifiers and feature sets, the better the combining output will be since they tend to capture different aspects of the classification task. Duin [1] summarized six combining strategies for multi-classifier system: (a) different initializations, (b) different parameter choices, (c) different architectures, (d) different classifiers, (e) different training sets, and (f) different feature sets. In a two stage classification model, the base classifiers are normally chosen to be significantly difference to each other. To train the base classifiers, Ting and Witten [5] developed the stacking algorithm where the training set is divided into several equal disjoint parts and each part plays as the test set once while the others play as the training set. The output of stacking-based algorithms is a set of fuzzy label [2], which gives the posterior probability that each observation belongs to an individual class according to each classifier. The set of

D.-S. Huang et al. (Eds.): ICIC 2014, LNAI 8589, pp. 33–43, 2014.
© Springer International Publishing Switzerland 2014

posterior probability of all observation is called Level1 data. Let N as the number of observations, K as the number of base classifiers and M as the number of classes $\{W_j\}$. For each observation X_i, $P_k(W_j \mid X_i)$ is the probability that X_i belongs to class W_j given by the kth classifier. The Level1 data of all observations, a $N \times MK$ - posterior probability matrix $\{P_k(W_j \mid X_i)\}$ $j = \overline{1,M}$ $k = \overline{1,K}$ $i = \overline{1,N}$, is defined by:

$$
\begin{bmatrix}
P_1(W_1 \mid X_1)... \; P_1(W_M \mid X_1) & ... & P_K(W_1 \mid X_1)...P_K(W_M \mid X_1) \\
P_1(W_1 \mid X_2)... \; P_1(W_M \mid X_2) & ... & P_K(W_1 \mid X_2)...P_K(W_M \mid X_2) \\
... & & ... \\
P_1(W_1 \mid X_N)...P_1(W_M \mid X_N) & ... & P_K(W_1 \mid X_N)...P_K(W_M \mid X_N)
\end{bmatrix}
\tag{1}
$$

The Level1 data of an observation X is defined by:

$$
Level1(X) := \begin{bmatrix}
P_1(W_1 \mid X) & ... & P_1(W_M \mid X) \\
\vdots & \ddots & \vdots \\
P_K(W_1 \mid X) & ... & P_K(W_M \mid X)
\end{bmatrix}
\tag{2}
$$

Based on stacking, several well-known combining classifiers algorithms were introduced. Kuncheva et al. [2] defined Decision Profile (denoted by $DP(X)$) of an observation X as equal to Level1(X) (eqn. 2) and Decision Template of ith class (denoted by DT_i) as average of Decision Profile of observations in training set where their labels are W_i. Based on measurement between DT_i ($i = \overline{1,M}$) and $DP(X)$, class label of X is predicted. Merz [11] proposed an algorithm called SCANN which first uses stacking result and the true label of learning set to build indicator matrix and applied Correspondence Analysis (CA) to that matrix to understand the relationship between observation and its class label. Then, K Nearest Neighbor is used to classify unlabeled observation on the Level1 data based on output of CA.

Kittler et al. [18] introduced fixed rules to combine outputs from classifiers. The difference between fixed rules and trainable combining algorithms is that fixed rule is pre-chosen and does not involve training on Level1 data. Six fixed combing rules evaluated in this paper are Sum, Product, Max, Min, Median and Majority Vote.

Two interesting questions related to multi-classifier system are whether there exist a subset of base classifiers and a subset of features that could achieve better classification results than the original sets? The first question is classifier selection problem. It means that in some situations, system with the elimination of some base classifiers may perform better than system with all base classifiers. The second question, which is the main subject of this paper, is feature selection. The purpose of this task is to reduce the number of features from feature set while still maintaining acceptable accuracy. In fact, some attributes of an object may have low discriminative power because of their nature and measurement bias. As a result, subset formed by removing these attributes may be more discriminative than its superset.

2   Recent Work

2.1   2-Stage Model

The idea of 2-stage model was published in [6] where K different feature vectors extracted from an object is used by K base classifiers respectively. However, extracting K feature vectors from an object is not always feasible so here we propose applying stacking algorithm to the 2-stage model as illustrated in Fig. 1. In the first stage, the dataset is classified by K base classifiers by stacking and their output is Level1 data (eqn. 1). Next, the Level1 data is classified by a new classifier (denoted by C in Fig.1). Level1 data can be viewed as "scaled results" given by K base classifiers (Level1 data is real value scaled in [0, 1]).

Our proposed model is very flexible in that the base classifiers can be chosen as different as possible (Strategy d) [1] while C is another arbitrary classifier. Here, the K base classifiers play the role of generating Level1 data from Level0 data by stacking, and then this new data is classified by C. Our aim is to gain lower error rate in comparison to when C is applied directly to Level0 data (classification directly by C), best result from all base classifiers, as well as popular combining classifier algorithms namely fixed rules, SCANN [11] and Decision Template [2].

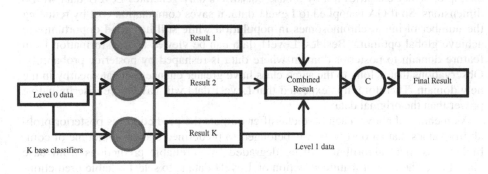

Fig. 1. Proposed 2-stage model based on stacking algorithm

2.2   GA in Multi-classifier System

Feature selection is an important technique in pattern recognition, data analysis and data mining [7]. Generally speaking, methods that transform features to a new domain with a reduction in the dimension of feature can be treated as feature selection. Therefore, strategy to solve this problem is very diverse, for instance, linear transformations, search techniques and Genetic Algorithm (GA). Here, we review literatures about GA for multi-classifier system.

Kuncheva et al. [4] introduced join and disjoin mechanism for GA approach. Firstly, chromosome was encoded by $\{0,1,...,K\}$ where k ($1 \le k \le K$) means that feature associated with this point is only used by k^{th} classifier and 0 is otherwise. In the second version, classifier encoding was added in the same chromosome with feature encoding, in which Venn diagram and integer values were used for feature and classifier encoding, respectively. In Kuncheva's experiment, the first version was not good while the second version is hard to implement since Venn diagram becomes more complicated with many classifiers. Nanni et al. [9] employed GA to improve SCANN algorithm [11] by building representations where each includes encoding of M classes. Gabrys et al. [8] tried to put classifier, feature and rule encoding in a single chromosome as a 3-dimensional cube. However, the above 2 approaches are hard to implement because of complicated crossover stage.

In our experiments, error rates of classification task on Level1 data of several data sources are not significantly improved compared with that of classification directly by C. It motivated us to apply GA on Level1data in order to achieve lower error rate. In the next section, we propose a novel chromosome encoding on Level1 data. After that, empirical evaluations are conducted on 18 UCI data files and CLEF 2009 medical image database to compare its performance with several existing algorithms. Finally, we conclude with some discussion about future developments.

3   2-Stage GA on Level1 Data

As mention above, Level1 data usually has fewer dimension than Level0 data, for example, binary classification by 3 base classifiers only generates Level1 data with 6 dimensions. So if GA is applied to Level1 data, it saves computation cost by reducing the number of initial chromosomes in population while still has good opportunity to achieve global optimum. Besides, Level1 data can be viewed as transformation from feature domain to posterior domain where data is reshaped by posterior probability. Observations that belong to the same class have greater chance to locate nearby in the new domain. Therefore, it is expected that Level1 data will have more discriminative power than the original data.

As mentioned above, each base classifier outputs the predictions as posterior probability values that an observation is belonged to predefined classes. Outcome of combining classifiers algorithm may be degraded by unreliable predictions from base classifiers. The idea of feature selection on Level1 data is to select reliable predictions of a base classifier for combining algorithms and discard unreliable predictions. As a result, the discriminant ability of new Level1 data is more power than the original Level1 data.

There are a numerous of feature selection methods introduced recently. They can be categorized in 3 main types namely filter, wrapper and embedded [17]. In this paper we focus on wrapper methods by introducing a GA-based approach to find an optimized subset from original Level1 data. Here, we propose the following chromosome structure. Each chromosome includes $M \times K$ genes due to the number of features of Level1 data. We use two elements $\{0,1\}$ to encode for each gene in a chromosome in which:

$$Gene(i) = \begin{cases} 1 & \text{if } i^{th} \text{ feature is selected} \\ 0 & \text{otherwise} \end{cases} \tag{3}$$

At crossover stage, we employ single splitter since the same single random point is selected on all pairs of parents. Each parent exchanges its head with the other while keeps its tail. After this stage, two new offspring chromosomes are generated. Next, based on mutation probability, we select random genes on two offspring chromosomes and change its values by $0 \rightarrow 1$ or $1 \rightarrow 0$. This mutation helps GA reach not only local extreme values but also global one. Here we select the accuracy of classifier C as fitness value of GA. Our GA approach is detailed below:

Algorithm 1: GA on Level1 data

Training procedure:
Input: Level0 data, K base classifiers, classifier C, TMax: number of interaction, PMul: mutation probability, L: population size
Output: optimal subset of Level1 feature and classification model.
Step1: Initialize a population with L random chromosomes
Step2: Compute fitness function of each initialized chromosome by cross validation method on Level1 data of training set and then classify by C.
Step3: Loop to select N best chromosomes.
Do
  • Withdraw with replacement L/2 pairs from population, conduct crossover and mutation (based on PMul) to generate new L chromosomes.
  • Add new L chromosomes to population.
  • Compute fitness value of new chromosomes
  • Select new population with L best fitness value chromosomes.
While (converge = true)
Step4: Select optimal chromosome from final population.
Step5: Build classification model based on encoding of best chromosome.

Test procedure:
Input: unlabeled observation XTest.
Output: predicted label of XTest.
Step1: Compute Level1 data of XTest
Step2: Select features from Level1 of XTest based on optimal chromosome encoding.
Step3: Classify selected features by C.

---

4   Experimental Results

We conducted experiments on several types of data namely UCI data files [13] and CLEF 2009 medical image database. Since only a single dataset was available for both training and testing, 10-fold cross validation was performed. The procedure was run 10 times so a total of 100 test results were obtained. In our assessment, we compared the error rates of 11 methods: six fixed rules (Sum, Product, Max, Min, Median and Majority Vote), best result from base classifiers, classification directly by C, 2-stage model, two well-known combining algorithms (Decision Template and SCANN) and our 2-stage model with Level1 feature selection by GA (2-stage GAFS). To initialize parameters of GA, we set mutation probability PMul = 0.015, population size L = 20. Three base classifiers namely Linear Discriminant Analysis (LDA), Naïve Bayes and K Nearest Neighbor (with K set to 5, denoted as 5-NN) are selected. The second classifier C is a Decision Tree. Finally, paired t-test was used (level of significance set to 0.05) on the results of our model and a benchmark algorithm for statistical significance.

4.1   UCI Data Files

We chose 18 data files with 2 classes (Bupa, Artificial, Sonar, etc...) to 6 classes (Dermatology). The number of attributes also changes in a wide range from only 3 (Haberman) to 60 (Sonar). The number of observations in each file also varies considerably, from small files like Fertility to quite big files such as Skin&NonSkin (Table1). Our purpose is to conduct an objective experiment to evaluate the advantage of our approach for a diversity of data sources. At the preprocessing step, we remove missing observations and encode attributes in several datasets to be available with Matlab environment. Experimental results on all 18 files are summarized in Table 2, 3, 4 and 5.

First, it is interesting to note that classification result from the 2-stage model is not significant better than classification directly by Decision Tree on Level0 data. There are 8 cases where direct classification by Decision Tree is better than the 2-stage model and 8 cases where performance is worse. The proposed 2-stage GAFS model, on the other hand, is significantly better than both Decision Tree and the 2-stage model. Our model obtained 12 wins and only 3 losses compared with the former and 11 wins and 0 loss compared with the latter (Table 6).

Next, we compare the 2-stage GAFS approach to the best result from three base classifiers. The 2-stage GAFS approach has superior performance in 6 files, namely Artificial (0.2313 vs. 0.2496), Balance (0.0938 vs. 0.1442), Fertility (0.125 vs. 0.155), Skin&NonSkin (4.15e-04 vs. 5e-04), Ring (0.1251 vs. 0.2374) and TicTacToe (0.1218 vs. 0.1557), while it has inferior performance in 3 files: Iris (0.036 vs. 0.02), Haberman (0.2748 vs. 0.2596) and Twonorm (0.0312 vs. 0.0217).

Then, two well-known combining classifiers algorithms, namely SCANN and Decision Template, are compared with our 2-stage GAFS model. Unfortunately, SCANN cannot be performed on Balance, Skin&NonSkin and Fertility files because the indicator matrix has columns with all 0 posterior probability values from the K base classifiers. As a result, its column mass will be singular and the standardized residuals is not available. These files are not included in the comparison with our model. SCANN outperforms 2-stage GAFS model on 4 files and underperforms on 6 files. Decision Template, in turn, obtains only 1 win and 7 losses among the 18 files. Significant results are Phoneme (0.1133, 0.1462 and 0.1229), Ring (0.1251, 0.1894 and 0.2150) and TicTacToe (0.1218, 0.1304 and 0.1680) for 2-stage GAFS, Decision Template and SCANN, respectively. Besides, comparing with 6 fixed combining rules, our model achieves 7 wins and 4 losses (Sum rules), 8 wins and 4 losses (Product Rules) and 9 wins and 3 losses (Max, Min, Median and Majority Vote rule).

Finally, the dimensions of Level0 data, Level1 data and data of 2-stage GAFS model are compared. From Fig. 2 we can see that GA helps reduce the number of features significantly while error rate is maintained at acceptable value (Table 5). There are only 4 files, Iris, Balance, Titanic and Page Blocks, where GAFS approach gives higher dimension than that of Level0 data; i.e. 1, 3, 2 and 1 higher, respectively. However, in the others files, the proposed GA outperforms Level0 data, especially on 4 files: Sonar (1 vs. 60), Artificial (1 vs. 10), Ring (1 vs. 20) and Twonorrn (2 vs. 20).

Table 1. UCI data files used in our experiment

	File name	# of attributes	Attribute Type	# of observations	# of classes	# of attribute on Level1 (3 classifiers)
1	Bupa	6	C,I,R	345	2	6
2	Artificial	10	R	700	2	6
3	Iris	4	R	150	3	9
4	Sonar	60	R	208	2	6
5	Phoneme	5	R	540	2	6
6	Haberman	3	I	306	2	6
7	Titanic	3	R,I	2201	2	6
8	Balance	4	C	625	3	9
9	Page Blocks	10	R,I	5473	5	15
10	Dermatology	34	C,I	358	6	18
11	Cleveland	13	C,I,R	270	5	15
12	Fertility	9	R	100	2	6
13	Skin&NonSkin	3	R	245057	2	6
14	Wdbc	30	R	569	2	6
15	Ring	20	R	7400	2	6
16	Twonorm	20	R	7400	2	6
17	Magic	10	R	19020	2	6
18	TicTacToe	9	C	958	2	6

Table 2. Error rates and variances of three base classifiers on UCI files

File name	LDA		Naive Bayes		5-NN		Best result from base classifiers	
	Mean	Variance	Mean	Variance	Mean	Variance	Mean	Variance
Bupa	0.3693	8.30E-03	0.4264	7.60E-03	0.3331	6.10E-03	0.3331	6.10E-03
Artificial	0.4511	1.40E-03	0.4521	1.40E-03	0.2496	2.40E-03	0.2496	2.40E-03
Iris	0.0200	1.40E-03	0.0400	2.30E-03	0.0353	1.50E-03	0.0200	1.40E-03
Sonar	0.2629	9.70E-03	0.3042	7.40E-03	0.1875	7.60E-03	0.1875	7.60E-03
Phoneme	0.2408	3.00E-04	0.2607	3.00E-04	0.1133	2.00E-04	0.1133	2.00E-04
Haberman	0.2669	4.50E-03	0.2596	4.40E-03	0.2829	3.80E-03	0.2596	4.40E-03
Titanic	0.2201	5.00E-04	0.2515	8.00E-04	0.2341	3.70E-03	0.2201	5.00E-04
Balance	0.2917	2.90E-03	0.2600	3.30E-03	0.1442	1.20E-03	0.1442	1.20E-03
Page Blocks	0.0641	1.11E-04	0.1136	2.17E-04	0.0432	6.15E-05	0.0432	6.15E-05
Dermatology	0.0363	9.00E-04	0.0422	1.10E-03	0.1145	1.90E-03	0.0363	9.00E-04
Cleveland	0.4589	4.90E-03	0.4688	5.10E-03	0.5558	4.70E-03	0.4589	4.90E-03
Fertility	0.3460	2.01E-02	0.3770	2.08E-02	0.1550	4.50E-03	0.1550	4.50E-03
Skin&NonSkin	0.0659	2.74E-06	0.1785	6.61E-06	0.0005	1.68E-08	0.0005	1.68E-08
Wdbc	0.0397	7.00E-04	0.0587	1.20E-03	0.0666	8.00E-04	0.0397	7.00E-04
Ring	0.2381	2.27E-04	0.2374	2.23E-04	0.3088	1.30E-04	0.2374	2.23E-04
Twonorm	0.0217	3.12E-05	0.0217	3.13E-05	0.0312	3.96E-05	0.0217	3.12E-05
Magic	0.2053	6.85E-05	0.2255	7.33E-05	0.1915	4.81E-05	0.1915	4.81E-05
TicTacToe	0.4170	2.95E-03	0.4152	2.98E-03	0.1557	9.09E-04	0.1557	9.09E-04

Table 3. Error rates and variances of Sum, Product, Max rule

File name	Sum rule		Product rule		Max rule	
	Mean	Variance	Mean	Variance	Mean	Variance
Bupa	0.3028	4.26E-03	0.3021	4.12E-03	0.2986	4.15E-03
Artificial	0.2230	2.06E-03	0.2193	2.05E-03	0.2450	2.57E-03
Iris	0.0387	2.59E-03	0.0407	2.39E-03	0.0440	3.13E-03
Sonar	0.2259	9.55E-03	0.2285	9.81E-03	0.2260	7.01E-03
Phoneme	0.1713	1.90E-04	0.1518	2.87E-04	0.1407	1.95E-04
Haberman	0.2392	2.39E-03	0.2424	3.08E-03	0.2457	3.18E-03
Titanic	0.2167	6.01E-04	0.2167	5.65E-04	0.2167	6.59E-04
Balance	0.1113	5.55E-04	0.1131	4.95E-04	0.1112	4.82E-04
Page Blocks	0.0513	5.00E-05	0.0463	4.89E-05	0.0489	4.41E-05
Dermatology	0.0352	9.18E-04	0.0447	9.81E-04	0.0327	8.60E-04
Cleveland	0.4280	3.05E-03	0.4453	3.53E-03	0.4346	2.65E-03
Fertility	0.1290	2.46E-03	0.1290	2.26E-03	0.1270	1.97E-03
Skin&NonSkin	0.0412	1.40E-06	0.0006	2.73E-08	0.0006	2.22E-08
Wdbc	0.0401	7.07E-04	0.0517	8.19E-04	0.0485	8.03E-04
Ring	0.2122	1.62E-04	0.2275	1.09E-04	0.2436	1.53E-04
Twonorm	0.0221	3.00E-05	0.0225	2.69E-05	0.0231	2.39E-05
Magic	0.1925	5.31E-05	0.1921	5.16E-05	0.1911	6.40E-05
TicTacToe	0.2376	9.79E-04	0.2288	6.47E-04	0.2090	9.74E-04

Table 4. Error rates and variances of Min, Median, Majority Vote rule

File name	Min rule		Median rule		Majority vote	
	Mean	*Variance*	*Mean*	*Variance*	*Mean*	*Variance*
Bupa	0.2970	4.89E-03	0.3428	4.46E-03	0.3429	4.04E-03
Artificial	0.2453	2.90E-03	0.3089	1.36E-03	0.3073	1.03E-03
Iris	0.0413	2.56E-03	0.0333	1.64E-03	0.0327	1.73E-03
Sonar	0.2298	9.32E-03	0.2104	1.00E-02	0.2079	8.16E-03
Phoneme	0.1417	2.10E-04	0.2254	2.71E-04	0.2257	2.73E-04
Haberman	0.2461	2.47E-03	0.2524	1.67E-03	0.2504	1.76E-03
Titanic	0.2167	5.00E-04	0.2216	5.25E-04	0.2217	4.61E-04
Balance	0.1232	4.99E-04	0.1155	4.93E-04	0.1261	4.63E-04
Page Blocks	0.0466	4.72E-05	0.0539	5.24E-05	0.0544	4.54E-05
Dermatology	0.0439	9.53E-04	0.0380	9.49E-04	0.0369	7.95E-04
Cleveland	0.4517	3.84E-03	0.4209	3.80E-03	0.4207	4.26E-03
Fertility	0.1280	2.02E-03	0.1330	2.81E-03	0.1310	2.34E-03
Skin&NonSkin	0.0006	2.13E-08	0.0528	2.23E-06	0.0528	1.90E-06
Wdbc	0.0522	7.71E-04	0.0395	5.03E-04	0.0406	6.47E-04
Ring	0.2437	1.33E-04	0.2368	1.93E-04	0.2365	2.00E-04
Twonorm	0.0233	3.92E-05	0.0217	2.74E-05	0.0216	2.82E-05
Magic	0.1905	5.72E-05	0.2004	5.81E-05	0.2006	7.49E-05
TicTacToe	0.2074	7.66E-04	0.3063	1.33E-03	0.3062	1.43E-03

Table 5. Error rates and variances of our model and the other approches on UCI files

File name	Decision Tree		2-stage model		2-stage GAFS		Decision Template		SCANN	
	Mean	Variance	Mean	Variance	Mean	Variance	Mean	Variance	Mean	Variance
Bupa	0.3514	6.10E-03	0.3748	6.30E-03	0.3467	4.41E-03	0.3348	7.10E-03	0.3304	4.29E-03
Artificial	0.2414	2.20E-03	0.2884	3.20E-03	0.2313	1.71E-03	0.2433	1.60E-03	0.2374	2.12E-03
Iris	0.0507	2.40E-03	0.0287	1.50E-03	0.0360	2.79E-03	0.0400	2.50E-03	0.0320	2.00E-03
Sonar	0.2866	6.20E-03	0.2374	9.70E-03	0.2042	7.00E-03	0.2129	8.80E-03	0.2128	8.01E-03
Phoneme	0.1298	2.00E-04	0.1548	3.00E-04	0.1133	1.56E-04	0.1462	2.00E-04	0.1229	6.53E-04
Haberman	0.3142	4.70E-03	0.3296	6.80E-03	0.2748	1.68E-03	0.2779	5.00E-03	0.2536	1.74E-03
Titanic	0.2101	3.00E-04	0.2149	3.00E-04	0.2158	5.60E-04	0.2167	6.00E-04	0.2216	6.29E-04
Balance	0.2107	2.10E-03	0.0960	1.00E-03	0.0938	1.36E-03	0.0988	1.40E-03	x	x
Page Blocks	0.0351	5.37E-05	0.0429	5.09E-05	0.0420	4.80E-05	0.0513	1.00E-04	0.0512	5.10E-05
Dermatology	0.0528	1.30E-03	0.0427	1.00E-03	0.0391	7.51E-04	0.036	7.00E-04	0.0372	8.29E-04
Cleveland	0.5055	6.30E-03	0.4649	4.60E-03	0.4483	4.78E-03	0.4432	3.40E-03	0.4120	3.70E-03
Fertility	0.1730	7.20E-03	0.2050	1.11E-02	0.1250	2.08E-03	0.4520	3.41E-02	x	x
Skin&NonSkin	4.16e-04	1.55E-08	0.0007	2.52E-08	4.15e-04	1.05E-08	0.0332	1.64E-06	x	x
Wdbc	0.0705	1.10E-03	0.0404	7.00E-04	0.0364	4.91E-04	0.0385	5.00E-04	0.0397	5.64E-04
Ring	0.2485	1.32E-04	0.1525	1.77E-04	0.1251	1.07E-04	0.1894	1.78E-04	0.2150	2.44E-04
Twonorm	0.0536	4.22E-05	0.0314	3.57E-05	0.0312	3.50E-05	0.0221	2.62E-05	0.0216	2.39E-05
Magic	0.1777	8.80E-05	0.2430	9.64E-05	0.1919	6.60E-05	0.1927	7.82E-05	0.2002	6.14E-05
TicTacToe	0.1490	1.53E-03	0.1480	1.34E-03	0.1218	6.91E-04	0.1304	8.55E-04	0.1680	1.27E-03

Table 6. Statistical test compares 2-stage GAFS with the benchmarks on UCI files

	Better	Competitive	Worse
2-stage GAFS vs. Sum rule	7	7	4
2-stage GAFS vs. Product rule	8	6	4
2-stage GAFS vs. Max rule	9	6	3
2-stage GAFS vs. Min rule	9	6	3
2-stage GAFS vs. Median rule	9	6	3
2-stage GAFS vs. Majority Vote rule	9	6	3
2-stage GAFS vs. Decision Template	7	10	1
2-stage GAFS vs. SCANN	6	5	4
2-stage GAFS vs. best result from base classifiers	6	9	3
2-stage GAFS vs. Decision Tree	12	3	3
2-stage GAFS vs. 2-stage model	11	7	0
Decision Tree vs. 2-stage model	8	2	8

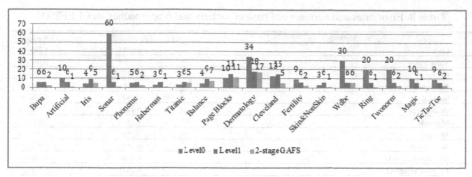

Fig. 2. Comparing dimensions of data on UCI files

4.2   CLEF2009

The CLEF2009 medical image dataset collected by Archen University includes 15363 medical images which are allocated to 193 hierarchical categories. Information about images in this dataset and their class is showed in Table 7. The images are first histogram equalized and then the feature vector of image is extracted using Histogram of Local Binary Pattern (HLBP) [12]. The base classifiers we used are LDA, Quadratic Discriminant Analysis (QDA) and Naïve Bayes, and C is 5-NN. By changing base classifiers and C, we want to illustrate the flexible characteristic of our model. Experimental results are illustrated in Table 8.

Again, the benefits of our 2-stage GAFS approach are obvious. It helps reduce both the dimensions of feature on Level1 data and the error rates on the 10-class dataset. Our 2-stage GAFS outperforms (paired t-test with level of significance set to 0.05) best result from base classifiers, 5 fixed rules (Sum, Max, Min, Median and Majority Vote) and 5-NN and is competitive with 2-stage model, Product rule, Decision Template and SCANN. Besides, the dimension of data is reduced from 32 to 30 and then 25 in the case of Level0 data, Level1data and 2-stage GAFS, respectively.

Table 7. Information of CLEF2009 medical images

Image					
Description	Abdomen	Cervical	Chest	Facial cranium	Left Elbow
Number of observation	80	81	80	80	69
Image					
Description	Left Shoulder	Left Breast	Finger	Left Ankle Joint	Left Carpal Joint
Number of observation	80	80	66	80	80

Table 8. Error rates and variances of base classifiers and 6 approaches on CLEF2009

Methods	Mean	Variance
LDA	0.1684	1.24E-03
Naïve Bayes	0.3694	3.10E-03
QDA	0.1732	1.78E-03
Best result from base classifiers	0.1684	1.24E-03
Sum rule	0.1525	1.47E-03
Product rule	0.1428	1.13E-03
Max rule	0.1572	1.63E-03
Min rule	0.1567	1.72E-03
Median rule	0.1656	1.25E-03
Majority Vote rule	0.1689	1.37E-03
Decision Template	0.1443	1.40E-03
SCANN	0.1528	1.62E-03
5-NN	0.2893	2.21E-03
2-stage model	0.1459	1.40E-03
2-stage GAFS	0.1440	1.39E-03

5   Conclusion and Future Work

In this paper, we have introduced a novel multi-classifier system based on stacking algorithm and GA feature selection in which GA feature selection is applied on Level1 data. Our aim is not only to build an effective classification model but also to find a subset of Level1 data which have better discriminative power than its superset. Extensive experiments on 18 UCI data files and CLEF2009 have demonstrated the benefits of our 2-stage GA feature selection approach on Level1 data. Although our approach is more time-consuming than other combining classifiers methods due to feature selection strategy by GA approach.

In the near future, we plan to improve our model by combining it with GA approach on Level0 data to solve both the feature and classifier selection problems in multi-classifier system. It is expected to result in significantly better combining classifiers model.

References

1. Duin, R.P.W.: The Combining Classifier: To Train or Not to Train? In: 16th International Conference on Pattern Recognition, vol. 2, pp. 765–770 (2002)
2. Kuncheva, L.I., Bezdek, J.C., Duin, R.P.W.: Decision Templates for Multi Classifier Fusion: An Experimental Comparison. Pattern Recognition 34(2), 299–314 (2001)
3. Kuncheva, L.I.: A theoretical Study on Six Classifier Fusion Strategies. IEEE Transactions on Pattern Analysis and Machine Intelligence 24(2) (2002)
4. Kuncheva, L.I., Jain, L.C.: Designing Classifier Fusion Systems by Genetic Algorithms. IEEE Transactions on Evolution Computation 4(4) (2000)
5. Ting, K.M., Witten, I.H.: Issues in Stacked Generation. Journal of Artificial in Intelligence Research 10, 271–289 (1999)
6. Lepisto, L., Kunttu, I., Autio, J., Visa, A.: Classification of Non-Homogeneous Texture Images by Combining Classifier. In: Proceedings International Conference on Image Processing, vol. 1, pp. 981–984 (2003)

7. Raymer, M.L., Punch, W.F., Goodman, E.D., Kuhn, L.A., Jain, A.K.: Dimensionality Reduction using Genetic Algorithms. IEEE Transactions on Evolutionary Computation 4(2) (2000)
8. Gabrys, B., Ruta, D.: Genetic Algorithms in Classifier Fusion. Applied Soft Computing 6, 337–347 (2006)
9. Nanni, L., Lumini, A.: A Genetic Encoding approach for Learning Methods for Combining Classifiers. Expert Systems with Applications 36, 7510–7514 (2009)
10. Seeward, A.K.: How to Make Stacking Better and Faster While Also Taking Care of an Unknown Weakness. In: Proceedings of the Nineteenth International Conference on Machine Learning, pp. 554–561 (2002)
11. Merz, C.: Using Correspondence Analysis to Combine Classifiers. Machine Learning 36, 33–58 (1999)
12. Ko, B.C., Kim, S.H., Nam, J.Y.: X-ray Image Classification Using Random Forests with Local Wavelet-Based CS-Local Binary Pattern. J. Digital Imaging 24, 1141–1151 (2011)
13. UCI Machine Learning Repository,
    http://archive.ics.uci.edu/ml/datasets.html
14. Rokach, L.: Taxonomy for characterizing ensemble methods in classification tasks: A review and annotated bibliography. Journal of Computational Statistics & Data Analysis 53(12), 4046–4072 (2009)
15. Zhang, L., Zhou, W.D.: Sparse ensembles using weighted combination methods based on linear programming. Pattern Recognition 44, 97–106 (2011)
16. Sen, M.U., Erdogan, H.: Linear classifier combination and selection using group sparse regularization and hinge loss. Pattern Recognition Letters 34, 265–274 (2013)
17. Saeys, Y., Inza, I., Larrañaga, P.: A review of feature selection techniques in bioinformatics. Bioinformatics 23(19), 2507–2517 (2007)
18. Kittler, J., Hatef, M., Duin, R.P.W., Matas, J.: On Combining Classifiers. IEEE Transactions on Pattern Analysis and Machine Intelligence 20(3) (1998)

Improved Bayesian Network Structure Learning with Node Ordering via K2 Algorithm[*]

Zhongqiang Wei, Hongzhe Xu, Wen Li, Xiaolin Gui, and Xiaozhou Wu

Shaanxi Key Lab of Computer Network, Xi'an Jiaotong University, Xi'an, China
{xuhz,leewhen,xlgui}@mail.xjtu.edu.cn

Abstract. The precise construction of Bayesian network classifier from database is an NP-hard problem and still one of the most exciting challenges. K2 algorithm can reduce search space effectively, improve learning efficiency, but it requires the initial node ordering as input, which is very limited by the absence of the priori information. On the other hand, search process of K2 algorithm uses a greedy search strategy and solutions are easy to fall into local optimization. In this paper, we present an improved Bayesian network structure learning with node ordering via K2 algorithm. This algorithm generates an effective node ordering as input based on conditional mutual information. The K2 algorithm is also improved combining with Simulated Annealing algorithm in order to avoid falling into the local optimization. Experimental results over two benchmark networks Asia and Alarm show that this new improved algorithm has higher classification accuracy and better degree of data matching.

Keywords: Bayesian Network Classifier, Structure Learning, Search Strategy, Conditional Mutual Information, K2 Algorithm.

1   Introduction

Pearl [1] put forward Bayesian network in 1980s, after that Bayesian network was widely applied to the classification process of data mining. The process of constructing Bayesian network classifier can be generally divided into three steps: (1)Learning Bayesian network topological structure from the training data which represents the dependent relationships between variables. (2)Learning Conditional Probability Table based on network structure. (3)Classifying and making decisions through finding the maximum posteriori probability. Among these three steps, Bayesian network structure learning is the key to the construction of Bayesian network model.

Bayesian network structure learning can be divided into two main categories: learning algorithms based-on dependent relationships [2, 3] and learning algorithms based-on scoring functions [4-6]. Learning algorithms based-on dependency analysis construct Bayesian network structure through the establishment of the conditional

---

[*] This research is supported by the central university basic scientific research business (XKJC2014008).

D.-S. Huang et al. (Eds.): ICIC 2014, LNAI 8589, pp. 44–55, 2014.

dependent relationships between nodes. The learning process of this kind of algorithm is more intuitive, conditional independence test and the search of network structure can be separated, but these algorithms are oversensitive with the error of condition independence test. Learning algorithms based-on scoring functions view learning as the optimization process, and select the best network structure which maximizes the scoring value. Generally network structures learned by this kind of algorithms have high precision, but structures are inclined to fall into local optimum.

K2 algorithm is a classical Bayesian network structure learning algorithm based-on scoring search proposed by Cooper and Herskovits [6], which combines Bayesian scoring method and hill-climbing search strategy. When using K2 algorithm, it has to determine the initial node ordering and maximal number of parents of each node, which are hard to obtain in many practical applications. In order to solve these problems, researchers have put forward many improved algorithms [7-9], the application effects of these algorithms are not very ideal. K2 algorithm is computationally efficient given an initial node ordering, which reduces the search space significantly. But, an improper node ordering may give poor results. So, it is of great importance to provide the K2 algorithm with an effective node ordering. David et al. [10] introduced polynomial algorithms for finding the highest scoring network structures in the special case where every node has at most k=1 parent. These kinds of algorithms limit parent number of each to great extent, consequently resulting network structures are close to a maximum spanning tree algorithm, which reflects the relationships between the nodes not well. Chen et al. [3] applied mutual information to determine node ordering and K2 algorithm to search node ordering space to learn network structure. But this algorithm didn't use class attribute information to obtain initial node ordering, which play an important role in node ordering. Simulated Annealing (SA) [11] algorithm is a search-score learning algorithm based-on the Monte Carlo approach. It has better global optimization ability, but its efficiency is low. Combining K2 algorithm and Simulated Annealing algorithm not only can be avoidance of the local optimum, but can also improve learning efficiency. In those algorithms based-on scoring search, researchers have proposed many scoring methods [12-15]. In this paper, we use the likelihood-equivalence Bayesian Dirichlet score metric [12] as the evaluation standard of network structures.

In this paper, the method of an effective node ordering is put forward based on conditional mutual information (CMI). Then an improved K2 algorithm called K2-SA algorithm is used to learn Bayesian Network structure and a heuristic Bayesian Network classification algorithm is proposed. Finally experimental results show the effectiveness of this algorithm.

This paper is organized as follows. Section 2 explains how to generate the initial node ordering based on conditional mutual information in detail. The improved algorithm K2-SA is given in Section 3. Experimental results of two benchmark network data sets with known structure are presented in Section 4. Finally, conclusion of this paper is given.

2   An Effective Node Ordering Based-On Conditional Mutual Information

Learning a Bayesian network structure is a time-consuming task. When providing an initial node ordering, K2 algorithm can improve learning efficiency by reducing search space. However, the performance of K2 algorithm is greatly affected by the initial node ordering. This paper presents a method to obtain an effective node ordering. It consists of three main components. Firstly, obtain a weighted undirected graph structure based on conditional mutual information. Secondly, assign orientations for every edge of the undirected structure. Thirdly, establish a maximal weight spanning tree (MWST) based on the directed graph.

2.1   Obtain a Weighted Undirected Graph Structure Based on CMI

Firstly, we obtain an undirected network structure based on conditional mutual information and the connectivity is established in the network. The values of CMI among pairs of random variables are assigned to edges in the network as the weight. Then a weighted undirected network structure can be obtained.

The degree of dependence between two random variables can be acquired by evaluating CMI value between these two variables. The high CMI value between variables X_i and $X_j (1 \le i, j \le n, n$ is the number of random variables) given variable Z represents variable X_i and X_j have a great degree of mutual dependence under the condition of giving variable Z. This dependence can be used to establish connections between variables.

The CMI between two random variables X_i and X_j given random variable Z, denoted by $I(X_i; X_j \mid Z)$, is mathematically defined as follows:

$$I(X_i; X_j \mid Z) = \sum_{k}^{q_z} \sum_{l}^{q_i} \sum_{m}^{q_j} P(x_l, y_m \mid z_k) \log \frac{P(x_l, y_m \mid z_k)}{P(x_l \mid z_k)P(y_m \mid z_k)}, i \ne j \quad (1)$$

where q_z represents the number of all the possible values of variable Z, z_k represents all the possible values of variable Z, q_i represents the number of all the possible values of variable X_i, x_l represents all the possible values of variable X_i, q_j represents the number of all the possible values of variable X_j, y_m represents all the possible values of variable X_j, $i \ne j$ represents random variables X_i and X_j cannot be the same variable, both variable X_i and X_j are different from variable Z. Importantly, the CMI is symmetric in nature, that is $I(X_i; X_j \mid Z) = I(X_j; X_i \mid Z)$.

To obtain an undirected graph structure, the CMI is computed between every two node pairs for all the nodes in the network we except to construct. Assuming that a given variable set $X = \{X_1, X_2, ..., X_n\}$ which has n variables or attributes, and a class attribute C. Then the CMI value between any two variables X_i and X_j can be computed as $I(X_i; X_j \mid C), 1 \le i, j \le n, i \ne j$. An edge is added between these two nodes or variables and the weight is assigned to $I(X_i; X_j | C)$. Because of the symmetric nature of CMI, the edges established in the graph structure are undirected. Compute CMI for each node or variable pairs, and add the undirected edges and the corresponding weights. After that, a weighted undirected complete graph structure can be obtained.

2.2   Assign Orientations for Every Edge of the Undirected Structure

This phase is to assign orientations for every edge of the undirected graph structure in order to obtain the initial node ordering information. The conditional relative average entropy (CRAE) [16] is used to determine orientations for edges. For two random variables X_i and X_j, their CRAE is defined as:

$$CRAE(X_j \to X_i) = \frac{H(X_i | X_j)}{H(X_i) \cdot | X_i |}, i \neq j \tag{2}$$

where $H(X_i) = -\sum_{x_l} P(x_l) \log P(x_l))$ is the entropy of random variable X_i, $H(X_i | X_j) = -\sum_{x_l, x_m} P(x_l, x_m) \log P(x_l | x_m)$ is the conditional entropy of random variable X_i given random variable X_j, $| X_i |$ represents the number of all possible values of X_i, x_l represents all possible values of X_i, x_m represents all possible values of X_j.

For two nodes X_i and X_j, if $CRAE(X_j \to X_i) > CRAE(X_i \to X_j)$ the orientation of edge between nodes X_i and X_j is $X_j \to X_i$. If $CRAE(X_i \to X_j) > CRAE(X_j \to X_i)$, the edge orientation between nodes X_i and X_j is $X_i \to X_j$. If $CRAE(X_i \to X_j) = CRAE(X_j \to X_i)$, the edge between those two nodes is not assigned the orientation at this point. On this occasion, we will use Bayesian score method to set the orientations for the remaining undirected edges after this step.

By using CRAE, most edges can be set the orientations and the number of edges whose orientations are remaining to be assigned is small. Thus, the exhaustive search combining with Bayesian score method is used to determine the remaining undirected edges' orientations.

After the above steps, a weighted directed graph structure can be acquired. In the next section, this directed graph is used to establish a maximal weight spanning tree (MWST) in order to obtain the initial node ordering.

2.3   Establish a MWST Based on the Weighted Directed Graph

This phase is to construct the MWST of the weighted directed graph. Assuming that the initial graph is $G(V, E)$ and the generated MWST is $T(U, D)$, the process is as follows:

- Initialize U and D as the empty set, that is $U \leftarrow \varnothing, D \leftarrow \varnothing$.
- Select a node u_0 randomly from V in the graph G and put this node into G. Then $U = \{u_0\}$ and $V = V - \{u_0\}$. A directed edge $<u_0, v>$ or $<v, u_0>$ that relates to node u_0 and has maximal weight is selected. The node v is added into U, then $U = U \cup \{v\}$ and $V = V - \{v\}$. This directed edge is added into D.
- Select a directed edge $<x, y>$ in the E from graph G, which satisfies the following conditions: 1) one of the two nodes associated with this edge is in V, the other one is in U; 2) the in-degree of the arc head of this edge is 0; 3) this edge has the maximal weight among all the directed edges which satisfy condition 1) and 2). The selected edge $<x, y>$ is added into D and the node that associates with this edge and is absent in U is added into U.
- Repeat the third step until U contains all the nodes.

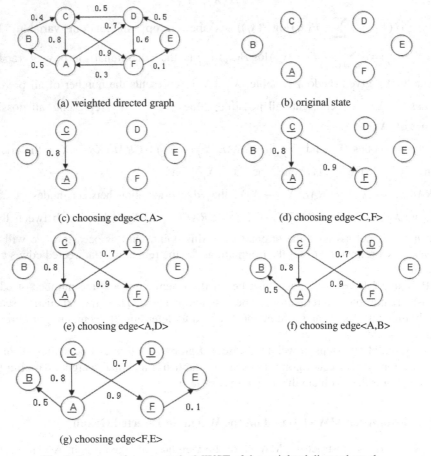

(a) weighted directed graph

(b) original state

(c) choosing edge<C,A>

(d) choosing edge<C,F>

(e) choosing edge<A,D>

(f) choosing edge<A,B>

(g) choosing edge<F,E>

Fig. 1. Process of construct the MWST of the weighted directed graph

After the above steps, a MWST based on the directed graph can be obtained. The initial node ordering can be determined according to this MWST. As shown in Fig.1, node A is selected as initial node randomly, and weights of directed edges are set as (a) in Fig.1. The resulting MWST is shown as Fig.1.(g), which the initial node ordering can be obtained according to it. For example, a topological sort algorithm can be used to obtain node ordering, which will generate many optional node ordering plans. Sorting directed edges according to weight, node ordering can be determined more accurate. In Fig.1, the node ordering is unique, that is CFABDE. This obtained node ordering will be used to learn the Bayesian network structure effectively in order to construct Bayesian network model.

3   K2-SA Algorithm

K2 algorithm with greedy search strategy can reduce the solution space significantly, but it often can't get the global optimal solution. SA algorithm is a random optimization algorithm, using the Metropolis criterion with probabilistic jumping properties to find the global optimal solution of the objective function randomly in the solution space. SA algorithm utilizes the similarity between annealing process and combination optimal problem to simulate the energy of physical system into the optimization problem of objective function, thereby search the global optimal value of optimization problem. The characteristic of SA algorithm can remedy the insufficiency of greedy search strategy that falls into local optimal solution easily. The algorithm combining K2 and SA algorithm (K2-SA) controls solving process to the optimization direction of minimum, and can escape from local extreme points with accepting inferior solutions at a certain probability. As long as the initial temperature is high enough and the annealing process is slow enough, this algorithm can converge to the global optimal solution. Therefore, K2 and SA algorithm can be combined to learn more optimal network structure.

When giving the node ordering, K2-SA algorithm searches the set of parents for each node orderly in order to establish the network structure. For any node X_i, only nodes appearing before node X_i in the node ordering can be the parents of node X_i. For example, node X_j comes prior to node X_i in the ordering, then node X_j can be a parent of node X_i, but node X_i cannot be a parent of node X_j.

The process of using K2-SA algorithm to choose optimal set of parents for each node is as following:

- Initialize the parent node set $Pa(X_i)$ of node X_i an empty set, which is $Pa(X_i) \leftarrow \varnothing$.
- Calculate evaluation function:

$$f(S) = CH(< X_i, Pa(X_i) >| D) = \sum_{j=1}^{q_i} [\log \frac{\Gamma(\alpha_{ijk})}{\Gamma(\alpha_{ij} + N_{ij})} + \log \frac{\Gamma(\alpha_{ijk} + N_{ijk})}{\Gamma(\alpha_{ijk})}] \qquad (3)$$

- where q_i represents all the configuration number of parents of node X_i, N_{ijk} is the sample number of the i th variable when the configuration of its parents is the j th and its value is the k th, N_{ij} is the sample of the i th variable when the configuration of its parents is the j th, that is $N_{ij} = \sum_k N_{ijk}$, α_{ijk} represents the hyper-parameter in the prior distribution.
- Pick the score function value minus and convert it to energy representation of simulated annealing, that is $E_{old} = -f(S)$. E_{old} represents the energy of simulated annealing, and lower energy value has higher probability of acceptance.
- Set initial temperature T_0, let current temperature $T_k = T_0$, and set minimum temperature T_g.
- Set the iteration counter $l = 0$ and the termination condition of iteration is $l_e = L(x)$, where $L(x)$ is variable step function.
- Under the current temperature T_k, perform the following actions:

— Step1: Sequentially select all nodes which are before node X_i and don't exist in the parent node set $Pa(X_i)$, calculate the scoring value for each such node, add the node which makes the scoring value maximal into the parent node set of node X_i, then produce a new adjacent set $Pa_{new}(X_i)$.
— Step2: Calculate the value of evaluation function $f(S')$. Pick minus the score function value and convert it to energy representation of simulated annealing, that is $E_{new} = -f(S')$.
— Step3: If $Pa(X_i)$ is an empty set, the new node set is accepted as the parent node set of node X_i. If not, the value of E_{old} and E_{new} will be compared. If $E_{new} < E_{old}$, the new parent node set will be accepted, that is $Pa(X_i) = Pa_{new}(X_i)$, $E_{new} = E_{old}$. If $E_{new} \geq E_{old}$, the new parent node will be accepted with probability P.
— Step4: If accepting the new set of parents, the iteration counter l is set to 0 and go to Step1. If not, set $l = l+1$ and go to the next step.

- Judge whether the iteration number l achieves the maximum iteration number or not. If $l < l_g$, go to the sixth step and continue to iterate. If not, algorithm has reached balance state under the current temperature and executes the next step.
- Determine whether current temperature achieves the minimum temperature or not. If $T_k > T_g$, carry on temperature decrease according to formula $T_{k+1} = r \cdot T_k$. The step size under each temperature is $L(x) = c \cdot C \cdot r \cdot P_{acc}$, where c is a constant, C is dimension of node states, r is decline coefficient, P_{acc} is the probability of acceptance. Then go to the fifth step. If not, continue to execute the next step.
- Output the parent node set $Pa(X_i)$.

With respect to K2-SA algorithm, there are some descriptions that need to explain:

- If initial temperature is set high enough, almost all of the adjacent can be accepted, and the algorithm can converge to global optimum. However, the resulting computational efficiency will be lower. Therefore, in the real case, the setting of initial temperature generally adopt a compromise with considering the quality and efficiency of optimization. In this paper, we produce several states (sets of parents of each node) firstly. Secondly, the difference values $|\Delta|$ of score functions between every two states are computed and select the maximal difference value as $|\Delta_{max}|$. Then the initial temperature T_0 is determined as $T_0 = -|\Delta_{max}|/\ln P_0$ where P_0 is the initial acceptance probability that needs to set manually.

- If algorithm has reached balance state under the current temperature, annealing action will be taken. The method used in this paper is $T_{k+1} = r \cdot T_k$ where $0.5 < r < 0.99$ represents annealing coefficient.

- The acceptance criterion for new solutions is the main difference between SA algorithm and the general search algorithm. Method of energy value computed in this paper is picking the Bayesian score function value minus, which is $E = -f(S')$.

  The specific approach of accepting the new state with probability P is that algorithm produces a random number γ between 0 and 1. If $P > \gamma$, $Pa_{new}(X_i)$ will be accepted as the new parent node set of node X_i and E_{old} is set as E_{new}. If not, keep the original state and parent node set unchanged.

- Generally, the termination condition is that several successive solutions are not accepted, at this time, the isothermal search at this temperature will be terminated. In this paper, a dynamic control method of iterations is employed and the step size is represented as $L(x) = c \cdot C \cdot r \cdot P_{acc}$. This control mode of iterations can guarantee to achieve the balance state at current temperature, reduce the redundant iterative number, and improve the speed of operation to a certain extent.

- K2 algorithm predefines the maximum number of parent nodes. When existing a node whose parent nodes exceed the maximum in the network structure, it needs to correct that parent node in order to make the parent node number of the current node no more than the maximum. The correction algorithm uses the mutual information between nodes and its parent nodes. Select a parent node which has minimum mutual information to delete sequentially until the condition of upper limit of number of parents is satisfied. The correction algorithm selects several nodes from the parent node set to delete so that the criterion about the maximum number of parent nodes can be satisfied.

According to the above method, we can obtain near optimal set of parents for each node. Bayesian network structure can be constructed by using these sets of parents. After learning conditional probability tables of network structure, Bayesian network model can be establish.

4   Experiment and Analysis

In order to validate the method proposed by this paper, experimental procedure uses 10-fold cross-validation method. Assume that all attributes are discrete and there is no missing value. Experiment tests one hundred times on each data set, selects the best result and average result respectively and calculates the standard deviation of results. Before experimenting, data need to be discretized. This paper utilizes conditional mutual information to determine the initial node ordering. Then K2-SA algorithm is used to learn Bayesian Network structure and conditional probabilities are also learned. Finally, the Bayesian Network classifier is constructed. All parameters in the experiment employ BDe [12] priori information and the global union tree algorithm is employed as inference algorithm. The priori value α_{ijk} is set as 1. The initial temperature of simulated annealing is set as $T_0 = 1$, the termination temperature is set as $T_g = 0.001$, the jumping probability P is set as 0.2 and the temperature decline coefficient is set as $r = 0.8$.

Parameters used for analysis and comparisons among several algorithms are given as follows:

1. **S (Score):** The BDe score value of the best individual network learned.
2. **CA (Classification Accuracy):** The value of classification accuracy of the best individual network learned.
3. **MR (Mean Result):** This is the averaged result over 100 runs of several algorithms on 100 different and independent datasets for network structures.
4. **SD (Standard Deviation):** It is the Standard Deviation of the results.
5. **BR (Best Result):** It is the best result.

Experimental results are carried with our method on two standard network data sets (Asia, Alarm). Other algorithms (TAN, Hill Climbing, and Random K2) are also performed to make a comparative study.

Asia [17]: The Asia network is a small Bayesian Network, and is used for a fictitious medical example. This example researches whether a patient has tuberculosis, lung cancer, or bronchitis, related to the chest clinic. It consists of eight nodes with eight edges connecting them. Each random variable (node) is discrete in nature and takes two discrete states (values).

Alarm [18]: The Alarm network is a medical diagnostic system for patient monitoring. This network consists of 37 with 46 edges connecting them. The random variables in the network are discrete in nature and can take two, three or four states.

Table 1. Experimental result comparison of network structure learned by these methods (Asia, Alarm)

Methods	MR		SD		BR	
	Asia	*Alarm*	*Asia*	*Alarm*	*Asia*	*Alarm*
Our method	97.55	97.54	16.4	17.4	98.51	98.65
TAN	93.64	94.77	15.8	16.1	94.98	95.73
Hill Climbing	97.42	97.55	19.2	19.0	98.21	98.60
Random K2	94.32	94.26	22.3	21.1	96.39	96.44

As can be seen from Table 1, both the presented method and other three algorithms have high classification accuracy, but classification accuracy of our method is higher than other three classification algorithms as a whole. In order to compare network structure performance of these methods, this paper compares network structures learned by these methods. Table 2 and 3 list the BDe score values of network structures generated by these methods.

Table 2. Score value comparison of network structures learned by these methods (Asia)

Methods	MR	SD	BR
Our method	-525.10	17.5	-523.21
TAN	-530.20	17.4	-526.43
Hill Climbing	-531.17	19.2	-526.33
Random K2	-529.84	22.7	-523.97

Table 3. Score value comparison of network structures learned by these methods (Alarm)

Methods	MR	SD	BR
Our method	-526.17	18.6	-524.22
TAN	-555.26	17.8	-553.74
Hill Climbing	-529.63	20.1	-525.72
Random K2	-532.38	22.6	-526.13

As can be seen from data in Table 2 and 3, BDe score value of our method is larger than compared with other three classification algorithms, which shows that the network structure learned by our method is more in line with real variable dependent relationships and has a higher degree of data fitting.

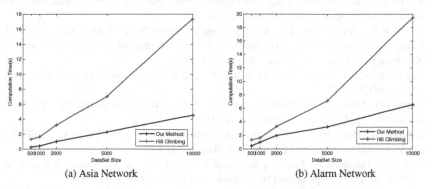

(a) Asia Network   (b) Alarm Network

Fig. 2. Computation time of learning Alarm network (seconds)

As can be seen from the above three forms, results gained by our method is similar to Hill Climbing algorithm. However, Fig.2 shows that the growth rate of computation time (seconds) of Hill Climbing is much higher than our method with the growth of dataset size.

Experimental results show that the novel method proposed by this paper has good classification performance and the network structure learned is more reasonable compared with the three algorithms. This network structure fits data better and can be more truly reflect the dependencies between variables.

5   Conclusion

In view of K2 algorithm needs to determine the initial node sequence and is easy to fall into local optimum, this paper presents a novel method which uses conditional mutual information to generate the initial node ordering and employs K2-SA algorithm to learn the Bayesian Network structure. Experimental results show the effectiveness of the algorithm. In the later study and research, we will use this algorithm for applications, research more data mining classification methods and classification-oriented scoring method.

Acknowledgment. This work was supported in part by NSFC under Grant 61172090, Scientific and Technological Project in Shaanxi Province under Grant 2012K06-30.Thanks to the condition and environment provided by Shaanxi Key Lab of Computer Network. Thanks to the guidance of Pro. Xu and Pro. Gui.

References

1. Pearl, J.: Fusion, Propagation, and Structuring in Belief Networks. Artificial Intelligence 29, 241–288 (1986)
2. Wong, M.L., Leung, K.S.: An Efficient Data Mining Method for Learning Bayesian Networks using an evolutionary algorithm based hybrid approach. IEEE Transactions on Evolutionary Computation 8(4), 378–404 (2004)
3. Chen, X.W., Anantha, G., Lin, X.: Improving Bayesian Network Structure Learning With Mutual information-based node ordering in the K2 algorithm. IEEE Transactions on Knowledge and Data Engineering 20(1), 1–13 (2008)
4. Darwiche, A.: A Differential Approach to Inference in Bayesian Networks. arXiv preprint arXiv:1301.3847 (2013)
5. Lerner, B., Malka, R.: Investigation of the K2 Algorithm in Learning Bayesian Network classifiers. Applied Artificial Intelligence 25(1), 74–96 (2011)
6. Cooper, G., Herskovits, F.E.: A Bayesian Method for The Induction of Probabilistic networks from data. Machine Learning 9(4), 309–347 (1992)
7. Lam, W., Bacchus, F.: Learning Bayesian Belief Networks: all Approach Based on the MDL Principle. Computational Intelligence 10(4), 269–293 (1994)
8. De Campos, L.M., Juan, M.F., José, A.G., et al.: Ant Colony Optimization for Learning Bayesian networks. Computational Intelligence 10, 269–293 (1994)
9. Chickering, D.M.: Optimal structure identification with Greedy Search. The Journal of Machine Learning Researeh 3, 507–554 (2002)
10. Chickering, D., Geiger, D., Heckerman, D.: Learning Bayesian Networks: Search methods and experimental results. In: Proceedings of Fifth Conference on Artificial Intelligence and Statistics, pp. 112–128 (1995)

11. Aarts, E., Korst, J., Michiels, W.: Simulated annealing. Search methodologies, pp. 187–210. Springer, US (2005)
12. Steck, H.: Learning the Bayesian Network Structure: Dirichlet prior Versus Data. arXiv preprint arXiv:1206.3287 (2012)
13. Rissanen, J.: Minimum description length principle. Encyclopedia of Machine Learning, pp. 666–668. Springer, US (2010)
14. Lam, W., Bacchus, F.: Using Causal Information And Local Measures to Learn Bayesian networks. In: Proceedings of the Ninth International Conference on Uncertainty in Artificial Intelligence, pp. 243–250. Morgan Kaufmann Publishers Inc. (1993)
15. Yun, Z., Keong, K.: Improved MDL Score for Learning of Bayesian Networks. In: Proceedings of the International Conference on Artificial Intelligence in Science and Technology, AISAT, pp. 98–103 (2004)
16. Jiang, J., Wang, J., Yu, H., Xu, H.: Poison Identification Based on Bayesian Network: A Novel Improvement on K2 Algorithm via Markov Blanket. In: Tan, Y., Shi, Y., Mo, H. (eds.) ICSI 2013, Part II. LNCS, vol. 7929, pp. 173–182. Springer, Heidelberg (2013)
17. Lauritzen, S.L., Spiegelhalter, D.J.: Local Computations with Probabilities on Graphical structures and their application to expert systems. Journal of the Royal Statistical Society, Series B (Methodological), 157–224 (1988)
18. Beinlich, I.A., Suermondt, H.J., Chavez, R.M., Cooper, G.F.: The ALARM monitoring system: A Case Study with Two Probabilistic Inference Techniques for Belief Networks, pp. 247–256. Springer, Heidelberg (1989)

Combining Multi Classifiers Based on a Genetic Algorithm – A Gaussian Mixture Model Framework

Tien Thanh Nguyen[1], Alan Wee-Chung Liew[1], Minh Toan Tran[2],
and Mai Phuong Nguyen[3]

[1] School of Information and Communication Technology, Griffith University, Australia
[2] School of Applied Mathematics and Informatics, Hanoi University of Science and
Technology, Vietnam
[3] College of Business, Massey University, New Zealand
tienthanh.nguyen2@griffithuni.edu.au

Abstract. Combining outputs from different classifiers to achieve high accuracy in classification task is one of the most active research areas in ensemble method. Although many state-of-art approaches have been introduced, no one method performs the best on all data sources. With the aim of introducing an effective classification model, we propose a Gaussian Mixture Model (GMM) based method that combines outputs of base classifiers (called meta-data or Level1 data) resulted from Stacking Algorithm. We further apply Genetic Algorithm (GA) to that data as a feature selection strategy to explore an optimal subset of Level1 data in which our GMM-based approach can achieve high accuracy. Experiments on 21 UCI Machine Learning Repository data files and CLEF2009 medical image database demonstrate the advantage of our framework compared with other well-known combining algorithms such as Decision Template, Multiple Response Linear Regression (MLR), SCANN and fixed combining rules as well as GMM-based approaches on original data.

Keywords: Stacking Algorithm, feature selection, Gaussian Mixture Model, Genetic Algorithm, multi-classifier system, classifier fusion, combining classifiers, ensemble method.

1   Introduction

In recent years, ensemble learning is one of the most active research areas in supervised learning [1, 2]. Ensemble methods can be divided into two categories [2]:

- Mixture of experts: Using a fixed set of classifiers and after that making decision from outputs of these classifiers.
- Coverage: Generating generic classifiers, which are classifiers from the same family but have different parameters. The classifiers are combined to reach a final decision.

In this paper, we focus on the first type of ensemble methods where decision is formed by combining outputs of different base classifiers. There are several combining strategies and among them, Stacking-based approaches are one of the most popular

D.-S. Huang et al. (Eds.): ICIC 2014, LNAI 8589, pp. 56–67, 2014.
© Springer International Publishing Switzerland 2014

ensemble methods. Stacking was first proposed by Wolpert [3] and was further developed by Ting and Witten [7]. In this model, the training set is divided into several equal disjoint parts. One part plays as the test set and the rest play as the training set during training. The outputs of Stacking are posterior probabilities that observations belong to a class according to each base classifier. Posterior probabilities of all observations form meta-data or Level1 data. The original data is denoted as Level0 data.

To apply Stacking to ensemble of classifiers, Ting and Witten [7] proposed Multiple Response Linear Regression algorithm (MLR) to combine posterior probabilities of each observation based on sum of weights calculated from linear regression functions. The idea of MLR is that each classifier sets a different weight on each class and combining algorithm is then conducted based on posterior probability and its associated weight. Kuncheva et al. [4] defined Decision Template for each class computed on posterior probability of observations in training set and their true class label and then detailed eleven measurements between posterior probability of unlabeled observation and each Decision Template [4] to output the prediction. Merz [9] combined Stacking, Correspondence Analysis (CA) and K Nearest Neighbor (KNN) into a single algorithm called SCANN. His idea was to form representation on a new space for outputs of base classifiers generated by Stacking and the true label of each observation. Finally, KNN is used as a classifier on the new space to obtain prediction for unlabeled observation. Recently, Szepannek et al. [10] developed idea from pairwise combining by finding which classifier is the best for a pair including class i and j ($i \neq j$) and used a pairwise coupling algorithm to combine outputs of all pairs to make posterior for each class. Zhang and Zhou [11] used linear programming to find weight that each classifier puts on a particular class. Sen et al. [12] introduced a method that was inspired by MLR which uses hinge loss function to the combiner instead of using conventional least square loss. By using new function with regularization, he proposed three different combination, namely weighted sum, dependent weighted sum and linear stacked generalization based on different regularizations with group sparsity.

Another popular and simple approach to combine outputs from base classifiers is using fixed rules. Kittler et al. [8] presented six fixed rules, namely Sum, Product, Vote, Min, Max and Average. The advantage of applying fixed rules for ensemble system is that no training based on Level1 data is needed.

In this paper, we propose a novel combining classifiers approach that exploits the distribution of outputs of base classifiers. Our approach combines Gaussian Mixture Model (GMM) classifiers and Genetic Algorithm (GA) based feature selection in a single framework. In the next two sections, we introduce the proposed framework in detail and then perform empirical evaluations on two popular datasets, namely UCI Machine Learning Repository [20] and CLEF2009 medical image database. Finally, we summarize and propose several future improvements.

2   Methodology

2.1   Classifier Fusion Based on GMM

Let us denote $\{W_j\}$ as the set of class label, N as the number of observations, K as the number of base classifiers, and M as the number of classes. For an observation

X_i, $P_k(W_j | X_i)$ is the probability that X_i belongs to class W_j given by the kth classifier. Level1 data of all observations, a $N \times MK$ - posterior probability matrix $\{P_k(W_j | X_i)\}$ $j = \overline{1,M}$ $k = \overline{1,K}$ $i = \overline{1,N}$, is in the form:

$$
\begin{bmatrix}
P_1(W_1 | X_1) \ldots P_1(W_M | X_1) \ldots P_K(W_1 | X_1) \ldots P_K(W_M | X_1) \\
P_1(W_1 | X_2) \ldots P_1(W_M | X_2) \ldots P_K(W_1 | X_2) \ldots P_K(W_M | X_2) \\
\ldots \quad\quad\quad\quad\quad\quad\quad\quad\quad\quad\quad \ldots \\
P_1(W_1 | X_N) \ldots P_1(W_M | X_N) \ldots P_K(W_1 | X_N) \ldots P_K(W_M | X_N)
\end{bmatrix}
\tag{1}
$$

The Level1 data of an observation X is defined as:

$$
Level1(X) := \begin{bmatrix}
P_1(W_1 | X) & \cdots & P_1(W_M | X) \\
\vdots & \ddots & \vdots \\
P_K(W_1 | X) & \cdots & P_K(W_M | X)
\end{bmatrix}
\tag{2}
$$

In our algorithm we employ Stacking Algorithm to generate Level1 data from the original training set. The most important distinction between our work and the previous work is we use GMM-based approach on Level1 data to construct combining classifiers model. In our knowledge, all previous GMM-based approaches were applied to Level0 data. Attributes in Level0 data are frequently different in nature, measurement unit, and type. As a result, GMM does not perform well when it is used to approximate the distribution of Level0 data. Level1 data, on the other hand, can be viewed as scaled result from feature domain to posterior domain where data is re-shaped to be all real values in [0, 1]. Observations that belong to the same class will have nearly equal posterior probabilities generated from base classifiers and would locate close together in the new domain. As a result, Level1 data is expected to be more discriminative than the original data, and GMM modeling on Level1 data will be more effective than on Level0 data. Besides, it is well known in literature that the higher the dimension of the data, the lower the effectiveness of GMM approximation. As Level1 data has lower dimension than Level0 data, GMM on Level1 data would be more effective.

Our proposed model is illustrated in Fig 1. Here training set is classified by K base classifiers based on Stacking to generate Level1 data. Because the label of all observations in a training set is known so we can group Level1 data into M classes such that observations belonging to the same class are grouped together. Next, GMM is employed as a statistical representation model for each class and then the class label of an observation is predicted by posterior probability using Bayes model.

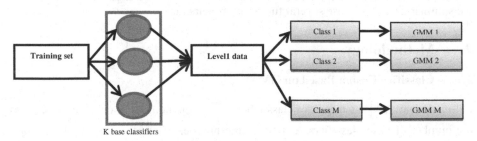

Fig. 1. GMM-based approach on Level1 data

During testing, unlabeled observation is classified by K base classifiers to generate Level1 data of that observation (eqn. 2). That meta-data is then gone through M-GMMs as input data and the final prediction is obtained through maximization among posterior probabilities corresponding with all classes.

For i^{th} class, we propose the prediction framework based on Bayes model

$$\underset{posteriror}{P(GMM_i \mid \mathbf{x})} \sim \underset{likelihood}{P(\mathbf{x} \mid GMM_i)} \times \underset{prior}{P(GMM_i)} \tag{3}$$

Here the likelihood function is modeled by GMM:

$$P(\mathbf{x} \mid GMM_i) = P(\mathbf{x} \mid \mathbf{\mu}_{ip}, \mathbf{\Sigma}_{ip}, \omega_{ip}) = \sum_{p=1}^{P_i} \omega_{ip} N\left(\mathbf{x} \mid \mathbf{\mu}_{ip}, \mathbf{\Sigma}_{ip}\right) \tag{4}$$

where

$$N\left(\mathbf{x} \mid \mathbf{\mu}_{ip}, \mathbf{\Sigma}_{ip}\right) = \frac{1}{(2\pi)^{MK/2} \left|\mathbf{\Sigma}_{ip}\right|^{1/2}} \exp\left\{-\frac{1}{2}\left(\mathbf{x} - \mathbf{\mu}_{ip}\right)^{\mathbf{T}} \mathbf{\Sigma}_{ip}^{-1}\left(\mathbf{x} - \mathbf{\mu}_{ip}\right)\right\} \tag{5}$$

P_i is the number of Gaussian components in GMM_i and $\mathbf{\mu}_{ip}, \mathbf{\Sigma}_{ip}$ are the mean and covariance of pth component in the model of ith class, respectively. The prior probability in (eqn. 3) of ith class is defined by:

$$P(GMM_i) = \frac{n_i}{N} \tag{6}$$

where n_i is the number of observations in i^{th} class and N is the total number of observations in the training set. To find parameters of GMM, we apply Expectation Maximization (EM) algorithm by maximize the likelihood function with respect to the means, covariances of components, and mixing coefficients [19].

Now with dataset $\mathbf{X} = \{\mathbf{x}_i\}$ $i = 1, N_i$ corresponding with i^{th} class, the question is how to find the number of component for GMM. Frequently, it is fixed by a specific number. Here, we propose applying Bayes Criterion Information (BIC) to find the optimal model [19] with an assumption that we have a set of model $\{F_j\}$ with parameters $\mathbf{\theta}_j$ where $\mathbf{\theta}_j$ denotes for all parameters of model. To find model by BIC, for i^{th} class ($i = \overline{1, M}$), we compute:

$$\ln P(\mathbf{X} \mid F_j) \approx \ln P(\mathbf{X} \mid F_j, \mathbf{\theta}_{MAP}) - \frac{1}{2}\left|\mathbf{\theta}_j\right| \ln n_i \tag{7}$$

where $\mathbf{\theta}_{MAP}$ is corresponding with the maximum of posterior distribution and $\left|\mathbf{\theta}_j\right|$ denotes the number of parameters in $\mathbf{\theta}_j$. In GMM modeling, $\{F_j\}$ are group in which each element is a GMM and $\{\mathbf{\theta}_j\}$ include three parameters namely means, covariances of Gaussian components, and mixing coefficients in mixture model and n_i is the number of observations in i^{th} class.

As Level1 data is obtained from K base classifiers, it conveys the posterior information from each classifier about how much support a classifier has for an observation to belong to a class. In some cases, there are columns in Level1 data in which $\exists k, m$, $P_k(W_m \mid X_i)$ is nearly constant for all i. Hence, the covariance matrix may be singular and EM is unable to solve for GMM. To overcome this problem, we randomly choose several elements in this column and perturb their values by a small quantity. This procedure only adds small value to some random elements in a column so it does not significantly affect the posterior probability or the covariance matrix.

Algorithm 1: GMM for combining classifiers
Training process:
Input: Training set (Level0 data), K base classifiers, PiMax: maximum number of Gaussian component for i^{th} class.
Output: GMM for each class.
Step1: Generate Level1 data from Level0 data.
Step2: Group observations with the same labels into M classes; compute $P(GMM_i)$ (eqn. 6), mean, and covariance for each class.
Step3:For i^{th} class
          For p=1 to PiMax
                • Apply EM algorithm to find GMM model corresponding with p components.
                • Compute BIC.
          End
     Select Pi that maximizes BIC, and GMM with Pi components.
     End
Testing process:
Input: Unlabeled observation XTest.
Output: predicted label of XTest.
Step1: Compute Level1 data of XTest.
Step2: For each class
        Compute $P(XTest \mid GMM_i)$ (eqn. 4) and posterior related to class (eqn. 3) as $P(GMM_i \mid XTest) \sim P(XTest \mid GMM_i) \times P(GMM_i)$
     End
Step3: $XTest \in W_t$ if $t = \arg\max_{i=1,M} P(GMM_i \mid XTest)$

End

2.2   GAGMM Framework

Recently, several GA-based approaches have been proposed to improve the accuracy of classifier fusion by solving both the classifier and feature selection problems [6, 15, 16]. For GMM, GA is only applied to improve EM algorithm [15]. Here, we employ

GA as a feature selection technique on Level1 data. It means that if columns $P_k(W_m \mid X_i)(i = \overline{1,n})$ for each m and k, which is posterior probability that X_i belong to class W_m given by k^{th} classifier, is not discriminative enough, its elimination from Level1 data could increase the classification accuracy of the combining algorithm.

To build the GAGMM framework on Level1 data, first, we propose the structure of chromosome in population. Each chromosome includes $M \times K$ genes due to dimension of Level1 data. We use two values {0,1} to encode for each gene in a chromosome in which:

$$Gene(i) = \begin{cases} 1 & \text{if } i^{th} \text{ feature is selected} \\ 0 & \text{otherwise} \end{cases} \tag{8}$$

At crossover stage, we employ single splitter since the same random point is selected on both parents. Each parent exchanges its head with the other while keeping its tail. After this stage, two new offspring chromosomes are created. Next, based on mutation probability, we select random genes from the offspring chromosomes in population and change their values by $0 \rightarrow 1$ or $1 \rightarrow 0$. Mutation helps GA reaches not only local extreme value but also global one. Here we use the accuracy of combination by GMM-based approach on Level1 data as fitness value of GA. Liu et al. [18] showed that when dimension of data is high, the effectiveness of GMM approximation is reduced. By using a subset of feature set, the dimension of Level1 data is reduced; as a result, the approximation of distribution by GMM is more precise than that on original Level1 data.

This framework can be viewed as a mechanism to learn an optimal subset of features from Level1 data of training set by using GMM based classifiers and therefore it can be classified to wrapper type among feature selection methods [21]. Since only a single training set is available, cross validation method is used to divide Level1 data of training set to several disjoint parts in which a part is selected as Level1 validation set while the rest constitute the Level1 training set. Attributes are selected according to the encoding of each chromosome. Fitness value is computed by averaging all accuracy values obtained from Level1 validation test. Due to the low dimensionality of Level1 data, GA-based approach can usually converge after a small number of interactions.

GAGMM framework is detailed below:

```
Algorithm 2: GAGMM

Training process:

Input: Training set (Level0), K base classifiers, mutation proba-
bility (PMul), population size (L).
Output: optimal chromosome encoding as an optimal subset of Level1
feature and GMMs for classes associated with that chromosome.
Step 1: Generate Level1 of training set.
Step 2: Initialize a population with L random chromosomes.
Step 3: Compute fitness of each chromosome.
Step 4: Loop to select L best chromosomes.
While(not converge)
```

- Draw with replacement L/2 pairs from population; conduct crossover and mutation (based on PMul) on them to generate new L chromosomes.
- Compute fitness of each offspring chromosome and add new L chromosomes to population.
- Select new population with L best fitness value chromosomes.

End

Step 5: Select and save the best chromosome from final population and GMMs for classes associated with that chromosome.

Testing process:

Input: unlabeled observation XTest.

Output: predicted label of XTest.

Call test process in Algorithm 1.

3   Experimental Results

We empirically evaluated our framework on two data sources namely UCI dataset and CLEF2009 medical image database. In our assessment, we compared error rates of our model with 6 benchmark algorithms: selecting the best results from fixed rules based on outcomes on test set, Decision Template (measure of similarity S_1 [4] is

used defined as $S_1(Level1(X), DT_i) = \dfrac{\|Level1(X) \cap DT_i\|}{\|Level1(X) \cup DT_i\|}$ where DT_i is Decision

Template of ith class and $\|\alpha\|$ is the relative cardinality of the fuzzy set α), MLR, SCANN, GMM on Level0 data and GMM on Level1 data. The comparison with GMM on Level0 data and GMM on Level1 data was that we wanted to demonstrate the performance of our model on Level 1 data compared to that on original data, and also the benefit of feature selection on Level1 data. Three base classifiers namely Linear Discriminant Analysis, Naïve Bayes and K Nearest Neighbor (with K set to 5) were chosen. The reason for choosing these diverse base classifiers is that they ensure diversity of the ensemble system. Parameters for GA were initialized by setting population size to 20 and mutation probability to 0.015.

To ensure objectiveness, we performed 10-fold validation and implemented the test 10 times to obtain 100 error rates result for each data file. To assess statistical significance, we used paired t-test to compare the classification results of our approach and each benchmark (level of significance set to 0.05)

3.1   Experiment on UCI Files

We chose 21 common UCI data files with number of classes ranging from 2 (Bupa, Artificial, etc…) to 26 (Letter). The number of attributes also changes in a wide range from only 3 attributes (Haberman) to 60 attributes (Sonar). The number of observations in each file also varies considerably, from small files like Iris, Fertility to big file such as Skin&NonSkin (up to 245057 observations) (Table 1). Experimental results of all 21 files are shown in Table 2.

From the paired t-test result in Table 3, it can be concluded that our framework is superior to the compared methods. First, there are 6 cases in which GAGMM outperforms the best result from fixed rules while on the other 13 files, both methods are competitive. Besides, GAGMM is better than Decision Template in 14 cases and is not worse in any cases.

Next, GAGMM has 18 wins and 1 loss whereas GMM on Level1 data has 16 wins and 1 loss with respect to GMM on Level0 data. This demonstrates that GMM on Level1 data significantly outperformed GMM on Level0 data. As mentioned in section 2, Level1 data is more uniform than Level0 data in which all components of feature vector are real data type and are scaled to [0, 1]. Therefore, GMM models the distribution of Level1 data better than the original data. For the one loss, we should point out that Ring is a simulated dataset generated from a multivariate Gaussian distribution [20], so GMM is expected to perform well on this dataset at Level0.

Comparing SCANN to GAGMM reveals that SCANN outperforms on just 1 file (Letter 0.063 vs. 0.0802) but has 8 losses. In our experiment, SCANN cannot be run on 3 files, namely Fertility, Balance and Skin&NonSkin, because the indicator matrix has columns with all 0 posterior probabilities from the K base classifiers. As a result, its column mass is singular and standardized residuals is not available. Here, we did not put these files in the comparison.

In the comparison between GAGMM and MLR, GAGMM obtained 6 wins and 2 losses. Significant improvements are on Ring (0.1108 vs. 0.17), Balance (0.0755 vs. 0.1225) and Tae (0.4313 vs. 0.4652).

Finally, GAGMM outperforms GMM on Level1 data, posting 5 wins and no loss. The five wins are Pima (0.2279 vs. 0.2432), Balance (0.0755 vs. 0.0839), Fertility (0.127 vs. 0.185), Wdbc (0.0321 vs. 0.0387) and Iris (0.0267 vs. 0.036). It shows that feature selection on Level1 data not only reduces the dimension of data but also improves classification performance.

We also assess the dimension of data in Level0 data, Level1 data, and the data resulted from GAGMM. Fig 2 shows that the dimension of data from GAGMM is significantly lower than those of Level0 and Level1. Significant results are observed on Fertility, Pima, Phoneme (with just 1 dimension) and Magic, Twonorm and Haberman (with just 2 dimensions). Hence, our GAGMM framework helps reduce the dimension of data while also improves the classification performance.

3.2   Experiment on CLEF2009 Database

The CLEF 2009 database is a large set of medical images collected by Archen University. It includes 15,363 images allocated to 193 hierarchical categories. In our experiment we chose 7 classes where each has different number of images (Table 4). Histogram of Local Binary Pattern (HLBP) is extracted from the image as feature vector. The results of the experiment on 7 classes are summarized in Table 5.

Table 6 shows the comparison between GAGMM and 6 benchmark algorithms, which are best result from fixed rules, MLR, Decision Template, SCANN, GMM on Level0 data, and GMM on Level1 data. GAGMM posts 6 wins compared with other state-of-art combining algorithms. GMM on Level1 data outperforms GMM on Level0 data (0.1469 vs. 0.2453) but underperforms GAGMM (0.1469 vs. 0.1164).

Table 1. UCI data files used in our experiment

File name	# of attributes	Attribute type	# of observations	# of classes	# of attributes on Level1 (3 classifiers)
Bupa	6	C,I,R	345	2	6
Pima	6	R,I	768	2	6
Sonar	60	R	208	2	6
Heart	13	C,I,R	270	2	6
Phoneme	5	R	540	2	6
Haberman	3	I	306	2	6
Titanic	3	R,I	2201	2	6
Balance	4	C	625	3	9
Fertility	9	R	100	2	6
Wdbc	30	R	569	2	6
Australian	14	C,I,R	690	2	6
Twonorm (*)	20	R	7400	2	6
Magic	10	R	19020	2	6
Ring (*)	20	R	7400	2	6
Contraceptive	9	C,I	1473	3	9
Vehicle	18	I	846	4	12
Iris	4	R	150	3	9
Tae	20	C,I	151	2	6
Letter	16	I	20000	26	78
Skin&NonSkin	3	R	245057	2	6
Artificial	10	R	700	2	6

R: Real, C: Category, I: Integer (*) Simulated data

Table 2. Comparing error rates of different combining algorithms

File name	MLR		Best results from fixed Rules		SCANN		Decision Template	
	Mean	Variance	Mean	Variance	Mean	Variance	Mean	Variance
Bupa	0.3033	4.70E-03	0.2970	4.89E-03	0.3304	4.29E-03	0.3348	7.10E-03
Artificial	0.2426	2.20E-03	0.2193	2.05E-03	0.2374	2.12E-03	0.2433	1.60E-03
Pima	0.2432	2.30E-03	0.2365	2.10E-03	0.2384	2.06E-03	0.2482	2.00E-03
Sonar	0.1974	7.20E-03	0.2079	8.16E-03	0.2128	8.01E-03	0.2129	8.80E-03
Heart	0.1607	4.70E-03	0.1570	4.64E-03	0.1637	4.14E-03	0.1541	4.00E-03
Phoneme	0.1136	1.75E-04	0.1407	1.95E-04	0.1229	6.53E-04	0.1462	2.00E-04
Haberman	0.2428	3.30E-03	0.2392	2.39E-03	0.2536	1.74E-03	0.2779	5.00E-03
Titanic	0.2169	4.00E-04	0.2167	5.00E-04	0.2216	6.29E-04	0.2167	6.00E-04
Balance	0.1225	8.00E-04	0.1112	4.82E-04	X	X	0.0988	1.40E-03
Fertility	0.1250	2.28E-03	0.1270	1.97E-03	X	X	0.4520	3.41E-02
Skin&NonSkin	4.79E-4	1.97E-08	0.0006	2.13E-08	X	X	0.0332	1.64E-06
Wdbc	0.0399	7.00E-04	0.0395	5.03E-04	0.0397	5.64E-04	0.0385	5.00E-04
Australian	0.1268	1.80E-03	0.1262	1.37E-03	0.1259	1.77E-03	0.1346	1.50E-03
Twonorm	0.0217	2.24E-05	0.0216	2.82E-05	0.0216	2.39E-05	0.0221	2.62E-05
Magic	0.1875	7.76E-05	0.1905	5.72E-05	0.2002	6.14E-05	0.1927	7.82E-05
Ring	0.1700	1.69E-04	0.2122	1.62E-04	0.2150	2.44E-04	0.1894	1.78E-04
Tae	0.4652	1.24E-02	0.4435	1.70E-02	0.4428	1.34E-02	0.4643	1.21E-02
Contraceptive	0.4675	1.10E-03	0.4653	1.79E-03	0.4869	1.80E-03	0.4781	1.40E-03
Vehicle	0.2139	1.40E-03	0.2645	1.37E-03	0.2224	1.54E-03	0.2161	1.50E-03
Iris	0.0220	1.87E-03	0.0327	1.73E-03	0.0320	2.00E-03	0.0400	2.50E-03
Letter	0.0427	1.63E-05	0.0760	3.94E-05	0.063	2.42E-05	0.1133	4.91E-05

Table 2. (*continued*)

File name	GMM on Level0		GMM on Level1		GAGMM	
	Mean	Variance	Mean	Variance	Mean	Variance
Bupa	0.4419	5.80E-03	0.3022	5.31E-03	0.2999	4.79E-03
Artificial	0.4507	8.00E-03	0.2374	2.40E-03	0.2423	2.64E-03
Pima	0.2466	2.40E-03	0.2432	2.60E-03	0.2279	1.95E-03
Sonar	0.3193	1.26E-02	0.2009	6.20E-03	0.2050	8.09E-03
Heart	0.1715	7.30E-03	0.1559	4.51E-03	0.1544	4.67E-03
Phoneme	0.2400	4.00E-04	0.1165	2.01E-04	0.1150	1.39E-04
Haberman	0.2696	2.00E-03	0.2458	3.36E-03	0.2461	3.96E-03
Titanic	0.2904	2.01E-02	0.2167	5.91E-04	0.2217	1.16E-03
Balance	0.1214	1.10E-03	0.0839	1.21E-03	0.0755	9.68E-04
Fertility	0.3130	7.47E-02	0.1850	1.05E-02	0.1270	2.40E-03
Skin&NonSkin	0.0761	2.21E-06	4.10E-4	1.53E-08	4.13E-4	1.98E-08
Wdbc	0.0678	1.10E-03	0.0387	5.98E-04	0.0321	5.25E-04
Australian	0.1980	1.80E-03	0.1222	1.30E-03	0.1210	1.60E-03
Twonorm	0.0216	2.83E-05	0.0219	2.78E-05	0.0220	2.72E-05
Magic	0.2733	5.06E-05	0.1921	8.34E-05	0.1918	6.03E-05
Ring	0.0209	2.20E-05	0.1131	1.16E-04	0.1108	1.09E-04
Tae	0.5595	1.39E-02	0.4365	1.36E-02	0.4313	1.58E-02
Contraceptive	0.5306	1.80E-03	0.4667	1.30E-03	0.4624	1.30E-03
Vehicle	0.5424	2.40E-03	0.2166	1.40E-03	0.2131	1.46E-03
Iris	0.0453	2.50E-03	0.0360	2.10E-03	0.0267	1.10E-03
Letter	0.3573	9.82E-05	0.0797	3.03E-05	0.0802	1.32E-05

Table 3. Compare GAGMM with the benchmarks among 21 UCI files

	Better	Competitive	Worse
GAGMM vs. MLR	6	13	2
GAGMM vs. SCANN	8	9	1
GAGMM vs. Decision Template	14	7	0
GAGMM vs. best result from fixed Rules	6	13	2
GAGMM vs. GMM Level0	18	2	1
GAGMM vs. GMM Level1	5	16	0
GMM Level1 vs. GMM Level10	16	4	1

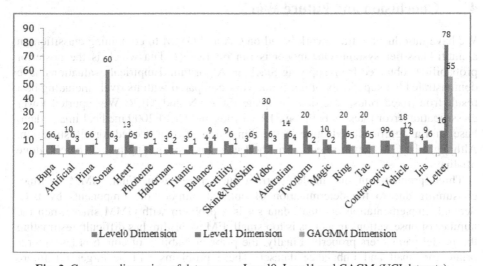

Fig. 2. Compare dimension of data among Level0, Level1 and GAGM (UCI datasets)

Table 4. Information of 7 classes CLEF2009 in our experiment

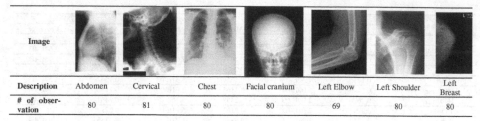

Image							
Description	Abdomen	Cervical	Chest	Facial cranium	Left Elbow	Left Shoulder	Left Breast
# of observation	80	81	80	80	69	80	80

Table 5. Error rates of different combining algorithms with CLEF2009

	Sum	Product	Max	Min	Median	Vote	MLR
Mean	0.1636	0.1831	0.1835	0.1978	0.1725	0.1851	0.1280
Variance	1.67E-03	2.18E-03	2.54E-03	2.35E-03	2.09E-03	1.87E-03	1.80E-03

	Decision Template	Best result from fixed Rules	SCANN	GMM Level0	GMM Level1	GAGMM	
Mean	0.1447	0.1636	0.1455	0.2453	0.1469	0.1164	
Variance	1.90E-03	1.67E-03	2.62E-03	3.49E-03	2.00E-03	2.10E-03	

Table 6. Comparing GAGMM with 6 benchmarks and GMM level1 with GMM level0 related to CLEF2009

	Better	Competitive	Worse
GAGMM vs. compared methods	6	0	0
GMM Level1 vs. GMM Level0	1	0	0

Table 7. Comparing dimension of feature related to CLEF2009

Data	# of dimension
Level0	32
Level1	21
Dimension from GAGMM	14

4   Conclusion and Future Work

We have introduced a framework based on GA and GMM to combining classifiers in a multi-classifier system. Our model is run on Level1 data which is the posterior probabilities obtained from applying Stacking Algorithm. Empirical evaluations have demonstrated the superiority of our framework compared with its rivals including best result from fixed rules, Decision Template, SCANN and MLR. We reported lower classification error rates on both the 21 UCI files and CLEF2009 medical image database, and a significant reduction in the dimension of data needed for classification. Although the GA feature selection method is a time-consuming task, it can be done off-line.

Three problems in our model warrant further research. First, GMM is time-consuming due to the determination of optimal number of components by BIC. Second, implementation on small data set is a problem with GMM since when the number of observations in a class is too small, EM algorithm has difficulty estimating the model parameters properly. Finally, the prior probability of eqn. 6 is less appropriate for small and imbalance dataset. These problems will be target for future improvement.

References

1. Rokach, L.: Taxonomy for characterizing ensemble methods in classification tasks: A review and annotated bibliography. Journal of Computational Statistics & Data Analysis 53(12), 4046–4072 (2009)
2. Ho, T.K., Hull, J.J., Srihari, S.N.: Decision Combination in Multiple Classifier Systems. IEEE Transactions on Pattern Analysis and Machine Intelligent 16(1), 66–75 (1994)
3. Wolpert, D.H.: Stacked Generalization. Neural Networks 5(2), 241–259 (1992)
4. Kuncheva, L.I., Bezdek, J.C., Duin, R.P.W.: Decision Templates for Multi Classifier Fusion: An Experimental Comparison. Pattern Recognition 34(2), 299–314 (2001)
5. Kuncheva, L.I.: A theoretical Study on Six Classifier Fusion Strategies. IEEE Transactions on Pattern Analysis and Machine Intelligence 24(2) (2002)
6. Kuncheva, L.I., Jain, L.C.: Designing Classifier Fusion Systems by Genetic Algorithms. IEEE Transactions on Evolution Computation 4(4) (2000)
7. Ting, K.M., Witten, I.H.: Issues in Stacked Generation. Journal of Artificial In Intelligence Research 10, 271–289 (1999)
8. Kittler, J., Hatef, M., Duin, R.P.W., Matas, J.: On Combining Classifiers. IEEE Transactions on Pattern Analysis and Machine Intelligence 20(3) (1998)
9. Merz, C.: Using Correspondence Analysis to Combine Classifiers. Machine Learning 36, 33–58 (1999)
10. Szepannek, G., Bischl, B., Weihs, C.: On the combination of locally optimal pairwise classifiers. Engineering Applications of Artificial Intelligent 22, 79–85 (2009)
11. Zhang, L., Zhou, W.D.: Sparse ensembles using weighted combination methods based on linear programming. Pattern Recognition 44, 97–106 (2011)
12. Sen, M.U., Erdogan, H.: Linear classifier combination and selection using group sparse regularization and hinge loss. Pattern Recognition Letters 34, 265–274 (2013)
13. Raymer, M.L., Punch, W.F., Goodman, E.D., Kuhn, L.A., Jain, A.K.: Dimensionality Reduction using Genetic Algorithms. IEEE Transactions on Evolutionary Computation 4(2) (2000)
14. Gabrys, B., Ruta, D.: Genetic Algorithms in Classifier Fusion. Applied Soft Computing 6, 337–347 (2006)
15. Pernkopf, F., Bouchaffra, D.: Genetic-Based EM algorithm for Learning Gaussian Mixture Models. IEEE Transactions on Pattern Analysis and Machine Intelligence 27(8) (2005)
16. Figueiredo, M.A.T., Jain, A.K.: Unsupervised learning of finite mixture models. IEEE Transactions on Pattern Analysis and Machine Intelligence 24(3), 381–396 (2002)
17. Moghaddam, B., Pentland, A.: Probabilistic Visual Learning for Object Representation. IEEE Transactions on Pattern Analysis and Machine Intelligence 19(7) (1997)
18. Liu, X.H., Liu, C.L.: Discriminative Model Selection for Gaussian Mixture Models for Classification. In: First Asian Conference on Pattern Recognition (ACPR), pp. 62–66 (2011)
19. Bishop, C.M.: Pattern Recognition and Machine Learning. Springer Press (2006)
20. UCI Machine Learning Repository,
    http://archive.ics.uci.edu/ml/datasets.html
21. Saeys, Y., Inza, I., Larrañaga, P.: A review of feature selection techniques in bioinformatics. Bioinformatics 23(19), 2507–2517 (2007)

A Precise Hard-Cut EM Algorithm for Mixtures of Gaussian Processes

Ziyi Chen[1], Jinwen Ma[1,*], and Yatong Zhou[1,2]

[1] Department of Information Science, School of Mathematical Sciences, Peking University,
Beijing, 100871, China
jwma@math.pku.edu.cn
[2] School of Information Engineering, Hebei University of Technology, Tianjin, 300401, China

Abstract. The mixture of Gaussian processes (MGP) is a powerful framework for machine learning. However, its parameter learning or estimation is still a very challenging problem. In this paper, a precise hard-cut EM algorithm is proposed for learning the parameters of the MGP without any approximation in the derivation. It is demonstrated by the experimental results that our proposed hard-cut EM algorithm for MGP is feasible and even outperforms two available hard-cut EM algorithms.

Keywords: Mixture of Gaussian process, Parameter learning, EM algorithm.

1   Introduction

Gaussian process (GP) is a powerful learning model for both regression and classification. Nevertheless, it cannot describe multimodality dataset and needs a large number of computations. To tackle these issues, Tresp [2] proposed the Mixture of GPs (MGP) in 2000, which was directly derived from the mixture of experts.

From then on, various versions of MGP models have been suggested. Most of them could be classified into the conditional model 'x→z→y' [2-6] and the generative model 'z→x→y' [1],[7-10] where x, y and z denote the input, output and the latent component indicator, respectively. In comparison with the first model, the second one has two advantages: (1). The missing features can be easily inferred from the outputs; (2). The influence of inputs on the outputs is more clear [8].

For the parameter learning or estimation of MGP, there are three kinds of learning algorithms: MCMC, variational Bayesian inference, and EM algorithm. As for MCMC approaches, Gibbs samples of the indicators, parameters or hyper-parameters were obtained in turn from their posteriors [8,9]. However, Nguyen & Bonilla [3] pointed out that the time complexity of the MCMC method is very high. As for the variational Bayesian algorithms, the main strategy is to approximate the posterior of parameters by a factorized and simplified form [5,6], but such an approximation may lead to a rather deviation from the true objective function.

---

* Corresponding author.

D.-S. Huang et al. (Eds.): ICIC 2014, LNAI 8589, pp. 68–75, 2014.

In general, EM algorithm is a popular and efficient choice for the parameter learning of mixture models. However, the posteriors of latent variables and Q function in the cases of MGP are rather complicated as the outputs are dependent. In order to alleviate this difficulty, some EM algorithms with the help of certain approximation mechanisms have already been proposed successively.

Tresp [2] firstly proposed the EM algorithm for MGP, in which the M-step integrated the posterior probability of each sample belonging to a GP component into the learning of each component. Along this direction, Stachniss et al. [4] developed a similar EM algorithm for the sparse MGP. However, the learning in the M-step was heuristic in [2] and [4] without maximizing Q function and the time complexity in [2] was still high as that in [1]. On the other hand, Yuan & Neubauer [1], Nguyen & Bonilla [3], Miguel et al. [5] and Sun & Xu [10] proposed some variational EM algorithms for MGP in which the posterior probabilities were approximated with certain factorized forms. Recently, Yang & Ma [7] also proposed the EM algorithm for MGP with the help of leave-one-out cross-validation (LOOCV) probability decomposition of the total likelihood.

Although the approximations or simplifications have been made for the learning in the M-step, these soft EM algorithms for MGP are still time-consuming. In order to reduce the time complexity, Nguyen & Bonilla [3] recently proposed a variational hard-cut EM algorithm for MGP under certain sparseness constraints, which actually partitions all the samples into these components according to the MAP criterion in the E-step. In fact, this hard-cut EM algorithm was more efficient than the soft EM algorithm since we could get the hyper-parameters of each GP independently in the M-step rather than maximize a complicated Q function. Moreover, in the same way, the soft EM algorithm for MGP with the LOOCV probability decomposition [7] can be easily turned into a hard-cut version of the EM algorithm for MGP, which is here referred to as the LOOCV hard-cut EM algorithm for convenience.

In this paper, we follow the generative MGP model and propose a precise hard-cut EM algorithm for MGP without any approximation used for the likelihood function or Q function. Actually, we further refine the MGP model to exclude extra priors from the main chain 'z→x→y' and according to this refined model, the hard-cut EM algorithm becomes more accurate than some popular algorithms since the posterior probability of each sample belonging to each component used in the algorithm is strictly derived, and the heuristic approximations used in [3] and [7] are avoided. It is demonstrated by the experimental results that our proposed hard-cut EM algorithm is feasible and even outperforms the two available hard-cut EM algorithms.

2   MGP Model

2.1   GP Model

For regression and prediction task, a GP is mathematically defined by

$$y = [y_1 \quad y_2 \quad \cdots \quad y_N]^T \sim N[m(X), K(X,X) + \sigma^2 I],$$ (1)

where $\{(x_t, y_t)\}_{t=1}^{N}$ denotes the set of given sample points or dataset, σ^2 is the intensity of noise, I is an $N \times N$ identity matrix, $m(X) = [m(x_1) \quad m(x_2) \quad \cdots \quad m(x_N)]^T$ and $K(X,X) = [K(x_i, x_j)]_{N \times N}$ are the means and kernel matrix, respectively. For simplicity, we set $m(X) = 0$. The most commonly used kernel is the SE kernel [11], being given by $K(x_i, x_j) = l^2 \exp(-0.5 f^2 \|x_i - x_j\|^2)$.

In order to learn the hyper-parameters $\theta = \{l, f, \sigma\}$, we use the commonly adopted approach------maximize likelihood estimation (MLE).

After the parameter learning process, the prediction of the output at a test input $x*$ is

$$\hat{y}* = K(x*, X)[K(X,X) + \sigma^2 I]^{-1} y \tag{2}$$

where $y = [y_1, y_2, \cdots, y_N]^T$ is the vector of training outputs, $K(X,X) = [K(x_i, x_j)]_{N \times N}$, and $K(x*, X) = [K(x*, x_1), K(x*, x_2), \cdots, K(x*, x_N)]$ denotes the kernel relationship vector of the training inputs to the test input.

2.2   Generative MGP Model

We adopt a full generative model ('z→x→y') due to its resistance to missing features and clear relationship between inputs and outputs [8].

At first, the latent indicators $\{z_t\}_{t=1}^{N}$ are generated by the Multinomial distribution:

$$\Pr(z_t = c) = \pi_c; c = 1 \sim C \; i.i.d \; for \, t = 1 \sim N \tag{3}$$

Given indicators, each input fulfills a Gaussian distribution:

$$p(x_t \mid z_t = c) \sim N(\mu_c, S_c); c = 1 \sim C \; i.i.d \; for \; t = 1 \sim N \tag{4}$$

After specifying $\{z_t, x_t\}_{t=1}^{N}$, the outputs of each component fulfill a GP, that is

$$p\left(\begin{bmatrix} y_{c,1} & y_{c,2} & \cdots & y_{c,N(c)} \end{bmatrix} \middle| \begin{bmatrix} x_{c,1} & x_{c,2} & \cdots & x_{c,N(c)} \end{bmatrix}\right) \sim N\left(\vec{0}, \left[K(x_{c,i}, x_{c,j} \mid \theta_c)\right]_{N(c) \times N(c)} + \sigma_c^2 I_{N(c)}\right) \; i.i.d. \, for \, c = 1 \sim C \tag{5}$$

where for the c-th component, $\{x_{c,i}, y_{c,i}\}_{i=1}^{N(c)}$ are the samples, $\theta_c = \{l_c, f_c, \sigma_c\}$ are the GP hyper-parameters and $K(x_{c,i}, x_{c,j} \mid \theta_c) = l_c^2 \exp(-0.5 f_c^2 \|x_i - x_j\|^2)$ is the SE kernel function.

The whole generative model can be completely defined by Eqs (3), (4) & (5). In fact, after the allocation of samples to these components, each GP can be learnt independently, as suggested in (4) and (5).

3   Precise Hard-Cut EM Algorithm

For this generative MGP model, we here adopt the EM algorithm framework to learn the whole parameters $\{\pi_c, \mu_c, S_c, l_c, f_c, \sigma_c\}_{c=1}^{C}$, taking $\{z_t\}_{t=1}^{N}$ as the latent variables.

Firstly, we derive the posterior probabilities of these indicators, i.e., z_t. Based on Eqs (3), (4) & (5), we get the following likelihood function:

$$p(z_t = c, x_t, y_t) = \pi_c N(x_t \mid \mu_c, S_c) N\left(y_t \mid 0, l_c^2 + \sigma_c^2\right) \tag{6}$$

According to the Bayesian formula, it can be derived that

$$p(z_t = c \mid x_t, y_t) = \pi_c N(x_t \mid \mu_c, S_c) N\left(y_t \mid 0, l_c^2 + \sigma_c^2\right) / \left\{ \sum_{c'=1}^{C} \pi_{c'} N(x_t \mid \mu_{c'}, S_{c'}) N\left(y_t \mid 0, l_{c'}^2 + \sigma_{c'}^2\right) \right\} \tag{7}$$

As the computational complexity of Q function is $O(C^N)$, it is more reasonable to construct a hard-cut version of the EM algorithm to reduce the computational burden. According to this idea, we propose a hard-cut EM algorithm as follows.

- Step 1   Initialization of indicators: cluster$\{(x_t, y_t)\}_{t=1}^{N}$ into C classes by the k-means clustering, and set $z_t \leftarrow$ *The indicator of the $t-$ th sample to the cluster*

- Step 2   M-step: calculate π_c, μ_c and S_c in the way of the Gaussian mixture model:

$$\pi_c \leftarrow \frac{1}{N} \sum_{t=1}^{N} I(z_t = c), \ \mu_c \leftarrow \frac{\sum_{t=1}^{N} I(z_t = c) x_t}{\sum_{t=1}^{N} I(z_t = c)}, \ S_c \leftarrow \frac{\sum_{t=1}^{N} I(z_t = c)(x_t - \mu_c)(x_t - \mu_c)^T}{\sum_{t=1}^{N} I(z_t = c)} \tag{8}$$

and obtain the GP parameters $\{l_c, f_c, \sigma_c\}_{c=1}^{C}$ by maximizing the likelihood (5).

- Step 3   E-step: classify each sample into the corresponding component according to the MAP criterion:

$$z_t \leftarrow \arg\max_{1 \le c \le C} p(z_t = c \mid x_t, y_t) = \arg\max_{1 \le c \le C} \pi_c N(x_t \mid \mu_c, S_c) N(y_t \mid 0, l_c^2 + \sigma_c^2) \tag{9}$$

- Step 4   If the indicators do not change any more, stop and output the parameters of MGP. Otherwise, return to Step 2.

After the convergence of the hard-cut EM algorithm, we have obtained the estimates of all the parameters of the MGP. For a test input x^*, we can classify it into the z-th component of MGP by the MAP criterion as follows:

$$z = \arg\max_{1 \le c \le C} p(z^* = c \mid x^*) = \arg\max_{1 \le c \le C} \pi_c N(x^* \mid \mu_c, S_c) \tag{10}$$

Based on such a classification, we can predict the output of the test input via the corresponding GP using (2).

4   Experimental Results

4.1   On a Typical Synthetic Dataset of MGP

In order to test the validity and feasibility of our proposed hard-cut EM algorithm, we begin to generate a typical synthetic dataset of MGP with 4 GP components. Actually, for each GP, we typically generate 500 training samples as well as 100 test samples. For the evaluation of the algorithm, we will compute the absolute error between the learned and true parameters as well as the root mean squared error (RMSE) for the output prediction. The true parameters of the four components are given in Table 1.

We implement our proposed hard-cut EM algorithm on this synthetic dataset. It is found by the experiments that the classification error rates on the training and test datasets are 0.30% and 0.50%, respectively. The running time for both the learning and prediction tasks is just 81.8680s with an Intel(R) Core(TM) i5 CPU and 4.00GB of RAM using Matlab R2013a, which is acceptable since we have 2000 training samples and 400 test samples in total. The true and estimated values of the parameters as well as the absolute error between them are listed in Table 1. From Table 1, we can observe the absolute errors are generally very small, and it can be found from Fig.1 that the predictive curve fits the test points very well. Moreover, the RMSE of the output prediction is only 0.4901. In sum, our proposed algorithm for MGP is demonstrated valid and feasible on the synthetic datasets.

Table 1. The true value (TV), estimated value (EV) and absolute error (AE) of the parameters for each GP component (C) on the typical synthetic dataset of MGP

C	Value	π_c	μ_c	S_c	l_c^2	σ_c^2	f_c^2
1	TV	0.2500	0.0000	0.1000	0.9000	0.1000	2.0000
	EV	0.2500	-0.0069	0.1006	0.7978	0.1041	0.3461
	AE	0.0000	0.0069	0.0006	0.2022	0.0041	1.6539
2	TV	0.25	3.0000	0.2000	1.0000	0.2000	3.0000
	EV	0.2495	3.0151	0.1916	1.3512	0.2007	3.1279
	AE	0.0005	0.0151	0.0084	0.3512	0.0007	0.1279
3	TV	0.2500	6.0000	0.3000	1.1000	0.3000	4.0000
	EV	0.2500	6.0219	0.2982	2.7321	0.2966	2.6416
	AE	0.0000	0.0219	0.0018	1.6321	0.0034	1.3584
4	TV	0.2500	9.0000	0.4000	1.2000	0.4000	5.0000
	EV	0.2505	9.0060	0.4085	0.5385	0.3790	7.7323
	AE	0.0005	0.0060	0.0085	0.6615	0.0210	2.7323

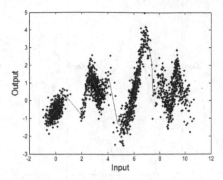

Fig. 1. The predictive curve (red solid line) and test sample points (black dots) of our proposed hard-cut EM algorithm on the typical synthetic dataset

4.2 Comparison with the LOOCV and Variational Hard-Cut EM Algorithms

We further compare our proposed hard-cut EM algorithm with the LOOCV and variational hard-cut EM algorithms on a toy dataset and a motorcycle dataset used in [7] and [8]. Actually, the toy dataset consists of four groups which are generated from 4 continuous functions. For each group, there are 50 training samples and 50 test samples. For the purpose of prediction, we certainly use the MGP with 4 components.

Motorcycle dataset is another popular one for the evaluation of the MGP methods [3],[7]. It consists of observations of accelerometer readings at 133 different times, belonging to three actual classes. Here, we use the 7-fold cross-validation, with the k-th fold being composed of $\{(x_t, y_t) : t = 7s + k, s = 0, 1, \ldots, 18\}$, where the inputs x_t are sorted in an ascending order. In this case, we use the MGP with 3 components.

We implement each of the three hard-cut EM algorithms five times on these two datasets under the same computational environment as above. The average prediction RMSEs and running times of the three algorithms on toy and motorcycle dataset are listed in Tables 2, respectively, while the corresponding predictive curves are also plotted in Figs. 2 & 3.

From Table 2, it can be found that our proposed hard-cut EM algorithm converges more accurately than the LOOCV and variational hard-cut EM algorithms. The reason may be that our proposed algorithm is more precise since it is strictly derived without

Table 2. The average prediction RMSEs and running times of the three hard-cut EM algorithms on toy and motorcycle dataset

The hard-cut EM algorithms	Toy dataset		Motorcycle dataset	
	Average RMSE	Average Time (s)	Average RMSE	Average Time (s)
Our proposed EM algorithm	16.5199	8.4513	19.1109	2.0401
The LOOCV EM algorithm	24.9874	43.4974	28.9551	10.5501
The variational EM algorithm	20.4238	57.3100	26.7883	62.0234

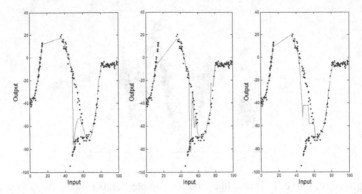

Fig. 2. The predictive curves (red solid line) and test points (black dots) of our proposed hard-cut EM algorithm (left), the LOOCV hard-cut EM algorithm (middle) and the variational hard-cut EM algorithm (right) on toy dataset

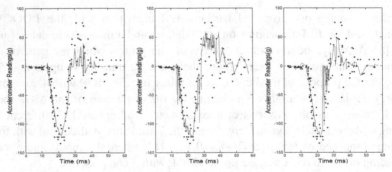

Fig. 3. The predictive curves (red solid line) and test points (black dots) of our proposed hard-cut EM algorithm (left), the LOOCV hard-cut EM algorithm (middle) and the variational hard-cut EM algorithm (right) on motorcycle dataset with 7 fold CV

any approximations like those used in the LOOCV decomposition and variational inference. Besides, our algorithm consumes much less time than the two EM algorithms due to its easy computation of the posterior probabilities. It can be also observed from Figs 2 & 3 that the predictive curves of our algorithm fit at least as well as the variational hard-cut EM algorithm, while these two hard-cut EM algorithms are smoother and fit better than those of the LOOCV hard-cut EM algorithm. On the whole, our proposed hard-cut EM algorithm clearly outperforms the LOOCV and variational hard-cut EM algorithm for prediction on both toy and motorcycle datasets.

5   Conclusion

We have investigated the learning problem of mixture of Gaussian processes (MGP) and proposed a precise hard-cut EM algorithm for it. In order to get this algorithm, the generative MGP model is redefined and the posterior probabilities of the latent indicators are strictly derived. In the algorithm design, the samples are partitioned in a hard-cut way according to the MAP criterion on their posterior probabilities obtained in E-step, while each GP component is learned independently in M-step. It is demonstrated by the experimental results that our proposed hard-cut EM algorithm is effective and efficient, and even outperforms the LOOCV and variational hard-cut EM algorithms.

Acknowledgements. This work was supported by the Natural Science Foundation of China for Grant 61171138. The authors would like to thank Dr. Yang Yan for her valuable discussions on the analysis and comparison of the LOOCV hard-cut EM algorithm for MGP.

References

1. Yuan, C., Neubauer, C.: Variational mixture of Gaussian process experts. In: Advances in Neural Information Processing Systems, vol. 21, pp. 1897–1904 (2009)
2. Tresp, V.: Mixtures of Gaussian processes. In: Advances in Neural Information Processing Systems, vol. 13, pp. 654–660 (2000)
3. Nguyen, T., Bonilla, E.: Fast Allocation of Gaussian Process Experts. In: Proceedings of The 31st International Conference on Machine Learning, pp. 145–153 (2014)
4. Stachniss, C., Plagemann, C., Lilienthal, A.J., et al.: Gas Distribution Modeling using Sparse Gaussian Process Mixture Models. In: Proc. of Robotics: Science and Systems, pp. 310–317 (2008)
5. Lázaro-Gredilla, M., Van Vaerenbergh, S., Lawrence, N.D.: Overlapping Mixtures of Gaussian Processes for the data association problem. Pattern Recognition 45, 1386–1395 (2012)
6. Ross, J., Dy, J.: Nonparametric Mixture of Gaussian Processes with Constraints. In: Proceedings of the 30th International Conference on Machine Learning, pp. 1346–1354 (2013)
7. Yang, Y., Ma, J.: An efficient EM approach to parameter learning of the mixture of Gaussian processes. In: Liu, D., Zhang, H., Polycarpou, M., Alippi, C., He, H. (eds.) ISNN 2011, Part II. LNCS, vol. 6676, pp. 165–174. Springer, Heidelberg (2011)
8. Meeds, E., Osindero, S.: An alternative infinite mixture of Gaussian process experts. In: Advances in Neural Information Processing Systems, vol. 18, pp. 883–890 (2006)
9. Sun, S.: Infinite mixtures of multivariate Gaussian processes. In: Proceedings of the International Conference on Machine Learning and Cybernetics, pp. 1–6 (2013)
10. Sun, S., Xu, X.: Variational inference for infinite mixtures of Gaussian processes with applications to traffic flow prediction. IEEE Trans. on Intelligent Transportation Systems 12(2), 466–475 (2011)
11. Rasmussen, C.E., Williams, C.K.I.: Gaussian processes for machine learning. In: Adaptive Computation and Machine Learning. The MIT Press, Cambridge (2006)

Comparison of EM-Based Algorithms and Image Segmentation Evaluation

Mei Niu, Qinpei Zhao[*], and Hongyu Li

School of Software Engineering, Tongji University, Shanghai, China
{sunnyniu1093,qinpeizhao,hongyuli.tj}@gmail.com

Abstract. Expectation-Maximization (EM) algorithm is used in statistics for finding maximum likelihood estimates of parameters in probabilistic models, where the model depends on unobserved latent variables. The idea behind the EM algorithm is intuitive and natural, which makes it applicable to a variety of problems. However, the EM algorithm does not guarantee convergence to the global maximum when there are multiple local maxima. In this paper, a random swap EM (RSEM) algorithm is introduced and compared to other variants of the EM algorithms. The variants are then applied to color image segmentation. In addition, a cluster validity criterion is proposed for evaluating the segmentation results from the EM variants. The purpose of this paper is to compare the characteristics of the variants with split and merge strategies and stochastic ways and their performance in color image segmentation. The experimental results indicate that the introduced RSEM performs better with simpler implementation than the other variants.

Keywords: EM algorithm, color image segmentation, clustering, evaluation.

1   Introduction

The usefulness of mixture models in any area which involves the statistical modeling of data (such as pattern recognition, computer vision, signal and image analysis, machine learning) is currently widely acknowledged. As a standard method for fitting finite mixture models, particularly normal mixture models, the use of the EM algorithm has been demonstrated for the analysis of data from a wide variety of fields. However the EM algorithm does not necessarily lead to globally optimal solution. If the likelihood function has several maxima and stationary points, convergence of the EM algorithm to global or local depends on the choice of starting points and the configuration of the data set.

To overcome the problems mentioned above, many variants of the EM algorithm have been proposed. As the final result will be affected by the initialization, some of the variants focus on choosing the starting values for the EM algorithm. Usually, EM starts with randomly assigning values to all the parameters to be estimated. A

---

[*] Corresponding author.

D.-S. Huang et al. (Eds.): ICIC 2014, LNAI 8589, pp. 76–86, 2014.

common way is to repeat the initialization with different random selections, and take the best selection. A more practical way is to first run hard clustering (e.g. k-means) on the data set and then interpret each cluster as a Gaussian component. Several initialization methods have been presented in [1, 2]. However, the methods improving the initialization step cannot solve the problem of local maxima. Thus, most of the other variants try to improve the algorithm itself. Among the variants, split and merge strategies and stochastic way are quite popular.

To address the inappropriate distribution of the components in data space, Ueda and Nakano proposed split and merge EM (SMEM) algorithm [3]. A more efficient greedy EM (GEM) algorithm was proposed in [4]. It reduces the problem of learning a k-component mixture model to a sequential learning of two-component model, and it offers a mechanism of dynamically allocating new components outside the over-populated center regions. GEM utilizes the split strategy, in which model selection method is employed as a stopping rule. To avoid the local trap of the EM algorithm, stochastic versions were proposed. These stochastic versions are the SAEM algorithm [5] and reversible jump MCMC [6]. Resampling methods and randomization are used in these versions.

Cluster validity is an important issue in clustering analysis. The problem comes up after clustering that how good the algorithm is. In modeling-based clustering, a number of model selection methods have been proposed. Criteria, such as Bayesian Information Criterion (BIC) [7] and minimum description length (MDL) [8], have been used for mixtures. In image segmentation problem, evaluating the segmentation result is a necessary study. We propose an evaluation criterion in this paper to estimate the image segmentation results from the EM-based algorithms.

The purpose of this paper is to compare the characteristics of the EM variants with split and merge strategies and stochastic methods. The RSEM algorithm as a stochastic method demonstrates the effectiveness and simpleness in the comparison. To display how the EM-based algorithm works in reality, the variants are applied in color image segmentation. The cluster validity problem thus forms into evaluating the segmented results with different algorithms.

2   EM Algorithm and Its Variants

2.1   Generalized EM Algorithm

We briefly review the main features of the EM algorithm [9]. The log-likelihood function for complete data $Z = \{X, Y\}$ is defined as:

$$L(\Theta \mid Z) = \sum_{n=1}^{N} \log\left(P\left(y_i \mid x_i\right)P(x)\right) = \sum_{n=1}^{N} \log\left(\alpha_{x_n} g_{x_n}\left(y_n \mid \theta_{x_n}\right)\right) \qquad (1)$$

where, Y is the observed data, X is the missing information. The log-likelihood function for incomplete data can be defined as:

$$L(\Theta \mid Y) = \log \prod_{n=1}^{N} p\left(y_n \mid \Theta\right) = \sum_{n=1}^{N} \log \sum_{m=1}^{M} \alpha_m g_m\left(y_n; \theta_m\right) \qquad (2)$$

where, N is the number of data points, M is the number of components, and α_m is the mixing proportion of the mth component. The $g_m(y_n ; \theta_m)$ is a d-dimensional density model corresponding to the mth component, where y_n is the nth observed data point, and θ_m is the parameters of the mth component. The α_m must satisfy the constraints:

$$\sum_{m=1}^{M}\alpha_m = 1, and, \alpha_m \geq 0, m = 1,...,M. \tag{3}$$

It is difficult to optimize Equation 2 because it contains the log of the sum and the log-likelihood $L(\Theta/Z)$ is unobservable. Hence the expectation function of $L(\Theta/Z)$ given Y is defined in the EM algorithm: $Q(\Theta; \Theta^{(k)}) = E\{ L(\Theta/Z)|Y, \Theta^{(k)}\}$.

Define the EM algorithm as follows:

1. Initialize the parameters;
2. E-step: calculate Q function $Q(\Theta; \Theta^{(k)})$;
3. M-step: choose Θ to be any value which maximizes $Q(\Theta; \Theta^{(k)})$;
4. Evaluate the log-likelihood, and stop until convergence.

It may not be numerically feasible to find the value of Θ that globally maximizes the Q function. That is, one choose $\Theta^{(k+1)}$ to increase the Q function $Q(\Theta; \Theta^{(k)})$ over its value at $\Theta = \Theta^{(k)}$, rather than to globally maximize it over all Θ. Under suitable regularity conditions, $\Theta^{(k)}$ converges to a stationary point of $L(\Theta)$. However, when there are several stationary points (local maxima, minima, saddle points), $\Theta^{(k)}$ does not necessarily converge to a significant local maximum of $L(\Theta)$ [9]. In practice, the EM algorithm has been observed to be extremely slow in some applications. As mentioned in [9], slow convergence appears when the proportion of missing information is high.

2.2   Variants

Split and merge strategies have been often used in image segmentation. An iterative split-and-merge algorithm is proposed to generate codebook in vector quantization [10]. Splitting is applied in the X-means algorithm [11] to improve the problems existing in K-means. With the split and merge (SM) operations in the EM algorithm, the criteria for choosing the candidates should be considered. There are a variety of ways to decide the criteria for splitting and merging based on the data set and algorithm itself. Consequently, finding a proper criterion is one of the constraints in the SM operations. For example, the split operation in SMEM is identified as an ill-posed problem [12].

The stochastic versions commonly utilize resampling method and randomization, which is heavy computationally involving. We introduce RSEM, a variant based on random swap, which has less time complexity than the usual stochastic variants. Two variants based on split and merge strategies are also discussed in the comparison. One is the SMEM algorithm [3] with unchanged number of components while the other is the GEM algorithm [4], which only employs split operation and the number of components increases in each successive iteration.

A random swap EM (RSEM) algorithm represents a random version in the comparison. The algorithm can be described as follows:

1. Initialization by several runs of k-means;
2. Perform a number of RSEM iterations.

In each RSEM iteration, the algorithm starts with a random swap. Then a conventional EM algorithm is applied to promise the algorithm convergence.

In each random swap, the algorithm randomly chooses one component (i) and relocates it to a position among the points (y_x). Update the Q function as follows:

$$Q^* = Q_i' + \sum_{j \neq i} Q_j \tag{4}$$

where, Q_i' is the updated Q function of the randomly selected component where the parameter $\Theta_i' = \{\alpha_i', u_i', \sum_i'\}$. Let the new parameters $u_i' = y_x$, $\alpha_i' = \sum_{j=1, j \neq i}^{m} (\alpha_j * P_j)$ and the covariance $\sum_i'$ is kept unchanged.

Run the conventional EM algorithm to converge, where Θ^* is the obtained parameters. If the log-likelihood improves, then accept the new solution, $Q = Q^*$, $\Theta = \Theta^*$; otherwise, continue the process.

To avoid the sensitivity of initialization, RSEM utilizes the randomization in the algorithm. The procedure of randomization breaks up the configuration of the previous step, which can overcome the problems of EM. Moreover, the underlying EM characteristic helps to find good estimates of the parameter in a comparatively small number of iterations. Finally, it is easy to implement.

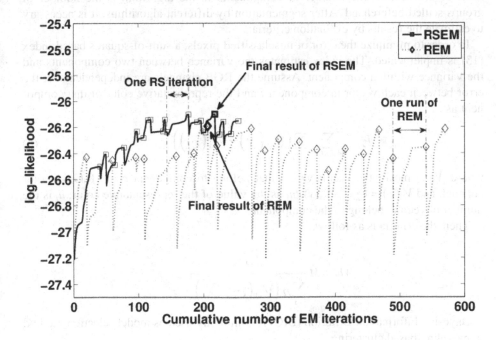

Fig. 1. An example run of repeated EM (REM) and the random swap EM (RSEM) on synthetic dataset S1. Number of repeats and number of swaps is set to 20. After each initialization (for REM) and after each swap (for RSEM) EM was iterated until convergence. Final log likelihood for REM is -26.19 and RSEM -26.10.

Differences in obtained log-likelihood and the amount of work spent, in terms of EM iterations, are seen in Fig. 1. We see that after each restart in REM, it takes many iterations for EM to reach a level that is comparable to previously converged solution. In contrast, RSEM needs only a few EM iterations to reach an even higher level of log-likelihood.

3   Color Image Segmentation Evaluation

Unsupervised methods are commonly accepted as the image segmentation approaches. Color image segmentation is the process of partitioning the features of colors in a digital color image to find the segments with similar colors. The color image segmentation problem can be solved by the EM-based algorithms as follows. Let N color features (e.g. RGB color space in this paper) of the image be the observed data y_n, M denotes the number of groups, the unobserved latent data is the nearest representative color $\{c_1, c_2, ..., c_M\}$ for each feature vector. In the EM-based algorithms, vector (n) belongs to component (m) with its probability $\{\alpha_{n1}, \alpha_{n2}, ..., \alpha_{nM}\}$. To produce partitions, the component with the highest probability will be taken as the representative label of the vector. The number of components in the algorithm is the number of groups settled beforehand. After segmentation by different algorithms, it is necessary to estimate the results by evaluation criteria.

In order to minimize the error of misclassified pixels, a sum-of-squares based index [13] is implemented. The index estimates the variance between two components and the variance within a component. Assume the RGB channels are independent, set the error between each vector in component i and the representative color of that component as:

$$E_i = \sum_{x \in \{R,G,B\}} \sum_{y \in C_i} \left(V_x(y) - V_x(c_i)\right)^2 \tag{5}$$

where, $V_x(y)$ means the vector values that belong to component c_i in $\{R,G,B\}$ color channel, and $V_x(c_i) = (\sum_{y \in i} V_x(y))/n_i$ is the value of the representative color, n_i is the number of vectors belong to the component.

Then, the criteria is as follows:

$$WbI = M \frac{\sum_{i=1}^{M} E_i}{\sum_{i=1}^{M} n_i \left(V_x(c_i) - \bar{V}\right)^2} \tag{6}$$

Bayesian Information Criterion (BIC) is commonly used as model selection method in modeling based clustering.

$$BIC = L \times N - \frac{1}{2} M (d+1) \sum_{m=1,...,M} \log(n_m) \tag{7}$$

where, L is the log-likelihood of the model, and d is the dimension of the data. A comparison of log-likelihood values between RSEM and EM is shown in Fig. 2.

$$\text{Combined} = \frac{BIC - MIN(BIC)}{MAX(BIC) - MIN(BIC)} + \frac{WbI - MIN(WbI)}{MAX(WbI) - MIN(WbI)} \qquad (8)$$

where, MIN returns the minimum value, while MAX returns the maximum one.

The WbI criterion has the statistical property which discovers the relationship of the segmented image and the original image, while BIC has the background of information theory, which can find the best model configuration. Thus the combined criterion has the advantage of both criteria. The smaller value of this combined criterion indicates better performance.

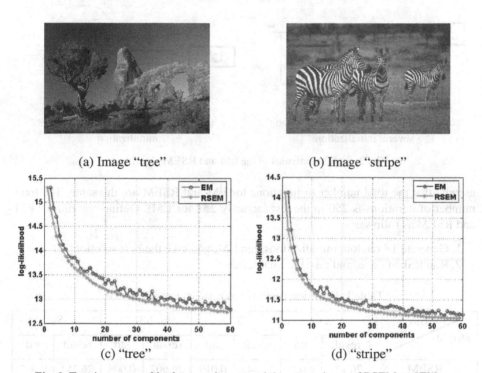

Fig. 2. Two images used in the experiments and the comparisons of RSEM and EM

4   Experiments and Results

In this paper, we employ both artificial and real data sets. The data sets S1 to S4[1] are generated with varying complexity in terms of spatial data distributions. Each set has 5000 vectors scattered around 15 predefined components with a varying degrees of overlap. Iris is a 3-dimensional real data set, which contains 3 classes of 50 instances each. One dimensional real data set "Galaxy" consists of the velocities of 82 distant

---

[1] http://sse.tongji.edu.cn/zhaoqinpei/Datasets/

galaxies. The image data sets in Fig. 2 are in RGB color space. In addition, the log-likelihood value mentioned in this section represents the value of the log-likelihood divided by the sample size.

4.1 Comparisons of the Variants

The comparison between RSEM with one initialization rather than EM with several randomly initial solutions is performed in the following experiment (Fig. 3). We

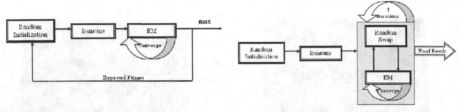

(a) EM: the best initial solution among
several initializations

(b) RSEM: the final result with one
initialization

Fig. 3. System diagrams of the EM and RSEM algorithms

assume that the total number of iterations for EM and RSEM are the same. The total number of iterations is 250 in the test (actually 252 for EM). Define one run for EM and RSEM as follows:

1. Generate 18 random initializations, run EM, and take the best solution;
2. Run RSEM once, and take the final result.

Table 1. Numerical comparisons between EM and RSEM

Method \ Data	S1		S2		S3		S4	
	result	std	result	std	result	std	result	std
RSEM	26.121	0.025	26.445	0.019	26.602	0.008	26.354	0.007
EM	26.140	0.041	26.451	0.030	26.616	0.010	26.392	0.010

To make it more reliable, 50 runs of both algorithms on data set S1-S4 are tested. The mean value and standard deviation are listed in Table 1. The average log-likelihood of RSEM is better than that of EM. The standard deviations also indicate that RSEM is more stable. The comparisons of EM and RSEM using different components in Fig. 5 demonstrate that RSEM achieves smaller log-likelihood, which means RSEM performs better than EM on the image data sets.

We find out that SMEM algorithm is not compatible with the log-likelihood framework, which is also mentioned in [14]. In this case, the "Galaxy" data is tested to visualize the results from different variants. The Gaussian mixture models by each variant are displayed in Fig. 4.

4.2   Evaluation Results

The prospective applications of RSEM could be different kinds of image and speech processing. We applied the proposed algorithm in color image segmentation, where images with RGB color space are tested. However, the problem of evaluating the results still remains. As Fig. 6 demonstrates, it is difficult to estimate the segmentation results by eyes with different number of segments. Thus it is necessary to employ the evaluation criteria to get a range of the number of segments. The Iris and S1-S4 data sets are tested first as the correct number of components is known. The results by the combined index (WBIndex and BIC) are shown in Fig. 5. It gives a good indication to 3 in the case of Iris, it also predicts a range from 13 to 16 for S1-S4 data sets. According to these results, we tested the criterion on two images: stripe and tree. As shown in Fig. 5, The segmentation results have little difference with the cluster number from 17 to 21 in image "stripe"; while the range from 23 to 28 is good for image "tree". The evaluation criteria could at least give a prior knowledge about the segment numbers in the segmentation results.

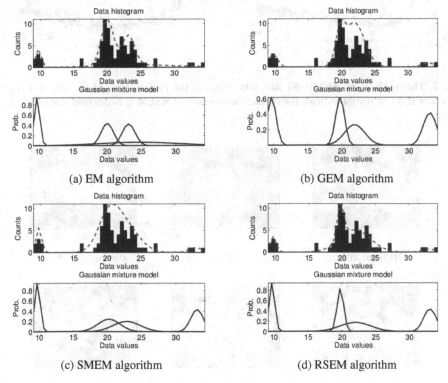

Fig. 4. The results (GMMs) from different variants on one dimensional data set "Galaxy"

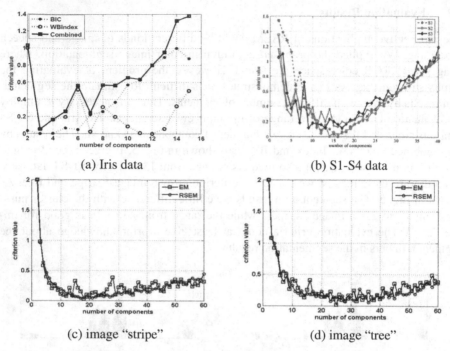

(a) Iris data   (b) S1-S4 data

(c) image "stripe"   (d) image "tree"

Fig. 5. The result of Iris and S1-S4 data sets indicates the performance of the criterion. The results of EM and RSEM indicate the segmentation result of RSEM is better than that of EM in general.

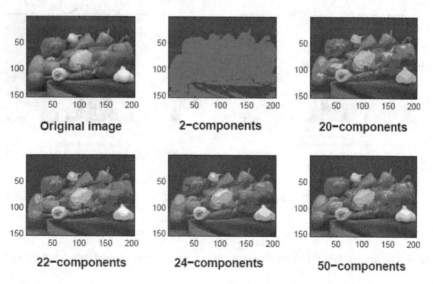

Original image   2-components   20-components

22-components   24-components   50-components

Fig. 6. The segmentation results by RSEM algorithm on image "pepper" under different number of clusters

5   Conclusion

In this paper, we proposed a random swap EM algorithm under the assumption of Gaussian distribution. With randomly perturbing the configuration of the results from the EM algorithm, the method achieved better performance with lower time complexity. Meanwhile, an image segmentation evaluation method for the EM-based algorithms' results is introduced. Experiments conducted on different data sets and different methods demonstrated that the proposed algorithm made a clear improvement over the conventional EM algorithm. Meanwhile, the try in color image segmentation also made it applicable. We believe that the proposed method also adapts to other distributions, and it could have a wider application.

Acknowledgement. The paper is supported by the Fundamental Research Funds for the Central Universities.

References

1. Biernacki, G., Celeux, C., Govaert, G.: Choosing Starting Values for The EM Algorithm for Getting The Highest Likelihood in Multivariate Gaussian Mixture Models. Computational Statistics and Data Analysis 41, 561–575 (2003)
2. Karlis, B.D., Xekalaki, E.: Choosing Initial Values for The EM Algorithm for Finite Mixtures. Computational Statistics and Data Analysis 41, 577–590 (2003)
3. Udea, N., Nakano, R., Gharhamani, Z., Hinton, G.: SMEM Algorithm for Mixture Models. Neural Computation 12, 2109–2128 (2000)
4. Verbeek, J., Vlassis, N., Krose, B.: Efficient Greedy Learning of Gaussian Mixture Models. Neural Computation 15(12), 469–485 (2003)
5. Wei, G.G.G., Tanner, M.A.: A Monte Carlo Implementation of The EM Algorithm and The Poor Man's Data Augmentation Algorithms. Journal of the American Statistical Association 85, 699–704 (1990)
6. Richardson, S., Green, P.J.: On Bayesian Analysis of Mixtures with An Unknown Number of Components. Journal of the Royal Statistical Society 59(4), 731–792 (2002)
7. Frayley, C., Raftery, A.: How Many Clusters? Which Clustering Method? Answers Via Model-based Cluster Analysis. Technical Report no. 329, Department of Statistics, University of Washington (1998)
8. Rissanen, J.: Stochastic complexity in statistical inquiry. World Scientific, Singapore (1989)
9. Mclachlan, G., Krishnan, T.: The EM Algorithm and Extensions. John Wiley & Sons (1996)
10. Kaukoranta, T., Fränti, P., Nevalainen, O.: Iterative splitand-merge algorithm for vector quantization codebook generation. Optical Engineering 37(10), 2726–2732 (1998)
11. Pelleg, D., Moore, A.: X-means: Extending K-means with efficient estimation of the number of clusters. In: Proceeding of the 17th Int. Conf. on Machine Learning, pp. 727–734 (2000)

12. Zhang, Z., Chen, C., Sun, J., Chan, K.: EM algorithms for Gaussian mixtures with split-and-merge operation. Pattern Recognition 36(9), 1973–1983 (2003)
13. Zhao, Q., Xu, M., Fränti, P.: Sum-of-squares based cluster validity index and significance analysis. In: Kolehmainen, M., Toivanen, P., Beliczynski, B. (eds.) ICANNGA 2009. LNCS, vol. 5495, pp. 313–322. Springer, Heidelberg (2009)
14. Minagawa, A., Tagawa, N., Tanaka, T.: SMEM algorithm is not fully compatible with maximum-likelihood framework. Neural Computation 14(6), 1261–1266 (2002)

Artificial Curiosity Driven Robots with Spatiotemporal Regularity Discovery Ability

Davood Kalhor and Chu Kiong Loo

Faculty of Computer Science, University of Malaya, Malaysia
kalhor@siswa.um.edu.my,
ckloo.um@um.edu.my

Abstract. Autonomous reinforcement learning (RL) robots usually need to learn from raw, high dimensional data generated by visual sensors and often corrupted by noise. These sorts of tasks are quite challenging and cannot be addressed without an efficient mechanism to encode and simplify the raw data. A recent study proposed an artificial curios robot (ACR) for this problem. However, this model is incapable of handling non-Markovian tasks and discovering spatiotemporal patterns in its milieu. This paper presents a method to solve this problem by extending ACR. A straightforward, but not efficient, solution is to keep recoding of previous observations which makes the algorithm intractable. We, instead, construct a perceptual context in a compact way. Using different environments, we show that the proposed algorithm can discover the regularity in its environment without any prior information on the task.

Keywords: Artificial curiosity, autonomous robots, developmental robotics, reinforcement learning, spatiotemporal discovery.

1 Introduction

In the real-world applications, autonomous reinforcement leaners are often faced with extremely hard problems due to the following reasons: They must process raw data coming from multiple sensory systems that are noisy and high dimensional. Their environments are generally dynamic and contain multiple still and moving objects (or goals), which some of them may even follow specific space-time patterns. The robot does not initially have any clue to where it must search, how long should continue exploration, it is better to wait or move in a particular time, it should revisit a state or not, and when is the best time for exploitation. Coping with these problems are the main sources of motivation for researchers in this interesting area.

Majority of works in the literature address RL problems with Markov properties, where the history of observations does not play any role in state transitions. Real problems are often more complex and they are non-Markovian. These tasks cannot be handled without a memory. There are a few approaches to address non-Markovian tasks, including recurrent neural networks [1], finite state automata [2], and long

D.-S. Huang et al. (Eds.): ICIC 2014, LNAI 8589, pp. 87–98, 2014.
© Springer International Publishing Switzerland 2014

short-term memory [3]. Despite promising results for these methods, they are evaluated only for tasks with a low-dimensional space (such as a T maze with size 10) which is far from problems that a robot for real-world application encounters.

The main objective of this paper is to develop an autonomous developmental robot that can learn discrete time series and sequential patterns from high-dimensional noisy observation. To the best of our knowledge this problem has not been addressed by previous studies. This work attempts to address this problem by extending the ACR through introducing perceptual context for encoding the history of observation.

1.1 Reinforcement Learning

In RL [4], an agent (or robot) is not told what to do. Instead, it must learn from its own experience which behaviors lead to the most reward. Learning an optimal policy is the core of any RL algorithms. These methods use either value iteration or policy iteration approaches, and they are divided into model-based and model-free methods [4], [5]. One of the well-known methods of RL is the Dyna architecture [6], which is based on dynamic programming and learns an environment model online.

Skill acquisition in humans is made through a hierarchical approach. This plays a crucial role in the problem solving competency of humans. The same strategy of learning has been employed by some researchers. For example, Vigorito and Barto [7] presented a framework for autonomous learning of a hierarchy of abstract skills using intrinsically motivated hierarchical RL [8]. In contrast with many of the previous works, they used factored Markov Decision Process (MDP) instead of standard MDP. This enables learning options with different state representation and leads to the state abstraction. They showed that the construction of policies of abstract skills can significantly reduce the computational burden compared with flat policy representations.

1.2 Artificial Curiosity

Classical RL algorithms are goal driven in this sense that they adapt their behavior to achieve more rewards, i.e., external motivation from a psychological point of view. A major problem of these algorithms is they lack any idea what to do in the absence of external goals. Neuropsychological evidence illustrates that there is also an internal mechanism, known as intrinsic motivation, deriving human behaviors even without any goal. One of early computational models of intrinsic motivation was introduced by Schmidhuber [9]; please refer to [10] for a summary on computational motivation systems. In the artificial curiosity (AC) framework [20], intrinsic reward is defined as a measure of the prediction learning progress. A group of researchers [11] made some comparison between the AC and active learning and suggested that the AC can be seen as an extension to active learning. The AC framework offers two advantages: 1) it enables a robot to continue exploration in the absence of external goals, and 2) it will not be confused to guide the robot in noisy environments. These advantages have been experimentally proven by previous studies [12-14]. For these reasons, we use the AC framework for computing reward.

1.3   Developmental Robotics

The work in this paper can be categorized under developmental robotics [15], which attempts to imitate the developmental learning of the human fetus and infants [16]. This class of learning is the result of self-discovery through interactions with environments in an open-ended, incremental, and self-organized manner. To build truly intelligent robots, it is crucial to equip them with developmental learning. However, this is very difficult in real configurations. For this reason, there are few attempts trying to address this challenge. For example, Ngo et al. [13] developed a developmental system for skill learning through an AC driven exploration. Upon receiving an instance of observation, a confidence-based approach is used to determine if the current situation is known or not. In a more recent work, Frank et al. [12], for the first time, implemented a curiosity-driven agent for realtime motion planning on a humanoid robot. Despite very promising results, these works are limited to Markovian tasks where an environment is modeled by MDP. In another work, Luciw et al. [14] introduced an artificial curios robot (ACR). This system maps noisy visual sensory data into some internal states using a perceptual module. Another module of this system learns to predict state transitions and reward for each state-action pair. It also generates some signals representing interestingness of each state transition. The last module of the system constructs a tabular map to relate a state-action pair to an internal value. Higher value pairs are supposed to lead to more rewards in the long run and this guides the robot to choose the best action at each time. Lack of memory is one of main disadvantages of ACR that limits its application for non-Markovian environments. The method in this paper extends this work by introducing memory to the perceptual module. Direct usage of the observation history is not feasible due to associated time complexity. Therefore, we use the concept of context to resolve this problem.

2   Developmental Autonomous Robots with Artificial Curiosity Driven Mechanism for Spatiotemporal Regularity Discovery

ACR-STRD transfers features of ACR from Markovian framework to non-Markovian. It also extends the spatial novelty detection of ACR to spatiotemporal regularity discovery. ACR-STRD, similar to original ACR, consists of three main modules: perceptual, cognitive, and value modules. The main difference of the two systems is the perceptual module, which is discussed in detail here. However, the two other modules are also briefly reviewed. ACR-STRD receives a noisy observation of its world. The main duty of the perceptual module is to encode this raw sensory data and construct a context for perception. The context carries important information on the history of observations and gives a perception memory to the robot. The cognitive module provides the value module with some estimations of possible outcome for each action. Using this information, the value module selects and performs the best possible action at each time step. An action potentially changes the world state, thereby leading to a new observation. After each action, the robot also receives an external reward describing the immediate value of the action. By continuing this process, the robot learns from its experience and improves its model of the world. An illustration of the architecture of ACR-STRD is depicted in Fig. 1. The ACR-STRD algorithm is presented in a modular form in Fig. 2.

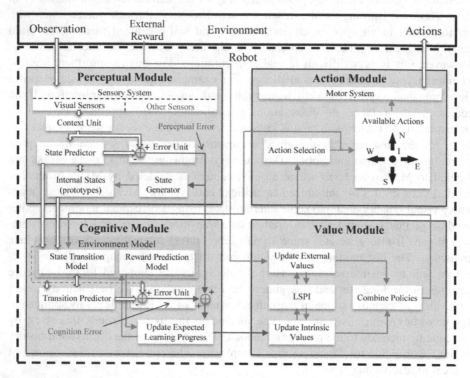

Fig. 1. The ACR-STRD architecture. The robot can perform five actions denoted by N, S, W, E, and I (immobile); each action (except I) is a step toward the corresponding direction.

2.1 Perceptual Module

In the real-world applications, a robot often needs dealing with problems that have an extremely large observation space due to the high-dimensionally nature of visual sensors. These observations represent physical (or external) states of the environment that are unknown for the robot. The state space is usually huge owing to noise and inherent complexity of the real-world problems. In addition, the real world is dynamic and comprises not only stationary objects, but also moving objects whose movements usually follow regular time-space patterns. This explains the importance of the robot ability to discover spatiotemporal regularities in its environment. High dimensional observations and huge state space are sufficient to impose a high computational burden on the robot. Adding the time dimension to this problem makes it even more intractable. In ACR-STRD, these difficulties are addressed as follows. The first problem is handled by building an internal model of the world to map each observation into an internal state. To resolve the second problem, observations are divided to several clusters, each of which is an internal state representing a group of observations. The third problem is addressed by using the idea of context to describe the history of observations.

An observation is expressed by $\mathbf{o}(t) = [o_1(t), \ldots, o_N(t)]^T$, where $\mathbf{o} \in \mathcal{O}$ (observation space, $\mathcal{O} \subset \mathbb{R}^N$). ACR-STRD, in contrast to ACR, uses a sequence of observations for inference. This enables the robot to handle non-Markovian environments. A sequence is defined as $\mathbf{O}(\leq t) = [\mathbf{o}(1), \ldots, \mathbf{o}(t)]$. Processing of such a huge data is not feasible for practical settings. To tackle the problem, we borrowed some ideas from Merge Growing Neural Gas (MGNG) [17], an algorithm initially developed for time series analysis, and combined it with adaptive vector quantization [18]. In ACR-STRD, a sequence of observations is represented by two vectors: current observation $\mathbf{o}(t)$ and global context $\mathbf{c}_g(t)$. In this way, sequence $\mathbf{O}(\leq t)$ that has an enormous dimension tN is represented with a substantially smaller vector of size $2N$. An internal state s is modeled as a node with two vectors $\mathbf{w}_s$, $\mathbf{c}_s \subset \mathbb{R}^N$. At each time step, a winner node is determined by finding the node (or state) with the smallest distance d_s

$$s^* = \operatorname{argmin}_s d_s(t), \tag{1}$$

$$d_s(t) = (1 - \alpha_c) \cdot \|\mathbf{o}(t) - \mathbf{w}_s(t)\|^2 + \alpha_c \cdot \|\mathbf{c}_g(t) - \mathbf{c}_s(t)\|^2, \tag{2}$$

where s^* is the winner node, parameter $\alpha_c \in [0,1]$ is used to control the importance of the current observation over the past, and $\mathbf{c}_g$ provides a compact representation of the history of observation into an N-dimensional vector defined as

$$\mathbf{c}_g(t + 1) = (1 - \beta_c) \cdot \mathbf{w}_{s^*}(t) + \beta_c \cdot \mathbf{c}_{s^*}(t), \tag{3}$$

where $\beta_c \in [0,1]$ is a weight parameter controlling the significance of the recent experiences over the far. If the distance is larger than a predefined threshold κ, then a new node is added to the network with the following prototypes:

$$\mathbf{w}_{s_{new}}(t) = \mathbf{o}(t), \quad \mathbf{c}_{s_{new}}(t) = \mathbf{c}_g(t). \tag{4}$$

Otherwise, the weights of the winner node are adjusted according to update rules:

$$\mathbf{w}_{s^*}(t + 1) = \mathbf{w}_{s^*}(t) + \eta_{s^*}^{per}(\mathbf{o}(t) - \mathbf{w}_{s^*}(t)), \tag{5}$$

$$\mathbf{c}_{s^*}(t + 1) = \mathbf{c}_{s^*}(t) + \eta_{s^*}^{per}(\mathbf{c}_g(t) - \mathbf{c}_{s^*}(t)), \tag{6}$$

where η_s^{per} is an adaptive learning rate to speed up the learning of new experiences, adopted from [19]; please see [14] for a good summary on this parameter.

Using the above mechanism, ACR-STRD adapts itself to its observations by either adding new prototypes or adjusting the existing ones. This process is incremental and growing-based, hence appropriate for online learning. Furthermore, ACR-STRD does not involve in costly computations, because its model of the environment is based on a sparse representation. This can be thought as an intelligent sampling of the important space-time information of the world. The density of the sampling can be controlled by parameter κ.

$\text{ACRSTRD}(\eta^{val}, \eta^{short}, \eta^{long}, \kappa, \nu, \omega, \tau, \gamma, \alpha_v, \beta_v, \alpha_c, \beta_c, \alpha_p, T)$

1. initialization
   $t := 1$; $s := 1$; $a := 1$; // time, state, and action
   $\mathbf{c}_g := \text{zeros}(N, 1)$;   // global context
   $\mathbf{c}_s := \text{zeros}(N, 1)$;   // 1st prototype's context
   $\mathbf{w}_s := \text{zeros}(N, 1)$;   // 1st prototype's weight
2. $\mathbf{o} := \text{sense}(world)$;   // 1st observation
3. repeat
   a) $t := t + 1$;
   b) $world := \text{action}(world, a)$; // execute action a
   c) $[\mathbf{o}', r^{ext}] := \text{sense}(world)$; // update observation
   d) $s' := \text{argmin}_s \, dist(\mathbf{o}', \mathbf{c}_g, \mathbf{w}_s, \mathbf{c}_s, \alpha_c)$; // find the closest state
                                        compute $dist$ using Eq. (2)
   e) $\mathbf{c}_g' := \text{updateGlobalContext}(\mathbf{c}_g, \mathbf{c}_{s'}, \beta_c)$; // for next time step
   f) $a' := \text{argmax}_i \, q_{s',i}$; // find the best action
   g) if $t > 1$
                   updatePereceptualModule(...);
                   updateCognitiveModule(...);
                   updateValueModule(...);
   h) end
   i) $\mathbf{c}_g := \mathbf{c}_g'$; $a := a'$; $s := s'$; $\mathbf{o} := \mathbf{o}'$;
4. until $t = T$ (robot dies)

updatePereceptualModule(...)

1. compute perceptual error
   $err_{per} := dist(\mathbf{o}', \mathbf{c}_g, \mathbf{w}_{s'}, \mathbf{c}_{s'})$; // compute $dist$ using Eq. (2)
2. if $err_{per} > \kappa$
      $N_s := N_s + 1$;
      $[\mathbf{w}_{N_s}, \mathbf{c}_{N_s}] := \text{newState}(\mathbf{o}', \mathbf{c}_g, N_s)$; // add a new state
      $err_{per} := 0$; // reset error for the new state
3. else
         // update the internal state & its age
         $\mathbf{w}_{s'} := \mathbf{w}_{s'} + \eta_{s'}^{per}(\mathbf{o}' - \mathbf{w}_{s'})$;
         $\mathbf{c}_{s'} := \mathbf{c}_{s'} + \eta_{s'}^{per}(\mathbf{c}_g - \mathbf{c}_{s'})$;
         $age_{s'} := age_{s'} + 1$;
4. end

Fig. 2. Pseudocode of ACR-STRD and its perceptual module. For details of the cognitive and value module please refer to [14].

2.2   Cognitive Module

A robot does not often have access to information on how the environment's states change, how these changes are influenced by its action, and possible reward for an action. This problem can be addressed by building a probabilistic model of the world. In the initial step, it is assumed the probability of transition from a state to any other state under a given action is equal. As soon as the robot acts upon its environment and observes new states, it adjusts its transition model using an update rule. ACR-STRD is based on the AC framework, and thus its behavior is driven by a combination of the external and intrinsic rewards. These signals are produced by the cognitive module. While the expected external reward is computed using an update rule similar to the state transition model, the intrinsic reward is defined as a measure of the expected learning progress [20] and calculated using short term and long term prediction errors. These two are also based on the total error which combines perceptual and cognitive errors. The cognitive error is the distance between the predicted state and the actual internal state after performing an action. For further details, please refer to [14].

2.3   Value Module

The value module is the center of decision making in ACR-STRD. An optimal plan should bring the most possible reward in long term. This requires a careful planning. This module builds a map by assigning an internal (or Q) value, to each state-action pair. Pairs with higher values lead to more reward in the long run and this guides the module to choose the best action at each state. Q-values are determined based on both intrinsic and external rewards. This combination is crucial for addressing exploration-exploitation trade-off which is a fundamental dilemma for a reinforcement learner. Q-values are learned using a mixture of SARSA [21] and LSPI [22]. A major problem of SARSA is the lack of convergence guarantees while this method is very fast and appropriate for online learning. On the other hand, LSPI is an offline, value approximation method that enjoys stability and convergence. Therefore, LSPI would be a good complement to SARSA. For fast adaptation, SARSA updates values at each step. But LSPI update is applied every τ steps to reduce the time complexity imposed by LSPI.

3   Experimental Setup and Results

3.1   Setup

For experiments, we created four new environments inspired by previous works [23], [14], but fundamentally different from them. The new environments are non-Markovian and include non-stationary goals whose their spatiotemporal patterns encode some regularities. The robot interacts with the environment via three sensorimotor channels. First channel enables the robot to get a visual observation of its world and itself. This input is a noisy, high-dimensional (225 dimensions) gray image (see Fig. 3). The environments are non-stationary and continuously change as shown in Fig. 4. The second channel allows the robot to perform five actions: one-step (i.e., 5 pixels) movement in any of four directions (up, down, left, or right) or standstill. The initial position of the robot is the upper left corner of the grid. The robot uses the last channel to collect information about the immediate value for the current state of the environment; this is known as external reward. Whenever the robot reaches a goal state, it receives a positive external reward. Other movements are penalized. Any attempt to move through the walls leads to no transition but causes a penalty. The robot does not have any prior information about the spatiotemporal behavior of environments and the visual features of itself or the goal.

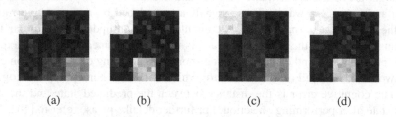

                (a)   (b)   (c)   (d)

Fig. 3. a) A sample (current) observation of environment b. b) The global context within which the observation was experienced. c) The reconstructed observation. d) The associated perceptual context. White square represents the robot and the gray one is the goal. The observation is corrupted by Gaussian noise.

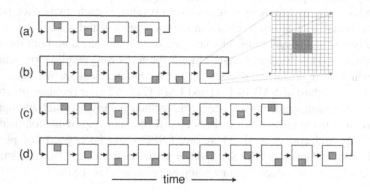

Fig. 4. Space-time changes of the four environments (a-d) used for experiments, excluding the robot. Each environment is a 15×15 grid as shown by the enlarged figure. The green regions indicate the position of goals, changing with time. Black arrows show the direction of changes. Each spatiotemporal pattern, which is repeated continuously, describes the dynamic of the corresponding environment.

There are several reasons to construct the new environments. First, the previous worlds, such as those introduced in [23] and [14], are Markovian and contain static goals. These sorts of milieus are not suitable to demonstrate the power of the proposed algorithm in this paper. Second, the new environments are more challenging since they pose problems with substantially larger state space to the robot and cannot be handled by memory-less robots. Third, a robot will not be able to cope with non-stationary worlds without action standstill, because sometimes the best action to not miss a goal is to keep its current position; agents in [23] and [14] are not equipped with this action. Fourth, conventional RL algorithms are generally intractable for visual observations, which are usually noisy and high-dimensional.

The parameters of the system were set as follows. The external reward (or penalty) for reaching a goal state, attempting to pass through a wall, and the other transitions were 30, -9, and -1, respectively. SARSA and LSPI were enabled during all tests. Other parameters to configure ACR-STRD were as follows: $\alpha_c = 0.5$, $\beta_c = 0.5$, $\kappa = 0.05$, $\omega = 0.05$, $\gamma = 0.9$, $\eta^{val} = 0.25$, $\eta^{short} = 0.2$, $\eta^{long} = 0.1$, $\alpha_v = 1, \nu = 1$, $\tau = 10$, $T = 10^4$.

We evaluated the performance of ACR-STRD in different scenarios. The first group of experiments was to show the ability of ACR-STRD in discovery spatiotemporal regularities in its world without any prior information. This test comprised two stages. First, the robot was given the opportunity to explore the world for 10000 time steps. In this stage, the system was set to follow a fully Intrinsically Motivated Policy (IntMP): $\beta_v = 0$. This setting represents fully curious robots which their mission is to discover spatiotemporal regularities (goal patterns) in its environment without any previous information about them. After gaining experience, the robot mission changed to obtain as much reward as possible. This requires a purely goal-driven action selection, or fully Externally Motivated Policy (ExtMP). During this phase (200 time steps), the robot was not allowed to gain any experience. The average of the total obtained rewards over several runs was calculated as the performance measure. To compare the performance of the proposed algorithm in this paper with ACR, we repeated the same test for all environments.

We also investigated the effect of choosing four different types of policies: fully Random Policy (RandP), IntMP, ExtMP, and Combined Policy (CombP) with $\beta_v = 0.1$. This group of experiments also consisted of two phases. In the first phase (lasted for 200 steps), the robot was allowed to gain experience. This was followed by a 100-step test phase in which learning was frozen and the robot used its experience based on ExtMP to collect reward. The robot continued the exploration-exploitation process for 4000 steps.

3.2   Results

Experimental results showed that ACR-STRD with ExtMP can obtain the maximum possible reward during the exploitation period for all the environments (see Table 1 for a summary of results). This will be possible only when the robot follows the goal pattern for all steps in the exploitation period. This implies the discovery of the space-time patterns in the exploration phase since the training was frozen in the exploitation period. On the other side, ACR was incapable of learning any of these patterns. It managed to collect some rewards by partially tracking some parts of the goal patterns. However, it always confused to make the right decision whenever it observed the environment states with multiple action options, such as the second frame of environment a, Fig. 4(a), and the second and third frames of environment b, Fig. 4(b). This was predictable because ACR is memory-less and does not have any mechanism to incorporate the state-action history into its decisions making process. In comparison, the perceptual module of ACR-STRD encodes the history of observations as the context for the robot decision making. This feature enables ACR-STRD to cope with non-Markovian tasks, such as those described in the previous section. For a better illustration of the performance of both systems on a single run, please see Fig. 5. This figure depicts the weakness of ACR in learning spatiotemporal patterns.

Table 1. Performance comparison for ACR and ACR-STRD

Env.	Avg. Obtained Reward (%)	
	ACR	ACR-STRD
1	60.73	100
2	64.09	100
3	58.92	100
4	54.53	100

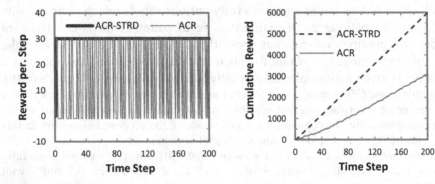

Fig. 5. Performance evaluation on a sample run for environment a. The left figure shows ACR-STDR obtains reward for all steps by tracking the goal spatiotemporal pattern but ACR fails this task. (b) Cumulative reward. Best viewed in color.

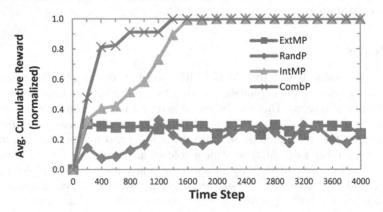

Fig. 6. Average cumulative reward obtained by ACR-STRD using four different policies. Best viewed in color.

Based on the experimental results, shown in Fig. 6, we found out that ACR-STRD with RandP cannot discover the existing spatiotemporal pattern in its environment. Similarly, ExtMP had a poor performance. Both IntMP and CombP can guide the robot to discover the environment spate-time pattern. However, CombP could manage to learn faster. These results support this idea that totally goal-driven and random behaviors are not effective policies. Instead, an artificially curiosity driven behavior can lead to discovery of spatiotemporal patterns in noisy environments.

4   Conclusions

In this paper, we proposed an artificial curiosity method for developmental robots with ability to autonomously discover spatiotemporal regularities in their milieus. To encode the role of history into a manageable data, we used the notion of perceptual context. This enables the robot to distinguish between two visual sequences with the same current observation but different history. The ACR-STRD behavior is guided by a hybrid motivation system that combines both intrinsic and external values. Experimental results showed that the proposed algorithm can learn spatiotemporal patterns in its environment without any teacher and previous knowledge of the environment.

There are a few directions for extending this work. We would like to explore the performance of the algorithm in larger environments and for more complex space-time patterns. Current version of ACR-STRD re-plans at fixed intervals. Planning is the most time consuming part of the algorithm and its high computational cost makes the exploration and learning very long. This can be modified by an algorithm that only re-plans whenever it is necessary.

Acknowledgements. This research was supported by UMRG Grant (RG115-12ICT) from University of Malaya.

References

1. Lin, L.-J., Mitchell, T.M.: Reinforcement learning with hidden states. In: Proc. 2nd Int. Conf. From Animals to Animats, pp. 271–280 (1993)
2. Chrisman, L.: Reinforcement learning with perceptual aliasing: The perceptual distinctions approach. In: Proc. 10th Nat. Conf. AI, pp. 183–188. AAAI Press (1992)
3. Bakker, B.: Reinforcement Learning with Long Short-Term Memory. In: Neural Information Processing Systems (NIPS), pp. 1475–1482 (2001)
4. Sutton, R.S., Barto, A.G.: Reinforcement Learning: An Introduction. MIT Press, Cambridge (1998)
5. Kaelbling, L.P., Littman, M.L., Moore, A.W.: Reinforcement Learning: A Survey. Journal of Artificial Intelligence Research 4, 237–285 (1996)
6. Sutton, R.S.: Integrated architectures for learning, planning, and reacting based on approximating dynamic programming. In: Proc. 7th Int. Conf. Machine Learning, pp. 216–224 (1990)

7. Vigorito, C.M., Barto, A.G.: Intrinsically Motivated Hierarchical Skill Learning in Structured Environments. IEEE Trans. Auton. Mental Develop. 2, 132–143 (2010)
8. Barto, A.G., Singh, S., Chentanez, N.: Intrinsically motivated learning of hierarchical collections of skills. In: Proc. 3rd Int. Conf. Development and Learning, La Jolla, CA, pp. 112–119 (2004)
9. Schmidhuber, J.: A possibility for implementing curiosity and boredom in model-building neural controllers. In: Meyer, J.A., Wilson, S.W. (eds.) Proc. Int. Conf. Simulation of Adaptive Behavior: From Animals to Animats, pp. 222–227. MIT Press (1991)
10. Oudeyer, P.-Y., Kaplan, F.: What is intrinsic motivation? A typology of computational approaches. Frontiers in Neurorobotics 1 (2007)
11. Lopes, M., Montesano, L.: Active Learning for Autonomous Intelligent Agents: Exploration, Curiosity, and Interaction. arXiv preprint arXiv:1403.1497v1 (2014)
12. Frank, M., Leitner, J., Stollenga, M., Förster, A., Schmidhuber, J.: Curiosity driven reinforcement learning for motion planning on humanoids. Frontiers in Neurorobotics 7 (2014)
13. Ngo, H., Luciw, M., Förster, A., Schmidhuber, J.: Confidence-based progress-driven self-generated goals for skill acquisition in developmental robots. Frontiers in Psychology 4 (2013)
14. Luciw, M., Graziano, V., Ring, M., Schmidhuber, J.: Artificial curiosity with planning for autonomous perceptual and cognitive development. In: 2011 IEEE Int. Conf. on Development and Learning (ICDL), pp. 1–8. IEEE Press (2011)
15. Asada, M., Hosoda, K., Kuniyoshi, Y., Ishiguro, H., Inui, T., Yoshikawa, Y., Ogino, M., Yoshida, C.: Cognitive Developmental Robotics: A Survey. IEEE Trans. Auton. Mental Develop. 1, 12–34 (2009)
16. Hruby, R., Maas, L.M., Fedor-Freybergh, P.G.: Early brain development toward shaping of human mind: an integrative psychoneurodevelopmental model in prenatal and perinatal medicine. Neuro. Endocrinol. Lett. 34, 447–463 (2013)
17. Andreakis, A., Hoyningen-Huene, N.v., Beetz, M.: Incremental unsupervised time series analysis using merge growing neural gas. In: Príncipe, J.C., Miikkulainen, R. (eds.) WSOM 2009. LNCS, vol. 5629, pp. 10–18. Springer, Heidelberg (2009)
18. Carpenter, G.A., Grossberg, S., Rosen, D.B.: Fuzzy ART: Fast stable learning and categorization of analog patterns by an adaptive resonance system. Neural Networks 4, 759–771 (1991)
19. Weng, J., Luciw, M.: Dually optimal neuronal layers: Lobe component analysis. IEEE Trans. Auton. Mental Develop. 1, 68–85 (2009)
20. Schmidhuber, J.: Formal Theory of Creativity, Fun, and Intrinsic Motivation. IEEE Trans. Auton. Mental Develop. 2, 230–247 (2010)
21. Singh, S.P., Sutton, R.S.: Reinforcement learning with replacing eligibility traces. Machine Learning 22, 123–158 (1996)
22. Lagoudakis, M.G., Parr, R.: Least-squares policy iteration. The Journal of Machine Learning Research 4, 1107–1149 (2003)
23. Lange, S., Riedmiller, M.: Deep auto-encoder neural networks in reinforcement learning. In: The 2010 Int. Joint Conf. Neural Networks (IJCNN), pp. 1–8. IEEE Press (2010)

The Equivalence Relationship between Kernel Functions Based on SVM and Four-Layer Functional Networks

Yongquan Zhou, Qifang Luo, Mingzhi Ma, and Liangliang Li

College of Information Science and Engineering,
Guangxi University for Nationalities, Nanning Guangxi, 530006, China
yongquanzhou@126.com

Abstract. This paper based on the concept of function interpolation, a function-al network interpolation mechanism was analyzed, the equivalent between functional network and kernel functions based SVM, and the equivalent relationship between functional networks with SVM is demonstrated. This result provides us a very useful guideline when we perform theoretical research and applications on design SVM, functional network systems.

Keywords: Functional network, interpolation representation, kernel functions, SVM.

1   Introduction

Functional Networks which is a new type of neural network model was proposed by Enrique Castillo in 1998 [1]. It is an extension of the Artificial Neural Network (ANN). Unlike ANN, connection between neurons of functional networks is without weights, and neuronal function is adjustable. We usually select some combination of given base functions (such as polynomial, trigonometric function, Fourier function) more satisfactory generalization performance size can be obtained with smaller network after training and learning; Literature [2] [3] studies have shown that functional network can not only solve the problems which can be solved by ANN, but also can solve the problems which can't be, the performance of functional networks is better than the artificial neural network in many aspects. Thus, the functional network with its unique structure and method for processing information, have been used in prediction of chaotic time series, solving differential or difference and functional equation, CAD, linear or nonlinear regression, nonlinear system identification and prediction [4], multidimensional data analysis [5], intelligent optimization calculation [6] [7] [8] [9], dynamic modeling [10], software reliability verification [11][12] and functional network computing [16] and other fields. In recent years, although the functional network made great success in the applications, the basic mathematical theories of functional network are not perfect, therefore, researchers need to continuously put forward the novel network structure which is more suitable for solving the problem and improve its mathematical theories to further expand its scope of application.

D.-S. Huang et al. (Eds.): ICIC 2014, LNAI 8589, pp. 99–106, 2014.

Support vector machine (SVM) is a new machine learning algorithm proposed by Vapnik ect. [13] , because of its excellent learning ability, SVM has received wide attention in the academic circle both domestic and overseas, and has acquired more and more applications in pattern recognition and function estimation [14]. Zhang Ling has proved Vapnik's kernel function based on SVM and three-layer feedforward neural networks, and proved kernel function existence theorems [15]. With his covering algorithm of neural network that put forward by him as a tool. The obtained results have guiding significance in studying the relationship between the functional network and SVM in this article.

The main work of this paper is based on the concept of interpolation function to study the equivalence between the functional network and SVM, prove the equivalence functional network and SVM. It is no doubt that the results obtained for improving the basic theory of the functional network, and studying the relationship between the functional network and based on SVM kernel function , respectively for improving the functional network and SVM has the important theory significance and the application value.

2   Functional Networks

In general, a functional network consists of the following elements: (1) Input units layer: this layer contains the input data. Input units are represented by solid circles with their corresponding names. (2) Output unit's layer, this layer contains the output data. Output units are also represented by solid circles with their corresponding names. (3) Neurons or processing units with one or several layers, each neuron is a computing unit. These units evaluate a set of input values, coming from the previous layer (of intermediate or input units) and deliver a set of output values to the next layer (of intermediate or output units). Processing units are connected with each other, and the output value of each neuron can be used as a part of another neuron or output unit data. Once the input value is given, the output is determined by the types of neurons, which is defined by a function. Neurons are represented by circles with the name of the corresponding function f_j . (4) A set of directed links. In general, the n layer network model is:

$$Y = f_n \bullet f_{n-1} \bullet \cdots \bullet f_1(X) \tag{1}$$

To illustrate the problem intuitively, we adopted a medical diagnosis example to show the modeling process of functional networks: Supposing a kind of disease d corresponding to three symptoms x, y and z , namely $d = D(x, y, z)$. Generally speaking, when doctors in the diagnosis of disease d , he can use the following three different orders to diagnose:

1) If the doctor diagnoses disease d by the sequence $x \rightarrow y \rightarrow z$, then:

$$\left.\begin{array}{c} x \\ y \\ z \end{array}\right\} P(x,y) \left.\right\} \quad F(P(x,y),z) = D(x,y,z) \qquad (2)$$

The corresponding functional network of the functional equation is

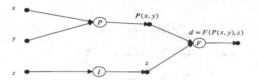

Fig. 1. The corresponding network structure by adopting the sequence

2) If the doctor diagnoses disease d by the sequence $y \rightarrow z \rightarrow x$, then:

$$\left.\begin{array}{c} y \\ z \\ x \end{array}\right\} Q(y,z) \left.\right\} \quad G(Q(y,z),x) = D(x,y,z) \qquad (3)$$

The corresponding functional network of the functional equation is

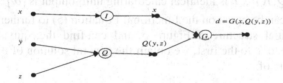

Fig. 2. The corresponding functional network structure by adopting the sequence

3) If the doctor diagnoses disease d by the sequence $x \rightarrow z \rightarrow y$, then:

$$\left.\begin{array}{c} x \\ z \\ y \end{array}\right\} R(x,z) \left.\right\} \quad H(R(x,z),y) = D(x,y,z) \qquad (4)$$

The corresponding functional network of the functional equation is

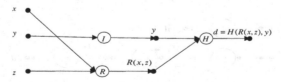

Fig. 3. The corresponding functional network structure by adopting the sequence

Then, we get the functional equation

$$d = D(x, y, z) = F(P(x, y), z) = G(Q(y, z), x) = H(R(x, z), y) \qquad (5)$$

Based on the above three conditions, and according to the Equation (5), the diagnosis of the disease d is corresponding to the functional network structure as follows:

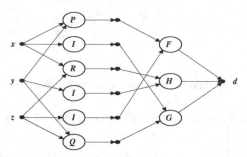

Fig. 4. Diseases d that corresponding to the functional network initial topological structure of the three symptoms x, y and z

The functional network's input is $X = \{x, y, z\}$, the intermediate calculation unit is $\{F, G, H, P, Q, R\}$, I is identical calculating unit; output is $\{d\}$.

The following can be based on the functional Equation (5) to further simplify functional network initial structure of Figure 5, and can find the equivalent functional networks structure to it, to the first; we obtain the general solution of functional Equations (5) in the form of:

$$F(x, y) = k[f(x) + r(y)] \; ; \quad P(x, y) = f^{-1}[p(x) + q(y)];$$

$$G(x, y) = k[n(x) + p(y)]; \quad Q(x, y) = n^{-1}[q(x) + r(y)]; \qquad (6)$$

$$h(x, y) = k[m(x) + q(y)]; \quad R(x, y) = m^{-1}[p(x) + r(y)].$$

By formula (5) and (6) we can obtain:

$$d = D(x, y, z) = k[p(x) + q(y) + r(z)] \qquad (7)$$

At this point, Figure 6 functional network's equivalent network structure is

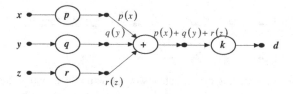

Fig. 5. Equivalent functional network structure to Fig. 5

Easy to see that Figure 4's functional network is much simpler than Figure 4, so that we can use Figure 5's functional network to complete diagnosis of this disease's training learning. To sum up, it can be seen from the topological structure of functional networks, what the functional network corresponds is some functional transformations; its topological structure describes a function transformation system.

3   Interpolation Principles of Functional Networks

3.1   The Three Layer Network Model with Single Input and Single Output

Fig. 6 shows a three layer network model with single input and single output

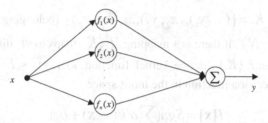

Fig. 6. Three layer functional network model with single input and single output

where $f_i(x), i = 1,2,...,n.$ The neurons and the output expression of network is

$$y = \sum_{i=1}^{n} f_i(x) \tag{8}$$

3.2   The Four Layer Network Model with Double Input and Single Output

Fig. 7 shows a four layer network model with double input and single output

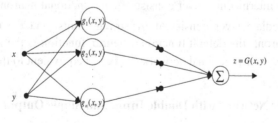

Fig. 7. Four layer functional network model with double input and single output

In Fig. 7, if the input data of network is $\{x, y\}$, and output data is $\{z\}$, then the network will get

$$z = G(x, y) = \sum_{j=1}^{m} g_j(x, y) \tag{9}$$

In Eq. (9), $g_j(x, y)$ is binary functional neuron. Particularly. when $g_j(x, y) = p_j(x)q_j(y)$, The output of this separable functional network is

$$z = F(x, y) = \sum_{i=1}^{n} f_i(x)g_i(y) \tag{10}$$

4   Based on the Kernel Function of SVM and the Equivalence of Functional Network

4.1   Based on the Kernel Function of SVM

Given a sample set $K = \{(x_1, y_1), (x_2, y_2), ..., (x_p, y_p)\}$ (belonging to the n dimension Euclidean space N), if there is a mapping F, K maps to m dimension Euclidean space M s subset $F(K)$, define kernel function $k(x, y) = <F(x), F(y)>$, and define a nonlinear decision function in the input space.

$$I[\mathbf{x}] = Sgn(\sum_i a_i k(x_i, \mathbf{x}) + b_0) \tag{11}$$

By the decision function (11) learning machine defined called SVM based on kernel $k(\mathbf{x}, y)$. Supposing the corresponding sample set K, the SVM machine has got its solution $(\mathbf{a}, b_0)$, where $\mathbf{a} = (a_1, a_2, ..., a_m)$. Makes

$$F(x) = (f_1(x), f_2(x), ..., f_m(x)), \tag{12}$$

As a functional neuron A_i, the corresponds functional neuron function is $f_i(x)$.

Shaped as Figure 8 with double input and single output four layer network $N(\mathbf{K})$: the first intermediate layer consist of m functional neuron $A_1, A_2, ..., A_m$, the second intermediate layer consist of m storage units, and the output layer has one functional neuron, the default functional neuron function is the identity function $I(x)$. Obviously, four functional network $N(\mathbf{K})$ is equivalent to the above SVM.

4.2   Four Layer Network with Double Input and Single Output

Given a sample set $\mathbf{K} = \{(x_1, y_1), (x_2, y_2)..., (x_p, y_p)\}$, supposing four layers functional network $N(K)$ (single output), the first intermediate layer is comprised of m functional neurons $A_1, A_2, ..., A_m$, the corresponding functional neuron is $f_i(x)$. The second intermediate layer is composed of m storage units, which are used to store the calculate results of the previous unit, the output layer is a functional neuron B, the corresponding functional neuron function is the identity function $I(x)$. and $N(\mathbf{K})$ can identify the sample set $\mathbf{K}$ correctly. $F(x) = (f_1(x), f_2(x), ... f_m(x))$, F is the mapping N to M , the kernel function $k(\mathbf{x}, y) = <F(\mathbf{x}), F(y)>$, then, SVM based on $k(\mathbf{x}, y)$ and the four layer functional networks $N(\mathbf{K})$ are equivalent. The conclusion from the above analysis as follow:

Theorem. The SVM based on kernel functions and the four layers functional networks are equivalent.

5   Discussion and Conclusion

In the section 4, the two most basic network models which is the single input-single output (Figure 6) and the double input-single output (Figure 7) model had studied the equivalence between the functional network and SVM. First of all, we introduce the definition as follows:

Definition 1. If the space of real numbers in n dimension and m dimension has function transformation $F : R^n \rightarrow R^m$, and $X \in R^n$, then functional equation is

$$F = F(f_1, f_2, ..., f_m) \tag{13}$$

where $X = (x_1, x_2, ..., x_n), f_i = f_i(x_1, x_2, ..., x_n); i = 1, 2, ..., m.$ at this time, the functional Equation (13) corresponds to the functional network model is the basic Multi-Input and Single Output FunNets (MIOFN). MIOFN network topology structure similar to Fig. 8. We can construct MIOFN according to the function transformation, and can simplify, the simplification of the initial MIOFN depends on the functional equation. For a given MIOFN, should analyse and determine whether there is corresponding to the input of the MIOFN are equivalent functional network with the same output. Then, we introduce two MIOFN is equivalent or equivalent class concept.

Definition 2. For any given set of input, If two MIOFN have the same output, we call the two MIOFN are equivalent. For a MIOFN, we can take all of the MIOFN which are equivalent to it form an equivalence class.

In practical application, we can find out the simplest MIOFN from the equivalence class to complete the proof of equivalence between the functional network and SVM by MIOFN equivalence and equivalence class.Because the network is a new neural network model proposed in recent years, some basics of theory and application are not perfect, researchers need to constantly put forward the novel network structure and improve the basic theory. In this paper, based on the concept of function interpolation, the equivalent between functional networks and SVM is revealed and demonstrated. I believe the results offered a new way to study the relationship and study the problems between SVM based on kernel function and functional network.

Acknowledgements. This work is supported by national science foundation of china under grant no. 61165015. Key project of guangxi high school science foundation under grant no. 20121ZD008, 201203YB072.

References

1. Castillo, E.: Function Networks. Neural Processing Letters 7, 151–159 (1998)
2. Castillo, E., Gutierrez, J.M.: Nonlinear Time Series Modeling And Prediction Using Functional Networks. Extracting Information Masked By Chaos. Physics Letters A 244, 71–84 (1998)
3. Castillo, E., Cobo, A., Gutierrez, J.M.: Working With Differential And Difference Equations Using Functional Networks. Applied Mathematical Modeling 23, 89–107 (1999)
4. Oscar, F.R., Enrique, C., Amparo, A.B.: A Measure of Fault Tolerance for Functional Networks. Neurocomputing 62, 327–347 (2004)
5. Castillo, E., Hadi, A., Locruz, B.: Semi-Parametric Nonlinear Regression And Transformating Using Functional Networks. Computational Statistics & Data Analysis 52, 2129–2157 (2008)
6. Zhou, Y.Q., Zhao, B., Jiao, L.C.: Approximate Factorization Learning Algorithm of Multivariate Polynomials Based On Functional Networks. Journal of Information and Computational Science 2, 205–210 (2005)
7. Zhou, Y.-Q., Jiao, L.-c.: Interpolation Mechanism of Functional Networks. In: Duch, W., Kacprzyk, J., Oja, E., Zadrożny, S. (eds.) ICANN 2005. LNCS, vol. 3697, pp. 45–51. Springer, Heidelberg (2005)
8. Zhou, Y.Q., Zhao, B., Jiao, L.C.: Universal Learning Algorithm of Hierarchical Function Networks. Chinese Journal of Computers 28, 1277–1286 (2005)
9. Zhou, Y.Q., Zhao, B., Jiao, L.C.: Serial Function Networks Method And Learning Algorithm With Applications. Chinese Journal of Computers 31, 1074–1081 (2008)
10. Qu, H.B., Hu, B.G.: Variational Learning for Generalized Associative Functional Networks in Modeling Dynamic Process of Plant Growth. Ecological Informatics 4, 163–176 (2004)
11. Dossevi, A., Cosmelli, D., Garnero, L., Ammari, H.: Multivariate Reconstruction of Functional Networks From Cortical Souces Dynamics in MRI. IEEE Transactions on Biomedical Engineering 55, 2074–2086 (2008)
12. Emad, A., El-Sebakhy: Software reliability identification using functional networks. A comparative study. Expert Syst. Appl. 36, 4013–4020 (2009)
13. Vapnik, V.N.: The nature of statistical learning theory. Springer, New York (1995)
14. Li, Y.G., Zhang, F., Liu, Z.Y.: Combining Position-Specific-Value Method and SVM for Remote Protein Classification. Computer Science 31, 44–50 (2008)
15. Zhang, L.: The Relationship Between Kernel Function Based SVM and Three-Layer Feedforward Neural Networks. Computer Science 25, 696–700 (2002)
16. Guo, P., Zhou, Y., Xiao, Q.: Expansion Type Functional Neuron Network Model and its Parameters to Directly Determine the Method. Journal of Software 8, 443–450 (2013)

Multiple Kernel Learning Based on Cooperative Clustering

Haiyang Du[1], Chuanhuan Yin[1], and Shaomin Mu[2]

[1] School of Computer and Information Technology, Beijing Jiaotong University, Beijing, China
{12120406,chyin}@bjtu.edu.cn, msm@sdau.edu.cn
[2] School of Computer and Information Engineering, Shandong Agricultural University, China

Abstract. In recent years, kernel methods with single kernel had been challenged by the big data because of its heterogeneousness. In order to exploit the advantages of kernel methods, multiple kernel learning was proposed several years ago. However, the time and space complexity of multiple kernel learning increases greatly due to great amount computation of multiple kernels. So far, the research on improving training efficiency for multiple kernel learning mostly focuses on reducing the complexity of solving the objective function, other than reducing the training set. In this paper, the time performance of multiple kernel learning is improved through shrinking the training sets by introducing cooperative clustering, which is a novel method based on k-means clustering. Applying cooperative clustering to multiple kernel learning problems is proposed to reduce the number of support vectors, and then reduce the time complexity of multiple kernel learning algorithms. Experimental results show that the new method improves the efficiency of multiple kernel learning greatly with a slight impact on classification accuracy.

Keywords: Support vector machine, Multiple kernel learning, Cooperative clustering.

1 Introduction

During the last several years, kernel methods have been widely used to solve classification problems [1]. However, traditional single kernel methods are not discriminative enough for heterogeneous datasets [2] or big datasets [3]. Developments of SVMs and recent publications [4] have shown that using multiple kernels instead of a single one can improve classifier performance. The decision function of multiple kernels has better interpretability due to multiple kernels and multiple feature representations used to train the classifier in this method. Therefore, multiple kernel learning (MKL) methods have higher classification accuracy than traditional kernel methods [5].

However, the efficiency of solving MKL problems is low due to great amount computations of multiple kernels and multiple feature representations. Recent publications show that a lot of research work has been done to improve the efficiency of solving MKL problems. Lanckriet [6] converted MKL problem into the semi

D.-S. Huang et al. (Eds.): ICIC 2014, LNAI 8589, pp. 107–117, 2014.

definite programming (SDP) problem to improve the solving efficiency. However, this method becomes rapidly intractable as the number of learning examples or kernels becoming larger. To solve this problem, Bach [7] proposed a dual formulation of quadratically constrained quadratic program (QCQP) as a second-order cone programming problem, in which the sequential minimal optimization (SMO) techniques can be used to solve the MKL problem. Sonnenburg [8] reformulates the MKL problem as a new algorithm which converts MKL problem into semi-infinite linear program (SILP). The advantage of this formulation is that the algorithm solves MKL problem by iteratively solving a classical SVM problem with a single kernel. Therefore, this method provides a more efficient solution for large scale MKL problems. Rakotomamonjy [9] proposed a new method called SimpleMKL, in which the MKL problem is solved by using a projected gradient method. Just as the name implies, SimpleMKL can speed up the procedure of solving MKL problem.

The above work about improving the efficiency of MKL is focusing on reducing the time complexity of solving the MKL problem. However, there is little work about shrinking the original training datasets to improve the training efficiency of MKL problems. In this paper, we propose a new method to improve the performance of MKL through shrinking of training sets by introducing cooperative clustering. Applying cooperative clustering to the MKL problems can reduce the number of support vectors, and then reduce the time complexity of MKL algorithms. Cooperative clustering is a novel clustering algorithm based on k-means clustering and coevolution [10]. It can find pairs of clustering centers crossing the boundary of two classes. In the following sections, we combine cooperative clustering algorithm and MKL to reduce the time complexity of MKL in training procedure.

2   Multiple Kernel Learning

Support vector machine (SVM) is a discriminative classifier for classification problems proposed by Vapnik, which is based on the theory of structural risk minimization [11]. Given a training set $\{(\mathbf{x}_i, y_i)\}$, where $\mathbf{x}_i \in R^d$ is the input feature vector, $y_i \in \{1, -1\}$ is the class label. SVM aims to find the best hyper-plane with the maximum margin between positive class and negative class in the feature space. The introduction of kernel function $k(\mathbf{x}_i \cdot \mathbf{x}_j) = \langle \Phi(\mathbf{x}_i) \cdot \Phi(\mathbf{x}_j) \rangle$ makes SVM much more discriminative for nonlinear classification problems. The classification function of learning problem can be written as $f(\mathbf{x}) = \sum_{i=1}^{n} \alpha_i y_i k(\mathbf{x}_i, \mathbf{x}) + b$ after introducing the kernel function, where coefficient α_i and b can be learned from the training set $\{(\mathbf{x}_i, y_i)\}$.

Compared with one single kernel learning methods like SVM, combination of multiple kernels is a more flexible model for complex classification problems. Multiple kernel functions are used to train the classifier to get higher classification accuracy in multiple kernel learning methods. Most of the existing MKL algorithms try to combine predefined kernels in an optimal way. The combination function of multiple kernels and its corresponding parameters can be written as

$$K(\mathbf{x}_i, \mathbf{x}_j) = f_\lambda(\{k_m(\mathbf{x}_i^m, \mathbf{x}_j^m)\}_{m=1}^P) \tag{1}$$

where $f_\lambda : R^P \to R$ is the combination function, it can be a linear or nonlinear function [12]. Kernel functions $\{k_m : R_i^{D_m} \times R_j^{D_m} \to R\}_{m=1}^P$ take P feature representations of samples: $\mathbf{x}_i = \{\mathbf{x}_i^m\}_{m=1}^P$ where $\mathbf{x}_i^m \in R^{D_m}$, m is the number of the corresponding feature representations, and P is the number of kernels in MKL algorithms. λ is the parameter of each kernel in the combination function.

In most situations, the convenient form of combination is to consider that the combination function $K(\mathbf{x}_i, \mathbf{x}_j)$ is actually a linear convex combination of basis kernels: $K(\mathbf{x}_i, \mathbf{x}_j) = \sum_{m=1}^P \lambda_m K_m(\mathbf{x}_i^m, \mathbf{x}_j^m)$, with $\lambda_m \geq 0, \sum_{m=1}^P \lambda_m = 1$. In such case, the performance of MKL methods mostly depends on the choice of the weight λ_m and kernels.

According to training method of MKL algorithms, they can be divided into two main groups: One-step methods and Two-step methods. In One-step methods, MKL calculates both the coefficient α_i and the weight λ_m in one single pass. Two-step methods update the λ_m and α_i in an alternating manner during each iteration [12].

MKL is inefficient in learning process compared with one single kernel learning method. The time complexity of MKL increases greatly due to the large amount computation of multiple kernels with multiple feature representations. What's more, MKL also needs much time to learn the weight λ_m of each kernel.

3   MKL Based on Cooperative Clustering

In this section, we try to use cooperative clustering method to preprocess training dataset used in the MKL algorithms. Using this method, we can reduce the size of training dataset to a small scale and archive the goal of improving efficiency of MKL problems. This method can also avoid the loss of classification information due to the reduction of samples used for training.

3.1   Method of Cooperative Clustering

K-means is a typical unsupervised learning clustering algorithm based on distance which is widely used in clustering problem [13]. The cooperative clustering is a novel algorithm based on k-means clustering. It has good performance in preprocessing the training data sets for SVMs [11] and multiclass problems [14]. This algorithm can quickly compute the clustering centers of the two classes and draw them towards the boundary of the two classes. Because these clustering centers locate at the boundary where support vectors used for training can be found here. Therefore, these clustering centers carry more classification information than the other samples in two classes. In this section, cooperative clustering algorithm is used to find clustering centers surrounding the boundary of given classes. The cooperative clustering is a practical method to select representative points, which are used to train multiple kernel classifiers.

Suppose that there are two-class data sets X^+ and X^-. With k-means algorithm, the two data sets are both divided into k clusters marked with $\{X_1^+, X_2^+, ..., X_k^+\}$ and $\{X_1^-, X_2^-, ... X_k^-\}$. Now let $v^+ = \{v_1^+, ..., v_k^+\}$ be the set of the clustering centers in class +, and $v^- = \{v_1^-, ..., v_k^-\}$ be the set of clustering centers in class -, and H represents the distance matrix between two the sets v^+ and v^-. The ij-th entry h_{ij} of matrix H can be computed as

$$h_{ij} = \left\| v_i^+ - v_j^- \right\|^2 \quad i, j = 1, ..., k \tag{2}$$

We take out each pair $\left(v_a^+, v_b^- \right)$ of clustering centers with the smallest distance from sets v^+ and v^- iteratively according to matrix H. Let r_a^+ be the average radius of cluster i in class + and r_b^- be the average radius of cluster j in class -. So we have

$$r_a^+ = \frac{1}{n_a^+} \sum_{x \in X_a^+} \left\| x - v_a^+ \right\| \tag{3}$$

$$r_b^- = \frac{1}{n_b^-} \sum_{x \in X_b^-} \left\| x - v_b^- \right\| \tag{4}$$

where n_a^+ and n_b^- are sample numbers in clusters X_a^+ and X_b^- respectively. Then each pair $\left(v_a^+, v_b^- \right)$ is updated as follows

$$v_a^+ = v_a^+ + \theta \frac{r_a^+}{r_a^+ + r_b^-} (v_b^- - v_a^+) \tag{5}$$

$$v_b^- = v_b^- + \theta \frac{r_b^-}{r_a^+ + r_b^-} (v_a^+ - v_b^-) \tag{6}$$

where $\theta \in (0,1)$ is a key parameter that controls the distance between v_a^+ and v_b^-.

The whole procedure of the cooperative clustering is as follows:

1. Partition X^+ and X^- into k clusters with k-means respectively, and get the clustering centers $v^+ = \{v_1^+, \ldots, v_k^+\}$ in class + and $v^- = \{v_1^-, \ldots, v_k^-\}$ in class -.

2. Set $V^s = \{\ \}$ and compute the matrix $H = \{h_{ij}\} 1 \leq i, j \leq k$ with equation (2).

3. Find a smallest element h_{ab} in H and add (v_a^+, v_b^-) into the set V^s

4. Delete the row i and the column j from matrix H.

5. If the number of pairs in set V^s is less than k, go to step 3.

6. Update each center pair (v_a^+, v_b^-) in V^s with equations (5) and (6).

7. Partition X^+ into clusters $\{X_1^+, X_2^+, \ldots, X_k^+\}$ and X^- into clusters $\{X_1^-, X_2^-, \ldots X_k^-\}$ with updated clustering centers, and compute new clustering centers $v^+ = \{v_1^+, \ldots, v_k^+\}$ in class + and $v^- = \{v_1^-, \ldots, v_k^-\}$ in class -.

8. If the partition is changed in step 7, go to step 2.

With the above procedure, we can get k pairs of clustering centers in V^s. Each pair of cluster centers crosses the boundary of the two classes. Because the training examples list near the boundaries between adjacent classes, they carry ample classification information. Therefore, they can be used for training MKL classifier.

3.2   CC-MKL

Using cooperative clustering method, we can get pairs of clustering centers near the boundaries between two classes, which carry ample classification information for training classifier. After cooperative clustering, the size of training datasets becomes much smaller than the original ones. It is worth trying to apply cooperative clustering to MKL with the purpose of improving the efficiency of MKL.

Suppose that there are P feature representations defined as $A_1 = \{(x_i^1, y_i)\}$ $A_2 = \{(x_i^2, y_i)\}$ $\cdots$ $A_P = \{(x_i^P, y_i)\}$, where $x_i^m \in R^d$ is a feature vector, $y_i \in \{1, -1\}$ is the class label.

To improve the efficiency of MKL training procedure, we use cooperative clustering as follows. Firstly, we combine these P feature representations as a whole one $A = \{(x_i, y_i)\}, x_i = x_i^1 x_i^2 \cdots x_i^P$. Then all the samples labeled 1 are assigned as positive class X^+ and the other samples labeled -1 as negative class X^-. The cooperative clustering procedure is run on these two classes simultaneously. After that, we can get k pairs clustering centers used for training MKL.

The training procedure of the CC-MKL classifier is described as follows:

1. Combine P feature representations $\{(x_i^m, y_i)\}$ as one $\{(x_i, y_i)\}$.

2. Partition $\{(\mathbf{x}_i, y_i)\}$ into class $+$ $X^+ = \{\mathbf{x}_i \mid y_i = 1\}$ and class $-$ $X^- = \{\mathbf{x}_i \mid y_i = -1\}$ according to label y_i.

2. Run cooperative clustering on X^+ and X^- simultaneously, then get 2*k clustering centers $\mathbf{v}^+ = \{\mathbf{v}_1^+, \dots, \mathbf{v}_k^+\}$ and $\mathbf{v}^- = \{\mathbf{v}_1^-, \dots, \mathbf{v}_k^-\}$.

3. Split these cluster centers into P feature representations according to the original number of attributes in each feature representation.

4. Use the new P feature representations as training dataset to train MKL classifier.

The new MKL method based on cooperative clustering is named as CC-MKL. Because the size of new training dataset becomes many times smaller than original dataset, the efficiency of CC-MKL will be improved significantly. As for the impact on accuracy, it will be validated by experiments in next section.

4   Experiments

To test and verify the higher efficiency of our new method, we apply it to two benchmark datasets composed of different feature representations. The chosen data sets PENDIGITS[1] and MFEAT[2] are both digit recognition multiple classes datasets.

The widely used Gaussian kernel and Polynomial kernel are chosen as the base kernels in our experiments. The important hyper parameter C is chosen among {0.01, 0.1, 1, 10, 100, 1000}. The one that has highest accuracy is used to train the final classifier in our experiments. We run every classification experiment for 10 times and compute average values as the final results.

The parameter k in the cooperative clustering procedure is used to control the time complexity of cooperative clustering and the accuracy of classifiers. If k is too large, the efficiency of cooperative clustering will be inefficient. On the contrary, the testing accuracy will be not high enough, if k is too small. After a series of experimental verification, we find that when k is fixed at 10% of average size of each class, the CC-MKL will has a good performance both in accuracy and time complexity. And parameter θ controls the distance between two cluster centers. In order to prevent two classes of cluster centers from drawing towards the boundary too fast, we choose a small θ like 0.25.

We compare the performance of different MKLs with the following methods:

MKL: The ordinary multiple kernel learning method.
SimpleMKL: A high efficient MKL algorithm.
CC-MKL: Multiple kernel learning based on cooperative clustering, its base learning algorithm is also the ordinary MKL.

---

[1] Available at http://mkl.ucsd.edu/dataset/pendigits
[2] Available at
http://archive.ics.uci.edu/ml/datasets/Multiple+Features

4.1   Pendigits Digit Recognition Experiments

Pendigits (PENDIGITS) is a multiple kernels dataset about pen-based digit recognition, which has 10,992 samples (7,494 for training and 3,498 for testing). It contains four different feature representations. And the detailed information of four feature representations is summarized in the following table 1.

Table 1. Feature representations of PENDIGITS dataset

Name	Attribute	Data Source
DYN	16	8 successive pen points on two-dimensional coordinate system
STA4	16	4×4 image bitmap representation
STA8	64	8×8 image bitmap representation
STA16	256	16×16 image bitmap representation formed by connecting the points in dyn representation with line segments

Pendigits is a multiclass digit recognition dataset. We use One-VS-All method to construct 10 binary classification problems by reclassifying the samples according to the label in the dataset. One digit is chosen as positive class and the other digits as negative class.

In experiments of CC-MKL, the parameter θ of cooperative clustering is fixed at $\theta = 0.25$, and parameter k is fixed at about 10% of the number of samples in positive class. The other parameters are kept the same in all three algorithms.

We just calculate the correct accuracy of positive class prediction in each experiment. The performance of each digit classification experiment was recorded in detail for comparison. Each digit classification experiment is run 10 times for computing the average values. The average performance values of each digit experiments using different methods are summarized in the following table 2 in detail.

Table 2. Performance of each digit in PENDIGITS dataset

Digit	Method	Accuracy(%)	Training Time(s)	Testing Time(s)
0	MKL	99.18	1453.18	0.87
	SimpleMKL	99.18	117.11	1.28
	CC-MKL	99.73	22.06 + 0.02	0.25
1	MKL	99.45	1378.15	1.51
	SimpleMKL	99.45	75.17	1.99
	CC-MKL	100	12.93 + 0.02	0.2
2	MKL	98.81	1459.05	1.28
	SimpleMKL	98.81	42.39	1.6
	CC-MKL	99.40	13.59 + 0.02	0.22
3	MKL	97.53	1446.05	0.67
	SimpleMKL	97.53	81.81	0.59
	CC-MKL	98.08	13.4 + 0.02	0.22

Table 2. (*continued*)

	MKL	97.61	1332.09	1.32
4	SimpleMKL	97.61	86.77	1.60
	CC-MKL	100	13.56 + 0.02	0.22
	MKL	99.40	1383.3	1.21
5	SimpleMKL	99.40	113.04	1.52
	CC-MKL	100	13.89 + 0.02	0.23
	MKL	89.29	1339.03	1.01
6	SimpleMKL	89.29	63.48	1.56
	CC-MKL	99.45	25.77 + 0.02	0.22
	MKL	99.40	1250.54	0.74
7	SimpleMKL	99.40	74.42	0.86
	CC-MKL	99.70	14.83 + 0.02	0.22
	MKL	97.02	1294.27	1.19
8	SimpleMKL	97.02	80.82	1.95
	CC-MKL	99.11	14.32 + 0.02	0.21
	MKL	95.04	1390.81	0.81
9	SimpleMKL	95.04	55.41	1.18
	CC-MKL	98.07	13.92 + 0.02	0.26

The training time of CC-MKL experiments is composed of cooperative clustering time and training time. The first part is cooperative clustering time and the latter part is classifier training time. According to table 2, we can find that the total training time of CC-MKL is dozens of times shorter than the ordinary MKL in every experiment. It is even several times faster than highly efficient SimpleMKL. What's more, the time of cooperative clustering dominates the overall value in training time. The actual training time of MKL can even be neglected. At the same time, the testing accuracy of CC-MKL is also higher than ordinary MKL and SimpleMKL. The reason for higher accuracy is that CC-MKL method can make the sample number of two classes become balance. In addition, the testing time of CC-MKL is also shorter than the ordinary MKL and SimpleMKL by a few times.

4.2   Multiple Features Digit Recognition Experiments

Multiple Features Digit dataset (MFEAT) consists of features of handwritten numerals ('0'-'9') from the UCI Machine Learning Repository, composed of six different feature representations. There are 2,000 samples in the MFEAT dataset without separating training set and testing set. Therefore, One-third samples are picked up from the original dataset randomly for testing and the rest samples are used for training. The detailed information of six feature representations is summarized in the following table 3.

Table 3. Feature representations of MFEAT data set

Name	Attribute	Data Source
mfeat-fou	76	Fourier coefficients of the character shapes
mfeat-fac	216	Profile correlations
mfeat-kar	64	Karhunen-Love coefficients
mfeat-pix	240	Pixel averages in 2 x 3 windows
mfeat-zer	47	Zernike moments
mfeat-mor	6	Morphological features

One-VS-All method is also used to construct 10 binary classification problems for this dataset. In experiments of CC-MKL, the parameter θ of cooperative clustering is fixed to $\theta = 0.25$, and parameter k is also fixed at 10% of the number of samples in positive class. The other parameters are kept the same in all three algorithms.

We also calculate the correct accuracy of positive class prediction in each experiment. And the performance of each digit classification experiment is recorded in detail for comparison. Each digit classification experiment is run 10 times for computing the average values. The average values of 10 experiments using different methods are summarized in the following table 4 in detail.

Table 4. Performance of each digit in MFEAT dataset

Digit	Method	Accuracy(%)	Training Time(s)	Testing Time(s)
0	MKL	100	25.79	0.07
	SimpleMKL	100	2.81	0.07
	CC-MKL	100	3.27 + 0.03	0.03
1	MKL	98.48	22.05	0.07
	SimpleMKL	98.48	1.95	0.07
	CC-MKL	100	2.51 + 0.02	0.06
2	MKL	96.97	23.13	0.05
	SimpleMKL	98.48	3.92	0.07
	CC-MKL	100	4.00 + 0.02	0.05
3	MKL	95.45	22.85	0.08
	SimpleMKL	95.45	3.69	0.09
	CC-MKL	100	3.85 + 0.02	0.05
4	MKL	98.48	24.09	0.06
	SimpleMKL	98.48	2.76	0.06
	CC-MKL	100	3.56 + 0.02	0.02
5	MKL	96.97	23.81	0.09
	SimpleMKL	98.48	1.32	0.08
	CC-MKL	100	2.75 + 0.02	0.04

Table 4. (*continued*)

6	MKL	93.94	23.12	0.05
	SimpleMKL	93.94	3.50	0.05
	CC-MKL	95.45	2.97 + 0.02	0.04
7	MKL	98.48	22.89	0.06
	SimpleMKL	98.48	2.49	0.07
	CC-MKL	100	3.80 + 0.03	0.04
8	MKL	100	23.82	0.04
	SimpleMKL	100	3.38	0.07
	CC-MKL	100	2.94 + 0.03	0.03
9	MKL	96.97	22.17	0.07
	SimpleMKL	96.97	3.20	0.07
	CC-MKL	98.48	4.13 + 0.02	0.04

The training time of CC-MKL is also composed of cooperative clustering time and training time. Table 4 shows that in the MFEAT dataset the performance of CC-MKL is still much better than the ordinary MKL. However, compared with highly efficient SimlpeMKL, our new method is slower in most experiments. Considering that the size of MFEAT is much too small (only 2,000 samples) and it takes more time in the procedure of cooperative clustering due to much more attributes in MFEAT. Therefore, the advantage of CC-MKL in MFEAT experiments is not outstanding as PENDIGITS dataset. The new method also has higher accuracy than ordinary MKL and SimpleMKL because of the imbalance of positive class and negative class.

5   Conclusion

In this paper, we proposed an efficient multiple kernel learning method based on cooperative clustering. Cooperative clustering is used to reduce the size of training datasets to a small scale. The new method reduces the number of support vectors used for training, and then reduce the time complexity of MKL algorithms. After cooperative clustering, pairs of cluster centers are found near the boundary of two classes. The representative points list near the boundaries between adjacent classes, so they carry more classification information than the samples in the center of class. This avoids reducing the accuracy of the classification due to the reduction of samples.

After a series of experiments with two benchmark datasets, we find that the efficiency of MKL based on cooperative clustering is much higher than the ordinary MKL and SimpleMKL, especially for big scale multiclass datasets. Because CC-MKL can make two classes of samples becomes balanced, it also has an advantage on accuracy for multiclass datasets. Therefore, it is meaningful to use cooperative clustering for improving the efficiency of MKL problems. In our future work, we plan to do the further research on selecting better clustering centers and further improve the efficiency of CC-MKL, especially the efficiency of cooperative clustering.

Acknowledgments. This work is supported by the National Natural Science Foundation of China (No. 61105056), the Fundamental Research Funds for the Central Universities, Shandong Provincial Natural Science Foundation, China (No. ZR2012FM024).

References

1. Müller, K., Mika, S., Rätsch, G., Tsuda, K., Schölkopf, B.: An introduction to kernel-based learning algorithms. IEEE Trans. on Neural Networks 12, 181–202 (2001)
2. Pavlidis, P., Weston, J., Cai, J., Grundy, W.N.: Gene functional classification from heterogeneous data. In: Proceedings of the 5th Annual International Conference on Computational Biology, pp. 242–248. ACM, Montreal (2001)
3. Rakotomamonjy, A., Bach, F.R., Canu, S., Grandvalet, Y.: More efficiency in multiple kernel learning. In: Proceedings of the 24th International Conference on Machine Learning, pp. 775–782. ACM, Corvalis (2007)
4. Lanckriet, G., Bie, T.D., Cristianini, N., Jordan, M., Noble, W.: A statistical framework for genomic data fusion. Bioinformatics 20, 2626–2635 (2004)
5. Lee, W.-J., Verzakov, S., Duin, R.P.W.: Kernel combination versus classifier combination. In: Haindl, M., Kittler, J., Roli, F. (eds.) MCS 2007. LNCS, vol. 4472, pp. 22–31. Springer, Heidelberg (2007)
6. Lanckriet, G., Cristianini, N., Bartlett, P., Ghaoui, L.E., Jordan, M.: Learning the kernel matrix with semidefnite programming. The Journal of Machine Learning Research 5, 27–72 (2004)
7. Bach, F.R., Lanckriet, G.R.G., Jordan, M.I.: Multiple kernel learning, conic duality, and the SMO algorithm. In: Proceedings of the 21st International Conference on Machine Learning, pp. 41–48. ACM, Banff (2004)
8. Sonnenburg, S., Ratsch, G., Schafer, C., Scholkopf, B.: Large scale multiple kernel learning. The Journal of Machine Learning Research 7, 1531–1565 (2006)
9. Rakotomamonjy, A., Bach, F.R., Canu, S., Grandvalet, Y.: SimpleMKL. The Journal of Machine Learning Research 9, 2491–2521 (2008)
10. Tian, S., Mu, S., Yin, C.: Cooperative clustering for training SVMs. In: Wang, J., Yi, Z., Żurada, J.M., Lu, B.-L., Yin, H. (eds.) ISNN 2006. LNCS, vol. 3971, pp. 962–967. Springer, Heidelberg (2006)
11. Vapnik, V.N.: Statistical Learning Theory. Join Wiley and Sons, New York (1998)
12. Gönen, M., Alpaydin, E.: Multiple Kernel Learning Algorithms. Journal of Machine Learning Research 12, 2211–2268 (2011)
13. Maulik, U., Bandyopadhyay, S.: Performance evaluation of some clustering algorithms and validity indices. IEEE Trans. Pattern Anal. Mach. Intell. 24, 1650–1654 (2002)
14. Yin, C., Mu, S., Tian, S.: Using Cooperative Clustering to Solve Multiclass Problems. In: Wang, Y., Li, T. (eds.) ISKE 2011. AISC, vol. 122, pp. 327–334. Springer, Heidelberg (2011)

Automatic Non-negative Matrix Factorization Clustering with Competitive Sparseness Constraints

Chenglin Liu and Jinwen Ma[*]

Department of Information Science, School of Mathematical Sciences and LMAM,
Peking University, Beijing, 100871, China
jwma@math.pku.edu.cn

Abstract. Determination of the appropriate number of clusters is a big challenge for the bi-clustering method of the non-negative matrix factorization (NMF). The conventional determination method may be to test a number of candidates and select the optimal one with the best clustering performance. However, such strategy of repetition test is obviously time-consuming. In this paper, we propose a novel efficient algorithm called the automatic NMF clustering method with competitive sparseness constraints (autoNMF) which can perform the reasonable clustering without pre-assigning the exact number of clusters. It is demonstrated by the experiments that the autoNMF has been significantly improved on both clustering performance and computational efficiency.

Keywords: Non-negative matrix factorization, Competitive learning, Clustering analysis.

1 Introduction

The Non-negative matrix factorization (NMF) can be used as a bi-clustering method, especially for high-dimensional data processing. Given a non-negative matrix $A \in R^{m \times n}$, NMF is to find out two non-negative matrices $W \in R^{m \times k}$ and $H \in R^{k \times n}$, so that

$$A \approx WH , \tag{1}$$

where k is a positive integer and $k < min(m, n)$.

One challenge of the NMF for clustering analysis is to determine the appropriate number of clusters. An improper one often leads to unstable or nonsense clusters. Previous methods address this problem by repetition test. They firstly perform NMF for a list of candidates, and then choose the optimal number that results in the best clustering performance according to a certain criterion [1, 2]. Moreover, some criteria also require multiple runs per a number for performance evaluation. Obviously, this kind of work requires a large computational cost.

---

[*] Corresponding author.

D.-S. Huang et al. (Eds.): ICIC 2014, LNAI 8589, pp. 118–125, 2014.

An important approach to determining the appropriate number of clusters is the competitive mechanism. In this paper, we try to incorporate such a competitive mechanism into the matrix factorization, and propose an automatic NMF clustering method with competitive sparseness constraints, being referred as autoNMF. Actually, it imposes certain competitive sparseness constraints to the factorized matrix, and performs an automatic clustering analysis by attenuating the elements of matrix corresponding to the fake clusters to zeroes.

The rest of this paper is organized as follows. In Section 2, we firstly review the basic NMF and sparse NMF, and then propose the autoNMF algorithm. In Section 3, we evaluate its clustering performances on 20 different synthetic datasets. We compare the autoNMF with the basic NMF and SNMF algorithm on both the clustering accuracy and computational time. We further discuss its robustness to the initialization of the number of clusters. Finally, we make a brief conclusion in Section 4.

2   Methods

2.1   Reviews of NMF and SNMF Algorithms

The model of NMF can be considered as the following optimization problem:

$$\min_{W,H} f(W,H) = \frac{1}{2} \| A - WH \|_F^2, \quad \text{s.t. } W, H \geq 0, \tag{2}$$

where $W \in R^{m \times k}$ is a basis matrix, $H \in R^{k \times n}$ is a coefficient matrix, and $k < min(m, n)$.

The alternating non-negativity constrained least squares (ANLS) method solves NMF based on a single right hand solution:

$$\arg \min_H \| A - WH \|_F \quad \text{and} \quad \arg \min_W \| A^T - H^T W^T \|_F \tag{3}$$

ANLS algorithm firstly solves H using the least squares algorithm, and separates the columns of H into active set Z (with all non-negative values) and inactive set P (containing negative values). The values of set P are modified by:

$$H_p \leftarrow H_p + \{\min_{k \in P} H_k / [H_k - (W^T A)_k]\}(W^T A)_P, \tag{4}$$

The details of this algorithm can be found in [3-5].

As to NMF clustering, the introduction of sparseness constraints to W and H improves the clustering performances by enhancing the differences of clusters and making the clustering more robust to noises. The SNMF method has properly balanced the sparseness of W and H by introducing a L-1 norm to the coefficient matrix [1]. The optimization function of SNMF is written as follows:

$$\min_{W,H} f(W,H) = \frac{1}{2} \{\| A - WH \|_F^2 + \eta \| W \|_F^2 + \beta \sum_{j=1}^{n} \| H(:,j) \|_1^2 \},$$

$$\text{s.t. } W, H \geq 0, \tag{5}$$

where $H(:,j)$ is the jth column vector of H, $\beta > 0$ controls the sparseness of H, $\eta > 0$ suppresses W to a proper scale.

The problem is solved by ANSL based on the following two least squares:

$$\min_H \| [\begin{matrix} W \\ \sqrt{\beta}e_{1\times k} \end{matrix}] H - [\begin{matrix} A \\ 0_{1\times n} \end{matrix}] \|_F^2, \quad \text{s.t. } H \geq 0,$$

$$\min_W \| [\begin{matrix} H^T \\ \sqrt{\eta}I_k \end{matrix}] W^T - [\begin{matrix} A^T \\ 0_{k\times m} \end{matrix}] \|_F^2, \quad \text{s.t. } W \geq 0,$$

(6)

where $e_{1\times k} \in R^{1\times k}$ is a row vector with all elements equal to one, and $0_{1\times n} \in R^{1\times n}$ is a zero vector, I_k is an identity $k \times k$ matrix, and $0_{k\times m}$ is a zero matrix.

2.2   The autoNMF Algorithm for Automatic Cluster Number Determination

In this paper, we propose the autoNMF algorithm for the automatic determination of the optimal number of clusters. In SNMF method, the sparseness degrees of W and H are fixed, which overlooks the differences between clusters. In fact, the mathematical precision increases monotonically with any increase of cluster number. In contrast, autoNMF tries to introduce a new cost function with different sparseness degrees to different columns of W. Columns with all small elements imply the fake clusters. Hence, large sparseness degrees are imposed on these columns to attenuate the elements to zeroes. On contrary, columns with a mount of relatively large elements indicate the true clusters, and smaller sparseness degrees are imposed.

Let $\vec{\lambda} = (\lambda_1, \lambda_2, \ldots \lambda_k)$ denotes the sparseness degrees of the columns of W. We formulate the cost function related to λ_r as follows:

$$J(\vec{\lambda}, W) = \frac{1}{2}(\eta \| W \|_F^2 + \eta \sum_{r=1}^{k} \| \lambda_r W(:,r) \|_F^2) - \theta \sum_{r=1}^{k} \lambda_r \log \lambda_r,$$

$$s.t. \sum_{r=1}^{k} \lambda_r = k$$

(7)

where θ is a trade-off between sparseness and variation of clusters, η is the to balance the sparseness between W and H.

Then, the objective function of the autoNMF is written as follows:

$$\min_{W,H} f(W,H) = \frac{1}{2}\{\| A - WH \|_F^2 + \eta \sum_{r=1}^{k} (1+\lambda_r^2)\| W(:,r) \|_F^2$$

$$+ \beta \sum_{j=1}^{n} \| H(:,j) \|_1^2\} - \theta \sum_{r=1}^{k} \lambda_r \log \lambda_r,$$

(8)

$$s.t. W, H \geq 0, \sum_{r=1}^{k} \lambda_r = k.$$

The autoNMF algorithm can solve W, H and $\vec{\lambda} = (\lambda_1, \lambda_2, \ldots \lambda_k)$ alternatively from the above objective function. Actually, W and H are solved by ANSL method. The least square rule for H is the same as Eq. (5). The rule for W has a small modification:

$$\min_W \| \begin{bmatrix} H^T \\ D \end{bmatrix} W^T - \begin{bmatrix} A^T \\ 0_{k \times m} \end{bmatrix} \|_F^2, s.t. W \geq 0, \tag{9}$$

where D is a $k \times k$ diagonal matrix with diagonal element equal to $\sqrt{\eta(1 + \lambda_r^2)}$.

As to λ_r , we employ the gradient learning algorithm to find the optimum sparseness degrees that can minimize the cost function Eq. (6). We denote $\vec{w}_r = \| W(:, r) \|_F^2$. Since $\sum_{r=1}^{k} \lambda_r = k$, we treat $\vec{\lambda}$ as $k - 1$ independent variables and $\lambda_c = k - \sum_{r=1, r \neq c}^{k} \lambda_r$. Based on mathematical deviation, the gradient of λ_r is written as:

$$\frac{\partial J(\vec{\lambda}, W)}{\partial \lambda_r} = \theta log(\frac{k - \lambda_r}{\lambda_r}) \frac{exp(\frac{\eta}{\theta} \lambda_r \vec{w}_r)}{\sum_{c \neq r} exp(\frac{\eta}{\theta} \lambda_c \vec{w}_c)}), r = 1, 2, \ldots, k , \tag{10}$$

We control the step sizes of λ_r to guarantee $\sum_{r=1}^{k} \lambda_r = k$ during the iterations.

2.3   The Convergence and Parameter Selections of the autoNMF Algorithm

As to W and H, the convergences have been proved by SNMF. Here, we only discuss the convergence of λ_r . We calculate the second derivatives of the cost function as to λ_r , and find that $J(\vec{\lambda}, W)$ has the minimum value if and only if $\eta / \theta > (\frac{1}{\lambda_r} + \frac{1}{\lambda_c})/(\vec{w}_r + \vec{w}_c)$. In the autoNMF algorithm, W and H are randomly initialized with non-negative values, and W is normalized so that $\| \vec{w}_r \|_F^2 = 1$. λ_r is initialized as one ($r = 1, 2, \ldots k$). η is set as the largest element of input matrix A, and

$$\theta = \min(\eta, \min_{1 \leq r, c \leq k, r \neq c} \eta(\vec{w}_r + \vec{w}_c)/(\frac{1}{\lambda_r} + \frac{1}{\lambda_c}), \tag{11}$$

The discussion about parameter β will be included in Section 3.

3   Experimental Results

In this section, we evaluate the clustering performances of the autoNMF algorithm on 20 synthetic data. Moreover, it is also compared with the NMF [6] and SNMF algorithms. We assign sample i to cluster r if $W(i, r) = \max W(i, :)$ and feature j to cluster r if $H(r, j) = \max H(:, j)$. We then evaluate the performance by matrices

$$Purity = \frac{1}{n}\sum_{q=1}^{l}\max_{1\le j\le k}n_q^j,$$

$$Entropy = -\frac{1}{nlog_2 l}\sum_{q=1}^{l}\sum_{j=1}^{k}n_q^j log_2 \frac{n_q^j}{n_q},$$

(12)

where k is the true class number, and l is the number generated by the algorithm, n is the size of data, n_q^j is the number of data in cluster q that belong to true class $j(1\le j\le k)$, and n_q is the size of cluster q. Combining the above two indexes together, we set the clustering performance index (CPI) as

$$CPI = Purity\times(1 - Entropy).$$

(13)

A larger CPI indicates a better clustering performance.

3.1   Synthetic Data Construction

The synthetic data are generated by 20 Gaussian mixtures. We set the number of cluster numbers as $k=6$, and the data size as 1800×100. We call the rows of data as samples, and columns as features.

- Randomly assign the 100 features to k clusters, and set their values $m\in M$.
- Set k covariance matrices $Cov_r \in R^{100\times100}$ as a diagonal matrix with diagonal value equal to d if feature j belongs to cluster r, $d\in D$.
- Assign the 1800 samples to k clusters according to the size rule S.
- Set the mix proportions as one if sample i belongs to cluster r, and t otherwise.
- Generate 1800 Gaussian mixtures based on the above parameters.
- Turns the negative values of the matrix to zeroes.

Table 1. Parameters of 20 data sets

I D	1	2	3	4	5	6	7	8	9	10
M	0.5	0.5	0.5	0.5	1	1	1	1	lm	lm
D	0.3	0.3	0.7	0.7	0.3	0.3	0.7	0.7	0.3	0.3
S	300	300	300	300	300	300	300	300	300	300
T	0.1	0.3	0.1	0.3	0.1	0.3	0.1	0.3	0.1	0.3
I D	11	12	13	14	15	16	17	18	19	20
M	lm	lm	lm	lm	lm	lm	lm	lm	lm	lm
D	0.7	0.7	lsd	lsd	0.3	0.3	0.7	0.7	lsd	lsd
S	300	300	300	300	lsize	lsize	lsize	lsize	lsize	lsize
T	0.1	0.3	0.1	0.3	0.1	0.3	0.1	0.3	0.1	0.3

$lm = \{0.5,1,1.5\}$, $lsize = \{300,500,200,400,100,300\}$, $lsd = \{0.1,0.3,0.5,0.7\}$.

In SNMF and autoNMF, η is set as the largest element of the input matrix A. To investigate the impact of β, we compare the results of autoNMF based on each $\beta \in \{0.1,0.3,0.5,0.7,0.9\}$. The initial number of clusters is set as $l = 12$. The results describe the average performances after 100 trials.

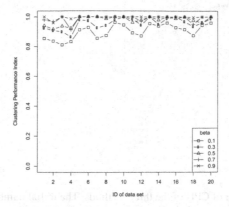

Fig. 1. CPIs of autoNMF based on different β in the 20 Gaussian mixture data sets clustering

3.2   Experimental Results on the Twenty Synthetic Datasets

As shown in Figure 1, autoNMF achieves good performances when $\beta \geq 0.5$. In most cases, it can find the true number of clusters when $\beta > 0.7$. However, too large β is not desirable because it may lead the algorithm overly emphasize the differences of clusters and imbalance the sizes of different clusters.

Next, we compare the performances of NMF, SNMF and autoNMF based on $\beta = 0.7$ and initialize the number of clusters as l = 12. To demonstrate the clustering accuracy of autoNMF, we also compare the algorithm with NMF and SNMF when the true number of clusters l = 6 is initialized, denoted as NMF* and SNMF*.

As shown in Figure 2(a), autoNMF, NMF* and SNMF* algorithm exhibit similar performances in most cases, and constantly better than the performances of NMF and SNMF. It indicates that autoNMF can obtain competitive clustering accuracy with NMF and SNMF algorithms when the initialization of l is correct, and much better performances when the initialization is larger than real.

The most advantage of autoNMF is the time efficiency due to the automatic section of number of clusters. To demonstrate this ability, we compare the time cost of autoNMF with NMF and SNMF. The initial number of clusters for autoNMF is set as $l = 12$. The candidate numbers of NMF and SNMF are set as $l \in \{4, 5, \ldots 12\}$. NMF and SNMF are performed once for each candidate number, and the best number is based on the inflection point in the residual sum of squares (RSS) curve [2]. β is set as 0.7. The time cost for single run NMF and SNMF are also included as contracts for the computational time cost. The comparisons are shown in Figure 2(b).

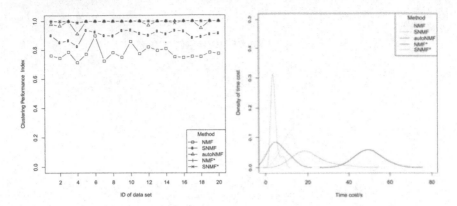

Fig. 2. (a)The comparison of CPIs of the three methods. The initial number of clusters of NMF, SNMF and autoNMF are set as 12, while those of NMF*, and SNMF* are set as 6 (true values). (b) Time costs of the NMF, SNMF and autoNMF. The lines are the density of time costs. NMF* and SNMF* are performed based on the candidate number list $l \in \{4,5,\ldots 12\}$. The initial cluster number of NMF, SNMF and autoNMF are 12.

Figure 2(b) describes the density of different ranges of time cost by each method. The autoNMF algorithm requires similar time cost with single run SNMF, and more efficient than basic NMF. By avoiding the repeated work like NMF and SNMF for the optimal number selection, our algorithm is significantly time efficient comparing to NMF* and SNMF* algorithm. Moreover, autoNMF avoids the case when the real number is not included in the candidate list.

Furthermore, we investigate the impact of the initial number of clusters l to the determination of the true number. Here, β is set as 0.7, and the initial number of clusters are set as $l \in \{8,10,\ldots,24\}$. We test if the algorithm can find the true number k, and compare the time costs based on different initializations.

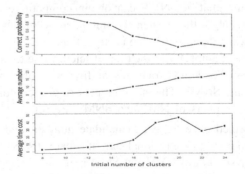

Fig. 3. The average performances of autoNMF based on 20 synthetic datasets across multiple choices of initial number of clusters: (top) the percentage of the correct number of clusters determined; (middle) the average number of clusters determined; (bottom) average time costs.

Figure 3 presents the average performances of autoNMF based on different number of clusters initializations across all the generated datasets. The three subfigures describe the probability that our algorithm determines the true number of clusters, the average number of clusters obtained, and the average computational time costs. It shows that autoNMF can find the true number of clusters when $l < 2k$. The number of clusters decreases when $l \geq 2k$, but they may not converge to the correct one. With the increase of l, the algorithm requires more computational time.

4   Conclusion

In this paper, we have constructed an efficient autoNMF algorithm for the bi-clustering analysis. It addresses the problem of automatic determination of cluster number by incorporating the competitive mechanism into the matrix factorization process. The experiments show that the autoNMF algorithm excels the state-of-art NMF method when the initial number of clusters is larger than the true value, and can obtain good performances when the provided number is not larger than twice of the actual number of clusters. Since it does not require repeated tests over different candidate numbers to determining a correct one, the autoNMF algorithm is significantly time and computational efficient for the NMF based bi-clustering analysis.

Acknowledgments. This work was supported by the Natural Science Foundation of China for Grant 61171138.

References

1. Kim, H., Park, H.: Sparse non-negative matrix factorizations via alternating non-negativity-constrained least squares for microarray data analysis. Bioinformatics 23, 1495–1502 (2007)
2. Brunet, J.-P., Tamayo, P., Golub, T.R., Mesirov, J.P.: Metagenes and molecular pattern discovery using matrix factorization. Proceedings of the National Academy of Sciences 101, 4164–4169 (2004)
3. Lawson, C.L., Hanson, R.J.: Solving least squares problems, vol. 161. SIAM (1974)
4. Kim, H., Park, H.: Nonnegative matrix factorization based on alternating nonnegativity constrained least squares and active set method. SIAM Journal on Matrix Analysis and Applications 30, 713–730 (2008)
5. Peharz, R., Pernkopf, F.: Sparse nonnegative matrix factorization with ℓ0-constraints. Neurocomputing 80, 38–46 (2012)
6. Lee, D.D., Seung, H.S.: Algorithms for Non-Negative Matrix Factorization. In: Advances in Neural Information Processing Systems, pp. 556–562 (2000)

Fusing Decision Trees Based on Genetic Programming for Classification of Microarray Datasets

KunHong Liu[1], MuChenxuan Tong[2], ShuTong Xie[3], and ZhiHao Zeng[1]

[1] Software School of Xiamen University, Xiamen, Fujian, P.R. China
lkhqz@xmu.edu.cn, zhihao5star@gmail.com
[2] Department of Electrical Engineering, Xiamen University, Xiamen, Fujian, P.R. China
demon386@gmail.com
[3] School of Computer Engineering, Jimei University, Xiamen, Fujian, P.R. China
shutong@jmu.edu.cn

Abstract. In this paper, a genetic programming(GP) based new ensemble system is proposed, named as GPES. Decision tree is used as base classifier, and fused by GP with three voting methods: min, max and average. In this way, each individual of GP acts as an ensemble system. When the evolution process of GP ends, the final ensemble committee is selected from the last generation by a forward search algorithm. GPES is evaluated on microarray datasets, and results show that this ensemble system is competitive compared with other ensemble systems.

Keywords: Microarray datasets, genetic programming(GP), ensemble system, decision tree.

1   Introduction

With the development of microarray technology, it is possible for ones to measure the expression level of thousands of genes simultaneously. Microarray brings new challenge for researchers because of the high dimension and small sample size problem. In general, for microarray classification, the difficulty of bias-variance dilemma [1] is mostly on the variance side.

Many researchers are devote to setting new learning models for classification of different diseases based on microarray data, and there are many successful applications to it. An effective way to reduce overfitting is ensemble learning. In general, ensemble learning is a technique that needs many classifiers, and then combines results from them to form final decisions instead of trusting only the best one. This technique has been proved to be able to boost accuracy significantly in many fields, and as pointed out in literatures, the improvement of ensemble learning is determined by a balance between the accuracy and diversity of base classifiers [2, 3].

For classification problem, we can denote the N training samples as $(x1, y1)$, $(x2, y2)$, $\cdots$, (xN, yN), where each x is a M-dimension vector, and each y is a scalar in $\{1, \cdots, c\}$ for a c-class problem. Diversity can be injected at the different levels: sample, feature, and base classifier. Researchers and practitioners usually combine

D.-S. Huang et al. (Eds.): ICIC 2014, LNAI 8589, pp. 126–134, 2014.
© Springer International Publishing Switzerland 2014

strategies at multiple levels to maximize the diversity: (1) For injecting diversity in sample, one doesn't need to alter the feature space of x, but the distribution of samples. Typical methods include Bagginng [4] and Boosting [5]. (2) For injecting diversity in feature, one maps the original M dimension feature space into a M' dimension space(M>M'). Ho [6] and Bay [7] trained multiple base classifiers with random subset of features, and this technique was named as random subspace[8]. (3) For injecting diversity at the level of base classifier, Maclin et al. used competitive learning to trained neural network with diverse initial weights [9]. The mixture of different levels can further improve the diversity in general.

After generating base classifiers, ensemble selection is the next key to producing powerful ensemble system by selecting a subset of the L base classifiers as the final committee in a static or dynamic way. When L is small, we usually just keep them all. However, when L becomes large, keeping all the available base classifiers may be redundant and even ineffective.

In this paper, we propose an Genetic programming(GP) based ensemble system(GPES for short). GP is a widely deployed evolutionary framework, and has been successfully applied to the classification of microarray data. Because each individual in the evolutionary group is a syntax tree, which can be used to produce a "yes/no" answer, it is easy to use GP to directly generate decision rules for classification of binary class problem. This approach has been applied in microarray classification with success. Langdon [10], as well as Yu et al. [11] used GP to do both gene selection and model generation. We extended the original GP framework by reconstruct the individuals in GP so as to form a sub-ensemble system, and make each individual deal with multiclass problem directly [12]. However, when using GP to generate classification rules, the computational complexity makes running GP based algorithm a time-consumption work. However, in this paper, the usage of GP is different from the previous works. It is used to fuse base classifiers so as to produce robust ensemble systems. Decision tree, a preferred model for many ensemble framework (Bagging, Random Forest, Rotation Forest, etc.), is used as base classifier because it's an unstable model[13]. And as pointed out in [14], different fusion strategies could make differences, so the individuals of GP could be of high diversity and low error rate in the evolution process. A number of individuals in the last generation of GP are selected with forward search techniques to form the final ensemble committee. It should be noted that although the base classifier of our ensembles (terminal in GP) is decision tree, other classifiers, such as SVM, neural network, can also be used in this framework.

2   Methods

2.1   Genetic Programming

Genetic programming (GP) is an evolutionary based optimization method. In GP, each individual expresses by a syntax tree. There are two kinds of nodes: terminal and nonterminal. Terminal is a leaf node without child nodes, while nonterminal is the inner node with child nodes. A child node can be terminal or nonterminal. Usually, terminals are primitive elements, and nonterminals are operators for combining these primitive elements.

Just like genetic algorithm (GA), in GP there is a pool of individuals to "compete" with each other. The initial population is usually generated randomly. In each generation, there are some individuals selected to survive, and others would be eliminated. The ones with higher fitness values would more likely survive. These fitness values are calculated by the criterion needing to be minimized (e.g. classification error rate). After the selection, two operators, crossover and mutation, are applied to survived individuals. The purpose of crossover operator is to create a chance of integrating strengths of different individuals. It is usually done by swapping some branches of two individuals (trees). The purpose of mutation operator is to inject randomness to avoid falling into local minima in evolutionary process. And it can be achieved by changing some terminals or nonterminals of some individuals. When crossover and mutation operations finish, some new individuals are produced and are picked to join the new generation. The whole evolutionary process leads the population to develop towards the global optimum rapidly. The algorithm keeps iterating until a criterion is met.

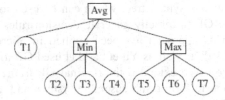

Fig. 1. An example of the individual of GP in the proposed algorithm

In our algorithm, each individuals in GP is an ensemble of decision trees. Each terminal is a single decision tree, while the nonterminal is one of the ensemble fusion operators: Average, Max and Min. For the binary class problem, let the label be -1 for negative class, and +1 for positive class. Then the Min operator prefers the negative class. That is, if a child node outputs a negative vote, the final decision of the Min operator is a negative label(-1) even when more children produce positive votes. And the Max operator works in a contrariwise way, and the output is positive class label if only one input is positive label(+1). The output of Average operator is hardened in the algorithm. That is, when the result is smaller than 0, then it is regarded as a negative label(-1); otherwise, it is a positive label. So the Average operator works in the same manner as the majority voting, and the final decision is the label got the most votes. In order to make these operators effective, we enforced each nonterminal to have three children, which can be either terminals or nonterminals. An example of the individuals is illustrated in Fig. 1. In this example, T1–T9 are decision trees; Avg, Min, Max are fusion operators. In this individul, if T1, T2, T5 produce negative votes for a sample, and other decision trees produce positive votes, the final output of the ensemble system is a negative label.

2.2   Generation of Decision Tree

As mentioned above, each terminal in GP is a single decision tree. In this section, the process of building decision trees is described.

1. Feature Standardization

Before training, each feature is standardized to zero mean and unit variance across training samples, and the same standardization step (fitted from the training set) will later be applied to the test set.

2. Feature subset selection

We apply the following four popular feature selection method:

F-Test: It is a classical technique used in one-way analysis of variance, to test whether the population means of different groups are equal. We use the implementation in scikit-learn library (0.12) [15] with the default setting.

RELIEF: It utilizes nonlinear classifier (nearest neighbor) for evaluating the usefulness of features[16]. We use the implementation in mlpy library (3.5.0) [17], setting the parameter sigma to 2.

Random Forest: The quality of the split in decision trees can be used for assessing the importance of the feature. Random Forest can be used for feature selection by summing the scores assigned by all decision trees. We use the implementation in scikit-learn library (0.12) with the default setting.

SVM-RFE: This is an embedded feature selection method. It iteratively trains SVM and removes a number of least important features. We use the implementation in scikit-learn library (0.12) with linear kernel, C=1, step=2.

Each of these 4 methods select 50 features, and only the unique features of all feature subsets are kept to form a pool of candidates. In this way, the dimension of datasets is reduced to a number from 50 to 200. After that, we adapt the idea of random feature subset selection so that each decision tree only sees a random projection of the candidate pool. For each tree, N f number of features are randomly selected, and N f obeys the distribution Int(N (5, 3)). By this means, the diversity among base classifiers is generated at the feature level.

2.3   Balanced Sub-Sampling

In building decision trees, we adapt the idea of Bagging to encourage diversity at the level of samples. Originally, Bagging uses sampling with replacement to introduce diversity in samples. However, in microarray data we usually don't have many samples (less than 100), thus sampling with replacement may not make enough differences.

Another concern is that sometimes microarray data is unbalanced. This is generally harmful for training classifiers. Considering these two factors, we used a balanced sub-sampling technique. For two-class classification problems, the number of samples in two classes are denoted as N1 and N2, respectively. The number of samples in the smaller class is Ns = Min(N1, N2). In building each classifier, Ns samples are sampled without replacement in each class, respectively. Thus, all the samples in the smaller class will be kept, while the samples in the larger class will be sub-sampled without replacement.

2.4   Evolutionary Process and Ensemble Selection

Based on the tree building process described above, 300 candidate decision trees are generated firstly. As random feature subset technique is used to generate trees, there would inevitably exist decision trees with unqualified performance. So only the trees above average accuracy of the 300 candidates are selected to construct the pool of base classifiers. And then the tree pool is used as the set of terminals in GP.

After that, a typical GP evolution schema is applied. Each individual is an ensemble classifier, and is generated with ramped methods. The max depth of each individual is restricted to 3, and each nonterminal is forced to have exactly three children, which can be terminals or nonterminals. The crossover rate is 0.8, and the mutation rate is 0.4. The population size is 80, and the number of maximum generation is 200. The fitness function for each individual is its accuracy in the validation set (See Sec. 2.4). The implementation of GP is based on the Pyevolve library (0.6rc1) [18].

Population in the last generation are reserved for ensemble selection. Since each individual is already an ensemble classifier, this further ensemble is a kind of meta-learning. First, like the decision tree building phase (Sec. 2.2), we select individuals above the average accuracy of all the individuals as the available individuals. Second, a forward search algorithm is applied: initially, the best individual is selected in the final committee. And then at each step, we iterated all pairs of the available individuals (not yet in the final committee) to find out the best pair that reduces the majority voting error (MVE) most. Usually this step stops when all the individuals are exhausted or no improvement could be found. However, because in microarray data we have so few samples, and each individual is usually a competent sub-ensemble system, it's found that this forward search process would cease after adding one or two pairs. As a result, we still take the risk of getting a low-bias and high-variance model. Based on this observation, in order to reduce the variance, it is necessary to contain more individuals in the final ensemble. We adapted this algorithm, and the forward search process stops when one of the following two conditions are satisfied: (1) the setback of MVE is observed, instead of no improvement; (2) all individuals are exhausted.

2.5   Cross Validation to Avoid Overfitting

For the forementioned classifiers building steps, we need to evaluate the accuracy in various phases:

✧   Selecting decision trees above the average accuracy of the 300 candidates.
✧   Evaluating the fitness value for each individual.
✧   Selecting accurate individuals in the population of the last generation, and using the forward search algorithm to select proper members for the final committee.

One common approach to avoid overfitting in training set is to split part of it as validation set. However, the algorithm can also overfit the validation set because of the samll sample size. To overcome it, in the implementation, we split the training set with 3-fold stratified cross validation, and feed different folds for these three phases, respectively. For 3-fold cross validation, the data is randomly and evenly split into 3 groups. The evaluation takes 3 iterations. In each iteration, one of the 3 groups is selected in turn as the validation set, and the remaining 2 groups are selected as the training set. And in our 3-fold stratified cross validation, the ratio of number of samples in two classes is kept in both training and validation set in each group.

3   Results and Analysis

To evaluate the effectiveness of GPES, we compare it with some other tree-based classifiers, including decision tree, Random Forest and Rotation Forest. They are evaluated on several binary and multiclass microarray datasets.

3.1   Implementations

For decision tree (both in standalone and GP) and Random Forest, the implementations in scikit-learn library (0.12) with the default settings are used. For Rotation Forest, the implementation in Weka with the default setting is used. To be fair, the standardization and the feature selection steps are applied in all the classifiers. In other words, all the classifiers receive the same reduced subset of features for a given training and test dataset.

3.2   Results and Analysis on Microarray Datasets

In this section, we present and analysis the results of GPES in five binary class microarray datasets. The detailed information about these datasets is listed in Table 1. Most of the original datasets are unbalanced. For example, for Lung dataset, the number of sample in two classes in training dataset is 134/15, so the ratio of two classes is about 1/9, which makes classification problem to be a hard job. As we use the sub-sample technique to obtain a balanced training set, the balanced training set consists of 15 samples for each class. Because of the under sub-sampling technique, the base classifier of different ensemble system could be of high diversity.

For the evaluation of fitness function in GP evolutionary process, the 10-fold stratified cross validation is applied. Each algorithm ran 20 times with the same 10-fold split, and we record the average accuracy and standard deviation on test sets in Table 2.

Table 1. Binary class datasets used in experiments

Datasets	No. of Genes	No. of samples of two classes	Reference
Ovarian	15154	162/91	[19]
Leukemia	7129	47/25	[20]
Colon	20000	40/22	[21]
2Lung	12533	150/31	[22]
Prostate	126000	77/59	[23]

Table 2. Average accuracy and standard deviation on the datasets

Datasets/Algorithms	GPES	DT	Random Forest	Rotation Forest
Ovarian	**0.997 ± 0.003**	0.984 ± 0.000	0.988 ± 0.006	**0.997 ± 0.003**
Leukemia	**0.961 ± 0.013**	0.863 ± 0.000	0.944 ± 0.016	0.934 ± 0.020
Colon	0.810 ± 0.024	0.743 ± 0.000	0.804 ± 0.030	**0.820 ± 0.025**
Lung	**0.992 ± 0.004**	0.967 ± 0.000	0.985 ± 0.005	0.990 ± 0.005
Prostate	0.902 ± 0.014	0.889 ± 0.000	0.889 ± 0.017	**0.912 ± 0.020**
Average	**0.932 ± 0.011**	0.889 ± 0.000	0.837 ± 0.020	0.930 ± 0.015

From the results shown in Table 2, we can see that both GPES and Rotation Forest wins three cases in all datasets. Random Forest and tree never achieves the best performance in all experiments. In detailed analysis, it can be found that when the base classifier is accurate enough, for example, in the case of Leukemia data, GPES can obtain 96.1% average accuracy, which is about 10% improvement compared with the accuracy of a single tree. In comparison of GPES and Rotation Forest, it should be noted that Rotation Forest uses PCA to transform the original features, so their inputs are different from the original feature subsets. But the highest average accuracy is still achieved by GPES with the lowest deviation. It is obvious that GPES is able to improve the accuracy of decision tree, and outperform Random Forest in most cases.

Table 3. Percentage of different operators in the final committee

Datasets	% of Min	% of Avg	% of Max
Ovarian	0.257	0.567	0.177
Leukemia	0.240	0.567	0.194
Colon	0.374	0.449	0.177
Lung	0.333	0.487	0.180
Prostate	0.178	0.582	0.240

Table 3 lists the average percentages of different operators appearing in the final ensemble committee for different datasets. In all the cases, the Average operator appears with the highest frequency. However, the distributions of different operators are different among different datasets. For example, the percentages of Min operators are relatively high in Colon and Lung datasets, while low in Prostate datasets. So the use of different fusion operators contributes to the effective of final ensemble committee.

4   Conclusion

In this paper, we propose a new ensemble method, GPES. The algorithm is based on GP, which is used to combine decision trees with three operators: Min, Max and Average. The evolutionary process makes the ensemble system adapted to better solve the classification problem.

The training process is carefully designed to inject diversity at feature, sample and base classifier levels. In this way, the final ensemble committee is composed of accurate and diverse base classifiers, so it can effectively avoid overfitting. The effectiveness of GPES is evaluated in five microarray datasets. It is found that the algorithm is proved to be adaptive to the characteristics of datasets. In addition, although the base classifier of this algorithm is decision tree, other classification models can also be used as base classifiers, such as neural networks and k-nearest neighbor. By using elaborate as base classifier, or applying other sampling techniques, the performance of GPES may be further improved.

Acknowledgment. This work is supported by National Science Foundation of China (No. 61100106, No. 61174161 and No. 61303080).

References

1. Geman, S., Bienenstock, E., Doursat, R.: Neural Networks And the Bias/Variance Dilemma. Neural Computation 4(1), 1–58 (1992)
2. Krogh, A., Vedelsby, J.: Neural Network Ensembles, Cross Validation, And Active Learning. In: Advances in Neural Information Processing Systems, pp. 231–238. MIT Press (1995)
3. Kuncheva, L.I., Whitaker, C.J.: Measures of Diversity in Classifier Ensembles and Their Rela-tionship with the Ensemble Accuracy. Machine Learning 51(2), 181–207 (2003)
4. Breiman, L.: Bagging Predictors. Machine Learning 24(2), 123–140 (1996)
5. Freund, Y., Schapire, R.: A Desicion-theoretic Generalization of On-line Learning And An Application to Boosting. In: Vitányi, P.M.B. (ed.) EuroCOLT 1995. LNCS, vol. 904, pp. 23–37. Springer, Heidelberg (1995)
6. Ho, T.K.: Random Decision Forests. In: Proceedings of the Third International Conference on Document Analysis and Recognition, vol. 1, pp. 278–282 (1995)
7. Bay, S.D.: Combining Nearest Neighbor Classifiers through Multiple Feature Subsets. In: Proceedings of the Fifteenth International Conference on Machine Learning, pp. 37–45 (1998)
8. Ho, T.K.: The Random Subspace Method for Constructing Decision Forests. IEEE Transactions on Pattern Analysis and Machine Intelligence 20(8), 832–844 (1998)
9. Maclin, R., Shavlik, J.W., et al.: Combining the Predictions of Multiple Classifiers: Using Competitive learning to initialize neural networks. In: International Joint Conference on Artificial Intelligence, pp. 524–531. Lawrence Erlbaum Associates Ltd. (1995)
10. Langdon, W.B., Buxton, B.F.: Genetic programming for mining DNA chip data from cancer patients. Genetic Programming and Evolvable Machines 5(3), 251–257 (2004)
11. Yu, J., Yu, J., Almal, A.A., Dhanasekaran, S.M., Ghosh, D., Worzel, W.P., Chinnaiyan, A.M.: Feature Selection and Molecular Classification of Cancer Using Genetic Programming. Neoplasia 9(4), 292 (2007)
12. Liu, K., Xu, C.: A Genetic Programming-based Approach to the Classification of Multiclass Microarray datasets. Bioinformatics 25(3), 331–337 (2009)
13. Breiman, L.: Heuristics of Instability and Stabilization in Model Selection. The Annals of Statistics 24(6), 2350–2383 (1996)
14. Kuncheva, L.I.: A Theoretical Study on Six Classifier Fusion Strategies. IEEE Transactions on Pattern Analysis and Machine Intelligence 24(2), 281–286 (2002)

15. Pedregosa, F., Varoquaux, G., Gramfort, A., Michel, V., Thirion, B., Grisel, O., Blondel, M., Prettenhofer, P., Weiss, R., Dubourg, V., Vanderplas, J., Passos, A., Cournapeau, D., Brucher, M., Perrot, M., Duchesnay, E.: Scikit-learn: Machine Learning in Python. Journal of Machine Learning Research 12, 2825–2830 (2011)

16. Sun, Y.: Iterative RELIEF for Feature Weighting: Algorithms, Theories, and Applications. IEEE Transactions on Pattern Analysis and Machine Intelligence 29(6), 1035–1051 (2007)

17. Albanese, D., Visintainer, R., Merler, S., Riccadonna, S., Jurman, G., Furlanello, C.: mlpy: Machine Learning Python. Tech. rep. (2012)

18. Perone, C.S.: Pyevolve: a Python Open-source Framework for Genetic Algorithms. SIGEVOlution 4(1), 12–20 (2009)

19. Petricoin III, E.F., Ardekani, A.M., Hitt, B.A., Levine, P.J., Fusaro, V.A., Steinberg, S.M., Mills, G.B., Simone, C., Fishman, D.A., Kohn, E.C., Liotta, L.A.: Use of Proteomic Patterns in Serum to identify ovarian cancer. The Lancet 359(9306), 572–577 (2002)

20. Golub, T., Slonim, D., Tamayo, P., Huard, C., Gaasenbeek, M., Mesirov, J., Coller, H., Loh, M., Downing, J., Caligiuri, M., Bloomfield, C.D., Lander, E.S.: Molecular Classification of Cancer: Class Discovery and Class Prediction by Gene Expression Monitoring. Science 286(5439), 531–537 (1999)

21. Alon, U., Barkai, N., Notterman, D.A., Gish, K., Ybarra, S., Mack, D., Levine, A.J.: Broad patterns of gene expression revealed by clustering analysis of tumor and normal colon tissues probed by oligonucleotide arrays. Proceedings of the National Academy of Sciences 96(12), 6745–6750 (1999)

22. Gordon, G.J., Jensen, R.V., Hsiao, L.-L., Gullans, S.R., Blumenstock, J.E., Ramaswamy, S., Richards, W.G., Sugarbaker, D.J., Bueno, R.: Translation of Microarray Data into Clinically Relevant Cancer Diagnostic Tests Using Gene Expression Ratios in Lung Cancer and Mesothelioma. Cancer Research 62(17), 4963–4967 (2002)

23. Singh, D., Febbo, P.G., Ross, K., Jackson, D.G., Manola, J., Ladd, C., Tamayo, P., Renshaw, A.A., D'Amico, A.V., Richie, J.P., Lander, E.S., Loda, M., Kantoff, P.W., Golub, T.R., Sellers, W.R.: Gene expression correlates of clinical prostate cancer behavior. Cancer Cell 1(2), 203–209 (2002)

A Study of Data Classification and Selection Techniques for Medical Decision Support Systems

Ahmed J. Aljaaf[1], Dhiya Al-Jumeily[1], Abir J. Hussain[1], David Lamb[1], Mohammed Al-Jumaily[1], and Khaled Abdel-Aziz[2]

[1] Applied Computing Research Group, LJMU, Liverpool, L3 3AF, UK
[2] Walton Centre Foundation Trust, Lower Lane, Fazakerley, Liverpool, L9 7LJ, UK
A.J.Kaky@2013.ljmu.ac.uk,
{d.aljumeily,a.hussain d.j.lamb}@ljmu.ac.uk,
khaled.abdel-aziz@thewaltoncentre.nhs.uk

Abstract. Artificial Intelligence techniques have been increasingly used in medical decision support systems to aid physicians in their diagnosis procedures; making decisions more accurate and effective, minimizing medical errors, improving patient safety and reducing costs. Our research study indicates that it is difficult to compare different artificial intelligence techniques which are utilised to solve various medical decision-making problems using different data models. This makes it difficult to find out the most useful artificial intelligence technique among them. This paper proposes a classification approach that would facilitate the selection of an appropriate artificial intelligence technique to solve a particular medical decision making problem. This classification is based on observations of previous research studies.

Keywords: Artificial intelligence, decision support system, Medical error.

1   Introduction

Naturally, intelligence is the capability to think, recognize and understand; instead of doing things by behavioural instinct or automatically [1, 2]. Although, it is possible to claim that there is no absolute standard definition of the term Artificial Intelligence (AI) [1]. According to Kurzweil's definition, AI is *"The art of creating machines that perform functions that require intelligence when performed by people"*. However, AI is *"The study of how to make computers do things at which, at the moment, people are better"* as reported by Rich and Knight [1].

Artificial intelligence has become widely used in health-related decision support systems [3], in order to increase the accuracy and effectiveness of disease diagnosis as well as to avoid or at least to minimize medical errors [7]. Various artificial intelligence techniques such as artificial neural networks [11], genetic algorithms [12], decision trees [13, 14], rule-based expert systems [15, 16] and fuzzy techniques [17, 18] have already been employed in clinical decision support systems to diagnose a wide range of health problems [3]. Systems based on intelligence strategies either employ a single technique to deal with certain problems, or utilise a combination of two or

D.-S. Huang et al. (Eds.): ICIC 2014, LNAI 8589, pp. 135–143, 2014.
© Springer International Publishing Switzerland 2014

more techniques to tackle complex problems that involve a level of uncertainty and ambiguity [4].

Within this paper, we aimed to a) discuss the circumstances leading to medical error and how medical-related decision support systems can deduce medical error rates, b) review previous studies which have employed AI techniques in medical decision support systems, c) discuss the importance of identifying a problem domain associated with a medical decision support system and d) propose a classification guide (CG) to classify problem domains within a medical decision support systems which would facilitate the selection of an appropriate AI technique to be used within such systems. This would give more efficient and reliable results; reflecting positively on patient safety, minimizing medical errors and overall costs.

2   Intelligent Decision and Medical Errors

Making decisions intelligently in the medical field are affected by two primary factors; (a) the quantity and validity of information that can be accessed and utilized combined with, (b) the use of intelligence techniques in the decision-making procedure [5, 8].

The information gathered by clinicians, such as a patient's symptoms, clinical examination and laboratory testing outcomes could potentially represent more than one possible disease. Thus, discriminating among the possibilities with complete certainty is often difficult. In such circumstances, non-specialist clinicians often seek the opinion of more experienced colleagues. Indeed, this uncertainty should not be thrown on the responsibility of some clinician with less experience. However, it reflects the uncertainty inherited in the information provided by the patients themselves, where the pathogens are invisible within the patient's body and must be inferred from insufficient external signs [5]. On the other hand, the necessary information needed to make an accurate decision seems to consist of more variables than the human mind can accommodate. There are persuasive evidences which indicate that the human ability to discover and understand complicated configuration relationships could be limited. There is no doubt that scientific genius exists, however a very little evidence shows that the majority of clinicians are geniuses [6]. Therefore, information processing with the limited ability of the human mind to manipulate a large quantity of information or variables in considering a complex problem possibly would lead to wrong decisions. The most prominent challenge is whether clinicians can be effective decision makers or not [6].

According to the foregoing facts that have been represented as (a) the uncertainty surrounding the decision, (b) a large quantity of information required to come up with the right decision, and (c) human mind limitations regarding to the analysis, evaluation and summarizing of a complex problem. Hence, making a decision with the existence of any such difficulties probably would lead to loss of the patient's life, or at least it might have a serious and life-altering consequence on the rest of his/her life.

Medical errors or *medical malpractice* is both costly and harmful. In the U.S. medical errors lead to a larger number of deaths annually than highway accidents, breast cancer, and AIDS combined [3]. Researchers concluded that from the inclusion of 37 million patient medical records, approximately 200 thousand patients in the U.S. died due to potentially preventable medical errors [3]. Another study conducted by the National Patient Safety Foundation showed that roughly less than half of 100 million subscribers within a phone survey thought that they personally have suffered from medical errors [7]. Consequently a reliable, robust and effective medical-related decision support system (MDSS) based on AI techniques are urgently required; such systems could potentially reduce medical error rates by informing decisions made by physicians, thus allowing earlier treatment, improving patient safety and survival rates, whilst reducing total cost [3, 7].

3   Medical Decision Support System

The growing availability of electronically stored medical records (EMRs) which is considered a great source of patient information and the use of computers to maintain and manage these records has led to a growing interest in the automation of medical decision making [8, 9]. Intelligent Medical Decision Support Systems (IMDSSs) are computer-based systems designed to aid clinicians by converting the raw medical-related data, documents, and expert practice into a set of sophisticated algorithms; applying numerical, logical and intelligent techniques in order to find out a proper solution to a medical problem as well as to make clinical decisions [7, 8].

IMDSSs are considered useful in circumstances where patients are suffering from complex conditions or there is lack of access to specialist doctors. Such systems can be used in aiding the doctors to take an appropriate decision and action in complex situations, in which any small error possibly may lead to loss of the patient's life [10]. Over the last years, several medical decision support systems based on Artificial Intelligence have been developed. This paper reviews and summarises a number of recent studies that employ a variety of Artificial Intelligence techniques in IMDSSs.

Artificial Neural Networks (ANNs) have already been successfully employed in a range of real world classification problem, particularly within medical diagnosis. An ANN is a collection of processing units, similar to neurons with weight connections among them. ANN mimics the human brain in order to make a prediction through based in its learning of patterns in a data set. Kumar, V., et al [11] have highlighted several uses of ANN in clinical diagnosis, medication development and image and signal analysis and interpretation. In this setting, ANNs were shown to excel in the classification of complicated data with high rate of accuracy (e.g. correctly classified 91.5% of presenting anticancer medications and knowledge-based ANN showed 87.36% in the prior knowledge of metabolic features of normal and cancerous breast tissues) [11]. This is because neural networks are nonlinear models, learning by example, which makes them more flexible, adaptable and highly effective in modelling complicated relationships within the medical diagnosis process. ANNs estimate posterior probabilities, providing the basis for creating classification rules and performing statistical analysis [19].

Sivasankar, E. and Rajesh, R. S. have used a Genetic Algorithm (GA) based feature selection approach for clinical decision support system; comparing the performance of the GA with two feature ranking algorithms. The objective is to utilise the GA as a per-classification stage in order to increase the classification accuracy [12]. This is mainly because the overall performance of the classifier can be affected by the selected features of the data that could be inherently noisy [20]. The ranking approaches Information Gain and Chi-Square algorithm, along with the GA's classification accuracy were measured using two methods; Bayesian and K-Nearest Neighbour respectively. GA showed a classification accuracy of 88.68% and 85.85% while Information Gain returned 83.96% and 81.13% and Chi-Square algorithm (83.96% and 81.13%) on the Appendicitis data set. On the Parkinson's dataset, the accuracy was as follows; GA (88.21% B, 89.74% K-NN), Information Gain (81.03% B, 84.10% K-NN) and Chi-Square algorithm (81.03% B, 84.10% K-NN) [12].

Shanthi, D., et al, have applied a decision tree classifier to identify the patient's post-operative recovery decision which include General Hospital, Go-Home or Intensive-care. It is a complicated classification problem concerning the prediction of the recovery area that the postoperative patients will be admitted to. The decision should be made depending on particular input vectors. These vectors usually are attributes corresponding generally to clinical observations (e.g. internal and surface temperature, blood pressure and oxygen saturation), where a rapid and correct decision in such circumstance is crucial [14].

The decision tree classification algorithm is a popular cross-domain classification method, utilised in problem domains such as marketing and customer retention, fraud detection and medical diagnosis [22]. The tree is constructed by recursively partitioning the training data until all members of every partition assign to the same class label or there are no further attributes remaining for partitioning, according to an appropriate algorithm [23].

The Gini splitting algorithm has been used to identify the patient's post-operative recovery section and the output generated as a binary tree structure, easy to follow and manipulate; the formulated model can be expressed as a set of decision rules. Building and executing decision trees are fast when compared to other learning methods. Additionally, the output model can easily be visually presented as a binary tree graphic, which is easy for non-technical domain experts to interpret [24].

Rule based clinical decision support systems have been used to provide quick diagnosis from Complete Blood Count (CBC) test results to present a probable list of diseases; assisting physicians to diagnose and recommend medicines to a patient with a haematological disease. The CBC test is the most commonly ordered blood test which is a calculation of the cellular (formed components) of blood [15]. The main inputs to the system comprise patient information (e.g., gender, age, altitude and pregnancy period) and the patient's CBC report from the laboratory. A list of probable haematological diseases will be suggested by the system based on these factors. The system follows a rule-based technique to reason on patient inputs with an internal knowledge base, including a set of haematological diseases combined with their symptoms. CBC is considered a basic method of testing; unable to provide an ultimate diagnosis, nevertheless the results are an extremely good starting point for further, more disease-specific tests to confirm the diagnosis [15].

The basic principle of a rule-based technique is pattern identification followed by a recommendation of what should be done in response. Therefore, the rule is a conditional statement that links the supplied conditions to actions or results. Ideally, rules are simple to express, providing knowledge in near-linguistic form; both human and machine interpretable and easily updatable as required. This facilitates the separation of knowledge from processing, in addition to allowing incomplete or uncertain knowledge to be expressed and bounded. Conversely, learning the format of a knowledge-based expression, in addition to navigating categorisations and relationships in a large knowledge base can be complicated and time consuming [25].

4   Results and Discussion

Previous studies on intelligent medical support system have highlighted the importance of identifying and understanding a problem domain that is related to a medical decision support issue [11-18]. They also focused on the selection mechanisms of an artificial intelligence technique that could be applied in each one of them, where an essential question has been identified and should be answered. The question that emerges is how to select an appropriate artificial intelligence technique to be employed in the medical decision support system.

It is possible to solve the same problem using more than one AI technique which may provide different outcomes. Each AI technique has strengths, functional advantages, and problem domain spaces to which it is well suited, that distinguish it from the others.

Accordingly, in this paper a classification guide is proposed to identify a problem domain within a MDSS. This would facilitate a selection approach of the best AI technique that could be adapted to solve a problem. As a result, more accurate and effective intelligent MDSSs can be developed; in turn reflecting positively on improving clinical treatment, minimizing medical errors and decreasing overall cost.

Many other studies [12, 13, 21] have aimed to examine and evaluate various AI techniques in medical-assisting systems through the use of the same data set to detect and identify the differences in terms of accuracy, cost, and time consumption. In contrast, the objective of this paper is to review the artificial intelligence techniques that have already been utilized in medical decision support (Table 1), with the intent of finding a classification guidance that best fits the problem domain. Use of a particular AI technique to solve a specific decision-related problem may have much better outcomes if compared with other techniques to solve exactly the same problem.

MDSSs based on AI techniques can be used to support many different activities, such as diagnosis, imaging and therapy. However, the performance and effectiveness of these systems are largely dependent on factors including system design, information quantity and validity as well as the AI technique used to generate a knowledge which significantly contributes to making a more accurate decision.

As such, the proposed classification guide (CG) categorizes the problem domain within the medical decision support systems under two main headings, namely; (a) the problem domain relevant to the classification, (b) the problem domain relevant to the selection, in addition to the uncertainty or ambiguity levels associated with both previous categories as illustrated in Figure 1. A classification problem domain arises when a set of symptoms grouped and assigned into a predefined group of diseases, according to a number of attributes associated with each disease, while a selection problem domain is the ability to choose or labelling one disease within the group of diseases, depending on a number of the observed symptoms.

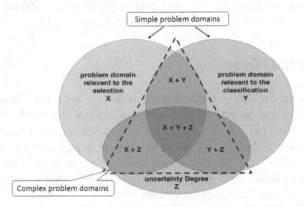

Fig. 1. Classification Guide (CG) of a Problem Domain Related to MDSSs

As the areas on both sides represent the simple problem domain, that can be solved using a single AI technique. However, the overlapping areas surrounded by a triangle in the middle reflects the complex problem domain, which requires more than a single AI technique to be used. The main purpose of the classification guide is in the pre-construction stages of the MDSS to assist in reflecting on the problem domain and desired system goals. This would facilitate the selection of the most appropriate AI technique (or set of techniques) in constructing the MDSS; thus increasing the accuracy and effectiveness of the system. For further illustration, the Classification Guide (CG) will be used to identify a problem domain of a decision-making related to headache diagnosis system via a set of symptoms. The system can assist the physician to diagnose the type of headache through a set of symptoms, as there are more than one type of headache.

As proposed in Figure 1, the headache diagnosing system would be classified under complex problems for the following underlying factors: First, the diagnosis of a specific headache type from other sorts of headache deemed within the Selection area. Second, some signs and symptoms might indicate more than one type of headaches, which indicates a degree of uncertainty surrounding the decision-making process. Therefore, the problem domain can be expressed according to the following hypothesis:

$$Problem\ Domain = Complex\ (X + Z)$$

Where, the above hypothesis can be clearly represented and derived from the full expression of the classification guide Figure 2, or through following the dotted path of the tree diagram shown in Figure 3, which describes the classification guide (CG).

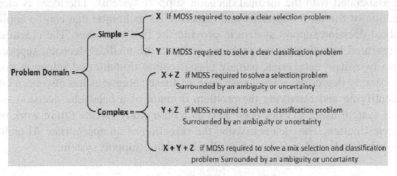

Fig. 2. Full expression of the Classification Guide (CG)

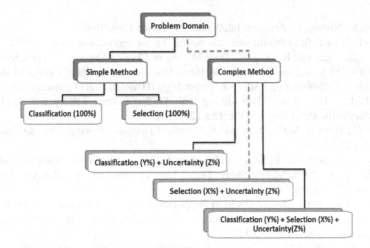

Fig. 3. Tree diagram of the Classification Guide (CG)

After identifying the problem domain, the selection of an appropriate AI technique is straightforward. For instance, the rules based technique could be used within the scope of selection and the fuzzy technique could be used within the scope of uncertainty, thus a combination of two AI technique (fuzzy rule-based technique) could be used to solve a specific complex problem related to medical decision support systems as well as to provide the most reliable and effective outcomes.

5   Conclusion

Medical decision support systems have utilised artificial intelligence (AI) techniques and methods to aid physicians and medical doctors in making more accurate decisions within a reasonable time period. The selection of an appropriate AI technique to be applied in such systems is extremely important and has a major role in the effectiveness and accuracy of the results. As results of investigating different research works, this paper proposes a classification guide (CG) to describe and identify the problem domain associated with the medical decision support systems. The ideas is facilitate the selection of the most beneficial and effective AI technique that can be applied in the medical decision support system to provide the best outcomes. The classification guide presented here divides the problem space in the medical decision support systems into two main categories, namely classification domain and selection domain. The uncertainty degree associated with each of these categories has also been considered. Identifying and clarifying the problem domain of a headache decision support diagnosis system using CG has been used as a case study for the future work related to this investigation. The idea is assisted the selection of an appropriate AI technique to be included and employed in such medical decision support system.

References

1. Russell, S., Norvig, P.: Artificial Intelligence. Prentice-Hall (1995)
2. Warwick, K.: Artificial Intelligence: the basics. Taylor and Francis Group (2012)
3. Amin, S.E., Agarwal, K., Beg, R.: Data Mining in Clinical Decision Support Systems for Diagnosis, Prediction and Treatment of Heart Disease. International Journal of Advanced Research in Computer Engineering & Technology (IJARCET) 2(1) (January 2013)
4. Rokach, L., Maimon, O.: Data Mining with Decision Tree: Theory and Applications. World Scientific Publishing Co. Pte. Ltd. (2008)
5. Sox, H.C., Higgins, M.C., Owens, D.K.: Medical Decision Making, 2nd edn. John Wiley & Sons, Ltd. (2013)
6. Moses, A., Lieberman, M., Kittay, I., Learreta, J.A.: Computer-Aided Diagnoses of Chronic Head Pain: Explanation, Study Data, Implications, and Challenges. Journal of Craniomandibular Practice (2006)
7. Jao, C.S., Hier, D.B.: Clinical Decision Support Systems: An Effective Pathway to Reduce Medical Errors and Improve Patient Safety. In: Jao, C.S., Hier, D.B. (eds.) Decision Support Systems, ch. 8. InTech (2012)
8. Berner, E.S. (ed.): Clinical Decision Support Systems: Theory and Practice, 2nd edn. Springer (2007)
9. Kwiatkowska, M., McMillan, L.: A Semiotic Approach to Data in Medical Decision Making. In: 2010 IEEE International Conference on Fuzzy Systems (2010)
10. Janghel, R.R., Shukla, A., Tiwari, R., Tiwari, P.: Clinical Decision support system for fetal Delivery using Artificial Neural Network. In: International Conference on New Trends in Information and Service Science. IEEE (2009)
11. Kumar, V., Ahmed, E., Kumar, D., Singh, R.K.: Artificial Neural Network Based Intelligent Decision Support System Model for Health Care Management. The Journal of Computer Science and Information Technology 6(1) (2007)
12. Sivasankar, E., Rajesh, R.S.: Design and Development of a Clinical Decision Support System for diagnosing appendicitis. Computing. In: Communications and Applications Conference (ComComAp). IEEE (2012)
13. Kumar, D.S., Sathyadevi, G., Sivanesh, S.: Decision Support System for Medical Diagnosis Using Data Mining. IJCSI International Journal of Computer Science Issues 8(3(1)) (May 2011)

14. Shanthi, D., Sahoo, G., Saravanan, N.: Decision Tree Classifiers to Determine the Patient's Post-Operative Recovery Decision. International Journal of Artificial Intelligence and Expert Systems (IJAE) 1(4) (2010)
15. Chen, Y.Y., Goh, K.N., Chong, K.: Rule Based Clinical Decision Support System for Hematological Disorder. In: 4th IEEE International Conference on Software Engineering and Service Science (ICSESS). IEEE (2013)
16. Al-Hajji, A. A.: Rule-Based Expert System for Diagnosis and Symptom of Neurological Disorders "Neurologist Expert System (NES)". ICCIT (2012).
17. Papageorgiou, E., Stylios, C., Groumpos, P.: A Combined Fuzzy Cognitive Map and Decision Trees Model for Medical Decision Making. In: Proceedings of the 28th IEEE EMBS Annual International Conference, USA (2006)
18. Levashenko, V., Zaitseva, E.: Fuzzy Decision Tree in Medical Decision Making Support System. In: 2012 Federated Conference on Computer Science and Information Systems (FedCSIS). IEEE (2012)
19. Zhang, G.P.: Neural Networks for Classification: A Survey. IEEE Transactions on Systems, Man, and Cybernetics, Part C: Applications and Reviews 30(4) (2000)
20. Mohamad, M.S., Deris, S., Yatim, S.M., Othman, M.R.: Feature Selection Method Using Genetic Algorithm for the Classification of Small and High Dimension Data. In: First International Symposium on Information and Communications Technologies, Putrajaya, Malaysia (2004)
21. Aziz, A.S.A., Azar, A.T., Salama, M.A.: Genetic Algorithm with Different Feature Selection Techniques for Anomaly Detectors Generation. In: 2013 Federated Conference on Computer Science and Information Systems. IEEE (2013)
22. Zhouab, Y., Tana, Y., LIb, H., Gub, H.: A Multi-Classifier Combined Decision Tree Hierarchical Classification Method. In: 2011 International Symposium on Image and Data Fusion (ISIDF). IEEE (2011)
23. Thangaparvathi, B., Anandhavalli, D., Mercy Shalinie, S.: A High Speed Decision Tree Classifier Algorithm for Huge Dataset. In: IEEE-International Conference on Recent Trends in Information Technology, ICRTIT, MIT, Anna University, Chennai. IEEE (2011)
24. Floares, A., Birlutiu, A.: Decision Tree Models for Developing Molecular Classifiers for Cancer Diagnosis. In: The 2012 International Joint Conference on Neural Networks (IJCNN). IEEE (2012)
25. Abraham, A.: Rule-based Expert Systems. In: Handbook of Measuring System Design. John Wiley & Sons, Ltd. (2005)

A Reduction SVM Classification Algorithm
Based on Adaptive AP Clustering Granulation

Xiuxi Wei

Information Engineering Department,
Guangxi International Business Vocational College, Nanning 530007
weixiuxi@163.com

Abstract. The classification speed of SVM is inversely proportional to the number of Support Vectors (SVs). Therefore, the less SVs means the more sparseness and the higher classification speed. In order to reduce the number of SVs but without losing of generalization performance, a new algorithm called Classification Algorithm of Support Vector Machine based on Adaptive Affinity Propagation clustering Granulation (CSVM-AAPG) is proposed, which employs Affinity Propagation (AP) clustering algorithm to cluster the original SVs and the cluster centers are used as the new SVs, then aiming to minimize the classification gap between SVM and CSVM-AAPG, a quadratic programming model is built for obtaining the optimal coefficients of the new SVs. Meanwhile, it is proven that when clustering the original SVs, the minimal upper bound of the error between the original decision function and the fast decision function can be achieved by AP. Finally, experiments show that compared with original SVs, the number of SVs decreases and the speed of classification increases using CSVM-AAPG, while the loss of accuracy is in the acceptable level.

Keywords: SVM, Adaptive, Affinity Propagation clustering, Granulation, Classification.

1   Introduction

Support Vector Machines (SVM) is known as a new generation learning system based on statistical learning theory[1]. Because of its profound mathematical theory, SVM delivers excellent performance in many real-world predictive data mining applications such as text categorization, medical and biological information analysis[2-4],etc. Generally, SVM is divided into two steps in the application process for classification. Firstly, the training samples with class labels are trained to determine a classification model. Secondly, the samples without class labels are to identify classification using the trained model. Correspondingly, the study on accelerating classification speed of SVM is also divided into two directions which contain how to speed up the training process of SVM and how to speed up the classification process of SVM.

At present, the study on accelerating the training speed of SVM has achieved many good results[5-8]. These methods are feasible and effective for improving the training speed of SVM. However, the study on the classification speed of SVM is rare. As we

D.-S. Huang et al. (Eds.): ICIC 2014, LNAI 8589, pp. 144–153, 2014.
© Springer International Publishing Switzerland 2014

know, the classification speed is very important to improve the performance of SVM in the practical applications, especially in the context of the rise of the current cloud computing and networking. Some applications such as the network intrusion detection, anomaly detection for time series and other online application business[9], whose training process can be completed by the background cloud with the strong computing power. Then the background cloud distributed the training business rules (classification model) to the various clients of the Trans-Link. However, how to use the classification model from the background cloud to quickly realize the classification will directly affect the experience of users who have the weak computing power client such as some mobile devices. Therefore, the study on the classification speed of SVM is an urgent need to address the real-world applications.

In order to improve the classification speed of SVM, a new algorithm called classification algorithm of support vector machine based on adaptive affinity propagation clustering granulation (CSVM-AAPG) is proposed in this paper. Where, Affinity Propagation (AP) clustering algorithm is a new clustering algorithm[10]. The main idea of CSVM-AAPG is that the AP clustering algorithm is used to cluster the original SVs and the cluster centers are used as the new SVs, which can improve the sparseness of SVs set. Meanwhile, based on the sparseness of SVs, it is proven that when clustering original SVs, the minimal upper bound of the error between the original decision function and the fast decision function can be achieved by AP clustering algorithm clustering the original SVs in input space. Experiments show that compared with original SVs, CSVM-AAPG has better classification speed.

2   The Method of CSVM-AAPG

2.1   The Main Idea of CSVM-AAPG

Support vector machine (SVM) is a new data mining method based on statistical learning theory and its basic idea is to find the optimal separating hyper-plane which can meet the classification requirements. The mathematical model of SVM is shown as follows:

Given training sample sets $(x_i, y_i), i = 1, 2, \cdots, l, x \in R^n, y \in \{\pm 1\}$ and the classification hyper-plane with kernel function is denoted by

$$f(x) = \sum_{i=1}^{s_1} \alpha_i y_i K(x_i, x) + b, \tag{1}$$

where, $x_i \in R^n$ are the support vectors, s_1 is the number of support vectors, $\alpha_i \in R$ is the Lagrange coefficient, b is a constant distance from the origin of coordinates, x is the test data, y_i are the corresponding class labels and $K(x_i, x)$ is the kernel function.

Seen visually from the equation (1), the classification speed of SVM depends on the size of s_1. As we know, the classification function of traditional SVM contains all of the support vectors, this result will affect its classification speed. Therefore, the key to improve the classification speed is to reduce the number of support vectors s_1. Based on the above analysis, we define a new fast classification function as follows:

$$g(x) = \sum_{j=1}^{s_2} \beta_j y_j K(x'_j, x) + b, \tag{2}$$

where, $s_2 < s_1$, $x'_j \in R^n$, and $\{x'_j\}_{j=1}^{s_2}$ is the new support vector set, β_j is the Lagrange coefficient. Meanwhile, the loss function is defined as follows:

$$L(x) = \|g(x) - f(x)\|^2 \tag{3}$$

For the SVM, the original training samples set $\{x_k\}_{k=1}^n$ are sparse as the support vectors set $\{x_i\}_{i=1}^{s_1}$, which is one of the reasons for SVM having better classification speed compared with other classification methods. Similarly, for the FCSVM-AAPCA, the support vectors set $\{x_i\}_{i=1}^{s_1}$ are sparse as the new supprot vectors set $\{x'_j\}_{j=1}^{s_2}$, which would improve the classification speed because of $s_2 < s_1$. Generally, s_2/s_1 is called the compression ratio. The smaller the value of s_2/s_1, the less the number of support vectors, resulting in the faster classification. Therefore, how to construct the new support vectors set $\{x'_j\}_{j=1}^{s_2}$ and the coefficient of the new classification function β_j without reducing the classification accuracy is the key of this paper. The methods of constructing $\{x'_j\}_{j=1}^{s_2}$ and β_j are as follows.

2.1.1   The Constructing Method of $\{x'_j\}_{j=1}^{s_2}$

In this paper, the main idea of constructing method is that the original support vectors are clustered into s_2 classes by the adaptive AP clustering algorithm and the cluster centers x'_j are used as the new support vectors. If the construing method is valid, we must prove that the loss function has the upper bound by using this constructing method. The proof process is as follows:

Theorem 1. If the original support vectors $\{x_i\}_{i=1}^{s_1}$ are clustered into s_2 classes accordance with $\{D_j\}_{j=1}^{s_2}$ as the division and $\{x'_j\}_{j=1}^{s_2}$ as the cluster centers, the upper bound of the loss function can be expressed as follows:

$$L(x) \le C \sum_{j=1}^{s_2} \sum_{i \in D_j} \left\| x_i - x'_j \right\|^2 , \tag{4}$$

where, C is a constant.

Proof: The loss function expressed by equation (3) can be rewritten as

$$
\begin{aligned}
L(x) &= \sum_{j=1}^{s_2} L_j(x) \\
&= \sum_{j=1}^{s_2} \left(\left(\sum_{i \in D_j} \alpha_i y_i K(x_i, x) \right) - \beta_j y_j K(x'_j, x) \right)^2 \\
&= \sum_{j=1}^{s_2} \left(\sum_{i \in D_j} \left(\alpha_i y_i K(x_i, x) - \frac{\beta_j}{|D_j|} y_j K(x'_j, x) \right) \right)^2
\end{aligned}
\tag{5}
$$

where, $|D_j|$ is the number of elements in the j th class. For each subkey $L_j(x)$ of $L(x)$, we can get the following by Cauchy-Schwarz inequality $\left(\sum_{i=1}^{n} a_i \right)^2 \le n \sum_{i=1}^{n} (a_i)^2$,

$$
\begin{aligned}
L_j(x) &= \left(\sum_{i \in D_j} \left(\alpha_i y_i K(x_i, x) - \frac{\beta_j}{|D_j|} y_j K(x'_j, x) \right) \right)^2 \\
&\le |D_j| \sum_{i \in D_j} \left(\alpha_i y_i K(x_i, x) - \frac{\beta_j}{|D_j|} y_j K(x'_j, x) \right)^2
\end{aligned}
\tag{6}
$$

There exists a constant c_{ij} which can make

$$\left(\alpha_i y_i K(x_i, x) - \frac{\beta_j}{|D_j|} y_j K(x'_j, x) \right)^2 = c_{ij} \left(K(x_i, x) - K(x'_j, x) \right)^2 \tag{7}$$

Let $z = \sqrt{\sigma} \| a - b \|$, if using the gaussian kernel function, we can get

$$K(a,b) = k(z) = \exp(-z^2) = \exp(-\sigma \| a - b \|^2) ,$$

$$k'(z) = -2z \cdot k(z), \quad k'(z) \le k'(1/\sqrt{2}) = \sqrt{2/e}$$

According to the lagrange mean value theorem, $\exists \xi$, make

$$\left(K(x_i,x)-K(x'_j,x)\right)^2 = \left(\sqrt{\sigma}k\,'(\xi)\left(\|x_i-x\|-\|x'_j-x\|\right)\right)^2$$

$$\leq (2\sigma/e)\left(\|x_i-x\|-\|x'_j-x\|\right)^2 \tag{8}$$

According to the triangle inequality theorem, we can get

$$\|x_i-x\|-\|x'_j-x\| \leq \|(x_i-x)-(x'_j-x)\| = \|x_i-x'_j\| \tag{9}$$

Substitute equation (6), (8) and (9) into equation (5)

$$L(x) = \left((2\sigma/e)|D_j|^{max}c_{ij}^{max}\right)\sum_{j=1}^{s_2}\sum_{i\in D_j}\|x_i-x'_j\|^2, \tag{10}$$

where, $|D_j|^{max} = \max(|D_j|)$, $c_{ij}^{max} = \max(c_{ij})$

Let $C = (2\sigma/e)\max(|D_j|)\max(c_{ij})$, the constant C can be obtained.

End.

2.1.2 The Constructing Method of β_j

In order to make the classification accuracy loss to reach the minimum, we must minimize the loss function. Thus the unconstrained optimization problem can be got as follows:

$$\min_{\beta_j}\sum_{k=1}^{n}\left\{\left\|\sum_{j=1}^{s_2}\beta_j y_j K(x_j,x_k)-\sum_{i=1}^{s_1}\alpha_i y_i K(x_i,x_k)\right\|^2\right\} \tag{11}$$

Solving equation (11) we can obtain the weighting coefficient $\{\beta_j\}_{j=1}^{s_2}$ under the minimum loss of classification accuracy.

In order to facilitate solving the above problem, each variable can be expressed as in the form of vector.

Let $\alpha = [\alpha_1,\alpha_2,\cdots,\alpha_{s_1}]^T$, $\beta = [\beta_1,\beta_2,\cdots,\beta_{s_2}]^T$, then we can get,

$$K_\alpha(x) = \begin{pmatrix} K(x_1,x) \\ K(x_2,x) \\ \vdots \\ K(x_{s_1},x) \end{pmatrix}_{s_1\times 1} \tag{12}$$

$$K_\beta(x) = \begin{pmatrix} K(x'_1, x) \\ K(x'_2, x) \\ \vdots \\ K(x'_{s_2}, x) \end{pmatrix}_{s_2 \times 1} \tag{13}$$

Here, we prove that equation (11) can be expressed in the form of quadratic programming, which can use Matlab to solve. Through such method, the value of β_j can be got.

Theorem 2. The minimum optimizaiton problem expressed by equation (11) can be expressed as the following form of quadratic programming problem.

$$\min_\beta \frac{1}{2} \beta^T H \beta + h^T \beta \tag{14}$$

Proof: Firstly analysis each general form x_k of equation (11):

$$\|g(x) - f(x)\|^2 = g(x)^2 - 2f(x)g(x) + f(x)^2 \tag{15}$$

Because the training set $\{x_k\}_{k=1}^n$ has been identified and $f(x)^2$ is constant, which don't affect solving equation (11). Therefore, we should consider the solving method of $g(x)^2$ and $f(x)g(x)$, which are solved as follows:

$$\begin{aligned} g(x)^2 &= \left(\sum_{j=1}^{s_2} \beta_j y_j K(x'_j, x) \right)^2 \\ &= \sum_{i=1}^{s_2} \sum_{j=1}^{s_2} \beta_i \beta_j K(x_i, x) K(x'_j, x) y_i y_j \\ &= \beta^T (K_\beta(x) K_\beta^T(x)) \beta \end{aligned} \tag{16}$$

$$\begin{aligned} f(x)g(x) &= \left(\sum_{i=1}^{s_1} \alpha_i y_i K(x_i, x_k) \right) \left(\sum_{j=1}^{s_2} \beta_j y_j K(x'_j, x) \right) \\ &= \sum_{i=1}^{s_1} \sum_{j=1}^{s_2} \alpha_i \beta_j K(x_i, x) K(x'_j, x) y_i y_j \\ &= \alpha^T (K_\alpha(x) K_\beta^T(x)) \beta \end{aligned} \tag{17}$$

Making equation (16), (17) into equation (11), we can see equation (11) can be expressed as follows:

$$\min_{\beta}\left(\beta^T \left(\sum_{k=1}^{n}(K_\beta(x)K_\beta^T(x)) \right)\beta - 2\left(\alpha^T\sum_{k=1}^{n}(K_\alpha(x)K_\beta^T(x)) \right)\beta \right) \quad (18)$$

Apparently, equation (18) is a standard form of quadratic programming and the quadratic coefficient matrix is as follows:

$$H = \sum_{k=1}^{n}(K_\beta(x_k)K_\beta^T(x_k)) \quad (19)$$

The vector of the one degree term coefficient is as follows:

$$h = -\left(\alpha^T\sum_{k=1}^{n}(K_\alpha(x)K_\beta^T(x)) \right)^2 \quad (20)$$

End.

2.2 The Steps of CSVM-AAPG

The specific steps of CSVM-AAPG is as follows:

- Step1: Extract the original support vectors. The training samples are trained to extract the original support vectors.
- Step2: Extract the new support vectors. AP algorithm is used to cluster the original support vectors and the cluster centers are treated as the new support vectors.
- Step3: Carry out the secondary training for SVM. The new support vectors are trained to get the new classification model.
- Step4: Test the classification ability of the new SVM model. The classification model in Step3 is used for testing on the testing samples.

2.3 The Time Complexity Analysis

Equation (16) is a quadratic programming problem whose variable scale is s_2. So the time complexity of constructing β_j is $O(s_2^3)$. Moreover, the time complexity of AP clustering algorithm is $O(s_1^2)$. Therefore, the time of carrying out the secondary training for SVM is $O(s_2^3)+O(s_1^2)$. As we know, the training time of SVM is $O(N^3)$, and $s_1 \ll N$, $s_2 \ll N$. Therefore, the training time of CSVM-AAPG is acceptable. Furthermore, the support vectors scale of CSVM-AAPG is s_2, the support vectors scale of SVM is n and $s_2 < n$, so the classification time of CSVM-AAPG is less than the classification time of SVM.

3   Experimental Results and Analysis

In order to test the performance of CSVM-AAPG, the experiments for five typical UCI data sets are carried out in Matlab7.11. The experiment results are compared with FD-SVM from [11] and SVM. In this paper, Gaussian kernel function is used as kernel function, penalty parameter C and kernel parameter σ is selected by using 10-fold cross-validation method. The convergence coefficient of AP clustering algorithm is set $\lambda=0.5$. Table 1 shows the average results of SVM, FD-SVM and CSVM-AAPG with 15 independent runs on ten benchmark datasets.

Table 1. Result comparisons of three algorithms on five UCI datasets

Dataset	CSVM-AAPG	FD-SVM	SVM
	Classification accuracy	Classification accuracy	Classification accuracy
	The number of support vectors	The number of support vectors	The number of support vectors
	Time (s)	Time (s)	Time (s)
Banana	89.32%	86.24%	89.34%
	92.3	107.5	150.7
	0.38	0.56	1.12
Breast Cancer	73.05%	70.45%	73.12%
	34.3	66.8	121.0
	0.05	0.12	0.41
Diabetis	76.28%	73.23%	76.42%
	205.8	286.9	394.8
	0.64	0.97	2.23
Flare-Solar	67.28%	64.69%	67.35%
	187.6	379.8	558.2
	0.87	1.36	3.43
German	74.68%	74.68%	74.68%
	84.6	123.6	205.7
	0.41	0.87	1.44

From Table 1, we can see that the number of support vectors have reduced a lot after compression. However, compared with the SVM, the classification accuracy of CSVM-AAPG doesn't reduce obviously, whose value is within the acceptable range. For example, for the Banana dataset, the number of support vectors is from 150.7 to 92.3 by compressing, but the classification accuracy is almost the same. Therefore, the classification speed will be increased substantially because of the decrease of support vectors. Compared with FD-SVM from [11], our algorithm can obtain higher classification accuracy in a shorter time. These results indicate that CSVM-AAPG is feasible and effective.

4   Conclusion

In this paper, in order to improve the classification speed of SVM under the premise of keeping the small loss of classification accuracy, the original support vectors are compressed. Under the above idea, it is proven that when clustering original SVs, the minimal upper bound of the error between the original decision function and the fast decision function can be achieved by AP clustering algorithm. Accordingly, a fast algorithm called CSVM-AAPG is proposed. The experiments of UCI data set s and the breast tumor diagnostic show that compared with original SVs, the number of SVs decreases and the speed of classification increases, while the loss of accuracy in acceptable level for CSVM-AAPG.

Acknowledgments. This work is supported by the Foundation of Guangxi Department of education research project (No.201010LX088), and the Foundation of Guangxi University of science and technology research project (2013YB326).

References

1. Bellocchio, F., Ferrari, S., Piuri, V.: Hierarchical Approach for Multiscale Support Vector Regression. IEEE Transactions on Neural Networks and Learning Systems 23(9), 1448–1460 (2012)
2. Kumar, M.M., Khemchandani, R., Gopal, M., Chandra, S.: Knowledge Based Least Squares Twin Support Vector Machines. Information Sciences 180(23), 4606–4618 (2010)
3. Ye, Q.L., Zhao, C.X., Chen, X.B.: A Feature Selection Method for TWSVM via a Regularization Technique. Journal of Computer Research and Development 48(6), 1029–1037 (2011)
4. Shao, Y.H., Zhang, C.H., Wang, X.B., et al.: Improvements on Twin Support Vector Machines. IEEE Transactions on Neural Networks 22(6), 962–968 (2011)
5. Platt, J.C.: Fast Training of Support Vector Machines Using Sequential Minimal Optimization. In: Advances in Kernel Methods-Support Vector Learning, California, USA, pp. 185–208 (1999)
6. Ye, Q.L., Zhao, C.X., Ye, N., et al.: Localized Twin SVM via Convex Minimization. Neurocomputing 74, 580–587 (2011)
7. Xu, Y.T., Wang, L.S., Zhong, P.: A Rough Margin-based V-twin Support Vector Machine. Neural Computing & Applications 21, 1307–1317 (2012)
8. Qi, Z.Q., Tian, Y.J., Shi, Y.: Twin Support Vector Machine with Universum Data. Neural Networks 36, 112–119 (2012)

9. Zhang, X.S., Gao, X.B.: Twin Support Vector Machines and Subspace Learning Methods for Micro-calcification Clusters Detection. Engineering Applications of Artificial Intelligence 25, 1062–1072 (2012)
10. Shao, Y.H., Deng, N.Y.: A Novel Margin-based Twin Support Vector Machine with Unitynorm Hyperplanes. Neural Computing & Applications 22, 1627–1635 (2013)
11. Zhang, Z.C., Wang, S.T., et al.: Fast Decision Using SVM for Incoming Samples. Journal of Electronics & Information Technology 33(9), 2181–2186 (2011)

Learning Automata Based Cooperative Student-Team
in Tutorial-Like System

Yifan Wang, Wen Jiang, Yinghua Ma, Hao Ge, and Yuchun Jing

Department of Electronic Engineering, Jiao Tong University, Shanghai, China
{wangyifan_1123,wenjiang,ma-yinghua,sjtu_gehao,
jingyc}@sjtu.edu.cn

Abstract. A novel learning automata (LA) based cooperative student-team in tutorial-like system is presented in this paper. The students in our system are modeled using LA. The new philosophy of a student is that he acquires knowledge not only from teacher, but also from team-workers. The self-examination indicator makes it possible for students to evaluate his learning outcomes. The below normal learner adopts the collective intelligence to improve himself. Experiments demonstrate the proposed method's convergence speed is comparable to the student-classroom interaction method [9] and even better when the environment becomes harder. Compared to the previous interaction method and the single operated student, our accuracy is significantly improved.

Keywords: Tutorial-like system, Learning Automata (LA), cooperative method, current state of Learning (*SoL*).

1 Introduction

Tutorial-like systems are becoming hotspot areas of research. Researchers endeavor to simulate every component of the system using appropriate learning models. What to teach (domain model), who to teach (student model) and how to teach (teacher model) are the main modules. In recent years, the combination of tutorial-like system and learning automata (LA) has been a new study direction. LA is useful in these cases, such as the simulation of learning process in student model [9], the classification of learning type in student model and the imitation of teaching skill improvement in teacher model. The aim of this paper is to present a new philosophy, which is trying to model the cooperative student-team learning process based on LA. In this innovative model, student acquires knowledge not only from teacher, but also from fellow team-workers.

1.1 Learning Automaton (LA)

Learning automaton (LA), a promising field of artificial intelligence, is a self-adaptive machine. By strategically letting a set of actions interact with environment, LA is able to track the optimal one. A wide range of learning automata applications are

D.-S. Huang et al. (Eds.): ICIC 2014, LNAI 8589, pp. 154–161, 2014.

published in areas such as cooperative spectrum sensing [1], adaptive polling scheme for clustered wireless ad-hoc networks [2], the congestion avoidance in wired networks [3] and multi-constraint assignment problems [4]. When the information is incomplete or the environment is noisy, LA is significantly superior to other methods.

Since the idea of learning automata is put forward, many algorithms are committed to accelerate the learning speed without compromise the accuracy. The concept of discretization [5] and estimation [6] are two landmarks in the development path of single LA. Stochastic estimator algorithm [7], which is the state-of-art of LA, takes the advantage of discretization and estimation idea, and adds a random perturbation with zero mean to the deterministic estimates. Bayesian Pursuit algorithm (BPST) [8], which using Bayesian estimator (BE) instead of maximum likelihood estimator (MLE), is the newly publish LA algorithm. The maturity of single LA theory provides the solid basis for application.

1.2   Related Work

A student-classroom interaction in tutorial-like system is proposed in [9] by Oommen and Hashem. This classroom consists of three types of students, they have different learning speed. Knowledge provider, knowledge seeker and independent learner are three possible statuses of a student. One of them is selected as the student's interaction status by *Tactic-LA*. If the student is a knowledge seeker, he can pick an interaction strategy to communicate with a knowledge provider. Such strategies are transferring no knowledge, embracing provider's knowledge, incorporating reliable knowledge and unlearning useless information after probation.

However, there are some unreasonable setups in this system. The preset reward probability leads the *Tactic-LA* to converge to one of its actions, which is controlled by this preprogrammed probability instead of the current state of learning. This setting does not take each student's individual learning style into account. Furthermore, at every interaction time, student can only communicate with one student.

In this paper, a cooperative student-team in tutorial-like system is proposed. Student can communicate with the whole team and utilize the team's knowledge to improve his. An index that reflects the current state of learning is proposed as a self-examination indicator. The team's average knowledge is performed as the self-examination standard. This index and this standard make it possible for the students to discover their weakness in learning. Experiments demonstrate that the proposed method's convergence speed is accelerated and the learning accuracy is also enhanced. This is the first published result that demonstrates the cooperative team-work effects among student-team.

2   Cooperative Student-Team in Tutorial-Like System

The cooperation of student-team in tutorial-like system is proposed in this section. The student, modeled by LA, is able to interact with the teacher and cooperate with other fellow team-workers. During the cooperation process, student does self-examination about his current state of learning and then utilizes the team's knowledge to improve his own. Knowledge, teacher and student modules are the three important components of tutorial-like system, the architecture of which is shown below.

2.1   Knowledge and Teacher Models

In the tutorial-like systems that we study, domain knowledge [10] is presented using a Socratic model via multiple choice questions. For each question, every choice is associated with a reward probability. $E = [e_1, e_2, ..., e_r]$, which has r options, is an instance of domain knowledge. e_i is the reward probability of the ith option. The higher the reward probability is, the more correct the choice is. The optimal answer is defined as the one with the highest reward probability. $optimal = \arg\max\{e_i\}$. The final goal of a student is to choose the optimal option with probability one.

Teacher attempts to present the domain material to students. Stochastic is the nature of this teacher model. That is to say, teacher doesn't pass on his knowledge directly to students, but replies to the students' choices through the heuristic approach. The option gets a "yes" or "no" response from teacher with a certain probability, which is equal to this choice's reward probability. Let β^i denotes the teacher's feedback to option i, and

$$\beta^i = \begin{cases} 1 & p = e_i \\ 0 & p = 1 - e_i \end{cases} \tag{1}$$

2.2   Student Model

Student does "try and error" work in learning process. At every time instant, student interacts with teacher. Firstly, student chooses one option from the candidate set. Then teacher gives him a response of this choice. This feedback helps the student to refine his decision strategy. This process cycles until student achieve the ultimate goal.

The simulator mimics a real-life student's learning process. In our model, the student simulator is a learning automaton (LA), which can be depicted as $\{A, B, P, T\}$.

$A = \{\alpha_1, \alpha_2, ..., \alpha_r\}$, a finite set of r actions. α_i is corresponding to the ith option in the multiple-choice question and A is the candidate set.

$B = \{0, 1\}$, a feedback set. $\beta \in B$ denotes the teacher's reaction to the present choice.

$P(t) = [p_1(t), p_2(t), ..., p_r(t)]$ is the probability distribution of the action set, α_i is selected on the basis of probability $p_i(t)$ at the time instant t. $P(t)$ simulates the strategy how a student chooses an option from the candidate.

T is the learning algorithm. Probability vector is updated based on the learning algorithm, selections and feedbacks. Through adjust $P(t)$ distribution, T simulates the optimization process of decision.

Different learning algorithms can present different convergence speed. As we discussed in the Introduction section, the epoch-making ideas of LA are discretization and estimation. Based on the experimental results presented in [7], [11], algorithms can be ranked according to their speed: Stochastic Estimator, Generalized Pursuit Algorithm (GPA), Pursuit Algorithm, No-estimator Variable Structure Stochastic Learning Automaton (VSSA) and Fixed Structure Stochastic Automaton (FSSA).

2.3   Cooperation Method in Student-Team

Various algorithms of LA can model students with different learning speed. However, fast is a relative concept. It is safe to say that faster learning algorithms do possess higher learning speed, but they may not always faster than others at any time of learning process. In our proposed cooperation method, student does self-examination to evaluate his learning outcomes. The below normal learner utilizes the team's knowledge to improve his. Cooperation occurs at an interval of M iterations.

Let SoL denotes the current state of learning. It is defined as the difference between the maximum probability and the second maximum probability. That is

$$SoL\{P(t)\} = \max\{P(t)\} - sub\max\{P(t)\} \tag{2}$$

The option with maximum probability and the option with the second maximum probability are the two most competitive candidates in the set. When SoL value is small, the difference between maximum and second maximum value is tiny. Student or LA is much confused when making a choice. The trend of convergence is not obvious and the current state of learning is not well. However, when the SoL is large, the trend of convergence is clear. Student is confident on this question and his current state of learning is well. The SoL works as a self-examination indicator.

In order to provide a standard to identify the SoL value of a student is small or large, the average of the team's probability vector is proposed. Average probability vector reveals the collective wisdom of the student-team. Consider the team has N students. We have an agreement that the superscript is the index of LA and the subscript is the index of actions. Then the average probability can be written as:

$$AvgP(t) = [Avgp_1(t), Avgp_2(t), ..., Avgp_r(t)] \tag{3}$$

$$Avgp_i(t) = \frac{1}{N}\sum_{j=1}^{N} p_i^j(t) \tag{4}$$

Then we can describe the cooperation process as follows. At an interval of M iterations, student calculates his SoL index. The situations are listed below.

If $SoL\{P^j(t)\} < SoL\{AvgP(t)\}$, which means the current observed index SoL of the jth student is lower than the SoL of average. The performance of the simulator is below normal standard. In our proposed cooperative method, this type of student is becoming a knowledge seeker. He uses the average probability vector to fully replace his knowledge.

If $SoL\{P^j(t)\} \geq SoL\{AvgP(t)\}$, which means the current observed index SoL of the jth student is equal or greater than the SoL of average. The performance of the simulator is above normal standard. In our proposed cooperative method, this type of student is becoming a knowledge provider. He offers help to others in an implicit way. He contributes to the team through enhancing the SoL of average probability.

Here to explain the difference between interaction with teacher and the cooperation among student-team. Teacher helps student to refine his decision strategy to achieve the ultimate goal. All the students can learn the correct answer from teacher, but the learning speed differs from others. However, the cooperation among student-team is to accelerate the learning process. There could be a case that a student has a strong preference of bad action, and *SoL* of this student is higher than the average baseline. He won't benefit from collective knowledge in this cooperation, but after sufficient interaction with teacher, student can gradually change this preference and benefit from the collective in the future cooperation.

3   Experimental Results

In this section, experimental results are presented to test the implementation of cooperative student-team. These results were obtained by performing numerous simulation experiments.

In all the experiments, a student is considered to have finished study if the probability of choosing an action is greater or equal to a threshold *Thr* ($0 < Thr < 1$). Moreover, a student is considered to have learned the correct answer if the probability vector converged to the action with the highest reward probability. We ran the experiment *NE* times and took the average as convergence speed. Accuracy is calculated as the ratio between correct convergence times and *NE*. The ratio between the reduction of iteration number and the baseline iteration is taken as iteration gain.

Firstly, table 1 shows a comparison between scenario 1 and 2. Scenario 1 is our proposed cooperative student-team system. Scenario 2 is the best situation in the student-classroom interaction system published in [9]. We adopt exactly the same environment setting and parameters as [9], that is three types of LA, each category has three instance, four environments, *M*=10, *Thr*=0.99 and *NE*=75.

Table 1. Comparison between cooperative student-team system and student-classroom interaction system on the convergence and accuracy

Env.	Student type	Single student		Scenario 1			Scenario 2		
		#iter.	#wro.	# iter.	%gain	#wro.	# iter.	%gain	#wro.
$E_{4,A}$	Fast	572	0	549	4	0	492	14	0
	Normal	996	0	535	46	0	476	52	0
	Below	1382	0	549	60	0	507	63	0
$E_{4,B}$	Fast	1482	0	731	51	0	972	34	0
	Normal	2201	0	683	69	0	879	60	9
	Below	2633	0	697	73	0	758	71	9
$E_{10,A}$	Fast	686	1	614	10	0	585	15	0
	Normal	1297	1	612	53	0	570	56	1
	Below	1804	0	629	65	0	586	68	1
$E_{10,B}$	Fast	1655	4	876	47	1	1053	36	1
	Normal	2114	4	783	63	1	780	63	13
	Below	2859	4	810	72	1	784	73	13

#iter.: the average number of iterations required for convergence

#wro.: the number of wrong convergences;

%gain: the iteration gain.

The results of this comparison are given in table 1. In scenario 1, all types of students benefited from the proposed strategy. The convergence speed is comparable to the scenario 2 and even better when the environment is becoming harder. For example, in environment $E_{4,B}$, all types of students got more improvement than what they got in scenario 2. Moreover, the accuracy is also improved. In scenario 2, nine out of NE times normal and below normal students study the wrong answer. While in our proposed system, all types of students converge correctly.

Next, experiments were performed to demonstrate the cooperative system is superior to the single operated LA in the aspect of speed and accuracy. Three estimator based LA algorithms: Stochastic Estimator reward-inaction (SEri), Discretized Pursuit reward-inaction (DPri) and Discretized Generalized Pursuit Algorithm (DGPA) were performed in this experiment. Environments are set to be the well-known benchmark in [7]. Thr=0.999, NE=250000, M=10 and "Best parameters" for each benchmark were used. "Best parameters" and single LA's performance were clearly published in [7].

Table 2. Comparison between cooperative student-team system and single-student on the convergence and accuracy

Env.	Student type	Single student		Cooperative team		
		#iteration	accuracy	# iteration	accuracy	%iter gain
E_1	SEri	426	0.997	416	0.999	2.34
	DPri	1086	0.995	438	0.999	59.7
	DGPA	880	0.997	440	0.999	50.0
E_2	SEri	834	0.996	749	0.999	10.2
	DPri	2500	0.994	788	0.999	68.5
	DGPA	1677	0.996	792	0.999	52.8
E_3	SEri	2540	0.995	2038	0.999	19.8
	DPri	9613	0.993	2044	0.999	78.7
	DGPA	5191	0.995	2097	0.999	59.6
E_4	SEri	325	0.998	323	0.999	0.615
	DPri	783	0.996	345	0.999	55.9
	DGPA	754	0.997	345	0.999	54.2
E_5	SEri	729	0.997	673	0.999	7.68
	DPri	2363	0.994	709	0.999	70.0
	DGPA	1445	0.997	710	0.999	50.9

It is obvious that all types of LA got improved both in terms of learning speed and accuracy. For instance, in E_1, SEri converged in 416 iterations instead of 426, which is a 2.34% improvement. DPri converged in 438 iterations instead of 1086, which is a 59.7% improvement. DGPA converged in 440 iterations instead of 880, which is a 50.0% improvement. When in the hardest environment E_3, SEri converged in 2038 iterations instead of 2540 iterations, which is a 19.8% improvement. DPri converged in 2044 iterations instead of 9613, which is 78.7% faster. DGPA converged in 2097 iterations instead of 5191, which is a 59.6% improvement. Furthermore, the accuracy of all types of LA achieved 0.999 under all environments. For example, in E_5, the accuracy of SEri is improved from 0.997 to 0.999, DPri is improved from 0.994 to 0.999 and DGPA is improved from 0.994 to 0.999.

From above data analysis, the cooperative student-team does well. Learning speed and accuracy are both benefited from the proposed strategy. This novel idea of cooperation provides another reference for the real-life students' learning, as well as explores a new research field in the intelligent tutorial systems (ITSs).

4   Conclusion and Future Work

This paper has presented a brand new class of cooperative student-team system, in which all LA simulated students working as a team and using collective wisdom to improve their learning. *SoL* is performed as a self-exam indicator, which is difference between the maximum probability and the second maximum probability. When the student discovers his weakness in learning, he adopts the average knowledge. The experiment evidently showed that our proposed methods convergence speed is comparable to the student-classroom interaction method and even better when the environment is becoming harder. The accuracy is clearly improved compared to the previous interaction method and the single operated student.

However, this paper doesn't aim to surpass all the proposed method, but to show the cooperative interaction can effectively serve as an attractive method to perform in tutorial-like system. For future work, we believe that a cooperative method can be used in the LA theoretical field to accelerate learning speed without compromising the accuracy. And the studying of students in tutorial-like system is still open.

Acknowledgement. This work is funded by National Science Foundation of China (61271316), 973 Program (2010CB731403, 2010CB731406, 2013CB329605) of China, Chinese National "Twelfth Five-Year" Plan for Science & Technology Support (2012BAH38 B04), Key Laboratory for Shanghai Integrated Information Security Management Technology Research, and Chinese National Engineering Laboratory for Information Content Analysis Technology.

References

1. Yuan, W., Leung, H., Cheng, W., et al.: Optimizing Voting Rule for Cooperative Spectrum Sensing through Learning Automata. IEEE Trans. Vehicular Technology 60(7), 3253–3264 (2011)
2. Torkestani, J.A.: LAAP: A Learning Automata-based Adaptive Polling Scheme for Clustered Wireless Ad-hoc Networks. Wireless Personal Communications 69(2), 841–855 (2013)
3. Misra, S., Oommen, B.J., Yanamandra, S., et al.: Random Early Detection for Congestion Avoidance in Wired Networks: a Discretized Pursuit Learning-automata-like Solution. IEEE Transactions on Systems, Man, and Cybernetics, Part B: Cybernetics 40(1), 66–76 (2010)

4. Horn, G., Oommen, B.J.: Solving Multiconstraint Assignment Problems Using Learning Automata. IEEE Transactions on Systems, Man, and Cybernetics, Part B: Cybernetics 40(1), 6–18 (2010)
5. Thathachar, M.A.L., Oommen, B.J.: Discretized reward-inaction learning automata. Cybern. Inf. Sci. 2(1), 24–29 (1979)
6. Thathachar, M.A.L., Sastry, P.S.: A new approach to the design of reinforcement schemes for learning automata. IEEE Transactions on Systems, Man and Cybernetics 1(1), 168–175 (1985)
7. Papadimitriou, G.I., Sklira, M., Pomportsis, A.S.: A New Class of ε-optimal Learning Automata. IEEE Transactions on Systems, Man, and Cybernetics, Part B: Cybernetics 34(1), 246–254 (2004)
8. Zhang, X., Granmo, O.-C., Oommen, B.J.: The Bayesian Pursuit Algorithm: A New Family of Estimator Learning Automata. In: Mehrotra, K.G., Mohan, C.K., Oh, J.C., Varshney, P.K., Ali, M. (eds.) IEA/AIE 2011, Part II. LNCS, vol. 6704, pp. 522–531. Springer, Heidelberg (2011)
9. Oommen, B.J., Hashem, M.K.: Modeling a Student–classroom Interaction in a Tutorial-like System Using Learning Automata. IEEE Transactions on Systems, Man, and Cybernetics, Part B: Cybernetics 40(1), 29–42 (2010)
10. Oommen, B.J., Hashem, M.K.: Modeling a Domain in a Tutorial-like System Using Learning Automata. Acta Cybern. 19(3), 635–653 (2010)
11. Agache, M., Oommen, B.J.: Generalized Pursuit Learning Schemes: New Families of Continuous and Discretized Learning Automata. IEEE Transactions on Systems, Man, and Cybernetics, Part B: Cybernetics 32(6), 738–749 (2002)

Clustering-Based Latent Variable Models for Monocular Non-rigid 3D Shape Recovery

Quan Wang[1], Fei Wang[1], Daming Li[2], and Xuan Wang[1]

[1] Xi'an Jiaotong University, Xi'an, Shaanxi Province, China
bhxspring@stu.xjtu.edu.cn, wfx@mail.xjtu.edu.cn,
xwang.cv@gmail.com
[2] Beijing Institute of Spacecraft System Engineering, Beijing, China
damli@163.com

Abstract. The difficulty of monocular non-rigid 3D reconstruction using statistical learning approaches is to get a model that can represent as many deformations as possible. Given a known dataset to learn a model, existing latent variable models (LVMs) fail to focus on how to attain labeled samples. In this paper, we propose novel clustering-based LVMs in which we automatically select representative samples to be the labeled ones. To this end, G-means algorithm is adopted to cluster latent variables and obtain the labeled samples. These labeled samples are corresponding to the latent variables closest to clustering centers. We learn the Gaussian Process Latent Variable Model (GPLVM) and the Constrained Latent Variable Model (CLVM) into which we introduce clustering in the context of monocular non-rigid 3D reconstruction, and compare them to those without clustering. The experimental results show that our clustering-based LVMs could perform better.

Keywords: latent variable models, G-means, non-rigid 3D reconstruction.

1   Introduction

Recovering the shape of 3D deformable surfaces from monocular images is a notoriously intractable problem because of its inherent ambiguities. However, it is rather interesting and attractive because this research could benefit many different fields, such as the entertainment industry, the sports, the medical field and so on. Therefore, varieties of approaches have been put forward to resolve the problem of non-rigid 3D reconstruction from monocular images. Physically-based approaches [1,2,3] aim at modeling the true behavior of an object, but they usually need to know some physical parameters which are hard to obtain and the non-linearity of the true physics is complicated to deal with. Structure from motion methods [4,5,6] resolve the ambiguities by tracking points across image sequences, but they depend on stronger constraints and a good initialization. Statistical learning approaches [7,8,9,10] have become a kind of favored methods to recover the shape of 3D deformable surfaces.

Actually, LVMs are commonly applied to learn the appropriate models from training data, since they can project the high-dimensional data into a low-dimensional

D.-S. Huang et al. (Eds.): ICIC 2014, LNAI 8589, pp. 162–172, 2014.
© Springer International Publishing Switzerland 2014

space and represent the unknown mappings by learning. In [9], the GPLVM [11,12] is used to learn local deformation models from data. [10] has proposed a new model CLVM, and made the GPLVM as a comparison. These LVMs have ignored that, what kind of the labeled samples to be chosen can influence the effectiveness of models. Although these LVMs perform well, we think that the resulting LVMs will be more effective if we are able to select those representative samples to be labeled samples.

Given a dataset with thousands of obtained samples, there is a need to select representative samples with an appropriate method as the labeled ones since there must be many whose intrinsic attributes are similar. If we just choose the labeled samples randomly, the probability will increase that some whose intrinsic attributes are similar are chosen as labeled samples. If all the labeled samples have similar attributes, which of course is the worst situation, the model will only adapt to 3D reconstruction within a very small range of deformations. Thus, clustering can be used to get the labeled samples. G-means assumes that the subsets of data follow the Gaussian distribution, which is applicative to our data in latent space. In fact, due to the advantage of low dimensionality in latent space, G-means expenses little additional cost of time. Therefore, it's a good choice to make use of the G-means to obtain the labeled samples.

More specifically, we make use of the G-means [13] to cluster the latent variables which may be obtained by the Principal Component Analysis (PCA) or other dimensionality reduction techniques. These latent variables closest to the clustering centers are taken to be labeled samples. We then learn the LVMs using these labeled samples. This method can ensure that labeled samples are as far different from each other as possible in the latent space. Consequently, the resulting model could represent a wider range of shapes as to the problem of monocular non-rigid 3D shape recovery.

By introducing the G-means to LVMs, we can improve the performance of LVMs. We have demonstrated the effectiveness of the clustering-based LVMs on synthetic data. The clustering-based LVMs do outperform their original LVMs without clustering in the context of non-rigid 3D shape recovery from monocular images.

2   Related Work

Non-rigid surface reconstruction from monocular images is a well-known challenging problem. Physically-based approaches are inspired by Mechanical Engineering techniques. The Finite Element Method, a classic physically-based method, is widely used in computer vision [1,2,3]. However, it's hard to obtain those very appropriate physical parameters. Structure from motion methods try to recover non-rigid 3D shapes with video sequences. [4] uses multiple factorizations, [5] introduces locally-rigid motion framework, and [6] presents a variational approach for dense 3D reconstruction. At the same time, statistical learning approaches develop well. [7] learns local deformation models with the GPLVM, [8] learns linear local models for surface patches, [9] learns GP mappings from intensity patterns to shapes of local surface patches, and [10] proposes to learn a novel model named CLVM.

Among statistical learning approaches in context of non-rigid 3D reconstruction, it can be seen that either simple linear models or more complex LVMs are learnt. As we know, the linear models are too simple to represent the problem in many occasions.

Hence, the LVMs are more commonly studied and used. The GPLVM has been explained in details in [11,12], and the subsequent study on the GPLVM has been presented by Urtasun in [14]. However, all of the studies have failed to consider the problem of the way to choose the labeled samples, which we will propose as follows.

3   Clustering-Based LVMs

3.1   Latent Variable Models

Monocular 3D reconstruction of deformable surfaces always involves handling data in high dimensionality. As an effective vehicle for dimensionality reduction, LVMs have drawn more and more attention. Naturally, it occurs to researchers that they could exploit the LVMs to reconstruct the shape of 3D deformable surfaces. GPLVM is used to deal with the trouble of high dimensionality in the space of possible deformations in [7]. CLVM [10] really gains improvements over GPLVM on the problem of non-rigid 3D reconstruction from monocular images.

Given a dataset full of varieties of available shapes Y, the first thing is to get their latent variables X. Here, Y, which is in a high dimensional space, is represented as triangulated meshes with 3D vertex coordinates; X, which is in a low dimensional latent space, is acquired by PCA. Then, n pairs of latent variables and their individual shapes are chosen as the labeled samples for both the GPLVM and the CLVM, and m latent variables as the unlabeled ones specially for the CLVM. Now we state the GPLVM and the CLVM, respectively.

GPLVM. Given n training pairs of the latent variables and their shapes $[(x_1, y_1), ..., (x_n, y_n)]$, our goal is to predict the output $y' = M(x')$ from a novel input x'. Let $Y_L = [y_1, ..., y_n]^T$, $X_L = [x_1, ..., x_n]^T$. The likelihood

$$p(Y_L \mid X_L, \Theta) = \frac{1}{\sqrt{(2\pi)^{ND} \mid K \mid^D}} \exp(-\frac{1}{2} tr(K^{-1} Y_L Y_L^T)), \tag{1}$$

where K is the kernel matrix whose elements are defined by the covariance function, $k(x_i, x_j)$. Here, we take this function to be the sum of a RBF and a noise term:

$$k(x_i, x_j) = \theta_1 \exp(-\frac{\theta_2}{2}(x_i - x_j)^T (x_i - x_j)) + \theta_3. \tag{2}$$

GPLVM is learnt by maximizing $p(Y_L \mid X_L, \Theta) p(\Theta)$ with respect to Θ.

At reference, given a latent variable x', the mean prediction can be expressed as

$$\mu(x') = Y_L^T K^{-1} k(x'), \tag{3}$$

where $k(x') = [k(x', x_1), ..., k(x', x_n)]^T$. y' is taken to be this mean prediction.

CLVM. The main idea of CLVM is explicitly imposing equality or inequality constraints on the model's output during learning. So, the CLVM is the problem of learning a LVM incorporating constraints on the generated outputs. As you may guess, those m unlabeled samples are used to impose constraints.

Let $X_U = [x_1^U, ..., x_m^U]$, representing the m unlabeled latent variables. The latent variable model can be written as the form

$$y' = W\phi(x').$$ (4)

The learning can be formulated as the constrained optimization problem

$$\min_W \frac{1}{2} \| W\phi(X_L) - Y_L \|_F^2 + \frac{\gamma}{2} \| W \|_F^2$$ (5)

$$s.t.\ C(W\phi(X_U)) = 0,\ or,\ D(W\phi(X_U)) \leq 0,$$

where $\phi(X_L) = [\phi(x_1), ..., \phi(x_n)]$.

By solving this problem with Lagrange multiplier method and introducing kernel function, we can finally get the prediction for an input x^* in closed-form as

$$y^* = AK_{:,*}$$ (6)

with

$$A = [M - \sum_u G_u^T \lambda_u K_{u,:}]B^{-1},\qquad Z = [\phi(X_L), \phi(X_U)],$$ (7)

$$B = K_{:,L}K_{L,:} + \gamma K_{:,:},\qquad\qquad M = Y_L K_{L,:},$$

where $K_{:,*} = [K_{*,L}, K_{*,U}]^T$, $K_{:,:} = ZZ^T = \begin{bmatrix} K_{L,L} & K_{L,U} \\ K_{U,L} & K_{U,U} \end{bmatrix}$.

3.2   The G-Means Algorithm

The G-means algorithm [13] circularly runs K-means with initial cluster centroid positions, whose data have undergone a statistical test, till all the subsets of data follow Gaussian distribution. Hence, at the heart of the G-means algorithm is the statistical test for the hypothesis that a subset of data follows Gaussian distribution.

The statistical test is a process in which decisions are made whether the cluster should be split into two sub-clusters or not. Anderson-Darling statistic test is used. Assuming that c is a center whose cluster is about to undergo the statistical test, the algorithm of the statistical test is as follows.

1. Initialize two centers with $c \pm m$, where $m = s\sqrt{2\lambda / \pi}$, which is obtained by doing PCA on X, data of the center c, s being the main principal component corresponding to eigenvalue λ.

2. Run k-means on the above two centers in X and get two new centers, c_1 and c_2.

3. Let vector $v = c_1 - c_2$ and then project X onto v: $x^{'} =< x,v > / \parallel v \parallel^2$.

Transform $X^{'}$ into Y so that it has mean 0 and variance 1.

4. Let $z = F(y)$, where F is the $N(0,1)$ cumulative distribution function. The Anderson-Darling statistic is

$$A^2(Z) = -\frac{1}{n}\sum_{i=1}^{n}(2i-1)[\log(z_i) + \log(1 - z_{n+1-i})] - n \tag{8}$$

and the corrected one is

$$A_*^2(Z) = A^2(Z)(1 + 4/n - 25/n^2) \tag{9}$$

5. If $A_*^2(Z)$ is lower than the critical value, keep the original center c; if not, keep c_1 and c_2 instead of c.

3.3   Clustering-Based LVMs for Non-rigid 3D Reconstruction

The main idea of clustering-based LVMs is to take advantage of the clustering method G-means to automatically obtain representative samples as the labeled samples for LVMs. The procedure of the clustering-based LVMs is as follows:

1. Given a dataset Y, randomly choose M test samples Y_T. Perform PCA on the rest samples Y_R to get all the latent variables X in the dataset. Let X_R be latent variables corresponding to Y_R. Let X_T be known test latent variables.

2. Cluster X_R with G-means according to subsection 3.2. Choose the latent variables closest to the clustering centers in X_R to be N labeled latent variables X_L, the corresponding Y_L to be labeled samples. Choose V unlabeled latent variables X_U randomly from X_R specially for the CLVM.

3. Learn GPLVM and CLVM according to Eq. (2) and Eq. (7) respectively.

4. Predict test samples according to Eq. (3) and Eq. (6) respectively. As a test, X_T can be known latent variables to predict their shapes $Y_T^{'}$. During simulating

reconstruction, optimized latent variables can be used to predict the shapes Y_T'. The mean of reconstruction errors between Y_T' and Y_T can judge the performance of models.

As is seen from subsection 3.1, latent variable models learn a model that gives a mapping from the latent space to the output space, such as 3D shapes. However, for an unknown shape, its latent variable is also unknown. So, the optimization is necessarily used to reconstruct. Now let's see the process of the optimization in brief.

First the latent variable is initialized with some knowledge or by intuition. Then we iteratively obtained the prediction of the 3D shape with the latent variable models and minimized the reprojection error. The optimization can be written as

$$\min_x \sum_i \| (u_i', v_i') - (u_i, v_i) \|_2^2 \qquad (10)$$

where (u_i', v_i') and (u_i, v_i) are the reprojection and the real location of the input image's i^{th} feature point. The details of solution to reprojection can be found in [15].

4   Monocular Non-rigid 3D Reconstruction

We verified the effectiveness of the clustering-based LVMs when applied to non-rigid 3D reconstruction from a single camera. We first conducted some experiments about the variation with the critical value in G-means, of the reconstruction errors and the constraint errors on the clustering-based LVMs using the known latent variables. Then an acceptable critical value was chosen to do some experiments on the clustering-based LVMs and LVMs without clustering respectively, using the known latent variables and the optimized latent variables respectively. For all the experiments, we made use of the cardboard dataset and the cloth dataset which had been employed in [10]. The two datasets separately contain 2299 and 1885 sample shapes represented by 81 and 63 3D vertex coordinates respectively.

4.1   Variation of Errors on Clustering-Based LVMs

For both datasets, we first chose M=300 test samples, then clustered the latent variables of the rest samples with G-means to attain the labeled samples, and finally chose V=50 unlabeled samples from the rest samples for CLVM. We learnt CLVM as well as GPLVM and predicted 300 shapes of the test samples with the known latent variables which were obtained by PCA.   Equality constraints were used here. We varied the critical value, and observed the changes of the mean reconstruction errors and the mean constraint errors over the 300 test samples. Fig.1 has shown the variation of those errors.

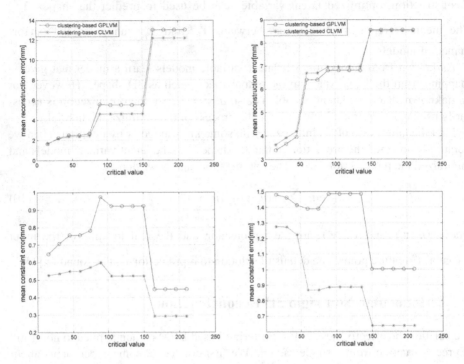

Fig. 1. Mean reconstruction errors and constraint errors' variation with the critical value. The left is for the cardboard dataset, and the right is for the cloth dataset.

From Fig.1, it can be observed that the mean reconstruction errors roughly showed a trend of increase with the critical value while the mean constraint errors did not show very significant changes, which meant that a lower critical value resulted in a better model. However, there was a truth during the experiments that the number of clusters decreased with the increasing critical value. That was to say, when the critical value increased, the number of labeled samples decreased, which was beneficial to lowering the computational burden. So there was a balance between the quality of models and the computation cost.

4.2 Comparison to LVMs without Clustering

Having known from subsection 4.1 that a balance had lain between the quality of models and the computation cost, we picked an appropriate critical value 30 to do the following experiments. We first demonstrated that the clustering-based LVMs were effective and useful using the known latent variables, and then we generated synthetic data to perform the reconstruction with the optimization method.

Table 1. Comparison between the LVMs without clustering and the clustering-based LVMs when predicting shapes for the cardboard from the known latent variables

Methods	Equality Constraints		Inequality Constraints	
	R.E.[mm]	C.E. [mm]	R.E. [mm]	C.E. [mm]
GPLVM	2.6824 ±0.1467	0.8575 ±0.0528	2.6824 ±0.1467	0.8448 ±0.0515
Clustering-Based GPLVM	**2.2282**	**0.7079**	**2.2282**	**0.6959**
CLVM	2.5558 ±0.1276	0.6093 ±0.0399	2.6756 ±0.1267	0.7284 ±0.0372
Clustering-Based CLVM	**2.1732**	**0.5369**	**2.2360**	**0.6472**

Table 2. Comparison between the LVMs without clustering and the clustering-based LVMs when predicting shapes for the cloth from the known latent variables

Methods	Equality Constraints		Inequality Constraints	
	R.E.[mm]	C.E. [mm]	R.E. [mm]	C.E. [mm]
GPLVM	3.9028 ±0.1012	1.5765 ±0.0277	3.8413 ±0.1112	1.1761 ±0.0692
Clustering-Based GPLVM	**3.6687**	**1.4621**	**3.6687**	**1.1086**
CLVM	4.0807 ±0.1034	1.2220 ±0.0444	3.7837 ±0.1046	1.0234 ±0.0617
Clustering-Based CLVM	**3.9224**	1.2739	**3.6434**	**0.9782**

Prediction with Known Latent Variables. A comparison was made between twokinds of original LVMs (GPLVM and CLVM) and the clustering-based LVMs (clustering-based GPLVM and clustering-based CLVM).

We still maintained $M=300$ test samples and $V=50$ unlabeled samples. For the clustering-based LVMs, indices of the labeled samples derived from the G-means with the critical value 30, and the number N was 59 for the cardboard and 56 for the cloth. For the original LVMs without clustering, we performed 24 times learning with only labeled samples randomly chosen every time and with the test samples and unlabeled samples maintained the same as those clustering-based LVMs, and predicted the test samples with the known latent variables every times, taking the mean value and the standard deviation over 24 times results of mean errors of the test samples. Equality and inequality constraints were imposed respectively. A clear comparison about the reconstruction and constraint errors was presented in Table 1 and Table 2.

The comparison clearly pointed out that reconstruction errors and constraint errors of the clustering-based LVMs were scarely outside the range of those of LVMs without clustering. In fact, nearly all the errors of the clustering-based LVMs were lower than the mean errors of LVMs without clustering. This revealed that the clustering-based LVMs could improve the performance of original LVMs in nature.

Table 3. Comparison between the LVMs without clustering and the clustering-based LVMs when reconstructing the shapes for the cardboard with the optimized latent variables. M=300, V=50, N=59. Reconstruction errors and constraint errors were presented.

Methods	Equality Constraints		Inequality Constraints	
	R.E.[mm]	C.E. [mm]	R.E. [mm]	C.E. [mm]
GPLVM	2.9868	0.9249	2.9868	0.8915
	±0.1441	±0.0228	±0.1441	±0.0254
Clustering-Based GPLVM	**2.5401**	**0.8187**	**2.5401**	**0.7873**
CLVM	2.6703	0.7409	2.9527	0.8072
	±0.1427	±0.0394	±0.1293	±0.0136
Clustering-Based CLVM	**2.2395**	**0.7139**	**2.5180**	**0.7533**

Table 4. Comparison between the LVMs without clustering and the clustering-based LVMs when reconstructing the shapes for the cloth with the optimized latent variables. M=300, V=50, N=56. Reconstruction errors and constraint errors were presented.

Methods	Equality Constraints		Inequality Constraints	
	R.E.[mm]	C.E. [mm]	R.E. [mm]	C.E. [mm]
GPLVM	4.9655	1.8372	4.9655	1.4516
	±0.0655	±0.0313	±0.0655	±0.0484
Clustering-Based GPLVM	**4.6727**	**1.7322**	**4.6727**	**1.3814**
CLVM	4.8743	1.7062	4.8630	1.3385
	±0.1476	±0.0401	±0.0619	±0.0441
Clustering-Based CLVM	**4.6321**	**1.6363**	**4.5864**	**1.2701**

Reconstruction with Optimized Latent Variables. We now generated the synthetic data for both datasets and simulated the process of optimization and reconstruction. 1000 random points were spread on each mesh-represented shape to be the 3D feature points. We specified their barycentric coordinates by setting 3 random numbers whose sum was 1. 2D locations of those 3D feature points could be created by projection relationship between them. We then added Gaussian noise with standard deviation 2 to these 2D locations. When we completed generating the synthetic data, we could use them to optimize the latent variables. For this part, we just performed 5 times learning for the LVMs without clustering since we took it into account that the process of optimization was rather time-consuming. The N labeled samples were chosen randomly for the LVMs without clustering. The M test samples and V unlabeled samples maintained the same as the clustering-based LVMs.

As observed from Table 3 and Table 4, errors of the clustering-based LVMs were all much smaller than the mean of errors of those without clustering. That was to say, the clustering-based LVMs gained improvements during the optimization which was a necessary process in the real situation. Therefore, we could draw a conclusion that the clustering-based LVMs had obvious advantages over their original ones. Fig. 2 and Fig. 3 showed shapes of 2 frames of deformations for the cardboard and the cloth respectively, including their real shapes and shapes reconstructed with GPLVM, clustering-based GPLVM, CLVM and clustering-based CLVM. The shapes were reconstructed using equality constraints in Fig. 2 and inequality constraints in Fig.3.

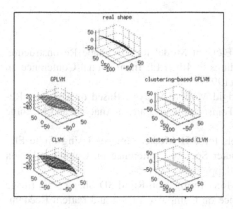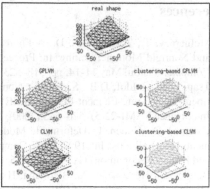

Fig. 2. Shapes of the 161th (left) and 1803th (right) deformations for the cardboard

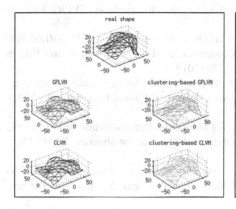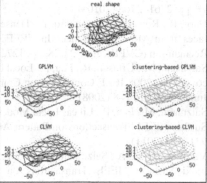

Fig. 3. Shapes of the 945th (left) and 1454th (right) deformations for the cloth

5   Conclusions

In this paper, we presented clustering-based LVMs in which we introduced G-means into LVMs. In theory, the clustering-based LVMs can predict a wider range of shapes, thus performing better. We conducted experiments on the cardboard dataset and the cloth dataset and compared the clustering-based LVMs with their original GPLVM and CLVM. The results demonstrated the effectiveness of the clustering-based LVMs in the context of monocular non-rigid 3D reconstruction. In future work, it is worth considering how to give a good initial latent variable during the optimization.

Acknowledgments. This work was supported by the Fund of National Natural Science Foundation of China (61231018) and the National Key Technology R&D Program (2013BAH62F03-3).

References

1. McInerney, T., Terzopoulos, D.: A Finite Element Model for 3D Shape Reconstruction and Nonrigid Motion Tracking. In: Proceedings of 4th IEEE International Conference on Computer Vision, May 11-14, pp. 518–523 (1993)
2. Tsap, L.V., Goldof, D.B., Sarkar, S.: Nonrigid Motion Analysis Based on Dynamic Refinement of Finite Element Models. IEEE Transactions on Pattern Analysis and Machine Intelligence PAMI-22(5), 526–543 (2000)
3. Cohen, L.D., Cohen, I.: Deformable Models for 3-D Medical Images Using Finite Elements and Balloons. In: 1992 IEEE Computer Society Conference on Computer Vision and Pattern Recognition (CVPR), June 15-18, pp. 592–598 (1992)
4. Bregler, C., Hertzmann, A., Biermann, H.: Recovering Non-Rigid 3D Shape from Image Streams. In: Proceedings of IEEE Conference on Computer Vision and Pattern Recognition (CVPR), June 13-15, pp. 690–696 (2000)
5. Taylor, J., Jepson, A.D., Kutulakos, K.N.: Non-Rigid Structure from Locally-Rigid Motion. In: IEEE Conference on Computer Vision and Pattern Recognition (CVPR), June 13-18, pp. 2761–2768 (2010)
6. Garg, R., Roussos, A., Agapito, L.: Dense Variational Reconstruction of Non-Rigid Surfaces from Monocular Video. In: IEEE Conference on Computer Vision and Pattern Recognition (CVPR), June 23-28, pp. 1272–1279 (2013)
7. Salzmann, M., Urtasun, R., Fua, P.: Local deformation models for monocular 3D shape recovery. In: 2008 IEEE Conference on Computer Vision and Pattern Recognition (CVPR), June 23-28, pp. 1–8 (2008)
8. Salzmann, M., Fua, P.: Linear Local Models for Monocular Reconstruction of Deformable Surfaces. IEEE Transactions on Pattern Analysis and Machine Intelligence, PAMI 33(5), 931–944 (2011)
9. Varol, A., Shaji, A., Salzmann, M., Fua, P.: Monocular 3D Reconstruction of Locally Textured Surfaces. IEEE Transactions on Pattern Analysis and Machine Intelligence, PAMI 34(6), 1118–1130 (2012)
10. Varol, A., Salzmann, M., Fua, P., Urtasun, R.: A constrained latent variable model. In: 2012 IEEE Conference on Computer Vision and Pattern Recognition (CVPR), June 16-21, pp. 2248–2255 (2012)
11. Lawrence, N.D.: Gaussian Process Latent Variable Models for Visualisation of High Dimensional Data. In: 17th Annual Conference on Neural Information Processing Systems, NIPS (December 8, 2003), In: Advances in Neural Information Processing Systems, NIPS 2004, vol. 16, pp. 329–336 (2004)
12. Lawrence, N.D.: Probabilistic non-linear principal component analysis with Gaussian process latent variable models. The Journal of Machine Learning Research 6, 1783–1816 (2005)
13. Hamerly, G., Elkan, C.: Learning the k in k-means. In: 17th Annual Conference on Neural Information Processing Systems, NIPS 2003 (December 8, 2003), In: Advances in Neural Information Processing Systems, NIPS 2004, vol. 16, pp. 281–288 (2004)
14. Urtasun, R., Fleet, D.J., Geiger, A., Popović, J., Darrell, T.J., Lawrence, N.D.: Topologically-constrained latent variable models. In: Proceedings of the 25th International Conference on Machine Learning, pp. 1080–1087. ACM, New York (2008)
15. Salzmann, M., Fua, P.: Deformable surface 3D reconstruction from monocular images. Synthesis Lectures on Computer Vision 2(1), 1–113 (2010)

Comments-Attached Chinese Microblog Sentiment Classification Based on Machine Learning Technology

Bo Yan, Bin Zhang, Hongyi Su, and Hong Zheng

Key Lab of Intelligent Information Technology, Beijing Institute of Technology, Beijing, CN
{yanbo,binz,henrysu,hongzheng}@bit.edu.cn

Abstract. Nowadays, with the rapid development of social networks, community-oriented Web sentiment analysis technology has gradually become a hot topic in the field of data mining. Being concise and flexible, Chinese microblog poses new challenges for sentiment analysis. This paper proposes an approach to classify Chinese microblog sentiments into positive and negative by the plain Naive Bayes (NB) and Support Vector Machine (SVM). Based on data preprocessing, sentiment lexicon construction, combining element of users' comments, this research posit this Comments-attached Microblog Sentiment Classification, which is a novel method of attaching microblog users' comments to the target microblog in order to improve the accuracy of sentiment classification. The experiment proves the vitality of this method and the advancement of the indecency from the way of language expressions.

Keywords: Chinese Microblog, sentiment classification, machine learning, user's comments.

1   Introduction

Microblog is a popular networking tool for obtaining and publishing information. Users can instantly share 140-Word text messages. With the development of microblog, there has been a lot of microblog with a perspective, so that there is an urgent need to analyze and get information on the Semantic Orientation (SO) of these. At popular events, the user's SO is far more dramatic. It is of great importance to study sentiment classification[1], for it is beneficial for monitoring, discovery and guidance of public opinions.

As Chinese microblog have a lot of differences from conventional texts. The sentiment analysis has its special feature, which is mainly that the microblog sentiment analysis is not limited to microblog text itself. Analysis for it can be implemented from various perspectives based on the characteristics of microblog product. As there are no stand expressions for the microblog texts, it is very difficult to judge the SO of the text. Compared with the traditional text, microblog has the following characteristics: short

---

[1] The sentiment of the text are divided the subjective texts into positive and negative, or positive, negative and neutral. The criteria used to measure the degree of preference for something as shown in the texts is semantic orientation.

D.-S. Huang et al. (Eds.): ICIC 2014, LNAI 8589, pp. 173–184, 2014.
© Springer International Publishing Switzerland 2014

length, informality (being colloquial), semi-structure, and a large number of ellipsis and reference which pose challenges on sentiment classification [11] [14].

In this paper, sentiments expressed by microblog are divided into positive and negative. Combining machine learning method, sentiment lexicon and a small-sized annotated corpora, the sentiment classification is studied. Considering the characteristics of microblog, we divide it into main bodies of messages and subdivided opinions. We propose a novel method, which regards the users' comments as a part of the additional features related to the features of the main bodies. We get the subjective microblog of hot topics for feature extraction and calculation of feature weight. Then we build classifiers to compare and analyze the performance of different classifiers under different conditions. Research results show that the method above can not only extract features effectively and accurately, but also improve the performance of classifier and get higher value of F1-Measure to the classification results.

2   Related Works

With regards to traditional text research, Pang et. al, (2002) applied different features election methods such as NB, Maximum Entropy (ME) and SVM to classify the movie comments. The experiment indicates that Boolean value as feature weights works yields more accurate results than term frequency. Among the three classifiers, SVM classifier obtain the best effects [1].Mullen et. al. (2004) accomplish more by taking the evaluative factor (good or bad, EVA), potency (strong or weak, POT) and activity (active or passive, ACT) values of adjectives, along with the SO value of the valuable phrases as features, based on Turney's [2] five combination model of polarity words [3].Employing CHI and information gain, Ni (2007) implement sentiment classification based on NB, SVM and Rocchio's algorithms [6].

In researches for Chinese language, Tang Huifeng et. al, (2007) studied the text features and sentiment classification based on supervised learning under the training set of various quantities and sizes [4]. It is found that sentiment classification can achieve better results with sufficient training set and features, applying Bigram features, information gain feature selection and SVM method. Considering the impact of negative adjectives and adverbs on semantic tendency, Xu Jun (2007), analyzed and classified the news comments corpora with NB method and ME model [5]. As to the feature weighting, the binary feature weighting is selected to improve the accuracy of classification. Combining the semantic features and the automatic polarity identification system on machine learning, Xu Wang calculated the lexical inclination by HowNet and analyzed the polarity of the text by SVM [7].

Presently, microblog sentiment classifications are mainly based on machine learning algorithms, such as the SVM, NB, ME. By learning training set features, a classifier model is constructed to be applied in the classification of the test set [10].The key of emotion classification based on machine learning is effectively extracting characteristic information [8, 9] [13]. With the development of training corpora and the introduction of the semantic features [11, 12], sentiment classification based on machine learning has been performing quite well.

3   Sentiment Classification Considering Users` Comments

Chinese grammar and language convention values the context. The microblog text contains more than one sentence in a specific context. Although it can accommodate up to 140 characters, the information that microblog contain still does not accurately describe the semantic orientation. We introduced the features related to the theme into the Chinese microblog, to explore whether it is conducive to classification. In order to fully cater to its special features, we have used the information of user's comments extended target microblog to increase classification accuracy.

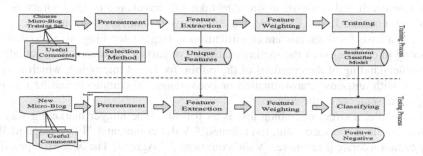

Fig. 1. Sentiment classification framework considering the comments

Fig.1shows basic framework of sentiment classification based on machine learning techniques consists of two similar phases: training and testing phase. These two stages include similar processes, such as selection process of user's comments, microblog pretreatment, feature extraction, feature weight calculation. Different machine learning methods are then used to train and get text classification model. In the classification process, it is used to predict a new microblog classification of polarity.

As shown in algorithm, users' comments are selected and processed. Then the valuable user's comments are attached. The whole algorithm processes as follows:

Algorithm. Microblog sentiment classification considering the comments.
Input: The training set of labeled samples for current batch, T_n; The testing set of unlabeled samples for current batch, U_n; The comment set of target microblog samples for current batch, C_n;
Output: The semantic orientation, $O(U_n)$;
Require: Classifiers for the classification, M_{nb} or M_{svm};
   1: Extracting the set of reliable negative or positive comments V_n from C_n with help of *Selection Method*;
   2: $V_n = V_{Tn}$ *cup* V_{Un}, for target microblog samples;
   3: Training the *model* on $T_n \cup V_{Tn}$, with help of classifiers
   4: Classifying testing samples in $U_n \cup V_{Un}$;
   5: **return** $O(U_n)$
Finally: Evaluation the capacity of classifiers;

3.1   Selection Process of Comments

Microblog Comments. Expressions of the sentence play a great role in deciding the sentiment of sentences. A challenge for sentiment classification is to make machines to understand these expressions. Compare two sentences: "the skin creams, suitable for people with very dry skin."" With a skin cream, skin will become very dry. "These two sentences using the word almost," skin cream, "" skin "and" very dry." But the first sentence is positive; the second sentence is likely to be negative. Another expression-related issue is "ironic". Many positive words are affected by the effects of Forum culture, they have transformed to negative trends, such as "you are so talented!"

After research and analysis, in order to take full advantage of the characteristics microblog possessed, we find that the subjective comments information can be viewed as a basis for the sentiment classification of target microblog, also a tool to fix the wrong interpretation of the sentiment. Effective part of the comment is actually a basic understanding of the emotions of the public for target microblog, which is synonymous with antisense transformation or conversion. So comments can be treating as a specific sentence in the context.

For example, a target microblog text is as follow: "The hunger marketing way of XiaoMi mobile is too successful, too talented!" Valid comment 1 "I lost it again! We really cannot tolerate it anymore. "Valid comment 2:" Agreed! The XiaoMi's practice hurts Mi-fans' feelings!!! "Mere analysis of target microblog may produce inaccuracy. Two valid comments are negative comments related to this microblog. So sentiment of the target microblog is likely to be negative.

Selection Process. After the analysis of microblog characteristics, it is found that user's comments, which contain the main characteristics are an important part of microblog. Two kinds of valuable information, one of which is holding the same point as target microblog, the other on the contrary to it are expected to be extracted from the comments collection intermixed with extraneous ones. We sequence the user's comments based on the relevancy (similarity), quality and time. Interception within a certain number of comments is perceived as valuable on the target microblog.

The first factor needs consideration is the relevance between the comments and the target microblog, which refers to the degree of similarity between these two. It can effectively filter out a lot of comment spam and advertising comments. However, relevance alone cannot guarantee the accuracy. Thus some peculiar characteristics of microblog production can be applied to delve in more significant issues, which often consist of the number of browsing users, number of "good" as well as the relevance between the latter and time. (This relevance is included in the formula 4) To further explain, a comment with more browsing users, which means more attention, obviously is of greater significance and worth for the research. A comment which gains a lot of "good" is more likely to own higher quality, with more approval from the users. Moreover, the earlier the comment is released, the more attention it tends to attract.

The calculation method based on the semantic similarities of HowNet[2] is introduced in this paper. The formula of the similarities between two sentences is as the following:

---

[2] Dong Zhendong & Dong Qiang. HowNet [EB/OL].http://www.keenage.com/

$$\text{Sim}(s_t, s_i) = \frac{\left(\sum\limits_{u=1}^{n} \max_{1 \le v \le m} \text{Sim}(w_{1u}, w_{2v})/n + \sum\limits_{v=1}^{m} \max_{1 \le u \le n} \text{Sim}(w_{1u}, w_{2v})/m\right)}{2} \tag{1}$$

In target microblog sentence (S_t), there are n words: $w_{11}, w_{12} \ldots w_{1n}$. In the comment sentence (S_i), there are m words: $w_{21}, w_{22} \ldots w_{2m}$. $Sim(w_{1u}, w_{2v})$ refers to the similarities between two sentences' words [15].

$$H_i = \frac{userful_i}{\sum\limits_{j=1, t=t_i}^{n, t_{now}} view_j} \tag{2}$$

$$Q_i = \frac{useful_i}{view_i} \tag{3}$$

$$T_i = 1 + \alpha \left(\frac{t_i - t_0}{t_{now} - t_0} \right) (\alpha \in 0,1) \tag{4}$$

The quality of the comments include: the quality of the heat (H) and quality of comments (Q) as well as the time dimension (T). It represents the number of user`s comments to be read within a certain time, and as well as the extent to which the public would find it helpful, it can objectively reflect influence of this microblog.

In the equation of H_i and Q_i, $userful_i$ represents that the users regard the comment as useful and $view_i$ means the times that the comment is browsed. As to T_i, t_i is the time at which the comment is published, t_{now} refers to the time when the comment is extracted and stands for a constant from zero to one.

Sort the evaluation score based on the relevance, quality and time. Formula is as follows:

$$rank_i = \alpha \times sim_i + (1 - \alpha) \times fun(T_i, H_i, Q_i, \beta) \tag{5}$$

$$fun(T_i, H_i, Q_i, \beta) = \beta * (Q_i + H_i) + (1 - \beta) * T_i \tag{6}$$

For $i=1 \ldots n$, where a, β are constants of 0 to 1. The $fun(T_i, H_i, Q_i, \beta)$ refers to the formula the quality of the $Comment_i$, which requires certain alteration based on different conditions. A simple method adopted in this paper.

It is often required to remove duplication in the data, which is no meaningful data, such as "right", "agrees LZ (Topic Creator)". But valid comments require some special words as properties of comments. These properties can tell the difference of emotional relationships between comments and target microblog, such as emoticons or expressions of support or opposition. Table 1 filters some microblog emoticons:

Table 1. Microblog emoticons

support		oppose	
	[good]		[bad]
	[cooperation]		[no]
	[powerful]		[spill]

After the above process, as shown in Fig. 2, valuable subjective comments of target microblog are attached to the target text as an extended text.

```xml
<?xml version="1.0"?>
- <project name="xsimilarity">
  - <weibo id="001">
    - <sent id="1">
        <include text="小米手机的营销方式太成功了、太有才了！"/>
      </sent>
    </weibo>
  - <comments>
    - <comment correlation="1" weight="1" target="001">
        <element message="我这次也没有抢到，这种营销方式真是让我们吃不消。"/>
      </comment>
    - <comment correlation="1" weight="0.9" target="001">
        <element message="小米公司的这种做法实在让米粉寒心！"/>
      </comment>
    </comments>
  </project>
```

Fig. 2. Structure of attached properties on Microblog

3.2 Pretreatment Works

Eliminate the Informal Language. The colloquial and subjective microblog texts and frequently appeared network expressions imposed higher requirements on sentiment classification. For example, 'I BS the quality of the computer at this price.' Since "BS" is the abbreviation of the negative word "contempt" (pronounced as "*bishi*" in Chinese), so this sentence expresses negative emotions. Under this circumstance, we need to make corresponding pretreatment on microblog corpus to eliminate the adverse effects brought by the informal language on the sentiment classification accuracy.

In this paper, internet terms and acronyms, which affect the effect of classification, have been added to the extended lexicon and to correspond to the replacement words, as shown in table 2 which has filtered out common Web words.

Table 2. Emotion-related Internet vocabulary

Internet Terms	Synonyms
56	boring
XI FAN	like, enjoy
DA XIA	swordsman
BS	despise
CAI NIAO	rookie
YA LI	pressure

Chinese Word Segmentation. Chinese text needs word segmentation before further processing. In this paper, we use NLPIR[3], which is designed by HuaPing Zhang. The next work is the removal of stop word in text and special symbols. These special symbols, including punctuation marks, numbers and letters as well as the inclusion in the text, which can have an impact on word segmentation. The privative will be removed from the stop word because it will change the semantic orientation of polarity words.

Features Extraction. Privative and polarity words require special handling. The objects of privative words can be viewed by setting the sliding window and then be added to the feature set; In this paper, we use "words set for sentiment analysis" provided by the HowNet to calculate the semantic orientation of the polarity words. The results of the calculation served as the principal evidence of classification. In case of multiple semantic orientations, the relationship between sentences needs to be considered when extracting the sentiment features of microblog. For instance, in the sentences like "although...but...", the emotion of the speakers are concentrated on the clause leaded by "but...", so more often than not, the latter part are the main evidences of extraction.

In every microblog, including those of the training set and testing set, the TF and TF-IDF score the highest before a Word as an attribute value is recorded as:

$$\text{features}(t) = \left\{ \omega_1, \omega_2, \cdots, \omega_N, \forall i \neq j, \omega_i \neq \omega_j \right\} \qquad (7)$$

For train sets, select features values in different classifications as the sum of the values of the features, namely as:

$$\text{features}(\text{category} Y) = \left\{ features(t) | \forall t, s.t. category(t) = Y \right\} \qquad (8)$$

For test set, on per microblog and corresponding valid comments sets from which the top n most frequently occurring words are selected as features' values.

Features Weight. The weights of features are mainly calculated in accordance with the size of the TF (Term Frequency) and size of TF-IDF (Term Frequency-Inverse Document Frequency). TF refers to the number that the feature appears in the document, whereas TF-IDF modified TF calculation method, taking the document with the feature into consideration in accordance with the following formula:

$$W(t,d) = tf(t,d) \times \lg\left(\frac{N}{n_i}\right) \qquad (9)$$

N represents the total number of training documentation set documents, n_i is the number of documents that contain the word t.

---

[3] NLPIR, Chinese word segmentation system http://ictclas.nlpir.org

3.3   Machine Learning Methods

After the texts are represented as a document-term matrixes use a machine learning method to train a sentiment classifier There are a lot of machine learning methods and the widely used methods are Support Vector Machine and Naïve Bayes.

Support Vector Machine. As a machine learning method based on structural risk minimization principle, SVM is a projections tool with good generalization capability. Considered an efficient method, SVM constructs a hyperplane or set of hyperplanes in a higher- or infinite-dimensional space, which can be used for classification, regression, or other tasks. Intuitively, a good separation is achieved by the hyperplane that has the largest distance to the nearest training data point of any class (so-called functional margin), since in general the larger the margin the lower the generalization of the classifier.Based on the feature vector sentiment classification divided the documents into two kinds, which are positive and negative. SVM method is similar to a solution to an optimization problem with one constraint condition. Classification function is as follows:

$$dx = \sum_{i=1}^{m} a_i y_i K(x_i, x) + b \tag{10}$$

The a_i and b are obtained by SVM learning algorithms, and k (x_i, x) represents kernel functions, mapping the samples to higher dimensions. The sample became "support vectors" when the corresponding a_i of x_i is not zero.

LIBSVM[4], an efficient software package designed by Lin ChihJen on SVM pattern recognition and regression is adapted to train and test the SVM classifier in this paper.

Naive Bayes. Bayes theorem is applied to predict the of an unknown category sample. Then the category with the most is selected as a sample. Within the given category y, under the polynomial Bayes algorithm, it is observed that joint W_1, W_2... Wn probability is as follow:

$$P(\omega_1, \omega_2, \cdots, \omega_N | Y) = \prod_{i=1}^{n} P(\omega_i | Y) \tag{11}$$

The microblog t belongs to a certain category of probability, located the eigenvalues of t as W_1, W_2, ..., W_N, m categories C_1, C_2, ..., C_M. M=2 in this case considering the duality classification of positive and negative sentiment. Based on Bayes formula, the probability of t, $P\{C_i/t\}$ belonging to the class of C_i is:

$$P\{C_i | t\} = \frac{P\{t | C_i\} P(C_i)}{P(t)} = \frac{\prod_{j=1}^{N} P\{\omega_j | C_i\} P(C_i)}{\sum_{j=1}^{M} \prod_{j+1}^{N} P\{t | C_i\} P(C_i)} \tag{12}$$

---

[4] C.- C. Chang and C.- J. Lin. LIBSVM : a library for support vector machines . Available at http://www.csie.ntu.edu.tw/~cjlin/libsvm/

$P(C_i)$ is obtained by dividing the number of microblog in C_i by the number in training set. $P\{w_j/C_i\}$ is obtained by dividing the frequency of W_j in feature of C_i by the feature in C_i. If $P\{w_j/C_i\}$ equals zero, then smooth processing is adopted, assigning as $P = \frac{1}{count(t)+1}$, in which *count (t)* represents the number of all the microblog in the training set.

4   Experiment Result Analysis

4.1   Data Collection

We utilize SINA[5]microblog API interface to develop program, which is used to collect microblog of hot topic. A *data sets* including 4,000 pieces of microblog is picked out and divided them into a training set and a testing set. They mark consistent results and an emotional tone. Training set contains 3,000 target microblog, with 1429 positive ones and 1571 negative ones; test set contains 1,000 records, with positive ones numbered 578 and negative at 422.

4.2   Evaluation Indicators

According to the evaluation scores, the method posited is analyzed. The performance evaluation indexes of classifier mainly consist of recall, precision, F1-Meatrue. The *recall* is defined as the percentage of correctly classified texts among all texts belonging to that category, and the *precision* is defined as the percentage of correctly classified texts among all texts that were assigned to the category by the classifier. And *F1-measure* to evaluate the performance of sentiment classifier. Evaluation Indicator as follow:

$$F1 = \frac{2 \times precision \times recall}{precision + recall} \times 100\% \tag{13}$$

4.3   Experiment   Process

Experiment 1: basic classification experiment. Original corpus is preprocessed. Pretreatment includes noise reduction, segmentation and POS tagging under different conditions in which are different classifiers, features and feature weight calculation. Better classifier and features are selected based on experimental results; the text is split into a number of clauses, and obtained sentence polarity in the test data. Finally, the whole microblog eventual polarity is obtained by adding up the polarity numbers of the sentence. Experiment 2: After processing, Users' comments are attached to the training set and testing set, with the same procedures as the experiment 1.

---

[5] Sina microblog open platform
http://open.weibo.com/wiki/%E9%A6%96%E9%A1%B5

4.4   Experiment Results

Basic Classification Results. Different from the classification based on theme, the one of sentiment is primarily dependent on polarity words, which have some effect on classification. In addition, people often use privative with polarity words to express negative emotions. These experiments applied the combination of three methods to describe the features, which are Unigram (U-gram), Polarity words (Pol), Privative stand for Negative words (Neg).

This experiment compared the effects of TF and TF-IDF, two methods of feature weighting on experimental results. Fig. 3 shows the classification performance of U-gram features classifier, which is based on NB classification algorithm and SVM algorithm, in different ways of weights calculation and different feature numbers. Table 3 shows the classification performance of the combination among different features weighting with classification algorithms under different features.

According to fig. 3, SVM overweigh NB in performance. In the initial stage, the error often occurred due to the lack of features. As the number of features increased, classification results raised at first then reduced. The reason of the reduction is the interference from certain irrelevant words as the number features became larger. Generally, using TF as the weight calculation method is more applicable.

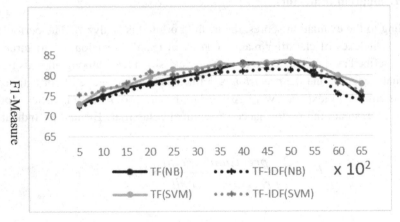

Fig. 3. Comparison of different weights under the NB / SVM algorithm

Table 3 indicates that, taking the classifier stability in various combinations in Fig. 4, into consideration, features selected number of 3,500. It is observed that adding polarity words and dealing with privative, can improve the accuracy.

Table 3. Result of Microblog-lever basic Classification (F1%)

weight feature	TF		TF-IDF	
	NB	SVM	NB	SVM
Unigram	**77.3**	*79.1*	**75.6**	*77.1*
U-gram + Polarity	**80.5**	*81.3*	**80.1**	*79.8*
U-gram + Pol +Neg	**82.7**	*82.6*	**82.0**	*82.2*

Classification Results of Attached Comments. Regarding the comments as single sentence in a specific context, this experiment adds the comments of the subjective microblog as the criteria. Table 4 shows that SVM classifier with TF feature weighing performs the best. It also performs better than the ones without users' comments information attached. It can be concluded that the attachment of comments can improve the accuracy.

Table 4. Result of Microblog -lever attached Comments (F1%)

feature \ weight	TF		TF-IDF	
	NB	SVM	NB	SVM
Unigram	*81.1*	*82.7*	*80.3*	*81.7*
U-gram + Polarity	*82.4*	*83.8*	*81.2*	*82.3*
U-gram + Pol +Neg	*83.7*	*84.5*	*83.2*	*84.1*

As the table 5 shows, precision of the comment reached at 85.6%, almost 3% higher than the sentence-level precision. So, one error in the sentence prediction on the whole has little effect, which also verifies the function of the attached comments to promote accuracy.

Table 5. Result of attached user Comments (%)

type	Recall		Precision	
	sentence	Microblog	sentence	Microblog
Pos	*80.3*	*83.4*	*82.7*	*85.6*
Neg	*83.1*	*83.6*	*82.8*	*84.2*

Experimental Summarize. To sum up, based on the SVM, NB methods of machine learning, as well as a variety of features, this paper implemented the sentiment classification of the Microblog data sets, and attached comments.

Experimental results show that SVM works better than NB, with the F1-measure of 84.5%. In terms of feature selection, unigram is the most important feature. With the unigram feature alone, SVM classifiers reached the F1-measure of 82.7%. And after adding sentiment polarity words and privative after two features, it rose by about 2%, to the degree of 84.5%. In terms of feature weight calculation, though simple, TF works much better. So attaching the processed comments, this paper obtained a precision rate of 85.6% and the F1-measure of 84.5%, which is a satisfactory result.

5   Conclusion

Chinese microblog sentiment analysis originates from the sentiment analysis of traditional text. But its peculiarities are obvious, worth an in-depth study. At present, a developed filtering technique is needed to filter out the spam interfering with the sentiment classification. This paper had accomplished some filiations of the spam in the comments. Due to the absence of network language corpus, the work of microblog

Chinese word segmentation is confronted with great difficulties. In order to relieve this problem, we build a few extended corpuses.

To optimize the classification results of machine learning, the microblog is divided into different categories. The best learning strategy would be adopted according to the characteristics of each category. In addition, sentiment classification of text should be combined with other text-mining technologies, which would dig out more valuable information. This information can enhance the value of classification.

References

1. Pang, B., Lee, L., Vaithyanathan, S.: Thumbs up? Sentiment Classification Using Machine Learning Techniques. In: Proceedings of the ACL 2002 Conference on Empirical Methods in Natural Language Processing, vol. 10, pp. 79–86. Association for Computational Linguistics (2002)
2. Turney, P.D.: Thumbs up or thumbs down? Semantic Orientation Applied to Unsupervised Classification of Comments. In: Proceedings of the 40th Annual Meeting on Association for Computational Linguistics, pp. 417–424. Association for Computational Linguistics (2002)
3. Mullen, T., Collier, N.: Sentiment Analysis using Support Vector Machines with Diverse Information Sources. In: EMNLP, vol. 4, pp. 412–418 (2004)
4. Tang, H., Tan, S., Cheng, X.Q.: Research on Sentiment Classification of Chinese Comments Based on Supervised Machine Learning Techniques. J. Chinese Information Processing 21(6), 88–94 (2007)
5. Xu, J., Ding, Y.X., Wang, X.L.: Sentiment Classification for Chinese News Using Machine Learning Methods. J. Chinese Information Processing 21(6), 95–100 (2007)
6. Ni, X., Xue, G.R., Ling, X.: Exploring in the Weblog Space by Detecting Informative and Affective Papers. In: Proceedings of the 16th International Conference on World Wide Web, pp. 281–290. ACM (2007)
7. Xu, L., Lin, H., Yang, Z.: Text Orientation Identification Based on Semantic Comprehension. J. Chinese Information Processing 1, 015 (2007)
8. Zhai, Z., Xu, H., Jia, P.: An Empirical Study of Unsupervised Sentiment Classification of Chinese Comments. J. Tsinghua Science & Technology 15(6), 702–708 (2010)
9. Zhai, Z., Xu, H., Kang, B., et al.: Exploiting Effective Features for Chinese Sentiment Classification. Expert Systems with Applications 38(8), 9139–9146 (2011)
10. Liu, Z., Liu, L.: Empirical Study of Sentiment Classification for Chinese Microblog Based on Machine Learning. J. Jisuanji Gongcheng yu Yingyong: Computer Engineering and Applications 48, 1–4 (2012)
11. Wang, S., Yang, A., Li, D.: Research on Sentence Sentiment Classification Based on Chinese Sentiment Word Table. J. Computer Engineering and Applications 45(24), 153–155 (2009)
12. Huo-song, X.I.A., Tao, M.: The Influence of Stop Word Removal on the Chinese Text Sentiment Classification Based on SVM Technology. J. Chinese Information Processing 30(4), 347–352 (2011)
13. Pei, Y., Wang, H.W., Zheng, L.J., et al.: Sentiment Classification of Chinese Online Comments: Analysing and Improving Supervised Machine Learning. J. Web Engineering and Technology 7(4), 381–398 (2012)
14. Zhang, J.F., Xi, Y.Q., Yao, J.M.: A Review towards Microtext Processing. J. Chinese Information Processing (2012)

Evaluation of Resonance in Staff Selection through Multimedia Contents

Vitoantonio Bevilacqua[1], Angelo Antonio Salatino[1], Carlo Di Leo[1], Dario D'Ambruoso[2], Marco Suma[2], Donato Barone[1], Giacomo Tattoli[1], Domenico Campagna[1], Fabio Stroppa[1], and Michele Pantaleo[2]

[1] Dipartimento di Ingegneria Elettrica e dell'Informazione, Politecnico di Bari
via Orabona 4 - Bari, Italy
[2] AMT Services s.r.l.
Viale Europa, 22 - Bari, Italy

Abstract. In this paper we present the results of an experimental Italian research project finalized to support the classification process of the two behavioural status (resonance and dissonance) of a candidate applying for a job position. The proposed framework is based on an innovative system designed and implemented to extract and process the non-verbal expressions like facial, gestural and prosodic of the subject, acquired during the whole job interview session. In principle, we created our own database, containing multimedia data extracted, by different software modules, from video, audio and 3D sensor streams and then used SVM classifiers that perform in terms of accuracy 72%, 79% and 63% respectively for facial, vocal and gestural features. ANN classifiers have also been used, obtaining comparable results. Finally, we combined all the three domains and then reported the results of this last classification test proving that the experimental proposed work seems to perform in a very encouraging way.

Keywords: Emotional Speech Classification, Facial Expression Recognition, Gestural Expression Recognition, SVM, ANN, Resonance, Dissonance, Emotion Recognition.

1   Introduction

This work describes the research activities and the final results of an Italian research project between AMT Services, an ICT company with a laboratory accredited by the Ministry of University and Italian Research, and Polytechnic of Bari, both located in Bari. Moreover, the company Adecco s.a., leader in staff selection processes, allows experimental tests providing several job interviews.

In literature [1-2], it is known that the resonance is one of the fundamental processes throughout the organization of life is structured. Every person, every event, every situation is associated with a particular resonance state. In other words, when we resonate with other people, we feel related and we communicate much better. The principle of the resonance brings people to identify and communicate with each other.

D.-S. Huang et al. (Eds.): ICIC 2014, LNAI 8589, pp. 185–198, 2014.

Often this communication takes place on the emotional and unconscious level, without awareness from rationality and conscious processes. According to the Theory of Vital Needs [3] the resonance plays a fundamental role in the most important choices of our life. The resonance represents the harmonic relationship between two or more people and it is related to a dynamic version of the behaviour. The resonance is some form of agreement or sympathy, among the emotions that people express during a conversation. This explains how resonance is strictly related to the emotions.

According to what is written above, we propose here a system for resonance/dissonance classification in staff selection through multimedia contents. As the resonance status is linked with the emotional status, an early study has been made to estimate the emotional status in the in the three different domains: facial, vocal and gestural. The algorithm for facial emotion recognition is able to detect the action units, the atomic actions which constitutes one or part of the facial expressions. It combines the action units and the Ekman work [4] to detect the 7 fundamental emotions. Our speech emotion recognition method is based on an extended version of our recent work [5]. The new module is able to process a speech signal, extract features and make a classification of the emotional status. The gestural emotion recognition algorithm is designed to recognize 13 different human gestural actions and map them into 7 emotions [6-8].

The purpose of this work is based on using the emotional information extracted to assess the resonance status between two people in a job interview. In Section 2 our data is described along with the methods used to assess the resonance. In Section 3 a detailed explanation of the whole system will be provided. In the last Section 4, an overview of the obtained results is given.

2   Materials

Our database consists of 120 job interviews. We used a HD Webcam Logitech C525 to acquire the facial video and the speech streams, while the Microsoft Kinect for Windows is used to acquire the skeleton of the candidate's body. The methods used to extract the features are described in Section 3. In order to assess the resonance/dissonance value for each interview, the psychologists designed a questionnaire that the recruiter and the candidate had to compile at the beginning and at the end of the interview. The first version of the questionnaire consisted of 83 questions divided in key variables, explanatory variables and structural variables. The key variables are necessary to determine the negative and positive resonance and they are based on the feeling of the future. Instead, the explanatory variables describe the emotions and the feelings related to the key variables of resonance. Finally, the structural variables contain personal data of the person who compiled the questionnaire. It was submitted to 160 people and by means of a correlation analysis it is figured out that only two key variables are necessary. Therefore, in order to obtain the final questionnaire, 10 most related items representing the first key variable and 10 representing the second one have been considered.

The ultimate version contains 20 questions, 10 positive items (cheerful, appreciated, pleased, determined, enthusiastic, joyful, happy, safe, satisfied and sociable) and 10 negative items (agitated, anxious, unhappy, insecure, useless, dissatisfied, lonely, sad, abdominal pain and tremors). The candidate and the recruiter had to answer each question giving a score between zero and ten.

Table 1 represents the spread of the multimedia content of the database. Moreover, information about the resonance and dissonance is given.

Table 1. Database summary

Domain	Total interviews	Resonant	Dissonant
Facial	116	56	60
Vocal	110	54	56
Gestural	111	54	57
Cross-domain	105	52	53

Fig. 1. Example of scenario during the interview

3   Methods

The aim of this work is to find a correlation between the features extracted to assess the emotional information and the resonance or dissonance status known from the questionnaire shown in Section 2. In this section a brief introduction of the techniques used by our system is given.

3.1   Facial

Previous works have faced facial emotion recognition by means of image processing techniques [9-12]. In those works we assumed to extract emotional information based on the variation of the area of the polygons built using the facial feature points. In this work we have increased the number of extracted facial feature points and calculated emotional information by means of the distances between them. The facial information related with resonance/dissonance will be given by the distances of the facial feature points along with the emotions detected per frame. Figure 2 summarizes the workflow used. First of all, a video is acquired using the camera device described in Section 2. Each frame is processed and a list of features is provided along with the emotions detected. In the first phase, the candidate's face is detected using the Viola Jones Algorithm [13].

Then, using the Region Of Interest (ROI) that contains the detected face, a new search is performed to detect areas that contain the eyes, the mouth, the eyebrows and the wrinkles using the Viola Jones Algorithm.

For each ROI, a set of points is detected: the pupil centre, the mouth corners, the upper and lower lip centres and the brow-corners. After that, the distances between each feature and the eyeline (line that link together the pupils) or the face bisector are calculated and normalized based on the distance between the eyes.

The Action Units (AUs) are minimal facial actions whose combination is used to compose the facial expressions. In this work, an AU is calculated as the distance described above. Figure 3 represents the AUs considered. Thus, six emotions are detected using the following equations, as defined in [14]:

$$Fear = \frac{AU1_L + AU1_R + AU2_L + AU2_R + AU20_L + AU20_R}{6} \tag{1}$$

$$Surprise = \frac{AU1_L + AU1_R + AU2_L + AU2_R + AU26}{5} \tag{2}$$

$$Hanger = \frac{AU4_L + AU4_R + AU24_L + AU24_R}{4} \tag{3}$$

$$Sadness = \frac{AU4_L + AU4_R + AU15_L + AU15_R}{4} \tag{4}$$

$$Disgust = \frac{AU4_L + AU4_R + AU10}{3} \tag{5}$$

$$Happiness = \frac{AU12_L + AU12_R}{2} \tag{6}$$

where R and L mean *right* and *left* respectively. So, for each frame we extract 28 features summarized in Table 2.

The result of this module is a set of N 28-dimensional observations, where N is the number of frames in which the face has been detected. In order to reduce the amount of information extracted, four statistical parameters (mean, standard deviation, skewness and kurtosis) are derived from each feature.

Using this mechanism, we have been able to reduce the data from $N \times 28$ to 1×112 (4×28).

Table 2. Facial features

Name	Description
$AU1_L$	Lifting of the internal corner of the left eyebrow
$AU1_R$	Lifting of the internal corner of the right eyebrow
$AU2_L$	Lifting of the external corner of the left eyebrow
$AU2_R$	Lifting of the external corner of the right eyebrow
$AU4_L$	Lowering of the internal corner of the left eyebrow
$AU4_R$	Lowering of the internal corner of the right eyebrow
AU10	Lifting of the centre point of the upper lip
$AU12_L$	Lifting of the left external corner of the lips
$AU12_R$	Lifting of the right external corner of the lips
$AU15_L$	Lowering of the left external corner of the lips
$AU15_R$	Lowering of the right external corner
$AU20_L$	Movement to left of the left external corner of the lips
$AU20_R$	Movement to right of the right external corner of the lips
$AU24_L$	Movement to right of the left external corner of the lips
$AU24_R$	Movement to left of the right external corner of the lips
AU26	Lowering of the centre point of the lower lip
R1	Frontal wrinkle
R2	Glabella
R3	Left eye wrinkle
R4	Right eye wrinkle
R5	Left nasolabial wrinkle
R6	Right nasolabial wrinkle
E1	Fear
E2	Surprise
E3	Hanger
E4	Sadness
E5	Disgust
E6	Happiness

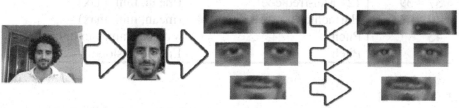

FaceDetection   ROI Selection   Feature Points Extraction

Fig. 2. Workflow for the feature detection

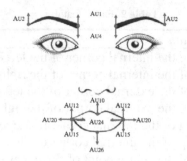

Fig. 3. Action units extracted

3.2   Vocal

In the vocal domain, we extract vocal features related with emotional information from the recorded speech based on our recent work on speech emotion recognition [5]. First the speech signal is split in a set of two-seconds signals. This value has been established experimentally to maximize the emotion classification. Each sub-signal is split into 197 overlapped frames whose length is 40 ms. For each frame, the pitch, the MFCCs and harmonic-to-noise ratio (HNR) are calculated. As these features have a great variability frame-by-frame, statistical features are extracted for each sub-signal. The features extracted are summarized in Table 3. To recognize the resonance or dissonance, at the end of the processing, an instance is retrieved every two seconds. In order to reduce the amount of information extracted, three statistical parameters (mean, standard deviation, range (max - min)) are derived from each feature. We have been able to reduce the data from $N \times 44$ to 1×132 (3×44).

Table 3. Descriptors and their statistics for the vocal domain

#	Feature	Statistic index
1 – 3	Pitch	(mean, min, max)
4 – 6	1st Mel-freq coef.	(mean, min, max)
. . . . . .	. . . . . .	. . . . . .
37 – 39	12th Mel-freq coef.	(mean, min, max)
40 – 42	Harmonic-to-noise ratio	(mean, min, max)
43	Pitch	Voiced frames in pitch
44	Pitch	Envelope slope

3.3   Gestural

In the gestural domain, we developed a system that is able to recognize 13 different human gestural actions and map them into 7 emotions [15]. Using the Kinect, we are able to process the depth map to perform features extraction and emotion recognition. The extracted features are distances between body joints or joint Euler rotation angles.

The joints detected (Fig. 4) are head (H), neck (N), torso (T), left shoulder (SL), right shoulder (SR), left elbow (EL), right elbow (ER), left hand (HL) and right hand (HR). Due to the low accuracy of both shoulders and hands joints, it has been necessary to enhance the detection of these points. Skin detection algorithm has been used to locate the skin regions in the image. The real hand will be searched within the skin regions on the line joining the elbow and the hand joints. The shoulder detection procedure finds the ROI surrounding the initial joint position. Then the edge detection is used to find the shoulder contours within the ROI cropped from the binary image containing the user silhouette.

To avoid false positive, only the biggest contour has been taken into account. After finding the highest point in the contour, we start a new search horizontally 3 pixels below until we find the first contour pixel. To clarify the *Features-Gestures* mapping, here we report an example.

The gesture G-AU2 ("Right hand – Head") is detected by the simultaneous satisfaction of the following conditions (in brackets the nomenclature used for a simple understanding):

— Distance between right hand joint and the head joint less than 200 mm. (HR_H < 200);
— Right shoulder rotation angle in relation to z-axis between -10 and 90 degrees (-10 < SR_z < 90);
— Right elbow rotation angle in relation to y-axis lesser than 75 degrees (ER_y < 75);

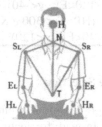

Fig. 4. Workflow for the feature detection

We developed an algorithm to process joint distances and angles to extract what we call gestural action units (G-AUs). In this context, we consider G-AUs as the basic elements for the construction of a complex gesture, or movement. The G-AUs recognized are illustrated in Table 4.

Each G-AUs variable can be zero or one according to the presence or not of the related action. If the variable related to the G-AU5 is set to 1, this means that the user's left hand is touching the left shoulder. For each frame, starting from the subject's skeleton, we determine if one or more AUs are present or not. In order to use this system for the purposes of our work, after a single processing we obtain a set of N instances, where N is the number of frames in which the subject's skeleton has been detected. As little AUs have high variance during the video, the Spectral Feature Selection (SPEC) method [16] has been used to select the set of AUs with maximum variance. We figured out that the most relevant features are AU9, AU10 and AU11. Then a set of statistical features is calculated as listed in Table 5.

Table 4. Action Units and connection with features for the gestural domain

#	AU	Description
G-AU1	Left hand - Head	HL_H < 200 && -90<SL_Z<10 && EL_y>-75
G-AU2	Right hand - Head	HR_H < 200 && -10<SR_Z<90 && ER_y<75
G-AU3	Left hand - Face	HL_H < 200 && HL_N > 130 && 30<SL_Z<100 && -60<EL_y<-20
G-AU4	Right hand - Face	HR_H < 200 && HR_N > 130 && -100<SR_Z<30 && 20<ER_y<60
G-AU5	Left hand - Left Shoulder	HL_SL < 150 && -30<SL_Z<90 && -50<EL_y<10
G-AU6	Right hand - Left Shoulder	HR_SL < 180 && HR_H > 200 && -150<SR_Z<-70 && 40<ER_y<80
G-AU7	Left hand - Right Shoulder	HL_SR < 180 && HL_H > 200 && 70<SL_Z<150 && -80<EL_y<-40
G-AU8	Right hand - Right Shoulder	HR_SR < 150 && -90<SR_Z<30 && -10<ER_y<50
G-AU9	Hands closed	(HL_HR < 200) OR (HL_ER<300 && HR_EL<300)
G-AU10	Tilt head to left	H_z > 7
G-AU11	Tilt head to right	H_z < -7
G-AU12	Left hand on the side	(HL_T < 300 && 10<SL_Z<50 && -120<EL_y<-40)
G-AU13	Right hand on the side	(HR_T < 300 && -50<SR_Z<-10 && 40<ER_y<120)

3.4   Classification and Feature Selection

For the classification stage, Support Vector Machines (SVM) and Artificial Neural Networks (ANN) have been used. Instead, for feature selection SVM-Recursive Feature Elimination (SVM-RFE) has been performed. The use of SVM classification technique is due to the tight relation with the SVM-RFE, instead the ANN has been chosen because of its advantages in modelling nonlinear problems. However, also other classification techniques have been performed even though there were not noteworthy results. A brief introduction of these methods will be given below.

Artificial Neural Network. An Artificial Neural Network (ANN) is an information processing paradigm that is inspired by the way of the biological nervous systems process information. The key element of this paradigm is the novel structure of the information processing system.

Table 5. Action Units for the gestural domain (n is number of frames)

#	Feature
1	$\dfrac{1}{n}\sum_{i=1}^{n} AU9_i$
2	$\dfrac{1}{n}\sum_{i=1}^{n} AU10_i$
3	$\dfrac{1}{n}\sum_{i=1}^{n} AU11_i$
4	$\dfrac{\sum_{i=1}^{n} AU9_i}{\sum_{i=1}^{n} AU9_i + AU10_i + AU11_i}$
5	$\dfrac{\sum_{i=1}^{n} AU10_i}{\sum_{i=1}^{n} AU9_i + AU10_i + AU11_i}$
6	$\dfrac{\sum_{i=1}^{n} AU11_i}{\sum_{i=1}^{n} AU9_i + AU10_i + AU11_i}$
7	$mean(4,5,6)$
8	$std(4,5,6)$

It is composed of a large number of highly interconnected processing elements (neurons) cooperating to solve specific problems. All connections among neurons are characterized by numeric values (weights) that are updated during the training. The ANN is trained by the supervised learning process of the Error Back Propagation algorithm [17].

Support Vector Machine. The SVM is a model for supervised and unsupervised learning for data analysis and pattern recognition [18]. The purpose of the SVM classifier is to find a linear separating hyper plane with the maximal margin using the data in the training set. The SVM uses a kernel to map samples in a higher dimensional space where it can handle the non-linear relation between the attributes and the class label. There are different kinds of kernel but in this case the RBF kernel has been used. The SVM classifier performance depends on two parameters: C and γ. Consequently, some kind of model selection (parameter search) must be performed.

SVM-Recursive Feature Elimination. The SVM-RFE [19-21], for features selection, has been carried out using Weka [22] in order to build a ranking of the features. The SVM-RFE has been initialized with the Weka default parameters. The output of the SVM-RFE is the list of the features ranked based on SVM classification results. Given the rank of the features, a Grid Search [22] has been performed for the identification of the best C and γ. An exponential growing of the values has been used during the search, so $C = 2^{-5}, 2^{-3}, ..., 2^{15}$ and $\gamma = 2^{-15}, 2^{-13}, ..., 2^{3}$.

3.5 Output

An SVM classifier has been trained for each domain. Then the results of the three classifiers are combined together to assess whether the two people involved in the interview are resonant or not. Considering *F(x)* as the result of the classification of a given instance *x* in the facial domain, while *V(x)* and *G(x)* are respectively the result obtained from the vocal and gestural classifiers, the output result *y(F,V,G)* will be a combination of the previous results as shown in the Eq. 7.

$$y(F,V,G) = \alpha \cdot F(x) + \beta \cdot V(x) + \delta \cdot G(x) \tag{7}$$

α, β and δ are the associated weights.

The set of values of F(x), V(x) and G(x) is {0,1} where the 0 value means dissonance and 1 value means resonance. These weights range from zero to one with $\alpha + \beta + \delta = 1$. The range of *y(F,V,G)* is a real value and it falls in [0,1]. To determine the status of the resonance or dissonance, a threshold σ has been set, in order to transform the real value in boolean one, as shown in Eq. 8.

$$\begin{cases} resonance & if\ y(F,V,G) \geq \sigma \\ dissonance & otherwise \end{cases} \tag{8}$$

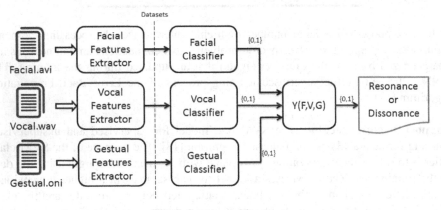

Fig. 5. System architecture

4 Results

In this section we present the classification results obtained for each single domain and the combination of them. Experiments have been performed using a ten-fold cross-validation method. The dataset is randomly partitioned into 10 equal size folds. The classification result is calculated as the average classification result of ten classification steps. At the *ith* iteration, the *ith* fold is used as test set while the remaining folds are used as training set. The accuracy metric is used to evaluate the classification performance, as defined in Eq. 9.

$$Accuracy = \frac{1}{N} \sum_{i=1}^{C} corrected_i \qquad (9)$$

where $corrected_i$ is the number of samples belonging to the ith class that have been classified as belonging to the ith class and N is the number of samples. Moreover resonance and dissonance classification rates have been considered as described respectively in Eq. 10 and Eq. 11.

$$Resonance\ Rate = \frac{correct_{resonant}}{correct_{resonant} + uncorrect_{resonant}} \qquad (10)$$

$$Dissonance\ Rate = \frac{correct_{dissonant}}{correct_{dissonant} + uncorrect_{dissonant}} \qquad (11)$$

4.1   Single Domain Result

In the facial domain, after the feature selection phase, the number of features has been reduced to 28, C has been set to 2^9 and γ has been set to 2^{-7}. Exploiting the SVM, the obtained accuracy is 73.3%, whilst $resonance_{rate} = 80.3\%$ and $dissonance_{rate} = 66.6\%$. With the ANN we obtained an accuracy = 72.4%, $resonance_{rate} = 71.4\%$ and $dissonance_{rate} = 73.4\%$. In Table 6 the confusion matrices for both the classifiers are represented.

In the vocal domain, experimental results have demonstrated that the minimum number of features to select is 24, with $C = 2^{11}$ and $\gamma = 2^{-7}$, while with the SVM the obtained accuracy is 79.1%, $resonance_{rate} = 81.5\%$ and $dissonance_{rate} = 76.7\%$ whereas with ANN we obtained accuracy = 72.7%, $resonance_{rate} = 75.9\%$ and $dissonance_{rate} = 69.6\%$. In Table 7 there are the confusion matrices for both the SVM and the ANN.

In the gestural domain, the minimum number of feature to keep is 2, with $C = 2^7$ and $\gamma = 2^3$. With the SVM, the obtained accuracy is 63%, $resonance_{rate} = 70.3\%$ and $dissonance_{rate} = 56.1\%$. Exploiting the ANN, the obtained accuracy = 63%, $resonance_{rate} = 66.6\%$ and $dissonance_{rate} = 59.6\%$. Table 8 shows the confusion matrices obtained in the gestural domain for both the classifiers.

Table 6. Confusion matrices of the facial domain with: a) SVM, b) ANN (**gt** stands for ground truth and **out** stands for test outcome)

	$Diss_{gt}$	Res_{gt}			$Diss_{gt}$	Res_{gt}
$Diss_{out}$	40	11		$Diss_{out}$	44	16
Res_{out}	20	45		Res_{out}	16	40
	(a)				(b)	

Table 7. Confusion matrices of the vocal domain with: a) SVM, b) ANN

	$Diss_{gt}$	Res_{gt}
$Diss_{out}$	43	10
Res_{out}	13	44
(a)		

	$Diss_{gt}$	Res_{gt}
$Diss_{out}$	39	13
Res_{out}	17	41
(b)		

Table 8. Confusion matrices of the gestural domain with: a) SVM, b) ANN

	$Diss_{gt}$	Res_{gt}
$Diss_{out}$	32	16
Res_{out}	25	38
(a)		

	$Diss_{gt}$	Res_{gt}
$Diss_{out}$	34	18
Res_{out}	23	36
(b)		

4.2   Cross-Domain Results

In the previous results it is clear that the results obtained from the SVM are higher than those of the ANN. For this reason the combination with the single domains has been performed using only the SVM classifiers. Table 9 summarizes the obtained classification results. An exhaustive search has been performed to set the weights of every domain classifier. The dataset is composed by the interviews that contain all the three domains. The threshold value σ has been set to 0.5. All the weights have been chosen having as an optimization criterion the classification results of the worst classified class between the classes considered. It can be clearly seen that, considering Dissonance and Resonance Rates, the configuration with facial and vocal data achieves the best balanced results.

Table 9. Results obtained on cross-domain tests

Facial	Vocal	Gestural	Accuracy	Dissonance Rate	Resonance Rate
0.5	0.3	0.2	70.5%	77.5%	63.46%
0.55	0.45	*	67.6%	67.92%	67.3%
0.75	*	0.25	67.6%	67.92%	67.3%
*	0.75	0.25	67.6%	64.15%	69.8%

5   Conclusion

In conclusion, the designed and developed system has acceptable performance in classifying resonance and dissonance using facial, vocal and gestural information. Experimental results show that facial and vocal domains achieve better performance when used separately or combined together, much better than the results achieved by the gestural domain. A depth analysis of the gestural data shows up that interviewees are very restrained so the number of detected gestures is very low. This explains why the number of features among the three domains is unbalanced. For these reasons, it makes sense to not consider the gestural information. Further studies can involve the chance to improve classification performance by using different kind of classifiers or using different features.

Acknowledgements. This study was supported by the Italian PON FIT Project called "Sviluppo di un sistema di rilevazione della risonanza (SS-RR) N° B01/0660/01-02/X17" - Politecnico di Bari and AMT Services s.r.l., Italy.The experimental tests were also conducted according to Adecco S.A., the world's leading provider of HR solutions. In particular we would thank to Nadia Cristofoli and Maria Luisa Bruno, Adecco Permanent Division, Rossana Rebuglio and Lucia Cosulich, Adecco Information Management Department, and Adecco Bari's branch.

References

1. Di Corpo, U.: Life Energy, Syntropy, Complementarity and Resonance. Syntropy Journal 2, 4–38 (2013)
2. Badinelli, R., Barile, S., Ng, I., Polese, F., Saviano, M., Di Nauta, P.: Viable service systems and decision making in service management. Journal of Service Management 23(4), 498–526 (2012)
3. Di Corpo, U., Vannini, A.: Anxiety, Depression and Anguish in the light of the Theory of Vital Needs (2011)
4. Ekman, P.: Emotion in the Human Face. Cambridge University Press, Cambridge (1982)
5. Bevilacqua, V., Guccione, P., Mascolo, L., Pazienza, P.P., Salatino, A.A., Pantaleo, M.: First progresses in evaluation of resonance in staff selection through speech emotion recognition. In: Huang, D.-S., Jo, K.-H., Zhou, Y.-Q., Han, K. (eds.) ICIC 2013. LNCS (LNAI), vol. 7996, pp. 658–671. Springer, Heidelberg (2013)
6. Roether, C.L., Omlor, L., Christensen, A., Giese, M.A.: Critical features for the perception of emotion from gait. Journal of Vision 9(6) (2009)
7. Coulson, M.: Attributing emotion to static body postures: Recognition accuracy, confusions, and viewpoint dependence. Journal of Nonverbal Behavior 28(2), 117–139 (2004)
8. Meijer, M.: The contribution of general features of body movement to the attribution of emotions. Journal of Nonverbal Behavior 13(4), 247–268 (1989)
9. Bevilacqua, V., Suma, M., D'Ambruoso, D., Mandolino, G., Caccia, M., Tucci, S., De Tommaso, E., Mastronardi, G.: A supervised approach to support the analysis and the classification of non verbal humans communications. In: Huang, D.-S., Gan, Y., Bevilacqua, V., Figueroa, J.C. (eds.) ICIC 2011. LNCS, vol. 6838, pp. 426–431. Springer, Heidelberg (2011)
10. Bevilacqua, V., D'Ambruoso, D., Mandolino, G., Suma, M.: A new tool to support diagnosis of neurological disorders by means of facial expressions. In: 2011 IEEE International Workshop on Medical Measurements and Applications Proceedings (MeMeA), pp. 544–549. IEEE (2011)
11. Bevilacqua, V., Casorio, P., Mastronardi, G.: Extending hough transform to a points' cloud for 3d-face nose-tip detection. In: Huang, D.-S., Wunsch II, D.C., Levine, D.S., Jo, K.-H. (eds.) ICIC 2008. LNCS (LNAI), vol. 5227, pp. 1200–1209. Springer, Heidelberg (2008)
12. Bevilacqua, V., Filograno, G., Mastronardi, G.: Face detection by means of skin detection. In: Huang, D.-S., Wunsch II, D.C., Levine, D.S., Jo, K.-H. (eds.) ICIC 2008. LNCS (LNAI), vol. 5227, pp. 1210–1220. Springer, Heidelberg (2008)

13. Viola, P., Jones, M.: Rapid object detection using a boosted cascade of simple features. In: Proceedings of the 2001 IEEE Computer Society Conference on Computer Vision and Pattern Recognition, CVPR 2001, vol. 1, pp. I–511. IEEE (2001)

14. Strupp, S., Schmitz, N., Berns, K.: Visual-based emotion detection for natural man-machine interaction. In: Dengel, A.R., Berns, K., Breuel, T.M., Bomarius, F., Roth-Berghofer, T.R. (eds.) KI 2008. LNCS (LNAI), vol. 5243, pp. 356–363. Springer, Heidelberg (2008)

15. Bevilacqua, V., D'Ambruoso, D., Suma, M., Barone, D., Cipriani, F., D'Onghia, G., Mastrandrea, G.: A new tool for gestural action recognition to support decisions in emotional framework. To appear in the IEEE Proceedings of INISTA (2014)

16. Alelyani, S., Tang, J., Liu, H.: Feature selection for clustering: A review. In: Aggarwal, C., Reddy, C. (eds.) Data Clustering: Algorithms and Applications, CRC Press

17. Bevilacqua, V.: Three-dimensional virtual colonoscopy for automatic polyps detection by artificial neural network approach: New tests on an enlarged cohort of polyps. Neurocomputing 116, 62–75 (2013)

18. Bevilacqua, V., Pannarale, P., Abbrescia, M., Cava, C., Paradiso, A., Tommasi, S.: Comparison of data-merging methods with SVM attribute selection and classification in breast cancer gene expression. BMC Bioinformatics 13(suppl. 7), S9 (2012), doi:10.1186/1471-2105-13-S7-S9

19. Cortes, C., Vapnik, V.: Support-vector networks. Mach. Learn. 20(3), 273–297 (1995)

20. Guyon, I., Weston, J., Barnhill, S., Vapnik, V.: Gene Selection for Cancer Classification using Support Vector Machines. Machine Learning 46(1-3), 389–422 (2002)

21. Chang, C.-C., Lin, C.-J.: LIBSVM: A library for support vector machines. ACM Transactions on Intelligent Systems and Technology 2, 27:1–27:27 (2011), http://www.csie.ntu.edu.tw/cjlin/libsvm

22. Hall, M., Frank, E., Holmes, G., Pfahringer, B., Reutemann, P., Witten, I.H.: The weka data mining software: an update. ACM SIGKDD Explorations Newsletter 11(1), 10–18 (2009)

Using Rough Set Theory and Decision Trees to Diagnose Enterprise Distress – Consideration of Corporate Governance Variables

Fu Hsiang Chen[1], Der-Jang Chi[1,*], and Chun-Yi Kuo[2]

[1] Department of Accounting, Chinese Culture University No. 55,
Hwa-Kang Road, Taipei, 11114, Taiwan
derjang@yahoo.com.tw
[2] Department of Accounting, Tamkang University No.151, Yingzhuan Rd.,
Danshui Dist., New Taipei City 25137, Taiwan

Abstract. This study discusses the key factors of financial distress warning models for companies using corporate governance variables and financial ratios as the research variables, sieving out influential variables based on the attribute simplification process of rough set theory (RST). Then, we construct some classification models for diagnosing enterprise distress based on RST, using a data mining technique of decision trees with the selected indicators and variables. The empirical results obtained from analysis of enterprise distress indicators, show that financial distress is not only affected by the traditional financial ratios, but also by corporate governance variables. In addition, enterprise distress diagnosis models constructed based on RST and decision trees can effectively diagnose firms in times of crisis. In particular, the RST models are more accurate. This study provides a reference for better understanding the symptoms that might lead to a company's financial crisis in advance and thus provide a valuable reference for investment decision making by stakeholders.

Keywords: data mining, enterprise distress diagnosis, financial ratios, corporate governance, rough set theory(RST), decision trees.

1   Introduction

Following the fall of the South-Eastern Asian stock markets and foreign exchange markets leading up to the financial storm that swept across the world, and as a result of the manipulation of speculators that became apparent in July 1997, a series of local-based companies in Taiwan have experienced financial distress in the late stages of the Asian Financial Crisis. A number of major companies, including Tong Lung Metals and the Kuangsan Group have encountered major financial problems, with bounced checks, high debt, and defaults on delivery, as a result of loss of investments and company assets by the responsible officers. In recent years, scandals involving

---

* Corresponding author.

D.-S. Huang et al. (Eds.): ICIC 2014, LNAI 8589, pp. 199–211, 2014.
© Springer International Publishing Switzerland 2014

Procomp, the Pacific Electric Wire & Cable Co. Ltd, Rebar Group and Chia Hsin Food have made the news, as well as constant reports about the crises faced by financial institutions at home and abroad. The occurrence of those events not only directly harms the rights of the company stakeholders, but also leads to a huge social burden. If business managers could identify risk warnings or problems earlier, they should be able to take corresponding corrective actions to prevent the occurrence or deterioration leading to such crises. For this purpose, effective enterprise distress diagnosis models are increasingly important.

In the 1960s, some studies began to construct company financial distress warning models mostly using different financial ratios, together with different statistical methods (e.g., discriminant analysis, logistic regression analysis, neural networks, etc.) and achieved good predictions. However, the focus in most past studies was solely on using financial ratios as indicator variables to construct the diagnosis models [1, 2, 3]. The concept of corporate governance generally refers to the company's management system and monitoring methods. Rajan and Zingales [4], and Prowse [5] conducted empirical studies in the East Asian region after the Asian Financial Crisis. They found that the Asian Financial Crisis was mainly caused by the over-concentrated shareholding structure of the companies in East Asia and lack of good corporate governance. Therefore, they concluded that "strengthening corporate governance mechanisms" facilitates a firm's capital structure adjustments, and operational performance. This has been widely recognized as a favorable weapon to protect enterprises against financial crisis [4, 6, 7]. Clearly, it is indeed an important task to strengthen corporate governance.

Enterprise distress diagnosis models have been a topic of concern for governmental agencies, financial institutions, enterprises and investors for a long time [8, 9]. Properly-structured enterprise distress diagnosis models can effectively detect signs of financial distress before problems occur. Banks can use these financial warning models to assess the real risks of the enterprise so as to lower their operating risks and costs. Investors, creditors and transaction partners can take early contingency measures in advance to avoid major losses due to a company's bankruptcy and closure. According to a study performed by Claessens et al. [10], poor corporate governance mechanisms comprised one of the major causes of the Asian Financial Crisis. In addition, dispersed ownership and separation of ownership and control are common characteristics of TSEC/GTSM Listed Companies. Claessens et al.[10] held a different viewpoint. Based on the method proposed by La Porta et al. [11], they studied a total of 2980 listed companies in nine countries in East Asia, including Taiwan, and found that more than half of the companies had a family member-based shareholding structure (48.2% in Taiwan).

Most of the past studies dealing with enterprise distress diagnosis models employed traditional statistical analysis methods, such as logistic regression and multivariate analysis in order to distinguish normal companies from those in financial distress [1, 2, 12, 3, 5, 4, 10, 13, 14]. These conventional statistical analysis methods have some restrictive assumptions such as the linearity, normality, and independence of the predictor or input variables, and have intrinsic limitations in terms of effectiveness and validity. However, following the development of artificial intelligence (AI) in recent years, advanced technologies including data mining have been employed for the construction of enterprise distress diagnosis models. The methods include artificial neural network (ANN) [15, 16, 17], decision tree (DT) [16, 18, 19], support vector machine (SVM) [18, 20], and rough set theory (RST)[21] techniques. Slowinski and Zopoudinis [22] first applied RST in financial forecasting, and suggested this to be a very useful method for decision makers to identify favorable features out of many, according to the financial ratios and related indicators of the enterprises. In practice, RST is able to the provide rules set by the enterprise and explain their decision-making process [21]. Meanwhile, the classification rules summarized by the decision tree method are presented in a tree-like structure which makes them easy to understand due to the simple knowledge structure [16, 18]. RST and decision tree analysis are applied to construct enterprise distress diagnosis models with or without the consideration of corporate governance indicators. The models are used to understand the operating status of those enterprises in financial distress.

2   Literature Review

2.1   Enterprise Distress

Beaver [1] first clearly defined company failure as when a company declares bankruptcy, defaults on debts, overdraws its bank accounts, or fails to pay preferred stock dividends. Lau [23] used five consecutive stages to describe financial status, and estimated the probability of the enterprise entering each stage. The five stages are: financial stability, halting or reduction of dividend payments, loan and technical default, subject to the protection of bankruptcy laws, declaration of bankruptcy, and finally liquidation. Ward and Foster [24] defined a company in financial distress as one that delays, lowers, or is unable to pay its debts and interest, as well as those under debt reconstruction. Pindado et al. [25] pointed out that the failure of a business in financial distress to meet its financial obligations does not inevitably lead to it filing of bankruptcy. Kim and Upneja [19] defined business financial distress as situations in which a firm cannot fulfill its financial obligations to its creditors, suppliers, and/or vendors. Most studies have adopted certain legal definitions for failure of a company, such as the trading of an enterprise, the termination of its listing on the stock exchange, taken from formal (legal) insolvency notices of debt servicing and bond or loan defaults, default swaps (Ericsson et al. [26]) or stock market suspensions.

2.2   Corporate Governance

It can be seen from the Enron and World Telecom scandals that healthy operation of an enterprise and the stability of the financial system are closely related to corporate governance. To strengthen corporate governance mechanisms, the US government promulgated the Sarbanes-Oxley Act of 2002, and considerably modified the Securities Act and Securities Exchange Act, giving topics related to corporate governance top reform priority. According to the definition of the Organization of Economic

Cooperation and Development (OECD), corporate governance generally refers to "the system by which business corporations are directed and controlled. The corporate governance structure specifies the distribution of rights and responsibilities among different participants in the corporation, such as the board, managers, shareholders and other stakeholders, and specifies the rules and procedures for making decisions on corporate affairs. By doing this, it also provides the structure through which the company objectives are set, and the means of attaining those objectives and monitoring performance." Specifically, corporate governance is supposed to manage and guide the enterprise through design mechanisms, improve management performance and monitor managers to ensure protection of the interests of company shareholders and other stakeholders [27].

In their empirical study, Claessens et al. [28] reported that poor corporate governance mechanisms comprised one of the causes of the Asian Financial Crisis. Although the impact of the 1997 Asian Financial Crisis on Taiwan was relatively mild, listed companies, including Procomp, Summit Computer Technology, Infodisc Technology, and Universal Abit have faced financial difficulties since 2004, highlighting the numerous agency problems as a result of inadequacy in laws and regulations in Taiwan. The most concerning case was the Rebar Group scandal at the end of 2006. Companies belonging to the Rebar Group invested in each other, were involved in insider trading, and received bank loans by virtue of mutual pledges of shares of affiliated companies, while their boards of directors and supervisors were controlled by family members who deliberately hid major information and presented false financial statements. Yeh et al. [29] categorized the functions of corporate governance into two types, those that promoted benefits and that should be abolished as harmful. The functions that promote benefits can enhance strategic management performance and ensure the implementation of company strategy in the right direction. One example of functions that should be abolished because they are harmful is that the company should have independent directors and supervisors to monitor management, ensuring transparency of information in real time and due reward for outside shareholders and creditors. Hsu et al. [30] surveyed 63 companies in financial distress and 126 companies with normal financial status during the period of 1998 to 2004. Based on this information they constructed 7 warning models employing logistic regression analysis, and discussed the predictive capacity of the various for forecasting the probability of financial distress based on earnings management indicators, accounting ratios and corporate governance variables. Their results indicated that the cash flow ratio has the most significant negative impact on financial distress probability among all the accounting information variables. As for the corporate governance variables, the percentage of pledged shares being held by directors and supervisors have the most significant positive impact on the probability of financial distress. Morellec et al. [7] predicted that a good corporate governance system would serve the best interests of shareholders by inducing managers to make more timely (upward) adjustments to capital structure deviations. Kim and Lu [31] indicated that corporate governance reforms had an important impact on corporate investments, and suggested that the gap in investor protection between capital exporting and importing countries distorts firm-level allocation of foreign capital inflows and reduces the benefits of globalization.

2.3   Enterprise Distress Diagnosis Models

Deakin [32] replaced linear discriminant analysis (LDA) with the quadratic discriminant function (QDA), and added the trends of financial ratios, variance and slope as independent variables in the models in order to establish a differential function for each year to improve the differentiation results. The fundamental assumptions must be satisfied for models established using the multivariate statistical method. The correctness and accuracy of prediction are questioned if such assumptions are not satisfied.

Compared with traditional statistical models, the artificial intelligence methods developed in recent years, such as data mining, decision tree, SVM, RS and ANN techniques, have considerably higher accuracy rates in enterprise distress diagnosis model applications. Xiao et al. [34] used a rough set (RS) method to determine the weight of each single prediction method and utilized the Dempster–Shafer evidence theory in a combination method. They carried out an empirical experiment using data for Chinese listed companies. The results indicated that the performance of the RS method is superior to those of the single classifier and other multiple classifiers. Lin et al. [35] proposed an integrated approach to feature selection for the financially distressed prediction problem that embeds expert knowledge with the wrapper method and found that the prediction model based on the feature set selected outperforms the traditional feature selection model in terms of prediction accuracy. Kim and Upneja [19] used a decision tree approach to predict financial distress for restaurants. Their results indicated that financially distressed restaurants rely more heavily on debt, with lower increasing rates of assets than others did, and as the net profit margin decreases with the lack of capital efficiency, the chance of financial distress increases.

3   Research Methodology

3.1   Rough Set Theory

Rough sets (RS) or rough set theory (RST) was first proposed by the Polish scholar, Z. Pawlak in 1982 [36]. RST is a new method of discrete data reasoning, and set theory is its mathematical basis. RST is used to deal with problems of vagueness and imprecise information. This method does not require any a priori data or additional information, such as the probability distribution or the membership function of fuzzy theory. It is able to reduce data and obtain the minimum expression of the knowledge according to the information provided by the data, while keeping key information from which to establish decision rules and discover hidden knowledge in the data set. Thus, RST can be used for knowledge mining in the data set to further produce knowledge of decision rules. RST can also be used to: (1) reduce attributes; (2) find out the hidden form of the data; and (3) produce decision rules [37, 38].

After years of research and development, RST has been applied in many fields, such as knowledge discovery from databases, expert systems and the pattern recognition of decision support systems. The RST algorithm is based on the following premise: reducing the accuracy of the information can make the data patterns more visible and the core of the RST is the knowledge of classification capability. In other words, it uses the classification patterns or a group of cases to deduce the patterns of the decision rules.

3.2   Decision Trees

DT is a powerful and widely used classification and prediction tool, which presents the rules in the form of a tree. This is a top-down approach that gradually divides specific object sets into smaller subsets along with the growth of the DT. The main structure of the DT includes the root node, node, branch, and leaf node. Different selection rules are used in cases with different input variables, and the data are repeatedly divided in order to establish different nodes. The repetitive procedure is applied during construction until dividing is impossible or the stop rule conditions are met to finally finish the leaf nodes. The dividing condition is to find the important variable selection rules for data classification. The structure formed by such a method is overly large and complex, leading to difficulty in analysis and understanding. The trimming of the DT is mainly done in order to solve the problem of over-learning and improve the judgment capacity of the DT. The trimming standards adopt the error estimation ratio as the judgment rule, and the trimming procedure is performed in the top-down direction. We employed DT models for enterprise distress prediction based on the following advantages. DT does not require any statistical assumption concerning the data in a training sample [19], and a DT model can handle incomplete and qualitative data [39].

4   Empirical Study

4.1   Research Variables

To discuss the effects of corporate governance indicators and financial ratio indicators on enterprise distress, and verify the effectiveness of the enterprise distress diagnosis models of the RST, this study uses the data from TSEC/GTSM Listed Companies from 2002-2007. This was the period when financial turmoil in the US led to a global financial crisis where most corporate financial performance was significantly reduced. To avoid this particular economic condition affecting the information as to normal financial status, data was selected for the sample only up to 2007 for empirical study. Data for normal companies from the same industry and with similar amounts of capital (30% below or above) to the companies in financial distress were used with the year before the financial crisis set as the current year. A total of 213 companies, including 71 with financial difficulties and 142 normal companies (proportion of 1:2 [1]) were selected in the sample. The data used in this study were sourced from the Taiwan economic journal (TEJ) data bank, and covered all the industries. They covered TSEC/GTSM listed companies with the exclusion of the financial sector and insurance sector due to their unique nature. The 5-fold cross-validation method [40] was used. The training data was used to make a prediction, or generalization estimate. Cross-validation is a useful statistical technique to determine the robustness of a model. The sample is randomly split into two subsamples, training and test sets. The training sample is used for model fitting and/or parameter estimation and the predictive effectiveness of the fitted model is evaluated using the test sample [41]. In this study, the sample dataset was randomly divided into five groups, each including 80% training groups and 20% test groups.

For the weighing of dependent variables, virtual variables are adopted, setting companies that may have financial distress as 1, and companies without financial distress as 0. With regard to the selection of research variables based on open information sources and TEJ, this study identifies 21 financial ratios that are not screened but have been commonly adopted by many scholars in the past and 7 corporate governance variables as the variable indicators for enterprise distress diagnosis models [1, 4, 6, 7, 12, 34, 35]. The related variables are defined as shown in Table 1.

Table 1. Definitions of research variables

Variable Type	Variable Code	Variable Name	Variable Definition
Financial variables	X1	Current Ratio	Current assets / current liabilities *100
	X2	Acid-test Ratio	(Current assets – inventory – prepaid expense – other current assets) / current liabilities * 100
	X3	Debt Ratio	Total liabilities / total assets * 100
	X4	Long term fund fitness	(Net worth + long term liabilities) / fixed assets * 100
	X5	Loan dependence	Long and short term borrowing / net worth * 100
	X6	Times-interest-earned ratio	Net income excluding income taxes and interest expense / interest expense
	X7	Assets turnover	Net sales / average total assets
	X8	Accounts receivable turnover	Net sales / average account receivables
	X9	Inventory turnover	Cost of goods sold / average inventory
	X10	Fixed assets turnover	Net sales / average fixed assets
	X11	Rate of return on total assets	(Net income after taxation + interest expense * (1-tax rate)) / average total assets * 100
	X12	Operating margin	Profit on operations / net sales * 100
	X13	Net profit margin before taxation	Net profit before taxation / net sales * 100
	X14	Net worth growth rate	(Net worth – previous period net worth) / previous period net worth * 100
	X15	Cash flow ratio	Cash flow from operations / current liabilities *100
	X16	Working capital ratio	(Current assets - current liabilities) / total assets * 100
	X17	Non-operating revenue and expenditure ratio	(Net non-operating revenue and expenditure / net sales) * 100
	X18	Cash reinvestment ratio	(Operating activity cash-cash dividends) / (fixed assets gross amount + long term investment + other assets + working capital) * 100
	X19	Net worth turnover	Net sales / average net worth
	X20	Cash and equivalent cash	Amount of cash and equivalent cash
	X21	Current liabilities	Amount of current liabilities

Table 1. (*continued*)

Corporate Governance variables	C1	Size of the Board of directors	Number of directors
	C2	Size of supervising committee	Number of supervisors
	C3	Shareholding ratio of major shareholders	Total shareholding ratio of major shareholders
	C4	Shareholding ratio of managers	Ratio of managers' shares from outstanding shares
	C5	Shares held by the company directors	Total number of shares held by company directors
	C6	Shareholding ratio of company directors and supervisors	The ratio of shares held by directors and supervisors of the total number of outstanding shares
	C7	Percentage of pledged shares held by company directors and supervisors	Percentage of shares held by company directors and supervisors being pledged.

Source: [1]; [4]; [6]; [7]; [12]; [34]; [35]

4.2   Empirical Analysis

In order to improve the effectiveness of the overall classification process and keep the original data after the selection of variables, this study uses the selected 21 financial ratios and 7 corporate governance variables after the attribute reduction by RST as the standard method of variable selection. Various groups of reduced attributes can be obtained from one decision table; however, the core attribute is the most important attribute after reduction. The Rosetta software is adopted to implement RST in this study. This is a suite of software based on RST that integrates such methods and obtains the attribute reduction results shown in Table 2. The most important attribute and the important indicators screened are the current ratio, debt ratio, times-interest-earned ratio, rate of return on total assets, operating margin, shareholding ratio of major shareholders and shareholding ratio of directors and supervisors. The screening results are as shown in Table 3.

Based on the important indicators, it can be found that in addition to financial ratios, the attributes of the shareholding ratio of major shareholders and the shareholding ratio of directors and supervisors, which appear in both the corporate governance variables and the major indicators, indicate that corporate governance variables do have effects on the enterprise distress diagnosis models.

Finally, the important indicators screened by RST are input into the RST and DT classification models. To underline whether the general classification accuracy of the enterprise distress diagnosis models has been significantly improved after adding the corporate governance indicators, this study individually judges the results of the model diagnosis both with and without corporate governance indicators. The diagnosis results of the RST are shown in Table 4.

Table 2. Attribute reduction results

#	Reduction	#	Reduction
1	{ X1, X3, C3, C6}	16	{ X3, X6, X11, X18, C1}
2	{X15, X19, X21, C4}	17	{ X3, X6, X12, X18, C1}
3	{X1, X10, X15, X19}	18	{X7, X11, X18, C1, C6}
4	{X6, X15, X17, C6}	19	{X1, X12, X18, C3, C5}
5	{X1, X3, X16, C4}	20	{X5, X6, X11, X18, C1, C6}
6	{X1, X10, X15, X21}	21	{X6, X11, X18, C1}
7	{X3, X16, X18, X20}	22	{X8, X11, X18, C1}
8	{X1, X5, X6, X12}	23	{X7, X11, X12,X18, C1, C2}
9	{X1, X15, X19, C2, C3, C6}	24	{ X1,X7, X11, X18, C1, C7}
10	{ X1,X3,X13, X20, X21, C6}	25	{ X4,X7, X11, X19, C1, C5}
11	{X3, X6, X17, C2, C3}	26	{X7, X11, X14, C1, C4, C7}
12	{X1, X6, X16, C4}	27	{ X2, X5, X7, X11, X18, C3}
13	{X2, X5, X15, X18}	28	{ X1, X6, X7, X11, X18, C6}
14	{X4, X5, X6, X17}	29	{X7, X11, X18, C1, C3, C6}
15	{X3, C3, C5, C7}	30	{X7, X11, X18, C2, C5, C6}

Table 3. Selected financial indicators

Variables	Indicators
X1	Current Ratio
X3	Debt Ratio
X6	Times-interest-earned ratio
X11	Rate of return on total assets
X12	Operating margin
C3	Shareholding ratio of major shareholders
C6	Shareholding ratio of directors and supervisors

Table 4. RST Diagnosis results

Group	RST accuracy (%)	
	With corporate governance variables	Without corporate governance variables
Group 1	83.4	77.5
Group 2	83.3	71.2
Group 3	83.4	76.5
Group 4	85	71.2
Group 5	83.3	76.5
Average	83.68	74.58

Similarly, the important indicators variables are selected into the classification models of the DT. The diagnosis results are shown in Table 5.

Table 5. DT Diagnosis results

Group	DT accuracy (%)	
	With corporate governance variables	Without corporate governance variables
Group 1	73.3	68.5
Group 2	73.4	71.2
Group 3	78.8	66.9
Group 4	80	71.6
Group 5	76.5	72.8
Average	76.4	70.2

The empirical study results show that the addition of corporate governance indicator variables does improve the accuracy of the enterprise distress diagnosis models, indicating that the establishment and effectiveness of corporate governance mechanisms has effects on financial distress prediction. The accuracy rate of the RST classification tool in the empirical study is 83.68%, indicating the achievement of a relatively high rate of accuracy for diagnosing companies in financial distress. The classification accuracy of the RST is better than the DT.

5   Conclusions and Contributions

The global economic environment has been changing rapidly in recent years. The financial storm that originated in the US in March 2008 swept across the world in the second half of 2008, and severely damaged the world economy, including the so-called "BRIC Countries"--- Brazil, Russia, India, and China. The economic recession is still in progress at present, causing considerable costs and potential financial distress to many companies. With the amount of enterprises in distress increasing, it is important to identify which businesses are most exposed to the risk of financial distress, because recognizing a potentially financially distressed business and identifying its problems provide the best chance for managers to take the necessary corrective actions to turn the firm around before it is too late. Therefore, it is an important topic for business and academia to establish effective enterprise distress diagnosis and prediction models. The ability to accurately predict enterprise distress is critical in financial decision making because making incorrect decisions is likely to lead to a financial crisis or bankruptcy. Enterprise distress diagnosis technologies include traditional statistical methods, non-parametric methods, and artificial intelligence methods. RST is a data mining method that has been developed in recent years. It requires no a priori or additional information about the data sets, and uses knowledge mining techniques on the data set itself to further produce decision rule knowledge to create a useful classification tool. In this study, we propose a hybrid model that combines RST and DT that not only enhances classification accuracy but also extracts meaningful rules for enterprise distress prediction.

This study compares enterprise distress diagnosis models established using the RST and DT methods. The attribute reduction function of the RST is used to select important indicator variables, in order to simplify the model and develop a faster and more accurate enterprise distress diagnosis model. Based on Tables 4-5, it can be seen that the models can be ranked in order of accuracy and that RST models are superior to the DT models. In addition, during the discussion of enterprise distress weighing indicators, this study also adds non-financial weighing indicators - corporate governance variables with reference to some financial ratios commonly used in past studies to obtain more diversified enterprise information in order to help investors assess the true value of the enterprise and make correct decisions. This study also shows that the models, including corporate governance variables and financial ratios, provide better classification results than the predictors which only use financial ratios. The process is helpful for company financial distress analysis, diagnosis and predictions, and can serve as a reference for management decision making and academic study.

References

1. Beaver, W.H.: Financial Ratios as Predictors of Failure. Journal of Accounting Research 4, 71–111 (1966)
2. Altman, E.I.: Financial Ratios Discriminate Analysis and the Prediction of Corporate Bankruptcy. Journal of Finance 23(4), 589–609 (1968)
3. Ohlson, J.: Financial Ratios and the Probabilistic Prediction of Bankruptcy. Journal of Accounting Research 18(1), 109–131 (1980)
4. Rajan, R.G., Zingales, L.: Which Capitalism? Lessons from the East Asian Crisis. Journal of Applied Corporate Finance 11(3), 40–48 (1998)
5. Prowse, S.: Corporate Governance: Emerging Issues and Lessons from East Asia, Responding to the Global Financial Crisis–World Bank Mimeo (1998)
6. Sueyoshi, T., Goto, M., Omi, Y.: Corporate Governance and Firm Performance: Evidence From Japanese Manufacturing Industries after the Lost Decade. European Journal of Operational Research 203(3), 724–736 (2010)
7. Morellec, E., Nikolov, B., Schürhoff, N.: Corporate Governance and Capital Structure Dynamics. Journal of Finance 67(3), 803–848 (2012)
8. Chen, W.S., Du, Y.K.: Using Neural Networks and Data Mining Techniques for the Financial Distress Prediction Model. Expert Systems with Applications 36(2), 4075–4086 (2009)
9. Cohen, S., Doumpos, M., Neofytou, E., Zopounidis, C.: Assessing Financial Distress where Bankruptcy is not an Option: An Alternative Approach for Local Municipalities. European Journal of Operational Research 218(1), 270–279 (2012)
10. Claessens, S., Fan, P.H.J., Djankov, S., Lang, H.P.L.: On Expropriation of Minority Shareholders: Evidence from East Asia. Available at SSRN 202390 (1999)
11. La Porta, R., Lopez-de-Silanes, F., Shleifer, A., Vishny, R.: The quality of government. Journal of Law, Economics, and organization 15(1), 222–279 (1999)

12. Altman, E.L., Edward, I., Haldeman, R., Narayanan, P.: Zetatm Analysis a new Model to Identify Bankruptcy Risk of Corporations. Journal of Banking and Finance 23(1), 29–54 (1977)
13. Johnson, S., Boone, P., Breach, A., Friedman, E.: Corporate Governance in the Asian Financial Crisis. Journal of Financial Economics 58(1-2), 141–186 (2000)
14. Aziz, M.A., Dar, H.A.: Predicting Corporate Bankruptcy: Where We Stand? Corporate Governance 6(1), 18–33 (2006)
15. Odom, M.D., Sharda, R.: A Neural Networks Model for Bankruptcy Prediction. In: IEEE International Joint Conference on Neural Networks, California, San Diego, pp. 163–167 (1990)
16. Tam, K.Y., Kiang, M.Y.: Managerial Applications of Neural Networks: The Case of Bank Failure Prediction. Management Science 38(7), 926–947 (1992)
17. Altman, E.I., Marco, G., Varetto, F.: Corporate Distress Diagnosis: Comparisons Using Linear Discriminant Analysis and Neural Networks (the Italian Experience). Journal of Banking and Finance 18, 505–529 (1994)
18. Kotsiantis, S., Koumanakos, E., Tzelepis, D., Tampakas, V.: Forecasting Fraudulent Financial Statements Using Data Mining. International Journal of Computational Intelligence 3(2), 104–110 (2006)
19. Kim, S.Y., Upneja, A.: Predicting Restaurant Financial Distress Using Decision Tree and AdaBoosted Decision Tree Models. Economic Modelling 36, 354–362 (2014)
20. Kirkos, E., Spathis, C., Manolopoulos, Y.: Data Mining Techniques for the Detection of Fraudulent Financial Statements. Expert Systems with Applications 32(4), 995–1003 (2007)
21. Bose, I.: Deciding the Financial Health of Dot-coms Using Rough Sets. Information and Management 43(7), 835–846 (2006)
22. Slowinski, R., Zopounidis, C.: Application of the Rough Set Approach to Evaluation of Bankruptcy Risk. International Journal of Intelligent Systems in Accounting. Finance and Management. 4, 27–41 (1995)
23. Lau, A.H.L.: A Five-state Financial Distress Prediction Model. Journal of Accounting Research 25(1), 127–138 (1987)
24. Ward, T.J., Foster, B.P.: An Empirical Analysis of Thomas's Financial Accounting Allocation Fallacy Theory in a Financial Distress Context. Accounting and Business Research 26(2), 137–152 (1996)
25. Pindado, J., Rodrigues, L., De la Torre, C.: Estimating Financial Distress Likelihood. Journal of Business Research 61, 995–1003 (2008)
26. Ericsson, J., Jacobs, C., Oviedo, R.: The Determinants of Credit Default Swap Premia. Journal of Financial and Quantitative Analysis 44, 109–132 (2009)
27. Salzman, J.: Decentralized Administrative Law in the Organization for Economic Cooperation and Development. Law and Contemporary Problems 68(3/4), 189–224 (2005)
28. Claessens, S.S., Djankov, S.J., Fan, P.H., Lang, L.H.P.: Corporate Diversification in East Asia: The Role of Ultimate Ownership Group Affiliation, World Bank, Research Paper 2089 (1999)
29. Yeh, Y.H., Lee, T.S., Ko, C.E.: Corporate Governance and Rating System. Sunbright Culture Eds, Taipei (2002)
30. Hsu, S.N., Ouyang, H., Chen, C.F.: Corporate Governance, Earnings Management, and the Construction of Financial Warning Models. Journal of Accounting and Corporate Governance 4(1), 85–121 (2007)
31. Kim, E.H., Lu, Y.: Corporate Governance Reforms around the World and Cross-border Acquisitions. Journal of Corporate Finance 22, 236–253 (2013)

32. Deakin, E.B.: A Discriminant Analysis of Predictors of Business Failure. Journal of Accounting Research, pp. 167–179 (1972)
33. Xiao, Z., Yang, X., Pang, Y., Dang, X.: The Prediction for Listed Companies' Financial Distress by Using Multiple Prediction Methods with Rough Set and Dempster–Shafer Evidence Theory. Knowledge-Based Systems 26, 196–206 (2012)
34. Lin, F., Liang, D., Yeh, C.C., Huang, J.C.: Novel Feature Selection Methods to Financial Distress Prediction. Expert Systems with Applications 41(5), 2472–2483 (2014)
35. Pawlak, Z.: Rough Sets. International Journal of Information and Computer Secience 11, 341–356 (1982)
36. Pawlak, Z., Slowinski, R.: Rough Set Approach to Multiattribute Decision Analysis. European Journal of Operational Research 72, 443–459 (1994)
37. Kusiak, A.: Rough Set Theory: a Data Mining Tool for Semiconductor Manufacturing. IEEE Transactions on Electronics Packaging Manufacturing 24(1), 44–50 (2001)
38. Kirkos, E., Spathis, C., Manolopoulos, Y.: Audit-firm Group Appointment: an Artificial Intelligence Approach. Intelligent Systems in Accounting, Finance and Management 17(1), 1–17 (2010)
39. Stone, M.: Cross-validatory Choice and Assessment of Statistical Predictions. Journal of the Royal Statistical Society 36, 111–147 (1974)
40. Zhang, G., Hu, M.Y., Patuwo, B.E., Indro, D.C.: Artificial Neural Networks in Bankruptcy Prediction: General Framework and Cross-validation Analysis. European Journal of Operational Research 116(1), 16–32 (1999)

Fuzzy Propositional Logic System and Its λ-Resolution

Jiexin Zhao and Zhenghua Pan

School of Science, Jiangnan University, Wuxi, China
ZHAOJX_JIANGNAN@126.com, Pan_zhenghua@163.com

Abstract. This paper researches resolution principle of the fuzzy propositional logic with contradictory negation, opposite negation and medium negation (FLCOM). In this paper, concepts of λ-satisfiable and λ-unsatisfiable are proposed under an infinite-valued semantic interpretation of FLCOM. The λ-resolution method of FLCOM is introduced. The λ-resolution deduction in FLCOM is defined and λ-resolution principle of FLCOM is discussed. Moreover, completeness theorem of the resolution method is proved.

Keywords: the fuzzy propositional logic system FLCOM, semantic interpretation, λ-resolution principle, completeness.

1 Introduction

Resolution principle is known to be the important foundation of automated theorem proving. Since the resolution principle of classical logic has been proposed by Robinson J. A. [1] in 1965, researchers try to introduce the resolution method into the fuzzy logic. In 1967, Slagle J. R. [2] put forward the principle of semantic resolution and proved its completeness. Then, people started to research the resolution method of fuzzy logic. And scholars have paid much attention to the research of the resolution method based on non-classical logic. In 1971, Lee and Chang [3] established Fuzzy Logic that values in [0, 1] and introduced a resolution method into it. In 1976, Morgan C. G. [4] made the earliest work referring resolution-based automated in many-valued logic. In 1980, the Fuzzy logic that values in lattice was established by Liu Xu-Hua et al. [5]. They also did research on the resolution method of this logic. Liu [6] proposed Operator Fuzzy Logic System (OFL) in 1985. In 1985, Yager R. R. [7] established resolution-based automated reasoning for a many-valued propositional logic system which was proposed by him independently. In 1989, Liu Xu-Hua [8], [9] put forward λ-resolution principle in the operator fuzzy logic. Based on the standard semantic interpretation of the Medium Logic, traditional resolution method was introduced into Medium Predicate Calculus System MF by Qiu Wei-De et al. [10] in 1990. From 2000 to 2001, Xu et al. [11], [12] established the α-resolution principle in lattice-valued propositional logic $LP(X)$ and lattice-valued first-order logic $LF(X)$, and applied them to uncertainty reasoning and automated reasoning. The corresponding reliability and completeness theorems were also proved by them. In 2003, Pan Zheng-Hua [13] proposed an infinite-valued semantic interpretation of Medium Predicate Logic System

D.-S. Huang et al. (Eds.): ICIC 2014, LNAI 8589, pp. 212–219, 2014.
© Springer International Publishing Switzerland 2014

MF. And he provided a λ-resolution method of MF. Then, Zhang ShengLi [14] provided an improved form of semantic interpretation and λ-resolution method of MF in 2012. But, these resolution principles do not established in the fuzzy logic with three different negations.

FLCOM is a fuzzy propositional logic system with contradictory negation, opposite negation and medium negation (three different negations), which was proposed by Pan Zheng-Hua [15] in 2013. Moreover, paper [15] gave an infinite-valued semantic interpretation of FLCOM and under this semantic interpretation the completeness of FLCOM is proved. Based on this infinite-valued semantic interpretation, in this paper, a λ-resolution method in FLCOM is discussed and its completeness theorems is verified. All the work in this paper can provide a new tool for automated reasoning.

In this paper, the symbols of form are quoted from FLCOM [15].

2   Basis

Paper [15] has defined a fuzzy propositional logic system with contradictory negation, opposite negation and medium negation (FLCOM). And FLCOM is logic foundation of the contradictory negation, the opposite negation and the medium negation in fuzzy knowledge. The definition of FLCOM is:

Definition 1 [15]. Let S be a set of fuzzy atom propositions, '$\neg$', '$\dashv$ ', '$\sim$', '$\rightarrow$', '$\wedge$' and '$\vee$' be conjunctions of fuzzy proposition logic. And let $\neg$, $\dashv$ and $\sim$ be formal symbols of the contradictory negation, the opposite negation and the medium negation respectively.

(I)   For any fuzzy atom propositions a_1, a_2, ..., $a_n \in S$, connecting them into a formula by $\neg$, $\dashv$, $\sim$, $\rightarrow$, $\wedge$ and $\vee$, such formula is called fuzzy well-formed formula (formula for short). $\mathfrak{I}$ denotes a set of all the fuzzy well-formed formulas in FLCOM.

(II)   For any $A, B \in \mathfrak{I}$, the following formulas are axioms:

(A1) $A \rightarrow (B \rightarrow A)$; (A2) $(A \rightarrow (A \rightarrow B)) \rightarrow (A \rightarrow B)$; (A3) $(A \rightarrow B) \rightarrow ((B \rightarrow C) \rightarrow (A \rightarrow C))$; (M1) $(A \rightarrow \neg B) \rightarrow (B \rightarrow \neg A)$; (M2) $(A \rightarrow \dashv B) \rightarrow (B \rightarrow \dashv A)$; (H) $\neg A \rightarrow (A \rightarrow B)$; (C) $((A \rightarrow \neg A) \rightarrow B) \rightarrow ((A \rightarrow B) \rightarrow B)$; ($\vee$1) $A \rightarrow A \vee B$; ($\vee$2) $B \rightarrow A \vee B$; ($\wedge$1) $A \wedge B \rightarrow A$; ($\wedge$2) $A \wedge B \rightarrow B$; (Y$\dashv$) $\dashv A \rightarrow \neg A \wedge \neg \sim A$; (Y$\sim$) $\sim A \rightarrow \neg A \wedge \neg \dashv A$.

(III)   The deduction rule MP (modus ponens): If $A \rightarrow B$ and A then B.

(I), (II) and (III) compose a formal system, which is called *Fuzzy propositional Logic System with Contradictory negation, Opposite negation and Medium negation*, FLCOM for short.

Definition 2 [15]. In FLCOM, the relation between contradictory negation $\neg$, opposite negation $\dashv$ and medium negation $\sim$ is

$$\neg A = \dashv A \vee \sim A_\circ \tag{1}$$

Definition 3 [15] (**λ-assignment**). Let A be a formula in FLCOM, $\lambda \in (0, 1)$. The mapping $\partial: \mathfrak{I} \rightarrow [0, 1]$ is called λ-assignment of $\mathfrak{I}$ if

(a) $\partial(A) + \partial(\dashv A) = 1,$ (2)

$$
\partial(\sim A) = \begin{cases}
\lambda - \dfrac{2\lambda - 1}{1 - \lambda}(\partial(A) - \lambda), & \lambda \ge \tfrac{1}{2} \text{ and } \partial(A) \in (\lambda,\ 1] & (3)\\[2ex]
\lambda - \dfrac{2\lambda - 1}{1 - \lambda}\partial(A), & \lambda \ge \tfrac{1}{2} \text{ and } \partial(A) \in [0,\ 1-\lambda) & (4)\\[2ex]
1 - \dfrac{1 - 2\lambda}{\lambda}(\partial(A) + \lambda - 1) - \lambda, & \lambda \le \tfrac{1}{2} \text{ and } \partial(A) \in (1-\lambda,\ 1] & (5)\\[2ex]
1 - \dfrac{1 - 2\lambda}{\lambda}\partial(A) - \lambda, & \lambda \le \tfrac{1}{2} \text{ and } \partial(A) \in [0,\ \lambda) & (6)\\[2ex]
\partial(A), & \text{otherwise} & (7)
\end{cases}
$$

(c) $\partial(A \vee B) = \max(\partial(A),\ \partial(B)),\ \ \partial(A \wedge B) = \min(\partial(A),\ \partial(B)),$ \hfill (8)

(d) $\partial(A \to B) = \Re(\partial(A),\ \partial(B)),$ \hfill (9)

where $\Re: [0, 1]^2 \to [0, 1]$ is a two binary function.

In the following, $A \in \Im$ denotes that A is a fuzzy propositional formula in FLCOM.

Definition 4. Let $A \in \Im$, $\lambda \in (0.5, 1)$. A is called λ-satisfiable if there is a λ-interpretation ∂ of $\Im$ such that $\partial(A) > \lambda$. A is called λ-unsatisfiable if $\partial(A) \le \lambda$ for any λ-interpretation ∂ of $\Im$.

Theorem 1. Let $A \in \Im$, $\lambda \in (0.5, 1)$. Then the following conclusions hold:

(i) $\partial(A) > \partial(\sim A) > \partial(\daleth A)$ if and only if $\partial(A) \in (\lambda, 1]$;

(ii) $\partial(\daleth A) > \partial(\sim A) > \partial(A)$ if and only if $\partial(A) \in [0, 1-\lambda)$.

Theorem 2. Let $A \in \Im$, $\lambda \in (0.5, 1)$. Then the following conclusions hold:

(a) $\partial(A) > \lambda$ if and only if $1 - \lambda \le \partial(\sim A) < \lambda$;

(b) $\partial(A) < 1 - \lambda$ if and only if $1 - \lambda < \partial(\sim A) \le \lambda$;

Proof. (a) According to the definition 3, if $\lambda > 0.5$, and $\partial(A) > \lambda$, then $\partial(\sim A) = \lambda - \dfrac{2\lambda - 1}{1 - \lambda}(\partial(A) - \lambda)$. Since $\lambda - \dfrac{2\lambda - 1}{1 - \lambda}(1 - \lambda) \le \lambda - \dfrac{2\lambda - 1}{1 - \lambda}(\partial(A) - \lambda) < \lambda - \dfrac{2\lambda - 1}{1 - \lambda}(\lambda - \lambda)$, that is

$1 - \lambda \le \lambda - \dfrac{2\lambda - 1}{1 - \lambda}(\partial(A) - \lambda) < \lambda$, then $1 - \lambda \le \partial(\sim A) < \lambda$. Conversely, when $1 - \lambda \le \partial(\sim A) <$

λ, since $\lambda > 0.5$ and $\partial(\sim A) = \lambda - \dfrac{2\lambda - 1}{1 - \lambda}(\partial(A) - \lambda) \in [1-\lambda, \lambda)$, then the result $\partial(A) > \lambda$ can

be obtained. So, $\partial(A) > \lambda$ if and only if $1 - \lambda \le \partial(\sim A) < \lambda$.

(b) According to the definition 3, we also can prove that $\partial(A) < 1 - \lambda$ if and only if $1 - \lambda < \partial(\sim A) \le \lambda$. Due to the limitation of space, we will not give the proofs here.

In [16], the relations among a fuzzy set A and its contradictory negation $A^\neg$, opposite negation $A^\daleth$ and medium negation $A^\sim$ are as follows: the relation between A and $A^\neg$ is

CFC [16], the relation between A and $A^{\daleth}$ is OFC [16], the relation between A and $A^{\sim}$ is MFC [16], and the relation between $A^{\daleth}$ and $A^{\sim}$ is MFC [16]. Corresponding to FLCOM, A and $\neg A$, A and $\daleth A$, A and $\sim A$ or $\daleth A$ and $\sim A$ all have the relation of negation.

The definition 2 demonstrates that $\neg A = \daleth A \vee \sim A$. Moreover, there is a relationship of negation between $\neg A$ and A. So, essentially the relation between $\neg A$ and A is equivalent of the relation between $\daleth A$ and A or $\sim A$ and A. Therefore, following definition can be provided:

Definition 5. In FLCOM, let $\{A, \daleth A, \sim A\}$ be a set of fuzzy atom formulas. Every two formulas in this set have the relation of negation, and one of them is called the negation of another one.

Theorem 3. The conjunction of every two formulas in $\{A, \daleth A, \sim A\}$ is λ-unsatisfiable.

Proof (Reduction to Absurdity). Suppose that there is a λ-interpretation ∂ such that $\partial(A \wedge \daleth A) > \lambda$, $\partial(A \wedge \sim A) > \lambda$ or $\partial(\daleth A \wedge \sim A) > \lambda$ when $\lambda \in (0.5, 1)$.

If $\partial(A \wedge \daleth A) > \lambda$, then $\partial(A) > \lambda$ and $\partial(\daleth A) > \lambda$. From the definition 3, we know that if $\partial(A) > \lambda$, there must has $\partial(\daleth A) < 1-\lambda \leq \lambda$, which is contrary to the conclusion that $\partial(\daleth A) > \lambda$ in the hypothesis. Similarly, we can prove that $\partial(A \wedge \sim A) > \lambda$ and $\partial(\daleth A \wedge \sim A) > \lambda$ don't hold.

So, the assumption is invalid and the conclusion in this theorem is true.

3   The Resolution Principle for FLCOM

Definition 6. A fuzzy propositional variable and its different negations are collectively called literals.

In fuzzy logic, we know any fuzzy propositional formula exits a equivalent conjunctive normal form (*CNF*). In FLCOM, the conjunctive normal form of any formula can be defined as follows:

Definition 7. Let $A \in \mathfrak{S}$, then the conjunctive normal form of A can be expressed as $A = (A_1^1 \vee A_2^1 \vee \ldots \vee A_r^1) \wedge \ldots \wedge (A_1^n \vee A_2^n \vee \ldots \vee A_r^n)$, where A_i^j are literals.

Lemma 4. $\forall A \in \mathfrak{S}$, A is λ-satisfiable *iff* its conjunctive normal form is λ-satisfiable.

Definition 8. In FLCOM, the conjunction of finite literals is called clause and is denoted by H. And a clause don't contain any literal is called empty clause, $\square$ in short. Empty clause is λ-unsatisfiable.

Definition 9. In FLCOM, let $H = C_1 \wedge C_2 \wedge \ldots \wedge C_n$, where C_i $(i = 1, 2, \ldots, n)$ are clauses. H is called clause set, if the logic symbol '$\wedge$' in H is rewritten as the logical symbol ',' and it has the form $H = \{C_1, C_2, \ldots, C_n\}$.

Because the empty clause is λ-unsatisfiable, so a clause set contains empty clause is λ-unsatisfiable.

Theorem 5. Let $A \in \mathfrak{I}$. A is λ-unsatisfiable *iff* its corresponding clause set H is λ-unsatisfiable.

Definition 10. Let p be a fuzzy propositional variable and let two different literals l, $l^c \in \{p, \neg p, \sim p\}$. We say there is a relation of negation between l and l^c. Moreover, l^c is called the negation of l.

Definition 11. Let C_1 and C_2 be two clauses, $C_1, C_2 \in H$, l_1 and l_2 be two literals of C_1 and C_2, respectively. If there is a MGU (most general unifier) σ, such that l_1^σ and l_2^σ have the relation of negation, then the clause $(C_1^\sigma - l_1^\sigma) \vee (C_2^\sigma - l_2^\sigma)$ is a dualistic λ-resolvent of C_1 and C_2, and is donated by $R(C_1, C_2)$. We call (l_1, l_2) a pair λ-resolution literals.

Theorem 6. Let C_1 and C_2 be two clauses. If $C_1 \wedge C_2$ is λ-satisfiable then $R(C_1, C_2)$ is λ-satisfiable, that is $C_1 \wedge C_2 \Rightarrow R(C_1, C_2)$.

Proof.   Let $C = R(C_1, C_2)$, and (l, l^c) be a pair of λ-resolution literals. We can assume that $C_1 = C_1' \vee l$ and $C_2 = C_2' \vee l^c$.

If $C_1 \wedge C_2$ is λ-satisfiable then there is a λ-assignment ∂ such that $\partial(C_1 \wedge C_2) > \lambda$, that is $\partial(C_1) = \partial(C_1' \vee l) > \lambda$ and $\partial(C_2) = \partial(C_2' \vee l^c) > \lambda$. Since $\partial(l) > \lambda$ and $\partial(l^c) > \lambda$ can't hold simultaneously, there must have $\partial(C_1') > \lambda$ or $\partial(C_2') > \lambda$. Then, there is a literal l' in C_1' or C_2' and satisfy $\partial(l') > \lambda$. Since C contains l', then $\partial(C) > \lambda$. Therefore, $R(C_1, C_2)$ (or C) is λ-satisfiable.

Definition 12 (λ-resolution deduction). Let H be a clause set in FLCOM. The λ-resolution deduction from H to the clause C is the following sequence with finite clauses: $C_1, C_2, \ldots, C_n$, where C_i $(i = 1, 2, \ldots, n)$ are clauses in H, or $C_i = R(C_j, C_k)$ $(j < i, k < i)$. In addition, let $C_n = C$. And $\{C_1, C_2, \ldots, C_n\}$ is also called λ-resolution sequence.

Theorem 7. Let H be a clause set in FLCOM, $\lambda \in (0.5, 1)$. If there is a λ-resolution deduction from H to $\square$, then H is λ-unsatisfiable.

Proof (Reduction to Absurdity). Suppose that H is λ-satisfiable, then there is a λ-interpretation ∂ such that $\partial(H) > \lambda$. That is for any clause $C \in H$, there exists $\partial(C) > \lambda$. For this and the theorem 6, it can be concluded that if there is a λ-resolution deduction from H to $\square$, then $\partial(\square) > \lambda$. However, $\square$ is λ-unsatisfiable. So the above assumption doesn't hold and H is λ-unsatisfiable.

Theorem 8. Let $H = \{C_1, C_2, \ldots, C_n\}$ be a clause set. H' is a clause set derived from removing all clauses contain l in H, and then removing l^c from the rest of clauses in H. If H is λ-unsatisfiable then H' is λ-unsatisfiable.

Proof. Let $H_1 = \{C_1, C_2, \ldots, C_i\}$, $C_p \in H$ and C_p contain l $(p = 1, 2, \ldots, i, i \leq n)$, $H_2 = \{C_{i+1}, C_{i+2}, \ldots, C_j\}$, $C_q \in H/H_1$ and C_q contain l^c $(q = i+1, i+2, \ldots, j, j \leq n)$, $H_3 = \{C_{j+1}, C_{j+2}, \ldots, C_n\}$, $C_r \in H$ and C_r don't contain l and l^c, $(r = j+1, j+2, \ldots, n)$, where H/H_1 is a clause set derived from removing H_1 from H. Let $H_2' = \{C'_{i+1}, C'_{i+2}, \ldots, C_j'\}$, where $C_k = C_k' \vee l^c$ $(k = i+1, i+2, \ldots, j; j \leq n)$.

In case H_3 is non-empty, then $H' = \{C'_{i+1}, C'_{i+2}, \ldots, C_j', C_{j+1}, C_{j+2}, \ldots C_n\}$, where $C_k = C_k' \vee l^c$ $(k = i+1, i+2, \ldots, j)$. Suppose that H is λ-unsatisfiable, then $\partial(H) \leq \lambda$ for any λ-interpretation ∂, that is for any clause $C \in H$, $\partial(C) \leq \lambda$ and $\partial(C_{j+1} \wedge C_{j+2} \wedge \ldots \wedge C_n) \leq \lambda$. Then, $\partial(C'_{i+1} \wedge \ldots \wedge C_j' \wedge C_{j+1} \wedge \ldots \wedge C_n) \leq \lambda$. Therefore, H' is λ-unsatisfiable.

In case H_3 is empty, then $H' = \{C'_{i+1}, C'_{i+2}, \ldots, C_j'\}$, where $C_k = C_k' \vee l^c$ $(k = i+1, i+2, \ldots, j; j = n)$. Suppose that H is λ-unsatisfiable. Then for any λ-interpretation ∂, $\partial(H) \leq \lambda$. So $\partial(C) \leq \lambda$ for all $C \in H$. So, there exists $\partial(C_k) = \partial(C_k' \vee l^c) \leq \lambda$, where $k = i+1, i+2, \ldots, j, j = n$. Then, $\partial(C_k') \leq \lambda$ $(k = i+1, i+2, \ldots, j; j = n)$. So, for any λ-interpretation ∂, $\partial(H') \leq \lambda$, that is H' is λ-unsatisfiable.

Theorem 9. Let H be a clause set in FLCOM, $\lambda \in (0.5, 1)$. If H is λ-unsatisfiable, then there is a λ-resolution deduction from H to $\square$.

Proof. Suppose that there are k fuzzy propositional variables in H. We adopt the inductive method to prove it.

There is only one fuzzy propositional variable p in H when $k = 1$. It is obvious that if there is $\square$ in H, then this theorem holds. But if $\square$ is not in H, then H contains p and $\neg p$, or p and $\sim p$, or $\neg p$ and $\sim p$ simultaneously. For $R(p, \neg p) = \square$, $R(p, \sim p) = \square$ and $R(\neg p, \sim p) = \square$, according to the definition 12, there is a λ-resolution deduction from H to $\square$.

Suppose that this theorem holds when $k < n$ $(n \geq 2)$. Next, this theorem can be proved when $k = n$.

Taking one of the fuzzy propositional variables from H arbitrarily and denoting it by p. It is known that p, $\neg p$ and $\sim p$ are literals. Let $l \in \{p, \neg p, \sim p\}$. And there has at least one clause contains l in H. Let l^c be the negation of l. H_1 denotes a set of clauses contain l in H. H_2 denotes a set of clauses contain l^c in H. H_3 denotes a set of clauses that are from H and don't contain l and l^c. Let H' be a clause set which is derived from removing all clauses contain l in H, and removing l^c from the rest of clauses. Let H'' be a clause set which is derived from removing all clauses contain l^c from H, and removing l from the rest of clauses. According to the theorem 8, if $H \wedge l$ is λ-unsatisfiable then H' is λ-unsatisfiable. And if $H \wedge l^c$ is λ-unsatisfiable then H'' is λ-unsatisfiable. Since H is λ-unsatisfiable, then $H \wedge l^c$ and $H \wedge l^c$ are all λ-unsatisfiable. So, H' and H'' are also λ-unsatisfiable. For the amounts of fuzzy proposition variables in H' and H'' are all less than n and according to the inductive assumption, there is a λ-resolution sequence of the deduction from H' to $\square$, that is $C_1, C_2, \ldots, C_i$ and there is a λ-resolution sequence of the deduction from H'' to $\square$, that is $D_1, D_2, \ldots, D_j$. Moreover, $C_i = \square$ and $D_j = \square$. If C_t $(1 \leq t \leq i)$ is a λ-resolvent of the clauses only from H_3, then C_t is unrelated to H_2', otherwise it has relation with H_2'. Similarly, the relation between D_t and H_1' can be defined. Next, we discuss two conditions as follows:

(1) If C_i is unrelated to H_2', or D_j is unrelated to H_1', then $\square$ can be derived from λ-resolution of clauses in H_3. So $\{C_1, C_2, \ldots, C_i\}$ is a λ-resolution sequence of the deduction from H to $\square$ when C_i is independent of H_2' or $D_1, D_2, \ldots, D_j$ is a λ-resolution sequence of the deduction from H to $\square$ when D_j is independent of H_1'.

(2) If C_i is relate to H_2', and D_j is relate to H_1', then for any $1 \le t \le i$, it can be defined as

$$C_t' = \begin{cases} C_t \vee l^c, & \text{when } C_t \text{ is relate to } H_2' \\ C_t, & \text{when } C_t \text{ is independent of } H_2'. \end{cases}$$

And for any $1 \le t \le j$, we define

$$D_t' = \begin{cases} D_t \vee l, & \text{when } D_t \text{ is relate to } H_1' \\ D_t, & \text{when } D_t \text{ is independent of } H_1'. \end{cases}$$

It can be easily concluded that $\{C_1', C_2', \ldots, C_i'\}$ and $\{D_1', D_2', \ldots, D_j'\}$ are all λ-resolution sequences of H. For $C_i' = l^c$, $D_j' = l$ and $R(C_i', D_j') = \square$, so $\{C_1', C_2', \ldots, C_i', D_1', D_2', \ldots, D_j'\}$ is a λ-resolution sequences of the λ-resolution deduction from H to $\square$. Therefore, this theorem holds when $k = n$.

According to the inductive method, this theorem is tenable.

Combing the theorem 7 and theorem 9, we can get the following conclusion:

Theorem 10 (Completeness). Let H be a clause set in FLCOM, $\lambda \in (0.5, 1)$. H is λ-unsatisfiable if and only if there is a λ-resolution deduction from H to $\square$.

4   Conclusion

(1) In a λ-interpretation of FLCOM [15], the fuzzy proposition and its different negations value in [0, 1], that shows the flexible procedure among fuzzy propositions with contradictory negation, opposite negation and medium negation. From the definition 4, we implicate a fuzzy formula λ-satiable when its true value is higher than λ and implicate it λ-unsatisfiable when its true value is less than or equal to λ. It reflects transit from true and false when the proposition values in the interval $(1-\lambda, \lambda)$.

(2) This paper introduces λ-resolution method in fuzzy propositional logic system FLCOM, defines λ-resolution deduction and proves completeness theorem of λ-resolution.

Acknowledgements. This work was supported by the National Natural Science Foundation of China (60973156, 61375004) and the Fundamental Research Funds for the Central Universities (JUSRP51317B).

References

1. Robinson, J.A.: A Machine-oriented Logic based on the Resolution Principle. J. Journal of the ACM 12(1), 46–54 (1965)
2. Slagle, J.R.: Automatic Theorem Proving with Renamable and Semantic Resolution. J. ACM 14, 687–697 (1967)
3. Chang, C.L., Lee, R.C.T.: Symbolic Logic and Mechanical Theorem Proving. Academic Press, New York (1973)
4. Morgan, C.G.: Resolution for many-valued logics. J. Logique et Analyse 19(74-76), 311–339 (1976)
5. Liu, X.H.: Generalized Fuzzy Logic and Lock Semantic Resolution Principle. J. Chinese Journal of Computers 2, 97–111 (1980)
6. Liu, X.H., Xiao, H.: Operator Fuzzy Logic and Fuzzy Resolution. In: Proc.of the 15th ISMVl, Kingston, Canada, vol. 5 (1985)
7. Yager, R.R.: Inference in a Multiple-Valued System. Int. J. Man-Machine Stud. 23, 27–34 (1985)
8. Liu, X.H., Xiao, H.: Operator Fuzzy Logic and l-Resolution. J. Chinese Journal of Computers 2, 81–91 (1989) (in Chinese)
9. Liu, X.H., An, Z.: an Improvement of Operator Fuzzy Logic and its Resolution Deduction. J. Chinese Journal of Computers 12, 890–899 (1990) (in Chinese)
10. Qiu, W.D., Zou, J.: Resolution Principle of the Medium Predicate Calculus System. J. Journal of Shanghai Polytechnic University 11(2), 156–161 (1990)
11. Xu, Y., Ruan, D., Kerre, E.E., Liu, J.: a-Resolution principle based on lattice-valued propositional logic LP(X). Int. J. Information Science 130, 195–223 (2000)
12. Xu, Y., Ruan, D., Kerre, E.E., Liu, J.: a-Resolution principle based on first-order lattice-valued logic LF(X). Int. J. Information Science 132, 221–239 (2001)
13. Pan, Z.H.: l-resolution of the Medium Predicate Logic System. J. Journal of Software 14(3), 345–349 (2004) (in Chinese)
14. Zhang, S.L., Pan, Z.H.: λ-resolution for medium predicate logic based on improved form of infinite-valued semantic interpretation. J. Journal of Shandong University (Natural Science) 02, 109–114+118 (2012) (in Chinese)
15. Pan, Z.: Three Kinds of Negation of Fuzzy Knowledge and Their Base of Logic. In: Huang, D.-S., Jo, K.-H., Zhou, Y.-Q., Han, K. (eds.) ICIC 2013. LNCS (LNAI), vol. 7996, pp. 83–93. Springer, Heidelberg (2013)
16. Pan, Z.H.: Three kinds of negation of fuzzy knowledge and their base of set. J. Chinese Journal of Computers 35(7), 1421–1428 (2012) (in Chinese)

Multi-level Linguistic Fuzzy Decision Network Hierarchical Structure Model for MCDM

Basem Mohamed Elomda[1], Hesham Ahmed Hefny[2], Maryam Hazman[1],
and Hesham Ahmed Hassan[3]

[1] Central Laboratory for Agriculture Expert Systems (CLAES),
Agriculture Research Center (ARC), Cairo, Egypt
{basem,m.hazman}@claes.sci.eg
[2] Institute of Statistical Studies and Research (ISSR), Cairo University, Cairo, Egypt
hehefny@cu.edu.eg
[3] Faculty of Computers and Information (FCI), Cairo University, Cairo, Egypt
h.hassan@fci-cu.edu.eg

Abstract. Linguistic Fuzzy Decision network (LFDN) method is an extension of Fuzzy Decision Map (FDM) for solving Multi-Criteria Decision Making problems (MCDM) in fuzzy environment having dependence and feedback among criteria. On the other hand, LFDN can't handle the complex decision making problem, particularly with the multi-level hierarchical structure model that consists of objectives, criteria, sub-criteria, etc. down to the bottom level (alternatives). The main objective of this paper is to develop the LFDN structure to be able to select a decision for multi-level structure problems. So the multi-level structure of LFDN is the general form of LFDN. Therefore, it can use for ranking alternatives and selecting the best one when the decision maker has multiple criteria. A case study was carried out to demonstrate the proposed model.

Keywords: Multi Criteria Decision Making, Soft Computing, Fuzzy Decision Map, Fuzzy Cognitive Map, Linguistic Fuzzy Decision Network.

1 Introduction

Multi-Criteria Decision-Making (MCDM) approaches have been developed during the mid-1960s; it has been an active area of research in decision theory, operations research, management science and system engineering [1]. MCDM is the study of methods and procedures by which concerns about multiple conflicting criteria can be formally incorporated into the management planning process and the best possible solution one can be identified from a set of alternatives, as defined by the International Society on MCDM [2]. Many different methods have been developed to solve MCDM problems. The Analytic Hierarchy Process (AHP) was first introduced by Thomas Saaty in the 1971 [3]. The Analytic Network Process (ANP) was developed by Saaty in 1996 to overcome the problem of dependence and feedback among

D.-S. Huang et al. (Eds.): ICIC 2014, LNAI 8589, pp. 220–229, 2014.
© Springer International Publishing Switzerland 2014

criteria in AHP method [4]. Fuzzy Decision Maps (FDM) method was proposed in 2006 [5] for handling MCDM problems with dependency and feedback to overcome the complexity drawback of ANP. In 2013, Linguistic Fuzzy Decision Networks (LFDN) is proposed to overcome the shortcoming of FDM [6, 7]. LFDN is proposed in order to facilitate the making of subjective assessment by the Decision Makers (DM) using the linguistic values in the form of triangular fuzzy numbers (TFN) [6, 7]. However, LFDN deals only with one level of criteria having dependence and feedback. Therefore the main purpose of this paper is to improve LFDN to address the need for considering the case of multi-level criteria before considering alternative. The enhanced LFDN method should take in consideration the multi-level hierarchical structure of objectives, criteria, sub-criteria and alternatives.

This paper is organized as follows: Section 2 presents Fuzzy Cognitive Map (FCM). Section 3 presents the LFDN method. The developed LFDN is explained in section 4. Case study is presented in section 5. Finally, the Conclusion is given in section 6.

2   Fuzzy Cognitive Map (FCM)

Fuzzy Cognitive Map (FCM) was originally introduced by Kosko in 1986 [8], as an extension of cognitive map model proposed by Axelrod in 1976 [9].   FCM in fuzzy environment using the concept of linguistic values is given in [6, 7]. It is applied in Linguistic Fuzzy Decision Network method (LFDN) to derive the fuzzy influence weight matrix among criteria, see [6, 7].   An example of a linguistic FCM is shown in Fig. 1. In Fig. 1, there are five concepts (or criteria) given by C1, C2, C3, C4, and C5. These are connected by weights. The relationships between two concepts Ci and Cj are described using a degree of influence i.e., Wij. Experts describe these degrees of influence using linguistic values by TFN as in table 1. After drawing the FCM using linguistic values to indicate the influence among criteria, fuzzy cognitive maps can be transformed into adjacency matrices.

Table 1. Linguistic values and corresponding fuzzy numbers for causal relationships [6]

linguistic values	TFN
No Influence (NI)	(0,0,0)
Extremely Low (EL)	(0,0.1,0.2)
Very Low (VL)	(0.1,0.2,0.3)
Low (L)	(0.2,0.3,0.4)
Medium Low (ML)	(0.3,0.4,0.5)
Medium (M)	(0.4,0.5,0.6)
Medium High (MH)	(0.5,0.6,0.7)
High (H)	(0.6,0.7,0.8)
Very High (VH)	(0.7,0.8,0.9)
Extremely High (EH)	(0.8,0.9,1)

The fuzzy influence relationship among criteria can be calculated using the following fuzzy updating equation [6, 7]:

$$\tilde{C}^{(t+1)} = F \left(\tilde{C}^{(t)} \cdot \tilde{W} \right), \quad \tilde{C}^{(0)} = \tilde{I}_{nxn} \tag{1}$$

Where $\tilde{W} = [W_{ij}]$ is $n \times n$ fuzzy weight matrix with TFNs, $\tilde{C}^{(t+1)}$ is the fuzzy state matrix at certain iteration (t+1), $\tilde{C}^{(0)}$ is the fuzzy identity matrix, $\tilde{C}^{(t)}$ is the fuzzy state matrix at certain iteration t, and F is a fuzzy transformation function with TFN. The fuzzy hyperbolic-tangent function was used as the fuzzy transformation function as in Eq. (2) [6-7], [10].

$$f(\tilde{x}) = \tanh(\tilde{x}) = \left(1 - e^{-x_l}/1 + e^{-x_l}, 1 - e^{-x_m}/1 + e^{-x_m}, 1 - e^{-x_u}/1 + e^{-x_u} \right), \quad \tilde{x} = (x_l, x_m, x_u) \tag{2}$$

Once the linguistic FCM has been created, it can be run directly using Eq. (1). In each step of cycling, the values of the concepts change according to Eq. (1). After very little iteration, it will reach to fuzzy steady state,) if one of the following situations occurs [11]:

- A fuzzy fixed point attractor: This case is arrived when the FCM state vector keeps fixed for successive.
- A fuzzy limit cycle: A series of FCM state vector remains repeating forming a cycle.

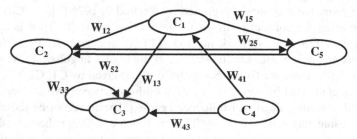

Fig. 1. An FCM as a graph

3   LFDN Method

It is an extension of linguistic FDM with fuzzy set theory to handle the uncertainty situations for solving MCDM. Now, we can summarize the steps of LFDN method to derive the priorities of criteria as follows [6, 7]:

Step 1: Derive the local fuzzy weight vector ($\tilde{L}$). In this step, perform the fuzzy preference matrix (or the fuzzy pair-wise comparison matrix) as in Eq. (3), and then apply the fuzzy eignvalue method to obtain the local fuzzy weight as given in Eq. (4). Finally, normalize the obtained local fuzzy weight vector.

$$\tilde{P} = \begin{bmatrix} \tilde{1} & \tilde{a}_{12} & \cdots & \tilde{a}_{1n} \\ 1/\tilde{a}_{12} & \tilde{1} & \cdots & \tilde{a}_{2n} \\ \vdots & \vdots & \ddots & \vdots \\ 1/\tilde{a}_{1n} & 1/\tilde{a}_{2n} & \cdots & \tilde{1} \end{bmatrix} \tag{3}$$

Where $\tilde{a}_{ij}=(l_{ij}, m_{ij}, u_{ij})$, $\tilde{a}_{ij}$-1 $=(1/u_{ij}, 1/m_{ij}, 1/l_{ij})$ for (i,j=1,.....,n and i ≠ j), $\tilde{1}$ = (1,1,1), $\tilde{P}$ represents n × n matrix with TFN and a_{ij} is the importance of criterion C_i w.r.t. criterion C_j according to the fuzzy importance scale for pair-wise comparison is shown in table 2.

$$\tilde{L}_i = \sum_{j=1}^{n} \tilde{a}_{ij} = \left(\sum_{j=1}^{n} l_{ij}, \sum_{j=1}^{n} m_{ij}, \sum_{j=1}^{n} u_{ij} \right), \quad i = 1,....,n \tag{4}$$

Table 2. Fuzzy preference scale for pair-wise comparison [6,7]

Intensity of fuzzy scale	Definition of linguistic values	Triangular fuzzy scale	Triangular fuzzy reciprocal scale
$\tilde{1}$	Equal importance (EI)	(1,1,1)	(1 , 1 , 1)
$\tilde{2}$	Equally Moderate importance(EMI)	(1, 2, 3)	(1/3, 1/2 ,1)
$\tilde{3}$	Moderate importance (MI)	(1,3,5)	(1/5, 1/3 ,1)
$\tilde{4}$	Moderately Strong importance(MSI)	(2, 4, 6)	(1/6, 1/4 ,1/2)
$\tilde{5}$	Strong importance (SI)	(3,5,7)	(1/7, 1/5,1/3)
$\tilde{6}$	Strongly Very Strong importance(SVSI)	(4, 6, 8)	(1/8, 1/6 ,1/4)
$\tilde{7}$	Very Strong importance (VSI)	(5,7,9)	(1/9,1/ 7,1/5)
$\tilde{8}$	Very Strongly Extreme importance(VSEI)	(6, 8, 9)	(1/9, 1/8 ,1/6)
$\tilde{9}$	Extreme importance (EXI)	(7,9,9)	(1/9,1/ 9,1/7)

Step 2: Derive the fuzzy influence weight matrix. In this step, draw the FCM with linguistic values, then transform FCM into adjacency matrix. The adjacency matrix with linguistic values changed into TFN using table 1. Therefore, run FCM model using the fuzzy updating Eq. (1) to determine the fuzzy steady-state matrix ($\tilde{C}^*$). Finally, normalize the obtained fuzzy steady-state matrix.

Step 3: Derive the global fuzzy weight vector ($\tilde{G}$). In this step, the global fuzzy weight vector is calculate using the fuzzy weighting equation as in Eq. (5). Then, normalize the obtained global fuzzy weight vector.

$$\tilde{G} = \tilde{L}_n + \tilde{C}_n^* \tilde{L}_n \tag{5}$$

Where ($\tilde{L}_n$) is the normalization of the local fuzzy weight vector and ($\tilde{C}_n^*$) is the normalization of the fuzzy steady-state matrix.

Since the final result is a vector of TFNs, then we need to use Eq. (6) (i.e., Center of Area (CoA) defuzzification method) for ranking such obtained fuzzy numbers to determine the criterion with highest priority.

$$D(\tilde{A}) = (1 + m + u)/3 \tag{6}$$

4   Developing LFDN for Solving Complex Decision Problems

The LFDN is a decision support tool which can be used in solving MCDM with two-level hierarchical structure (i.e., objectives and criteria). However, for solving complex decision problem, we developed LFDN method. The developed LFDN model, based on the identified criteria, sub criteria and sub sub-criteria. The lowest level of the hierarchy contains of the alternatives. In the multi-level LFDN, the decision problem is structured hierarchically at different levels, each level consisting of a finite number of elements. Where, level 1 is the goal of the analysis, and level 2 is multi criteria that consist of several factors. You can also add several other levels of sub criteria and sub-sub criteria. In this paper, we add only one sub criteria level. The last level (level 4 in Fig. 2) is the alternative choices. The general form for multi-level hierarchical structure of LFDN can be depicted as in Fig .2. LFDN with multi-level hierarchical structure steps can be described as follows:

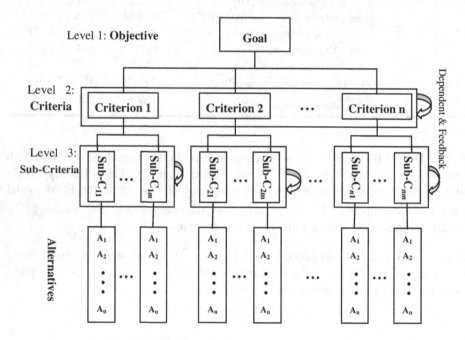

Fig. 2. LFDN with multi-level hierarchical structure

- **Step 1: Define the nature of the problem definition to be solved**
- **Step 2: Construct a hierarchy model of system elements for its evaluation**

This phase involves building the LFDN hierarchy model. An LFDN hierarchy is consists of an overall goal, a group of criteria (or factors) that relate to the goal, a group of alternatives (or options) that relate to the alternatives for reaching the goal. The criteria can be further broken down into sub-criteria, sub-sub criteria, and so on, in as many levels as the problem requires. After the hierarchy is built as Fig.2, calculating the weights among elements (criteria, sub-criteria, and alternative) at each level of hierarchy is as follows:

- **Step 3: Derive the Fuzzy Global Weight (FGW) for criteria in the second level**

  Calculating FGW regard to improve the LFDN process is done as the following:

  -The decision maker (DM) performs fuzzy pairwise comparisons matrix (FPCM) at level 1 of the hierarchy with respect to their relative importance with first level (Goal). FPCM in the LFDN assume that the DM can compare any two elements Ci, Cj at the same level of the hierarchy and provide a linguistic value aij for the ratio of their importance. The fuzzy preference scale used for LFDN is given as table 2. After build FPCM, we used the fuzzy eigenvalue method as in Eq. (4) to derive the fuzzy local weight vectors (FLW).

  -The DM draws FCM among criteria. DM used table 1 to indicate the influence degree among criteria. Then, apply Eq. (1) to get the influence among criteria. Therefore, derive the fuzzy global weight for all criteria using Eq. (5).

- **Step 4: Derive the fuzzy global weight for each group of sub-criteria in the third level**

  In this step, we can derive FGW for all sub-criteria as follows:

  -DM will be compared each group of the sub-criteria with regard to parent criteria in the upper level i.e., perform a set of FPCM, where each element in a criteria level is used to compare the elements immediately below with regard to it. This process will be continuous till all elements in level 3 are compared. Once all FPCM are created in level 3 i.e., FPCM among (Sub-C11,....., Sub-C1m) w.r.t criterion1, and until FPCM among (Sub-Cn1,....., Sub-Cnm) w.r.t criterion n, we used the fuzzy eigenvalue method as in Eq. (4) to derive the FLW for each group of sub- criteria.

  -DM will be drawn a FCM for each group of the sub-criteria i.e., draw a FCM among elements (Sub-C$_{11}$,....., Sub-C$_{1m}$), and until for (Sub-C$_{n1}$,....., Sub-C$_{nm}$). Then, apply Eq. (1) for each one of FCM to get the influence among sub-criteria. Therefore, derive the FGW for each group of the sub-criteria using Eq. (5).

- **Step 5: Derive the fuzzy weight for alternative**

  Calculating FLW for alternatives is done as the following:

  DM will be constructed FPCM among alternatives with regard to each sub-criteria in the upper level i.e., perform a set of FPCM among alternatives, where each element in a sub-criteria level is used to compare the elements (alternatives) immediately below with regard to it. For example, concerning the importance w.r.t elements (Sub-C11) we performed a n × n FPCM containing our comparison element (A1,..., An). This process will be continuous until all alternatives are compared w.r.t each sub-criteria. After build all FPCM among alternatives, we use the fuzzy eigenvalue method as in Eq. (4) to derive the FLW for the alternatives with respect to each sub-criteria.

- **Step 6: Derive the final fuzzy weight for alternative**

In the final step, the obtained FLW vectors for the alternatives with respect to each sub-criterion are given in matrix form as in Eq. (7), and the obtained final priority weights for sub-criteria are given as in Eq. (8). Then, the final fuzzy weight vector for alternative is obtained as in Eq. (9).

$$
\delta = \begin{array}{c} {\scriptstyle \text{Sub-}C_{11} \quad \cdots \quad \text{Sub-}C_{1m} \quad \text{Sub-}C_{21} \quad \cdots \quad \text{Sub-}C_{2m} \quad \cdots \quad \text{Sub-}C_{n1} \quad \cdots \quad \text{Sub-}C_{nm}} \\ \begin{array}{c} A_1 \\ A_2 \\ \vdots \\ An \end{array} \begin{bmatrix} \tilde{\omega}(A_1) & \cdots & \tilde{\omega}(A_1) & \tilde{\omega}(A_1) & \cdots & \tilde{\omega}(A_1) & \cdots & \tilde{\omega}(A_1) & \tilde{\omega}(A_1) \\ \tilde{\omega}(A_2) & \cdots & \tilde{\omega}(A_2) & \tilde{\omega}(A_2) & \cdots & \tilde{\omega}(A_2) & \cdots & \tilde{\omega}(A_2) & \tilde{\omega}(A_2) \\ \vdots & \cdots & \vdots & \vdots & \cdots & \vdots & \cdots & \vdots & \vdots \\ \tilde{\omega}(A_n) & \cdots & \tilde{\omega}(A_n) & \tilde{\omega}(A_n) & \cdots & \tilde{\omega}(A_n) & \cdots & \tilde{\omega}(A_n) & \tilde{\omega}(A_n) \end{bmatrix} \end{array}
\tag{7}
$$

$$
\beta = [\tilde{\omega}(C_{11}) \quad \cdots \quad \tilde{\omega}(C_{1m}) \quad \tilde{\omega}(C_{21}) \quad \cdots \quad \tilde{\omega}(C_{2m}) \quad \cdots \quad \tilde{\omega}(C_{n1}) \quad \cdots \quad \tilde{\omega}(C_{nm})]^T
\tag{8}
$$

$$
\varpi = \delta \cdot \beta
\tag{9}
$$

Since the final result (or the final fuzzy weight vector for alternative is a vector of TFNs representing the priority of each alternative, then using Eq.(6) for ranking such obtained fuzzy numbers to determine the alternative with highest priority.

5   Case Study

In this section, the developed LFDN model is implemented to choose the best car w.r.t. several criteria.   The following show how the case study was implemented.

- **Step1: the problem definition**

DM tries to purchase a car according to the following four criteria including Price (P), Durability (D), Robustness (R), and Repair Cost (C), for choosing the best alternative car among available cars namely: Passat, Corolla, Cerato, and Ménage.

- **Step2: Construct the hierarchy structure of the problem.**

Fig. 3 shows an illustrative 3-level hierarchy for the car selection problem. Level 1 includes the goal, level 2 contains criteria, and level 3 contain the alternatives.

- **Step3: Calculate the weight of criteria**

In this step, DM compares the importance among criteria using the fuzzy importance scale given in Table 2. Therefore, the FPCM with linguistic values is given in Table 3. The local fuzzy weights vector of each criterion is obtained using Eq. (4).

DM Draw the FCM with linguistic values to indicate the influence among criteria using Table 1. Therefore, The FCM for the considered case study is shown in Fig. 4. The adjacency matrix obtained from FCM with linguistic values ($\tilde{E}$) is shown in Table 4.

Run FCM model using the fuzzy updating Eq. (1) to obtain the fuzzy steady state-matrix ($\tilde{E}$) for the convergent state.

We can derive the global fuzzy weights ($\tilde{G}$) for criteria in level 1 using Eq. (5).

- **Step4: Compute the alternative evaluations**

In this step, DM compares the importance among alternatives with respect to the corresponding elements in the higher level (i.e., for each criterion in level 2), obtaining a FPCM. Therefore, the following tables 5, 6, 7 and 8 show FPCM among alternatives w.r.t price, durability, robustness and repair cost criteria respectively. Now we are going to get the fuzzy local weights vector for alternative with regard to parent criteria in the upper level using Eq. (4).

- **Step 5: Compute the final weight for alternative (i.e., final ranking)**

In this step, we need to obtain the final fuzzy weight vector (ϖ) for alternative as in Eq. (9). So, we can rank the fuzzy weights vector using COA method given in Eq. (6). Table 9 illustrates the final ranking of the obtained fuzzy weight vector for alternative. Thus, from tables 9 it is found that the selection criterion with highest priority is **Passat**, while the criterion with lowest priority is **Cerato**.

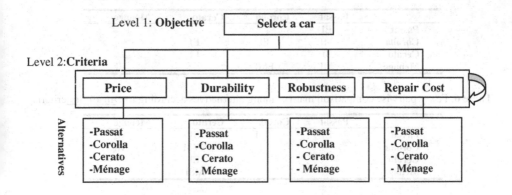

Fig. 3. LFDN hierarchical structure model for car selection problem

Table 3. Fuzzy pairwise comparison matrix among criteria with Lingusic values

	Price	Durability	Robustness	Repair Cost
Price	EI			
Durability	MI	EI		
Robustness	SI	MI	EI	EMI
Repair Cost	MI	EMI		EI

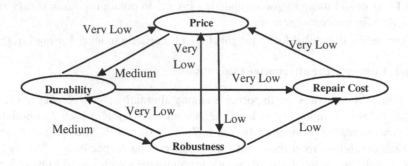

Fig. 4. FCM with linguistic values among criteria for car selection problem

Table 4. Adjacency matrix with linguistic values (Ẽ) obtained from FCM in Fig. 4

	Price	Durability	Robustness	Repair Cost
Price	No influence	Medium	Low	No influence
Durability	Very Low	No influence	Medium	Very Low
Robustness	Very Low	Very Low	No influence	Low
Repair Cost	Very Low	No influence	No influence	No influence

Table 5. Fuzzy pairwise comparison matrix among alternatives according to **Price** criteria

	Passat	Corolla	Cerato	Ménage
Passat	EI			
Corolla	MI	EI	EI	
Cerato	SI		EI	
Ménage	MI	EMI		EI

Table 6. Fuzzy pairwise comparison matrix among alternatives according to **Durability** criteria

	Passat	Corolla	Cerato	Ménage
Passat	EI	MSI	SI	MI
Corolla		EI	SVSI	MI
Cerato			EI	
Ménage			MSI	

Table 7. FPCM among alternatives according to **Robustness** criteria

	Passat	Corolla	Cerato	Ménage
Passat	EI	VSI	VSEI	SI
Corolla		EI	SI	
Cerato			EI	
Ménage		MI	SI	EI

Table 8. FPCM among alternatives according to **Repair Cost** criteria

	Passat	Corolla	Cerato	Ménage
Passat	EI			MI
Corolla	MSI	EI		MSI
Cerato	MSI		EI	SI
Ménage		EI		EI

Table 9. The ranking of the final fuzzy weight vector (ϖ) for alternative

The final weight (ϖ)	Fuzzy value	COA value	Ranking
Passat	(0.0704, 2.1204, 77.2815)	26.49	1
Corolla	(0.0481, 1.7375, 73.419)	25.07	3
Cerato	(0.0366, 1.1267, 48.2505)	16.47	4
Ménage	(0.0348, 1.5578, 73.7466)	25.11	2

6   Conclusion

Although Fuzzy decision maps method (LFDN) has been proposed to deal with Multi-Criteria Decision-Making (MCDM) problems, it uses fuzzy set theory instead of dealing with crisp numbers to simulate the real life situations. Unfortunately, it can't solve complex problems for ranking the actions (alternatives) to select the appropriate action. Therefore, we developed LFDN to deal with complex problems. The proposed LFDN model with multi-level structure is practical and effective for solving MCDM problems. Currently, the authors are preparing several applications in different area for the proposed model to solve such MCDM problems. A case study was implemented to evidence that the proposed model is capable of select the best alternative regard to a multiple criteria.

References

1. Zhenghai, A.: A new TOPSIS with triangular fuzzy number and uncertain weight based on cosines similar degree. In: Eighth International Conference on Computational Intelligence and Security (CIS), pp. 17–21. IEEE Computer Society, Washington, DC (2012)
2. Kou, G., Miettinen, K., Shi, Y.: Multiple criteria decision making: challenges and advancements. Journal of Multi-Criteria Decision Analysis 18, 1–4 (2011)
3. Saaty, T.L.: The analytic hierarchy process. McGraw-Hill, New York (1980)
4. Saaty, T.L.: Decision making with dependence and feedback: The Analytic Network Process. RWS Publications, Pittsburgh (1996)
5. Yu, R., Tzeng, G.H.: A Soft Computing Method for Multi-criteria Decision Making with dependence and feedback. In: Applied Mathematics and Computation, vol. 180, pp. 63–75 (2006)
6. Elomda, B.M., Hefny, H.A., Hassan, H.A.: An Extension of fuzzy Decision Maps for Multi-criteria Decision-making. Egyptian Informatics Journal 14, 147–155 (2013)
7. Elomda, B.M., Hefny, H.A., Hassan, H.A.: MCDM method Based on Improved Fuzzy decision map. In: IEEE International Conference on Electronics, Circuits, and Systems (ICECS 2013), Abu Dhabi, UAE, December 8-11, pp. 225–228 (2013)
8. Kosko, B.: Fuzzy cognitive maps. International Journal on Man–Machine Studies 24(1), 65–75 (1986)
9. Axelrod, R.: Structure of decision: The Cognitive Maps of Political Elites. Princeton University Press, Princeton (1976)
10. Elomda, B.M., Hefny, H.A., Hassan, H.A.: Fuzzy Cognitive Map with Linguistic Values. In: IEEE International Conference on Engineering and Technology (ICET 2014), Cairo, Egypt, April 19-20 (2014)

An Edge Sensing Fuzzy Local Information C-Means Clustering Algorithm for Image Segmentation

Xinning Wang, Xiangbo Lin[*], and Zhen Yuan

Faculty of Electronic Information and Electrical Engineering,
Dalian University of Technology, Dalian, China
linxbo@dlut.edu.cn

Abstract. In this paper, we present a variation of fuzzy local information c-means (FLICM) algorithm for image segmentation by introducing a novel tradeoff factor and an effective kernel metric. The proposed tradeoff factor utilizes both local spatial and gray level information in a new way, and the Euclidean distance in FLICM algorithm is substituted by Gaussian Radial Basis function. By the novel factor and kernel metric, the new algorithm has edge identification ability and is insensitive to noise. Experiments result on both synthetic and real world images show that the proposed algorithm is effective and efficient, providing higher segmenting accuracy than other competitive algorithms.

Keywords: Fuzzy clustering, image segmentation, edge sensing, local information constraints, kernel function.

1   Introduction

Image segmentation is a very crucial procedure in image understanding and computer vision, many methods have been proposed [1,2]. It is a technique that divides an image into certain non-overlapping regions and extracts interest targets. Due to introducing the fuzzy belongingness of each pixel, fuzzy theory based methods have been used to segment image successfully [3]. Fuzzy C-Means (FCM) algorithm [4,5] is famous and has been researched widely among these fuzzy clustering methods. By lack of considering any spatial information, FCM is very sensitive to noise. A lot of improved algorithms based on FCM have been presented to remedy this drawback.

D.L. Pham proposed RFCM algorithm [6], he changed the cost function of standard FCM algorithm to include spatial constraints. RFCM was proved to be more robust for noisy image segmentation. M.N. Ahmed *et al* presented FCM_S algorithm [7,8] for medical image segmentation. It modified objective function of FCM to allow the labeling of a pixel to be influenced by the labels in its immediate neighborhood. To eliminate the intensity non-uniformity (INU) in MR images, Ahmed introduced bias field model and changed the FCM_S to BCFCM. After that, many methods based on this bias field model were proposed to correct INU in MR images [9,10,11].

---

[*] Corresponding author.

D.-S. Huang et al. (Eds.): ICIC 2014, LNAI 8589, pp. 230–240, 2014.

However BCFCM and FCM_S had high computational complexity and were sensitive to noise. To FCM_S, S. Chen *et al* proposed FCM_S1, FCM_S2 and the corresponding kernel function based version KFCM_S1 and KFCM_S2 [12]. Kernel metric was more robust to noise and had been utilized to substitute Euclidean distance in some fuzzy clustering methods [13,14,15].

B. Caldairou made use of the non-local mean denoising method and modified RFCM algorithm to get better segmentation performance [16]. F. Zhao in [17] proposed fuzzy c-means clustering algorithm with self-tunning non-local spatial information. These methods utilizing non-local information were robust to noise and were studied widely recent years [10],[18,19,20], however they are time consuming.

To reduce the calculation time of standard FCM class algorithm, L. Szilagyi *et al* proposed a quick FCM algorithm named EnFCM algorithm [21]. The clustering was performed on the gray level histogram of the linearly weighted sum image. W. Cai *et al* modified the linearly weighted sum image and get another fast FCM algorithm version FGFCM [22] which was more robust to noise than EnFCM. But these methods need adjustable parameters to control the tradeoff between denoising performance and details preservation, and parameter selection had significant influence on the segmentation performance. S. Krinidis proposed the FLICM algorithm [23] which was free of any parameter selection. However this approach had some weakness in identifying the class boundary pixels. Moreover, when the images are contaminated by severe noise, the performance of this method will degrade. Based on FLICM, M. Gong *et al* presented the KWFLICM algorithm [24] to improve noise immunity. To better classify the boundary pixels, we present an edge sensing FLICM algorithm named ESFLICM. It can not only solve the boundary pixels classification problem, but also has more accurate clustering performance in low SNR situations.

The rest of this paper is organized as follows. In second section, we will analyze the weakness of FLICM in boundary pixels classification. In third part, we will introduce the whole framework of ESFLICM. In section four, we will present the experimental results with different methods. The conclusions will be drawn in the last part.

2   Problem Analysis

Based on classical FCM, the robust fuzzy local information c-means (FLICM) algorithm [23] introduced a fuzzy factor G_{ki} to define its objective function. This fuzzy factor incorporates local spatial information and gray level information defined as

$$G_{ki} = \sum_{\substack{j \in N_i \\ i \neq j}} \frac{1}{d_{ij} + 1} \left(1 - u_{kj}\right)^m \left\| x_j - v_k \right\|^2 \tag{1}$$

where N_i is the set of neighborhood pixels centering around the ith pixel. d_{ij} presents the spatial Euclidean distance between the ith pixel and jth pixel. u_{kj} is the membership degree of the jth pixel in the kth class. The parameter m stands for the weighting exponent and controls the fuzziness of the clustering results, we set $m=2$ usually. $\left\| \bullet \right\|$

denotes Euclidean norm, x_j is the gray level of the jth pixel, v_k is the prototype of the center of the kth class. Thus the objective function of FLICM algorithm is defined as

$$J_m = \sum_{i=1}^{N} \sum_{k=1}^{c} \left[u_{ki}^{m} \|x_i - v_k\|^2 + G_{ki} \right] \tag{2}$$

The minimization of the objective function J_m is performed in an iterative way. The update equations of membership u_{ki} and class center v_k are given as follows

$$u_{ki} = \cfrac{1}{\sum_{j=1}^{c} \left(\cfrac{\|x_i - v_k\|^2 + G_{ki}}{\|x_i - v_j\|^2 + G_{ji}} \right)^{1/m-1}} \tag{3}$$

$$v_k = \frac{\sum_{i=1}^{N} u_{ki}^{m} x_i}{\sum_{i=1}^{N} u_{ki}^{m}} \tag{4}$$

By introduction of the new factor G_{ki}, FLICM enhances noise insensitivity and outlier's resistance which can be proved by the two examples in [23]. Moreover FLICM is free of any parameters selection which contributes to automatically image segmentation. However, FLICM is not always effective. There is one possible case to prove FLICM's weakness. As shown in Fig. 1, improper clustering result is obtained in a simple example.

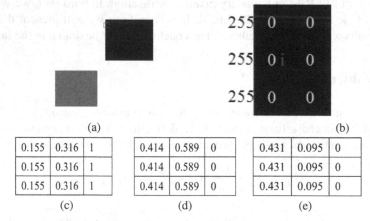

255	0	0
255	0	0
255	0	0

(a)   (b)

0.155	0.316	1
0.155	0.316	1
0.155	0.316	1

0.414	0.589	0
0.414	0.589	0
0.414	0.589	0

0.431	0.095	0
0.431	0.095	0
0.431	0.095	0

(c)   (d)   (e)

Fig. 1. Membership degree of edge pixels. (a) original image. (b) gray levels of pixels marked in rectangle in the original image. (c) Membership degree of the pixels in the 1th class. (d) Membership degree of the pixels in the 2th class. (e) Membership degree of the pixels in the 3th class.

This image has three classes, and it is not contaminated by any noise. Let's analyze boundary pixels between the black square region and white background. The size of the analysis window is 3×3. Calculate the equation (3) and (4) iteratively. After 4 iterations FLICM algorithm gets to the convergence condition. The centers of the 3 different classes are 0, 91 and 255, and the membership degrees of the 9 pixels are shown in Fig. 1(c), (d) and (e). It shows that the three pixels lied in the boundary between black square region and white backgrounds which belong to the black class are classified to the second gray class incorrectly.

This false clustering is related to the definition of the fuzzy factor G_{ki}. Let's study the influence of G_{ki} to the ith pixel in the above mentioned example. The gray level of x_i is 0, so x_i should be classified into the first class (the black region). That is to say, u_{1i} should be the largest membership degree. However, the range of $1/(d_{ij}+1)$ is from zero to one. Those pixels whose gray level is 255 belong to different class with respect to the ith pixel, therefore $(1-u_{1j})$ is relative large, and $\|x_j - v_i\|^2$ is a rather high value. As a result G_{1i} has a greater impact on the membership u_{1i}. This results in a smaller u_{1i} with respect to u_{2i}. So u_{2i} is the largest membership degree and x_i is classified to the gray region. We called this false clustering as *multi-class edge problem*. It refers to that at the edge of two different classes if there are the other classes whose gray level is between these two classes, the edge pixels will be possibly classified into the wrong class. Motivated by the above problem, an Edge Sensing Fuzzy Local Information C-Means (ESFLICM) algorithm is proposed. ESFLICM algorithm uses Gaussian Radial Basis Function as the similarity measure and defines a novel factor G_{ki} to regularize the influence of local information more effectively.

3   Edge Sensing Fuzzy Local Information C-Means (ESFLICM) Algorithm

Kernel functions can transform the nonlinear vectors in original feature space into linear vectors in a higher dimensional (possibly infinite) feature space. The complex nonlinear problems can be turned into more easier linear issues by this kind transformation. In this new algorithm ESFLICM, we use the kernel induced distance instead of the traditional Euclidean metric.

A kernel in feature space can be described as K [25,26]

$$K(x,y) = (\phi(x),\phi(y))$$ (5)

where $\phi(\bullet)$ is an nonlinear mapping. $(\phi(x),\phi(y))$ is the inner product operation.

In general, typical Radial Basis Function (RBF) kernel [12], [24] is common used. The mathematical formulation of RBF kernel is

$$K(x,y) = \exp\left(-\left(\sum_{i=1}^{d}|x_i - y_i|^a\right)^b \middle/ \sigma^2\right)$$ (6)

Where d is the dimension of vector x, $a \geq 0$ and $1 \leq b \leq 2$. In this paper, we use Gaussian Radial Basis Function (GRBF) with a=2 and b=1.σ refers to the kernel bandwidth. Its value has an important influence on the accuracy of the clustering result. In [24], Gong presented a method to choose proper σ whose effectiveness has been proved by some convincing experiments. So in ESFLICM algorithm we adopt this method to determine σ.

With above descriptions and through GBRF kernel substitution, the proposed kernel FLICM algorithm is defined as follows

$$J_{m}^{'} = \sum_{i=1}^{N}\sum_{k=1}^{c}\left[u_{ki}^{m}\left(1 - K\left(x_{i}, v_{k}\right)\right) + \sum_{\substack{j \in N_i \\ j \neq i}}\frac{1}{d_{ij}+1}\left(1 - u_{kj}\right)^{m}\left(1 - K\left(x_{j}, v_{k}\right)\right)\right] \quad (7)$$

where

$$K(x_{i}, v_{k}) = \exp(-\frac{\left|x_{i} - v_{k}\right|^{2}}{\sigma^{2}}) \quad (8)$$

The introducing of GRBF kernel can remedy too large influence of factor G_{ki} to cost function. Nevertheless, when images are contaminated by noise, *multi-class edge problem* cannot be solved effectively only depending on kernel. In the following, we present a novel tradeoff factor $G_{ki}^{'}$ to improve the effectiveness of local information. It is shown in the following

$$G_{ki}^{'} = u_{ik}^{m}\sum_{\substack{j \in N_i \\ j \neq i}} w_{ij}\frac{\alpha}{d_{ij}+1}\left(1 - u_{kj}\right)^{m}\left(1 - K\left(x_{j}, v_{k}\right)\right) \quad (9)$$

where

$$w_{ij} = \left(1 + \frac{e^{-\|x_{j}-x_{i}\|^{2}}}{\sum_{j \in N_i} e^{-\|x_{j}-x_{i}\|^{2}}}\right)\left(1 + e^{-\frac{\mathrm{var}(x_{j})}{\overline{\mathrm{var}(x_{i})}}}\right) \quad (10)$$

where x_i refers to the gray level of the ith pixel and so does x_j. $\mathrm{var}(x_j)$ is the intensity variance of pixels located in the neighbor window centering around the jth pixel, and $\overline{\mathrm{var}(x_i)}$ is the mean of $\mathrm{var}(x_j)$. The value of α is from 0.5 to 5 depending on noise level, and is easy to select.

w_{ij} reflects local gray level information. The first item in equation (10) depicts the intensity similarity between the ith pixel and the pixels in its neighbor window. If x_j is very close to x_i, the value of this item will be large. It indicates x_j and x_i may belong to the same class, the influence of $G_{ki}^{'}$ will increase, and vice versa. The second item of equation (10) denotes the intensity homogeneity degree in the local window of pixel j. When those pixels in the local window are located in the homogenous regions or not contaminated by noise, the role of the second item will be large; and the influence of the tradeoff factor will increase. The general objective function of ESFLICM algorithm is written as follows

$$J_m = \sum_{i=1}^{N} \sum_{k=1}^{c} \left[u_{ki}^{\ m} \left(1 - K\left(x_i, v_k\right)\right) + G_{ki}^{'} \right] \tag{11}$$

The membership function and prototypes of the class center can be attained by the Lagrange multiplier method. They are shown as follows

$$u_{ki} = \cfrac{1}{\displaystyle\sum_{j=1}^{c} \left(\cfrac{\left(1 - K\left(x_i, v_k\right)\right) + \displaystyle\sum_{\substack{j \in N_i \\ j \neq i}} w_{ij} \dfrac{\alpha}{d_{ij}+1} \left(1 - u_{kj}\right)^m \left(1 - K\left(x_j, v_k\right)\right)}{\left(1 - K\left(x_i, v_j\right)\right) + \displaystyle\sum_{\substack{j \in N_i \\ j \neq i}} w_{ij} \dfrac{\alpha}{d_{ij}+1} \left(1 - u_{jj}\right)^m \left(1 - K\left(x_j, v_j\right)\right)} \right)^{1/m-1}} \tag{12}$$

$$v_k = \cfrac{\displaystyle\sum_{i=1}^{N} u_{ki}^{\ m} K\left(x_i, v_k\right) x_i}{\displaystyle\sum_{i=1}^{N} u_{ki}^{\ m} K\left(x_i, v_k\right)} \tag{13}$$

With above modifications the new algorithm ESFLICM is not only able to solve the *multi-class edge problem*, but also more robust for noisy image segmentation than FLICM and KWFLICM.

4   Experimental Results and Analysis

In this section, we apply the proposed algorithm ESFLICM, FLICM [23], KWFLICM [24] and standard FCM to synthetic and real world images to compare the clustering performance of these methods. In our experiments, the convergence threshold is set to be 0.001, the maximal iteration number is 100 and the size of the local window (N_i) is 3×3.

The synthetic image includes four classes. The clustering results of these four algorithms on the image contaminated by Gaussian noise (0 mean and 0.02 variance) are shown in Fig.2. We set α=5.0 in ESFLICM algorithm and use intensity 0, 60, 120 and 180 to represent the four classes. Visually, in Fig. 2(c), FCM cannot resist the influence of noise. In Fig. 2(d), FLICM removes most of the noise however as Fig.2(g) shows, FLICM can not solve the *multi-class edge problem*. In Fig. 2(e), KWFLICM is still influenced by noise to some extent. In Fig. 2(f), ESFLICM removes almost all the noise and get clear class boundary which is proved in Fig. 2(i).

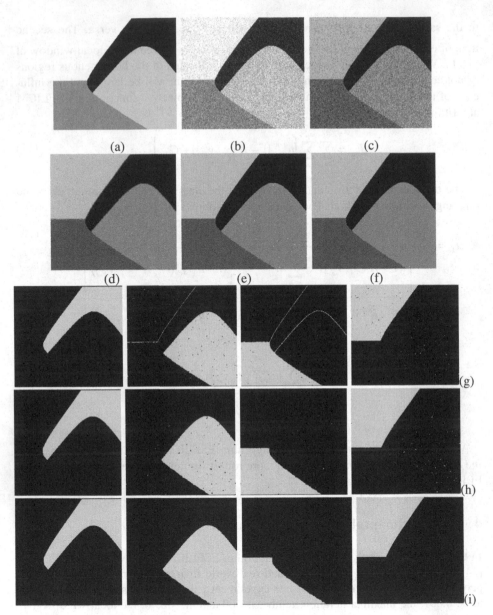

Fig. 2. Clustering results on image contaminated by Gaussian noise. (a) Original image. (b) Noisy image. (c) FCM result. (d) FLICM result. (e) KWFLICM result. (f) ESFLICM result. (g) Each class of FLICM. (h) Each class of KWFLICM. (i) Each class of ESFLICM.

Fig. 3 illustrates the segmentation performance of the synthetic image contaminated by Salt & Pepper noise (density 0.25). We set α=5.0 in ESFLICM algorithm. Fig. 3(c) shows FCM cannot remove any noise. FLICM still has a proportion of noise and *multi-class edge problem*. KWFLICM remains small part of noise, while ESFLICM achieves the most satisfactory result from both denoising performance and identifying ability to boundary pixels.

Table 1 presents the segmentation accuracy (SA) [8] to compare the denoising performance of different methods. SA is defined as the number of correctly classified pixels divided by the total number of pixels. As Table 1 shows, the proposed algorithm ESFLICM can get more accurate results than the other competitive algorithm in the present of high level noise.

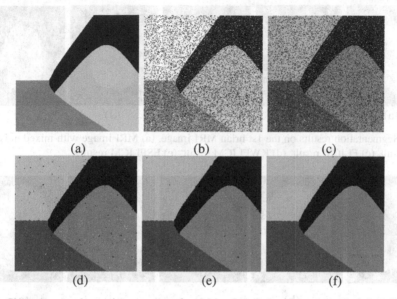

Fig. 3. Clustering results on image contaminated by Salt & Pepper noise. (a) Original image. (b) Noisy image. (c) FCM result. (d) FLICM result. (e) KWFLICM result. (f) ESFLICM result.

Table 1. SA (SA%) of four algorithms on synthetic images contaminated by noise

	FCM	FLICM	KWFLICM	**ESFLICM**
Gaussian	76.01	98.01	99.14	99.37
Salt & pepper	86.07	98.16	99.13	99.25

In addition, the proposed algorithm ESFLICM is used to segment brain MRI images comparing with the other three algorithms. Brian MRI images are generally classified into four classes, white matter (wm), gray matter (gm), cerebrospinal fluid (csf) and background.αis set to be 0.5 in ESFLICM algorithm. The first brain image contaminated by mixed noise is shown in Fig. 4(a), which is provided by [23]. Obviously, FLICM still has some noise in gm, csf and background. KWFLICM smoothes some details of csf. ESFLICM delineates sulcal csf more exactly. Another brain MRI image is from Brianweb [27]. We choose MRI image with T1 weighted phantom, 1mm slice thickness, and 9% Rician noise. The visual segmentation results are shown in Fig. 5.

Table 2 gives the quantitative evaluation measure Dice Similarity Coefficient (known as KI) [16],[28] to compare the classification performance. KI measures the overlap between the classification result using the algorithm and the standard segmentation. $0 \leq KI \leq 1$, and the closer to 1 KI is, the more similar the two results are.

As can be seen from the Fig.5 and Table 2, the new proposed algorithm ESFLICM achieves the most satisfactory segmentation result. This good performance benefits from the novel factor G'_{ki} utilizing the local spatial and intensity information effectively and the more robust GRBF kernel.

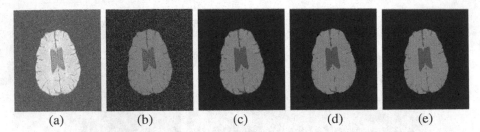

(a)   (b)   (c)   (d)   (e)

Fig. 4. Segmentation results on the 1st brian MRI image. (a) MRI image with mixed noise. (b) FCM result. (c) FLICM result. (d) KWFLICM result. (e) ESFLICM result.

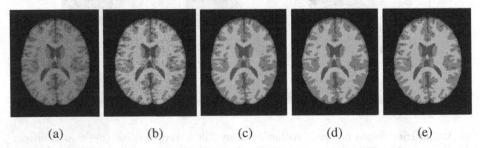

(a)   (b)   (c)   (d)   (e)

Fig. 5. Segmentation results on the 2nd brain MRI image. (a) MRI image with 9% Rician noise. (b) FCM result. (c) FLICM result. (d) KWFLICM result. (e) ESFLICM result.

Table 2. KI (%) of four algorithms on the 2nd brian MRI image

	FCM	FLICM	KWFLICM	**ESFLICM**
wm	88.66	93.92	92.95	93.78
gm	81.66	91.54	90.39	91.91
csf	83.01	88.73	85.43	89.58

5   Conclusion

This paper presents an improved algorithm of FLICM for accurate image segmentation. We make a reasonable analysis of FLICM algorithm to reveal the existed *multi-class edge problem*. To solve this problem, we propose a novel tradeoff weighted factor G'_{ki} and utilize GRBF kernel to substitute the Euclidean metric. The factor G'_{ki} combines local spatial and intensity information in a more reasonable and simpler manner, rendering the new proposed algorithm removing *multi-class edge problem*, more robust to noise type and guaranteeing more details. Compared to FLICM, the proposed ESFLICM has no significant increase in computational cost. In our future works, we will study the strategy to segment images influenced by low-frequency interference.

Acknowledgments. The authors would like to thank Dr. Krinidis for providing the code of FLICM, and the support of National Natural Science Foundation of China (61101230).

References

1. Pham, D.L., Xu, C., Prince, J.L.: Current Methods in Medical Image Segmentation. Annual Review of Biomedical Engineering 2(1), 315–337 (2000)
2. Zhang, H., Fritts, J.E., Goldman, S.A.: Image Segmentation Evaluation: A survey of Un supervised Methods. Computer Vision and Image Understanding 110(2), 260–280 (2008)
3. Naz, S., Majeed, H., Irshad, H.: Image Segmentation Using Fuzzy Clustering: A survey. In: 6th International Conference on Emerging Technologies, pp. 181–186. IEEE Press, Islamabad (2010)
4. Bezdek, J.C.: Pattern Recognition with Fuzzy Objective Function Algorithms. Bibliometrics, New York (1981)
5. Bezdek, J.C., Hall, L.O., Clarke, L.P.: Review of MR Image Segmentation Techniques Using Pattern Recognition. Medical Physics-Lancaster PA 20, 1033 (1993)
6. Pham, D.L.: Spatial Models for Fuzzy Clustering. Computer Vision and Image Understanding 84(2), 285–297 (2001)
7. Ahmed, M.N., Yamany, S.M., Farag, A.A., Moriarty, T.: Bias Field Estimation and Adaptive Segmentation of MRI Data Using A Modified Fuzzy C-Means Algorithm. In: IEEE Computer Society Conference on Computer Vision and Pattern Recognition, vol. 1, pp. 250–255. IEEE Press, Fort Collins (1999)
8. Ahmed, M.N., Yamany, S.M., Mohamed, N., Farag, A.A., Moriarty, T.: A Modified Fuzzy C-Means Algorithm for Bias Field Estimation and Segmentation of MRI Data. IEEE Transactions on Medical Imaging 21(3), 193–199 (2002)
9. Chen, W., Giger, M.L.: A Fuzzy C-Means (FCM) Based Algorithm for Intensity Inhomogeneity Correction and Segmentation of MR Images. In: IEEE International Symposium on Biomedical Imaging: Nano to Macro, pp. 1307–1310. IEEE Press (2004)
10. Chen, Y., Zhang, J., Wang, S., Zheng, Y.: Brain Magnetic Resonance Image Segmentation Based on an Adapted Non-local Fuzzy C-Means Method. IET Computer Vision 6(6), 610–625 (2012)
11. Szilágyi, L., Szilágyi, S.M., Benyó, B.: Efficient Inhomogeneity Compensation Using fuzzy C-Means Clustering Models. Computer Methods and Programs in Biomedicine 108(1), 80–89 (2012)
12. Chen, S., Zhang, D.: Robust Image Segmentation Using FCM with Spatial Constraints Based on New Kernel-Induced Distance Measure. IEEE Transaction on Syst. Man. Cybern 34, 1907–1916 (2004)
13. Thamaraichelvi, B., Yamuna, G.Y.: A Novel Efficient Kernelized Fuzzy C-means with Additive Bias Field for Brain Image Segmentation. In: 2013 International Conference on Communications and Signal Processing, pp. 68–72. IEEE Press, Melmaruvathur (2013)

14. Kaur, P., Lamba, I.M.S., Gosain, A.: Kernel Type-2 FCM Cluastering Algorithm in Segmentation of Noisy Medical Image. In: Recent Advances in Intelligent Computational Systems(RAICS), pp. 493–498. IEEE Press, New Delhi (2011)

15. Kannan, S.R., Ramathilagam, S., Devi, R., Sathya, A.: Robust Kernel FCM in Segmentation of Breast Medical Images. Expert Systems with Applications 38(4), 4382–4389 (2011)

16. Caldairou, B., Passat, N., Habas, P.A., Studholme, C.: A Non-local Fuzzy Segmentation Method: Application to Brain MRI. Pattern Recognition 44(9), 1916–1927 (2011)

17. Zhao, F.: Fuzzy Clustering Algorithms with Self-tuning Non-local Spatial Information for Image Segmentation. Neurocomputing 106, 115–125 (2013)

18. Wang, P., Wang, H.: A Modified FCM Algorithm for MRI Brain Image Segmentation. In: 8th International Seminar on Future BioMedical Information Engineering, pp. 26–29. IEEE Press, Wuhan (2008)

19. Shi, Z., Lihuang, S., Li, L., Hua, Z.: A Modified Fuzzy C-means for Bias Field Estimation and Segmentation of Brain MR Image. In: 25th Chinese Control and Decision Conference (CCDC), pp. 2080–2085. IEEE Press, Guiyang (2013)

20. Vansteenkiste, E., Philips, W.: Spatially Coherent Fuzzy Clustering for Accurate and Noise-robust Image Segmentation. IEEE Signal Processing Letters 20(4), 295–298 (2013)

21. Szilagyi, L., Benyo, Z., Szilágyi, S.M., Adam, H.S.: MR Brain Image Segmentation Using an Enhanced Fuzzy C-means Algorithm. In: Proceedings of the 25th Annual International Conference of the Engineering in Medicine and Biology Society, vol. 1, pp. 724–726. IEEE Press (2003)

22. Cai, W., Chen, S., Zhang, D.: Fast and Robust Fuzzy C-means Clustering Algorithms incorporating Local Information for Image Segmentation. Pattern Recognition 40(3), 825–838 (2007)

23. Krinidis, S., Chatzis, V.: A Robust Ruzzy Local Information C-means Clustering Algorithm. IEEE Transactions on Image Processing 19(5), 1328–1337 (2010)

24. Gong, M., Liang, Y., Shi, J., Ma, W., Ma, J.: Fuzzy C-means Clustering with Local Information and Kernel Metric for Image Segmentation. IEEE Transactions on Image Processing 22(2), 573–584 (2013)

25. Muller, K., Mika, S., Ratsch, G., Tsuda, K., Scholkopf, B.: An Introduction to Kernel-based Learning Algorithms. IEEE Transactions on Neural Networks 12(2), 181–201 (2001)

26. Girolami, M.: Mercer Kernel-based Clustering in Feature Space. IEEE Transactions on Neural Networks 13(3), 780–784 (2002)

27. BrainWeb: Simulated Brain Database,
    http://brainweb.bic.mni.mggill.ca/brainweb/

28. Crum, W.R., Camara, O., Hill, D.L.: Generalized Overlap Measures for Evaluation and Validation in Medical Image Analysis. IEEE Transactions on Medical Imaging 25(11), 1451–1461 (2006)

Application of Genetic Algorithm and Fuzzy Gantt Chart to Project Scheduling with Resource Constraints

Yu-Chuan Liu[1,*], Hong-Mei Gao[2], Shih-Ming Yang[3], and Chun-Yung Chuang[3]

[1] Dept. of Information Management, Tainan Univ. of Technology, Tainan, 71002, Taiwan
t00258@mail.tut.edu.tw
[2] Dept. of Economic Management, Tianjin Agricultural Univ., Tianjin, 300384, China
[3] Dept. of Aeronautics and Astronautics, National Cheng Kung Univ., Tainan, 70101, Taiwan

Abstract. Project scheduling with resource constraints is one of the most challenging optimization problems because of the complexity in estimating the resource requirement. This work aims at the application of fuzzy Gantt chart (FGC) and genetic algorithm (GA) to calculate optimal activity in project scheduling. Activity durations are considered adjustable for the optimal resource assignment under the constraints. GA determines not only the activity priority but also the activity duration within resource constraints. Numerical results of an example show that this application can effectively reduce the maximum resource input from 94 to 40 men with similar project makespan.

Keywords: project scheduling, resource constraints, genetic algorithm, fuzzy Gantt chart.

1 Introduction

Project scheduling is subject to considerable constraints and uncertainties that often lead to schedule disruptions. Jiang and Shi [1] recently presented a branch-and-cut procedure for minimizing the makespan of a construction project under resource constraints. Lu and Lam [2] evaluated the effect of multiple resource calendars on the makespan in a constructive project. El-Rayes and Jun [3] also developed two resource-leveling to minimize the impact of resource fluctuations in constructive project. Genetic algorithms (GA) have been applied to project scheduling as an effective tool to search for optimal solution in construction management [4], aircraft maintenance [5], and program-evaluation-and-review-technique [6], and many other project scheduling [7-10], multi-objective optimization [11], minimized cost [12], financial management [13], repetitive project [14], and ad hoc method [15]. In practice, however, many projects face not only resource constraints but also schedule uncertainties and state variable variations. Fuzzy set has been known suitable to describe the uncertainties. Chen and Huang [16] presented an activity time evaluation method by fuzzy logic. Yakhchali and Ghodsypour [17] also analyzed the start time

---

* Corresponding author.

D.-S. Huang et al. (Eds.): ICIC 2014, LNAI 8589, pp. 241–252, 2014.
© Springer International Publishing Switzerland 2014

of the activities in a project by fuzzy logic. However, both did not consider any resource constraints. Lin and Shyu [18] used fuzzy logic to determine the arrival schedule at an airport. Liang [19] developed a two-phase fuzzy approach to solve multi-objective project in linear systems. A recent work proposed the concept of fuzzy Gantt chart (FGC) to evaluate the activity duration [20]. This work aims at the applications of FGC with to project scheduling by using GA to meet the resource constraints. Numerical results show project scheduling under resource constraints can be solved effectively and efficiently by FGC with GA.

2   Fuzzy Gantt Chart for Resource Requirement

In project scheduling, accurate resource requirement of every activity is preferable, if not necessary. The resource requirements can be regarded as the product of activity duration and resource input. Liu [21] proposed FGC to determine activity duration by considering the uncertainty characteristics using fuzzy quality function deployment (FQFD) and fuzzy analytic hierarchy process (FAHP), where the degree of fuzziness for every project activity is calculated. The uncertainties are measured by the risk level of project characteristics such as project time limit, project activity start time, project budget, manpower, technological difficulty, and facility requirements. When applying FQFD to project management, the risk level of such project characteristics (PCs) are considered as customer attributes, while the project activities (PAs) are defined as engineering characteristics in FQFD. The application of FQFD for project management is shown as Fig. 1. PCs are translated into PAs by linguistic expressions {very low, low, median, high, very high} at different technical rankings, where VL, L, M, H, and VH represent very low, low, median, high, and very high, respectively. Figure 2 shows the triangular membership function of the five fuzzy linguistic spaces.

The element of relationship matrix R_{ij} indicates the relative risk level of j-th PA to ith PC, and W_j represents the total risk of the j-th PA calculated under the consideration of PCs. W_j is defined as the degree of fuzziness for each PA:

$$W_j = \sum_i^n I_i R_{ij}, \ j = 1, 2, \ldots, m \tag{1}$$

where I_i represents the importance factor of the i-th PC, n is the number of project characteristics, and m is the number of project activities. In conventional FQFD, the importance factor I_i and the element of relationship matrix R_{ij} are determined by experts. For more reliable value, I_i and R_{ij} are preferable to be determined by FAHP, if necessary [21]. One should note that the calculated W_j is in triangular fuzzy membership function represented by W_j (c_1, c_2, c_3), where c_1, c_2, c_3 are the location of the corners of triangle. The defuzzified value w_j or the degree of fuzziness W_j (c_1, c_2, c_3), can be determined by the center of area method as:

$$w_j = \frac{c_1 + c_2 + c_3}{3} \tag{2}$$

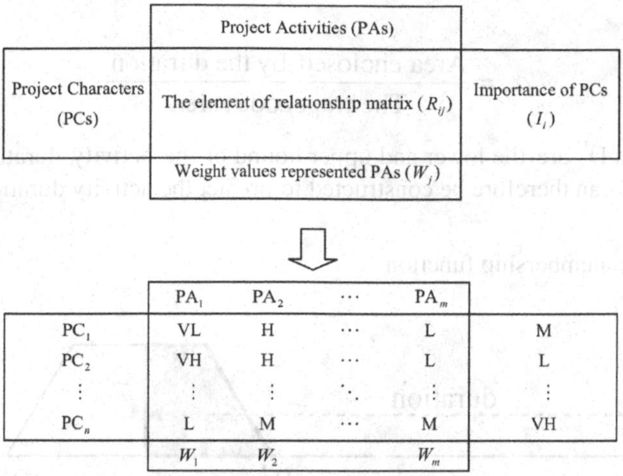

Project Characters (PCs)	Project Activities (PAs)			Importance of PCs (I_i)
	The element of relationship matrix (R_{ij})			
	Weight values represented PAs (W_j)			

	PA$_1$	PA$_2$	$\cdots$	PA$_m$	
PC$_1$	VL	H	$\cdots$	L	M
PC$_2$	VH	H	$\cdots$	L	L
$\vdots$	$\vdots$	$\vdots$	$\ddots$	$\vdots$	$\vdots$
PC$_n$	L	M	$\cdots$	M	VH
	W_1	W_2		W_m	

Fig. 1. The example of fuzzy quality function deployment (FQFD) with five linguistic spaces {very low, low, median, high, very high}

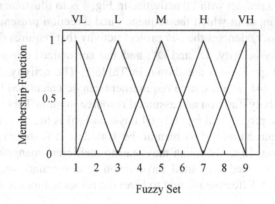

Fig. 2. The membership function of triangular fuzzy numbers

In FQFD, the importance factor I_i is translated into W_j by the experiences of experts, and each W_j is defuzzified as w_j by Eq. (2). One can normalize the degree of fuzziness by:

$$\alpha_j = \frac{w_j}{\overline{w} + 3\sigma}$$

(3)

where $\overline{w}$ is the mean and σ is the standard deviation of all w_j. The normalized degree of fuzziness α_j is a crisp index to represent how vague or unclear an activity would be. Hence, α_j can be applied to estimate the activity duration. For the fuzzy logic with trapezoidal membership function to describe the probability distribution of activity duration, the duration is determined as illustrated in Fig. 3 by the area ratio [21]:

$$\alpha_j = \frac{\text{Area enclosed by the duration}}{\text{The trapezodal area}} \tag{4}$$

where D^L and D^U are the lower and upper bound of the activity duration by assumption. An FGC can therefore be constructed to predict the activity duration.

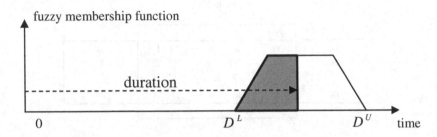

Fig. 3. The example of activity duration estimated by the degree of fuzziness

An example of a project with 20 activities in Fig. 4 is to illustrate the activity duration and resource input in which the sequence and direction presented by nodes and arrows [21]. PAi (d, r) denotes the *i-th* project activity that requires d days and r men. The range of every activity, D^L and D^U, and the normalized degree of fuzziness α_j determined from Eqs. (1)~(3) are shown in Table 1. The activity duration is determined by with Eq. (4). The resource requirement can be calculated by the product of the determined activity duration and assumed resource input in Table 1. Figure 5 (a) is the FGC where the project makespan is 64 days, and (b) is the resource diagram with the maximum resource input of 89 men at the 10th day. It is shown that the resource input fluctuates from 15 to 89 which may lead to inefficient manpower management. In practice, project scheduling shall have resource constraints such as manpower, budge, or equipment. Effective method to level the resource input is necessary.

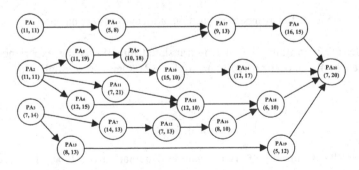

Fig. 4. The example of a project network, where PAi (d, r) denotes the ith project activity with d days and r men to complete the task

Table 1. The example of a project of 20 activities with the degree of fuzziness [21]

Project Activity	D^L	D^U	α_j	Duration	Resource Input	Precedence Activities
PA_1	5	18	0.381	11	11	-
PA_2	9	14	0.499	11	11	-
PA_3	3	10	0.653	7	14	-
PA_4	3	10	0.394	5	8	PA_1
PA_5	5	14	0.563	11	19	PA_2
PA_6	10	19	0.238	12	15	PA_2
PA_7	13	21	0.158	14	13	PA_3
PA_8	15	19	0.101	16	15	PA_{17}
PA_9	7	12	0.634	10	18	PA_5
PA_{10}	8	22	0.592	15	10	PA_6
PA_{11}	6	15	0.141	7	21	PA_2
PA_{12}	5	13	0.375	7	13	PA_7
PA_{13}	6	11	0.504	8	13	PA_3
PA_{14}	10	20	0.641	12	17	PA_{10}
PA_{15}	9	19	0.388	12	10	PA_{11}, PA_6
PA_{16}	5	11	0.569	8	10	PA_{12}
PA_{17}	7	17	0.232	9	13	PA_4, PA_9
PA_{18}	4	11	0.162	6	10	PA_{15}, PA_{16}
PA_{19}	4	9	0.097	5	12	PA_{13}
PA_{20}	3	9	0.649	7	20	PA_8,P A_{14}, PA_{18}, PA_{19}

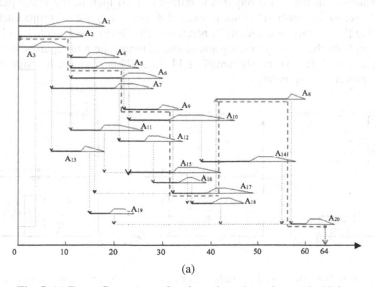

(a)

Fig. 5. (a) Fuzzy Gantt chart of project where the makespan is 64 days

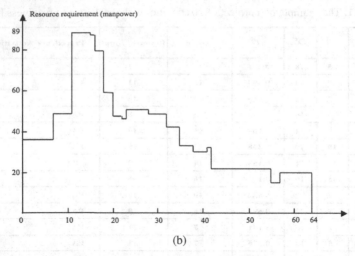

(b)

Fig. 5. (b) the resource diagram where the manpower fluctuates from 15 to 89 during the project period

3 Project Scheduling with Resource Constraints

Resource leveling can be achieved by rescheduling the project activities and adjusting the resource input for every activity. One can adjust the resource input to shorten the total makespan as illustrated in Fig. 6. Consider a project with 4 activities PAa, PAb, PAc, and PAd, in which (PAa, PAb) and (PAc, PAd) are conducted in sequence. The original schedule in Fig. 6 (a) requires maximum of 16 men in the makespan of 10 days. If the resource constraint is not to exceed 8 men a day for all project activities, rescheduling the activity priorities as illustrated in Fig. 6 (b) can complete in 13 days. By adjusting both the activity priority and resource input, the makespan of the project as shown in Fig. 6 (c) can be shortened in 11 days. This resource leveling helps to better manpower management.

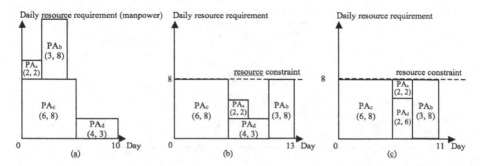

Fig. 6. The example of shortening the makespan by adjusting the priorities and resource inputs of the activities without violating the resource requirement: (a) original scheduling (b) rescheduling only by adjusting the priorities (c) rescheduling by adjusting the priorities and durations simultaneously.

As the resource input is inverse proportional to the activity duration, GA may be effective to find the priority and resource input of the activities. The chromosome of GA can be constructed as the priority and duration for every activity as shown in Fig. 7, where the first half of the chromosome string consists of the activity duration and the second half the activity priority. The adjustment of resource input, i.e., the activity duration can be described as

$$D_i^L \le D_i \le D_i^U, i = 1,2,...,n \qquad (5)$$

where D_i is the ith activity duration, D_i^L and D_i^U are the lower and upper bound, and n is number of activities in a project. The physical meaning of individual chromosome can be presented by

$$D_i = D_i^L + X_i \left(\frac{D_i^U - D_i^L}{2^l - 1} \right) \qquad (6)$$

where X_i is used to determine the ratio of the range between D_i^L and D_i^U, $X_i = \sum_{b=0}^{l-1} x_i(b) \cdot 2^b$, l is the length of binary code of a gene for one activity, and $x_i(b)$ is the binary value 0 or 1.

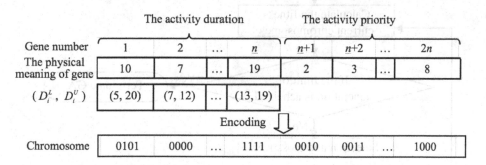

	The activity duration				The activity priority			
Gene number	1	2	...	n	$n+1$	$n+2$	...	$2n$
The physical meaning of gene	10	7	...	19	2	3	...	8
(D_i^L, D_i^U)	(5, 20)	(7, 12)	...	(13, 19)				
				Encoding				
Chromosome	0101	0000	...	1111	0010	0011	...	1000

Fig. 7. Binary representation of chromosome for the project activities

With the formulation for the activity duration Di in Eq. (6), the objective of minimizing project makespan within the resource constraints can be formulated as:

$$\text{minimize} \quad \max_i \{ s_i + D_i \} \qquad (7)$$

subjected to

$$s_j \ge s_i + D_i, \text{ for all } (i, j) \in H, i = 1, 2,...,n \qquad (8)$$

$$\sum_{G_t} R_{ik} \le R_k^T, k = 1, 2,...,m \qquad (9)$$

where si is the start time of activity i, H is the set of pairs of activities with precedence relationships, G_t is the set of in-progress activities at time t, R_{ik} is the resource input for *i-th* activity, R_k^T is the resource constraint, and m is the numbers of resource.

For optimal project scheduling, the activity priority and duration presented by the chromsomes are evaluated by GA. The initial population of the chromesomes is created randomly. The fitness function of GA is defined to minimize the total makespan of project as shown as Eq. (7). Shorter makespan means better fitness and more likely to be selected to create the offspring in the next generation by crossover. The chromosomes with higher fitness tend to be reserved for optimal solution. Mutation is the operator to enhance variation of chromosome and to avoid being trapedin local optimal solution. The calculation repeats for next generation and the calculation procedure can be summarized Fig. 8.

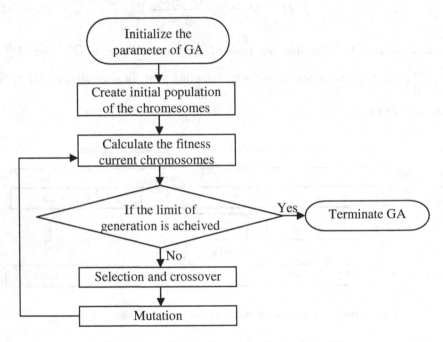

Fig. 8. The calculation procedure for applying GA

For the project in Fig. 4, the number of activities n = 20 and the parameters in GA are: population size equals to 35, the generation number is 200, and crossover and mutation rates are set as 0.8 and 0.05, respectively. Large population size and generation number often result in better solution, but they also increase computation time and memory requirement. Appropriate crossover and mutation rates result in chromosome diversity in the population. Table 2 shows the comparison for the scheduling results of FGC and GA including scheduling algorithm, chromosome definition, adjusted duration and resource inputs, project makespan, and maximum daily manpower. Case 1 is scheduled by FGC without considering any resource constraint which results in the maximum manpower requirement of 89 men for the project makespan of 64 days. If the manpower is limited to 40, $R_k^T = 40$, case 2 is the scheduling results of GA in optimizing the activity priority. The maximum manpower is 38 men for the entire project period and the project makespan is increased to 80 days. For GA optimization of both activity duration and priority, case 3 shows the project makespan can be significantly reduced to 69 days, which is very close to the FGC results of 64 days with manpower 89. The rescheduling result is shown in Fig. 9. The solid line in resource diagram is the resource requirement of case 3, and the dotted line is the resource requirement of case 1 in Fig. 5 (b). It shows that the fluctuation of resource diagram of case 3 is significantly less than case 1.

Table 2. The comparison for scheduling results with different methods and conditions

Scheduling Algorithm	Case 1	Case 2	Case 3
	FGC	FGC+ GA (with priority)	FGC+ GA (with priority and duration)
PA_1 (day, man)	(11, 11)	(11, 11)	(10, 12)
PA_2	(11, 11)	(11, 11)	(11, 11)
PA_3	(7, 14)	(7, 14)	(6, 16)
PA_4	(5, 8)	(5, 8)	(3, 13)
PA_5	(11, 19)	(11, 19)	(8, 26)
PA_6	(12, 15)	(12, 15)	(12, 15)
PA_7	(14, 13)	(14, 13)	(13, 14)
PA_8	(16, 15)	(16, 15)	(15, 16)
PA_9	(10, 18)	(10, 18)	(8, 23)
PA_{10}	(15, 10)	(15, 10)	(14, 11)
PA_{11}	(7, 21)	(7, 21)	(6, 25)
PA_{12}	(7, 13)	(7, 13)	(5, 18)
PA_{13}	(8, 13)	(8, 13)	(7, 15)
PA_{14}	(12, 17)	(12, 17)	(13, 16)
PA_{15}	(12, 10)	(12, 10)	(10, 12)
PA_{16}	(8, 10)	(8, 10)	(7, 11)
PA_{17}	(9, 13)	(9, 13)	(7, 17)
PA_{18}	(6, 10)	(6, 10)	(4, 15)
PA_{19}	(5, 12)	(5, 12)	(4, 15)
PA_{20}	(7, 20)	(7, 20)	(4, 35)
Project Makespan (day)	64	80	69
Maximum Daily Manpower (man)	89	38	40

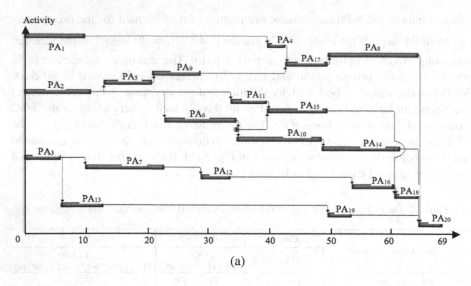

(a)

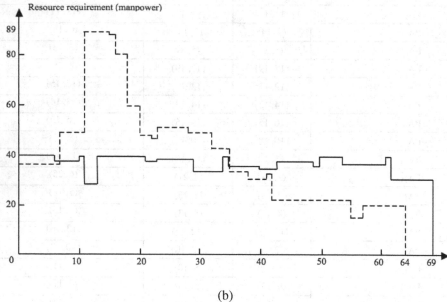

(b)

Fig. 9. (a) The Gantt chart and (b) resource diagram of case 3, and the dotted line is the resource requirement of case 1

4   Summary and Conclusion

In project scheduling facing uncertainties, precise resource requirement of every activity can be determined by using FGC. When facing both project uncertainties and resource constraints, hence, FGC is futile and GA is thus needed to determine the activity's priority and duration. The project example with 20 activities shows that scheduling by FGC without considering resource constraint requires 64 days to complete with the maximum resource of 89 men. If the resource constraint is limited to 40 men, GA shows rescheduling of the activity priority can complete the project in 80 days with the maximum resource of 38 men. When GA is further applied to determine the activity priority and duration, the project makespan is significantly shortened to 69 days with maximum manpower of 40. This resource leveling can improve the utilization of resources.

Acknowledgment. This work was financially supported by the Spark Program 2013 of the Ministry of Science and Technology of PRC, Tianjin Municipal Science and Technology Commission (13ZLZLZF04400), and Tianjin Administration of Foreign Experts Affairs project no.: B2014030.

References

1. Jiang, G., Shi, J.: Exact Algorithm for Solving Project Scheduling Problems under Multiple Resource Constraints. Journal of Construction Engineering and Management 131(9), 986–992 (2005)
2. Lu, M., Lam, H.C.: Critical Path Scheduling under Resource Calendar Constraints. Journal of Construction Engineering and Management 134(1), 25–31 (2008)
3. El-Rayes, K., Jun, D.H.: Optimizing Resource Leveling in Construction Projects. Journal of Construction Engineering and Management 135(11), 1172–1180 (2009)
4. Kim, J.L., Ellis, R.D.: Permutation-Based Elitist Genetic Algorithm for Optimization of Large-Sized Resource-Constrained Project Scheduling. Journal of Construction Engineering and Management 134(11), 904–913 (2008)
5. Hsu, C.L., Wang, T., Chen, S.H.: The Study of Optimum Engine Shop Visit Scheduling via Genetic Algorithm Method. Journal of Aeronautics, Astronautics and Aviation, Series A 42(2), 141–148 (2010)
6. Azaron, A., Perkgoz, C., Sakawa, M.: A Genetic Algorithm Approach for the Time-cost Trade-off in PERT Networks. Applied Mathematics and Computation 168(2), 1317–1339 (2005)
7. Chen, P.H., Weng, H.J.: A Two-phase GA Model for Resource-constrained Project Scheduling. Automation in Construction 18(4), 485–498 (2009)
8. Proon, S., Jin, M.Z.: A Genetic Algorithm with Neighborhood Search for the Resource-Constrained Project Scheduling Problem. Naval Research Logistics 58(2), 73–82 (2011)
9. Jaskowski, P., Sobotka, A.: Scheduling Construction Projects Using Evolutionary Algorithm. Journal of Construction Engineering and Management 132(8), 861–870 (2006)
10. Hegazy, T., Petzold, K.: Genetic Optimization for Dynamic Project Control. Journal of Construction Engineering and Management 129(4), 396–404 (2003)
11. Zeng, Q., Yang, Y., Liang, X.D., Ma, J.Q.: Multi-Objective Optimization for Equal Batch Splitting Flexible Job-Shop Scheduling Problem with Multiple Process Flows. Advanced Science Letters 4(4-5), 1808–1813 (2011)
12. Liu, Z., Wang, H.: GA-based Resource-Constrained Project Scheduling with the Objective of Minimizing Activities' Cost. In: Huang, D.-S., Zhang, X.-P., Huang, G.-B. (eds.) ICIC 2005. Part I. LNCS, vol. 3644, pp. 937–946. Springer, Heidelberg (2005)

13. Fathi, H., Afshar, A.: 'GA-based Multi-objective Optimization of Finance-based Construction Project Scheduling. Journal of Civil Engineering 14(5), 627–638 (2010)
14. Long, L.D., Ohsato, A.: A Genetic Algorithm-based Method for Scheduling Repetitive Construction Projects. Automation in Construction 18(4), 499–511 (2009)
15. Hegazy, T., Kassab, M.: Resource Optimization Using Combined Simulation and Genetic Algorithms. Journal of Construction Engineering and Management 129(6), 698–705 (2003)
16. Chen, C.T., Huang, S.F.: Applying Fuzzy Method for Measuring Criticality in Project Network. Information Sciences 177(12), 2448–2458 (2007)
17. Yakhchali, S.H., Ghodsypour, S.H.: Computing Latest Starting Times of Activities in Interval-valued Networks with Minimal Time Lags. European Journal of Operational Research 200(3), 874–880 (2010)
18. Lin, C.E., Shyu, F.Y.: Arrival Scheduling for Terminal Control Area Using Fuzzy Logic. Transactions of the Aeronautical and Astronautical Society of the Republic of China 31(2), 141-149。(1999)
19. Liang, T.F.: Fuzzy Multi-objective Project Management Decisions Using Two-phase Fuzzy Goal Programming Approach. Computers & Industrial Engineering 57(4), 1407–1416 (2009)
20. Liu, Y.C., Yang, S.M., Lin, Y.T.: Fuzzy Finish Time Modeling for Project Scheduling. Journal of Zhejiang University-Science A 11(12), 946–952 (2010)
21. Liu, Y.C.: Buffer Sizing Technique for Fuzzy Critical Chain Scheduling. In: Proc. of 2nd Intl. Conf. on Uncertainty Theo., Tibet, China, August 5-10 (2011)

Dynamic Output Feedback Guaranteed Cost Control for T-S Fuzzy Systems with Uncertainties and Time Delays[*]

Guangfu Ma, Yanchao Sun, Jingjing Ma, Chuanjiang Li, and Shuo Sun

Harbin Institute of Technology, Department of Control Science and Engineering, Harbin, China
{magf,sunyanchao,lichuan}@hit.edu.cn

Abstract. This paper introduces a dynamic output feedback (DOF) guaranteed cost controller for systems with uncertainties and time delays, which are usually described by Takagi–Sugeno (T-S) fuzzy models. We model the nonlinear system directly by parallel distributed compensation (PDC) rules. The uncertainties and time delays caused by modeling errors and parameter sensitivity of devices are considered. Since states are not always available, we choose dynamic output feedback and design guaranteed cost controller based on linear matrix inequality (LMI). The upper bound of performances and calculation method for controller parameters are given. A numerical simulation is provided to demonstrate the availability and effectiveness of our method.

Keywords: T-S fuzzy model, uncertainties, time delay, dynamic output feedback, guaranteed cost control.

1 Introduction

Constructing an accurate mathematical model is an important premise for systematic study. But practically it is difficult or impossible to avoid modeling errors. And considering the parameter sensitivity of sensor, actuators and other devices, there always exist uncertainties and time delays in physical system [1-4]. They are also important factors which cause systems to be nonlinear. Robust control is a powerful tool to process uncertainties. In recent years, guaranteed cost control of systems with uncertainties and time delays based on robust analysis and synthesis has received wide attention.

References [5-7] analyze uncertainties and time delays in linear systems without considering nonlinearity. However, nonlinear problem exists widely in physical systems. It can not only influence the performances of systems, but also undermine the stability of systems. Fuzzy control can successfully process different kinds of complicated nonlinear problems. Among them, the systems which are designed based on T-S fuzzy models draw people's attention widely.

---

[*] This work is supported by National Natural Science Foundation (NNSF) of China under Grant (No. 61304005, 61174200); Research Fund for the Doctoral Program of Higher Education (RFDP) of China (NO. 20102302110031).

D.-S. Huang et al. (Eds.): ICIC 2014, LNAI 8589, pp. 253–264, 2014.

T-S fuzzy model was first put forward by Takagi and Sugeno in 1985 [8]. Its idea is that decomposing nonlinear model under different states into several subsystems according to different rules. Then, controllers are designed for each subsystem. At last, a controller is designed applicable to the whole system via certain rules such as PDC rules. T-S fuzzy model converts many complex nonlinear problems into linear problems, so it establishes a bridge between linear system and nonlinear system. Reference [9] proposes T-S fuzzy controllers for nonlinear interconnected systems with multiple time delays. Reference [10] proposes dynamic output feedback stabilization controllers for T-S fuzzy system based on an observer. Reference [11] studies dynamic output feedback control problem for a T-S fuzzy system based on LMI. However, they only analyze the stabilization of the system and do not guarantee the performance of the system. People always have certain requirements for performances of systems in practical applications. Reference [12] researches robust non-fragile guaranteed cost controller design for a class of distributed delay uncertain systems. However, it only deals with linear systems. In [13], guaranteed cost control for uncertain singular time-delay system is studied. It considers the state feedback case. When states cannot be measured or have no definite physical meaning, it is difficult to approach physical realizability. So it is necessary to study the dynamic output feedback case. Although in [14] dynamic output feedback is considered, it does not consider uncertainties. Reference [15] discussed dynamic output feedback guaranteed cost control for uncertain stochastic system based on LMI. However, it ignores time delays.

In view of the current research situation, this paper put forward dynamic output feedback guaranteed cost control for T-S fuzzy systems with uncertainties and time delays. We combine robust control and fuzzy control and consider uncertainties caused by modeling errors and parameter variation. We model nonlinear system directly via T-S method. Dynamic output feedback is adopted because states are not always easily measured in practical system. In order to ensure the system have better convergence rate and energy saving properties, we design the guaranteed cost controller. The study in this paper is based on LMI. Theoretical analysis and simulation results both prove the effectiveness of our method.

2   System Description

The nonlinear system with uncertainties and time delays in this paper is described by r rules of T-S fuzzy model as follows [16]:

$$R_i : \text{if } Z_1 \text{ is } M_{i1} \text{ and},\cdots,Z_g \text{ is } M_{ig}; \text{ then}$$
$$\dot{x}(t) = [A_i + \Delta A_i]x(t) + [A_{id} + \Delta A_{id}]x(t-d) + [B_i + \Delta B_i]u(t)$$
$$x(t) = \varphi(t) \quad t \in [-d,0], i = 1,2\cdots r \tag{1}$$
$$y(t) = [C_i + \Delta C_i]x(t)$$

where R_i denotes the i-th rule of T-S fuzzy model, also known as fuzzy subsystem. $Z_1 \cdots Z_g$ denote the premise variables of fuzzy rules, M_{ij} denotes Fuzzy set, d denotes delay time and $d>0$. And

$$\Delta A_i = DFE_{i1}, \Delta B_i = DFE_{i2}, \Delta A_{id} = DFE_{id}, \Delta C_i = D_1 FE_{i1} \qquad (2)$$

When considering global fuzzy model, the fuzzy model can be expressed as follows:

$$\dot{x}(t) = \sum_{i=1}^{r} h_i \{[A_i + \Delta A_i]x(t) + [A_{id} + \Delta A_{id}]x(t-d) + [B_i + \Delta B_i]u(t)\}$$
$$y(t) = \sum_{i=1}^{r} h_i \{[C_i + \Delta C_i]x(t)\} \qquad (3)$$
$$x(t) = \varphi(t), t \in [-d, 0]$$

We will design dynamic output feedback controller under each corresponding rule, namely:

$$\dot{\hat{x}}(t) = A_{ci}\hat{x}(t) + B_{ci}y(t)$$
$$u(t) = C_{ci}\hat{x}(t) \qquad (4)$$

and form a global fuzzy model:

$$\dot{\hat{x}}(t) = \sum_{i=1}^{r} h_i \{A_{ci}\hat{x}(t) + B_{ci}y(t)\}$$
$$u(t) = \sum_{i=1}^{r} h_i \{C_{ci}\hat{x}(t)\} \qquad (5)$$

The global fuzzy model can make primary system asymptotically stable and form a corresponding guaranteed cost controller, where the performance index is given as follows.

$$J = \int_0^{\infty} [x^T Q x + u^T R u] dt \qquad (6)$$

3   Controller Introduction

First, rewrite the nonlinear system above to another form [17].

$$\dot{x}(t) = [A + \Delta A]x(t) + [A_d + \Delta A_d]x(t-d) + [B + \Delta B]u(t)$$
$$y(t) = [C + \Delta C]x(t) \qquad (7)$$

where

$$A = \sum_{i=1}^{r} h_i A_i \qquad A_d = \sum_{i=1}^{r} h_i A_{id} \qquad B = \sum_{i=1}^{r} h_i B_i$$

$$\Delta A = \sum_{i=1}^{r} h_i \Delta A_i \quad \Delta A_d = \sum_{i=1}^{r} h_i \Delta A_{id} \quad \Delta B = \sum_{i=1}^{r} h_i \Delta B_i$$
$$= DFE_1 \qquad\qquad = DFE_d \qquad\qquad = DFE_2$$

$$E_1 = \sum_{i=1}^{r} h_i E_{i1} \qquad E_d = \sum_{i=1}^{r} h_i E_{id} \qquad E_2 = \sum_{i=1}^{r} h_i E_{i2}$$

The corresponding controller is:

$$\begin{cases} \dot{\hat{x}}(t) = A_c\hat{x}(t) + B_c y(t) \\ u(t) = C_c\hat{x}(t) \end{cases} \tag{8}$$

where

$$A_c = \sum_{i=1}^{r} h_i A_{ci}, B_c = \sum_{i=1}^{r} h_i B_{ci}, C_c = \sum_{i=1}^{r} h_i C_{ci}$$

Rewrite the model and controller in closed-loop form. Suppose $\bar{x} = [x^T \; \hat{x}^T]^T$, then

$$\begin{aligned} \dot{\bar{x}} &= \bar{A}\bar{x} + \bar{A}_d x_d \\ x_d &= x(t-d) \end{aligned} \tag{9}$$

where

$$\bar{A} = \begin{bmatrix} A+\Delta A & BC_c + \Delta BC_c \\ B_c C + B_c \Delta C & A_c \end{bmatrix}, \bar{A}_d = \begin{bmatrix} A_d + \Delta A_d \\ 0 \end{bmatrix}$$

The performance index function (6) can be rewritten as [18]

$$J = \int_0^\infty (x^T Q x + \hat{x}^T C_c^T RC_c \hat{x})dt = \int_0^\infty \bar{x}\bar{C}^T \bar{C}\bar{x}dt \tag{10}$$

where

$$\bar{C} = \begin{bmatrix} Q^{1/2} & 0 \\ 0 & R^{1/2}C_c \end{bmatrix}$$

4   Guaranteed Cost Controller Design

Definition 1. For system (1) and corresponding performance function (6), if there exists a control law u and a positive constant J^*, such that the closed-loop system is asymptotically stable and $J \leq J^*$ under all uncertainties. Then we call u the guaranteed cost control law of networked control systems and J^* the performance bound of the systems [18].

   Lemma 1. (Schur Complement)
   If partitioned matrix A is real symmetric, i.e.,

$$A = \begin{bmatrix} A_{11} & A_{12} \\ A_{12}^T & A_{22} \end{bmatrix} \tag{11}$$

then the necessary and sufficient condition to make $A < 0$ is

$$A_{11} < 0, A_{22} - A_{12}^T A_{11}^{-1} A_{12} < 0 \quad \text{or} \quad A_{22} < 0, A_{11} - A_{12} A_{22}^{-1} A_{12}^T < 0 \tag{12}$$

Lemma 2. [19] Let M, D, F, and E are fitness-dimensional matrices, where M is symmetric and $F^T F \leq I$, then $M + DFE + (DFE)^T < 0$ if and only if there exists constant $\varepsilon > 0$ making $M + \varepsilon DD^T + \varepsilon^{-1} E^T E < 0$.

First, we demonstrate the sufficient conditions by the following theorem for the dynamic output feedback guaranteed cost controller dealing with models with uncertainties and time delays.

Theorem 1. For system (1) and its corresponding performance function (6), there exist symmetric positive definite matrices $P \in R^{2n \times 2n}$, $P_1 \in R^{n \times n}$ and A_c, B_c, C_c making the following inequality for arbitrary matrix F.

$$\psi = \begin{bmatrix} \overline{A}^T P + P\overline{A} + \hat{P} + \overline{C}^T \overline{C} & P\overline{A}_d \\ \overline{A}_d^T P & -P_1 \end{bmatrix} < 0 \tag{13}$$

where the definitions of correlative matrices are the same as above. Furthermore, a corresponding upper bound of the performance is given by:

$$J^* = V(0) = \overline{x}^T(0) P\overline{x}(0) + \int_{-d}^{0} x^T(\tau) P_1 x(\tau) d\tau \tag{14}$$

Proof

Select a Lyapunov function as

$$V(\overline{x}, t) = \overline{x}^T P\overline{x} + \int_{t-d}^{t} x^T(\tau) P_1 x(\tau) d\tau \tag{15}$$

It is easy to show that

$$\dot{V} = \dot{\overline{x}}^T P\overline{x} + \overline{x}^T P\dot{\overline{x}} + x^T P_1 x - x_d^T P_1 x_d$$

where (9) is utilized. Furthermore,

$$\dot{V} = \overline{x}^T \overline{A}^T P\overline{x} + x_d^T \overline{A}_d^T P\overline{x} + \overline{x}^T P\overline{A}\overline{x} = \overline{x}^T P\overline{A}_d x_d + \overline{x}^T \hat{P}\overline{x} - x_d^T P_1 x_d \tag{16}$$

Let $\hat{P} = \begin{bmatrix} P_1 & 0 \\ 0 & 0 \end{bmatrix}$, we can arrive at

$$\dot{V} = \begin{bmatrix} \overline{x}^T & x_d^T \end{bmatrix} \Phi_0 \begin{bmatrix} \overline{x} \\ x_d \end{bmatrix} \tag{17}$$

where

$$\Phi_0 = \begin{bmatrix} \bar{A}^T P + P\bar{A} + \hat{P} & P\bar{A}_d \\ \bar{A}_d^T P & -P_1 \end{bmatrix}$$

Because $\psi = \Phi_0 + \begin{bmatrix} \bar{C}^T\bar{C} & 0 \\ 0 & 0 \end{bmatrix}$ and $\begin{bmatrix} \bar{C}^T\bar{C} & 0 \\ 0 & 0 \end{bmatrix} \geq 0$, from (13), we can have that

$\Phi_0 < 0$, such that the augmented system of $[\bar{x}^T \ x_d^T]^T$ is asymptotically stable. Then we will prove that the guaranteed cost control is satisfied.

From(13) and (17), it follows that

$$\dot{V}(t) = \begin{bmatrix} \bar{x}^T & x_d^T \end{bmatrix} \Phi_0 \begin{bmatrix} \bar{x} \\ x_d \end{bmatrix} < -\begin{bmatrix} \bar{x}^T & x_d^T \end{bmatrix} \begin{bmatrix} \bar{C}^T\bar{C} & 0 \\ 0 & 0 \end{bmatrix} \begin{bmatrix} \bar{x} \\ x_d \end{bmatrix} \tag{18}$$

When integrating the above formula from 0 to ∞, we obtain the following result, $V(\infty) - V(0) < -J$, namely, $J < V(0) - V(\infty)$. And as $T \to \infty, V(\infty) \to 0$, it follows that $J < V(0)$. Furthermore, we can get that

$$J^* = V(0) = \bar{x}^T(0)P\bar{x}(0) + \int_{-d}^{0} x^T(\tau)P_1 x(\tau)d\tau$$

which is a corresponding upper bound of the system performance. From Definition 1, the guaranteed cost control is satisfied. ∎

The conditions we give in Theorem 1. contain uncertain matrix F and are only sufficient conditions. Solution method for A_c, B_c, C_c is not given. We will give the solution method for A_c, B_c, C_c in Theorem 2.

Theorem 2. For system (1) and its corresponding performance function (6), if there exist scalars ε_1, ε_2, matrices A_c, B_c, C_c and symmetric positive definite matrices S, R, P_1 making the following inequalities simultaneously.

$$\begin{bmatrix} J_{11} & J_{12} \\ * & J_{22} \end{bmatrix} < 0, \quad \begin{bmatrix} R & I \\ I & S \end{bmatrix} > 0 \tag{19}$$

where

$$J_{11} = \begin{bmatrix} AR + RA^T + & A + \hat{A}^T & A_d \\ \hat{C}^T B^T + B\hat{C} & & \\ * & A^T S + SA + & SA_d \\ & C^T B^T + \hat{B}C & \\ * & * & -P_1 \end{bmatrix} \quad J_{12} = \begin{bmatrix} D & RE_1^T + & D & 0 & R & RQ^{1/2} & \hat{C}^T R^{1/2} \\ & \hat{C}^T E_2^T & & & & \\ SD + & E_1^T & SD & 0 & I & Q^{1/2} & 0 \\ \hat{B}D_1 & & & & & \\ 0 & 0 & 0 & E_d^T & 0 & 0 & 0 \end{bmatrix}$$

$$J_{22} = \text{diag}\left\{ -\varepsilon_1^{-1}I \quad -\varepsilon_1 I \quad -\varepsilon_2^{-1}I \quad -\varepsilon_2 I \quad -P_1^{-1} \quad -I \right\}$$

Then using a set of feasible solution ε_1, ε_2, A_c, B_c, C_c, S, R and P_1, we get invertible matrices M and N via singular value decomposition on I-RS. The dynamic output feedback guaranteed cost controller parameters of system (1) can be obtained from the following matrix equations.

$$\hat{A} = NA_c M^T + NB_c CR + SBC_c M^T + SAR$$
$$\hat{B} = NB_c, \ \hat{C} = C_c M^T \tag{20}$$

Proof

We divide the coefficient matrices of augmented system of $[\bar{x}^T \ x_d^T]^T$ into determining parameter matrices and uncertain parameter matrices, that is

$$\bar{A} = \bar{\mathbb{A}} + \Delta\bar{A}, \bar{A}_d = \bar{\mathbb{A}}_d + \Delta\bar{A}_d, \psi = \Phi_1 + \Phi_1'$$
$$\Phi_1 = \begin{bmatrix} \bar{\mathbb{A}}^T P + P\bar{\mathbb{A}} + \hat{P} + \bar{C}^T \bar{C} & P\bar{\mathbb{A}}_d \\ * & -P_1 \end{bmatrix}, \Phi_1' = \begin{bmatrix} \Delta\bar{A}^T P + P\Delta\bar{A} & P\Delta\bar{A}_d \\ * & 0 \end{bmatrix} \tag{21}$$

Let

$$\bar{\mathbb{D}} = \begin{bmatrix} D \\ B_c D_1 \end{bmatrix}, \bar{\mathbb{E}} = [E_1 \ E_2 C_c], \bar{D} = \begin{bmatrix} D \\ 0 \end{bmatrix} \tag{22}$$

then, we obtain

$$\Phi_1' = \Phi_{11}' + (\Phi_{11}')^T + \Phi_{12}' + (\Phi_{12}')^T \tag{23}$$

where

$$\Phi_{11}' = \begin{bmatrix} P\bar{\mathbb{D}} \\ 0 \end{bmatrix} F [\bar{\mathbb{E}} \ 0], \Phi_{12}' = \begin{bmatrix} P\bar{D} \\ 0 \end{bmatrix} F [0 \ E_d]$$

Using Lemma 2, we can get that

$$\Phi_1 + \Phi_{11}' + (\Phi_{11}')^T + \Phi_{12}' + (\Phi_{12}')^T < 0$$
$$\Leftrightarrow \Phi_1 + \Phi_{12}' + (\Phi_{12}')^T + \varepsilon_1 \begin{bmatrix} P\bar{\mathbb{D}} \\ 0 \end{bmatrix} \begin{bmatrix} P\bar{\mathbb{D}} \\ 0 \end{bmatrix}^T + \varepsilon_1^{-1} [\bar{\mathbb{E}} \ 0]^T [\bar{\mathbb{E}} \ 0] < 0 \tag{24}$$

From Lemma 1, (24) is equivalent to the following inequality

$$\Phi_4 = \begin{bmatrix} \bar{A}^T P + P\bar{A} & P\bar{A}_d & P\bar{D} & \bar{E}^T & P\bar{D} & 0 & \tilde{E} & \bar{C}^T \\ * & -P_1 & 0 & 0 & 0 & E_d^T & 0 & 0 \\ * & * & -\varepsilon_1^{-1}I & 0 & 0 & 0 & 0 & 0 \\ * & * & * & -\varepsilon_1 I & 0 & 0 & 0 & 0 \\ * & * & * & * & -\varepsilon_2^{-1}I & 0 & 0 & 0 \\ * & * & * & * & * & -\varepsilon_2 I & 0 & 0 \\ * & * & * & * & * & * & -P_1^{-1} & 0 \\ * & * & * & * & * & * & * & -I \end{bmatrix} < 0 \qquad (25)$$

Let

$$P = \begin{bmatrix} S & N \\ N^T & W \end{bmatrix}, P^{-1} = \begin{bmatrix} R & M \\ M^T & U \end{bmatrix}, \phi_1 = \begin{bmatrix} R & I \\ M^T & 0 \end{bmatrix}, \quad \phi_2 = \begin{bmatrix} I & S \\ 0 & N^T \end{bmatrix} \qquad (26)$$

then, it is easy to show that

$$P\phi_1 = \phi_2, \ \phi_1^T P\phi_1 = \phi_1^T \phi_2 = \begin{bmatrix} R & I \\ I & S \end{bmatrix}$$

We use left-multiplication matrix $\mathrm{diag}\{\phi_1^T \quad I \quad \cdots \quad I\}$ and right-multiplication matrix $\mathrm{diag}\{\phi_1 \quad I \quad \cdots \quad I\}$ methods on Φ_4, and let

$$\hat{A} = NA_c M^T + NB_c CR + SBC_c M^T + SAR$$
$$\hat{B} = NB_c, \hat{C} = C_c M^T$$

then the following inequality follows

$$\Phi_5 = \begin{bmatrix} J_{11} & J_{12} \\ * & J_{22} \end{bmatrix} < 0 \qquad (27)$$

where the definition of partitioned matrices are the same as Theorem 1 and the correlative controller parameters can be obtained from (20). ∎

It is noteworthy that P and its inversion appear in (26). So we cannot get the results directly via linear matrix inequality. Considering this, let $P = \delta I$, $\delta > 0$ [19]. Then we can get the solution conveniently by selecting parameter δ.

Considering $\hat{A}$, $\hat{B}$, $\hat{C}$ and other uncertain items are all global variables, now we give the solution method of dynamic output feedback controller parameters A_{cj}, B_{cj}, C_{cj}, $j=1, 2 \ldots r$ in each T-S fuzzy subsystem.

Before moving on, we define that

$$\hat{A}_{ij} = NA_{cj}M^T + NB_{cj}C_iR + SB_iC_{cj}M^T + SA_iR$$
$$\hat{B}_j = NB_{cj}, \hat{C}_j = C_{cj}M^T \tag{28}$$

and

$$J_{11ij} = \begin{bmatrix} A_iR + RA_i^T + \\ \hat{C}_j^T\hat{B}_i^T + B_i\hat{C}_j & A_i + \hat{A}_{ij}^T & A_{id} \\ \hline * & \begin{matrix} A_i^TS + SA_i + \\ C_i^T\hat{B}_j^T + \hat{B}_jC_i \end{matrix} & SA_{id} \\ \hline * & * & -P_1 \end{bmatrix} \quad J_{12ij} = \begin{bmatrix} D & \begin{matrix} RE_{1i}^T + \\ \hat{C}_j^TE_{2i} \end{matrix} & D & 0 & R & RQ^{1/2} & \hat{C}_j^TR^{1/2} \\ \hline \begin{matrix} SD + \\ \hat{B}_jD_1 \end{matrix} & E_{1i}^T & SD & 0 & I & Q^{1/2} & 0 \\ \hline 0 & 0 & 0 & E_{di}^T & 0 & 0 & 0 \end{bmatrix}$$

$$J_{22} = \mathrm{diag}\left\{ -\varepsilon_1^{-1}I \quad -\varepsilon_1 I \quad -\varepsilon_2^{-1}I \quad -\varepsilon_2 I \quad -P_1^{-1} \quad -I \right\}$$

Let

$$\psi_{ij} = \begin{bmatrix} J_{11ij} & J_{12ij} \\ * & J_{22} \end{bmatrix} \tag{29}$$

then, it follows that

$$\Phi_5 = \sum_{i=1}^{r} h_i \sum_{j=1}^{r} h_j \psi_{ij} = \sum_{i=1}^{r} h_i h_i \psi_{ii} + \sum_{i<j}^{r} h_i h_j (\psi_{ij} + \psi_{ji}) < 0 \tag{30}$$

In the following section, we will give the sufficient conditions which make Theorem 2 hold, and give the solution method for $A_{cj}, B_{cj}, C_{cj}, j=1, 2 \ldots r$.

Theorem 3. For system (1) and its corresponding performance function (6), if there exist scalars $\varepsilon_1, \varepsilon_2$, matrices $\hat{A}$, $\hat{B}$ and $\hat{C}$ and symmetric positive definite matrices S, R and P_1 making the following inequalities hold simultaneously

$$\psi_{ii} < 0, i = 1 \quad 2 \quad \cdots \quad r$$
$$\psi_{ij} + \psi_{ji} < 0, i < j \le r \tag{31}$$
$$\begin{bmatrix} R & I \\ I & S \end{bmatrix} > 0$$

Then the parameters of dynamic output feedback controller in each T-S fuzzy subsystem, $A_{cj}, B_{cj}, C_{cj}, j=1, 2 \ldots r$, can be obtained based on the method of Theorem 2.

The proof is similar to Theorem 8-8 of [20] and therefore omitted.

5   Simulation Example

In order to prove the effectiveness of our method, we give a simulation example [20].

$\dot{x}_1(t) = -0.1x_1^3(t) - 0.0125x_1(t-d) - 0.02x_2(t) - 0.67x_2^3(t) - 0.1x_2^3(t-d) - 0.005x_2(t-d) + u(t)$

$\dot{x}_2(t) = x_1(t)$

where

$$x_1(t) \in [-1.5, 1.5] \ , x_2(t) \in [-1.5, 1.5], d = 4s \ ,$$

$$M_{11}(x_2(t)) = 1 - \frac{x_2^2(t)}{2.25} \ , M_{12}(x_2(t)) = 1 - M_{11}(x_2(t)) .$$

The nonlinear system above can be expressed as T-S fuzzy models with uncertainties and time delays.

Rule1: if $x_2(t)$ is M_{11}, then $\dot{x}(t) = [A_1 + \Delta A_1]x(t) + A_{1d}x(t-d) + B_1u(t)$

Rule2: if $x_2(t)$ is M_{12}, then $\dot{x}(t) = [A_2 + \Delta A_2]x(t) + A_{2d}x(t-d) + B_2u(t)$

where

$$x(t) = [x_1(t) \quad x_2(t)]^T$$

$A_1 = \begin{bmatrix} -0.1125 & -0.02 \\ 1 & 0 \end{bmatrix}$, $A_{1d} = \begin{bmatrix} -0.0125 & -0.005 \\ 0 & 0 \end{bmatrix}$, $A_2 = \begin{bmatrix} -0.1125 & -1.527 \\ 1 & 0 \end{bmatrix}$,

$A_{2d} = \begin{bmatrix} -0.0125 & -0.23 \\ 0 & 0 \end{bmatrix}$, $B_1 = B_2 = [1 \ 0]^T$, $\Delta A_1 = DF(t)E_{11}$, $\Delta A_2 = DF(t)E_{21}$,

$D = [-0.1125 \ 0]^T$, $D1 = 0$, $E_{11} = E_{12} = [1 \ 0]$, $E_{d1} = E_{d2} = [0 \ 0]$, $E_{21} = E_{22} = 0$

,

$$C_1 = C_2 = [-1 \ 0], F^T(t)F(t) \leq I .$$

Take $\varepsilon = 0.1$, $R = 0.06$, $Q = \begin{bmatrix} 0.056 & 0 \\ 0 & 0.056 \end{bmatrix}$, $\varepsilon_1 = 80$, $\varepsilon_2 = 100$. From the methods proposed in this paper, we can get that

$A_{c1} = \begin{bmatrix} 1.5 & -205.1 \\ 10 & -1395 \end{bmatrix}$, $A_{c2} = \begin{bmatrix} 1.5 & -205.6 \\ 13.3 & -1862.3 \end{bmatrix}$, $B_{c1} = [-1424.8 \ -9688.4]^T$,

$B_{c2} = [-1428 \ -12957]^T$, $C_{c1} = [0.0172 \ -2.4015], C_{c2} = [0.0172 \ -2.4004]$.

The system state trajectory without any control is shown in Fig. 1. It is obvious that this system is a strong nonlinear divergence system.

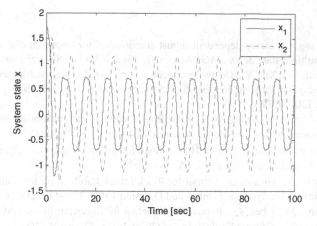

Fig. 1. Trajectories of the system states without control

When the designed controller of our paper is applied, the state trajectory varies. As shown in Fig. 2. We can see that it can guarantee the state to be stable and to converge to the origin in a period of time.

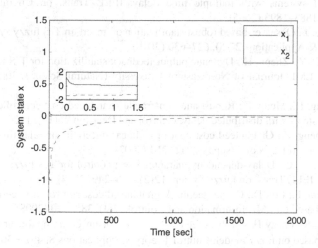

Fig. 2. Trajectories of the system states under the proposed DOF controller

6   Conclusions

This paper focuses on nonlinear system with uncertainties and time delays based on T-S fuzzy models. We design the guaranteed cost controller based on dynamic output feedback considering the immeasurability and impracticability of states. So the system can maintain better performance on the premise of stability. The simulation results of a nonlinear divergence system demonstrated the effectiveness of the proposed approach.

References

1. Gao, H., Wang, C.: Delay-dependent robust stabilization for uncertain discrete-time systems with multiple state delays. Acta Automatica Sinica 30(5), 789–795 (2004)
2. Jiang, Z., Gui, W., Xie, Y.: Delay-dependent decentralized robust stabilization for interconnected singular large-scale system with uncertainties. Control Theory & Application 26(11), 1303–1308 (2009)
3. Lee, Y.S., Moon, Y.S., Know, W.H., Park, P.G.: Delay-dependent robust $H\infty$ control for uncertain systems with a state-delay. Automatica 40(1), 65 – 72 (2004)
4. Fang, M.: Delay-dependent Robust $H\infty$ control for uncertain singular systems with state delay. Acta Automatica Sinica 35(1), 65 – 70 (2009)
5. Zheng, L., Liu, X., Huang, G.: Output feedback robust stabilization for a class of uncertain linear time-delay systems. Control and Decision 16(4), 339–442 (2001)
6. Sun, X., Jiang, M., Chen, Q.: Robust $H\infty$ control for uncertain linear systems with dependent time-delay. Control Engineering of China 16(4), 423 – 425 (2009)
7. Wang, J., Zhang, X.: Robust control of a uncertain generalized systems based on the LMI. Control Engineering of China 20(4), 691–693 (2013)
8. Takagi, T., Sugeno, M.: Fuzzy identification of systems and its application to control. IEEE Trans. on Systems, Man, Cybernetics 15(1), 116–132 (1986)
9. Hsiao, F.H., Chen, C.W., Liang, Y.W., Xu, S.D.: T-S fuzzy controllers for nonlinear interconnected systems with multiple time delay. IEEE Trans. on Circuits and Systems 52(9), 1883–1893 (2005)
10. Qi, L., Yang, J.: Observer-based robust stabilization of uncertain T-S fuzzy systems. Control Theory & Applications 27(5), 627–630 (2010)
11. Chen, Z., Jin, Y., Zheng, H.: Dynamic output feedback stabilization for T-S fuzzy control systems via LMI. Journal of Northeastern University (Natural Science) 30(5), 617–620 (2009)
12. Li, T., Zhang, H., Meng, F.: Robust and non-fragile guaranteed cost controller design for uncertain systems with distributed delay. Control and Decision 26(10), 1520–1529 (2011)
13. Feng, J., Cheng, Z.: Guaranteed cost control of linear uncertain singular time-delay systems. Control and Decision 7(suppl.), 711–714 (2002)
14. Guan, X., Chen, C.: Delay-dependent guaranteed cost control for T-S fuzzy systems with time delays. IEEE Trans. on Fuzzy System 12(2), 236–249 (2004)
15. Gao, Z., Qian, F., Liu, D.: Output feedback guaranteed cost control of uncertain stochastic systems based on LMI. Information and Control 37(4), 385–390 (2008)
16. Liu, Y., Hu, S.: Fuzzy $H\infty$ robust feedback control for uncertain nonlinear system with time-delay based on fuzzy model. Control Theory & Applications 8, 497 – 502 (2003)
17. Zhang, G.: Fuzzy Control of Nonlinear Systems based on T-S Model. Doctor Thesis of Xidian University, Xi'an (2003)
18. Li, Y.: Robust Dissipative Control for T-S Fuzzy Systems and its Application. Doctor Thesis of Harbin Institute of Technology, Harbin (2010)
19. Yu, L.: Robust Control-Linear Matrix Inequality approach. Tsinghua University Press, Beijing (2002)
20. Wu, Z.: Robust Control and Application of Nonlinear Systems. Machinery Industry Press, Beijing (2005)

Shape and Color Based Segmentation Using Level Set Framework

Xiang Gao, Ji-Xiang Du, Jing Wang, and Chuan-Min Zhai

Department of Computer Science and Technology, Huaqiao University, Xiamen 361021
rockey_gao@live.cn, jxdu77@gmail.com

Abstract. We propose a level set based variational approach that incorporates shape and color prior into Local Chan-Vese model for segmentation problem. Object detection and segmentation can be facilitated by the availability of a reference object. In our model, besides the level set function for segmentation, we introduce another labelling level set function to indicate the regions on which the prior shape and color should be compared. The active contour is able to find boundaries that are similar in shape and color to the prior, even when the entire boundary is not visible in the image. The experimental results demonstrate that the proposed model can efficiently segment the objects.

Keywords: Level Set Framework, shape prior, color prior, segmentation.

1 Introduction

Image segmentation is the process of dividing images into meaningful subsets that correspond to surfaces or objects. Numerous approaches have done great efforts and proposed a wide variety of methods for image segmentation. Recent works address that prior knowledge on the shape of interest can significantly facilitate segmentation processes [5, 7, 8]. Prior knowledge on the shape of interest can significantly facilitate these processes, particularly when the object boundaries missing data or occlusions. Color images carry much more information than gray-level ones, in many pattern recognition and computer vision applications, the color information can be used to enhance the image analysis process and improve segmentation results.

In this approach, our aim is to combine existing successful active contour and shape based image segmentation methods in segmentation problems. We use a novel level set model called Local Chan-Vese model, presented in [6] which utilize both global image information and local image information for image segmentation. By using the local image information, the images with intensity inhomogeneity can be efficiently segmented in limited iterations. Our contribution in this paper is using shape prior information which constrains the active contour to get a shape of interesting, concurrently with the segmentation we compare the current segmented image with the prior image's color similarity. Combined shape and color information will make the segmentation more robust and accurate.

D.-S. Huang et al. (Eds.): ICIC 2014, LNAI 8589, pp. 265–270, 2014.

2   Local Chan–Vese Model

Now, we present local Chan–Vese model (LCV). Traditional level set methods used either the image gradient or the global information to drive the evolving curve towards the true boundaries. The LCV model combined both global and local statistical information to overcome the inhomogeneous intensity distribution. By using the local image information, the images with intensity inhomogeneity can be efficiently segmented in limited iterations. The time-consuming re-initialization step in level set methods can be avoided by introducing a new penalizing energy.

The overall energy functional in LCV model can be described as:

$$E^{LCV} = \alpha \cdot E^G + \beta \cdot E^L + E^R \tag{1}$$

The global term E^G is defined based on the global properties, i.e., the averages of u_0 inside C and outside C, as follows:

$$E^G(c_1, c_2, C) = \int_{inside(C)} |u_0(x, y) - c_1|^2 dxdy + \int_{outside(C)} |u_0(x, y) - c_2|^2 dxdy \tag{2}$$

The local term E^L uses the local statistical information help to improve the segmentation capability of model.

$$E^L(d_1, d_2, C) = \int_{in(C)} |g_k * u_0(x, y) - u_0(x, y) - d_1|^2 dxdy + \int_{out(C)} |g_k * u_0(x, y) - u_0(x, y) - d_2|^2 dxdy \tag{3}$$

where g_k is a averaging convolution operator with k×k size window. d_1 and d_2 are the intensity averages of difference image ($g_k * u_0(x, y) - u_0(x, y)$) inside C and outside C.

We add to the regularization term E^R a length penalty term L(C) which is defined related to the length of the evolving curve C and a penalty term that can force the level set function to be close to a signed distance function. Then the regularization term E^R should be composed of two terms:

$$E^R(\phi) = \mu \cdot \int_\Omega \delta(\phi(x, y)) |\nabla \phi(x, y)| dxdy + \int_\Omega \frac{1}{2} (|\nabla \phi(x, y)| - 1)^2 dxdy \tag{4}$$

Where μ is the parameter which can control the penalization effect of length term.

So, the overall energy functional (2, 3, 4) can be further described as follows:

$$\begin{aligned} E^{LCV}(c_1, c_2, d_1, d_2, \phi) = &\int_\Omega (\alpha \cdot |u_0(x, y) - c_1|^2 + \beta \cdot |g_k * u_0(x, y) - u_0(x, y) - d_1|^2) H_\varepsilon(\phi(x, y)) dxdy \\ &+ \int_\Omega (\alpha \cdot |u_0(x, y) - c_2|^2 + \beta \cdot |g_k * u_0(x, y) - u_0(x, y) - d_2|^2)(1 - H(\phi(x, y))) dxdy \\ &+ \mu \cdot \int_\Omega \delta_\varepsilon(\phi(x, y)) |\nabla \phi(x, y)| dxdy + \int_\Omega \frac{1}{2} (|\nabla \phi(x, y)| - 1)^2 dxdy \end{aligned} \tag{5}$$

Where $H_\varepsilon(z)$ and $\delta_\varepsilon(z)$ are the regularized approximation of Heaviside function and Dirac delta function as follows:

$$H_\varepsilon(z) = \frac{1}{2} \left| 1 + \frac{2}{\pi} \arctan \left| \frac{z}{\varepsilon} \right| \right| \qquad \delta_\varepsilon(z) = \frac{1}{\pi} \cdot \frac{\varepsilon}{\varepsilon^2 + z^2} \tag{6}$$

Where C_1 and C_2 evolved concurrently with the evolution of $\phi(x, y)$ by:

$$c_1(\phi) = \frac{\int_\Omega u_0(x, y) H_\varepsilon(\phi(x, y)) dxdy}{\int_\Omega H_\varepsilon(\phi(x, y)) dxdy} \qquad c_2(\phi) = \frac{\int_\Omega u_0(x, y)(1 - H_\varepsilon(\phi(x, y))) dxdy}{\int_\Omega (1 - H_\varepsilon(\phi(x, y))) dxdy} \qquad (7)$$

3   Prior Shape Model and Color Consistency Constraint

3.1   Prior Shape Model

There is a serious practical problem with level set method that if the boundary has 'leak' or 'gap', standard level set method do not have any information about how the gaps are to be bridged. One solution to the problem is to incorporate into the energy functional some prior information about the expected shape of the boundary. Shape prior information can aid the segmentation problems involving missing data or occlusions. The shape prior can be defined by level set function. This description can naturally consistent with the level set framework of active contours.

Then, by the given prior image, prior shape term can be described as E^p. The shape term incorporated in the energy functional measures the non-overlapping areas between the prior shape and the zero level-set of ϕ. Let ϕ_p be a labeling function of the shape in the prior image. Thus, the shape term takes the form:

$$E^p(\phi, \phi_p) = \int_\Omega (H(\phi(x, y)) - H(\phi_p(x, y)))^2 dxdy \qquad (8)$$

3.2   Color Consistency Constraint

We adopt the color feature in this paper because it is usually the most dominant and distinguishing visual feature and the level set method is just apply on binary image, lacks other meaningful features-such as color and texture.

Color is perhaps the most dominant and distinguishing visual feature. Color histogram is the most widely used color descriptor in content based retrieval research. A color histogram captures global color distribution in an image. Color histograms are easy to compute, and they are invariant to the rotation and translation of image content. The RGB space has been widely used due to the availability of images in the RGB format from image scanners. Regardless of the color space, color information in an image can be represented either by three separate 1-D histograms. A suitable normalization of the histograms also provides scale invariance. Let H(i) be a histogram of an image. Then, the normalized histogram I is defined as follows:

$$I(i) = \frac{H(i)}{\sum_i H(i)} \qquad (9)$$

Euclidean distance is used as a metric to compute the distance between the feature vectors. Let P_R, P_G and P_B be the normalized color histograms of prior image and let I_R, I_G and I_B be the normalized color histograms of the image. The similarity between the prior image and the current segmented image $D(P, I)$ is given by the following equation:

$$D(P,I) = \sqrt{\frac{\sum_r (P_R - I_R)^2 + \sum_r (P_G - I_G)^2 + \sum_r (P_B - I_B)^2}{2 \times 3}} \tag{10}$$

4   Proposed Framework and Algorithm

We now describe our proposed framework. By combining shape term (8) with LCV energy functional, our energy functional can be written as:

$$E^{PLCV} = \alpha \cdot E^G + \beta \cdot E^L + \upsilon \cdot E^P + E^R \tag{11}$$

Or explicitly:

$$
\begin{aligned}
E^{PLCV}(c_1, c_2, d_1, d_2, \phi, \phi_p) &= \int_\Omega (\alpha \cdot |u_0(x,y) - c_1|^2 + \beta \cdot |g_k * u_0(x,y) - u_0(x,y) - d_1|^2) H_\varepsilon(\phi(x,y)) dxdy \\
&+ \int_\Omega (\alpha \cdot |u_0(x,y) - c_2|^2 + \beta \cdot |g_k * u_0(x,y) - u_0(x,y) - d_2|^2)(1 - H(\phi(x,y))) dxdy \\
&+ \mu \cdot \int_\Omega \delta_\varepsilon(\phi(x,y)) |\nabla \phi(x,y)| dxdy + \int_\Omega \frac{1}{2} (|\nabla \phi(x,y)| - 1)^2 dxdy \\
&+ \int_\Omega (H(\phi(x,y)) - H(\phi_p(x,y)))^2 dxdy
\end{aligned} \tag{12}
$$

c_1 and c_2 calculate in (7), d_1 and d_2 are given by:

$$d_1(\phi) = \frac{\int_\Omega (g_k * u_0(x,y) - u_0(x,y)) H_\varepsilon(\phi(x,y)) dxdy}{\int_\Omega H_\varepsilon(\phi(x,y)) dxdy}, d_2(\phi) = \frac{\int_\Omega (g_k * u_0(x,y) - u_0(x,y))(1 - H_\varepsilon(\phi(x,y))) dxdy}{\int_\Omega (1 - H_\varepsilon(\phi(x,y))) dxdy} \tag{13}$$

Then we can deduce the associated Euler–Lagrange equation for ϕ .The minimization of (12) can be done by introducing an artificial time variable $t \geq 0$, and moving ϕ in the steepest descent direction to a steady state. The associated gradient equation for ϕ is then:

$$
\begin{aligned}
\frac{\partial \phi}{\partial t} &= \delta(\phi)[-(\alpha(u_0 - c_1)^2 + \beta(g_k * u_0(x,y) - u_0(x,y) - d_1)^2) + (\alpha(u_0 - c_2)^2 + \beta(g_k * u_0(x,y) - u_0(x,y) - d_2)^2)] \\
&+ [\mu \delta_\varepsilon(\phi) \text{div}(\frac{\nabla \phi}{|\nabla \phi|}) + (\nabla^2 \phi - \text{div}(\frac{\nabla \phi}{|\nabla \phi|}))] + 2\delta(\phi)\left[H(\phi(x,y)) - H(\phi_p(x,y)) \right]
\end{aligned} \tag{14}
$$

Then, we can summarize the proposed algorithm for segmentation:

1. Choose an initial level-set function $\phi_{t=0}$ that determines the initial segmentation contour.
2. Set the value of parameters. In practice, $\Delta t = 0.1, \varepsilon = 1$, set the window size k of averaging convolution operator in local term. Set the values of the controlling parameter of global term ∂, the controlling parameter of local term β, and length controlling parameter μ .
3. Update ϕ using the gradient descent equation according to (14).
4. Repeat steps 3 until convergence.
5. Compare the color similarity between segmentation result and prior image.

5   Experiments and Results

In this Section, we present the experimental results of our model. In all experiments we set: $\Delta t = 0.1$, $\varepsilon = 1$, k=15. The contributions of the first three terms are normalized to [−1, 1], iterations are specified by the user. It can be found that we were able to get good segmentation results on a wide range of images.

We firstly considered the LCV model's advantage in less number of iterations and no sensitive to initial contour. Fig.1 (a) is the original image, 1(b) is the initial contour for level set function, (c) and (d) is the segmentation result of CV model and LCV model, total consumption of time and iterations shown in table 1. Next, we change initial contour that do not include the object, (e) shows the changed initial contour. (f) and (g) shows the segmentation result of CV model and LCV model.

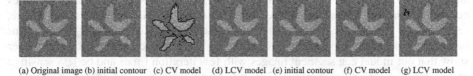

(a) Original image (b) initial contour   (c) CV model   (d) LCV model   (e) initial contour   (f) CV model   (g) LCV model

Fig. 1. Test image and the segmentation result of CV model and LCV model

Table 1. Iteration number and processing time of CV model and LCV model

	Iteration number	Processing time (s)
CV	800	16.8949
LCV	100	6.4740

Next, consider the object boundaries missing data or occlusions. Fig. 2a is the prior image. (b) is the image to segment which has a gap. Segmentation result using LCV model is shown in (c). (d) shows Successful Segmentation using our model.

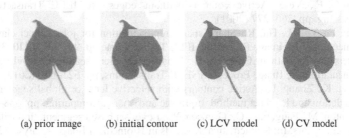

(a) prior image   (b) initial contour   (c) LCV model   (d) CV model

Fig. 2. Segmentation results of LCV model and proposed model

We next consider a color image. Fig. 3a shows prior image, it is a ripe banana, we want to split ripe banana from an image. (b) is the image to segment. (c) shows segmentation without prior knowledge of LCV model, all bananas are split out. The desired segmentation result using our proposed method is shown in Fig. 3d.

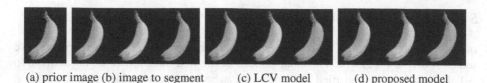

(a) prior image (b) image to segment   (c) LCV model   (d) proposed model

Fig. 3. Segmentation results of proposed framework

6   Conclusion

In the present paper, we introduce the modified Local level set framework used local information instead of edge alignment constraint. And on this basis we proposes a novel more robust shape-based segmentation approach based on level set techniques depending on both color and shape prior information. Given a reference image, segmentation process is carried out concurrently with a comparison of the prior instance of the object to the image to segment. The approach proposed in our paper show that our model can segment images with few iteration times and be less sensitive to the location of initial contour. The following work we can using an efficient algorithm extraction the prior shape and color information from the existing image database. More work is needed to address these issues.

Acknowledgments. This work was Supported by the Grant of the National Science Foundation of China (No.61175121, 61370006, 61100106), the Grant of the National Science Foundation of Fujian Province (No.2013J06014), the Promotion Program for Young and Middle-aged Teacher in Science and Technology Research of Huaqiao University (No.ZQN-YX108, ZQN-PY116).

References

[1] Chan, T.F., Vese: Active contours without edges. In: IEEE Transaction on Image Processing, pp. 266–277 (2001)

[2] Chen, F., Yu, H., Hu, R.: Shape sparse representation for joint object classification and segmentation. In: Image Processing IEEE Transactions, pp. 992–1004 (2013)

[3] Andersson, T., Lathen, G., Lenz, R.: Modified gradient search for level set based image segmentation. In: Image Processing IEEE Transactions, pp. 621–630 (2013)

[4] Zhang, K., Zhang, L.: Active contours with selective local or global segmentation: A new formulation and level set method. In: Image and Vision Computing, pp. 668–676 (2010)

[5] Leventon, M., Faugeraus, O., Grimson, W.: Level set based segmentation with intensity and curvature priors. In: Mathematical Methods in Biomedical Image Analysis (2000)

[6] Wang, X.F., Huang, D.S., Xu, H.: An efficient local Chan-Vese model for image segmentation. In: Pattern Recognition, pp. 603–618 (2010)

[7] Cremers, D., Sochen, N., Schnorr, C.: Towards recognition-based variational segmentation using shape priors and dynamic labeling. In: Griffin, L.D., Lillholm, M. (eds.) Scale-Space 2003. LNCS, vol. 2695, pp. 388–400. Springer, Heidelberg (2003)

[8] Riklin-Raviv, T., Kiryati, N., Sochen, N.: Prior-Based Segmentation by Projective Registration and Level Sets. In: Int'l Conf. Computer Vision, pp. 204–211 (2005)

An Integrated NRSFM Approach for Image Sequences with Small Size[*]

Ya-Ping Wang[1], Zhan-Li Sun[1,**], and Li Shang[2]

[1] School of Electrical Engineering and Automation,
Anhui University, Hefei, China
[2] Department of Communication Technology, Electronic Information Engineering College,
Suzhou Vocational University, Suzhou, China
zhlsun2006@126.com

Abstract. In this paper, a sub-sequence based integrated algorithm is proposed to deal with the non-rigid structure from motion (NRSFM) with small sequence size. In the proposed method, multiple sub-sequences are first extracted from the original sequence. Then, the extracted sub-sequences are used as the inputs of a recently reported NRSFM algorithm with rotation invariant kernel (RIK-NRSFM). Finally, the 3D coordinates estimated by the RIK-NRSFM algorithm are integrated to obtain the final estimation results. Experimental results on two widely used image sequences demonstrate the effectiveness and feasibility of the proposed algorithm.

Keywords: Sub-sequence, small sequence size, weaker estimator.

1 Introduction

Non-rigid structure from motion (NRSFM) is the process of recovering the time varying 3D coordinates of the feature points on a deforming object by utilizing the corresponding 2D points located on a sequence of 2D images [1]. In the NRSFM model, the objects generally undergo a series of shape deformations and pose variations. Thus, in the absence of necessary prior knowledge on shape deformation, reconstructing non-rigid structure from motion becomes a difficult ill-posed problem.

Currently, the large majority of works in NRSFM are the variants of the standard matrix factorization method [1]. As NRSFM is an under-constrained problem, recent research works have attempted to define some effective constraints to make NRSFM more tractable [2, 3]. In [4, 5], smoothness constraints are enforced on camera motion

---

[*] The work was supported by two grants from National Natural Science Foundation of China (Nos. 61370109, 61373098), a grant from Natural Science Foundation of Anhui Province (No. 1308085MF85), and 2013 Zhan-Li Sun's Technology Foundation for Selected Overseas Chinese Scholars from department of human resources and social security of Anhui Province (Project name: Research on structure from motion and its application on 3D face reconstruction).

[**] Corresponding author.

D.-S. Huang et al. (Eds.): ICIC 2014, LNAI 8589, pp. 271–276, 2014.

and object deformation. In order to compress the shape space of possible solutions, the kernel trick is applied in the NRSFM to complement the low-rank constraint by capturing non-linear relationships in the shape coefficients of the linear model [6]. Further, to deal with the data lacking temporal ordering or with abrupt deformation, a novel NRSFM with rotation invariant kernel (RIK-NRSFM) is proposed in [7] by utilizing the spatial variation constraint.

In practice, the quantity of available high quality images may be limited in many cases, such as the face images in a monitor system, etc. In this paper, a sub-sequence based integrated algorithm is proposed to deal with the small sequence problem. In the proposed method, the 3D coordinates of each frame is estimated one by one. For a test frame, except for itself, a few frames are first randomly extracted from the original sequence. Then, the extracted frames are combined with the test frame to form a sub-sequence, and used as the input of RIK-NRSFM. Similar to the classifier committee learning, the sub-sequence and the estimation process of RIK-NRSFM constitute a weaker estimator. Finally, the z-coordinates obtained by multiple weaker estimators are integrated and used as the final estimation of the test frame. Experimental results on two sequences demonstrate the effectiveness and feasibility of the proposed method.

The remainder of the paper is organized as follows. The proposed method is presented in Section 2. Experimental results are given in Section 3. . Finally, conclusions are made in Section 4.

2   Methodology

2.1   Sub-sequence Extraction

The first step of our proposed method is to extract the sub-sequences from the original sequence. Assume that $[x_{t,j}, y_{t,j}]^T (t = 1, 2, ..., n)$ is the 2D projection of the jth 3D point observed on the tth image, the n 2D point tracks of F images can be represented as a $2F \times n$ observation matrix $\mathbf{W}$. For the tth image, the observation $\mathbf{w}_t$ is a $2 \times n$ matrix,

$$\mathbf{w}_t = \begin{pmatrix} x_{t,1} & x_{t,2} & \cdots & x_{t,n} \\ y_{t,1} & y_{t,2} & \cdots & y_{t,n} \end{pmatrix}. \tag{1}$$

The observations $\mathbf{w}_t (t = 1, ..., F)$ of F images consist of the original sequence.

When the 3D coordinates of the tth image are to be estimated, a sub-sequence $\mathbf{W}_{s,j}$ can be constructed by randomly selecting $F_s - 1$ observations from $\mathbf{W}_r = [\mathbf{w}_1^T, ..., \mathbf{w}_{t-1}^T, \mathbf{w}_{t+1}^T, \mathbf{w}_F^T]^T$, and merging them with $\mathbf{w}_j$. When the process is repeated for N times, N sub-sequences can be obtained.

2.2   RIK-NRSFM Based Weaker Estimator

For each test frame, we can extract N sub-sequences $\mathbf{W}_{s,j}$. Further, we input $\mathbf{W}_{s,j}$ to the RIK-NRSFM algorithm [7] to restructure the 3D coordinates of the sub-sequence. According to the linear subspace model, $\mathbf{W}_{s,j}$ is factorized as:

$$\mathbf{W}_{sj} = \mathbf{MS} = \mathbf{D}(\mathbf{C} \otimes \mathbf{I}_3)\mathbf{S}, \tag{2}$$

where the matrices $\mathbf{D} \in R^{2F \times 3F}$, $\mathbf{C} \in R^{F \times K}$, $\mathbf{I}_3$ represents a block-diagonal rotation matrix, a shape coefficient matrix and a 3×3 identity matrix, respectively. $\mathbf{S}$ is the K basis shapes. Let $\mathbf{c}_t^T$ be the tth row of $\mathbf{C}$, and the 3D shape of tth image is modeled as

$$S(\mathbf{c}_t^T) = (\mathbf{c}_t^T \otimes \mathbf{I}_3)\mathbf{S} = \sum_{k=1}^{K} c_{t,k} \hat{\mathbf{S}}_k \tag{3}$$

$\mathbf{S}_k \in \mathbb{R}_{3 \times n}$ is the 3D basis shape, and K is the number of 3D basis shapes. In [7], c_t^T is represented as a nonlinear mapping of each 2D shape $\mathbf{w}^t$ via the kernel function,

$$c_{t,k} = f_k(\mathbf{w}^t) = \sum_{i=1}^{d} \kappa(\mathbf{w}^t, \mathbf{w}_b^i)x_{ik}, \tag{4}$$

After computing a complete kernel matrix $\mathbf{K}_{ww_b}$ and its eigenvector matrix $\mathbf{V}$ associated with the d largest eigenvalues in the diagonal matrix $\mathbf{\Lambda}$, each observation $\Phi(\mathbf{W})^T$ is projected on to the eigenfunction subspace. A new basis matrix $\mathbf{B}$ of $\mathbf{C}$ is defined as:

$$\mathbf{B} = \mathbf{K}_{ww}\mathbf{V}\mathbf{\Lambda}^{-1/2}. \tag{5}$$

Correspondingly,

$$\mathbf{M} = \mathbf{D}(\mathbf{BX} \otimes \mathbf{I}_3). \tag{6}$$

According to (4),

$$\mathbf{S} = \mathbf{M}^\dagger \mathbf{W} \tag{7}$$

where $\mathbf{M}^\dagger$ denotes the Moore-Penrose pseudo-inverse of $\mathbf{M}$. In terms of (4) and (7) the 3D shape of the tth image can be given by

$$S(\mathbf{c}_t^T) = S(f(\mathbf{w}^t)) = (\mathbf{c}_t^T \otimes \mathbf{I}_3)\mathbf{M}^\dagger \mathbf{W}. \tag{8}$$

2.3   Integration of Weaker Estimators

For the jth test frame, we can see from Section 2.2 that one set of estimated $\mathbf{z}_{sj}(j = 1, \dots, N)$ can be obtained for each sub-sequence $\mathbf{W}_{sj}$. Therefore, each input

$\mathbf{W}_{sj}$ and the reconstruction model can be considered as a weaker estimator. Further, in order to integrate the results obtained by N weaker estimators, the mean $\hat{\mathbf{z}}_t$ of $\mathbf{z}_{t1},...,\mathbf{z}_{tN}$ is computed as

$$\hat{\mathbf{z}}_t = \frac{\mathbf{z}_{t1} + ... + \mathbf{z}_{tN}}{N}, \tag{9}$$

and used as the final estimated z-coordinates of the tth test image.

3   Experimental Results

We evaluate the performance of our proposed method on 2 widely used sequences: stretch [7] and the Bosphorus database [8]. As the problem addressed in this paper is the small size sequence, we first select 15 images from the stretch sequence and use them as the experimental data. In experiments, F_s is set as 6, and the trials are repeated for 10 times, i.e. N=10. To evaluate the estimation accuracy, the correlation coefficient between the true z-coordinates z and the estimated z-coordinates $\hat{z}$ is used as the performance index. Moreover, to verify the performance of our sub-sequence based integrated RIK-NRSFM algorithm (denoted as SSI-RIK-NRSFM), we compare it to the original RIK-NRSFM method [7].

Table 1. The correlation coefficients of two algorithms on the stretch sequence when the sample size is 15

	RIK-NRSFM	SSI-RIK-NRSFM
1	0.1700	0.9668
2	0.2425	0.9655
3	0.3295	0.9822
4	0.4602	0.9850
5	0.6325	0.9943
6	0.8696	0.9896
7	0.9205	0.9950
8	0.9497	0.9965
9	0.9601	0.9943
10	0.9551	0.9928
11	0.9207	0.9905
12	0.9010	0.9857
13	0.8933	0.9832
14	0.8843	0.9804
15	0.8711	0.9785
μ	0.7307	0.9854
σ	0.0813	9.3e-5

Table 1 shows the correlation coefficients of two algorithms on the stretch sequence. In Table 1, the numbers 1 to 15 denote the 1th to the 15th frame. The last two rows show the mean (μ) and the standard deviation (σ) of the 15 frames. We can see from Table 1 that the performance of SSI-RIK-NRSFM is significantly improved compared to RIK-NRSFM.

Table 2. The correlation coefficients between the true and estimated z coordinates on different sample sizes of one individual of the Bosphorus database

	7	8	9	10	11	12
RIK-NRSFM	0.6047	0.4209	0.1721	0.3673	0.4356	0.4780
SSI-RIK-NRSFM	0.6992	0.7883	0.7478	0.7791	0.7738	0.8114
	13	14				
RIK-NRSFM	0.6219	0.6750				
SSI-RIK-NRSFM	0.8170	0.8196				

Further, we present the experimental results for the real Bosphorus database. In experiments, the z -coordinates of the frontal-view image are estimated. Table 2 shows the correlation coefficients when the sequence sizes are varied from 7 to 14 for one individual. It can be seen that the proposed method achieves a better performance than the RIK-NRSFM method.

4   Conclusions

In this paper, a sub-sequence based RIK-NRSFM algorithm is proposed for NRSFM with small size sequence. The experimental results on the artificial data and the real data verified the effectiveness and feasibility of the proposed method.

References

1. Bregler, C., Hertzmann, A., Biermann, H.: Recovering Non-Rigid 3D Shape from Image Streams. In: Proceedings of IEEE Computer Vision and Pattern Recognition, pp. 690–696 (2000)
2. Akhter, I., Sheikh, Y., Khan, S.: In Defense of Orthonormality Constraints for Non-Rigid Structure from Motion. In: Proceedings of IEEE Computer Vision and Pattern Recognition, pp. 1534–1541 (2009)
3. Paladini, M., Del Bue, A., Stosic, M., Dodig, M., Xavier, J., Agapio, L.: Factorization for Non-Rigid and Articulated Structure Using Metric Projections. IEEE Transactions on Pattern Analysis and Machine Intelligence, 2898 - 2905 (2009)

4. Gotardo, P.F.U., Martinez, A.M.: Computing Smooth Time-Trajectories for Camera And Deformable Shape in Structure from Motion with Occlusion. IEEE Transactions on Pattern Analysis and Machine Intelligence, 2051–2065 (2011)
5. Akhter, I., Sheikh, Y.A., Khan, S., Kanade, T.: Nonrigid Structure from Motion in Trajectory Space. Neural Information Processing Systems (2008)
6. Gotardo, P.E.U., Martinez, A.M.: Kernel Non-Rigid Structure from Motion. In: Proceedings of IEEE International Conference on Computer Vision, pp. 802–809 (2011)
7. Hamsici, O.C., Gotardo, P.F.U., Martinez, A.M.: Learning Spatially-Smooth Mappings in Non-Rigid Structure from Motion. In: Fitzgibbon, A., Lazebnik, S., Perona, P., Sato, Y., Schmid, C. (eds.) ECCV 2012, Part IV. LNCS, vol. 7575, pp. 260–273. Springer, Heidelberg (2012)
8. http://bosphorus.ee.boun.edu.tr/

Image Super-Resolution Reconstruction
Based on Two-Stage Dictionary Learning

Li Shang[1] and Zhan-li Sun[2]

[1] Department of Communication Technology,
College of Electronic Information Engineering,
Suzhou Vocational University, Suzhou 215104, Jiangsu, China
[2] Department of Automation, College of Electrical Engineering and Automation,
Anhui University, Hefei 230601, Anhui, China
sl0930@jssvc.edu.cn, zhlsun2006@yahoo.com.cn

Abstract. The general image super-resolution reconstruction (SRR) methods based on sparse representation utilizes the one-stage high and low resolution dictionary pairs to reconstruct a high resolution image, and this method can not restore much image detail information. To solve this detect, two-stage high and low resolution dictionaries are explored here. The goal of exploiting the two-stage dictionaries is to reconstruct the difference image between the original high resolution image and the reconstructed image obtained by using the one-stage dictionaries. In learning two-stage dictionaries, the difference image is used as the high resolution (HR) image, and the first-order and second-order gradient feature images of the one-stage reconstructed images are used as the low resolution (LR) images. Then, the two-stage dictionaries are learned by K-singular value decomposition (K-SVD) method. In test, an artificial and a real LR image are used, and simulation results show that, compared with other learning-based methods, our method proposed has remarkable improvement in PSNR and visual effect.

Keywords: Super-resolution reconstruction (SRR), Sparse representation, K- singular value decomposition (K-SVD), One-stage dictionaries, Two-stage dictionaries, Low resolution, High resolution.

1   Introduction

At present, image super-resolution reconstruction (SRR)methods based on learning have been the research hot [1-2]. The latest learning based method is that proposed by Yang et al [3-4] denoted SC-SRR method, here SC is the abbreviation of sparse coding. The theory of SC-SRR is that the coefficients of high resolution (HR) images can be represented by using those of low resolution (LR) images [5-7]. This method can restore certain high frequency details of images, but the reconstructed results have certain block effect to some extent. To restore much image details and improve the quality of reconstructed images, on the basis of SC-SRR, the two-stage HR and LR dictionary learning is proposed here. The one-stage HR and LR dictionaries are

D.-S. Huang et al. (Eds.): ICIC 2014, LNAI 8589, pp. 277–284, 2014.

learned by SC-SRR method. Utilizing the one-stage reconstructed image as the HR features, as well as the first order and the second order gradient features of the one-stage reconstructed image as LR features, the two-stage HR and LR dictionaries can be learned by the K- singular value decomposition (K-SVD) method [8-10]. In test, an artificial and a real millimeter wave (MMW) image are utilized. Compared with other learning-based methods, simulation results show that our method has great improvement in image structural similarity and visual effect.

2   Sparse Representation Based SRR Method

Let the matrix $D \in \mathfrak{R}^{N \times K}$ ($n << K$) be an over-complete dictionary of K prototype atoms, and supposed a vector $x \in \mathfrak{R}^N$ can be represented as a sparse linear combination of these atoms. Thus, this signal x can be approximately written as

$$\min_s \|s\|_p, \quad s.t. \quad \|x - Ds\|_2^2 \le \varepsilon \quad . \tag{1}$$

where $s \in \mathfrak{R}^K$ represents the sparse coefficient vector [5], $\|\cdot\|_p$ and $\|\cdot\|_2$ denote respectively the l_p and l_2 norm. Assumed that x_h and x_l denote respectively a HR and LR image patch vector, according to Eqn. (1), x_h can be approximated by the equation of $x_h \approx D_h s$, and x_l can be approximated by $x_l = T x_h \approx T D_h s$, here T is the mapping matrix. Thus, the LR dictionary D_l can be calculated by $D_l = T D_h$. Namely, the HR image patch x_h can be restored by using the equation of $x_h = D_h s$. Commonly, for an image I , to improve the quality of reconstructed images, it is randomly sampled L times with $p \times p$ patch, denoted by matrix $X = \{x_1, x_2, \cdots, x_L\} \in \mathfrak{R}^{N \times L}$, thus, the dictionary D is learned from X , and the optimized problem is described as

$$\{D, S\} = \arg \min_{D,S} \|X - DS\|_F^2 + \lambda \|S\|_1 \quad . \tag{2}$$

subject to $\|d_k\|_2^2 \le 1$ ($k = 1, 2, 3, \cdots, K$), d_k is the $k th$ atom of the dictionary D , and S denotes the sparse coefficient matrix.

3   Training Two-Stage Dictionaries

3.1   SC- SRR Based One-Stage Dictionary Pairs Learning

The one-stage HR and LR dictionary pairs are trained here by Yang's SC-SRR method [4-5]. This problem   by generalizing the basic SC scheme as follows [4-5]

$$\min_{D_h, D_l, \{s_i^{xh/xl}\}_{i=1}^L} \sum_{i=1}^L \left\{ \|x_{hi} - D_h s_i\|_2^2 + \|x_{li} - D_l s_i\|_2^2 \right\} + \lambda \|s_i\|_1 \quad . \tag{3}$$

Further, to guarantee the compatibility between adjacent patches, Yang et al modified the optimization problem of Equation (3), which is written as follows [6-7]

$$\min_{s}\left\{\frac{1}{2}\left\|\begin{bmatrix}Fx_{li}\\ \lambda x_{hi}\end{bmatrix}-\begin{bmatrix}FD_l\\ \lambda PD_h\end{bmatrix}\cdot s_i\right\|_2^2+\lambda\|s_i\|_1\right\}=\left\{\frac{1}{2}\left\|\tilde{y}-\tilde{D}s_i\right\|_2^2+\lambda\|s_i\|_1\right\}. \tag{4}$$

Where Eqns. (3) and (4) are subjected to $\|D_h(:,k)\|_2 \leq 1$ and $\|D_l(:,k)\|_2 \leq 1$ ($k=1,2,3,\cdots,K$). Parameter P extracts the region of overlap between current target patch and previously reconstructed high-resolution image, and F is a feature extraction operator. Given the optimal solution s^*, the HR patch can be reconstructed as $\hat{x}_h = D_h s^*$, and the patch $\hat{x}_h$ is put into the HR image $\hat{X}_{h1}$ set.

3.2   Two-Stage Dictionary Pairs Learning

Assumed that the original HR image is denoted by I_h, and it can be represented as $I_h = \hat{I}_1 + I_e$, where I_e is the difference image between the HR image I_h and the one-stage reconstructed image $\hat{I}_1$. The goal of training two-stage high and LR dictionaries is to reconstruct the difference image and make it appreciate I_e. Here, the training of two-stage dictionaries are mainly generalized as follows:

Step 1. Ascertaining the HR image patch set X^{h2}. The difference image I_e is used as the HR image. Randomly sampled I_e image L times with $d \times d$ pixels, the HR image patch set $X^{h2} = \{x_1^{h2}, x_2^{h2}, \cdots, x_L^{h2}\}$ with the size of $d^2 \times L$ can be obtained.

Step 2. Ascertaining the LR image patch set X^{l2}. Because the high frequency information of LR images has important effect on predicting detail information of HR images, for the one-stage reconstructed image $\hat{I}_1$, four one dimension gradient filters [19] are used to improve the prediction precision of HR images' details. Utilizing four gradient filters, the first order and second order gradient information of $\hat{I}_1$, distributing in vertical and horizontal direction, can be obtained. Therefore, the four gradient feature images obtained are used as the LR features of the two-stage LR dictionary. For each gradient feature image, the image patch random sampling method is the same as that utilized in Step 1 so as to obtain a LR image patch set with $d^2 \times L$ size, denoted by $X_{rg}^{l2} = \{x_{rg1}^{l2}, x_{rg2}^{l2}, \cdots, x_{rgL}^{l2}\}$ $(r=1,2,3,4)$, which represents respectively the first order horizontal and vertical gradient feature image patch set (i.e., X_{1g}^{l2} and X_{2g}^{l2} set), and the second order horizontal and vertical gradient feature image patch set (i.e., X_{3g}^{l2} and X_{4g}^{l2} set), . And then, connected each gradient feature image patch set in order, the total LR image patch training set $X^{l2} = \{X_{1g}^{l2}; X_{2g}^{l2}; X_{3g}^{l2}; X_{4g}^{l2}\}$ can be obtained.

Step 3. Selecting the high frequency features. Because the human visual system (HVS) is sensitive to image edge features, the high frequency features are selected as cluster features. For an original HR image I_h, its high frequency features are obtained by using contourlet transform method [20-21]. Here, the number of contourlet decomposition layer is set as 2, thus each layer has four high pass sub-band images. For each layer, the fusion image of four high pass sub-band images is first considered, and each layer's fusion result is denoted by I_h^{f1} and I_h^{f2}. And then, the fusion between I_h^{f1} and I_h^{f2} are considered again, and this high pass sub-band images' fusion result between each layer is denoted by I_h^f. Further, I_h^f image is segmented into M image patches by using the fixed $d \times d$ sub-window, and the cluster sample set of high frequency features $X_{hf} = \{x_{hf1}, x_{hf2}, \cdots, x_{hfM}\}$ with $d^2 \times M$ size are obtained.

Step 4. Defining classified marking vector and cluster center of high frequency fusion feature set X_{hf}. Set the threshold Δ, for $\forall x_{hfi} \in X_{hf} = \{x_{hf1}, x_{hf2}, \cdots, x_{hfM}\}$, if the variance of x_{hfi}, denoted by $\text{var}(x_{hfi})$, is smaller than Δ, then the set satisfying the condition of $\text{var}(x_{hfi}) < \Delta$ is defined as the smooth set X_{hfs}, conversely the remaining x_{hfi} elements comprise the feature set X_{hff}. Utilized K-means method, the sample set of K classes and the K cluster centers can be obtained.

Step 5. Classifying the HR and LR set X'^{h2} and X'^{l2}. The class label of the HR and LR vector x_i^{h2} and x_i^{l2} are determined by ϕ_i. Thus, the $K+1$ class HR and LR sample set can be obtained, denoted respectively by $X'^{h2} = \{X'_1^{h2}, X'_2^{h2}, \cdots, X'_K^{h2}, X'_{K+1}^{h2}\}$ and $X'^{l2} = \{X'_1^{l2}, X'_2^{l2}, \cdots, X'_K^{l2}, X'_{K+1}^{l2}\}$.

Step 6. Training the two-stage HR and LR dictionary pairs. The two-stage LR dictionary D^{l2} corresponding to X'^{l2} is trained by K-SVD algorithm. Utilized the LR coefficient matrix S_K and the HR training set X'^{h2}, the two-stage HR dictionary D_k^{h2} can be learned by $D_k^{h2} = \arg\min \|X'^{h2}_k - D_k^{h2} S_k\|_F^2$. Combined D_k^{h2} and D_k^{l2}, the two-stage dictionary pairs $\{D_k^{h2}, D_k^{l2}\}$ can be obtained.

4   Experimental Results and Analysis

4.1   SRR Results Using One-Stage Dictionary Pairs

In test, five natural images with 512×512 size were used. Each original image was sampled randomly with 8×8 image patch 10000 times, thus the HR image patch set denoted by X^{h1} with 64×50000 size was obtained. For each HR image, using a point spread function (PSF) filter and the down-sampling method, the corresponding LR image can be obtained. For example, a test image called Elaine and its LR version was shown in Fig. 1(a) and Fig.1(b). At the same time, the real LR image, namely the

MMW image, was also shown in Fig.1(d). Using the same image patch sampling method, the LR image patch set denoted by X^{l1} with 64×50000 size could be obtained. Let X^{h1} and X^{l1} to be the training samples of SC-SRR, it can be obtained the one-stage HR and LR dictionary pairs, denoted by D^{h1} and D^{l1} with 256 atoms.

(a) The original Elaine image   (b) The LR image

(c) The original   toy gun image   (d) MMW image

Fig. 1. The original images and corresponding low-resolution versions

And for given LR dictionary D^{l1} and the LR image patch feature vector $\{x_i^{l1}\}$, the LR coefficient vector $\{s_i^{l1}\}$ can be learned by the following form

$$s_i^{l1} = \arg\min_s \frac{1}{2}\left\| x_i^{l1} - D^{l1}s_i^{l1} \right\|_2^2 + \gamma \left\| s_i^{l1} \right\|_1 .\tag{5}$$

Utilizing the LR coefficient s_i^{l1} and the HR dictionary D^{h1}, the reconstructed HR image patch $\hat{x}_i^{h1}$ can be solved by $\hat{x}_i^{h1} = D^{h1}s_i^{l1}$, further, replaced $\hat{x}_i^{h1}$ to the corresponding location in $\hat{f}_1$, $\hat{f}_1$ can be reconstructed. Here, considering the paper's length, only the reconstructed examples of Elaine LR version and MMW image were given shown in Fig.2. Clearly, LR images' contour has been reconstructed, however, LR images' many details are not still restored, at the same time, the restored image edge is very blurred from visual appeal. Especially, for the MMW image restored by the one-stage reconstructed process, the background noise is not reduced well.

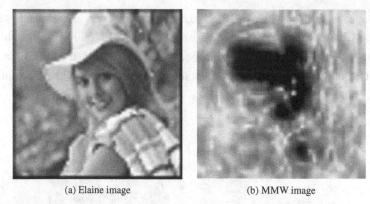

(a) Elaine image   (b) MMW image

Fig. 2. SRR images obtained by using one-stage high and LR dictionary pairs. (a) Reconstructed Elaine image. (b) Reconstructed MMW image.

4.2   SRR Results Using Two-Stage Dictionary Pairs

In this stage, for each reconstructed image $\hat{f}_1$, its high frequency feature set $\{x_{hfi}\}$ can be obtained by contourlet transform with 2 layers and 4 directions, then the class of x_{hfi} can be determined by K-means cluster method. And the four order gradient images of $\hat{f}_1$ are used as the LR feature set $\{x_i^{l2}\}$, and its class is defined by that of $\{x_{hfi}\}$. Then, according to the threshold Δ, the high frequency feature set $\{x_{hfi}\}$ is divided into the feature set $\{x_{hffi}\}$ and the smooth set $\{x_{hfsi}\}$. The class of image patches in the set $\{x_{hfsi}\}$ is marked as the $K+1$ class, and the corresponding LR feature image patch can be represented by LR dictionary D_{K+1}^{l2}. For feature set $\{x_{hfi}\}$, the class is defined by the form of $k_i = \arg\min_k \|x_{hfi} - c_k\|_2$. Thus, the set x_i^{l2} can be represented by using the $k_i - th$ dictionary $D_{k_i}^{l2}$. Further, the LR sparse coefficient vector s_i^{l2} can be obtained by the equation of $s_i^{l2} = \arg\min_s \frac{1}{2}\|x_i^{l2} - D_{k_i}^{h2}s_i^{l2}\|_2^2 + \gamma\|s_i^{l2}\|_1$,., the two-stage reconstructed image $\hat{f}_2$ can be computed by the following form:

$$\hat{f}_2 = \hat{f}_1 + \left[\sum_i (R_i)^T R_i\right]^{-1}\left[\sum_i (R_i)^T D_{k_i}^{h2} s_i^{l2}\right]. \tag{6}$$

where R_i is the ith image patch of the extracted image, and it is converted a column vector in solving the optimization of Eqn.(6).

And utilized Eqn. (6), the two-stage reconstructed images could be obtained. In the same way, considering the paper's length, only the reconstructed results of Elaine LR image and MMW image were given as shown in Fig.5 by using our method.In the same way, the reconstructed results of Elaine LR image and MMW image were given

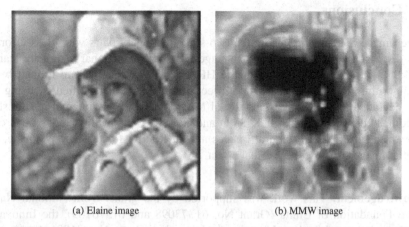

(a) Elaine image   (b) MMW image

Fig. 3. SRR images obtained by using the two-stage high and LR dictionary pairs. (a) Reconstructed Elaine image. (b) Reconstructed MMW image.

as shown in Fig.3 by our method. Compared Fig.3 and Fig.2, from the visual effect, the edge of the reconstructed Elaine image shown in Fig.3 (a) is clearer than those shown in Fig.2 (a), and much detail information is restored. And for reconstructed MMW image, much background noise in Fig.3 (b) has been reduced greatly.

In order to further differentiate the performance of different SRR algorithms, for reconstructed natural images, the PSNR measurement criterion is used to estimate the equality of reconstructed results. For the reconstructed MMW image, it is measured by using the relative SNR (RSNR) criterion. The calculated values of PSNR with different algorithms were listed in Table 1. And for the MMW image, the computed RSNR values were shown in Table 2.

Table 1. PSNR values of restored images using different algorithms (LR images were processed by PSF filter with Gaussian variance 2 and downsampling with the extracted scale 3)

Images \ Algorithms	Our method	SC-SRR	K-SVD	Noise images
Elaine	20.53	19.26	18.14	4.97
Lena	21.55	19.37	18.21	5.66
Boat	20.78	18.46	17.72	5.34
House	22.62	21.53	19.17	7.83

Table 2. RSNR values of restored MMW images using different algorithms

Images \ Algorithms	Our method	SC-SRR	K-SVD	MMW image
Reconstructed MMW image	15.36	14.84	14.63	12.37

5   Conclusions

A novel image SRR method using the two-stage dictionary pairs with class information is discussed in this paper. The one-stage HR and LR dictionary pairs are learned by SC-SRR method. And the two-stage HR and LR dictionaries are trained by K-SVD algorithm. In test, the restoration effect of our method is testified by using five highly degraded natural images and a real MMW image. Experimental results show that, compared with methods of K-SVD and SC-SRR, our method indeed has clear improvement in PSNR for natural images and in RSNR for the MMW image, and has better image structures and visual effect.

Acknowledgement. This work was supported by the grants from National Nature Science Foundation of China (Grant No. 61373098 and 61370109), the Innovative Team Foundation of Suzhou Vocational University (Grant No. 3100125),the grant from Natural Science Foundation of Anhui Province (No. 1308085MF85).

References

1. Elad, M.: Sparse and redundant representation: From theory to applications in signal and image processing, pp. 228–237. Springer, New York (2010)
2. Li, X., Orchard, M.T.: New edge-directed interpolation. IEEE Transactions on Image Processing 10(10), 1521–1527 (2001)
3. Yang, J.C., Wright, J., Ma, Y., Huang, T.: Image super-resolution as sparse representation of raw image patches. In: Proceedings of the 2008 IEEE Computer Society Conference on Computer Vision and Pattern Recognition, pp. 1–8. IEEE Press, Anchorage (2008)
4. Yang, J.C., Wright, J., Huang, T., Ma, Y.: Image super-resolution via sparse representation. IEEE Transactions on Image Processing 19(11), 2861–2873 (2010)
5. Zeyde, R., Elad, M., Protter, M.: On single image scale-up using sparse-representations. In: Boissonnat, J.-D., Chenin, P., Cohen, A., Gout, C., Lyche, T., Mazure, M.-L., Schumaker, L. (eds.) Curves and Surfaces 2011. LNCS, vol. 6920, pp. 711–730. Springer, Heidelberg (2012)
6. Freeman, W.T., Pasztor, E.C., Carmichael, O.T.: Example-based super-resolution. IEEE Computer Graphics and Application 22(2), 56–65 (2002)
7. Huang, D.S., Horace, H.S., lp, Law Ken, C.K., Chi, Z.R.: Zeroing polynomials using modified constrained neural network approach. IEEE Trans. on Neural Networks 16(3), 721–732 (2005)
8. Huang, D.S., Horace, H.S., lp, Ken, C.K., Chi, Z.R., Wong, H.S.: A new partitioning neural network model for recursively finding arbitrary roots of higher order arbitrary polynomials. Applied Mathematics and Computation 162(3), 1183–1200 (2005)
9. Aharon, M., Elad, M., Bruckstein, A.: K-SVD: An algorithm for designing overcomplete dictionaries for sparse representation. IEEE Transactions on Signal Processing 54(11), 4311–4322 (2006)
10. Huang, D.S., Horace, H.S., lp, et al Zheru, C.: Dilation method for finding close roots of polynomials based on constrained learning neural networks. Physics Letters A 309(5-6), 443–451 (2003)

Position Accuracy Improvement of Robots having Closed-Chain Mechanisms

Hoai-Nhan Nguyen[1], Jian Zhou[1], Hee-Jun Kang[1,*], and Tien-Dung Le[2]

[1] School of Electrical Engineering, University of Ulsan, 680-749 Ulsan, Korea
[2] Department of Electrical Engineering, The University of Da Nang - UST, Viet Nam
nhan.nguyenhoai@yahoo.com, freesoulzhou@hotmail.com,
hjkang@ulsan.ac.kr, dung.letien@gmail.com

Abstract. Improving the position accuracy of the closed chain robots needs a global robot kinematic model. Their closed loop constraints should be included in the global model via the specific passive joints of the robot. This paper proposes a simple but effective method to derive a mathematical formula for identification of all parameters of the robots having multiple closed chain mechanisms. An experimental calibration is carried out on a Hyundai HP160 robot to validate the correctness and effectiveness of the proposed method.

Keywords: kinematic calibration, parameter identification, robot having closed mechanism.

1   Introduction

Recently, robotic manipulators are widely used for advanced industrial manufacturing lines. Improving the position accuracy of these robots is really necessary. Manipulators are designed in the form of multiple closed loop mechanisms to increase the overall stiffness of robot structures. Calibration for serial robots was well defined in various previous works [1-5]. But, calibration for manipulators containing many closed loops will be challenging and difficult. Some works [6, 7], had assumed that the closed chain mechanism were perfect, so the authors treated the robots like the simple serial linkages. However, there were some other researches focusing on calibration for robot containing one or multiple closed loops [8, 9]. In the study [8], calibration for robot having closed loops was carried out by replacing an optimization problem imposed constraints with an optimization problem without constraints by means of using Lagrange multipliers. In the studies [9], the authors first found solutions of the equations of constraints explicitly and then integrated the solutions to the kinematic model of the robot main open chain to make the robot global kinematic model. The proposed model in [10, 11] did not select correctly the main open kinematic chain of the robot having one closed link mechanism.

---

* Corresponding author.

D.-S. Huang et al. (Eds.): ICIC 2014, LNAI 8589, pp. 285–292, 2014.

This paper presents a simple but effective method for the formulation of parameter identification equations of a robot having multiple closed chain mechanisms. First, a robot model is developed by combining kinematics of the main open loop and of the closed loops as in [9]. The global differential relationship is obtained by combining the differential relations of the closed loops and of the main open chain. An experimental calibration is carried out on Hyundai HP160 manipulator (as shown in Fig.1 (left)) to verify the correctness and effectiveness of the proposed method. Section 2 presents the modeling of robot kinematics. Section 3 formulates the identification equations. Section 4 presents experimental calibration on the HP160 robot. Section 5 withdraws some conclusions.

2   Model of HP160 Robot

A four degree of freedom (d.o.f) Hyundai HP160 robot (shown in Fig. 1 (left)) consists of one main open kinematic chain, which is connected by four revolute joints θ1, θ2, θ3p, θ4p, θ4a, and three closed chain mechanisms ABCD, AEFD, DGHK. The link frames are fixed on the joint axes, which are elements of the main open kinematic chain, based on the Danevit-Hartenberg (D-H) method [12] (shown in Fig. 1 (left)). Nominal D-H parameters of the open chain and the link lengths of the closed loops (Fig. 1(right), Fig. 2 (left and right)) are listed in Table 1.

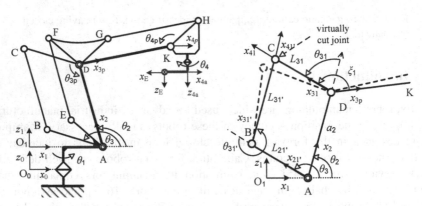

Fig. 1. Hyundai HP160 robot and its link frames assignment (left); closed loop ABCD, dashed line for degenerated loop (right)

A homogeneous transformation matrix from robot base frame {0} to end-effector link frame {E} is calculated by following equation:

$$_E^0T = {}_1^0T(\theta_1)\,{}_2^1T(\theta_2)\,{}_{3p}^2T(\theta_{3p})\,{}_{4p}^{3p}T(\theta_{4p})\,{}_{4a}^{4p}T(\theta_{4a})\,{}_E^{4a}T \,, \tag{1}$$

where ${}_i^{i-1}T$ is a transformation matrix from frame {i-1} to the frame {i}:

$$_i^{i-1}T = Rot(x_{i-1},\alpha_{i-1})Tr(x_{i-1},a_{i-1})Tr(z_i,d_i)Rot(z_i,\theta_i) \,, \tag{2}$$

$Rot(\cdot)$ and $Tr(\cdot)$ are (4×4) matrices of rotation about and translation along an axis, respectively, $i = 2, \ldots, 5$; the robot base and tool transformations matrix are:

$$_1^0T = Rot(x_0,\alpha_0)Tr(x_0,a_0)Rot(y_0,\beta_0)Tr(y_0,b_0)Rot(z_1,\theta_1)Tr(z_1,d_1), \tag{3}$$

$$_E^{4a}T = Tr(x_5,a_5)Tr(y_5,a_5)Tr(z_6,d_6), \tag{4}$$

where the matrices $_{3p}^2T$ and $_{4p}^{3p}T$ are computed with passive angles θ_{3p}, θ_{4p}.

Table 1. Nominal D-H parameters of HP160 robot (units: length [m], angle [°]; --: not exist)

			Nominal D-H parameters of the main open chain			
i	α_{i-1}	a_{i-1}	β_{i-1}	b_{i-1}	d_i	θ_i
1	0	0	0	0	0.6915	θ_1
2	90	0.300	--	--	0	θ_2
3	0	1.250	0	--	0	θ_{3p}
4	0	1.350	0	--	0	θ_{4p}
5	90	0.250	--	--	0.17	θ_{4a}
6	--	0.1	--	0.1	0.1	--
			Link lengths of closed chain mechanisms			
$L_{21'}$	0.5614		$L_{31'}$	1.2500	L_{31}	0.5614
$L_{22'}$	0.5300		$L_{32'}$	1.2500	L_{32}	0.530
$L_{33'}$	0.520		$L_{43'}$	1.350	L_{43}	0.520

The position constraint equations of closed loop ABCD in Fig. 1 (right) is as follows [13]:

$$a_2c\theta_2 + L_{31}c\theta_{2,31} - L_{21'}c\theta_3 + L_{31'}c\theta_{3,31'} = 0,$$
$$a_2s\theta_2 + L_{31}s\theta_{2,31} - L_{21'}s\theta_3 + L_{31'}s\theta_{3,31'} = 0, \tag{5}$$

where $c\theta_i$, $s\theta_i$ and $\theta_{i,j}$ are short form of $\cos(\theta_i)$, $\sin(\theta_i)$ and $\theta_i + \theta_j$, respectively.

This closed mechanism is a parallelogram, so $L_{21'} = L_{31}$ and $L_{31'} = a_2$. The passive joint angle in (5) is computed by (6):

$$\theta_{31} = \theta_3 - \theta_2, \ \theta_{31'} = 360 - \theta_3 + \theta_2, \tag{6}$$

The position constraints of closed loop AEFD as in Fig. 2 (left) is as follows:

$$a_2c\theta_2 + L_{32}c\theta_{2,32} - L_{22'}c\gamma - L_{32'}c(\gamma + \theta_{32'}) = 0,$$
$$a_2s\theta_2 + L_{32}s\theta_{2,32} - L_{22'}s\gamma - L_{32'}s(\gamma + \theta_{32'}) = 0, \tag{7}$$

This mechanism also has parallelogram structure, so $L_{22'} = L_{22}$ and $L_{32'} = a_2$ ($\gamma = 148.11$ [°]). Given a value of the joint angle θ_2, the values of passive joint positions of closed loop AEFD in (7) are computed by (8),

$$\theta_{32} = \gamma - \theta_2, \ \theta_{32'} = 360° + \theta_2 - \gamma. \tag{8}$$

Similarly to the both closed loops above, position constraint equations of closed kinematic loop DGHK (note $L_{33'} = L_{43}$ and $L_{43'} = a_3$) in Fig. 2 (right) are:

$$a_3 c\theta_{33} + L_{43} c\theta_{33,43} - L_{33'} c\theta_{33'} - L_{43'} c\theta_{33',43'} = 0,$$
$$a_3 s\theta_{33} + L_{43} s\theta_{33,43} - L_{33'} s\theta_{33'} - L_{43'} s\theta_{33',43'} = 0, \tag{9}$$

The joint position values need to be obtained in (7) and (9) are: $\theta_{33} = \theta_3 - \xi_1$ and $\theta_{33'} = \gamma - \xi_2$. Constant angles $\xi_1 = 175.9144$ [°], $\xi_2 = 112.892$ [°] and $\xi_3 = 35.218$ [°]. The values of passive joint positions of closed loop DGHK are:

$$\theta_{43} = \gamma + \xi_1 - \xi_2 - \theta_3, \ \theta_{43'} = 360° + \xi_1 + \xi_2 - \theta_3 - \gamma. \tag{10}$$

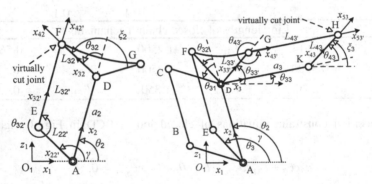

Fig. 2. The closed loop mechanism AEFD (left); Connection of the closed loops DGHK, ABCD and AEFD (right)

By solving (5), (7), and (9), then the passive joint positions such as θ_{3p} and θ_{4p} in (1) can be computed by the following formulas:

$$\theta_{3p} = \theta_{31} - \xi_1 = \theta_3 - \theta_2 - \xi_1, \ \theta_{4p} = -\theta_{43} + \xi_3 = -\theta_3 + \gamma + \xi_1 - \xi_2 + \xi_3. \tag{11}$$

The real link lengths of the closed mechanism of the physical robot deviate from their nominal values, so the actual values of the passive angles θ_{3p} and θ_{4p} should be computed by solving (5), (7) and (9) with the actual current link length parameters. The HP160 manipulator has three consecutive parallel joint axes, the transformation in (1) is modified as follows:

$$\substack{2\\3p}T = Rot(x_2,\alpha_2)Tr(x_2,a_2)Rot(y_2,\beta_2)Rot(z_{3p},\theta_{3p}),$$

$$\substack{3p\\4p}T = Rot(x_3,\alpha_3)Tr(x_3,a_3)Rot(y_3,\beta_3)Rot(z_{4p},\theta_{4p}),$$

(12)

where β_2 and β_3 are the link twist parameters about the axes y_2 and y_3.

3   Mathematical Formulation for Robot Error Identification

A differential homogeneous transformation of the main open kinematic chain could be obtained by differentiating (1) as follows [14]:

$$\Delta x = [J_{\alpha i}\, J_{ai}\, J_{\beta i}\, J_{di}\, J_{\theta i}]\Delta p = J\,\Delta p,$$

(13)

where Δx is a (3×1) vector of three differential positions of the robot end-effector. Δp is a (20×1) vector of the robot kinematic parameters. J is a (3×20) error propagation matrix which relates the vectors Δx with Δp. The column vectors of the matrix J are numerically computed as follows [14]:

$$J_{\alpha i} = [x_{i-1} \times p_{i-1}],\ J_{ai} = [x_{i-1}],\ J_{\beta i} = [y_{i-1} \times p_{i-1}],\ J_{di} = [z_i],\ J_{\theta i} = [z_i \times p_i], (14)$$

where x_{i-1}, y_{i-1}, z_i are directional vectors of link frames $\{i\text{-}1\}$ and $\{i\}$; p_i is a vector represents the position of robot end-effector with respect to the link frame $\{i\}$.

The positions of the passive joints of the closed loops could be expressed by functions of related parameters as follows

$$\theta_{3p} = f(q,L_1),\ \theta_{4p} = g(q,L_1,L_2,L_3),$$

(15)

where   vectors $q = [\theta_2,\theta_3]^T$, $L_1 = [L_{31},L_{21'},L_{31'},a_2]^T$,

$L_2 = [L_{32},L_{22'},L_{32'},a_2]^T$, $L_3 = [L_{43},L_{33'},L_{43'},a_3]^T$, f and g are mathematical functions. These functions are derived by combining (5), (7) and (9) for robot current parameter values. The direct differentiate of the functions f and g in (15) are:

$$\Delta\theta_{3p} = \frac{\partial f}{\partial q}\Delta q + \frac{\partial f}{\partial L_1}\Delta L_1,\ \Delta\theta_{4p} = \frac{\partial g}{\partial q}\Delta q + \frac{\partial g}{\partial L_1}\Delta L_1 + \frac{\partial g}{\partial L_2}\Delta L_2 + \frac{\partial g}{\partial L_3}\Delta L_3.$$

(16)

The joint C, F, H of the corresponding closed loops ABCD, AEFD, DGHK are assumed virtually cut off. The closed loop ABCD in Fig. 1 (right) comprises of two open kinematic chains ADC and ABC which are connected at joint C. The differential relationship of individual open chains are described as follows,

$$\Delta x_C = J_1\Delta h_1,\ \Delta x_C = J_2\Delta h_2,$$

(17)

where J_1 and J_2 are Jacobian matrices of the chains ADC and ABC, respectively. Δx_C is a vector of position error of endpoint C. The vectors of link parameter errors of the chains ADC and ABC are $\Delta h_1 = [\Delta\theta_2 \ \Delta\theta_{31} \ \Delta a_2 \ \Delta L_{31}]^T$ and $\Delta h_2 = [\Delta\theta_3 \ \Delta\theta_{31'} \ \Delta L_{21'} \ \Delta L_{31'}]^T$, respectively. Equaling the both right sides of the two equations in (17), the following equation is obtained:

$$J_1\Delta h_1 = J_2\Delta h_2 . \tag{18}$$

By solving (18) for the passive joint position $\Delta\theta_{3p}$ in term of the other link errors, the differential relation of the closed loop ABCD is as follows:

$$\Delta\theta_{3p} = \Delta\theta_{31} = J_c^1.\Delta h, \tag{19}$$

where $J_c^1 = [J_{\theta 2} \ J_{\theta 3} \ J_{a2} \ J_{L31} \ J_{L31'} \ J_{L21'}]^T$ ('c' means 'closed') is a Jacobian matrix and $\Delta h = [\Delta\theta_2 \ \Delta\theta_3 \ \Delta a_2 \ \Delta L_{31} \ \Delta L_{31'} \ \Delta L_{21'}]^T$ is a vector of the closed loop link errors.

Applying the singular value decomposition method [15] for evaluating of identifiable parameter, a reduced Jacobian matrix and a vector of closed chain errors are:

$$J_c^1 = [J_{\theta 3}^1 \ J_{L31'}^1 \ J_{L21'}^1]^T, \quad \Delta h = [\Delta\theta_3 \ \Delta L_{31'} \ \Delta L_{21'}]^T . \tag{20}$$

Similarly, a differential relation of the closed loop AEFD shown in Fig. 2 (left) is described in the following equation,

$$\Delta\theta_{32} = J_c^2.\Delta k, \tag{21}$$

where a reduced Jacobian matrix and a reduced vector of closed loop link errors are as follows: $J_c^2 = [J_{\theta 2}^2 \ J_{a2}^2 \ J_{L32}^2 \ J_{L32'}^2 \ J_{L22'}^2]^T$ and $\Delta k = [\Delta\theta_2 \ \Delta a_2 \ \Delta L_{32} \ \Delta L_{32'} \ \Delta L_{22'}]^T$.

The closed loop DGHK is connected with the two closed loops ABCD and AEFD as shown in Fig. 2 (right). Note that $\Delta\theta_{33} = \Delta\theta_{31}$, $\Delta\theta_{33'} = \Delta\theta_{32}$, and $\Delta\theta_{4p} = \Delta\theta_{43}$. The differential relationship of the closed loop DGHK in Fig. 2 (right) is:

$$\Delta\theta_{43} = J_c.\Delta w, \tag{22}$$

where a reduced Jacobian matrix and a reduced vector of link errors of the closed loop DGHK are:

$$J_c = [J_{\theta 33} \ J_{\theta 33'} \ J_{a3} \ J_{L43} \ J_{L33'} \ J_{L43'}]^T, \text{ and } \Delta w = [\Delta\theta_{33} \ \Delta\theta_{33'} \ \Delta a_3 \ \Delta L_{43} \ \Delta L_{33'} \ \Delta L_{43'}]^T .$$

Substituting (19) and (21) into (22) via joint error variables $\Delta\theta_{33}(\Delta\theta_{33} = \Delta\theta_{31})$ and $\Delta\theta_{33'}(\Delta\theta_{33'} = \Delta\theta_{32})$, we have (23):

$$\Delta\theta_{43} = \Delta\theta_{4p} = J_c^3 . \Delta g, \tag{23}$$

where a reduced Jacobian matrix and vector of closed loop errors are as follows,

$$J_c^3 = [J_{\theta 3}^3 \; J_{a2}^3 \; J_{a3}^3 \; J_{L21}^3 \; J_{L31}^3 \; J_{L22}^3 \; J_{L32}^3 \; J_{L33}^3 \; J_{L43}^3]^T,$$
$$\Delta g = [\Delta\theta_3 \; \Delta a_2 \; \Delta a_3 \; \Delta L_{21} \; \Delta L_{31} \; \Delta L_{22} \; \Delta L_{32} \; \Delta L_{33} \; \Delta L_{43}]^T. \tag{24}$$

Substituting (19) and (23) into (13) via passive joint error variables $\Delta\theta_{3p}$ and $\Delta\theta_{4p}$ to form the global identification equation of the robot kinematic parameters,

$$\Delta x = [J'_{\alpha i} \; J'_{ai} \; J'_{bi} \; J'_{\beta i} \; J_{di} \; J'_{\theta i}] = J' \Delta p', \tag{25}$$

where J' is a global Jacobian matrix, $\Delta p'$ is the global vector of errors.

4   Experimental Calibration and Results

The proposed method for derivation of the robot global differential relationship is applied to the practical calibration procedure of the Hyundai HP160 robot (as shown in Fig.1 (left)). The residual position errors after calibration at 25 measurement points are shown in Fig. 3. The Table 2 summarized the calibration results. The experimental calibration results demonstrate that model 2 (the proposed model) results in a better accuracy than model 1 [6, 7].

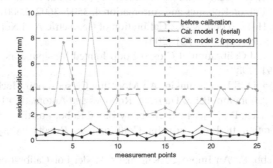

Fig. 3. Calibration results for robot HP160 with models 1 and 2

Table 2. Residual position error of models 1 and 2 (experimental calibration)

Robot models	Average error [mm]	Maximum error [mm]
Nominal model	3.5316	9.6538
Model 1 (serial)	0.6554	1.2682
Model 2 (proposed)	0.4426	0.6848

5   Conclusion

This paper proposed a formulation method for parameter identification. The global differential relationship is obtained by integrating differential relationships of the closed loops and the main open chain. The results of the experimental calibration demonstrated the correctness and effectiveness of the proposed method.

Acknowledgments. The authors would like to express appreciation for the financial support from the Ministry of Knowledge Economy under Robot Industry Core Technology Project.

References

1. Judd, R.P., Knasinski, A.B.: A Technique To Calibrate Industrial Robots With Experimental Verification. IEEE Transactions on Robotics and Automation 6(1), 20–30 (1990)
2. Park, I.-.W., et al.: Laser-Based Kinematic Calibration of Robot Manipulator Using Differential Kinematics. IEEE/ASME Transactions on Mechatronics, 1–9 (2011)
3. Everett, L., Driels, M., Mooring, B.: Kinematic Modelling For Robot Calibration. Proc. IEEE Int. Conf. on Robotics and Automation 4, 183–189 (1987)
4. Hayati, S., Tso, K., Roston, G.: Robot Geometry Calibration. Proc. IEEE Int. Conf. on Robotics and Automation 2, 947–951 (1988)
5. Everett, L.J., Suryohadiprojo, A.H.: A Study of Kinematic Models For Forward Calibration of Manipulators. Proc. IEEE Int. Conf. on Robotics and Automation 2, 798–800 (1988)
6. Newman, W.S., et al.: Calibration of A Motoman P8 Robot Based On Laser Tracking. Proc. IEEE Int. Conf. on Robotics and Automation 4, 3597–3602 (2000)
7. Joon, H.-J., Soo, H.-K., Yoon, K.-K.: Calibration of Geometric And Non-Geometric Errors of An Industrial Robot, pp. 311–321 (2001)
8. Louis, J.E.: Forward Calibration of Closed-Loop Jointed Manipulators. Int. J. of Robotics Research 8, 85–91 (1989)
9. Schröer, K., Albright, S.L., Lisounkin, A.: Modeling Closed-Loop Mechanisms In Robots For Purposes of Calibration. IEEE/Trans. on Robotics and Automation 13, 218–229 (1997)
10. Alici, G., Shirinzadeh, B.: Laser Interferometry Based Robot Position Error Modelling For Kinematic Calibration. In: Proc. IEEE Int. Conf. on Intel. Robots and Sys., pp. 3588–3593 (2003)
11. To, M., And Webb, P.: An Improved Kinematic Model For Calibration of Serial Robots Having Closed-Chain Mechanisms. Robotica, 1–9 (2011)
12. Hartenberg, R.S., Denavit, J.: A Kinematic Notation For Lower Pair Mechanisms Based On Matrices. Trans. ASME Jour. of App. Mech. 77, 215–221 (1955)
13. Siciliano, B., Sciavicco, L., Villani, L.: Robotics Modeling, Planning And Control, pp. 70–72. Springer, Berlin (2009)
14. Bennett, D.J., et al.: Autonomous Calibration of Single-Loop Closed Kinematic Chains Formed By Manipulators With Passive Endpoint Constraints. IEEE Trans. on Rob. and Aut. 5, 597–606 (1991)
15. Staton, D., And Mayer, J.R.R.: Robot Calibration Within Cim-Serach/I. In: Bernhardt, Albright (eds.) Robot Calibration, pp. 62–69. Chapman & Hall, London (1993)

GPU-Based Real-Time Range Image Segmentation

Xinhua Jin[1], Dong Joong Kang[2], and Mun-Ho Jeong[1]

[1] Robot Vision, Intelligence Lab Kwangwoon University Seoul, Korea
yipianqihei@gmail.com, mhjeong@kw.ac.kr
[2] Mechanical Engineering Pusan National University Seoul, Korea
djkang@pusan.ac.kr

Abstract. In this paper proposed a GPU-based parallel processing method for real-time image segmentation with neural oscillator network. Range image segmentation methods can be divided into two categories: edge-based and region-based. Edge-base method is sensitive to noise and region-based method is hard to extracting the boundary detail between the object. However, by using LEGION (Locally Excitatory Globally Inhibitory oscillator networks) to do range image segmentation can overcome above disadvantages. The reason why LEGION is suitable for parallel processing that each oscillator calculate with its 8-neiborhood oscillators in real time when we process image segmentation by LEGION. Thus, using GPU-based parallel processing with LEGION can improve the speed to realize real-time image segmentation.

Keywords: Range image segmentation, CUDA, LEGION, GPGPU.

1 Introduction

Recently, due to the popularity of 3D- camera (KINECT, Xtion) rang images are widely used. In order to the scene understanding of range image, need to do range image segmentation. The requirements of environmental recognition in real time are increasing these days.

X. Liu and D. Wang proposed use LEGION to do range image segmentation in [1]. It solved shortages are sensitive to noises and hard to extract detailed boundary between two objects in region-based and edge-based. However, it still remained a problem is hard to implement range image segmentation due to complexity of LEGION's structure and CPU's sequential execution.

In this paper, we realize real-time range image segmentation using GPU-based parallel processing techniques with LEGION. Section 2 and 3 are introduced the existing LEGION model and range image segmentation using LEGION. Section 4 is shown the contents of range image segmentation using GPU parallel processing with LEGION and conclusion also shown in section 5.

D.-S. Huang et al. (Eds.): ICIC 2014, LNAI 8589, pp. 293–297, 2014.

2   LEGION Model

LEGION algorithm was proposed with the object of keeping flexibility of Neural Oscillatory Network and simple the calculation.

$$\dot{x}_i = 3x_i - x_i^3 + 2 - y_i + I_i + S_i + \rho \tag{1}$$

$$\dot{y}_i = \varepsilon(\alpha(1 + \tanh(x_i / \beta)) - y_i) \tag{2}$$

$$S_i = \sum_{k \in N(i)} W_{ik} H(x_k - \theta_x) - W_z H(z - \theta_z) \tag{3}$$

The Eq. (1), (2) is showing LEGION's constitute elements. i is a single oscillator, x_i is excitability unit and y_i is inhibitory unit. In Eq. (1), (2) unit of excitability and unit of inhibitory defined as derivatives by time. Eq. (3) show the coupling strength of i oscillator from its 8 neighbors [2], [3]

3   Range Image Segmentation Using LEGION

X. Liu apply LEGION algorithm to range image segmentation, W_{ik} was defined as $W_{ik} = (\ 255\ /\ \Psi(i, k)\ +\ 1\ )$ in Eq. (3), set $\Psi(i, k)$ as an input value and calculated by measured value of feature vector (z, n_x, n_y, n_z, C, G). Z is the depth value and $(\ n_x,\ n_y\ ,\ n_z\ )$ is the surface normal. C is mean Gaussian curvature and G is Gaussian curvature. A surface normal in a point are represented as

$$(n_x, n_y, n_z) = \frac{\dfrac{\partial z}{\partial x} + \dfrac{\partial z}{\partial y}}{\| \dfrac{\partial z}{\partial x} + \dfrac{\partial z}{\partial y} \|}, \tag{4}$$

The surface curvature at p point in the direction of q point is shown as

$$k_{p,q} = \begin{cases} \dfrac{\| n_p - n_q \|}{\| p - q \|}, & \text{if } \| p - q \| \le (n_p + p) - (n_q + q) \| \\[4mm] -\dfrac{\| n_p - n_q \|}{\| p - q \|}, & \text{otherwise} \end{cases} \tag{5}$$

In Eq. (5), p and q refer to the 3-D coordinate vectors of the corresponding pixels, and n_p is unit normal vector at point p, n_q unit normal vector at point q.

The input value of LEGION algorithm is calculated as follows

$$\Psi(i, k) = \begin{cases} K_z |z_i - z_k| + K_n \|n_i - n_k\| & \text{if } |C_i| \leq T_c \\ K_c |C_i - C_k| + K_G |G_i - G_k| & \text{otherwise} \end{cases}. \tag{6}$$

K_Z, K_n, K_c and K_G are weighted value and T_c is a threshold value.

4   Gpu-Based Real-Time Range Image Segmentation

From Fig. 1, we can know with the development of the GPU graphics card, more and more powerful, the calculation has gone beyond the general CPU. Such a powerful chip if only as graphics will result in wasted computational capability. So NVIDIA launched the CUDA, lets graphics card can be used for purposes other than the image rendering.

We utilized parallel processing technique of CUDA (Computer Unified Device Architecture) in GPU offered from NVIDIA in this paper.

CUDA consists of two parts: Host and Device. Generally speaking, Host stands for CPU, Device stands for GPU. In CUDA structure, main programming is also executed by CPU while it need process part of data, CUDA would compile program GPU is able execute then send to GPU. And this program is called core in CUDA.

Actually, when execute the program CUDA would engender a lot of threads could execute in GPU; every thread would be to execute the program, although there is only one program, the reason why we calculate with different data is that there are different indexes.

To be executed on the GPU thread, there are three types (1 dimension, 2 dimensions and 3 dimensions) according to the most effective data sharing to set a block. These threads in the same block use the identical sharing internal memory. In addition, there is a limit to the size of every block, so could not put all of threads into one block (there are a lot of threads in GPGPU program). At this time can use the same dimension and size of block (need the same core) to construct a network (Grid) for batch processing (GPU need to compile it only one time).

Our machine environment in GPU parallel programming uses Intel Core i7-2600@3.40GHz, 8GB DDR3 ram, 1333MHz, GeForce GTX 580, 1536MB DDR5, NVidia driver version 335.23.

As Fig. 2 below every oscillator have some connection with its neighbor 8 oscillators, so we can regard one oscillator as a thread in parallel processing of GPU. It means that the structure of LEGION is suitable for parallel processing in GPU.

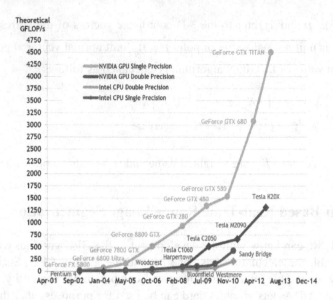

Fig. 1. Floating-point Operations per Second for the CPU and GPU (form [4])

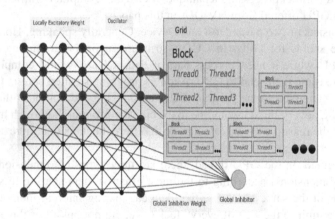

Fig. 2. 2D LEGION network and GPU thread

When implementing parallel processing of GPU get image distance from KINECT is 480 * 640, one oscillator (pixel) correspond one thread. To optimize the availability of GPU, set per block as 786, so in every SM, there are 1536 threads (2 blocks) would be processed in every time. Set the block dimension as 2D because input image is 2D. Fig. 3 shows the image segmentation result.

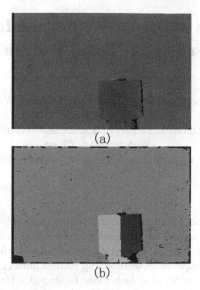

Fig. 3. Segmentation result of the LEGION network for the range image: (a) is the range image gets from KINECT, (b) is the image segmentation result and each color part represents a segmented region

5   Conclusion

In this paper, we use GPU-based parallel processing techniques to range image segmentation. Because of LEGION is suitable for parallel processing, so we get a significantly speed improvement than CPU-based serial processing. Finally, we realize the real-time range image segmentation.

References

1. Liu, X., Wang, D.L.: Range Image Segmentation Using An Oscillatory Network. IEEE Trans. Neural Networks 10, 564–573 (1999)
2. Terman And, D., Wang, D.L.: Global Competition And Local Cooperation in A Network of Neural Oscillators. Physicu D 82, 148–176 (1995)
3. Wang, D.L., Terman, D.: Image Segmentation Based On Oscillatory Correlation. Neural Computation (in Press); See Also Technical Report 19, Center for Cognitive Science. The Ohio State University (1996)
4. CUDA C PROGRAMMING GUIDE http://Docs.Nvidia.Com/Cuda/Cuda-C-Programming-Guide/Index.Html

Recognition of Human Action and Identification Based on SIFT and Watermark

Khawlah Hussein Ali and Tianjiang Wang

Huazhong University of Science and Technology,
Wuhan, Hubei, China
Khawlahussein@yahoo.com, tjwang@hust.edu.cn

Abstract. This paper presents a fast and simple method for action recognition and identity at the same time. A watermark embedding as a 2-D wavelet in the training data at the first step to identify the identity of who makes the action. The proposed technique relies on detecting interest points using SIFT (scale invariant feature transform) from each frame of the video for action recognition. More specifically, we propose an action representation based on computing a rich set of descriptors from 2D-SIFT key points. Since most previous approaches to human action recognition typically focus on action classification or localization, these approaches usually ignore the information about human identity. A compact yet discriminative semantics visual vocabulary was built by a K-means for high-level representation. Finally a multi class linear Support Vector Machine (SVM) is utilized for classification. Our algorithm can not only categorize human actions contained in the video, but also verify the person who performs the action. We test our algorithm on three datasets: the KTH human motion dataset, Weizmann and our action dataset. Our results reflect the promise of our approach.

Keywords: action recognition, water mark, SIFT, K-mean, PCA.

1 Introduction

Applications such as surveillance, video retrieval and human-computer interaction require recognizing human actions from distance and identity in various scenarios. The benefit of knowing the identity of the person who is certain actions of private security and control of sensitive places or security nature of the state task. But only few papers researched this problem and most papers only focus on recognizing or localizing human action, which usually cannot recognize the identity that makes the action at same time [4, 5]. Since most of the previous methods utilize the detailed human information (e.g. the face information) that is acquired from small distance to recognize the human identity. As these views a person appears as a small patch. So under above discussed conditions, it is a difficult problem in computer vision to capture such detailed information from such distance.

In this paper, we propose a framework to address the problems for recognizing actions and identity at the same time. Our framework can be summarized as follows

D.-S. Huang et al. (Eds.): ICIC 2014, LNAI 8589, pp. 298–309, 2014.

1) we propose an identity representation based on embedded a watermark as 2-D wavelet. 2) Detecting interest points for action recognition by using 2-D SIFT. 2-D SIFT has advantage which is limited size of the features vectors, which consumes less computation time than other techniques such as 3D descriptors[8], [18], and has another advantage which does not need to use topic modelling methods such as pLSA and LDA, where a separate topic model is learned for each action class and new samples are classified by using the constructed action topic model [17]. 3) Principle Component Analysis (PCA) is applied to the features for dimension reduction without losing information. The idea to use PCA is due to the fact that, in most of the action datasets, there is a large amount of redundant information unnecessary for action recognition. 4) The K-means clustering algorithm is utilized to construct visual words. 5) A linear multi class SVM [24] is trained to classify identity and action at the same time. Action representation can be categorized as: flow based approaches [25], spatio-temporal shape template based approaches [26,27], tracking based approaches [28] and interest points based approaches [29]. In flow based approaches optical flow computation is used to describe motion[10], it is sensitive to noise and cannot reveal the true motions. Spatio-temporal shape template based approaches treat the action recognition problem as a 3D object recognition problem and extracts features from the 3D volume[18,19]. The extracted features are very huge so the computational cost is unacceptable for real-time applications. Tracking based approaches suffer from the same problems. Interest points based approaches have the advantage of short feature vectors; hence low computational cost. They are widely used and are adopted in this work.

Recent work has focused on bag of spatial-temporal local interest point feature [3, 4], [6], [11]. With our best knowledge very few works of recognize actions and identity like [1, 2, 3]. In this paper SIFT [9] is used for detecting interest points where the extracted features are invariant to scale, location and orientation changes. In addition, the accuracy is better than all (to our knowledge) previous work in this field.

2 Related Works

In the work related [1], in their paper, they propose a framework to address the problems for recognizing actions and identity. They propose an action and identity representation based on computing rich descriptors from ASIFT key point trajectories which capture more global spatial and temporal information. And they used the object categories depicted using a model developed in the statistical text literature: Latent Dirichlet Allocation [1] (LDA), and combined action recognition and identification in a whole framework. They used trajectory extraction, which is may be not always useful for action recognition. Their method requires more computation and time for recognition and identification. Another paper [2], he motivated by the view – invariance issue in the gaitID problem. Address the general problem of classifying the "content" of human motions of unknown "style" and used bilinear modeling to improve the recognition when the test motion is performed in an unknown style. In [3] they have proposed a framework that performs action recognition and identity maintenance of multiple targets simultaneously. In the current paper we proposed a framework that performs identity as a watermark based 2-D wavelet transform on the video, and extract features using SIFT algorithm. To reduction the size of the dimension, we use PCA. The whole framework as shown in Figure 1, takes less time and little computation for action recognition and identity as compared to other researches.

3   Proposed Approach

We adopt the notion of watermark from [13] as identity to the training data. In our method we enter the watermark, which represents identity of human to the video we want training and then divided the video into short clips and each clip consists of frames of 0.5 second, the distribution of a video clip reveals the temporal variation of the pixel's intensity. The goal of digital watermarking is to hide a watermark (some sort of identification data) within the image, so that it is not perceptually visible to the human eye, but also it is robust enough to withstand the various kinds of transformations.

One of the many advantages over the wavelet transform is that it is believed to be more accurately model aspects of the HVS as compared to the FFT or DCT. This allows us to use higher energy watermarks in regions that the HVS is known to be less sensitive to, such as the high resolution detail bands {LH, HL, HH} [4]. Using discrete wavelet transform (DWT) is motivated by good time-frequency features and well-matching with human visual system directives. These two combined elements are important in building an invisible and robust watermark. In this paper, the watermark is embedded into high frequency DWT components of a specific sub-image and it is calculated in correlation with the image features and statistic properties.

3.1   Watermark Embedding

Let C be the matrix associated to block to be watermarked. C has order n power of 2. In the watermark embedding step DWT decomposition is applied to C to obtain the four sub-matrices which have the DWT coefficients of the decomposition level as elements .Only the entries of high frequencies detail matrices {HL,LH,HH} are modified by watermark. The number of DWT wavelet decomposition depends on the order of sub-image matrix associated to blocks. Typically we apply three DWT decomposition levels on a block with associated matrix of order 256, and two DWT decomposition levels on each block with associated matrix of order 128. Watermark embedding is calculated with the following formulas [15]:

$$\tilde{C}_k^\theta(i, j) = C_k^\theta(i, j) + \omega * \alpha(i, j), \tag{1}$$

where i,j=1,2,...,n_k and i,j=1,2,...,n_k and $\alpha(i,j)$ is the generic element of a matrix of order equal to the order of C_k^θ $\alpha(i,j) \in \{-1,0,1\}$ and its value is computed depending on the belonging of the corresponding DWT coefficient to an interval, more details can be found in [15]. Watermark detection applies a re-synchronization between the original and watermarked image. The watermark is embedded in a training video data set only, so that when we recognize the action, we also recognize the identity at the same time in the training data set, because in the training data sets, we know the identity by detecting a watermark.

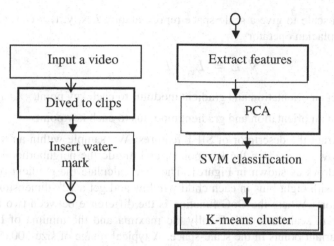

Fig. 1. Flow chart of training proposed algorithm

3.2   Feature Description Methods

A video sequence is represented as a collection of spatial-temporal words by extracting space-time interest points, after the localization of a space-time interest point in the video sequence; a video patch is extracted around the interest point location and described using one of the description methods. The main advantage of using such SIFT is the simplicity due to the unsupervised nature. The main idea is to extract features through a staged filtering that identifies stable points in the scale space: 1) select candidates for features by searching for peaks in the scale space of the difference – of – Gaussians (DoG) function. 2) localize each feature using measures of its stability, and 3) assign orientations based on local image gradient directions.   Provision of an interest point detection schema from three different planes along the spatial and the temporal axes; Formulation of the space-time detector that is scale and location invariant; Application of the developed ST-SIFT on a human action classification task, with comparison to other state-of-the-art approaches.   The 2D SIFT detector maps the special content of an image to a coordinate of scale, location and orientation invariant feature. This is achieved using a scale space kernel function such as the Gaussian, which is a continuous function to capture stable features in different scales. The feature extraction procedure of SIFT can be described as follows.

STEP1. Construct the DOG scale-space:

$$D(x, y, \sigma) = (G(x, y, k\sigma) - G(x, y, \sigma)) * I(x, y) = L(x, y, k\sigma) - L(x, y, \sigma) \qquad (2)$$

$$G(x, y, \sigma) = \frac{1}{2\pi\sigma 2} e^{-(x^2 + y^2)/2\sigma^2} \qquad (3)$$

Where $I(x, y)$ is the original image, and $L(x, y, \sigma) = G(x, y, \sigma) * I(x, y)$

STEP2. Get the key points: We get the key points at the scale space extreme in the difference of Gaussian function convolved with the image.

At a certain scale to give a scale space representation $L(x, y, t) = G(x, y, t) * I(x, y)$, then the Laplacian operator

$$\nabla^2 L = L_{xx} + L_{yy.} \tag{4}$$

STEP3. Assign an orientation and gradient modulus to each key point.

STEP3. Assign an orientation and gradient modulus to each key point.

STEP4. Construct the descriptor of SIFT features: We sample within an 8*8 neighborhood window centered on the key points, and divide the neighborhood into four 4*4 child windows as shown in Figure1. Then we calculate the gradient orientation histogram [9] with eight bins in each child window and get a 128-dimensional vector called descriptor. Where the DoG function is the difference between two neighbors with the constant scale factor K. Finally the maxima and the minima of $D(x, y, \sigma)$ give scale-invariant points in the scale-space. A typical image of size 500x500 pixels will give rise to about 2000 stable features (although this number depends on both image content and choices for various parameters). In our implementation the descriptor is about 2929 key points with 128 feature vectors. The size of this descriptor is reduced with PCA. The covariance matrix for PCA is estimated on 39,000 image patches collected from various images of actions. The 128 largest eigenvectors are used for description. In Table 1 shows the value of parameters that is used in SIFT algorithm.

Table 1. Parameters of SIFT key points detection

parameter	value
Number of octaves	4
First octave index	-1
Number of scale levels per octave	2
Edge threshold	0.8
Peak threshold	10

3.3   Feature Matching and Indexing

Indexing consists of storing SIFT keys and identifying matching keys from the new image. Lowe [9] used a modification of the k-d tree algorithm called best-bin-first search method that can identify the nearest neighbors with high probability using only a limited amount of computation. The best candidate match for each key point is found by identifying its nearest neighbor in the database of key points from training images. The probability that a match is correct can be determined by taking the ratio of distance of the second closest. In [9] rejected all matches in which the distance ratio is greater than 0.8.

3.4   Features Description

The SIFT feature vector consists of 128 elements, the coordinates of each point (the x and y location in the frame) are made use of to enhance the results as inspired by Lai et al. [16], so the new feature vector becomes 130 elements (the old 128 elements vector + x coordinate of the interest point + y coordinate of the interest point). One of the reasons to use 2-D SIFT (beside that it is invariant to scale, location and orientation changes) is its short feature vector which does not need to use topic modeling methods as pLSA and LDA, where a separate topic model is learned for each action class and new samples are classified by using the constructed action topic models.

3.5   Building a Visual Vocabulary

The objective here is to vector quantize the descriptors into clusters which will be the visual 'words' for action retrieval. Then when a new frame of the action is observed each descriptor of the frame is assigned to the nearest cluster, and this immediately generates matches for all frames throughout the action. The vocabulary is constructed from a subpart of the video, and its matching accuracy and expressive power. The vector quantization is carried out here by K-means clustering histogram.

3.6   Action Recognition

Various supervised learning algorithms can be employed to train an action pattern recognizer. Support vector machine (SVM) [12] is used in our approach. SVM has been successfully applied to a wide range of pattern recognition and classification problems. The advantages of SVM over other methods consist of: 1) providing better prediction on unseen test data, 2) providing a unique optimal solution for a training problem, and 3) containing fewer parameters compared with other methods. The whole recognition process can be divided into two phases: the training phase and the testing phase. During the training phase, as the flow chart shown in Figure 1, the interest points are extracted by SIFT detector from the training sequences, and then descriptor of each sequence is generated by the structural distribution of interest points. Descriptors from all training sequences are gathered together for further clustering by K-means which uses Euclidean distance as the clustering metric. The cluster centers are represented as the video words and they constitute the codebook. Each feature descriptor is assigned to a unique video word based on the distance between the descriptor and cluster centers. And the codebook membership of each feature descriptor is utilized to create a model representing the characteristics of each class of the training sequences [14]. All sequences were divided with respect to the subjects into a training set (8 persons), a validation set (8 persons) and a test set (9 persons). The classifiers were trained on a training set while the validation set was used to optimize the parameters of each method. The presented recognition results were obtained on the test set. Representations of motion patterns in terms of local features have advantages of being robust to variations in the scale, the frequency and the velocity of the pattern Whereas local features have been treated independently , the spatial and the temporal relations between features provide additional cues that could be used to improve the results of recognition.

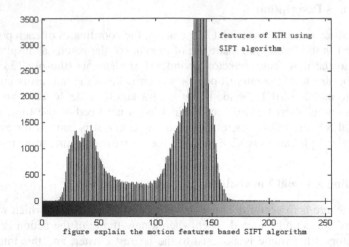

Fig. 2. Motion features based SIFT algorithm

4   Datasets

For our action recognition experiments, we chose to use the KTH human action dataset [1]. It contains six types of human actions: walking, jogging, running, boxing,
hand waving and handclapping. Each action class is performed several times by 25
subjects in different scenarios of outdoor and indoor environment. The camera is not
static and the video sequences contain scale changes. In total, the dataset contains 600
sequences. We divide the dataset into two parts: 16 people for training and 9 people
for testing. We limit the length of all video sequences to the first 300 frames. And
Weizmann datasets The human action dataset recorded at the Weizmann institute
contains 10 actions (walk, run, jump, gallop sideways, bend, one-hand wave, two-
hands wave, jump in place, jumping jack and skip), each performed by 10 persons.
The backgrounds are static and foreground silhouettes are included in the dataset.

Fig. 3. Sample images of our dataset

5   Experimental Results

We tested our approach on the KTH dataset, and our own action dataset for identity and action recognition. SVM with is chosen as the default classifier. In our experiment, we empirically used   128-bin SIFT histogram. We divide the dataset into two parts for training and testing respectively instead of leave-one-out classification. The results are shown in Figure 4, 5, 6 and table 2, 3. For matching SIFT features, we use ROC curve that computing true positive (TP) number of detected matches that are correct, and false positive (FP) number of detected matches that are incorrect, maximize area under the curve (AUC).

	Walk	Jog	Run	Box	Hand-wave	Hand-clap
Walk	94.8	5.2	0.0	0.0	0.0	0.0
Jog	2.1	94.3	3.6	0.0	0.0	0.0
Run	0.0	16.9	83.1	0.0	0.0	0.0
Box	0.0	0.0	0.0	100.0	0.0	0.0
Hand-wave	0.0	0.0	0.0	0.0	93.5	6.5
Hand-clap	0.0	0.0	0.0	1.6	0.0	98.4

97.8	bend	jack	jump	pjump	run	side	skip	walk	wave1	wave2
bend	100.	0.0	0.0	0.0	0.0	0.0	0.0	0.0	0.0	0.0
jack	0.0	100.	0.0	0.0	0.0	0.0	0.0	0.0	0.0	0.0
jump	0.0	0.0	100.	0.0	0.0	0.0	0.0	0.0	0.0	0.0
pjump	0.0	0.0	0.0	100.	0.0	0.0	0.0	0.0	0.0	0.0
run	0.0	0.0	0.0	0.0	90.0	0.0	10.0	0.0	0.0	0.0
side	0.0	0.0	0.0	0.0	0.0	100.	0.0	0.0	0.0	0.0
skip	0.0	0.0	0.0	0.0	10.0	0.0	90.0	0.0	0.0	0.0
walk	0.0	0.0	0.0	0.0	0.0	0.0	0.0	100.	0.0	0.0
wave1	0.0	0.0	0.0	0.0	0.0	0.0	0.0	0.0	100.	0.0
wave2	0.0	0.0	0.0	0.0	0.0	0.0	0.0	0.0	0.0	100.

Fig. 4. Confusion matrix for the KTH   and Weizmann datasets

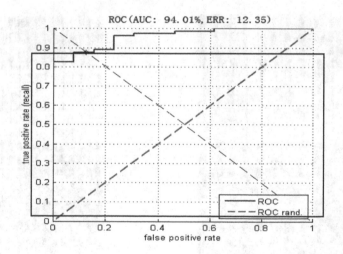

Fig. 5. Accuracy error of KTH dataset

Table 2. Comparison of method and Conventional detectors

detector	descriptor	Describing an identity	KTH	Weizmann	Our data
2D SIFT	2D HOG	Water mark as 2D wavelet	94.01%	90.14%	82.25%
ASIFT	2D HOG	Visual words	80.74%	78.21%	80.13%

In particular, this means that the comparison of two algorithms on a dataset does not always produce an obvious order. As the results show there is a significant increase in accuracy of recognition of human action and identity as shown in Table 2, 3. We recognize the action and identity at the same time without any extra computations, so the previous methods in table 3 recognize only actions. Our method is appropriate for real time applications.

Table 3. Shows a comparison between our method and a group of other previously proposed systems that use leave-one-out setup. The results show that for the KTH dataset our result is the best of them.

method	KTH	Weizmann
Neibles et al. [7]	83.3	90
Schuldt et al.[20]	71.72	---
Dollar [21]	81.17	85.2
Klaser[22]	91.4	84.3
Zhang[23]	91.33	92.89
Proposed method	94.01	90.14

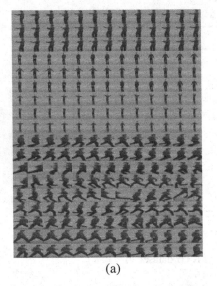
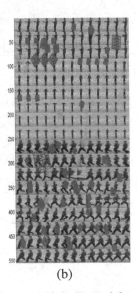

(a)   (b)

Fig. 6. (a) frames of KTH datasets (b) Scale-invariant interest points detected from a grey-level image using scale-space extrema of the Laplacian. The squares illustrate the selected detection scales of the interest points. Red squares indicate bright image features with $\nabla 2L < 0$. Extract features from KTH using SIFT detectors (c) The green arrows illustrate the orientation estimates obtained from peaks in local orientation histograms around the interest points, matching between training (embedding watermark) and testing (without watermark) using SIFT matching.

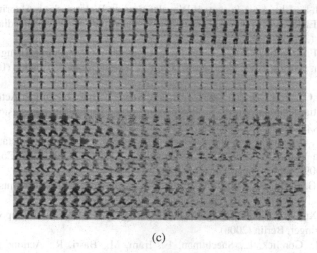

(c)

Fig. 6. (*continued*)

6   Conclusion

The main contribution of this paper can be summarized as follows: a watermark is embedded in the training data set to identify the identity of the human that makes an action, so we can recognize action and identity at same time without any extra computations, the only post processing is the watermark detection.   The results of the experiment suggest that the watermark can be embedded to a video as a human identity, in a simple and efficient manner by 2-D wavelet transform. Then we can extract action features by using 2-D SIFT descriptors.   We applied the algorithm on several data sets such as KTH and Weizmann and our datasets of our lab.   The results demonstrate the accuracy and speed of implementation and discrimination compared to other methods that are take high computed complexity. Future work includes applying the proposed system on different complex datasets, such as: sports and real actions ones. These datasets are more complex than the ones used here and the system may need some improvements to achieve acceptable recognition rate.

References

[1] Zhang, Z., Jia, L.: Recognizing Human and Identity Based on Affine-SIFT. In: IEEE Symposium on Electrical & Electronics Engineering (2012)

[2] Fabio, C.: Using Bilinear Models for View-invariant Action and Identity Recognition. In: Proceedings of the Conference on Computer Vision and Pattern Recognition (CVPR 2006), New York, NY, vol. 2 (June 2006)

[3] khamis., S., Morariu, V.I., Larry, S.D.: A Flow Model for Joint Action Recognition and Identity Maintenance. In: IEEE (2012)

[4] Langelaar, G., Setyawan, I., Lagendijk, R.: Watermarking Digital Image and Video Data. IEEE Signal Processing Magazine 17, 20–43 (2000)

[5] Meerwald, P., Uhl, A.: A Survey of Wavelet-domain Watermarking Algorithms. In: Proc. of SPIE, Electronics Imaging, Security and Watermarking of Multimedia Contents III, CA, USA (2001)

[6] Wong, S.F., Cipolla, R.: Extracting Spatiotemporal Interest Points Using Global Information. In: Proc. of IEEE International Conference on Computer Vis ion (ICCV), pp. 1–8 (2007)

[7] Niebles, J.C., Wang, H., Fei-Fei, L.: Unsupervised learning of Human Action Categories using Spatial-temporal Words. In: Proceedings of the British Machine Vision Conference, vol. 79(3), pp. 299–318 (2008)

[8] Duchenne, O., Laptev, I., Sivic, J., Bach, F., Ponce, J.: Automatic Annotation of Human Actions in Video. In: Proceedings of the International Conference On Computer Vision (ICCV 2009), Kyoto, Japan (September 2009)

[9] Lowe, D.G.: Distinctive Image Features from Scale Invariant Key Points. J. Computer Vision, IJCV 60(2), 91–110 (2004)

[10] Zhu, G., Xu, C., Gao, W., Qingming, H.: Using Optical Flow and Support Vector Machine. Springer, Berlin (2006)

[11] Blank, M., Gorelick, L., Shechtman, E., Irani, M., Basri, R.: Actions as Space-time Shapes. In: ICCV (2005)

[12] Enrique, F.M., Angel, S., Jose, V.: Support Vector Machines Versus Multi-layer Perceptron for Efficient Off-line Sig-nature Recognition. In: Biometry and Artificial Vision Group(GAVAP), Madrid, Spain,

[13] Xiaojun, Q.: An Efficient Wavelet-Based Watermarking Algorithm, Xiaojun.Qi@usu.edu

[14] Gao, R.: Dynamic Feature Description in Human Action Recognition. M.Sc. Computer Science (2009)

[15] Agreste, S., Guido, A., Daniela, P., Luigia, P.: An Image Adaptive, Wavelet-based Watermarking of Digital Images. J. of Computational and Applied Mathematics 210 (2007)

[16] Lai, K.-T., Hsieh, C.-H., Lai, M.-F., Chen, M.-S.: Human Action Recognition Using Key Points Displacement. In: Elmoataz, A., Lezoray, O., Nouboud, F., Mammass, D., Meunier, J. (eds.) ICISP 2010. LNCS, vol. 6134, pp. 439–447. Springer, Heidelberg (2010)

[17] Moussa, M.M., Fayek, M.B., Heba, A., Nemr, E.: An Enhanced Method for Human Action Recognition. Journal of Advanced Research (2013)

[18] Lian, Z., Godil, A., Sun, X.: Visual Similarity Based 3D Shape Retrieval Using Bag-of-Features. In: IEEE Proceedings of the Shape Modeling International Conference (2010)

[19] Sipiran, I., Bustos, B.: A Robust 3D Interest Points Detector Based on Harris Operator. Presented at the Proc. Euro-graphics Workshop on 3D Object Retrieval (3DOR 2010) (2010)

[20] Schuldt, C., Laptev, I., Caputo, B.: Recognize Human Actions Local SVM Approach. In: IEEE Int. Conf. ICPR, vol. 3, pp. 32–36 (2004)

[21] Dollar, P., Rabaud, V., Cottrell, G., Belongie, S.: Behavior Recognition via Sparse Spatio-temporal Features. In: IEEE International Workshop on Visual Surveillance and Performance Evaluation of Tracking and Surveillance, pp. 65–72 (2005)

[22] Klaser, A., Marszaek, M., Schmid, C.: A Spatio-temporal Descriptor Based on 3D-Gradients. In: British Mach Vision Conf. BMVC (2008)

[23] Zhang, Z., Hu, Y., Chan, S., Chia, L.-T.: Motion Context: A New Representation for Human Action Recognition. In: Forsyth, D., Torr, P., Zisserman, A. (eds.) ECCV 2008, Part IV. LNCS, vol. 5305, pp. 817–829. Springer, Heidelberg (2008)

[24] Chang, C., Lin, C.: LIBSVM: A Library for Support Vector Machines. ACM Trans. Intell. Syst. Technol. TIST (2011)

[25] Fathi, A., Mori, G.: Action Recognition by Learning Mid-level Motion Features. In: Conference of CVPR IEEE (2008)
[26] Blank, M., Gorelick, L., Shechtman, E., Irani, M., Basri, R.: Actions as Space-time Shapes. In: ICCV IEEE (2005)
[27] Ke, Y., Sukthanka, R., Hebert, M.: Efficient Visual Event Detection Using Volumetric Features. In: ICCV IEEE (2005)
[28] Sheikh, Y., Sheikh, M., Shah, M.: Exploring the space of a human action. In: ICCV IEEE, pp. 144–149 (2005)
[29] Chen, M.Y., Hauptmann, A.: MoSIFT: Recognizing human actions in surveillance videos. In: Technological Report. Carnegie Mellon University (2009)

Augmented Reality Surveillance System
for Road Traffic Monitoring

Alexander Filonenko[1], Andrey Vavilin[2], Taeho Kim[2], and Kang-Hyun Jo[1]

[1] Graduate School of Electrical Engineering,
University of Ulsan, Ulsan, 680-749, Republic of Korea
[2] FuturIST, Republic of Korea
{alexander,acejo}@islab.ulsan.ac.kr
andrey.korea@gmail.com
thkim@futist.co.kr

Abstract. This paper introduces the augmented reality surveillance system which evaluates the density of the traffic on roads and displays information in an easy to understand form over the video stream and a map. A mutual dependence between the real world, global coordinates and the position of the pixel in the image is explained. The way to find the real size of an object by knowing its dimension in the image is introduced. An operator can decide what points on the map it is required to survey, and the camera will know how to rotate to those points by mapping of global coordinates to pan and tilt angles. The density of the traffic is evaluated by processing video data and applying the knowledge about real width and length of cars.

Keywords: Traffic monitoring, Car detection, PTZ camera, Camera measurements.

1   Introduction

Our system allows mounting a camera at any part of a road with different altitude and position. The system automatically connects the visual information from the camera with data from a map. Overlays on the image and on the map show knowledge about density of the road traffic presented both by number of objects and their color which is easy to understand. Ways of getting the real sizes of objects were introduced in this work. It allows deciding whether there are cars presented in the image according to the real size of objects. In the third chapter of this paper the mappings from the real world to image coordinates and vice versa are introduced. The fourth chapter explains how to detect cars using background subtraction and find out the density of the traffic.

2   Related Works

Some researchers use microphones [1] to say whether there are many motors working at the same time. This approach cannot tell whether both sides of the road are

D.-S. Huang et al. (Eds.): ICIC 2014, LNAI 8589, pp. 310–317, 2014.

occupied. Another approach is detection of the traffic density where each car checks its surrounding and builds its own picture of the environment [2]. In developing countries movement of cars is not well organized. Special algorithms which take in account features of traffic of a particular country are developed [3]. It becomes more and more popular to use support vector machines and histogram of oriented gradients (HOG) in the process of detection vehicles [4]. Although detection results can be good, processing time for HOG with sliding window is not sufficient for the real-time processing. Some researchers try to use a simple method to achieve real-time performance by sacrificing detection accuracy [5]. When the camera is shaking, there is still a solution to detect moving objects [6], but the objects are not classified in this work.

3   Camera to Real World Projection

In this work the pan-tilt-zoom (PTZ) camera Samsung SNP-3370 was used which can rotate about its vertical and horizontal axes.

Relation between angle and the sensor size is shown in (1):

$$\alpha = 2 \cdot \tan^{-1}\left(\frac{d}{2 \cdot F}\right) \tag{1}$$

Then the sensor size can be estimated:

$$d = 2 \cdot F \cdot \tan\left(\frac{\alpha}{2}\right) \tag{2}$$

where F is the focal length and α is the angle.
Horizontal angle of view for different focal lengths (f) for Samsung SNP-3370 is

$$\alpha_H(f) = 2 \cdot \tan^{-1}\left(\frac{3.6}{2 \cdot f}\right) \tag{3}$$

Vertical angle of view for different focal lengths for Samsung SNP-3370 is

$$\alpha_V(f) = 2 \cdot \tan^{-1}\left(\frac{2.7}{2 \cdot f}\right) \tag{4}$$

The spherical model is considered in this work. In general, the angular resolutions do not distribute evenly on the sensor since the sensor is a plane not a spherical surface. However, in surveillance task mostly the narrow field of view is required. Thereby, distortions caused by the chosen model are inessential.

Fig. 1(a) represents a model of the real world where C is the position of the camera, p is a point in the real world. Fig. 1(b) shows the side view of the model. Fig. 2 shows the vertical view of a model of scene. Image resolutions in degrees are:

$$D_V = \frac{\alpha_V(f)}{R_V} \quad (5)$$

where R_V is the height of the image in pixels.

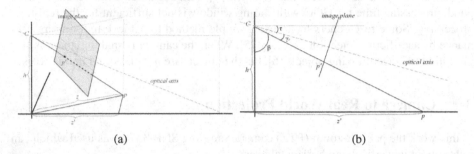

(a)   (b)

Fig. 1. (a) 3D model of scene; (b) Side view of a model of scene

$$D_H = \frac{\alpha_H(f)}{R_H} \quad (6)$$

where R_H is the width of the image in pixels.

Vertical and horizontal angles between p and the optical axis of the camera are:

$$\gamma_V = \left(y - \frac{R_V}{2} \right) \cdot D_V \quad (7)$$

$$\gamma_H = \left(y - \frac{R_H}{2} \right) \cdot D_H \quad (8)$$

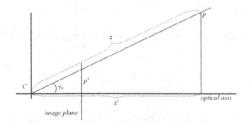

Fig. 2. Vertical view of a model of scene

$$\beta = 90 - \tau - \gamma_V \quad (9)$$

$$z' = h \cdot \tan(\beta) \quad (10)$$

$$z = \frac{z'}{\cos(\gamma_H)} \tag{11}$$

If two points are selected on the ground plane, then distance between them can be found using law of cosines.

$$d_p = \sqrt{z_1^2 + z_2^2 - 2 \cdot z_1 \cdot z_2 \cdot \cos(\rho)} \tag{12}$$

where ρ is the angle between two points, z_1 and z_2 are distances to the first and the second points respectively.

When pan and tilt equal to zero, the optical axis is directed to the north and parallel to the ground plane respectively. The GPS coordinates of the point are:

$$lat_2 = \arcsin\left(\sin(lat_1) \cdot \cos\left(\frac{z}{R}\right) + \cos(lat_1) \cdot \sin\left(\frac{z}{R}\right) \cdot \cos(\theta) \right) \tag{13}$$

where lat_1 and lat_2 are latitudes of the camera and the point respectively, z is found in (11), θ is the bearing angle, R is the mean radius of the Earth (6371 km).

$$lon_2 = lon_1 + \arctan\left(\frac{\sin(\theta) \cdot \sin\left(\frac{z}{R}\right) \cdot \cos(lat_1)}{\cos\left(\frac{z}{R}\right) - \sin(lat_1) \cdot \sin(lat_2)} \right) \tag{14}$$

where lon_1 and lon_2 are longitudes of the camera and the point respectively.

To show some exact position given in GPS coordinates, it is required to know pan and tilt angles to rotate to (as in [7]). The tilt angle is:

$$T = \left(\frac{\pi}{2} - \arctan\left(\frac{z_G}{h}\right) \right) \cdot \frac{180}{\pi} \tag{15}$$

$$z_G = \arccos(\sin(lat_1) \cdot \sin(lat_2) + \cos(lat_1) \cdot \cos(lat_2) \cdot \cos(lon_2 - lon_1)) \cdot R \tag{16}$$

Camera pan angle should be calculated with additional conditions.

$$P = \left(\arccos\left(\frac{z_1}{z_G}\right) \right) \cdot \frac{180}{\pi}, if\,(lat_1 < lat_2 \;\&\; lon_1 < lon_2) \tag{17}$$

where z_1 is the distance from the camera to the support point l which we introduced in the Mercator coordinate system (Fig. 3(a)).

$$P = \left(\pi - \arccos\left(\frac{z_1}{z_G} \right) \right) \cdot \frac{180}{\pi}, if\,(lat_1 > lat_2 \,\&\, lon_1 < lon_2) \tag{18}$$

$$P = \left(\pi + \arccos\left(\frac{z_1}{z_G} \right) \right) \cdot \frac{180}{\pi}, if\,(lat_1 > lat_2 \,\&\, lon_1 > lon_2) \tag{19}$$

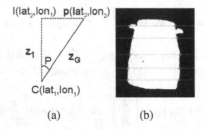

(a)   (b)

Fig. 3. (a) Support point used to find the pan angle of a camera; (b) Width estimation of the blob

$$P = \left(2 \cdot \pi - \arccos\left(\frac{z_1}{z_G} \right) \right) \cdot \frac{180}{\pi}, if\,(lat_1 < lat_2 \,\&\, lon_1 > lon_2) \tag{20}$$

4   Traffic Density Estimation

The region of interest is set as boundaries of the roads found by the lane detection algorithm based on vanishing point detection ([8]). To detect the moving cars on the road the very first step is building a background model. The scoreboard algorithm ([9]) which is a tradeoff between accuracy and computational cost is used in this work. After the background acquired, the vehicles are segmented from the background and the image is binarized by using the threshold introduced in [5].

If the camera is not fixed firmly, then the image quality will suffer from the shaking making the background in the image move in random directions. This becomes a serious problem with the narrow field of view. In this case another subtraction technique should be applied. Xinyi Cui et al. [10] proposed to analyze the behavior of background using the low rank constraint when the foreground satisfies the group sparsity constraint. The preliminary step of generating points and tracking their trajectories using GPU [11] processes less than 0.5 frames per second. So, if this approach is used, the camera position should be selected such that the car will appear in the sequence of frames during more than 4 seconds.

The connected components labeling algorithm is then applied. After that for each blob the average width is calculated using the following approach: for some random vertical coordinates in the image which consist the blob pixels, the horizontal scan lines are built and the distance in pixels between the leftmost and the rightmost pixels is found as shown in Fig. 3(b). After that the average value of these distances is chosen as the width of the blob. Delete the blobs which are not in (21):

$$Car \in \left[W_c, \sqrt{W_c^2 + L_c^2} \right] \tag{21}$$

where W_c and L_c are the real possible width and length of the car. W_c and L_c should be chosen suchwise they cover sizes of both small passenger cars and trucks.

After the car blobs are found, the road area is divided into 2 parts according to detected lanes (to represent the density of traffic in opposite directions) and for each part the number of blobs is calculated. The more blobs found the higher the density of the traffic.

In the final step these blobs should be shown on a map. Using equations (13) and (14) the coordinates of the center of these blobs is found and displayed on the map. According to the density of the traffic (number of cars), a colored layer is overlaid for each part of the road.

5   Experimental Results

To test the pan and tilt angles calculation the camera was mounted on the bridge above the road. GPS coordinates of the camera were collected during 3 minutes using 2 GPS receivers and the average value was decided as the real one to decrease the error. We aimed the camera at the first landmark, and set the pan angle to zero. The difference between the north direction and the orientation of the camera with zero pan value was taken into account for the all next measurements. Manual measurements showed that the average error of the orientation was about 3 meters by 100 meters of the distance. This error is permissible when rotating the camera to the point marked on the map. Using the comparison with a background (Fig. 4 (a)), the cars within 100 meters to the camera were detected correctly (Fig. 4 (b)).

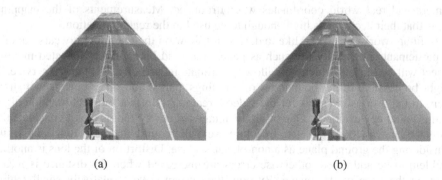

(a)   (b)

Fig. 4. Car detection: (a) Background of the road scene with ignored non-road regions; (b) Road scene with detected cars

Detection of the real sizes of objects was performed both indoors and outdoors. For the latter case (Table 2) the error of the automatic measurement is higher than indoors because a pixel covers bigger area in the real world due to its discrete nature. Distance from the camera to objects was 150 meters. In the second test (10.5 meters) the field of view was wider than during other tests that directly impact on the accuracy of measurements because of the camera model used.

During tests it was found out that different layers of the Google Maps service do not match (Fig. 5).

Table 1. Measurement of real size of objects using camera

Real size (m)	Average error (m)	Error (%)
5.5	0.1	1.8
10.5	0.7	6.6
20	0.2	1
25	0.3	1.2

Fig. 5. Positions of cars shown on different maps

6   Conclusions

The road traffic monitoring system was developed with augmented reality layers which make it easier to understand the density of the traffic. The mapping between camera and real world coordinates was introduced. Measurements of the mapping show that their accuracy is high enough to be used in the real application.

In future works we would like to detect and show on the map not only cars but other participants of the city life such as pedestrians and bikes. Every detected moving object will be tracked which will allow to estimate its velocity. Using map services it might be possible to get boundaries of buildings, parks, and other objects and show them to the user as overlays for the video stream. To do that the vectorization algorithm should be developed. One of disadvantages of our system is that it considers the ground plane as totally planar. In future we will rebuild mapping functions while considering the ground plane as a non-planar surface. Distortion of the lens is another problem to be addressed; otherwise errors are increased when the distance is measured on the edge of the image. By now the system covers relatively small region which is not enough to build a complete picture of traffic along the road. In the future we will mount a camera on a multicopter and will move it along roads which will raise new problems to solve, such that compensation of ego motion in image processing.

Acknowledgment. This work was supported by 2014 Research Funds of Hyundai Heavy Industries for University of Ulsan.

References

1. Tyagi, V.: Vehicular Traffic Density State Estimation Based on Cumulative Road Acoustics. IEEE Transactions on Intelligent Transportation Systems 13(3), 1156–1166 (2012)
2. Mao, R.X.: Road traffic density estimation in vehicular networks. In: Wireless Communications and Networking Conference, Shanghai, pp. 4653–4658 (2013)
3. Padiath, A., Vanajakshi, L., Subramanian, S.C., Manda, H.: Prediction of traffic density for congestion analysis under Indian traffic conditions. In: 12th International IEEE Conference on Intelligent Transportation Systems, St. Louis (2009)
4. Purusothaman, S.B.: Vehicular traffic density state estimation using Support Vector Machine. In: International Conference on Emerging Trends in Computing, Communication and Nanotechnology, Tirunelveli, pp. 782–785 (2013)
5. Royani, T., Haddadnia, J., Pooshideh, M.R.: A simple method for calculating vehicle density in traffic images. In: Machine Vision and Image Processing, 6th Iranian, Isfahan (2010)
6. Kim, T., Jo, K.H.: Real-time detection of moving objects from shaking camera based on the multiple background model and temporal median background model. Domestic Journal of Institute of Control, Robotics and Systems 16(3) (2010)
7. Filonenko, A., Jo, K.H.: Visual surveillance with sensor network for accident detection, Industrial Electronics Society. In: IECON 2013 - 39th Annual Conference of the IEEE, Vienna, pp. 5516–5521 (2013)
8. Kim, J.W., Kim, T.H., Jo, K.H.: Traffic road line detection based on the vanishing point and contour information. In: 2011 Proceedings of SICE Annual Conference (SICE), Tokyo, pp. 495–498 (2011)
9. Lai, A.H.S., Yung, N.H.C.: A fast and accurate scoreboard algorithm for estimating stationary backgrounds in an image sequence. In: Proceedings of the 1998 IEEE International Symposium on Circuits and Systems, ISCAS 1998, Monterey, vol. 4, pp. 241–244 (1998)
10. Cui, X., Huang, J., Zhang, S. N., Metaxas, D.N.: Background subtraction using low rank and group sparsity constraints. In: Fitzgibbon, A., Lazebnik, S., Perona, P., Sato, Y., Schmid, C. (eds.) ECCV 2012, Part I. LNCS, vol. 7572, pp. 612–625. Springer, Heidelberg (2012)
11. Sundaram, N., Brox, T., Keutzer, K.: Dense point trajectories by GPU-accelerated large displacement optical flow. In: Daniilidis, K., Maragos, P., Paragios, N. (eds.) ECCV 2010, Part I. LNCS, vol. 6311, pp. 438–451. Springer, Heidelberg (2010)

Effective Palm Tracking with Integrated Tracker and Offline Detector

Zhibo Yang, Yanmin Zhu, and Bo Yuan

Intelligent Computing Lab, Division of Informatics
Graduate School at Shenzhen, Tsinghua University, Shenzhen 518055, P.R. China
{yangzhibo450,zhuym111}@gmail.com, yuanb@sz.tsinghua.edu.cn

Abstract. In this paper, we propose a vision based palm tracking method with three inherently connected components: i) an offline palm detector that locates all possible palm-like objects; ii) a SURF-based tracking module that identifies the tracked palm's location using historical information; iii) an adaptive skin color model and a patch similarity calculation module. The outputs from the last component can effectively eliminate false detections and decide which palm is under tracking and also provide updated information to the first two modules. In summary, our work makes the following contributions: i) an effective offline palm detector; ii) a benchmark dataset for training and testing palm detectors; iii) an effective solution to tackling the challenges of palm tracking in adverse environments including occlusions, changing illumination and lack of context. Experiment results show that our method compare favorably with other popular tracking techniques such as Camshift and TLD in terms of precision and recall.

Keywords: Palm Tracking, Detection, SURF, Skin Color Model, Integrator.

1 Introduction

In recent years, vision based dynamic hand gesture recognition has received growing interests from the computer vision community. Due to its ease of use and power of metaphor, hand gesture has long been serving as an essential communication tools since the beginning of human society and has great potential in human computer inte-raction (HCI) [1]. Palm tracking is one of the most important components of dynamic hand gesture recognition. Typical applications include vision based remote control, sign language recognition and palm localization in palm print recognition. However, palm tracking in unconstrained environments still remains as a challenging task due to factors such as changing illumination, occlusions, deformable palm shape and the lack of body or arm context information.

One of the key steps in palm tracking is finding a proper set of features so that the palm region can be extracted as accurately as possible. Recent studies [2] show that, for hand detection, color is an effective feature compared to other texture features such as LBP (Local Binary Pattern), Haar-like [3] and gradient features. However, the color feature can fail when the illumination condition is poor. In this paper, we will investigate the possibility of adopting non-color features such as LBP and normalized

D.-S. Huang et al. (Eds.): ICIC 2014, LNAI 8589, pp. 318–327, 2014.
© Springer International Publishing Switzerland 2014

HOG (Histogram of Oriented Gradient) in the training of offline detectors while using the color feature as complementary information.

Another issue in palm tracking is how to use historical cues to increase the precision of tracking. In complex situations, offline detectors may often produce false or multiple detections and it is also difficult to obtain satisfactory tracking results consistently as the feature points may be unstable. To address this issue, there is a need to build a link between the detector and the tracker. In our work, we will show how a SURF (Speeded Up Robust Features) based tracking module can be used to propagate the information of accurately tracked palm on a frame to frame basis.

Furthermore, a representative human palm dataset is essential in the research of palm tracking. Unfortunately, there is a lack of publicly available datasets that are well suited for this purpose. As a result, a fully annotated palm dataset containing samples from various sources is introduced, as shown in Fig. 1.

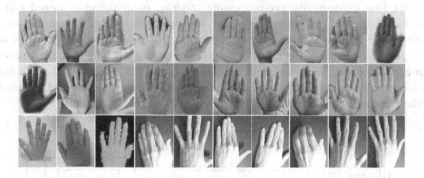

Fig. 1. Some positive hand samples in our benchmark dataset

2   Related Work

The issue of segmenting hands from images or videos in constrained environments has been extensively studied in the literature. However, in complicated real-world scenarios, methods independently using skin detection [4-6] or cascade detectors based on wavelet features [7, 8] often fail in tracking. Some approaches treat human pictorial structure as the spatial context for locating hand positions [9, 10]. However, these methods require the location information of certain human body parts (e.g., head and arms), which are not always visible in the image, especially when the user is close to the camera. In [11], a novel approach to hand segmentation is based on creating hand regions from contours, but extra information is still required in this method. There are also some studies in which different techniques are combined to achieve improved performance. For example, in [12], three detectors and two segmentation methods were used to locate the hand in images. However, this method cannot be extended to hand tracking in video due to its high computational complexity. By combining traditional tracking methods with the skin color features, it is possible to achieve fast tracking but the system is not robust enough against the interference of hand-like objects [13, 14].

Other well known tracking methods such as block tracking and the state-of-the-art TLD algorithm [15] have been tested in our preliminary experiments. The block tracking can fail to track the palm for many reasons. For example, the face can be accidently tracked after the hand moves across it. The failure of TLD is because there is not sufficient training data for the online detector and the deformation of palm can produce many problematic patches that may compromise the performance of detector. Motivated by the methods mentioned above, the proposed method not only detects the palm using multiple descriptors but also integrates the detections and tracking results, resulting in significantly better tracking performance.

3   Framework

This section gives an overview of our palm tracking method, as shown in Fig. 2. There are four components: pre-processing module, palm detector trained offline, SURF-based tracker and integrator. The pre-processing stage including image smoothing and color space transformation can effectively decrease noises, which may otherwise be falsely treated as tracking points. The task of tracking is actually to track the foreground and a foreground detector can be helpful [16]. In our method, multiple cues are used to detect the foreground and both the offline detector and the tracker take the preprocessed frame and the feedback of the integrator as inputs.

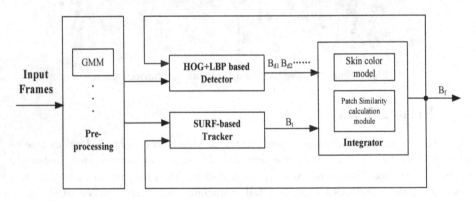

Fig. 2. The framework of the proposed tracking process

If there are multiple palm-like objects in the image, the detector will recognize all of them and output their bounding boxes B_{d1}, B_{d2}, $\cdots$. We noticed that classifiers trained offline were unable to detect the palm with large rotation degrees (e.g., more than ± 15 degrees). In practice, the palm is unlikely to rotate sharply between two frames. Therefore, it is possible to inject the rotation information into the detector by analyzing the bounding box in the last frame. The SURF-based tracker takes as input two consecutive frames and the bounding box in the last frame. A number of SURF points within the bounding box are tracked and outliers are filtered out. Remaining points are used to estimate the motion of the box region and the location of the new tracking box is equal to the location of the last bounding box plus the motion offset.

4   Implementation Details

4.1   Offline Palm Detector

We collected 500 palm images as positive training samples, together with their left-right reflections (1000 images in all), as shown in Fig. 1. Some of the images were generated by our lab members while others were retrieved from Cambridge Hand Gesture Data Set [17]. Negative training samples were based on the images used in a face detection project [18] and all images containing hand-like objects were removed, resulting in 6465 normalized images.

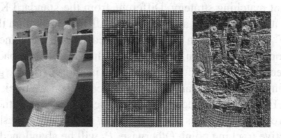

Fig. 3. A palm (left) and its HOG (middle) and LBP (right) reconstruction images

For feature extraction, we tested the discriminative power of different features: HSV, RGB, YCbCr, Gabor, Harr-like, HOG, BRIEF and LBP. In our results, color features often failed when illumination was poor, and the last three features ranked higher than Gabor and Harr-like features. Since HOG describes the shape and contour of palm while LBP is invariant to grey value and rotation, when combined together, they may generate better results. Therefore, HOG and LBP descriptors were selected to form combined feature vectors (Fig. 3).

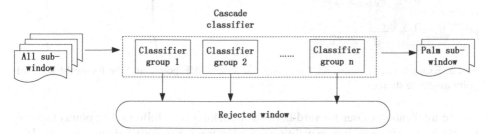

Fig. 4. The workflow of cascade classifiers

Cascade classifiers were used in the classifier training process, as shown in Fig. 4. The number of classifier stages depends on the specific training process. In each stage, the GAB (Gentle AdaBoost) algorithm boosted a strong classifier consisting of a group of decision trees (weak classifiers). The maximum tolerable false positive rate FPR (50% in our experiments) and minimum tolerable true positive rate TPR (99.9% in our experiments) influenced the number of weak classifiers in each stage. During testing, all sub-windows went through stages one by one, and more than half of the remaining negative sub-windows were rejected at each stage. In our experiments, the cascade had 20 stages and the number of weak classifiers was between 5 and 28. The overall false positive rate was $1 - 0.5 - 0.5^2 \cdots - 0.5^{20} \approx 0.0001\%$ and the overall true positive rate reached $0.999^{20} \approx 98\%$. Finally, the cascade classifiers identified all palm-like objects and generated a bounding box for each of them.

4.2   SURF-Based Tracker

There are two key factors for achieving reliable tracking: the selection of feature points and the point matching strategy. Different from the standard KLT tracker [19], we used SURF feature points [20] and, for each SURF point within the bounding box B_f in the frame F^t, its location in F^{t+1} was tracked. Next, the same procedure was conducted in the opposite direction, which means that we tracked backward the point that has already been tracked in F^{t+1}. For example, in Fig. 5 (right), given P_0 in frame F^t, its corresponding point P_1 in frame F^{t+1} is located. When we track P_1 backward, P_2 is identified. The Euclidean distance between these two points is then calculated. If the distance between P_1 and P_2 is smaller than a predefined threshold, P_1 will be treated as an effective tracking point. Otherwise, P_1 will be abandoned.

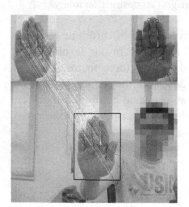 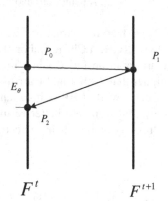

Fig. 5. Examples of extended KLT tracking based on SURF points and the forward-backward point matching strategy

The implication of our forward-backward tracking is as follows. For points that are visible in both frames, it is usually easy for the tracker to locate them correctly. If a point is invisible in the next frame, the tracker may locate a different point, leading to false tracking. Fig. 5 (left) gives an example of tracking SURF feature points, which are labeled as blue circles over the palm in the upper-right corner.

In summary, the tracking precision of the standard KLT tracker can be improved due to the use of SURF tracking points and the specific point matching strategy. The SURF points are invariant to rotations and scale change while falsely tracked points can be effectively removed by tracking them back to the last frame. The details of the error estimation of this bi-directional tracking can be found in [21].

4.3   Integrator

The integrator combines the outputs of the offline detector and tracker. It provides the rotation information to the detector and the object to be tracked in the next frame. In the following, our work is conducted in the perspective of patch. A single instance of the object's appearance is represented by an image p called patch, which is either described by color statistics or texture statistics. We generated all possible patches by shifting an initial box with the following parameters: scale step = 1.2, horizontal step size = 10% of the width, vertical step size = 10% of the height and minimal box size: 20×30 pixels for a 640×480 image. Finally, all these grids were normalized to create uniform patches with 40×60 pixels.

patch set(p+, p-)

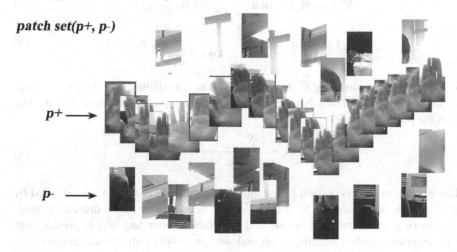

$p+ \longrightarrow$

$p- \longrightarrow$

Fig. 6. An example of the patch set. Positive patches consist of correctly tracked palms in the history while negative patches are areas without the palm.

A patch set M includes both positive patch p^+ and negative patch p^-, as illustrated in Fig. 6. Next, the skin color model extracts the color information from the positive patches while the patch similarity calculation module mainly considers the texture information of both positive and negative patches.

Skin Color Model can work well or badly depending on the illumination condition. To initialize the skin color model in the beginning of tracking, a face detector is employed to locate the human face, which provides the required skin color information. If there is no face in the frames, empirical threshold values are used. The model is then used to reject the inputs (bounding boxes) if they violate the skin color information. This model can also increase *recall* by finding skin regions outside the input bounding box. During the tracking process, the model keeps updating itself by extracting skin color information from positive patches in the latest frame.

Patch Similarity Calculation Module takes the outputs from the skin color model. As mentioned above, we define positive patches as those tracked correctly in the history and their background neighborhoods are regarded as negative patches (sampled from regions close to the trajectory). The similarity $S(p_i, p_j)$ between two patches is defined as (1).

$$S(p_i, p_j) = 0.5(NCC(p_i, p_j) + 1), \tag{1}$$

In (1), NCC is the Normalized Correlation Coefficient. Given an arbitrary patch p and a patch set M represented by (p_i^+, p_i^-), the similarity between p and negative patches is defined as (2). A large value of S^- means that the patch shares similar appearance with the background. The similarity between p and the last 10% positive patches is given by (3), where m is the total number of patches. In general, it is more accurate when considering only the last 10% patches as older patches of the palm may differ a lot from the newly generated ones.

$$S^-(p, M) = \max_{p_i^- \in M} S(p, p_i^-), \tag{2}$$

$$S_{10\%}^+(p, M) = \max_{p_i^+ \in M \wedge i > \frac{9m}{10}} S(p, p_i^+), \tag{3}$$

The conservative similarity S^c ranging from 0 to 1 is defined in (4). A large value of S^c implies high confidence that the patch resembles appearance observed in the last 10% of the positive patches.

$$S^c = \frac{S_{10\%}^+}{S_{10\%}^+ + S^-}, \tag{4}$$

The overlap rate O^θ is defined as the intersection area of two rectangles divided by their total area. According to (5), if the O^θ value between one of the detected bounding boxes (n is the number of boxes) and the tracked bounding box is greater than 50%, it implies a perfect tracking result and we can simply output the average of the two boxes. Otherwise, the integrator calculates the conservative similarity S^c between each input patch and the patch set M, and outputs the one with the maximum S^c.

$$B_f = \begin{cases} average_{i \in n}(B_d^i, B_t) & if \quad O_\theta > 0.5 \\ \max(S^c(p, M)) & else \end{cases} \tag{5}$$

If none of the tracker, the detector and the skin color model is able to generate a bounding box, the palm is declared as being invisible or a system failure is reported. In this situation, the skin color model will be re-initialized.

5   Experiments

We implemented the proposed tracking system in C++ (single thread), and tested it on a PC running Windows 7 32-bit version with Intel Core i5 CPU and 4GB RAM. Three video sequences named *Gao*, *Huang* and *Yang* were captured in normal lab working environments with different levels of difficulty. The environment in *Huang* was the simplest, without occlusion and other palms. In *Gao*, the tracked palm was occluded and out of frame sometimes. In *Yang*, there were a variety of challenging factors including occlusion, interference by other palms and discontinuity.

Qualitative Studies. In Fig. 7, the top row of images demonstrates the performance of the offline detector in recognizing palm and the back of the hand with rotation and deformation. An example with naked arms (middle of the first row) was also included to show that the exposure of arm skin has no negative effect on our approach. The middle row is an ordinary tracking example without occlusion.

To demonstrate the advantages of the proposed tracking system on dealing with complicated background and topology changes, we compared our system with the standard TLD, as shown by the sequence in the bottom row in Fig. 7. The standard TLD performed poorly when an identical object (e.g., another palm) passed over the object under tracking. For example, the blue bounding circle of TLD was distracted by the appearance of the left palm and stayed with it, losing the real target (right palm). By contrast, due to the patch comparison module in use, our method consistently tracked the right palm (red bounding box) in all frames.

Fig. 7. An illustration of the detection (top) and tracking (middle) performance of our approach and the comparison with TLD (bottom)

Quantitative Studies. In the three benchmark video sequences, the position of palm was manually annotated. The tracking performance was evaluated using the *precision* and *recall* measures: *precision* is the number of frames with correctly tracked palm divided by the number of frames with a bounding box; *recall* is the number of frames with correctly tracked palm divided by the number of frames that should be tracked according to the ground truth.

Table 1. Precision and recall of different tracking technqiues

Sequences	Number of frames	Camshift		Standard TLD		Mittal's [12]		Our Method	
		P	R	P	R	P	R	P	R
Huang	3368	0.42	0.37	0.81	0.63	0.98	0.60	0.94	0.85
Gao	4108	0.22	0.16	0.38	0.22	0.95	0.51	0.94	0.87
Yang	5884	0.10	0.08	0.22	0.14	0.97	0.45	0.92	0.80
Average	4453	0.25	0.20	0.47	0.33	0.967	0.52	0.93	0.84

In Table 1, the Mittal's method is a sophisticated hand segmentation method [12]. Its *precision* value was very high but it failed to detect the palm in many frames, leading to a low score in *recall*. The standard TLD performed well in simple situations but was not robust against occlusion or interferences.

As to time complexity, for a 640×480 frame, it generally takes our method 90 ms to 150 ms to process, depending on the number of detections produced by the offline detector. The average processing times on the three video sequences are presented in Table 2. The standard TLD often took significant amount of time in learning new samples especially at the beginning of tracking and the time reported was when the tracking was relatively stable. Note that the Mittal's method took much longer time compared to other tracking techniques.

Table 2. Average computation time of each frame (miliseconds)

Sequences	Camshift	Standard TLD	Mittals' [12]	Our Method
Huang	11	134	31856	99
Gao	9	118	31974	113
Yang	13	115	32196	135
Average	11	123	32008	116

6   Conclusions

In this paper, we presented a hybrid method for robust palm tracking in complex scenes. An offline palm detector with HOG and LBP descriptors based on cascade classifiers was trained to extract palm-like objects from the background. Furthermore, the traditional KLT tracker was extended with SURF features and an effective point matching strategy was employed to reduce false tracking. At last, a novel integrator containing the skin color model and the patch comparison module was proposed. It not only combines the offline detector and the tracker but also provides vital feedback information such as rotation degrees.

Experiments on three video sequences showed that our technique achieved better tracking performance in terms of *precision* and *recall*, compared to Camshift and standard TLD, especially in challenging environments. It also features reasonable computational complexity, making it suitable for conducting real-time tracking. A major direction of future work is to further improve the effectiveness of the offline detector so that it can better handle the deformation and rotation of palms.

References

1. Zhu, Y.M., Yang, Z.B., Yuan, B.: Vision Based Hand Gesture Recognition. In: 2013 International Conference on Service Science, pp. 260–265 (2013)
2. Li, C., Kitani, M.K.: Pixel-Level Hand Detection in Ego-Centric Videos. In: 2013 IEEE Conference on Computer Vision and Pattern Recognition, pp. 3570–3577 (2013)

3. Chen, Q., Georganas, N.D., Petriu, E.M.: Real-Time Vision-Based Hand Gesture Recognition Using Haar-like Features. In: 2007 IEEE Instrumentation and Measurement Technology Conference, pp. 1–6 (2007)
4. Wu, Y., Huang, T.S.: View-Independent Recognition of Hand Postures. In: 2000 IEEE Conference on Computer Vision and Pattern Recognition, vol. 2, pp. 88–94 (2000)
5. Wu, Y., Liu, Q., Huang, T.S.: An Adaptive Self-Organizing Color Segmentation Algorithm with Application to Robust Real-Time Human Hand Localization. In: 2000 Asian Conference on Computer Vision, pp. 1106–1110 (2000)
6. Dadgostar, F., Sarrafzadeh, A.: An Adaptive Real-Time Skin Detector Based on Hue Thresholding: A Comparison on Two Motion Tracking Methods. Pattern Recognition Letters 27(12), 1342–1352 (2006)
7. Viola, P., Jones, M.: Rapid Object Detection Using a Boosted Cascade of Simple Features. In: 2001 IEEE Conference on Computer Vision and Pattern Recognition, vol. 1, pp. 511–518 (2001)
8. Ong, E.J., Bowden, R.: A Boosted Classifier Tree for Hand Shape Detection. In: 6th Automatic Face and Gesture Recognition, pp. 889–894 (2004)
9. Karlinsky, L., Dinerstein, M., Harari, D.: The Chains Model for Detecting Parts by Their Context. In: 2010 IEEE Conference on Computer Vision and Pattern Recognition, pp. 25–32 (2010)
10. Kumar, M.P., Zisserman, A., Torr, P.H.: Efficient Discriminative Learning of Parts-Based Models. In: 12th International Conference on Computer Vision, pp. 552–559 (2009)
11. Arbelaez, P., Maire, M., Fowlkes, C., Malik, J.: From Contours to Regions: An Empirical Evaluation. In: 2009 IEEE Conference on Computer Vision and Pattern Recognition, pp. 2294–2301 (2009)
12. Mittal, A., Zisserman, A., Torr, P.H.: Hand Detection Using Multiple Proposals. In: 22nd British Machine Vision Conference, 75.1-75.11 (2011)
13. Kolsch, M., Turk, M.: Fast 2D Hand Tracking with Flocks of Features and Multi-Cue Integration. In: IEEE Workshop on Real-Time Vision for Human-Computer Interaction, pp. 158–165 (2004)
14. Bretzner, L., Laptev, I., Lindeberg, T.: Hand Gesture Recognition Using Multi-Scale Color Features, Hierarchical Models and Particle Filtering. In: 5th IEEE International Conference on Automatic Face and Gesture Recognition, pp. 423–428 (2002)
15. Kalal, Z., Krystian, M.K., Matas, J.: Tracking-Learning-Detection. IEEE Transactions on Pattern Analysis and Machine Intelligence 34(7), 1409–1422 (2012)
16. Rittscher, J., Tu, P.H., Krahnstoever, N.: Simultaneous Estimation of Segmentation and Shape. In: 2005 IEEE Conference on Computer Vision and Pattern Recognition, vol. 2, pp. 486–493 (2005)
17. http://www.iis.ee.ic.ac.uk/~tkkim/ges_db.htm
18. http://tutorial-haartraining.googlecode.com/svn/trunk/data/negatives/
19. Lucas, B.D., Kanade, T.: An Iterative Image Registration Technique with an Application to Stereo Vision. In: 7th International Joint Conference on Artificial Intelligence, pp. 674–679 (1981)
20. Bay, H., Tuytelaars, T., Van Gool, L.: SURF: Speeded up robust features. In: Leonardis, A., Bischof, H., Pinz, A. (eds.) ECCV 2006, Part I. LNCS, vol. 3951, pp. 404–417. Springer, Heidelberg (2006)
21. Kalal, Z., Mikolajczyk, K., Matas, J.: Forward-Backward Error: Automatic Detection of Tracking Failures. In: 20th International Conference on Pattern Recognition, pp. 2756–2759 (2010)

Monocular 3D Shape Recovery of Inextensibility Deformable Surface by Using DE-Based Niching Algorithm with Partial Reinitialization

Xuan Wang[1], Fei Wang[1], and Lei Chen[2]

[1] Xi'an Jiaotong University, Xi'an, China
xwang.cv@gmail.com, wfx@mail.xjtu.edu.cn
[2] Beijing Institute of Spacecraft System Engineering, Beijing, China
chenleibit@gmail.com

Abstract. Template-based deformable surface shape recovery is a well-known challenging problem for its compatible local minima and high degree of freedom. The gradient-based optimization method often converges to the local minimum, the premature convergence also occurs even using the evolution strategies which are highly effective in locating a single global minimum. Meanwhile, exploration in a high dimensional space is often time exhausted. To avoid these difficulties, a two-step method was proposed. The projections of vertices of a mesh were estimated firstly. Then the 3D positions of the vertices were estimated via estimating the depth along the sightlines calculated according to the given projections. While the depth of vertices was estimated, the problem was regarded as a multimodal optimization. A DE-based niching algorithm was used to solve it, and the partial reinitialization was used to keep the diversity of the population. The effectiveness of our method was demonstrated on both synthetic data and real images.

Keywords: Monocular shape estimation, Crowding differential evolution, Multimodal optimization.

1 Introduction

Recovering the 3D shape of a deformable surface in monocular images is a well-studied but still open field. In the particular template-based setup, there is a reference 3D view of the surface, called template. 3D reconstruction is then carried out from 3D to 2D correspondences established between the template and an input image of the object surface. This problem has been treated as an ill-posed problem. Only the reprojection information is not sufficient to well-constrain the problem, because lots of different surfaces project at the same positions on image. That is to say, the problem is highly ambiguous. In order to constrain the problem, most works assume the surface to deform isometrically or to be developed. Furthermore, learning-based deformation models [1-3] were proposed to restrict the possible deformations. In these works, the possible deformations were restricted in a scope corresponding to the

D.-S. Huang et al. (Eds.): ICIC 2014, LNAI 8589, pp. 328–338, 2014.

training samples. But the ambiguities still existed. The work in [4] tried to resolve this problem by using a covariance matrix adaptation (CMA) evolution strategy. It demonstrated the effectiveness of the evolutionary computation method in this field, but it could only process the learned deformations in a relative small range, because the global linear deformation models were used in this work. The most recent progress presented in [5] proved that the problem about template-based 3D reconstruction of an isometrically deformed surface has unique solution. It means that the problem can be regarded as a minimizing problem which locating the unique global minimum in the space of optimization variables.

For a global optimization problem, the evolutionary computation methods were demonstrated effective. In most situations, many nature-inspired optimization algorithms, in particular the CMA evolution strategy, showed excellent search abilities, but only when they were applied to some 30-100 dimensional problems [6]. But the degree of freedom (Dof.) of the deformable surface represented as a triangulation mesh was often larger than 100. Although the problem had the unique solution, it just meant that the objective function had the unique global minimum. But lots of local minima still existed and had comparable fitness value to the global one. To deal with this kind of problems, the premature convergence often occurred when using the original CMA evolution strategy or other evolutionary computation method.

In this paper, the problem about template-based 3D reconstruction of an inextensibility surface was divided into two sub-problems. The first one was estimating the projections of the vertices of the mesh. The second one was regarded as a multimodal optimization problem which locating the best 3d position of vertices via moving the vertices along the sightlines. For this, a variant of crowding differential evolution algorithm was proposed which originated from the works in [7]. In the DE niching algorithm, neighborhood mutation was used to improve the local exploitation abilities of the algorithm, and the partial reinitialization is used to keep the high diversity of the population.

The rest of this paper was organized as follows. Section 2 reviewed differential evolution (DE) for niching. Section 3 discussed the proposed method and showed how the method recovered the shape of an inextensibility surface. In section 4, the proposed method was applied on test set including both synthetic data and real image, and the performance was shown and discussed. Finally, this paper was concluded in section 5.

2   DE-Based Niching Algorithm

2.1   Differential Evolution

The DE algorithm was proposed by Storn and Price in [8]. It was a simple but efficient algorithm in solving global optimization problems [6, 9-10]. DE was a population-based stochastic global optimization technique. It contained four main steps, i.e. initialization, mutation, crossover and selection. Many different strategies were suggested. The original strategy DE/rand/1 was used in this paper, which represented as follows:

$$v_i = x_{r1} + F \cdot \left(x_{r2} - x_{r3} \right) \tag{1}$$

where, v_i is the mutation vector, $r1$, $r2$ and $r3$ are random and mutually different integers drawn from the current target vector x_i, and F is a scale factor used for scaling the differential vector.

Crossover was applied after mutation process to obtain the trial vector $u_{i,j}$,

$$u_{i,j} = \begin{cases} v_{i,j}, & \text{if} \quad rand_j \leq CR \quad \text{or} \quad j = k \\ X_{i,j}, & \text{otherwise} \end{cases} \tag{2}$$

where CR is a control parameter of DE that decides in a comparison with a random number $rand_j$ in the rang $[0,1]$ whether components are copied from v_i or x_i, respectively, into trial vector u_i. The selection process is based on a simple competition between the corresponding parent and offspring in the case of single objective global optimization.

2.2   Crowding DE Algorithm with Neighborhood Mutation

Several DE-based niching algorithms were proposed to solve the multimodal optimization [12-13]. Relating to our works, the crowding DE (CDE) algorithm was proposed in [11] by Thomsen. In CDE, when an offspring was generated, it would only compete with the most similar (measured by Euclidean distance) individual in the current population. The offspring would replace this individual if it had a better fitness value. The details were as follows:

— Randomly generate NP number of initial trial solutions.
— Loop: for $i = 1$ to NP
  • Produce an offspring u_i using the standard DE.
  • Calculate the Euclidean distance of u_i to the other individuals in the DE population.
  • Compare the fitness of u_i with the most similar individual and replace it if the u_i has a better fitness value.
— Stop if a termination criterion is satisfied. Otherwise, go to Loop.

In literature [7], the neighborhood-based CDE (NCDE) was proposed. In the NCDE algorithm, only one more parameter m was added, which was the neighborhood size. It was often chosen according to the population size. And the flow of the NCDE algorithm was as follows:

— Randomly generate NP number of initial trial solutions.
— Loop: for $i = 1$ to NP

- Find the parametrically most similar (the similarity is measured in terms of Euclidean distances of parameter space) m individuals of solution i to form a subpopulation $subpop_i$.
- Produce an offspring u_i using DE within $subpop_i$, i.e. pick $r1, r2, r3$ from the ith subpopulation.
- Calculate the Euclidean distance of u_i to the other individuals in the entire population.
- Compare the fitness of u_i with the most similar (in Euclidean distance) individual and replace the most similar individual if the u_i has a better fitness value.
— Stop if a termination criterion is satisfied. Otherwise go to Loop.

In the following section, it showed how the NCDE algorithm could be used to recover the shape of an inextensibility surface. According to the properties of the problem, a partial reinitialization mechanism was added to the NCDE algorithm to keep the diversity of the population. Besides, a different measurement replaced the Euclidean distance was used to measure the similarity between individuals.

3   Template-Based Shape Recovery of the Inextensibility Surface

3.1   Problem Statement

In the template-based setup, a 3D reference and an input image captured by a calibrated camera were given. Then the 3D-to-2D correspondences between them were established. According to the common way, the shape of the surface was represented as a triangulation mesh denoted by $M = \{V, E, F\}$, where $V = \{v_m | m = 1...N_v\}$ was the set of vertices, $E = \{(m,n) | m,n = 1...N_v\}$ was the set of edges recording all the pairs of vertices which were connected by an edge, and $F = \{(m,n,l) | m,n,l = 1...N_v\}$ was the set of facets recording all the triple of vertices belonging to the same facet.

Fig. 1. Template and input image

In the template-based setup, a 3D reference $M^{(T)} = \{V^{(T)}, E, F\}$ was given. Recovering the shape of the deformed surface in input image was a problem equaling to estimating the 3D position of the vertices v_I corresponding to the deformed surface. In this paper, two kinds of information were available. One was the reprojection information from the correspondences between template and input. The other one was invariance of the edges length known from the inextensibility assumption.

Assuming there were N_c feature points $Q = \{q_j | j = 1...N_c\}$ in 3D template and their corresponding point $Q^c = \{q_j^c | j = 1...N_c\}$ could be found in input image. Considering a specific feature point q_j , its 3d coordinates could be converted to barycentric coordinates, i.e. represented as a linear combination of coordinates of the three vertices of the facet where the feature point was located. Then, when the coordinates of vertices were computed, the reprojection of the feature point could be calculated, and the reprojection error could be evaluated by calculating the difference between the positions of its reprojection and its corresponding position q_j^c in input image. Now the problem could be regarded as a minimizing problem:

$$V^{(T)} = \arg\min_V \left(reproj\_err\left(V, V^{(T)}, F, Q_j, Q_j^C\right) + iso\_err\left(V, V^{(T)}, E\right)\right) \tag{3}$$

3.2   Estimation of the Projections

Because the degree of freedom of the mesh was often too large, minimizing the equation (3) directly led to a global optimization problem with relative large scale. In order to decrease the dimension of the problem, the problem was divided into two sub-problems.

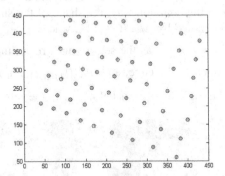

Fig. 2. Estimation of the projections of vertices with green stars as estimations and blue circles as projections of the ground-truth

First, the projections of vertices V^* needed to be estimated. Considering the ambiguities in the non-rigid reconstruction problem, the ambiguities were due to that the different points at the same sightline would be projected on the image at the same position. Conversely, There was no ambiguity in the process that only estimating the projections of vertices.

We assumed no deformation occurred on a single facet, and then the facets could be assumed flat during the deformation process. For each facet, 2D projective homographic transformation between the facet plane and its projections in image could be estimated. Through minimizing the residuals of all the homographic matrices corresponding to each facet, the projections $\tilde{v}$ of vertices $v^{\{l\}}$ were calculated.

3.3   3D Reconstruction of the Surface

Given the estimated projections of vertices, the sightlines through the projection of each vertex could be computed by equation (4).

$$s_m = A^{-1} \cdot v_m^* \cdot \left\| A^{-1} \cdot v_m^* \right\|^{-1} \tag{4}$$

Where A is the intrinsic matrix of camera which is calibrated, v_m^* is the homogeneous coordinates of the projection of vertex m and s_m is the sightline through v_m^*. Since the projections of vertices were fixed, the shape of surface only depended on the depths $D = \{d_m\}$ for each vertex. For the same reason, in the 3D reconstruction step, the reprojection errors could be ignored and only the iso_err needed to be minimized. Then the objective function was rewritten as equation (5).

$$D^{\{l\}} = \arg\min_{D} \sum_{(m,n) \in E} \left\| \left\| v_m^{\{T\}} - v_n^{\{T\}} \right\| - \left\| d_m s_m - d_n s_n \right\| \right\| \tag{5}$$

Before the reconstruction of the deformed surface, the comparison between this problem and a standard multimodal optimization problem is essential. Both problems had the objective function with lots of comparable local minima, but the standard multimodal optimization tried to locate all the local minima and our goal was to find the best one. According to the inherent properties of this problem, even the NCDE algorithm would prematurely converge to a subset of local minima. Since only the global minimum was really wanted in this case, a partial reinitialization mechanism is added to the traditional NCDE algorithm which can keep the diversity of the population. When every g generations passed, the current survival individuals would be clustered by KNN algorithm, the individual would be chosen as the cluster center who had the best fitness value in its own cluster. Then the clusters were sorted by the fitness value of their centers. The best c clusters were saved to the next generation and the other clusters would be removed and reinitialized.

— Randomly generate NP number of initial trial solutions.
— Loop: for $i = 1$ to NP

Find the parametrically most similar (the similarity is measured by equation (6)) m individuals of solution i to form a subpopulation $subpop_i$.

• Produce an offspring u_i using DE within $subpop_i$, i.e. pick $r1, r2, r3$ from the ith subpopulation.

- Calculate the Euclidean distance of u_i to the other individuals in the entire population.
- Compare the fitness of u_i with the most similar (in Euclidean distance) individual and replace the most similar individual if the u_i has a better fitness value.
— If every g times of iterations passed, clustering all the individuals and re-initializing the individuals not belong to the best c clusters. Go to Loop.
— Stop if a termination criterion is satisfied. Otherwise go to Loop.

Except for adding the partial reinitialization mechanism, the similarity between two meshes (individuals) was not calculated by a Euclidean distance (this was used in traditional NCDE algorithm) between their coordinates of vertices. Because not only the distances between corresponding vertices, but also the difference of degree between corresponding facets indicated the similarity between two meshes. For the two given meshes $V^{(1)}$ and $V^{(2)}$, the similarity between them was shown in equation (6).

$$sim\left(V^{(1)},V^{(2)}\right)=\sum_{m=1}^{N_v}\left\|v_m^{(1)}-v_m^{(2)}\right\|+\sum_{(m,n,l)\in F}nor\left(v_m^{(1)},v_n^{(1)},v_l^{(1)}\right)^{T}\cdot nor\left(v_m^{(2)},v_n^{(2)},v_l^{(2)}\right) \tag{6}$$

Where, *nor* is the function calculating the normal vector of the facet. As what was to be shown in section 4, the proposed method could yield good results on both synthetic data and real images.

4   Results

We first used the two motion capture datasets to generate the synthetic data: one was captured from cardboard and the other one was captured from cloth. Both of them were used in the literature [3] and available online. To generate the synthetic data, we sampled the barycentric coordinates of the ground-truth meshes and projected the resulting 3D points with a known camera. Then the Gaussian noise with deviant 1 was added. Then the proposed algorithm was applied on it. Thanks to the motion capture dataset, we could evaluate the accuracy of the proposed algorithm explicitly.

The Fig.3 and Fig.4 displayed the result from our method and comparisons. The effectiveness was illustrated by comparing with the two methods in [3]. In the Fig.3 and Fig.4, our method displayed as green curves, then the comparison, constrained latent variables model (CLVM) and Gaussian process latent variables model (GPLVM) were displayed as blue and red curves. For cardboard dataset, our method reached the comparable reconstruction error and our result was much more stable. Nearby the frame 30, CLVM and GPLVM reached the larger errors, since the object in these frames deformed strongly. This was because CLVM and GPLVM methods were both learning based method which capacity was restricted by the training samples. For the cloth dataset, the same situation is more obvious because the cloth could deform much more freely that the training samples cannot contain all the possible deformations. At the beginning and the end of the synthetic data sequence, the deformations rarely occurred on the object surface. Since our method used no more priors but inextensibility assumption, for example the smoothness regularization, our result reached the relative larger but acceptable errors at the ends of the data sequence. But when the obvious deformations occurred, our method was illustrated more accuracy and stable than the comparisons.

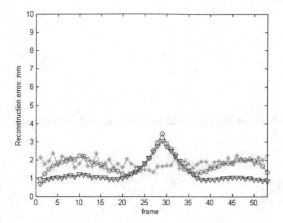

Fig. 3. 3D reconstruction errors on the cardboard dataset with our method in green, CLVM in blue and GPLVM in red

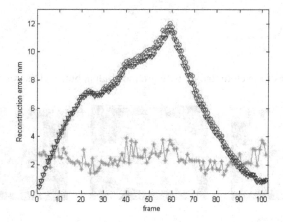

Fig. 4. 3D reconstruction errors on the cloth dataset with our method in green, CLVM in blue and GPLVM in red

Then we applied our proposed algorithm on the real images set in which the object surface was a piece of bend paper. The data could be found on the homepage of EPFL's CVLab, the data was used in Salzmann's works in [14].

In Fig.5, we could find that some reconstructed surfaces were not smooth. That is because we did not use any regularization terms in the objective function. It could be solved easily if any kinds of regularization were added to the objective function.

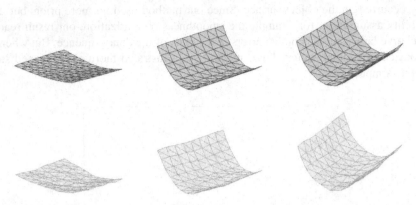

Fig. 5. Examples from cardboard dataset with ground-truth in blue and the results in green

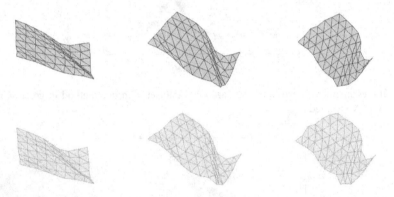

Fig. 6. Examples from cloth dataset with ground-truth in blue and the results in green

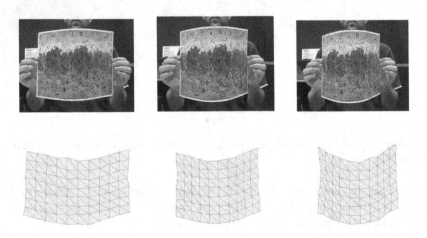

Fig. 7. Examples from real images with reprojection in the top and results in the bottom

5   Conclusions

In this paper, we proposed a two-step method for template-based shape recovery of inextensibility surface from the monocular images. The two-step strategy divided the original problem into two sub-problems with relative low dimensions and the partial reinitialization mechanism was used to keep the diversity of population. The proposed method was demonstrated effective on both synthetic data and real images.

In future work, we plan to use the evolutionary computation method on the non-rigid structure from motion case, since the template is often unavailable in many practical cases.

Acknowledgements. This work is supported by National High Technology Research and Development Program of China (2013AA014601) and National Natural Science Foundation of China (61231018).

References

1. Salzmann, M., Pilet, J., Ilic, S., Fua, P.: Surface Deformation Models for Nonrigid 3d Shape Recovery. IEEE Transaction on Pattern Analysis and Machine Intelligence 29(8) (2007)
2. Salzmann, M., Fua, P.: Linear Local Models forMonocular Reconstruction of Deformable Surfaces. IEEE Transaction on Pattern Analysis and Machine Intelligence 33(5) (2011)
3. Varol, A., Salzmann, M., Fua, P., Urtasun, R.: A Constrained Latent Variable Model. In: IEEE Conference on Computer Vision and Pattern Recognition (2012)
4. Moreno Noguer, F., Fua, P.: Stochastic Exploration ofAmbiguities for Non-Rigid Shape Recovery. IEEE Transaction on Pattern Analysis and Machine Intelligence 35(2) (2013)
5. Bartoli, A., Gerard, Y., Chadebecq, F., Collins, T.: on Template-Based Reconstruction From A Single View: Analytical Solutions And Proofs ofWell-Posedness forDevelopable, Isometric And Conformal Surfaces. In: IEEE Conference on Computer Vision And Pattern Recognition (2012)
6. Das, S., Suganthan, P.N.: Differential Evolution: A Survey ofThe State-Of-The-Art. IEEE Transaction on Evolutionary Computation 15(1) (2011)
7. Qu, B.Y., Suganthan, P.N., Liang, J.J.: Differential Evolution With Neighborhood Mutation forMultimodal Optimization. IEEE Transaction on Evolutionary Computation 16(5) (2012)
8. Storn, R., Whigham, P.: Differential Evolution: A Simple And Efficient Heuristic for-Global Optimization Over Continuous Spaces. J. Global Optimization 11(4) (1995)
9. Qin, V.K., Huang, V.L., Suganthan, P.N.: Differential Evolution Algorithm With Strategy Adaption forGlobal Numerical Optimization. IEEE Transaction on Evolutionary Computation 13(2) (2009)
10. Qu, B.Y., Suganthan, P.N.: Multiobjectives And Diversified Selection. Information Science 180(17) (2010)

11. Thomsen, R.: Multimodal Optimization Using Crowding-Based Differential Evolution. IEEE Congress on Evolutionary Computation (2004)
12. Rumpler, J., Moore, F.: Automatic Selection ofSubpopulations And Minimal Spanning Distances forImproved Numerical Optimization. IEEE Congress on Evolutionary Computation (2001)
13. Rigling, B., Moore, F.: Exploitation of Subpopulation In Evolutionary Strategies forImproved Numerical Optimization. In: Proc. 11th MAICS (1999)
14. Salzmann, M., Hartley, R., Fua, P.: Convex Optimization forDeformable Surface 3-D Tracking. In: IEEE International Conference on Computer Vision (2007)

Computer Vision Based Traffic Monitoring System for Multi-track Freeways

Zubair Iftikhar, Prashan Dissanayake, and Peter Vial

School of Electrical Computer and Telecommunications Engineering,
University of Wollongong, North Wollongong, NSW, Australia
zi770@uowmail.edu.au

Abstract. Nowadays, development is synonymous with construction of infrastructure. Such road infrastructure needs constant attention in terms of traffic monitoring as even a single disaster on a major artery will disrupt the way of life. Humans cannot be expected to monitor these massive infrastructures over 24/7 and computer vision is increasingly being used to develop automated strategies to notify the human observers of any impending slowdowns and traffic bottlenecks. However, due to extreme costs associated with the current state of the art computer vision based networked monitoring systems, innovative computer vision based systems can be developed which are standalone and efficient in analyzing the traffic flow and tracking vehicles for speed detection. In this article, a traffic monitoring system is suggested that counts vehicles and tracks their speeds in realtime for multi-track freeways in Australia. Proposed algorithm uses Gaussian mixture model for detection of foreground and is capable of tracking the vehicle trajectory and extracts the useful traffic information for vehicle counting. This stationary surveillance system uses a fixed position overhead camera to monitor traffic.

Keywords: computer vision, surveillance system, Gaussian mixture model. Vehicle Trajectory.

1   Introduction

Computer vision is effectively used in manufacturing industry for assembling electronics control in vehicles by robots. Quality control in multi-Billion dollar electronics industry is maintained by computer vision where involvement of the human eye is unheard of during the past decade. Large infrastructure projects are too vast to monitor by humans along. More and more computer vision based systems are developed for security and maintenance. Computer vision has been successfully used in the UK for monitoring traffic to avoid traffic bottlenecks. However, such systems cost tax payers millions of dollars every year and are far from economical in the current economic slowdown of the world. So, it is very desirable to develop new surveillance systems which are stand along and inexpensive that making use of the latest hardware developments even for smaller freeways. In a traffic monitoring system, vehicles can be detected and tracked for their speed for a short period of time

D.-S. Huang et al. (Eds.): ICIC 2014, LNAI 8589, pp. 339–349, 2014.

using a live video stream. This can be used for counting the traffic density of any track or section of the freeway and could potentially identify the traffic jams due to certain kind of breakdown or accidents. The propose can detect the event before the authorities are notified of any event. Major applications of such systems include intelligent traffic monitoring, counting vehicle on road, traffic rule violation detection, classification of E-TAG (Electronic tag) system. In recent years, there have been few instances of intelligent vehicles with autonomous driving. However, there are many challenges for such a system in practice. Major challenges in vehicle detection algorithm are weather conditions, poor road illumination, occlusions from other vehicles, view point orientation and the challenges posed by variety of vehicles with trailers. A lot of work has been attempted to handle some of these issues.

Vehicle detection can be classified as camera based systems or optical sensor based detection and tracking [1]. Computer vision has been successfully implemented in many vision-related scenarios with increasing reliability [2-11]. Camera quality and its sensing system have been improved a lot in recent times and same level of improvement can be seen in computational power of computer system. These factors are enabling researchers to enhance vehicle detection algorithm's accuracy. Number of sensors and quality of sensing device are key factors in computer vision application. Major goal of many current research activities regarding vehicle monitoring is geared towards improving accuracy with reduction in complexity to handle different weather conditions and occlusion problems.

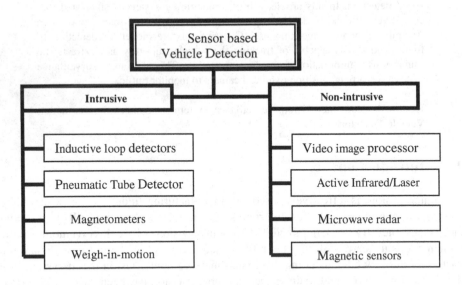

Fig. 1. Categorization of Sensor based Vehicle Detection Methods

1.1   Sensor Based Vehicle Detection

Sensor based vehicle detection system is a non-computer vision based system. It is based on the physical interaction or presence of vehicles. It is categorized in two major types as shown in Fig. 1. Intrusive techniques are based on the physical interaction of vehicles with sensors. In these approaches, embedded sensor detects the existence of vehicle through weight or metal body of vehicle. Second major type is non-Intrusive; it is based on the vehicle presence.

1.2   Vision Based Vehicle Detection

Out of all the existing technologies of traffic surveillance, computer vision-based systems are one of the latest and most considerable. Numerous research work have been carried out in computer vision-based Intelligent Transportation Systems (ITS) due to its various features as compare to others such as easy installation and maintenance, quick response, simple operation and maintenance, high flexibility, low cost, and their ability to monitor and analyze of wide areas [2-11]. Major issues in vision based system are occlusion and illumination. The common approach in vision based algorithms is to apply preprocessing such as filtering to remove sensor noise followed by detecting the foreground. Then, segmentation of vehicle position is attempted through connected component analysis. Finally, detected vehicle is tracked. In literature, vision based multi vehicles detection algorithm are classified as Background Detection [12], Color Cluster method [13], Graph Axis method [14], Optical flow [15] and Space Vector Differencing.

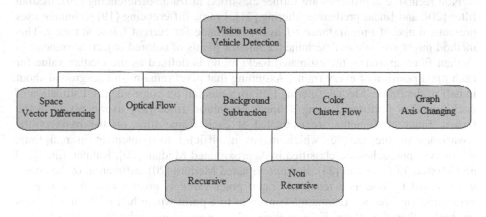

Fig. 2. Categorization of Vision based Vehicle Detection Methods

Background modeling approaches [16][17] are very popular and widely used because these techniques supply the most complete feature data for accurate detection of moving objects. Background modeling methods develop the background model by observing the scene for a particular time period. After background modeling, subtraction of current frame from the background frame is performed for detecting the moving objects. Accurate building of background model is a key challenge especially in dynamic outdoor environment. In addition, it is necessary to adapt to new conditions of weather, illumination and shadows and update the model. Background modeling approach has a very straightforward mechanism. This approach assumes that background scene will not change with sequence of frames.

Firstly, difference D_k from the current frame f_k and the background model frame b_k is determined by using equation (1) [18].

$$D_k(x, y) = | f_k(x, y) - b_k(x, y) | \qquad (1)$$

Then threshold the difference D_k according to equation (2) [18]:

$$R_k(x, y) = \begin{cases} 1, D_k(x, y) > T \\ 0, \text{otherwise} \end{cases} \qquad (2)$$

If R_k has value '1' then this pixel is foreground pixel. Selection of threshold value T has direct impact on the quality of background modeling. Too high T affects as broken occurrence on region of targeted moving object. Whereas, too low value of T introduces a lot of noise. Generally, threshold is chosen through gray histogram by finding double peaks or more and then selecting the bottom value between two peak values [18]. Foreground detection is an essential step in vision based detection. In literature, recursive and non-recursive approaches are presented to detect foreground.

Non recursive approaches are further classified in frame differencing [19], median filter [20], and linear predictive filtering [21]. Frame differencing [19] technique uses previous frame of time instance $t-1$ as the reference for current frame at time t. This method might not able to determine the interior pixels of colored object movement. In median filter approach, the estimated background is defined as the median value for each pixel position of every frame. Assuming that pixel remains in background about in half of the frame. Median filtering [20] uses median for color images. Estimation of background is performed using a linear prediction filter [21] to the pixels in the buffer. Filter coefficients are evaluated at each frame time on the basis of the covariance of the sample, which makes it difficult to implement in real time. Recursive approaches are classified in Approximated Median [20], Kalman filter [22] and Mixture of Gaussian [23]. In Approximated Median [20], estimation of the center is increased by one as a result of input pixel value is greater than the estimated value, and vice versa. Estimation converges to a point where half of the input pixels are higher than the others. Kalman filter [22] is a recursive technique for tracking of linear dynamical systems with Gaussian noise. Kalman filter estimates the global illumination change, and noise variance. Also, it performs the management of structures pixel. In mixture of Gaussian approach, pixel values that are not belonged to background are assumed as foreground. This is a famous technique for segmentation ofmoving regions at real-time. Gaussian model are updated using K-means approximation method. Each Gaussian distribution is appointed to represent the background or a moving object in the model of adaptive mixture. Each pixel is then evaluated and classified.

Color cluster method [13] combines the information of different sources, such as structure and color. The information structure is the census transforms and color information which is obtained from color histograms. Color image is divided into method clusters. Objects of interest are grouped based on color contrast so object shape is reflected by the respective color shape. To this end, it must have the prior knowledge and data of individual objects.

In graph axis method [14], pixels of an image are modified according to the x-axis and the y-axis. There is no such concept as a fixed background and variable background. This method can detect the moving objects by finding the change in position of pixels.

Optical flow method [15] is based on a relative movement, rather than the absolute motion, as in the case of motion vector search method. This technique uses the change trend of gray scale of each image point. Optical flow finds the image changes which are dependent on motion during a time interval. Optical flow field represents the speed of object movement of 3D points in a 2D image. However, this method is quite complicated and requires a higher material. So, as a consequence, it is not suitable for real time processing.

Space vector difference method [24] uses the difference in vector space to obtain a card unlike current video and the modeling of background frame. Then, an adaptive threshold is automatically calculated by analyzing the characteristic of difference map histogram. This method adopts the concept of technical RGB color. Space vector difference is applied to color images and gives good results, but it cannot detect the background objects if the color of the background and the object is the same.

2   Vehicles Detection

In proposed algorithm, Gaussian mixture model [23] is used to detect the foreground of video. There is need to provide some initial frames in Gaussian mixture model and then it computes probability density function (PDF) of each frame and then it measure the change in entrain intensity value of all frames. If change is greater than threshold intensity then value become part of foreground otherwise it assumes as background. A 0 or 1 is assigned to each pixel intensity value based on change in PDF in all initial frames. The concept of histogram equalization is observed to assign values. Fig. 3. shows the system flow diagram.

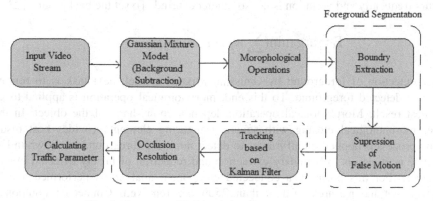

Fig. 3. Flowchart of Vision based Detection, Tracking

Gaussian Mixture model which is a background subtraction method is applied on a video stream shown in Fig. 4a, for foreground detection that is shown in Fig. 4b. In this Gaussian Mixture modeling, illumination variations in background are automatically adopted. The recent history $\{X_1, ...,X_t\}$ of a pixel is modeled through mixture of K (normally, 3 to 5) Gaussian distributions. The probability of current pixel is [23]:

$$P(X_t) = \sum_{i=1}^{K} w_{i,t} * \eta(X_t, \mu_{i,t}, \Sigma_{i,t}) \tag{3}$$

K defines the number of distributions, $\mu_{i,t}$ is the mean value of the i_{th} Gaussian in the mixture at time t, $w_{i,t}$ is an estimate of the weight of the i_{th} Gaussian in the mixture at time t, $\Sigma_{i,t}$ is the covariance matrix at time t for the i_{th} Gaussian in the mixture, and η is a Gaussian probability density function [23]:

$$\eta(X_t, \mu, \Sigma) = \frac{1}{(2\pi)^{\frac{n}{2}} |\Sigma|^{\frac{1}{2}}} e^{-1/2(X_t - \mu_t)^T \Sigma^{-1}(X_t - \mu_t)} \tag{4}$$

If the current pixel value does not match with any of the K distributions than the least predictable distribution takes the place of a distribution that has current value as its mean value, low prior weight, and an initially high variance. The prior weights $\omega_{k,t}$ at time t for the K distributions [23]:

$$\omega_{k,t} = (1 - \alpha)\omega_{k,t-1} + \alpha(M_{k,t}) \tag{5}$$

where $M_{k,t}$ is '1' for the matched model and '0' for the remaining models, and α is the learning rate. In addition, weights are renormalized after this approximation.

A surface is deemed to be background with higher probability (lower subscript k) if it occurs frequently and variation is not so much occurred. To set the background [23]:

$$B = \arg\min_b (\sum_{k=1}^{b} w_k > T) \tag{6}$$

After detection of foreground by Gaussian Mixture model, next task is to remove noise in detected foreground. To this end, morphological operation is applied to get the best result. Morphological operation depends upon shape of the object. In this case, morphological closing with square structuring element provides best result. Same process is applied on all frames to detect noiseless foreground as shown in Fig. 4c. The next step is to boundary extraction and reduction of falsely detected objects (vehicle). For this purpose, a connected component analysis is performed and all objects that having area of less than 2500 are removed. Connected component analysis is applied on all detected vehicles.

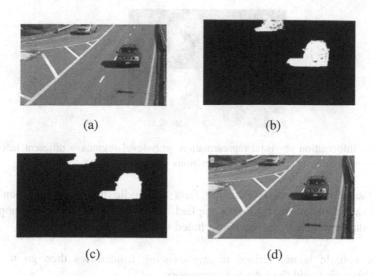

(a)   (b)

(c)   (d)

Fig. 4. (a).Original Video Frame, (b).Detected Foreground, (c).Morphological Operation (d).Detected Vehicles

3   Vehicles Tracking and Occlusion Resolution

Once true objects are detected as shown in Fig. 5b then at next step, Kalman filter is applied for vehicle tracking and centroid of every detected vehicle is also determined. Tracking process can be disrupted if vehicles are occluded at time of entrance to frame. Generally, vehicles occlusion is seen on side by side in horizontal direction across adjacent lanes. To solve this issue, a prior occlusion detection method [25] is proposed which is based on lane information. Firstly, a lane mask $L(x, y)$ is made with values:- -1 (neglected area), 0 (separated area), 1 (first lane), and so on as shown in Fig. 5(b). A label m with a label image $f(x, y)$ is assigned to each vehicle after connected-component labeling. After that, vehicles are monitored according to Eq.(7) to get a histogram $G(m,h)$ as regards to lane h. Eq.(8) is used with an adequate threshold $Thrsh = 5$ to detect the occluded region. At the end, occlusion is removed through splitting the vehicles with the help of reference of $L(x, y)$.

$$f(x, y) = m \text{ and } L(x, y) \neq -1 \text{ and } L(x, y) \neq 0$$

$$G(m, L(x, y)) = G(m, L(x, y)) + 1 \tag{7}$$

$$\text{if } \| G(m, h) - G(m, h+1) \| < \text{Thrsh, then} \tag{8}$$

Resolve the occlusion

Fig. 5. Lane information in visual representation, gray-level regions = different lanes, white color regions = separation areas, black color regions = neglected area

In case, the aforementioned technique fails to handle the occlusion then a post occlusion detection method [25] is applied. The following steps are applied to determine that a vehicle is shaped by occluded vehicles or not.

1. If vehicle is not related to any existing trajectories then go to step 2. Otherwise, add the vehicle to trajectory.
2. If an offset which is calculated through centers of last two nodes of trajectory and vehicle region are a superset of last trajectory node then go to step 3. Otherwise, a new vehicle trajectory is created.
3. Object is split into two moving objects.

4   Traffic Parameters

A virtual line is drawn to count vehicle. When centroid touches virtual line, then algorithm checks the last five position of centroid and takes the decision for counting. Algorithm keeps track of centroid value and stores center value of detected object for further computation. Traffic parameters are derived from tracking trajectories. However, in this proposed method, each trajectory is classified to a lane ID for calculating traffic parameters. The technique [25] which is used for classification a trajectory to lane ID is stated in Eq. (9). Way to get $G(m,h)$ in Eq.(9) is the same as Eq.(7) and l is $m(n, t-1)$ at time instance $(t-1)$ which show the n-th trajectory.

$$h*(n,t) = \arg \max_{h} G(m(n,t-1),h) \qquad (9)$$

Following equations [25] listed in Table 1 are used to calculate the traffic parameters.

5   Experimental Results

In these experiments, many common traffic video sequences are used which are available on different online data sources. However, we mostly focused on freeway traffic. In that, fixed position camera was located on a 1.5 meters long pole over a pedestrian bridge. Overall height of camera from the road is 6.5 meters, as observed in Fig. 6a. Experimental video was observed in a sunny day and consists of three minutes since *01:17pm* to *01:20 pm*.

Table 1. Traffic Parameters Equations

Traffic Parameters	Equations
Speed(h*)	$Speed(h*(n,t)) = 1/8\,Speed(h*(n,t)) + 7/8\dfrac{\lvert (C_{m(n,t-1)} - C_{m(n,t-N(n,t-1))}\rvert}{N(n,t-1)\,FramesperHour}\dfrac{0.005}{\bar{P}(n)}$ N(n,t-1) is number of nodes at at time t-1 in n-th trajectory , C is vehicle center, $\bar{P}(n)$ is nodes average width
Quantity(h*)	If N(n,t-1), Quantity (h*(n,t) = Quantity (h*(n,t) + 1
Volume (h*)	$Volume(h*(n,t)) = \dfrac{Quantity(h*(n,t))\,FramesperHour}{t}$

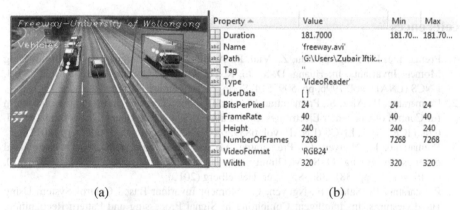

Property ▲	Value	Min	Max
Duration	181.7000	181.70...	181.70...
Name	'freeway.avi'		
Path	'G:\Users\Zubair Iftik...		
Tag	''		
Type	'VideoReader'		
UserData	[]		
BitsPerPixel	24	24	24
FrameRate	40	40	40
Height	240	240	240
NumberOfFrames	7268	7268	7268
VideoFormat	'RGB24'		
Width	320	320	320

(a)   (b)

Fig. 6. (a) Counting and Tracking; a case of prior occlusion (b) Video Information

(a)   (b)

Fig. 7. (a) Speed 98.4km/hr, Quantity 06, Volume 1383/hr, the 1000-th Frame; (b) Speed 105km/hr, Quantity 35, Volume 1802/hr, the 5000-th Frame

Information regarding to frame rate, data rate and siz of video can be seen in   6b. The proposed system is developed on Windows 7 platform with a Pentium® Dual-Core *2.1GHz CPU, 3GB Ram.*

6   Conclusion and Future Work

In this paper, different techniques of vehicle detection and tracking are discussed which are broadly categorized as sensor and vision based approaches. Researchers are taking interest in vision based approaches due to increasing quality of sensing device, computation power of computers, accuracy, human resource reduction and cost minimization. In proposed algorithm, Gaussian mixture model is used to detect foreground and Kalman filter is applied for tracking. In addition, a technique for occlusion handling is described. At the end, traffic parameters are determined. Future work is to enable system to handle serious occlusion and weather condition, and classification approach to differentiate the different types and sizes of vehicles.

References

1. Premaratne, P., Yang, S., Zou, Z., Vial, P.: Australian Sign Language Recognition Using Moment Invariants. In: Huang, D.-S., Jo, K.-H., Zhou, Y.-Q., Han, K. (eds.) ICIC 2013. LNCS (LNAI), vol. 7996, pp. 509–514. Springer, Heidelberg (2013)
2. Premaratne, P., Ajaz, S., Premaratne, M.: Hand Gesture Tracking and Recognition System for Control of Consumer Electronics. In: Huang, D.-S., Gan, Y., Gupta, P., Gromiha, M.M. (eds.) ICIC 2011. LNCS (LNAI), vol. 6839, pp. 588–593. Springer, Heidelberg (2012)
3. Premaratne, P., Nguyen, Q., Premaratne, M.: Human computer interaction using hand gestures. In: Huang, D.-S., McGinnity, M., Heutte, L., Zhang, X.-P. (eds.) ICIC 2010. CCIS, vol. 93, pp. 381–386. Springer, Heidelberg (2010)
4. Premaratne, P., Safaei, F., Nguyen, Q.: Moment Invariant Based Control System Using Hand Gestures. In: Intelligent Computing in Signal Processing and Pattern Recognition, vol. 345, pp. 322–333. Springer, Heidelberg (2006)
5. Premaratne, P., Premaratne, M.: Image Matching using Moment Invariants. Neurocomputing 137, 65–70 (2014)
6. Premaratne, P., Ajaz, S., Premaratne, M.: Hand Gesture Tracking and Recognition System Using Lucas-Kanade Algorithm for Control of Consumer Electronics. For Neurocomputing Journal 116(20), 242–249 (2013)
7. Premaratne, P., Nguyen, Q.: Consumer electronics control system based on hand gesture moment invariants. IET Computer Vision 1, 35–41 (2007)
8. Yang, S., Premaratne, P., Vial, P.: Hand Gesture Recognition: An Overview. In: 5th IEEE International Conference on Broadband Network and Multimedia Technology (2013)
9. Zou, Z., Premaratne, P., Premaratne, M., Monaragala, R., Bandara, N.: Dynamic hand gesture recognition system using moment invariants. In: ICIAfS, IEEE Computational Intelligence Society, Colombo, Sri Lanka, pp. 108–113 (2010)
10. Herath, D.C., Kroos, C., Stevens, C.J., Cavedon, L., Premaratne, P.: Thinking head: Towards human centred robotic. In: 2010 11th International Conference on Control, Automation, Robotics and Vision (ICARCV), Singapore, pp. 2042–2047 (2010)

11. Minge, E.: Evaluation of Non-Intrusive Technologies for Traffic Detection. Minnesota Department of Transportation, Office of Policy Analysis, Research and Innovation, SRF Consulting Group, US Department of Transportation, Federal Highway Administration (2010)
12. Cheung, S.-C., Kamath, C.: Robust techniques for background subtraction in urban traffic video. In: Proc. EI-VCIP, pp. 881–892 (2004)
13. Heisele, B., Kressel, U., Ritter, W.: Tracking non-rigid, moving objects based on color cluster flow. In: Proc. of IEEE Conference on Computer Vision and Pattern Recognition, San Juan, pp. 257–260 (1997)
14. Housni, K., Mammass, D., Chahir, Y.: Moving objects tracking in video by graph cuts and parameter motion model. International Journal of Computer Applications 40(10), 20–27 (2012), Published by Foundation of Computer Science, New York, USA (February 2012)
15. Ching, O.H.: Optical flow-based vehicle detection and Tracking. B.I.S. Thesis, University Tunku Abdul Rahman, Perak, Malaysia (2010)
16. Messelodi, S., Modena, C.M., Segata, N., Zanin, M.: A Kalman Filter Based Background Updating Algorithm Robust to Sharp Illumination Change. In: Roli, F., Vitulano, S. (eds.) ICIAP 2005. LNCS, vol. 3617, pp. 163–170. Springer, Heidelberg (2005)
17. Alatan, A., Onural, L., Wollborn, M.: Image sequence analysis for emerging Interactive multimedia services: The European COST 211 framework. IEEE Trans. Circuit Syst. Video Technol. 8(7), 802–813 (1998)
18. Xiao, L., Li, T.: Research on Moving Object Detection and Tracking. In: FSKD 2010, vol. 2, pp. 2324–2327 (2010)
19. Ahmad, K.A., et al.: Improvement moving vehicle detection using RGB removal shadow segmentation, pp. 22–26 (2011)
20. Gangodkar, D., et al.: Robust segmentation of moving vehicles under complex outdoor conditions. IEEE Transactions on Intelligent Transportation Systems 13, 1738–1752 (2012)
21. Bharti, T.T.: Background Subtraction Technique Review. International Journal of Innovative Technology and Exploring Engineering (IJITEE) 2(3), 166–168 (2013)
22. Chauhan, A.K., Krishan, P.: Moving Object Tracking Using Gaussian Mixture Model And Optical Flow. International Journal of Advanced Research in Computer Science and Software Engineering 3(4), 243–246 (2013)
23. Stauffer, C., Grimson, W.E.L.: Adaptive background mixture models for real-time tracking. In: Proceedings of the IEEE Computer Society Conference on Computer Vision and Pattern Recognition, vol. 2, pp. 246–252 (1999)
24. Qian, Z., et al.: Moving objects detection based on space vector difference, pp. 646–651 (2011)
25. Lin, S.P., et al.: A real-time multiple-vehicle detection and tracking system with prior occlusion detection and resolution, and prior queue detection and resolution, pp. 828–831 (2006)

Robust Pose Estimation Algorithm
for Approximate Coplanar Targets

Haiwei Yang[1], Fei Wang[1], Lei Chen[2], Yicong He[1], and Yongjian He[3]

[1] Xi'an Jiaotong University, Xi'an, Shaanxi, China
[2] Beijing Institute of Spacecraft System Engineering, Beijing, China
[3] Xi'an Communication Institute, Xi'an, China
{yanghw.2005,heyicong2013}@stu.xjtu.edu.cn,
wfx@mail.xjtu.edu.cn,
{chenleibit,heshifu}@gmail.com

Abstract. To uniquely determine the position and orientation of a calibrated camera from a single image, pose estimation algorithms have been developed. However, the presented algorithms usually encounter pose ambiguity problem when process approximate coplanar targets, which can be defined as that a majority of object points on these targets belongs to a plane while some others distract outside the plane. Based on a more comprehensive explanation for pose ambiguity from the influence of 3D object configuration, we propose a robust pose estimation algorithm. The approximate coplanar points are divided into coplanar points and non-coplanar points. When two candidate solutions are calculated by coplanar points, final pose is determined using non-coplanar points. Simulation results and experiments on real images prove the effectiveness of our proposed pose estimation algorithm.

Keywords: pose ambiguity, object configuration, pose estimation.

1 Introduction

The aim of pose estimation is to determine the position and orientation between a calibrated camera and an object. There are many applications of pose estimation in computer vision, such as hand-eye robot systems, augmented reality, photogrammetry and so on.

In computer vision literature, several pose estimation approaches have been proposed [1, 2, 3, 4, 5]. Most of them work for arbitrary 3D targets, while some of them are developed especially for planar objects [6, 7, 8]. In theory, pose can be estimated from four or more non-collinear points [9, 10, 11, 12, 13], if the intrinsic parameters of the camera are known. However, in some special applications, people often observe that pose calculated by these algorithms is not very robust, which results in significant jitter, pose jumps and gross pose outliers. We compare several state-of-the-art pose algorithms [1], [13], [14] and observe several pose jumps, which should not occur.

D.-S. Huang et al. (Eds.): ICIC 2014, LNAI 8589, pp. 350–361, 2014.
© Springer International Publishing Switzerland 2014

Pose estimation algorithms usually encounter pose ambiguity problem when process planar targets, which has been realized by most people. Oberkampf et al. [16] first mentioned pose ambiguity for planar points. They gave a straightforward interpretation and developed an algorithm based on scale orthogonally projection. However, this approach does not discuss the perspective situation. Schweghofer and Pinz [17] analyzed the pose ambiguity problem for planar points using the general case of perspective projection. They chose the correct pose with minimum re-projection error from two possible minima. However, with noise increasing, the two possible minima may be very close to each other and the correct rate of pose estimation will decrease regardless of the selected strategy. What's more, the algorithm cannot be applied for arbitrary 3D targets because of the hypothesis of planar object configuration.

Among these state-of-art pose estimation algorithms designed for arbitrary 3D targets, pose ambiguity also exists according to the analysis in our paper. Fiore's method [14] is a linear algorithm, which makes real-time performance possible for large number of object points. He initially constructs a set of linear equations from which the rotation and translation parameters from world to camera are eliminated, allowing the direct recovery of the point depths without considering the inter-point distances. Unfortunately, ignoring nonlinear constraints produces poor results in the presence of noise. Lu et al.'s [1] is one of the fastest and the most accurate algorithms. It minimizes an error expressed in 3D space, unlike many earlier methods that attempt to minimize re-projection residuals. However, the algorithm tends to converge fast but may get stuck in an inappropriate local minimum if incorrectly initialized. Our experiments reveal the results of Lu et al.'s would get less accurate for planar and approximate coplanar targets. Lepetit et al. [13] proposed a non-iterative solution, called EPnP, whose computational complexity grows linearly with the number of object points. The central idea is to write the coordinates of the 3D points as a weighted sum of four virtual control points. This reduces the problem to estimating the coordinates of the control points in the camera referential, which can be done in $O(n)$ time by expressing these coordinates as weighted sum of the eigenvectors of a 12×12 matrix and solving a small constant number of quadratic equations to pick the right weights. Unfortunately, the approach also causes problems when extended to planar or approximate coplanar configurations.

In this paper, we demonstrate that pose ambiguity also exists among non-planar targets, especially for approximate coplanar points and traditional pose estimation algorithms may produce large error in some cases. Through analyzing the cause of pose ambiguity and a more comprehensive explanation for pose ambiguity from arbitrary 3D objects, we point out that pose ambiguity has a compact relationship with the configuration of object points and the influence of planar points and non-planar points in 3D targets should be dealt with respectively. Based on this, a robust algorithm for approximate coplanar targets is presented. The approximate coplanar points are firstly divided into coplanar points and non-coplanar points. When two possible solutions are calculated by the coplanar points, final pose is determined using the remaining non-coplanar points.

2   Analysis of Pose Ambiguity

In this section, a more comprehensive explanation for pose ambiguity from arbitrary 3D objects is provided. We point out that pose ambiguity is an inherent phenomenon relating to the configuration of targets. Especially, pose ambiguity is obvious for both planar targets and approximate coplanar object points. Based on this, our novel robust pose estimation algorithm for approximate coplanar targets is presented in next section.

Assuming that the coordinates of corresponding 3D object points and normalized image points are $p_i = [X_i \ Y_i \ Z_i]^T$ and $v_i = [x_i \ y_i]^T$, for $i = 1, 2, \cdots, N$. The perspective projection camera model can be described as

$$l_i \begin{bmatrix} v_i \\ 1 \end{bmatrix} = s(Rp_i + T) \quad , \ i = 1, 2, \cdots, N \ , \tag{1}$$

where l_i and s are scale factors. R and T are the rotation matrix and translation vector describing the relationship between the camera and a 3D target. Pose estimation is then to seek the optimal rotation matrix R and translation vector T that best satisfy the N equations in (1). We define the data matrices

$$P = \begin{bmatrix} p_1, p_2, \cdots, p_n \\ 1 \ , \ 1 \ , \cdots, \ 1 \end{bmatrix}, \quad D = \begin{bmatrix} v_1, \ v_2, \cdots, v_n \\ 1 \ , \ 1 \ , \cdots, \ 1 \end{bmatrix}.$$

The null space of P can be calculated as

$$W = null(P).$$

We therefore have that

$$\begin{bmatrix} l_1 \begin{bmatrix} v_1 \\ 1 \end{bmatrix} & l_2 \begin{bmatrix} v_2 \\ 1 \end{bmatrix} & \cdots & l_N \begin{bmatrix} v_N \\ 1 \end{bmatrix} \end{bmatrix} W = s[R \ T]PW = 0 \tag{2}$$

We isolate the third row in (2) and let $L = (l_1, l_2, \cdots, l_N)^T$:

$$W^T L = 0$$

This means that L is determined by the null space of W^T. Noting that the matrix $W = (w_{ij})_{N \times (N-4)}$ is spanned by P, we therefore have that

$$L = P^T \alpha, \tag{3}$$

for an unknown 4×1 vector α. This allows us to rewrite the equations in (2) as

$$C\alpha \triangleq \begin{bmatrix} w_{1,1}\begin{bmatrix} x_1 \\ y_1 \end{bmatrix} & \cdots & w_{N,1}\begin{bmatrix} x_N \\ y_N \end{bmatrix} \\ \vdots & & \vdots \\ w_{1,N-4}\begin{bmatrix} x_1 \\ y_1 \end{bmatrix} & \cdots & w_{N,N-4}\begin{bmatrix} x_N \\ y_N \end{bmatrix} \end{bmatrix} P^T\alpha = 0 \tag{4}$$

The best α can then be found from the SVD of matrix C by choosing the right singular vector corresponding to the minimum singular value of C. Alternatively, we can pre-multiply by C^T in the two sides of (4) and seek the eigenvector corresponding to the minimum eigenvalue of the matrix $C^T C$.

Once α is determined, the parameter $L=(l_1,l_2,\cdots,l_N)^T$ will be calculated using (3) and the pose estimation in (1) can be changed to the absolute orientation. Denoting the left-sides of (1) as

$$b_i = l_i \begin{bmatrix} v_i \\ 1 \end{bmatrix}, \quad i = 1,2,\cdots,N$$

Then, (1) becomes

$$b_i = s(Rp_i + T) \quad, \quad i = 1,2,\cdots,N \tag{5}$$

Let $b_0 = \dfrac{1}{N}\sum_{i=1}^{N} b_i$ and $p_0 = \dfrac{1}{N}\sum_{i=1}^{N} p_i$ are the centroids of the two data sets. Two new data matrices can be decided as

$$\begin{cases} B = \begin{bmatrix} \tilde{b}_1, & \tilde{b}_2, \cdots, \tilde{b}_N \end{bmatrix} \\ A = \begin{bmatrix} \tilde{p}_1, & \tilde{p}_2, \cdots, \tilde{p}_N \end{bmatrix} \end{cases}$$

where $\tilde{b}_i = b_i - b_0$ and $\tilde{p}_i = p_i - p_0$.

This problem is known as the Orthogonal Procrustes Problem, whose solution is given by Golub and Van Loan [18] as

$$s = \frac{\sum_{i=1}^{N} \left\| \tilde{p}_i^T \tilde{b}_i \right\|}{\sum_{i=1}^{N} \left\| \tilde{p}_i \right\|^2}, \quad R = V_R U_R^T, \quad T = b_0 - sRa_0 \tag{6}$$

where U_R and V_R are the left and right singular vectors of the SVD of matrix AB^T.

Based on the above analysis, the process of determining α in (4), which solution is the eigenvector corresponding to the minimum eigenvalue of the matrix $C^T C$, is the key procedure of pose estimation. However, considering different distribution of 3D object points and image points, the minimum eigenvalue of matrix $C^T C$ may be different. What's more, different minimum eigenvalue will give different parameters l_i, resulting in different pose estimation results, especially when the minimum eigenvalue cannot be determined easily or the number of minimum eigenvalue is more than one. A sample simulation experiment illustrates this problem clearly.

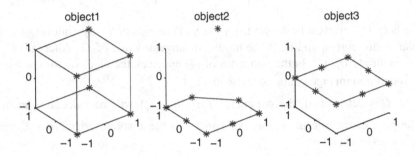

Fig. 1. Three characteristic objects: object1 is a cube; object2 contains a plane and a point outside the plane; object3 is a planar target.

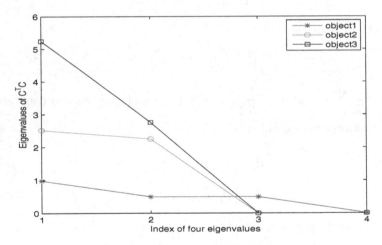

Fig. 2. Four eigenvalues of matrix $C^T C$ for each object

Fig.1 gives three characteristic objects with 8 points. The first one is a cube, denoting arbitrary uniform 3D object configuration. The second can be called approximate coplanar 3D object, while most points distribute in a plane with a small number of non-coplanar points (only one in this special case). The third is a planar configuration because all the points belong to a plane. The eigenvalues of the 4×4 matrix $C^T C$ are calculated for each object according to (4), whose results are shown in Fig.2.

As illustrated in Fig.2, object2 and object3 both have two minimum eigenvalues, leading to two different pose for each object configuration, which is called pose ambiguity. In [17], Schweghofer and Pinz propose an interpretation for the pose ambiguity of planar object points. The results above lead us to demonstrate that pose ambiguity also exists among non-planar targets. Meanwhile, we can conclude that pose ambiguity has a close relationship with the configuration of 3D object points and pose ambiguity may arise when object points are planar or approximate planar.

3   Robust Pose Estimation Algorithm

Based on above analysis, we realize that the influences of planar points and non-planar points in approximate coplanar targets are different and both the influences should be considered comprehensively when estimating pose. On the basis of this idea, we propose a robust pose estimation algorithm for approximate coplanar points.

The data set of approximate coplanar points can be written as follow

$$P = \begin{bmatrix} p_1, p_2, \cdots, p_n \\ 1, 1, \cdots, 1 \end{bmatrix} = P_{cop} + P_{out},$$

where P_{cop} is the coplanar object points and P_{out} is the remaining points outside the plane. The two subsets can be conveniently distinguished by using SVD method. According the segmentation in object coordinate system, the corresponding image points also can be rewritten as

$$D = \begin{bmatrix} v_1, v_2, \cdots, v_n \\ 1, 1, \cdots, 1 \end{bmatrix} = D_{cop} + D_{out},$$

where D_{cop} and D_{out} are image points corresponding to P_{cop} and P_{out} respectively.

As we know, there are two pose solutions for planar object points and a robust pose estimation algorithm from planar targets (RPP for short) has been presented by [20]. Two candidate pose solutions can be calculated from the selected coplanar points

$$[R_1, T_1, R_2, T_2] = RPP(P_{cop}, D_{cop}) \tag{7}$$

Once we get the two possible pose estimation results, a method should be found to choose the correct one by using the selected non-planar points. In general, there are two kinds of re-projection error function because of different projection methods. One is image space error function:

$$E_{is}(R,T) = \sum_{i=1}^{N_{out}} [(x_i - \frac{r_1 \cdot p_i + t_x}{r_3 \cdot p_i + t_z})^2 + (y_i - \frac{r_2 \cdot p_i + t_y}{r_3 \cdot p_i + t_z})^2] \tag{8}$$

And another is object space error function:

$$E_{os}(R,T) = \sum_{i=1}^{N_{out}} \|(I - V_i)(Rp_i + T)\|^2 , \quad V_i = \frac{v_i v_i^T}{v_i^T v_i}.$$ (9)

where N_{out} is the number of non-planar points ; I is the identity matrix; R and T are denoted as follow

$$R_{3\times3} = \left[r_1^T, r_2^T, r_3^T \right] , \quad T_{3\times1} = [t_x, t_y, t_z]^T .$$

The final and correct pose, which has the lowest re-projection error, can be determined easily:

$$pose = \begin{cases} (R_1, T_1) , & \text{if } E_{is/os}(R_1, T_1) < E_{is/os}(R_2, T_2) \\ (R_2, T_2) , & \text{if } E_{is/os}(R_1, T_1) \geq E_{is/os}(R_2, T_2) \end{cases}.$$ (10)

According to the above algorithm, pose parameters can be uniquely determined and pose ambiguity problem of approximate coplanar targets can be avoided. The central idea is to treat all the points in a target as two data subsets. One is coplanar points which belong to a plane and the other is non-coplanar points which distract outside the plane. Then, pose estimation can be done using the coplanar points, which usually leads to two possible solutions because of planar pose ambiguity. Finally, the final correct pose can be determined from the two candidate solutions, using the non-coplanar points. Our robust pose estimation algorithm is summarized as follows:

1) Selecting the majority of object points residing in a main plane using SVD method;
2) Calculating the two candidate solutions using the algorithm in [17] for the selected planar points from (7);
3) Computing the re-projection error in both image space and object space using (8) and (9);
4) Choosing the correct pose corresponding to the minimum re-projection error using the criterion in (10).

4   Experiment Results

In this section, we compare the effect of our approach against that of the state-of-the-art ones both on synthetic and real world data.

4.1   Synthetic Experiments

For all the synthetic experiments in this subsection, we use the following setup:

1) We produce synthetic 3D-to-2D correspondences in a 640×480 image acquired using a virtual calibrated camera with an effective focal length of $f_u = f_v = 800$ and a principal point at $(u_c, v_c) = (320, 240)$.

2) There are 8 object points, where 7 points are coplanar and the object shape is the same as object2 in Fig.1.

3) Given the true camera rotation matrix R_{true} and translation vector T_{true}, we then can compute the relative error of the estimated rotation matrix R by $E_R = \|R - R_{true}\| / R_{true} \times 100\%$. Similarly, the relative error of the estimated translation T is determined by $E_T = \|T - T_{true}\| / T_{true} \times 100\%$. Then we define a pose estimation solution is right if both $E_R < threshold$ and $E_T < threshold$.

4) The corresponding image points are calculated through rotation matrix R_{true} and translation vector T_{true}, and the different level Gaussian noise is added for all the image points.

For each noise level, we run 1000 independent simulations for our proposed algorithm and three state-of-art ones (Lu's algorithm, Fiore's algorithm and EPnP). Fig.3 shows the rate of finding the correct solution.

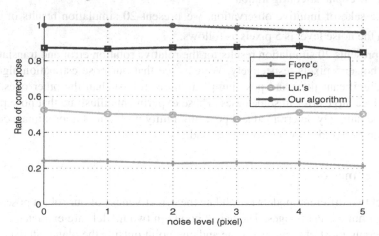

Fig. 3. Rate of correct pose for different algorithms in each noise level

In Fig.3, we can see that for different noise levels, Lu's algorithm has a correct rate about 50 percent, which is because that Lu's algorithm gives only one pose even if there are two possible solutions for approximate coplanar points. Fiore's algorithm is a linear algorithm and the correct rate is very low, which means Fiore's algorithm is extremely unstable for approximate coplanar points. EPnP algorithm has a great improvement in the correct rate. The slightly less robust results (about 15 percent) can be explained in this way: in a larger range, the result can be optimized to the correct pose; however, in some range, it will reach to the other wrong pose because of pose ambiguity. The correct rate of our algorithm is almost 100 percent for all the noise

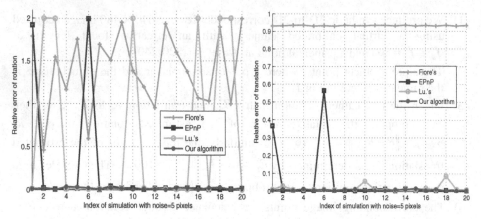

Fig. 4. Relative errors of rotation matrix and translation vector for different algorithms with noise=5 pixels

level, which demonstrates the effectiveness of our novel algorithm, which considers both the influences of planar points and non-coplanar points, especially the targets are approximate coplanar configuration.

For the sake of intuitive observation, we present 20 simulation results of E_R and E_T with the noise level is 5 pixels as follows.

Fig.4 presents 20 simulation results for the relative rotation error and translation error with noise=5 pixels respectively. We can see that our pose estimation algorithm, without significant jitter and pose jumps, is more robust than the other ones, which may produce large error in some cases. These experiments illustrate that pose parameters can be uniquely determined and pose ambiguity problem of approximate coplanar targets can be avoided by our robust method.

4.2   Real Images

We select two different models to validate the effectiveness of our robust pose estimation algorithm on real images. The object points in two models are extracted manually while keeping most of them in a plane and one point outside the plane. Statistic data is analyzed in Table 1. We can see obviously from Table 1 that the correct rate of our proposed algorithm is higher than others.

Table 1. Rate of correct pose estimation for different models

	Total Frames	Fiore's	Lu's	**Our algorithm**
Model 1	93	26 (28.0%)	58 (62.4%)	**93 (100%)**
Model 2	436	91 (20.9%)	294 (67.4%)	**429 (98.4%)**

Some pose estimation frames on two video sequences of the two different models are depicted in Fig.5 and Fig.6.

Fig. 5. Experiment for real images (pose in frame # 37 of model 1): top-left: model and approximate coplanar object points (red dots); top-right: our algorithm; bottom-left: Lu's algorithm; bottom-right: Fiore's algorithm.

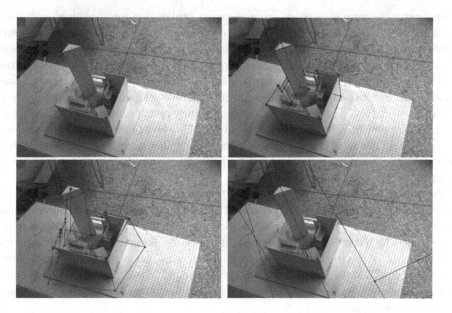

Fig. 6. Experiment for real images (pose in frame # 402 of model 2): top-left: model and approximate coplanar object points (red dots); top-right: our robust algorithm; bottom-left: Lu's algorithm; bottom-right: Fiore's algorithm.

The selected object points on the two models are approximate coplanar points, which means pose ambiguity exists according to the analysis in section 2. All the points are used to estimate pose in each frame, and then the model is re-projected to the image frame, which can confirm the effectiveness and precision intuitively and the rate of correct pose estimation can also be determined. In Fig.5 and Fig.6, approximate coplanar points are shown in the top-left images as red dots and re-projection of the model using three algorithms is depicted by thin blue lines respectively. We can see that Fiore's algorithm has significant deviation and the rate of correct pose estimation is very low, which means Fiore's algorithm is not suitable for the approximate coplanar configuration. Lu's algorithm has a certain bias, though the re-projection of planar points seems all right. To some extent, this situation proves that planar points and non-planar points have different effects on pose estimation and the algorithm will encounter problems if treating all the object points in the same way. Our algorithm matches the model very well and reaches a high rate in all the image frames of the two video sequences, by considering the influences of planar points and non-planar points comprehensively. Meanwhile pose calculated by our algorithm is robust and precise, without jitter and obvious bias. The real image results demonstrate the effectiveness of our robust algorithm.

5   Conclusions

In this paper, we provide a more comprehensive explanation for pose ambiguity from arbitrary 3D objects. We point out that pose ambiguity is an inherent phenomenon relating to the configuration of targets. Especially, pose ambiguity is obvious for both planar targets and approximate coplanar targets. Based on this, a robust pose estimation method for approximate coplanar targets, considering both the influences of coplanar points and non-coplanar points respectively, is developed and pose ambiguity is avoided effectively. The central idea of our robust pose estimation algorithm is to treat all the points in a target as two data subsets. One is coplanar points which belong to a plane and the other is non-coplanar points which distract outside the plane. When two candidate solutions are calculated by coplanar points, final and correct pose is determined using non-coplanar points.

According to the analysis, when we need to estimate pose in practice, the configuration of object points should be considered to avoid pose ambiguity. When designing the configuration of 3D targets, it would be better for the uniform distribution in all directions, which can decrease the probability of pose ambiguity and then increase the robustness of pose estimation. If the approximate coplanar configuration is encountered, our proposed algorithm can be adopted to improve the robustness of pose estimation. This robust algorithm should be relevant for many applications in AR and autonomous navigation.

Acknowledgements. This work was supported by the Fund of National Natural Science Foundation of China (61231018) and National High Technology Research and Development Program of China (2013AA014601).

References

1. Lu, C., Hager, G., Mjolsness, E.: Fast and Globally Convergent Pose Estimation from Video Images. IEEE Trans. Pattern Analysis and Machine Intelligence 22(6), 610–622 (2000)
2. Ansar, A., Daniilidis, K.: Linear pose estimation from points or lines. IEEE Transactions on Pattern Analysis and Machine Intelligence 25(5), 578–589 (2003)
3. Gordon, I., Lowe, D.G.: What and where: 3D object recognition with accurate pose. In: Ponce, J., Hebert, M., Schmid, C., Zisserman, A. (eds.) Toward Category-Level Object Recognition. LNCS, vol. 4170, pp. 67–82. Springer, Heidelberg (2006)
4. Martinec, D., Pajdla, T.: Robust rotation and translation estimation in multi-view reconstruction. In: CVPR (2007)
5. Caron, G., Dame, A., Marchand, E.: Direct model based visual tracking and pose estimation using mutual information. Image and Vision Computing 32, 54–63 (2014)
6. Agarwal, A., Jawahar, C., Narayanan, P.: A survey of planar homography estimation techniques. Technical Reports. International Institute of Information Technology, Hyderabad (2005)
7. Kumar, D.S., Jawahar, C.V.: Robust homography-based control for camera positioning in piecewise planar environments. In: Kalra, P.K., Peleg, S. (eds.) ICVGIP 2006. LNCS, vol. 4338, pp. 906–918. Springer, Heidelberg (2006)
8. Lourakis, M., Zabulis, X.: Model-based pose estimation for rigid objects. In: Chen, M., Leibe, B., Neumann, B. (eds.) ICVS 2013. LNCS, vol. 7963, pp. 83–92. Springer, Heidelberg (2013)
9. Joel, A., Stergios, H., Roumeliotis, I.: A Direct Least-Squares (dls) Solution for PnP. In: Int. Conf. on Computer Vision, Barcelona, Spain, November 6-13 (2011)
10. Wu, Y., Hu, Z.: PnP problem revisited. Journal of Mathematical Imaging and Vision 24(1), 131–141 (2006)
11. Nist´er, D.: An efficient solution to the five-point relative pose. IEEE PAMI 26(6), 756–770 (2004)
12. Gao, X., Hou, X., Tang, J., Cheng, H.: Complete Solution Classification for the Perspective-Three-Point Problem. IEEE Trans. Pattern Analysis and Machine Intelligence 25(8), 930–943 (2003)
13. Lepetit, V., Moreno-Noguer, F., Fua, P.: EPnP: An accurate O (n) solution to the PnP problem. Int. Journal of Computer Vision 81(2), 155–166 (2008)
14. Fiore, P.D.: Efficient linear solution of exterior orientation. IEEE Transactions on Pattern Analysis and Machine Intelligence 23(2), 140–148 (2001)
15. Gold, S., Rangarajan, A., Lu, C., Pappu, S., Mjolsness, E.: New algorithms for 2D and 3D point matching: Pose estimation and correspondence. Pattern Recognition 31(8), 1019–1031 (1998)
16. Oberkampf, D., De Menthon, D.F., Davis, L.S.: Iterative Pose Estimation Using Planar Feature Points. Computer Vision and Image Understanding 63(3), 495–511 (1996)
17. Schweighofer, G., Pinz, A.: Robust Pose Estimation from a Planar Target. IEEE Trans. Pattern Analysis and Machine Intelligence 28(12), 2024–2030 (2006)
18. Golub, G.H., Van Loan, C.F.: Matrix Computations, 3rd edn. Johns-Hopkins Univ. Press (1996)

A BP Neural Network Predictor Model for Stock Price

Li Huo[1], Bo Jiang[2], Tao Ning[1], and Bo Yin[3]

[1] Software Technology Institute, Dalian Jiaotong University, Dalian, China
[2] School of Information Science and Technology, Dalian Maritime University, Dalian, China
[3] Fundamental Course, Dalian Air force Communication NCO Academy, Dalian, China
alicia1900@163.com

Abstract. This paper presents a stock price forecast model using LM-BP algorithm, which has adopted three-layer back propagation neural networks. This model is more fast convergence rate and overcomes redundancy and noise of the samples. The proposed model is simulated in MATLAB platform and the experimental results of stocks price prediction show that the LM-BP model attains high accuracy of predicting the stock price in short-term. This paper provides also the comparison of actual value and forecast value, which proves that the model is effective and promising.

Keywords: BP neural networks, stock price prediction, LM-BP algorithm.

1   Introduction

The stock price forecast is one of the hottest fields of research recently due to its commercial applications owing to the high stakes and the kinds of attractive benefits that it has to offer. Unfortunately, stock market is essentially dynamic, non-linear, complicated, nonparametric, and chaotic in nature [1]. In addition, stock price is affected by many macro-economical factors such as political events, policies of firms, general economic conditions, commodity price index, bank rate, movements of other stock market, psychology of investors, etc [2]. The technical method for stock price prediction mainly includes statistical methods and artificial intelligent methods [3, 4]. The time series analysis method is one of the primary method based on statistics and the common models includes Random Walk model, ARMA model and Markov Chain model etc[5].The artificial intelligent, especially neural network, have achieved great development in recent years. Lots of research scholars have carried out the study of the stock forecasting based on neural network technique. Artificial neural network has the ability of self-organization and self-adaptability, self-study, high-nonlinear mapping etc. Thereby it is suitable for stock price forecasting. The BP neural network has a wide application in this field [6]. There are defects such as slow convergence speed and generalization ability weak in the traditional BP algorithm. Many scholars have improved BP algorithm [6]. The Approach to forecast stock price based on LM-BP algorithm has given in this paper. Furthermore, we have verified effectiveness of this forecasting method combining with the actual data of a stock market.

D.-S. Huang et al. (Eds.): ICIC 2014, LNAI 8589, pp. 362–368, 2014.

2   Back Propagation Neural Network

Back propagation neural network (BP-NN) has been proposed by Rumelhart and McClelland [7]. The BP-NN model is a well known a multi-layer feed forward training model. They are effective in many applications of fields such as pattern recognition, risk evaluation and self-adaptive control [10].

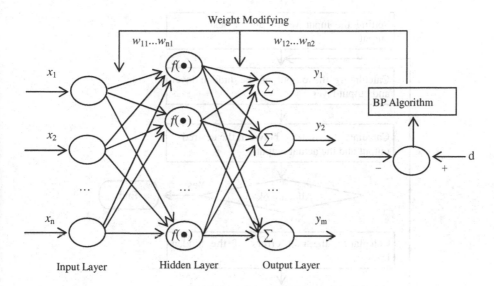

Fig. 1. Algorithm and training chart of BP network

The most common BP-NN is three-layer. The structure of three-layer BP-NN model is depicted in Fig. 1. There are three layers including input layer, hidden layer, and output layer. Two nodes of each adjacent layer are directly connected, which is called a link. Each link has a weighted value presenting the relational degree between two nodes [8]. Assume that there are n input neurons, many hidden neurons, and m output neurons. The outputs of all neurons are calculated by Eq. 1.

$$y_j = f\left(\sum_i w_{ji}x_i - \theta_j\right) = f(net_j) \tag{1}$$

Here y_j is the output value and net_j is the activation value of the j node. $f()$ is called the activation function, which is usually a Sigmoid function. θ_j is threshold value of j node and x_i is the input value, w_{ji} is connection weight value.

The BP-NN training process can be divided into two steps: information forward propagation and error back propagation. The flow of BP training is showed in Fig 2. The BP-NN training can reach any small approximation using the steepest gradient descent method [10].

The BP-NN possesses so strong fault tolerant ability that solves the noise problem in stock prediction. The structure of multi-layer forward feed can solve nonlinear problem. The BP-NN is a kind of supervised learning neural network. Therefore BP-NN is suitable to forecast stock price.

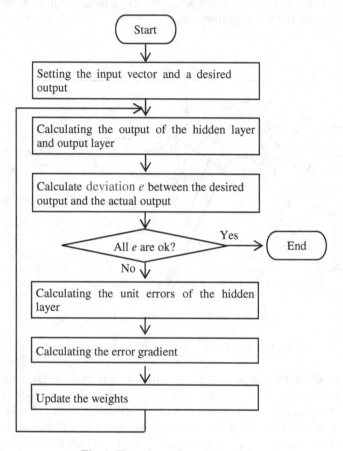

Fig. 2. Flow chart of network training

3   Parameter Selection

3.1   BP Model

This paper adopted three layers BP training model of $10\times12\times1$, in which has 10 input variables in the input layer, one hidden layer including 12 nodes and 1 output node in the output layer. The one output node is the closing price of next trading day.

We selected 15 technical indicators, which play a key role in the short-term stock price forecast. At last 10 technical indicators, which acted as input variables, have been retained by the preprocessing of attribute reduction and normalization.

The selection of hidden layer nodes is a complicated problem, which is directly related to the number of input and output layer nodes. In practical applications, the node number of hidden layer can be determined by Eq.(2) .

$$h = \sqrt{m*n} + a \tag{2}$$

Here h is the number of hidden layer node, n is the number of input layer node, m is the number of output layer node and a is a regulator constant ranging from 0 to 10 [9].

We adopted 12 nodes of the hidden layer by the method of trial and error, which acquired the satisfactory results. The selected parameters for the forecasting of stock price are showed in Table.1.

3.2   LM-BP Algorithm

We have implemented the simulation prediction by means of MATLAB 7.0. The neural network toolkit of MATLAB contains many customizable BP algorithms. We selected training function "trainlm", which is a Levenberg-Marquadt (LM) learning algorithm. The advantage of LM algorithm is fast convergence rate and so it can overcome slow speed of training process. LM can skip local minimum, thereby solve redundancy and noise of the samples [11].

Table 1. Parameters information of the BP_NN

Layer	Parameters	Value
Input layer	Concomitant variable 1	Trading volume
	Concomitant variable 2	MA5
	Concomitant variable 3	MA10
	Concomitant variable 4	MA20
	Concomitant variable 5	WR1
	Concomitant variable 6	WR2
	Concomitant variable 7	PSY
	Concomitant variable 8	PSYMA
	Concomitant variable 9	MACD
	Concomitant variable 10	KDJJ
	Number	10
	Rescale method	normalization
Hidden layer	The Number	1
	The node number	12
	Activation function	Hyperbolic tangent
Output layout	Dependent variable	Closing quotation
	The number	1
	Rescale method	normalization
	Activation function	Hyperbolic tangent
	Error function	Quadratic sum

4   Simulation and Results

4.1   Experimental Data

The raw data for the stock prediction experiment has been collected from one stock market in china. The total number of samples for the stock indices is 120 trading days from 5th January 2010 to 2th July 2010. Each sample consists of the closing price and trading volume. In order to reflect fully change of the stock market, the technical indicators are added to the BP network. A portion of raw data and technical indicators are shown in table 2. The data was divided into two groups, 70% - 80% of which was used for training sample, 20% -30% was used for testing sample.

We selected 100 samples for training and 40 samples for test in our research. The function *logsig* is selected as transfer function between input layer and hidden layer and the function *purelin* is selected as transfer function between the hidden layer and output layer.

Table 2. The original data

date	Closing price	Trading volume	MA. MA5	MA. MA20	WR. WR1	PSY	MACD	KDJ. J
01/04	3243.76	122446259	3236.8	3211.066	20.171	58.333	16.221	105.71
01/05	3282.18	142274964	3255.488	3208.581	5.129	66.667	21.758	109.67
01/06	3254.21	135131301	3263.975	3206.456	16.368	66.667	20.517	99.123
01/07	3192.78	145228703	3250.012	3204.119	46.698	58.333	10.773	59.305
01/08	3196	114602542	3233.788	3201.206	59.546	66.667	4.439	34.27
01/11	3212.75	161502182	3227.587	3199.478	59.535	66.667	2.29	29.383
01/12	3273.97	159128744	3225.938	3198.028	20.782	66.667	8.548	58.624
01/13	3172.66	185213142	3209.637	3192.941	85.012	66.667	-1.016	23.952
01/14	3215.55	151986663	3214.188	3190.959	57.82	66.667	-1.589	28.695
01/15	3224.15	131414089	3219.813	3193.209	52.368	66.667	-0.869	36.542
01/18	3237.1	145703540	3224.688	3199.372	44.158	66.667	1.18	47.648
01/19	3246.87	144950161	3219.275	3205.569	37.964	66.667	3.546	58.312
01/20	3151.85	172555629	3215.1	3210.631	97.778	66.667	-7.395	16.943
01/21	3158.86	132639664	3203.762	3214.887	82.23	66.667	-13.065	9.014
01/22	3128.59	156427043	3184.65	3213.644	72.98	66.667	-19.862	16.776
01/25	3094.41	93994068	3156.113	3211.297	85.05	66.667	-27.462	9.664

4.2   Prediction Results

The training result is showed in Fig.3. It shows that the error curve is declining and becomes smooth and steady with the increase of iterations (>200). After 500 iterations, the total error is closes to 0.0000145. The result has reached the desired precision and the mean square error is 0.0078. This indicated that the LM-BP has quite convergence and stability and fast training speed.

This paper provides the comparison of actual value and forecast value, which is showed in Fig.4. We selected samples of 20 trading days and implemented the forecast using trained LM-BP model. This shows that the accuracy of prediction attains 56% and the accuracy of trend prediction attains 76%. The result indicates that the LM-BP model can forecast the closing quotation considerably accurately.

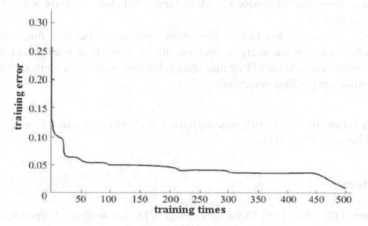

Fig. 3. Training Results

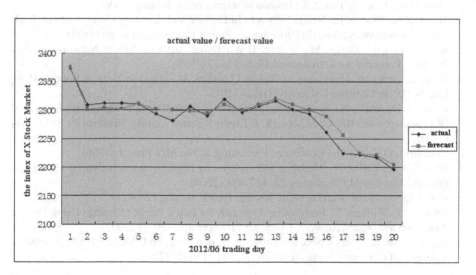

Fig. 4. The comparison of actual value and forecast value

5   Conclusion

The results demonstrated that the operation of the stock market in China has its inherent law. The optimized LM-BP model, which have fast convergence rate, can forecast the change of short-term stock price effectively. The predictions are quite accurate. The LM-BP algorithm is effective and promising.

The initial input variables comprise as many as possible technical indicators, which influence stock price. The initial variables were recombined use attribute reduction and normalization and consequently short-term prediction accords with the actual stock market regular.

However, the model has certain limitation without considering macro-economic factors, fundamental value analysis, movements of other stock market factors and so on. The further research should be one comprehensive prediction method, which will comprise different prediction technology.

Acknowledgements. This work was supported by National Natural Science Foundation of China (No. 61173034).

References

1. Fishman, M.B., Barr, D.S.: Using Neural Nets in Market Analysis. J. Technical Analysis of Stocks & Commodities 9(4), 135–138 (1991)
2. Ling, J.G., Zheng, J., Tou, Z.X.: Economic science press, Beijing (2006)
3. Trippi, R.R., Turban, E.: Neural Networks in Finance and Investing Using Artificial Intelligence to Improve Real-world Performance. Probus Company, Chicago (1993)
4. Kryzanowski, L., Galler, M., Wright, D.W.: Using Artificial Neural Networks to Pick Stocks. J. Financial Analysts Journal 49(2), 21–27 (1993)
5. Ning, T.: Study of Application of Hybrid Quantum Algorithm in Vehicle Routing Problem. D. Dalian Maritime University, Dalian (2013)
6. Zhang, Y.D., Wu, L.N.: Stock market prediction of S&P 500 via combination of improved BCO approach and BP neural network. J. Expert Systems with Applications 36, 8849–8854 (2009)
7. Rumelhart, D.E.: Parallel Distributed Processing. J. The MIT Press,1 (1986)
8. Basma, A.A., Kallas, N.: Modeling soil collapse by artificial neural networks. J. Geotechnical and Geological Engineering 22, 427–438 (2004)
9. Han, L.Q.: Artificial neural network tutorial. BUPT, Beijing (2006)
10. Chou, S.H., William, T.: Forecasting Approach for Initial Public Offerings Using Genetic Algorithm and Neural Network. Computer Engineering 33, 9–11 (2007)
11. Xiao, J., Pan, Z.L.: Stock price short-time prediction based on GA-LM-BP neural network. J. Journal of Computer Applications 32(s1), 144 –150 (2012)

Plant Leaf Recognition Using Histograms of Oriented Gradients

Qing Xia[1], Hao-Dong Zhu[2], Yong Gan[2], and Li Shang[3]

[1] College of Electronics and Information Engineering
Tongji University Shanghai, China
[2] School of Computer and Communication Engineering
ZhengZhou University of Light Industry
Henan Zhengzhou, 450002, China
[3] Department of Communication Technology,
College of Electronic Information Engineering,
Suzhou Vocational University, Suzhou 215104, Jiangsu, China

Abstract. Leaves from plants are proved to be a feasible source of information used to identify plant species [1]. In this paper, we present a method to recognize plant leaves employing Histograms of Oriented Gradients (HOG) as the feature descriptor. For better robustness to illumination, shadow, quality degradation, etc., five vital factors of original HOG algorithm are discussed to evaluate the respective effects in different configurations. Experimental results show that this method achieves excellent performance in recognition rate.

Keywords: Leaf recognition, Histograms of Oriented Gradients, ICL database.

1 Introduction

Plant information obtained from leaf images are proved to be sufficient for the identification of plant species [1]. For classification of images, the selection of a proper feature descriptor and a suitable classifier is significantly important for final accuracy rate and execution speed. In this paper, we will focus on the former.

We implement HOG algorithm to extract features for plant leaf recognition. HOG, short of Histograms of Oriented Gradient were proposed as an algorithm of image feature extraction by Navneet Dalal and Bill Triggs [2] for the purpose of human detection. Among kinds of different feature descriptors, the HOG descriptor gets some of the better performance for its invariance to 2D rotation and its capability of representing local structures as well as object contours. However, few publications can be found which successfully implement HOG to the classification of plant leaves. In order to get a robust feature descriptor which is insensitive to illumination and perform high execution speed, several crucial variables in the HOG algorithm will be discussed in this paper.

In 2005, Ling et al. [11] proposed using the inner-distance between landmark points to build shape descriptors while the inner-distance was defined as the length of

D.-S. Huang et al. (Eds.): ICIC 2014, LNAI 8589, pp. 369–374, 2014.

the shortest path between landmark points within the shape silhouette. Experiments showed that the inner-distance was articulation insensitive and more effective at capturing complex shapes with part structures than Euclidean distance. Besides, Manh et al. [9] make efforts for weed leaf identification using an algorithm of fitting a parametric model to leaf outlines by minimizing an energy term related to internal constraints of the model and salient features. Wang et al. [10] extract eight geometric features from leaves contours shape, meanwhile they propose a moving median hyper sphere classifier to achieve a better result. This algorithm has just been used to recognize 20 kinds of plants leaves. Du et al. [12] enhance and detail the moving median hyper sphere classifier based on a constructed leaf database. Those methods are applicable for different certain plant groups.

2   The HOG Descriptor for Leaf Recognition

The accordance of the original HOG algorithm, is that the appearance and shape of local object can often be characterized rather well by the distribution of light intensity gradients and directions, even without precise knowledge of the corresponding gradient or edge positions.

Features of HOG are extracted on the basis of evaluating well-normalized histograms of gradient orientations in the detection window. In detail, gradient values and orientations are firstly computed on each image pixels while the leaf image is divided into cells which are dense grids of uniformly adjacent and spaced regions with certain size. And then oriented gradient histograms are counted on cells by dividing orientations into fixed number of orientation bins. Aimed at achieving characteristics of invariance and robustness, oriented gradient histograms are locally normalized on blocks which are overlapping and spatially connected regions grouped by cells. In practice the size of cells, blocks, block overlapping strides and orientation bins should be analyzed and discussed according to experiment results. Following parts of this section will detail the procedures of extracting HOG from pant leaf images.

2.1   Computing Gradients and Orientation Angles

Considering pre-processing functions make little impact on the final performance of HOG features, only graying is applied into images before computing gradients. The leaf appearance can be outlined by the pixels which have large gradient values. Meanwhile gradient orientations show the tendency of gradients. If a pixel in leaf images is described as $p(x, y)$, where (x, y) is respectively the horizontal and vertical coordinates of the pixel, the gradient value can be given by:

$$g_p = g(x, y) = \sqrt{\Delta x^2 + \Delta y^2} \tag{1}$$

And the gradient direction is given by:

$$\theta_p = \theta(x, y) = arctan\frac{\Delta x}{\Delta y} \tag{2}$$

where Δx and Δy can be given by kinds of algorithm, such as Sobel operator, etc.

2.2   Counting Histograms of Oriented Gradient on Cells

Counting Histograms of Oriented Gradient

Gradient magnitudes, gradient orientations and relations between them contain sufficient informations of leaf appearances. Histograms of oriented gradient which count
the gradient magnitudes by diving orientation into separated bins are a direct structure
to represent those informations. So the dimensions of a histogram is same as the number of separated bins.

In HOG, gradient magnitudes are taken as weight of pixels and contribute to two
adjacent orientation bins according to the bilinear interpolation algorithm. If gradient
orientation range over $0° - 180°$ and are divided into m bins, the gradient angle of
θ_p could be represented by:

$$\delta_p = \frac{\theta_p * m}{180} - 0.5f \tag{3}$$

Meanwhile, the contribution of gradient magnitudes to two adjacent bins according to
the bilinear interpolation algorithm is given as v_1 and v_2:

$$\omega_p = \delta_p - FLOOR(\delta_p) \tag{4}$$

$$v_1 = g_p * (1 - \omega_p) \tag{5}$$

$$v_2 = g_p * \omega_p \tag{6}$$

Segment Images into Cells

The histogram computed on the whole image represents global structures, but is short
of local patterns. As a Supplement, computing histograms on image segments, called
cells, could represent local structures in a good manner. In summary, histograms on
cells represent local leaf features and the combination of those histograms represent
global leaf features.

If size of a leaf image is presented as (X, Y) and the size of cells is presented by
(C_x, C_y) in units of pixels, the number of spaced cells would be calculated by:

$$CN_x = \frac{X}{C_x} \tag{7}$$

$$CN_y = \frac{Y}{C_y} \tag{8}$$

where CN_x and CN_y are respectively the cell number in horizontal direction and the
cell number in vertical direction. Then the whole image are divided into $CN_x * CN_y$
spaced cells of same size and the number of histograms on cells are $CN_x * CN_y$ as
well.

2.3   Normalization Histograms on Overlapping Blocks

Features on cells are simple and efficient. But there is a potential drawback of
those cells features that is those features are tend to be very sensitive to impurity of

illumination. Images of plant leaves are inevitably to avoid illumination whether acquired through photographs or scanners. Normalizing the cell histograms will eliminate the impact of light and compensate for robustness by accumulating a measure of local histogram energy over the somewhat larger connected regions and using the results to normalize all cells in the large regions. The overlapping large regions grouped by cells are called blocks. Normalization on overlapping blocks make a cell express more than once in the final features. We choice rectangular block for leaf image recognition on account of its faster computing speed and almost the same effect compared with circular blocks.

If the size of blocks and block strides are respectively (B_x, B_y) and (D_x, D_y) in units of pixels, then the block size in units of cells is $(\frac{B_x}{C_x}, \frac{B_y}{C_y})$ and the number of blocks $BN_x \times BN_y$ is given by:

$$BN_x = 1 + \frac{X - B_x}{D_x} \tag{9}$$

$$BN_y = 1 + \frac{Y - B_y}{D_y} \tag{10}$$

Normalization is implemented on each block histogram combined by $\frac{B_x}{C_x} \times \frac{B_y}{C_y}$ embedded cell histograms.

3   Experiments

3.1   Database

We tested our method on the ICL public database [17] established by the Intelligent Computing Laboratory. All images in the ICL database are taken by cameras or scanners in a white background under vary illumination conditions after plucked leaves from plants. Two sides of a leaf are respectively taken to images. Images taken through this process are tend to be in a uniform size and colorful. To ensure a plain background, only one leaf is put in an image. The ICL database includes 200 species of plants. Each species includes 30 samples of leaf images (15 per side). Hence there are totally 6000 images.Fig.1 shows some samples from the ICL database in which the top images are the frontal side of plant leaves and the bottom images are the other side.

Fig. 1. Two sides of leaf images from the ICL database

3.2   Experimental Results

To evaluate the effect of variables in our method and detect the best value set of variables for accuracy rate, a series experiments by different schemes are carried out on the ICL database. We select randomly 50 species in our experiments. For each species, 20 images are used as training data and another 10 images are used as testing data. Variables including image size, block size, cell size, block stride, as well as number of orientation bins are taken into experiments to detect their effects on final recognition rate. In the experiment scheme for one variable, other variables are set to fixed values to observe the effect of the single variable.

Following Fig.2-Fig.5 lists the experimental results by different schemes, from which we could make the following findings: Firstly, the accuracy rate achieves best at 0.97 when block size is 2-4 times of cell size, which means one block contains 4-16 cells inside. Secondly, 9-11 is the best number of orientation bins and the accuracy rate changes little while increasing the number of orientation bins. More number of bins means slower execution speed. So we could choose 9 number of orientation bins without decreasing accuracy. Thirdly, for image size, it keeps a relation with the combination of block size, cell size and bins number, generally it performs best when one image could be divided into 64 cells. Though smaller size of block stride will increase dimensions of the final feature descriptor, the accuracy rate does not increase persistently while reducing the block stride.

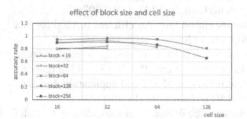

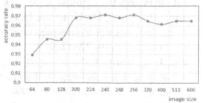

Fig. 2. Effect of block size and cell size   **Fig. 3.** Effect of number of orientation bins

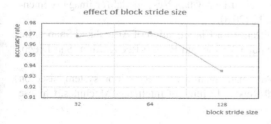

Fig. 4. Effect of block stride size   **Fig. 5.** Effect of image size

4   Conclusion

A leaf recognition method using HOG to represent leaf appearance features is proposed in this paper. The experimental results demonstrate that our approach significantly outperforms when all variables are set in proper values. Our work will focus on the improvement of HOG for plant leaf recognition.

Acknowledgments: This work is supported by the Science and Technology Innovation Outstanding Talent Plan Project of Henan Province of China under Professor De-Shuang Huang (No.134200510025), the National Natural Science Foundation of China (Nos. 61373105, 61373098 & 61133010).

References

1. Guyer, D.E., Miles, G.E., Schreiber, M.M., Mitchell, O.R., Vanderbilt, V.C.: Machine vision and image processing for plant identification. Trans. ASAE 29(6), 1500–1507 (1986)
2. Dalal, N., Triggs, B.: Histograms of oriented gradients for human detection. In: CVPR 2005 (2005)
3. Albiol, A., Monzo, D., Martin, A., Sastre, J., Albiol, A.: Face recognition using HOG-EBGM. Pattern Recognition Lett. 29(10), 1537–1543 (2008)
4. Bertozzi, M., Broggi, A., Rose, M.D., Felisa, M., Rakotomamonjy, A., Suard, F.: A pedestrian detector using histograms of oriented gradients and a support vector machine classifier. In: Proc. Intelligent Transportation Systems Conf., pp. 143–148 (2007)
5. Li, B., Huang, D.S.: Locally linear discriminant embedding: An efficient method for face recognition. Pattern Recognition 41(12), 3813–3821 (2008)
6. Zhao, Z.-Q., Huang, D.S., Sun, B.-Y.: Human face recognition based on multiple features using neural networks committee. Pattern Recognition Letters 25(12), 1351–1358 (2004)
7. Huang, D.S.: Radial basis probabilistic neural networks: Model and application. Int. Journal of Pattern Recognit., and Artificial Intell. 13(7), 1083–1101 (1999)
8. Huang, D.S., Du, J.-X.: A constructive hybrid structure optimization methodology for radial basis probabilistic neural networks. IEEE Trans. Neural Networks 19(12), 2099–2115 (2008)
9. Manh, A.G., Rabatel, G.: In-Field Classification of Weed Leaves by Machine Vision Using Deformable Templates. In: 3rd European Conference Precision Agriculture, ECPA 2001, Montpellier, June 18-28, pp. 599–604 (2001)
10. Wang, X.F., Huang, D.S., Xu, H.: An efficient local Chan-Vese model for image segmentation. Pattern Recognition 43(3), 603–618 (2010)
11. Ling, H., Jacobs, D.: Using the Inner Distance for Classification of Articulated Shapes. In: IEEE International Conference on Computer Vision and Pattern Recognition, San Diego, CA, USA, pp. 719–726 (2005)
12. Du, J.-X., Wang, X.-F., Huang, D.S.: Automatic plant leaves recognition system based on image processing techniques, Technical Report, Institute of Intelligent Machines, Chinese Academy of Sciences (October 2004)
13. ICL database of leaf images,
   http://www.intelengine.cn/dataset/index.html

An Implementation of an Intelligent PCE-Agent-Based Multi-domain Optical Network Architecture

Ying Xu, Tiantian Zhang, and Renfa Li

Key Laboratory for Embedded and Network Computing of Hunan Province,
College of Computer Science and Electronic Engineering,
Hunan University, Changsha, China, 410082
hnxy@hnu.edu.cn

Abstract. This paper presents a new optical network architecture based on the PCE (Path Computation Element) by adding a PCE-Agent to intelligently control PCEs in each layer of the conventional multi-domain optical network. The PCE-Agent uses a weighted sum policy to evaluate the performance of each domain managed by a PCE in each layer for the selection of the next hop domain when establishing a P2P (Point to Point) or P2MP (Point to Multiple Points) routing path. A PA-BRPC (PCE-Agent Backward Recursive PCE-based Computation Algorithm) routing algorithm is then proposed by extending the BRPC (Backward Recursive PCE-based Computation Algorithm) algorithm in the PCE-Agent-based multi-domain optical network architecture. Compared with the traditional BRPC algorithm, our PA-BRPC algorithm based on the improved PCE-Agent architecture obtains better performance in terms of both the utilization rate of resources and the load balance of the optical network.

Keywords: Path Computation Element, Multi-domain networks, PCE-Agent, BRPC.

1 Introduction

With the development of the optical network communication, the multi-domain optical network topology is becoming more and more complex, and the transfer requirement of large capacity communication is increasing. The traditional optical network architecture becomes unable to meet the needs of calculating the optimal paths in optical network. Therefore, how to calculate routing paths in the multi-layer multi-domain optical network has become an important research focus in the future development of optical networks.

In order to solve these problems, some protocols for the interconnection of multiple optical network domains have been proposed including Optical Border Gateway Protocol (OBGP), an optical network control protocol based on the border gateway protocol [1], External Network-to-Network Interface (E-NNI) [2], and the Path Computation Element (PCE) [3-5] and so on. At present, the main two types of routing mechanisms are the Generalized Multi-Protocol Label Switching (GMPLS) based multi-domain hierarchical routing [6-7] and the multi-domain hierarchical

D.-S. Huang et al. (Eds.): ICIC 2014, LNAI 8589, pp. 375–384, 2014.
© Springer International Publishing Switzerland 2014

routing based on PCE. PCE, proposed by the Internet Engineering Task Force (IETF), has gained wide attention due to its efficiency and flexibility for computing the constrained routing paths. PCE is a functional entity in the optical network to perform the path computation. The optimal path satisfying certain constraints is calculated according to the request of the path computation client. When many service requests to the same PCE for the centralized processing result in large processing delay, so, for distributed multi-layer and multi-domain network architecture, the traditional PCE-based optical network architecture is not an ideal solution.

In RFC5441, based on the PCE, IETF defines a Backward Recursive PCE-based Computation (BRPC) algorithm, which can support the multi-domain path computation based on backtracking [8]. With the known domain sequence, the algorithm of BRPC has an advantage of low complexity and can also provide the shortest path. However the algorithm does not guarantee a successful wavelength assignment. When the congestion and physical injury situation occur in the network, the algorithm of BRPC cannot timely and effectively avoid these situations. This will result in some cases that the network is blocked while there are still available resources in the network.

To handle the above problem, we improve the multi-domain optical network architecture by adding a PCE-Agent to intelligently help the selection of the domain sequence by evaluating the performance of each domain controlled by a PCE for each layer. Based on the implementation of our proposed architecture, a new routing algorithm named PA-BRPC is proposed (PCE-Agent-based Backward Recursive PCE-based Computation Algorithm) by extending the BRPC algorithm. In each layer, a PCE-Agent is designed to estimate the performance of each domain when choosing the best next hop domain to build P2P or P2MP routing paths. The performance of each domain, such as the free wavelength, the available bandwidth, the average transmission delay and the blocking probability, etc., is weighted by calculating the weights of boundary nodes according to a measurement strategy. The simulation results show that our proposed algorithm can improve the resource utilization rate of the whole optical network and decrease the network blocking rate.

The rest of the paper is organized as follows. In section 2, we present the relate work. Section 3 presents the proposed multi-domain optical network architecture based on PCE-Agent. Section 4 describes the proposed PA-BRPC routing algorithm. We evaluate our algorithm by experiments and summarize the obtained results in section 5. Finally, section 6 concludes this paper.

2   Related Work

From the BRPC algorithm that we know how to get a good domain sequence is a crucial problem to be solved. The manner by which the sequence of traversed domains is selected has been studied in both management [9,10] and Border Gateway Protocol (BGP) [11 – 13] based networks. The BGP-based scheme, which has the inter-domain link traffic engineering information, can dynamically perform domain sequencing. In a BGP-based scheme, however, the lack of knowledge on intra-domain resource information leads to a high blocking probability, especially in networks where intra-domain and inter-domain topologies are both complicated. So a k random path (KRP)-based inter-domain routing algorithm in a hierarchical PCE architecture has been proposed [14]. The network traffic blocking performance improves because of the introduction of factor k. As k increases, the proposed algorithm achieves lower blocking probability. Compared with the traditional scheme, the proposed algorithm achieves lower blocking probability and more efficient resource utilization while preventing the increase in PCEP signaling overhead. But the KRP algorithm also has some questions such as for different large optical networks how to set the value of k to make the optimal path is still worth discussing.

IETF has proposed a backtracking algorithm based on the PCE multi-domain optical network architecture. The algorithm calculates the shortest path from the source node to the destination node for the given domain sequence. The procedure is first to calculate the Virtual Shortest Path Tree（VSPT）from the destination node to the boundary nodes of the PCE that contains the destination node, then send the VSPT to the upper reaches of the PCE. The PCE of the upstream domain calculates the VSPT from the entrance boundary node to the outlet boundary nodes, reserves the optimal path, and removes the suboptimal path which does not meet the conditions. The upstream domains repeat the above process repeatedly, until the source node's domain is found and the optimal path from the source node to the destination node is thus obtained.

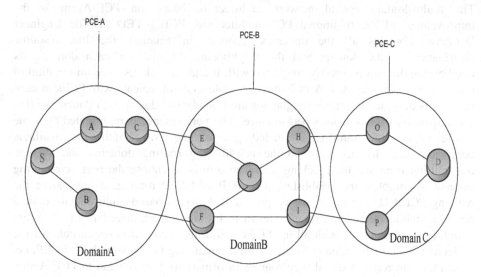

Fig. 1. The multi-domain optical network topology with three domains

Figure 1 shows an example calculation of a virtual shortest path from the source node S to the destination node D using BRPC. The process is as follows: the known domain sequence is Domain-A - Domain-B - Domain-C. The destination node D lies in Domain-C and the source node S lies in Domain-A.A path calculation request message is delivered from PCE-A, PCE-B and PCE-C in turn. The virtual shortest path tree is calculated by PCE-C from the entrance gateway and sent the results to PCE-B.PCE-B calculate the virtual shortest path tree of the Domain-B's import gateway nodes to the export gateway nodes. Combining the results sent by PCE-C, the shortest path tree from the import gateway nodes to the destination node .Then the results are sent to the upstream PCE-A. PCE-A calculate the virtual shortest path tree from the source node S to the adjacent gateway nodes between Domain-A and Domain-B. Combining the results from PCE-B, the shortest path tree from the source node to the destination node is thus obtained.

As mentioned above, the traditional BRPC algorithm assumes the domain sequence is known and has a low complexity. However, the algorithm does not guarantee a wavelength assignment for success when the congestion or physical injury situation occurs in the network. In order to balance the utilization of the network resources, this paper improves the traditional multi-layer multi-domain optical network architecture by designing a PCE-Agent-based multi-domain optical network architecture. Then a new PA-BRPC algorithm based on the designed architecture is proposed to test the effectiveness and efficiency of the new architecture.

3   The Multi-domain Optical Network Architecture Based on PCE-Agent

The multi-domain optical network architecture based on PCE-Agent is the improvement of the traditional PCE architecture. PCE's TED (Traffic Engineer Database) contains all the domain's resource information, the link resource information of the domain and the neighboring domain. for each domain, its neighboring domain is directly connected with it and the link resource information of other domains is unknown. A PCE-Agent is able to communicate with PCEs in each layer. It is designed to set the weights for the boundary nodes of each domain so that it can quantify each domain's performance. The performance is represented by some traffic engineering information including the free wavelength, the available bandwidth, the distance between the neighboring routing domains, the average transmitting time and the blocking rate. The aim is to choose the best neighboring domain to complete the establishment of P2P or P2MP routing. By extending the existing PCEP [15] protocol, we propose a PCE-Agent-based multi-domain optical network architecture as an example shown in Figure 2. Each layer has a PCE-Agent which can communicate with all the PCEs within this layer. It is responsible for the collection and collaboration of traffic information among PCEs through the PCEP, i.e. each PCE can report the real time routing information of its domain to PCE-Agent. The multi-domain optical network architecture based on PCE-Agent can be implemented as the extension of the traditional optical network layered routing model, where a PCE-Agent is added to intelligently determine the routing paths for the multi-domain multi-layer optical networks. When a path computation request arrives, PCE-Agent will weight each domain's performance and select the next neighboring domain to form a better domain sequence. The selected domain's PCE calculates the shortest path using the BRPC algorithm.

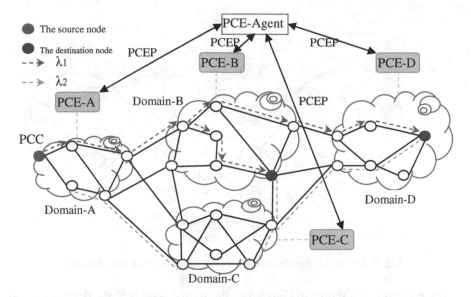

Fig. 2. The PCE-Agent-based multi-domain optical network architecture

The PCE-Agent-based multi-domain optical network architecture is effective with the increasing number of network domains. In Figure 2, there are four domains in the network, the source node's domain is Domain-A and the destination node is located in Domain-D. When calculating the routing path from Domain-A to Domain-D, which inter domain, i.e., Domain-B or Domain-C will be selected as the neighboring domain will be decided by the PCE-Agent considering the current traffic engineering information returned by PCE-B and PCE-C. A weighted sum policy is defined as a weighted combination W of cost, delay and packet loss rate each domain, the value of W is calculated according to the formula 1. Where C represents the network cost; D represents the delay; P represents the packet loss rate; x, y, z represent the weighs of C, D, P respectively. A domain with smaller W value has a better overall performance in terms of the current traffic information. The cost is the cost of the whole building of the link. The delay refers to the transit time of sending a message or packet from one node to another node. The packet loss rate is defined as to the percent of lost packets to the packets sent.

$$W = \begin{cases} C*x + D*y + P*z, x+y+z=1 \\ \infty \end{cases} \tag{1}$$

4   The PA-BRPC Algorithm

he PA-BRPC is an extension of the traditional BRPC algorithm based on the proposed multi-domain optical network architecture based on the PCE-Agent. The domain sequence is selected according to the current traffic condition. PA-BRPC is more flexible for large scale multi-domain optical network.

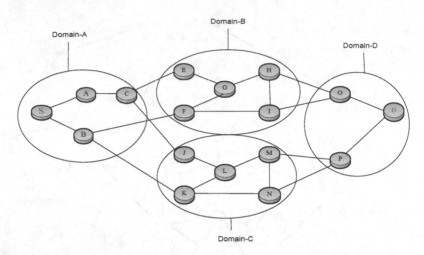

Fig. 3. The Multi-domain optical network topology with four domains

For example， as shown in Figure 3， the source node is S, the destination node is G, where the source node S lays is Domain-A and the destination node lays in Domain-B. Domain-A and Domain-B have the same boundary nodes E and F, which means Domain-A and Domain-B are neighboring domains. The shortest path tree from boundary nodes to the source node S is(S-B-F), (S-A-C-E) calculated by PCE-A, the shortest path tree from the source node S to the destination node is calculated as (S-B-F-G).

By using the defined weighted sum policy as in Formula 1, considering the impact of the routing delay, the packet loss rate and the energy consumption simultaneously. Domain-B's value of w is smaller than that of Domain-C. PCE-Agent thus concludes that Domain-B has better performance than that of Domain-C, and then the domain sequence is selected as Domain-A - Domain-B - Domain-D.

The same BRPC procedure is applied to calculate the shortest path from the source node S to the destination node D as follows:

1, The known domain sequence is Domain-A - Domain-B - Domain-D. The destination node D lies in Domain-D and the source node S lies in Domain-A. A path calculation request message is delivered from PCE-A, PCE-B and PCE-D in turn. The virtual shortest path tree is calculated by PCE-D from the entrance gateway O and P to the destination node D, i.e. O-D, and the results are sent to PCE-B.

2, PCE-B calculates the virtual shortest path tree of the Domain-B's import gateway E and F to the export gateway nodes O and P: E-G-H-O, F-G-H-O, and F-I-O. Combining the results sent by PCE-D, the shortest path tree from the import gateway nodes to the destination node is E-G-H-O-D and F-I-O-D. Then the results are sent to the upstream PCE-A.

3, PCE-A calculate the virtual shortest path tree from the source node S to the adjacent gateway nodes E and F between Domain-A and Domain-B is S-A-C-E and S-B-F. Combining the results from PCE-B, the shortest path tree from the source node to the destination node is S-B-F-I-O-D in this example.

5   The Analysis of Simulation Results

Based on the proposed PCE-Agent-based multi-domain optical network architecture, we use Matlab as the simulation software to build the experimental platform to test the performance of our proposed PA-BRPC algorithm. The network topology with 17 random nodes the simulation model is built in a square area of the 100*100. Each node's location coordinate is recorded, based on the geographical location, the network topology is divided based on K-means clustering method (MacQueen, 1967) is one of the simplest unsupervised learning algorithms that solve the well-known clustering problem. The procedure follows a simple and easy way to classify a given data set through a certain number of clusters (assume k clusters) fixed a priori. The main idea is to define k centroids, one for each cluster. These centroids should be placed in a cunning way because of different location causes different results [16]. The link bandwidth and the node energy are randomly generated in a certain range. The delay value represents the distance between two points divided by 2/3 speed of light, packet loss rate is randomly selected between 0 ~ 0.01. Three parameters (x, y, z) are used to describe the characteristics of each network topology as defined in formula (1). In order to investigate the effectiveness of the PA-BRPC algorithm, three different types of multi-domain optical network topologies as shown in Figure 5, Figure 6 and Figure 7 have been used as the test instances. We compare three algorithms including the traditional BRPC algorithm, the KRP algorithm and our proposed PA-BRPC algorithm based on the designed multi-domain optical network simulation platform. All the experiments are repeated 30 times on each network topology for each algorithm. The experimental results are shown in Figure 7, 8, 9, 10, 11 and 12.

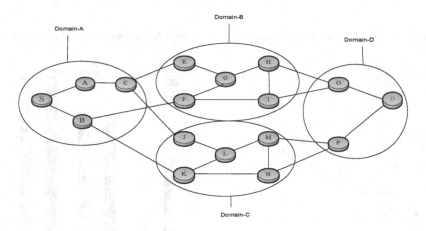

Fig. 4. The network topology A

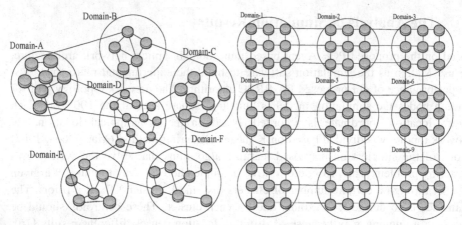

Fig. 5. The network topology B   **Fig. 6.** The network topology C

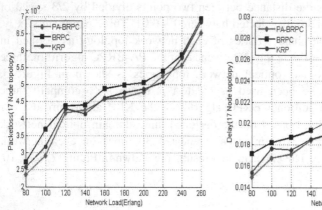

Fig. 7. The Packet Loss on the topology A   **Fig. 8.** The Delay on the topology A

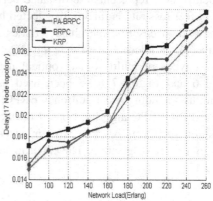

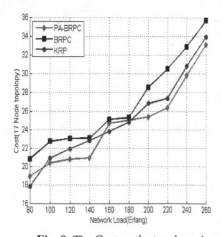

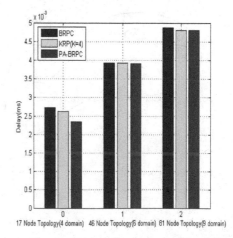

Fig. 9. The Cost on the topology A   **Fig. 10.** The Delay on three topologies

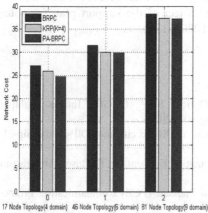

Fig. 11. The Cost on three topologies   **Fig. 12.** The Packet Loss rate on three topologies

The simulation results show that the adding of PCE-Agent in the multi-domain optical network architecture improves the utilization rate of resources in the optical networks. From Figure 7, 8 and 9, we can see that for most cases, the average delay, packet loss rate, cost obtained by the proposed PA-BRPC algorithm is better than those of the other two algorithms on the network topology Λ. The comparisons of three network parameters (i.e. the average delay, the packet loss rate and cost) on the three network topologies are shown in Figure 10, 11 and 12. Experimental results show that the PA-BRPC algorithm has the best overall performance than both the traditional BRPC algorithm and the KRP algorithm in terms of the average delay, the packet loss rate and cost on the three network topologies.

6   Conclusions

In this paper, we propose a new multi-domain optical network architecture based on the Path Computation Element (PCE) by adding a PCE-Agent to intelligently control PCEs in each layer of the conventional multi-domain optical network. In order to quantify each domain's network performance, the PCE-Agent is designed to intelligently set the weights for the boundary nodes of each domain, which can be helpful for choosing the best neighboring domain when building the P2P or P2MP routing paths in the multi-domain optical networks. Then, a PA-BRPC routing algorithm is proposed based on the PCE-Agent-based multi-domain optical network architecture. The simulation results indicate that network performance is improved because of the introduction of PCE-Agent. Compared with the traditional BRPC algorithm and the KRP algorithm, the proposed algorithm achieves lower blocking probability and more efficient resource utilization. This work can be further extended for solving the multicast routing problem in multi-domain multi-layer optical networks.

Acknowledgment. This research is supported by Natural Science Foundation of China (NSFC project No. 61202289) and the project of the support plan for young teachers in Hunan University, China (Ref. 531107021137).

References

1. Blanchet, M., Parent, F., St-Arnaud, B.: Optical BGP (OBGP): Interas Lightpath Provisioning. IETF Draft, Ietf-Draft-Parent-Obgp-01 (2001)
2. Ong, L.Y.: Intra-Carrier E-NNI Signaling Specification. OIF Specification OIF-E-NNI-Sig-01.0 (2004)
3. Farrel, A., Vasseur, J.P., Ash, J.: A Path Computation Element (PCE)-Based Architecture. RFC 4655 (2006)
4. Ash, J., Le Roux, J. (eds.): Path Computation Element (PCE) Communication Protocol Generic Requirements.RFC4657 (2006)
5. Le Roux, J.L.: Requirements for Path Computation Element (PCE) Discovery (2006)
6. Bernstein, G., Cheng, D. Pendarakis. D.: Domain to Domain Routing Using GMPLS, OSPF Extension V1. 1 (Draft). OIF2002 23 (2002)
7. Shiomoto, K., et al.: Requirements for GMPLS-Based Multi-Region And Multilayer Networks (MRN/MLN). RFC5212 (2008)
8. Vasseur, J.P., et al.: A Backward-Recursive PCE-Based Computation (BRPC) Procedure to Compute Shortest Constrained Inter-Domain Traffic Engineering Label Switched Paths. RFC 5441 (2009)
9. Vasseur, J.P., Roux, J.L.: Path Computation Element (PCE) Communication Protocol (PCEP). RFC5440 (2009)
10. Yannuzzi, M., Masip-Bruin, X., Bonaventure, O.: Open Issues In Interdomain Routing: A Survey. IEEE Network 19(6), 49–56 (2005)
11. Cugini, F., Paolucci, F., Valcarenghi, L., Castoldi, P., Weiin, A.: PCE Communication Protocol for Resource Advertisemen. In: Multidomain BOP-Based Networks, In: IEEE/OSA OFC, OWL3 (2009)
12. Yannuzzi, M., Masip-Bruin, X., SÁNchez-LÓPez, S., MarÍN-Tordera, E.: OBGPC: An Improved Path-Vector Protocol for Multi-Domain Optical Networks. Optical Switching and Networking 6(2), 111–119 (2009)
13. Buzzi, L., Bardellini, M.C.: Hierarchical Border Gateway Protocol (HBGP) for PCE Based Multi-Domain Traffic Engineering. In: 2010 IEEE International Conference on Communications, ICC, pp. 1–7 (2010)
14. Shang, S., Zheng, X., Zhang, H., Hua, N., Zhang, H.: A Hierarchical Path Computation Element (PCE)-Based Routing Algorithm In Multi-Domain WDM Networks. In: Asia Communications And Photonics Conference And Exhibition (P. 798903). Optical Society of America, Shanghai (2010)
15. Vasseur, J.P., Roux, J.L.: Path Computation Element (PCE) Communication Protocol (PCEP). RFC5440 (2009)
16. Macqueen, J.: Some Methods for Classification And Analysis Of Multivariate Observations. In: Proceedings of the Fifth Berkeley Symposium on Mathematical Statistics and Probability, vol. 1(281-297), p. 14 (1967)

A Heuristic Virtual Network Mapping Algorithm

Xiao-guang Wang[1,2], Xiang-wei Zheng[1,2], and Dian-jie Lu[1,2]

[1] School of Information Science and Engineering,
Shandong Normal University, Jinan 250014, China
[2] Shandong Provincial Key Laboratory for Distributed Computer Software Novel Technology,
Jinan 250014, China
samxiaoguang2008@163.com

Abstract. Virtual Network Mapping Problem (VNMP) is one of the key problems in network virtualization, which is proved as a non-deterministic polynomial hard (NP-Hard) problem. Considering the relevance of nodes and links, a virtual network mapping algorithm based on biogeography optimization algorithm is proposed in this paper. In order to reduce the cost of the substrate network embedding, it is designed as a one-stage algorithm and solved with heuristic algorithm. The experimental results suggest that the proposed algorithm increases the average operating income of the virtual networks and reduces the cost of substrate network comparing with other two-stage mapping algorithms.

Keywords: network virtualization, virtual network mapping, biogeography-based optimization.

1 Introduction

With the rapid increase of Internet users, Internet ossification gradually appears and network virtualization has become key technology to solve this problem. Network virtualization means that multiple independent virtual networks share all kinds of resources in substrate networks, and service providers can provide different services to various users according to their requirements [1]. Virtual network mapping problem is that the infrastructure providers distribute the substrate network resources to the requested virtual network with node and link resources constraints, which is a non-deterministic polynomial hard (NP-Hard) problem [2].

Generally speaking, the study of virtual network mapping problem is under certain constraints [3, 4]: (1) all requests of the virtual network are known beforehand; (2) the node and link resources of substrate network are infinite; (3) the limitations on position requirement are ignored.

Some efficient virtual network mapping problem algorithms have been proposed to solve different variants of virtual network mapping problem in the past years [2, 3, 4, 5, 6, 7]. Ref. [2] presented two stages mapping algorithm, handling the node mapping in the first stage and doing the link mapping in a second stage. In paper [3], the author studied virtual network assignment without reconfiguration. Ref. [4] studied the problem of mapping diverse virtual networks onto a common physical substrate. Ref. [5]

D.-S. Huang et al. (Eds.): ICIC 2014, LNAI 8589, pp. 385–395, 2014.

proposed a distributed and autonomic framework. Ref. [6] proposed a virtual network mapping algorithm based on subgraph isomorphism detection, which maps nodes and links during the same stage. Ref. [7] tried to solve the problem considering data rate constraints.

In the paper, the above restrictions on the virtual network mapping problems are not assumed, and a binary model of virtual network mapping problem with no assumption that substrate network needs to support path splitting is constructed [8, 9]. A new virtual network mapping algorithm based on biogeography-based optimization (VNM-BBO) is proposed. In order to reduce the substrate network cost, we adopted a one stage mapping method by fully considering the correlation between the nodes and links. Therefore, the virtual network mapping problem is modeled as global optimization and solved with intelligent algorithm. Experimental results demonstrated that the average operating income of substrate is effectively improved and the cost of substrate network is reduced.

The remainder of this paper is organized as follows. Section II describes the virtual network mapping problem and Section III describes model of VN mapping problem. Section IV introduces improved biogeography-based optimization (BBO) and details application of BBO to virtual network embedding problem. Section V presents the simulation experimental results and discussion. Section VI concludes the paper and points out future work.

2   Virtual Network Mapping Problem

This section briefly introduces the virtual network mapping problem.

Definition 1: Substrate network (SN). We define the substrate network by an undirected graph $G_S=(N_S,E_S,A^N_S,A^E_S)$, where N_S and E_S refer to the set of substrate node and the set of substrate link. A^N_S and A^E_S describe the attribute set of substrate node ns and the attribute set of substrate link e_s. A^N_S include the node computing capacity and physical location, while A^E_S refer to the remaining bandwidth resources of link e_s. P_S is denoted as the set of all paths in the substrate network.

Definition 2: Virtual network (VN). We also define the virtual network by an undirected graph $G_V=(N_V,E_V,B^N_V,B^E_V)$. Among, N_V and E_V refer to the set of virtual node request and the set of virtual link request. B^N_V and B^E_V describe the attribute set of virtual node request and the attribute set of virtual link request. B^N_V include the node computing capacity of virtual node request and status requirement, while A^E_S refer to the bandwidth resources of link request. We use Q (G_V, D, t, life, hop) to describe a virtual network request. Where D refers to the maximum limitation distance from a virtual node mapping to a substrate node, t refers to the time that the virtual network request arrived, life is defined to the life time of the virtual network and hop describe as the hop limit of the substrate path.

Figure 1(b) depicts the substrate network, the numbers near the nodes refer to the current computing resources, and the numbers nears the links denote the current bandwidth.

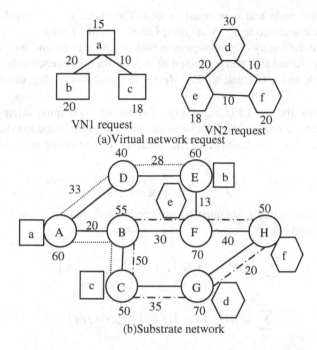

Fig. 1. The virtual network requests mapping scheme

Definition 3: Virtual network mapping problem (VNMP).

Virtual network mapping includes two stages: node mapping and link mapping. The node mapping problem is described as Fn : $(N_V, B^N_V) \rightarrow (N_S, A^N_S)$, where mapping the virtual node to the substrate node in certain constrain. And link mapping problem is defined as Fe : $(E_V, B^E_V) \rightarrow (E_S, A^E_S)$, where mapping the virtual link to substrate path which is meeting the bandwidth constraints.

As shown in figure 1(b), the node mapping scheme of the virtual network request VN1 and VN2 are $\{a \rightarrow A, b \rightarrow E, c \rightarrow C\}$ and $\{d \rightarrow G, e \rightarrow F, f \rightarrow H\}$. While the link mapping scheme are $\{$ (a, b) $\rightarrow \{(A,D), (D,E)\}$, (a, c) $\rightarrow \{(A,B), (B,C)\}\}$ and $\{$ (d, e) $\rightarrow \{(G,C),(C,B),(B,F)\}$,(d, f) $\rightarrow \{(G,H)\}$,(e, f) $\rightarrow \{(F,H)\}$ $\}$. Substrate nodes and links must satisfy the constraints of resources that virtual nodes and links request.

3   The Binary Model of Virtual Network Mapping

Different from paper [10], this paper uses intelligent algorithm to strengthen the correlation between the nodes and links, combining with a binary model of virtual network mapping problem with no assumption that substrate network needs to support path splitting, to reduce the cost of virtual network mapping.

Let *Ma* denote the binary node-mapping matrix with *m* rows and *n* columns. Each row represents one virtual node and *m* is equal to $|N_V|$. Meanwhile, each column represents a substrate node and n is equal to $|N_S|$. Therefore, if the virtual node n^v_i is mapped on the substrate node n^s_j, $Ma(i,j)=1$; Otherwise, $Ma(i,j)=0$.

Let Me denote the binary edge assignment matrix with p rows and q columns. Each row refers to one virtual link and p is equal to $|E_v|$. Furthermore, each column refers to one substrate link and q is equal to $|E_s|$. $Me(i,j)$ is equal to 1 if the substrate link e^s_y $\in P_S$.

For node constraint, the CPU capability of substrate node must satisfy the corresponding virtual node request, and the substrate node must be in the maximum limitation distance that the virtual allowed. The constrain condition of node is defined in (1).

$$\forall i \in N_V, \forall j \in N_S \begin{cases} Ma(i,j) \times cpu(i) \le cpu(j) \\ Ma(i,j) \times Dis(l(i),l(j)) \le D \end{cases} \tag{1}$$

In addition, each virtual node must be assigned to one and only one physical node as defined in (2) and (3).

$$\sum_i^m Ma(i,j) \le 1, j \in \{1,2,\cdots,n\} \tag{2}$$

$$\sum_j^n Ma(i,j) = 1, i \in \{1,2,\cdots,m\} \tag{3}$$

For link mapping, the virtual link e^α_v between virtual node n^i_v and n^j_v are mapped to the path set $P_S(i,j)$ between substrate node n^i_s and n^j_s. The available bandwidth of all substrate links in $P_S(i,j)$ must satisfy the bandwidth request of virtual link e^α_v, the constrain condition as described in (4).

$$Me(i,j) \times bw(e^\alpha_v) \le bw(e^\beta_v)$$
$$Me(i,j) \in \{0,1\}, i \in \{1,2,\cdots,m\}, j \in \{1,2,\cdots,n\} \tag{4}$$

The resources of substrate nodes assigned for virtual nodes is certain for each virtual network request, but the resources of substrate links assigned for virtual links will be different for the choose of path, therefore, we use the bandwidth resources of virtual links assigned by substrate links to measure the cost of mapping. The objective function is the summation of minimized link resources.

$$\min(\sum_{\alpha=1}^p \sum_{\beta=1}^q Me(\alpha,\beta)bw(e^\alpha_v)) \tag{5}$$

4   A Heuristic Virtual Network Mapping Algorithm

4.1   Introduction to BBO

Biogeography-Based Optimization (BBO) is a new heuristic algorithm proposed by Dan Simon in IEEE Transactions on Evolutionary Computation in 2008 [11]. Habitat

Suitability Index (HSI) and Suitable Index Vector (SIV) are defined in the paper [12].

In BBO, the algorithm is composed of n habitat and they realize information exchange between each other by immigration and emigration. The migration of habitat was judged by HSI, the mutation was defined to prevent the algorithm falling into local optimum.

4.2   Definitions and Algorithm

To solve the problem of the virtual network mapping problem, we redefine the habitat characteristic vector, habitat migration rate, immigration operation and emigration operator. The definitions are presented as follows.

Definition 4: Habitat characteristic vector: $x_i=(x_{i1},x_{i2},...,x_{iD})$ was defined as the I possible mapping scheme. Among, D is the number of virtual network nodes in the virtual network request, the value of x_{iD} indicates the substrate node number that the virtual node D mapped.

Definition 5: Habitat migration rate: the immigration λ_i and emigration μ_i are defined as the adjustment strategy of mapping scheme, the defined of immigration and emigration are described in (6). λ_i and μ_i is the number between 0 and 1. The greater of the value of λ_i, the more possible for the mapping scheme to adjust.

$$\lambda_i = I(1 - f(x_i))$$
$$\mu_i = E \times f(x_i) \tag{6}$$

Here
$$f(x) = \frac{\sum\limits_{e_v \in E_V} bw(e_v)}{\min(\sum\limits_{\alpha=1}^{p} \sum\limits_{\beta=1}^{q} Me(\alpha,\beta)bw(e_v^\alpha))} \tag{7}$$

$\sum\limits_{e_v \in E_V} bw(e_v)$ is the summation of link resources that the virtual network requests.

Definition 6: immigration operation and emigration operator: We define x_{ik} as the immigration operation of the habitat vector x_i, meanwhile, x_{jk} is defined as the emigration operator of habitat vector x_j, $i \neq j$, when proceeding the operation of immigration we use x_{jk} to replace the x_{ik} on the corresponding dimension.

The virtual network mapping algorithm based on biogeography-based optimization (VNM-BBO) is briefly described in the following.

We mark f(x) as the suitability function in VNM-BBO, x is defined as the possible mapping scheme.

Step 1 Initialize parameters. Initialize max immigration rate I, max emigration rate E, max mutation rate m_{max}, max iteration times MG. Initialize a random set of mapping scheme.

Step 2 Calculate $f(x_i)$ of each mapping scheme, immigration probability λ and emigration probability μ.

Step 3 Probabilistically use immigration and emigration to modify each mapping scheme.

Step 4 Mutation process. Compute mutation probability mp. Select a mapping scheme with mutation probability mp. Mutate selected mapping scheme with randomly generated SIVs.

Step 5 If immigration, emigration and mutation for population of current iteration are not finished, Go to Step 3.

Step 6 Go to step 2 for the next iteration. Otherwise, output final solutions.

5   Simulation Experiments

5.1   Evaluation Metrics

Two evaluation metrics are employed in our study and some definitions are described first in the following.

$R(G_v)$ is defined as the revenue of substrate network when accepting a virtual network request, including the computing resources and bandwidth resources of the nodes in the virtual network requests.

$$R(G_v) = \sum_{n_v \in N_V} cpu(n_v) + \sum_{e_v \in E_V} bw(e_v) \qquad (8)$$

$C(G_V)$ is defined as the cost of substrate networks when accepting a virtual network request, including the computing resources and bandwidth resources of the substrate nodes allocated for the virtual networks.

$$C(G_V) = \sum_{n_v \in N_V} cpu(n_v) + \sum_{e_v \in E_V} \sum_{e_s \in E_S} bw(e_v) \qquad (9)$$

The main evaluation metrics of the virtual network include:

1)   Average revenue of substrate network

The average revenue of substrate network is defined in (10).

$$\overline{R} = \frac{\sum R(G_V)}{T} \qquad (10)$$

2)   Average cost of substrate network

The average cost of substrate network is defined in (11).

$$\overline{C} = \frac{\sum C(G_V)}{T} \qquad (11)$$

5.2   Simulation Setting

GT-ITM tool is used to generate the substrate network topology and virtual network request. The substrate network topology is configured to own 100 nodes. Each pair of substrate nodes is randomly connected with probability 0.5. The CPU resources at nodes and the link bandwidths at links follow a uniform distribution ranging from 0 to 100.

In a virtual network request, the number of virtual network nodes is randomly determined following a uniform distribution between 2 and 10. The CPU requirements of the virtual nodes are real numbers uniformly distributed between 0 to 20 and the bandwidth requirements of the virtual links are uniformly distributed between 0 and 25. The arrivals of virtual network requests are modeled by a Poisson process with mean 5 reqs. per time window. We run all of our simulations for 10000 time windows, which corresponds to about 100 requests on average in one instance of simulation.

In VNM-BBO, the values of the parameters are set as follows: population size is 15, max iteration times is 10, max immigration rate I is 1, max emigration rate E is 1, and max mutation rate m_{max} is 0.005.

5.3 Discussion

In the following experiments, VNM-BBO is contrasted with greedy node mapping with Shortest Path Based Link Mapping (G-SP), which employs greedy strategy to select nodes with maximum residual resources.

Table1 shows simulation results of average revenue and cost of substrate network, which are both generated by using the VNM-BBO to dispose 100 virtual networks. Wherein, C/R means the ratio of the cost and revenue. The changing curves of revenue and cost are depicted in Fig.2 and Fig.3.

Table 1. Simulation results of revenue and cost of substrate network

VNs	1	2	3	4	5
revenue	1.08E+02	1.22E+02	1.30E+02	1.19E+02	1.14E+02
cost	2.53E+01	1.79E+01	1.68E+01	1.74E+01	1.53E+01
C/R	2.35E-01	1.46E-01	1.29E-01	1.46E-01	1.34E-01
VNs	6	7	8	9	10
revenue	1.30E+02	1.24E+02	1.30E+02	1.31E+02	1.32E+02
cost	1.58E+01	1.69E+01	1.60E+01	1.48E+01	1.65E+01
C/R	1.22E-01	1.37E-01	1.24E-01	1.13E-01	1.25E-01
VNs	11	12	13	14	15
revenue	1.32E+02	1.33E+02	1.27E+02	8.27E+01	5.28E+01
cost	1.65E+01	1.60E+01	1.61E+01	1.08E+01	6.77E+00
C/R	1.25E-01	1.21E-01	1.27E-01	1.31E-01	1.28E-01
VNs	16	17	18	19	20
revenue	4.32E+01	4.18E+01	4.30E+01	4.34E+01	4.03E+01
cost	5.97E+00	6.04E+00	5.71E+00	5.93E+00	5.58E+00

Table 1. (*continued*)

C/R	1.38E-01	1.44E-01	1.33E-01	1.36E-01	1.39E-01
VNs	21	22	23	24	25
revenue	4.02E+01	3.92E+01	3.79E+01	3.83E+01	3.69E+01
cost	5.54E+00	5.45E+00	5.30E+00	5.29E+00	5.41E+00
C/R	1.38E-01	1.39E-01	1.40E-01	1.38E-01	1.47E-01
VNs	26	27	28	29	30
revenue	3.75E+01	3.68E+01	3.74E+01	3.80E+01	3.91E+01
cost	5.42E+00	5.26E+00	5.39E+00	5.25E+00	5.34E+00
C/R	1.45E-01	1.43E-01	1.44E-01	1.38E-01	1.37E-01
VNs	31	32	33	34	35
revenue	4.01E+01	4.05E+01	4.05E+01	4.07E+01	4.16E+01
cost	5.51E+00	5.77E+00	5.85E+00	5.77E+00	5.82E+00
C/R	1.37E-01	1.43E-01	1.44E-01	1.42E-01	1.40E-01
VNs	36	37	38	39	40
revenue	4.20E+01	4.30E+01	4.37E+01	4.46E+01	4.49E+01
cost	5.86E+00	5.80E+00	6.02E+00	6.01E+00	6.34E+00
C/R	1.39E-01	1.35E-01	1.38E-01	1.35E-01	1.41E-01
VNs	41	42	43	44	45
revenue	4.52E+01	4.48E+01	4.58E+01	4.59E+01	4.64E+01
cost	6.35E+00	6.47E+00	6.61E+00	6.55E+00	6.48E+00
C/R	1.40E-01	1.44E-01	1.44E-01	1.43E-01	1.40E-01
VNs	46	47	48	49	50
revenue	4.66E+01	4.64E+01	4.62E+01	4.53E+01	4.45E+01
cost	6.81E+00	6.66E+00	6.65E+00	6.69E+00	6.66E+00
C/R	1.46E-01	1.44E-01	1.44E-01	1.47E-01	1.50E-01
VNs	51	52	53	54	55
revenue	4.44E+01	4.37E+01	4.41E+01	4.41E+01	4.44E+01
cost	6.67E+00	6.56E+00	6.53E+00	6.56E+00	6.60E+00
C/R	1.50E-01	1.50E-01	1.48E-01	1.49E-01	1.49E-01
VNs	56	57	58	59	60
revenue	4.46E+01	4.44E+01	4.47E+01	4.40E+01	4.40E+01
cost	6.66E+00	6.65E+00	6.65E+00	7.11E+00	7.03E+00
C/R	1.49E-01	1.50E-01	1.49E-01	1.62E-01	1.60E-01
VNs	61	62	63	64	65
revenue	4.38E+01	4.43E+01	4.29E+01	4.30E+01	4.21E+01
cost	7.13E+00	6.99E+00	6.75E+00	6.69E+00	7.05E+00

Table 1. (*continued*)

C/R	1.63E-01	1.58E-01	1.58E-01	1.55E-01	1.68E-01
VNs	66	67	68	69	70
revenue	4.10E+01	4.09E+01	4.24E+01	4.23E+01	4.18E+01
cost	6.86E+00	6.82E+00	6.89E+00	6.84E+00	6.78E+00
C/R	1.67E-01	1.67E-01	1.63E-01	1.62E-01	1.62E-01
VNs	71	72	73	74	75
revenue	4.19E+01	4.22E+01	4.24E+01	4.23E+01	4.24E+01
cost	6.86E+00	6.81E+00	6.80E+00	6.80E+00	6.76E+00
C/R	1.64E-01	1.61E-01	1.60E-01	1.61E-01	1.59E-01
VNs	76	77	78	79	80
revenue	4.14E+01	4.07E+01	3.95E+01	3.89E+01	3.86E+01
cost	6.61E+00	6.48E+00	6.30E+00	6.23E+00	6.09E+00
C/R	1.60E-01	1.59E-01	1.59E-01	1.60E-01	1.58E-01
VNs	81	82	83	84	85
revenue	3.87E+01	3.88E+01	3.90E+01	3.91E+01	3.95E+01
cost	6.13E+00	6.14E+00	6.14E+00	6.17E+00	6.19E+00
C/R	1.58E-01	1.58E-01	1.58E-01	1.58E-01	1.57E-01
VNs	86	87	88	89	90
revenue	3.95E+01	4.00E+01	3.97E+01	3.97E+01	3.98E+01
cost	6.30E+00	6.35E+00	6.27E+00	6.27E+00	6.29E+00
C/R	1.59E-01	1.59E-01	1.58E-01	1.58E-01	1.58E-01
VNs	91	92	93	94	95
revenue	3.94E+01	3.93E+01	3.89E+01	3.87E+01	3.86E+01
cost	6.23E+00	6.21E+00	6.12E+00	6.08E+00	6.05E+00
C/R	1.58E-01	1.58E-01	1.58E-01	1.57E-01	1.57E-01
VNs	96	97	98	99	100
revenue	3.88E+01	3.87E+01	3.87E+01	3.87E+01	3.88E+01
cost	6.08E+00	6.07E+00	6.07E+00	6.06E+00	6.05E+00
C/R	1.57E-01	1.57E-01	1.57E-01	1.57E-01	1.56E-01

In Fig. 2, the average revenue of the substrate network is shown in which 100 virtual networks arrives in limited time. The average revenue of the substrate network employing VNM-BBO is also higher than that employing G-SP.

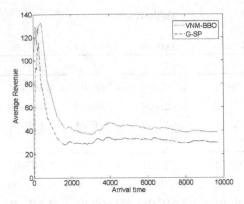

Fig. 2. Revenue of the substrate network

Fig.3 illustrates the average cost of the substrate network, at the beginning, the average cost of G-SP is lower than that of VNM-BBO, with more virtual networks arrive, the cost of the substrate network employing G-SP surpasses the VNM-BBO.

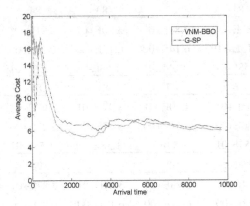

Fig. 3. Cost of the substrate network

6   Conclusions and Future Work

This paper presents a new virtual network mapping algorithm based on biogeography-based optimization. We redefine the habitat characteristic vector, habitat migration rate, immigration operation and emigration operator and define the cost of mapping as the fitness function. Simulation experiments demonstrate that the average revenue of substrate network is competitive in contrast with the G-SP algorithms. The experimental results suggest our approach increases the average operating income of the virtual network and reduces the cost of substrate network comparing with some other two-stage mapping algorithm.

Our work is primary and there are many problems that remain unresolved. In the future, we will continue our work from the following. On one hand, the virtual network mapping algorithm will be elaborated and analyzed. On the other hand, more simulation experiments will be designed and compared with other existing algorithms.

Acknowledgments. We are grateful for the support of the National Natural Science Foundation of China (61373149), the Promotive Research Fund for Excellent Young and Middle-aged Scientists of Shandong Province (BS2010DX033) and a Project of Shandong Province Higher Educational Science and Technology Program (J10LG08).

References

1. Chowdhury, N., Boutaba, R.: Network Virtualization: State of the Art and Research Challenges. IEEE Communications Magazine 47(7), 20–26 (2009)
2. Yu, M., Yi, Y., Rexford, J., et al.: Rethinking Virtual Network Embedding Substrate Support for Path Splitting and Migration. ACM SIGCOMM Computer Communication Review 38(2), 17–29 (2008)
3. Zhu, Y., Ammar, M.: Algorithms for Assigning Substrate Network Resources to Virtual Network Components. In: INFOCOM, pp. 1–12. IEEE, Barcelona (2006)
4. Lu, J., Turner, J.: Efficient Mapping of Virtual Networks onto a Shared Substrate. St.Louis: Department of Computer Science and Engineering, Washington University (2006)
5. Houidi, I., Louati, W., Zeghlache, D.: A Distributed Virtual Network Mapping Algorithm. In: Proc. of the IEEE Int'l Conf. on Communications, pp. 5634–5640 (2008)
6. Lischka, J., Karl, H.: A Virtual Network Mapping Algorithm Based on Subgraph Isomorphism Detection. In: Proc. of the 1st ACM SIGCOMM Workshop on Virtualized Infrastructure Systems and Architectures, pp. 81–88 (2009)
7. Fan, J., Ammar, M.H.: Dynamic Topology Configuration in Service Overlay Networks: A Study of Reconfiguration Policies. In: INFOCOM (2006)
8. Cheng, X., Zhang, Z.B., Su, S., Yang, F.C.: Virtual Network Embedding Based on Particle Swarm Optimization. Acta Electronica Sinica 39(10), 2240–2244 (2011)
9. Zhu, Q., Wang, H.Q., Lv, H.W., Wang, Z.D.: VNE-AFS: Virtual Network Embedding Based on Artificial Fish Swarm. Journal on Communications 33(Z1), 170–177 (2012)
10. Chowdhury, N.M.M.K., Rahman, M.R., Boutaba, B.: ViNEYard: Virtual Network Embedding Algorithms with Coordinated Node and Link Mapping. IEEE/ACM Transactions on Networking 20(1), 206–219 (2012)
11. Simon, D.: Biogeography-based Optimization. IEEE Trans Evolutionary Computation 12(6), 702–713 (2008)
12. Wesche, T., Goertler, G., Hubert, W.: Modified Habitat Suitability Index Model for Brown Trout in Southeastern Wyoming. North Amer J Fisheries Manage 7, 232–237 (1987)

Fault-Tolerant Control, Fault Diagnosis and Recovery in Runtime of Business Docking Service Composition Flow in the Cloud Environment

Jianhua Han[*], Silu He, Hengxin Li, and Jianping Huang

Faculty of Computer Science, Guangdong University of Technology, Guangzhou, China
hanjianhua@gdut.edu.cn, siluh@foxmail.com,
{5555xin,jphuang86}@163.com

Abstract. In the design and remote maintenance of district and city petition business docking platform, loose coupling environment of service composition flow, and autonomy of distributed resources can easily lead to failures of inter-process interaction, the platform running will gradually changes form trouble-free operation to operation with fault, so we design a docking platform with fault-isolation and fault-tolerance control based on services choreography and orchestration, and a remote monitoring system based on fault-diagnosis by fussy inference. More than two years' operation and remote maintenance verify the effectiveness of the fault-tolerant control mode in the process of business docking services composition. Achieve the goal, that the improving of fault-recovery mechanism and the changing of business docking requirements won't affect each other. Accordingly, it reduces the complexity of maintenancing services composition software, provides application value in business docking service mode for service computing and cloud computing, and SaaS-level monitoring and repair method.

Keywords: Fault-tolerant control, Fault diagnosis, Business docking service composition flow, Services choreography and orchestration.

1 Introduction

With the development and application of the Internet technology, the Internet has become a network computing platform with abundant resources because of its extensive connectivity, heterogeneity and self-organization. Innovative service modes, such as service computing and cloud computing, developed on the basis of Services-oriented Architecture which provides standards for the inter-connectivity, inter-communication, and inter-operability of network software, have greatly contributed to service-oriented software engineering [1] by incorporating the open-source software service resources into softwares. While services composition improves the inter-operability among business partners by combining services with greater range and

---

[*] Corresponding author.

D.-S. Huang et al. (Eds.): ICIC 2014, LNAI 8589, pp. 396–408, 2014.

further integrated resources, constant input or output flow [2] of network computing nodes are formed, because business collaborations involving multiple independent processes need long life cycle and peer-to-peer message interaction among participating services automatically. It is easy to lead failure of service interactions, not only because of the association between the service composition process and network execution environment, but also the autonomy of distributed dynamic resource. Consequently, it's necessary for service users to measure and control quality features of software , and enhance fault-tolerant control of service composition to realize the accurate fault localization and diagnosis along with self-repairing of failure entities , so as to make operate entity continuous run and improve under the cloud environment, and meet constant changing business requirement for the sake of sustainable development [3] under the constant changing external environment, when service composition software's design, operation and remote-maintenance.

2   Relevant Researches

As to set up web service system with fault-tolerance, reference [4] puts forward a method to enhance the reliability of service composition taking advantage of software fault-tolerant mode. Reference [5] has designed a fault-tolerant framework in engine performance to evaluate fault conditions according to service response time so as to choose different solutions for fault tolerance. However, various elements affect the normal operation of the service and its handling strategy has a weak pertinence.

As to monitor and repair service composition software, reference [6-7] has made a research on runtime monitoring and recovery services composition's architecture. However, they have to pre-select the monitoring point and prepare repair behavior in advance, but it's so difficult to predict faults in the future, and to adapt to the changing business needs of services composition when operating. Reference [8] gives the monitoring and recovery architecture of different levels (SaaS, PaaS, IaaS) service models in cloud computing environment, yet it does not mention the solution of monitoring and recovery.

In terms of the problems that networked software (service composition) faces on design and application, this article adopts fault-tolerant models of business docking web service composition with fault isolation and fault-tolerant control in petition businesses docking platform design according to the requirement of petition business docking between the district and city systems, leading to a platform in application being able to skip fault process by fault isolation so as not to affect other petition letters business automatic docking (advancing) flow just because of interaction failure of certain petition letter business docking flow; during the platform runtime and the process of remote maintenance, the fault diagnosis and remote monitoring system which is based on fuzzy reasoning was designed, in order to separate the processing logic of business service composition process and fault recovery mechanism, and to eliminate the deficiencies caused by intruding troubleshooting into[6-7] service composition process. By constantly improving fault diagnosis and repair abnormal process, and by enhancing the autonomic computing functions on self-localization, self-diagnosis and self-recovery, the complexity of continuingly operating the platform can be greatly reduced.

3   The Models of Services Choreography and Orchestration of the District and City Petition Businesses Docking Platform

3.1   The Model of Services Choreography of the District and City Petition Businesses Docking

The petition system was put into use in the district government since June, 2005. In the inter-connection project between the district and city petition systems at the beginning of 2012, one of these business demands is to deal with the skip-level petitions (which are called transferred petitions or just petitions for short in this article) of the city system by the district system.

The services choreography model of the district and city petition business docking is shown in Fig. 1 (the middle part) where the information interaction sequence (described from a district-operation point of view) agreed upon by the district and city systems includes: reading the list of transferred petitions, reading the data of transferred petitions and their attachments, confirming the reception of transferred petitions, turning back transferred petitions, registering accepted and disposed transferred petitions, reporting to the city petition system the treatment results of transferred petitions and their relevant attachments. Each interaction procedure in the interaction sequence possesses the request and response behavior. The interaction operation appears to be exclusive between turning back transferred petitions and accepting and disposing transferred petitions. That is, each petition treatment instance will normally terminate when it has implemented the message exchange between city and district systems of the turning back transferred petitions or reporting to the city system the information of disposed petition cases.

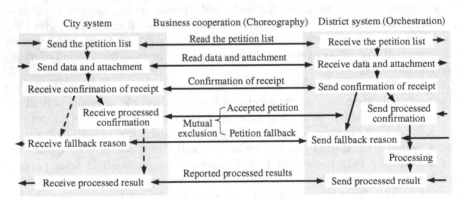

Fig. 1. The flow model of services choreography and orchestration in the district and city business docking

3.2   Service Choreography and Orchestration Methods of the District and City Businesses Automatic Docking Flow

Services choreography and orchestration offer an approach to design web services composition from whole to parts. Choreography defines the sequence [9] of interactions between cooperating web services (global-based interaction rules), and then the global rules are mapped to participants' local interaction rules (the process of each participant partner is formulated through services orchestration). When some active driving points of the processes in the parallel executions of all participants' handling processes bring about interactions between participants, a business cooperation model (similar to the one shown in Fig. 1) for composite services from a global view is established.

In district petition system, the flow routing and host departments are constituted by the petitions handling process, both of which are in dynamic change. Each petition letter processing instance may be handled by N different departments. From petitions acceptance to their settlements, the process may take days or even months. During the docking process, there are at least 3 to 4 times interactions between the district and city (receive petitions and attachments, accept or rollback, report closing results and attachments), rendering the service composition process of dynamic change, multiple participants and long life-cycle. It's difficult to supervise each process operation and locate fault points of anomalous instances when random various process messages exchange between multiple petition business flow and the city system. From this point of view, Java language has realized the district and city business automatic docking platform based on web services via service choreography and orchestration method on the strength of interaction protocol state machine, which effectively controls all operations of business docking process instances between district and city from a global view.

3.3   The Operating Principle of the Interaction Protocol State Machine

Interaction protocol state machine drives service combination by using business processes [10], which realizes automatic docking of the district and city petitions business by combining message exchange sequence (services choreography) with interaction rules (services orchestration, shown in Fig. 1) of the district system.

Interaction protocol is the logical realization of data exchange function completely accomplished between the district and city systems.

To improve the scalability of dynamic work flow of the district system and simplify the design of local interaction rules, the activity participated by department in charge ought to be chosen as the driving point of interaction protocol, or transfer some driving points to active points participated by department in charge. The process of the local virtual interaction rules of the district system with only one participant is expressed in the Petri nets in Fig. 2 [11]. Every active driving point in the figure corresponds to a protocol state value relative to the processing progress of transferred district's petition cases. With the progressing of the petition treatment, the active driving point which reflects the course of the interaction protocol change and the corresponding state value will also change, thus a virtual process composed of active driving points is formed, which reflects the advancement of transferred district's petitions is consistent with the progressing process of the interaction protocols, bearing the conditions of protocol change.

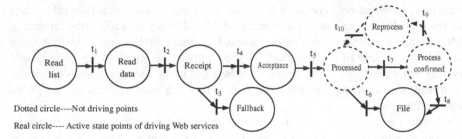

Fig. 2. The progressing process of the local virtual interaction rules in the district system

As is shown in Fig. 2, the interaction protocol state machine takes the competent department of district system as the participant, to progress the interaction protocol sequence of both cooperative parties which based on the local virtual interaction rules. And the inference rules, that used to progress process, is described in a triple [11]:

$$IR_i = (Spre_i , T_i , Spost_i) \tag{1}$$

If interaction protocol sequence is progressed as the sequence structure shown in Fig. 2, there is the relationship as $Spost_i = Spre_{i+1}$ between IR_i and IR_{i+1}. In the triple, $Spre_i$ symbolizes the pre-state value of the current active point, T_i the conditions to be met when the change from $Spre_i$ to $Spost_i$ happens (in Fig. 2, each active driving point and protocol state value only represent the order and procedures of progressing inte-raction protocol, some processing activities that haven't been marked between the driving point and its follow-up, is needed to meet the   conditions of driving $Spost_i$ interaction protocols), $Spost_i$ the post-state value of the active points; if $Spre_i$ is void of subsequent interaction protocols, $Spost_i$ is null, and the state is terminated.

The protocol state machine monitors the processing progress of every transferred petition instance in distric system and maps it onto the corresponding active driving point of the protocol state value; when it also meets the pre-state value of the current active point called $Spre_i$ (implying that $Spre_i$ interactive operation has been imple-mented) and the T_i vicissitude condition of driving the $Spost_i$ protocol, the inter-operation of the $Spost_i$ protocol is implemented (according to the virtual flow struc-ture of Fig. 2, the driving point of $Spre_i$ changes to the subsequent driving points of $Spost_i$). In this way, the interactive operation of every transferred petition instance is progressed according to inference rules in Eq.1, until the end of the task. The dotted circles in Fig. 2 is the processing activity of common petitions and important petitions before the case closed.

4   Fault-Tolerant Model of Service Composition Flow in the District and City Petition Business Docking

4.1   Reliability Analysis of Business Docking Service Composition Flow

Reliability is mathematically defined to an exponential distribution: $R = e^{-\lambda t}$, in which λ is failure rate, and t represents continuous running time but discrete call times to software [12].

The reliability of service composition flow needs to allow for web service functional reliability R_f, service binding reliability R_b, service connecting reliability R_c, and process logic reliability R_p in service composition flow realized by service users.

In the business docking platform, the reliability of process binding belongs to web service offered by the city system, if without wrong binding, $R_b=1$. Reliability of business docking process can be described as

$$R = R_f R_c R_p \tag{2}$$

Therefore, in the execution of business docking process, it will record information of three aspects : the (interactive) abnormal log of calling services, the process logs of the business docking platform, and the functions execution log of city system.

The exception of business docking service composition when running, is been classified as Table 1.

4.2   Fault-Tolerant Modes and Handling Strategies of Business Docking Flow

Software fault tolerance is the key technology to ensure the reliability and availability of the system through redundancy, so as to work normally even if system partial failure happens. Usually there are three fault-tolerant modes [4]:

Recovery block mode, N version program design mode, 1-out-of-N mode.

Four types of strategies are likely to be adopted when there is anomalous service composition: Retry, skip, reconfiguration (substitution), partial reconfiguration.

Fault-tolerant control is closely associated with fault diagnosis that the former can provide early-warning for the latter by logs of fault-tolerant control behavior, and offer important information of process instance's bug symptom for remote maintenance in time. Appropriate fault-tolerant mode and handling strategies can raise fault-tolerance and self-recovery of business docking process. Accordingly, we should pay attention to the following rules in designing each interaction protocol of the district and city business docking flow:

(1) Recovery block mode and skip processing strategy to achieve transactional integrity of interaction protocols: It is essential for protocol design to do with service atomicity (basic service), so as to lessen (decoupling) associations between information and operation (such as separate basic information and attachments of petitions). When there is any abnormities of either interaction part, record the relational abnormal logs, abandon the current instances, achieve similar operations as transaction processing, and make the process returns to the state before executing this interaction protocol, namely using skip processing strategy to achieve the fault-tolerant control of recovery block mode.

(2) Retry after interaction protocol failure: Because of the hardware temporary abnormity (such as full database space, server overload) or unstable operating environment (such as call service failure, network delay), the platform abandons the current operation when the executing certain interaction protocol goes wrong, polling mode can be used to retry this protocol after a period of time which means, using the retry processing strategy to achieve improved 1-out-of-N mode fault-tolerant control.

Table 1. The anomaly in business docking process

Anomaly classification	Fault causes	Fault involving cooperative parties	Handling strategies
Anomalous Web services	Incompatible service interface before and after the change	√	①Skip、②Retry
	Unavailable service	√	①、②
	Anomalous service implementation		①、②
	Anomalous transaction	√	①、③Partial reconfiguration
Anomalous service composition	Interaction protocol logic error		①、②
	Service interface parameter anomaly		①、②、③
	Unmatched data forma		①、②
Anomalous communication	Network connection failure	√	①、②
	Network delay		①、②
Anomalous hardware	Server fault of services providers	√	①、②
	Server fault of service composition users		①、②
	Insufficient data space		①、②

4.3   Fault Localization and Isolation of Business Docking Service Composition Process

(1) Polling and retry of automatic progressing interaction protocol sequence: Protocol state machine will touch off the execution of the docking process in accordance with specified intervals: First, the network is monitored, and then detect processing schedule T_i of every transferred petition instance and business docking process schedule $Spre_i$ to advance subsequent interaction protocol meeting Eq. 1, until the end of docking or the consistency of process instances progress and business docking. If the network is abnormal, trigger this mission after a period time.

(2) Fault localization: Inference rules of business docking flow (such as Eq. 1) are that interaction operation of $Spost_i$ protocol will be implemented if process instances of certain transferred petition meet the condition of $Spre_i \land T_i$, which reflects the protocol state value of progressing order being changed into $Spost_i$ from $Spre_i$. However, if operation has to quit owing to the fault (that is fault-tolerant control of recovery block mode by skip) when performing $Spost_i$ agreement, the state value of process instance is remained in $Spre_i$ (in Eq. 3), interaction protocol corresponding to state value $Spost_i$ ($Spre_{i+1}$) is decided as fault point by what is described (conditions) in Eq. 3, 4, 5.

$$Spre_i \land T_i = \text{true, but the protocol state value is still } Spre_i. \tag{3}$$

$$T_{i+1} = \text{true,} \quad Spre_{i+1} \neq Spost_i. \tag{4}$$

Fault detecting time > fresh schedule accomplished time of transferred petition + intervals of polling implementation interaction protocol. $\hspace{2em}$ (5)

(3) Fault isolation: As Eq. 1 describes, inference rules not only restrict the process order of business docking flows, but also play an important role in fault isolation instances. If a process instance still stays in a certain interaction protocol process of business docking process (in Eq. 3) because of faults, and it meets the transferred condition of subsequent interaction protocol, $T_{i+1} = \text{true}$, but does not meet its combination

condition, $Spre_{i+1} \wedge T_{i+1} \neq true$ (in Eq. 6). Subsequent interaction will not be promoted if precursor interaction protocol have not ran , protocol state machine will skip over process instances with fault point automatically refer to Eq.3 and 6.

$$T_{i+1} = true, \quad Spre_{i+1} \wedge T_{i+1} \neq true. \tag{6}$$

The model of fault-tolerant control process of interaction protocol based on Petri nets is shown in Fig. 3 and Table 2.

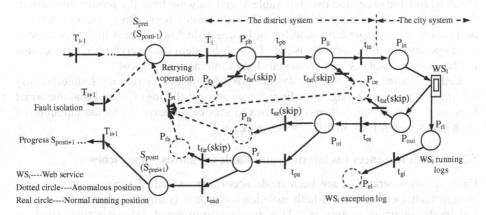

Fig. 3. The fault-tolerant control process model of the interaction protocol

Table 2. Transitions and places in interaction protocol fault-tolerant control process model

Places	Position	Transitions	Operation
S_{prei}	Protocol i pre-state	T_i	Protocol i operation
P_{pb}	Front protocol parameter	t_{pb}	Protocol front operation
P_{fa}	Fault points	t_{far}	Anomalous runtime (skip)
P_{ii}	WS_i requests service message	t_{in}	Runtime services
P_{in}	Input parameters	t_{es}	End of service
P_{out}	WS_i response message	t_{se}	Anomalous services (skip)
P_{oi}	Return results	t_{pa}	Protocol subsequent operation
P_r	Process results	t_{end}	End of protocol
P_{ce}	Communication faults	t_{fat}	Fail transmission (skip)
P_{rl}	Running logs	t_{gl}	Access logs
P_{el}	Exception log	t_{rt}	Retry or time out
S_{posti}	Protocol i post-state	T_{i+1}	Protocol i+1 operation

5   Fault Diagnosis Based on Fuzzy Inference and Remote Monitoring System

Fault diagnosis based on fuzzy inference and remote monitoring system are comprised of remote monitoring system with monitoring platform running and recovering fault, and remote testing system of petition businesses interface, whose supervision object is all business docking service composition process instances.

As is shown in Fig. 4, fault-tolerant control, fault diagnosis and recovery structure in the district and city business docking flow can be classified as monitoring, analysis, decision and self-recovery. In this case, the business docking platform with fault-free operation is made to realize the self-localization and self-isolation of fault by automatic computing [13], as well as fault diagnosis and recovery by monitoring system.

5.1   Monitoring Modes of Business Docking Platform Operating Status

By analyzing the exception listed in Table 1, and viewing from the service interaction, the combination service process logic and the execution logs of city system's service, on-line monitors progressing state log of business docking process instances in platform operation, calling anomalous log of the city system (web services) interface, also implement operation log and anomalous log of interaction protocol.

Petition business interface test system provides means for remote troubleshooting when interface goes wrong, including: detect and get the logs (normal and abnormal logs) of the city system, when it responses to service requests; detect the input/output changes of Web service interface when docking.

5.2   Fault Instances Localization and Fault Diagnosis Approaches

Fault recovery strategies are been made according to fault diagnosis consequences though fault localization and fault isolation have been realized by fault-tolerant control mode (referring to Section 4.2) of interaction protocol, inference rules (such as Eq.1) of interaction protocol sequence and timeout mechanism of interaction protocol advancing schedule (such as Eq. 3, 4, 5, 6).

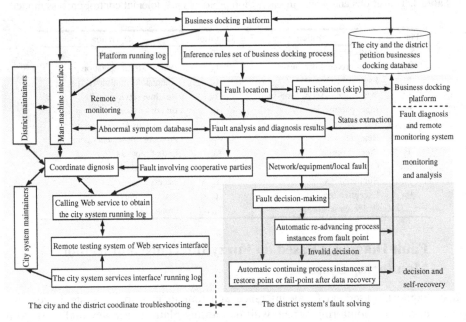

Fig. 4. Fault-tolerant control, fault diagnosis and recovery structure diagram in the district and city business docking flow

We can draw a conclusion from Table 1, it is unreasonable to deal with anomalies by using traditional two-value logic, because, those anomalous symptoms of platform operation log and fault causes corresponding to different fault points in protocol implementation are fuzzy with uncertainty and incomplete information.

More than two years' platform operation and remote maintenance (At the end of July 2013, the district petition system, the district and city petition docking platform, remote monitoring system were moved to operate in the cloud environment) have enabled us to collect more than 20 kinds of symptom information from platform's anomalous logs and the anomalous logs of interaction between city system and local district. We summarize the symptoms and causes of the failure (shown in Table 1) by experience from fault diagnosis, fault troubleshooting. We are still improving the method of fault diagnosis based on fuzzy reasoning [14] , and providing supplementary troubleshooting information to the remote maintenance.

Suppose there are m kinds of key anomalous symptoms of business docking process, described as x_i (i=1, 2, ..., m), and n possible fault causes, described as y_j (j=1, 2, ..., n). Then causality matrix C between key anomalous symptom information and fault causes is a m × n matrix:

$$C = \begin{bmatrix} c_{11} & c_{12} & \cdots & c_{1n} \\ c_{21} & c_{22} & \cdots & c_{2n} \\ \vdots & \vdots & \vdots & \vdots \\ c_{m1} & c_{m2} & \cdots & c_{mn} \end{bmatrix} = \left(c_{ij} \right)_{m \times n} \tag{7}$$

c_{ij} represents membership of ith anomalous symptom belonging to j^{th} fault cause, $0 \leq c_{ij} \leq 1$, i=1, 2, ..., m, j=1, 2, ..., n.

Fuzzy relational equation of key anomalous symptom info and fault causes is:

$$Y = X \circ C \tag{8}$$

X is fuzzy vector of key anomalous symptom information; Y is fuzzy vector of fault causes; $\circ$ is fuzzy operator.

The remote off-line fault diagnosis method in the system is listed as follows:

(1) Check progressing state log of every transferred petition in on-line docking process, to failure monitoring by Eq. 3,4,5 with the goal of accurately locate the fail-point (interaction protocol) of fault flow.

(2) According to abnormal classification listed in table 1, starting from the completion time (Assuming it is t) of interaction protocol without fault before fault flow, the system searches the anomaly information of procedure-related and time-related from the abnormal service interaction logs after time t, abnormal protocol processing logic logs, and web service function logs of city system.

(3) Take abnormal information (as string), which was got from step (2), with each value of anomalous symptom vector X (as keywords or sub-string), check their match case until each value of X has been checked.

(4) One or more membership fault causes can be found by Eq. 8, that can narrow the scope of troubleshooting; if values in vector X do not match the symptom value of

anomaly information, the causality matrix C (sparse arrays) and the anomalous symptom matrix X can be replenished by the troubleshooting results and constantly improve the fuzzy reasoning algorithm.

5.3   Troubleshooting and Fault Decision-Making

Abnormities listed in Table 1 show that some faults are temporary (such as network delay) while some need manual checking (such as software bugs, hardware faults). The causes of those faults involving coordinating parities, or invalid process instances (n faults) in which the faults got corrected and failure process repaired, will continue to be solved with the help of city maintainers. Processing strategy will be implemented in view of the fault's effect on failure process after it is been cleared:

(1) Promoting failure caused by abnormity, but the process has not change the docking state of city, district business, the retry strategy ("②" in Table 1) can be applied for self-repairing.

(2) The failure process that affect the transactional integrity of the city, district business docking, can be restored by partial reconfiguration strategy ("③" in Table 1), namely, give up the protocol operation that has already been done, return to the former status point which the collaboration parts are consistent with each other, and re-execute the interaction protocol which has been troubleshoot and modified to reach the original target state point.

5.4   System's Self-Repairing Principle and Its Applications

There are always certain faults listed in Table 1 after the platform has ran for over two years. Some abnormal process needed to repair instances by using partial reconfiguration strategy ("③" in Table 1). Such as:

(1) Protocol of turning back transferred petitions: As a result of successful responses of service suppliers though sending parameters including empty turn-back causes, Therefore, the district and city systems should be restored to the former protocol status without fault, and then the district system will re-execute the protocol of turning back transferred petitions.

(2) Reporting to the city petition system the treatment result of transferred petition and its relevant attachments: when the treatment result of transferred petition and its relevant attachments are reported to the city petition system, it happened that the attachment is too large to upload totally. However, the corresponding result of service supplier is successful. After the program bug has been modified, the district system will be responsible for submitting the attachment only.

While other anomalies in process implementation are able to realize self-repairing just by retry mechanism ("②" in Table 1) after the fault are eliminated. Such as:

(1) Business docking process instance (long life cycle) which executes before and after the changes of service interface, will go wrong because of the incompatibility of the service interface changes. After modifying the agreement which causes incompatibility, retry strategy can help it start from the failure point (protocol status points) to restore the abnormal process instance.

(2) Abnormalities can be caused by full database space, communication error, and hardware problems. In this case, retry strategy can be adopted to resume execution process instance after troubleshooting.

When some bugs are hard to troubleshoot, we will get the function logs of the city system; if the causes are still difficult to analyze and troubleshoot from the city system's logs, we will coordinate with the city system maintenance personnel.

6   Conclusions

Aiming at the stable operation of the district and city petitions businesses docking service composition flow, business docking platform based on services choreography and orchestration came into being with functions of fault isolation and fault-tolerant control, so that self-localization and self-isolation of fault flow can be realized during the platform's operation to promote the petition business docking flow. Also, a system for fault diagnosis and remote monitoring came into being based on the fuzzy inference. With the help of fault localization and analysis of its causes, the monitoring system can go through a recovery process based on the interaction protocol granularity. The experience of platform's continuous and stable operation and remote maintenance for over two year shows that, the automatic business docking service combination for fault-tolerance control mode and SaaS-level monitoring and repair method, which is mentioned in this paper is far more competitive than the service combination frame and monitoring method mentioned in reference [6-7].

(1) The automatic business docking platform combines the advantages of the service choreography (an overall view) and the service orchestration (partial view). When the processing mode of automatic business docking service integrates with the mode of fault-tolerance control, both the interactive orders determined by both collaborations and the requirements of promoting the district system's service can be taken into account. As a result, by monitoring all the online service composition processes which differ in automatic docking progress, it is easier to realize the automatic localization and isolation of failure process, as well as the adoption of skip and retry strategies when abnormality happens in this process.

(2) The business logic of the docking service composition flow and fault recovery can be separated, so that when design business logic , we only need to record relevant logs (the business automatic docking log and the information of fault-tolerant control behavior), and don't need to consider possible exception types and strategies for fault recovery. Owing to the performance of business docking platform and actual results of troubleshooting, the remote monitoring system can aim at improving the methods of fault diagnosis and recovery. Besides, business docking demands become variable while fault diagnosis and fault recovery mechanism get improved, and won't affect each other.

At the same time, the research and practical work in this paper have provided application value in business docking service mode for service computing and cloud computing, and SaaS-level monitoring and repair method.

Acknowledgements. This work was supported by the National Natural Science Foundation of China (No. 61142012). Thanks for the assistance and support by the staff from Guangzhou South Talent Information Industry Co., Ltd during the remote maintenance of the district and city petition business docking platform.

References

1. Li, D.Y.: Requirement Engineering—a Basic Study on Software Engineering of Complex Systems. China Basic Science (2), 4–8 (2009)
2. Li, D.Y., Zhang, H.S.: Internet Computation Beyond Turing Machines. Communication of the China Computer Federation(CCFF) 5(12), 8–16 (2009)
3. Zhang, L., Bai, X.Y., Wu, J.: Quality Outlook of Networked Software. Communication of the China Computer Federation(CCFF) 5(12), 36–43 (2009)
4. Zhong, D.H., Qi, Z.C.: Using Software Fault Tolerance to Increase the Reliability of Web Service Composition. Computer Engineering & Science 30(3), 28–31 (2008)
5. Zou, F., Gao, C.M.: Fault-tolerant Framework at Runtime of Web Service Composition. Computer Engineering 34(18), 89–92 (2008)
6. Huang, T., Wu, G.Q., Wei, J.: Runtime Monitoring Composite Web Services through Stateful Aspect Extension. Journal of Computer Science and Technology 24(2), 294–308 (2009)
7. Liu, A., Li, Q., Huang, L.S., Xiao, M.J.: Facts: a Framework for Fault-tolerant Composition of Transactional Web Services. IEEE Transactions on Services Computing 3(1), 46–59 (2010)
8. Wu, Y.G., Ren, J.L.: Monitoring and Recovery of System Security in Cloud Computing. Communication of the China Computer Federation(CCFF) 8(7), 8–14 (2012)
9. Papazoglou, M.P.: The Principle and Technology of Web Services. China Machine Press, Beijing (2010); translated by Gong, L., Zhang, Y.T.
10. Li, B., Huang, Y.F.: Demand Service Network Software Development. Communication of the China Computer Federation(CCFF) 5(12), 17–26 (2009)
11. Han, J.H., Luo, Y., Huang, J.P.: A study on the Scalable Flow Model of Web Services Choreography and Orchestration based on Dynamic Workflow. International Journal of Information and Communication Technology 5(3/4), 307–320 (2013)
12. Wang, L.J., Bai, X.Y., Chen, Y.N., Zhou, L.Z.: Data-driven Approach to Dynamic Reliability Evaluation of SOA Applications. Tsinghua University (Science & Technology) 49(10), 1729–1732, 1736 (2009)
13. Liao, B.S., Li, S.J., Yao, Y., Gao, J.: Conceptual Model and Realization Methods of Autonomic Computing. Journal of Software 19(4), 779–802 (2008)
14. Hu, L.M., Cao, K.Q., Xu, H.J., Dong, X.M.: Fault Diagnosis and Control Technology Based on Support Vector Machine. National Defence Industry Press, Beijing (2011)

Ambiguous Proximity Distribution

Quanquan Wang[1,2,3,*] and Yongping Li[1]

[1] Shanghai Institute of Applied Physics,
Chinese Academy of Sciences, Shanghai, 201800, China
[2] University of Chinese Academy of Sciences, Bejing 100049, China
[3] Provincial Key Laboratory for Computer Information Processing Technology,
Soochow University, Suzhou 215006, China
{wangquanquan,ypli}@sinap.ac.cn

Abstract. Proximity Distribution Kernel is an effective method for bag-of-featues based image representation. In this paper, we investigate the soft assignment of visual words to image features for proximity distribution. Visual word contribution function is proposed to model ambiguous proximity distributions. Three ambiguous proximity distributions is developed by three ambiguous contribution functions. The experiments are conducted on both classification and retrieval of medical image data sets. The results show that the performance of the proposed methods, Proximity Distribution Kernel (PDK), is better or comparable to the state-of-the-art bag-of-features based image representation methods.

Keywords: Bag-of-Features, Proximity Distribution Kernel, Soft Assignment.

1   Introduction

Due to the rapid growth of modern digital imaging and video technologies, the volume of visual data is also increasing significantly. The management and retrieval of visual information has attracted much attention of the research community of computer vision, machine learning, and scientific computing [1-10]. Recently, Content-Based Image Retrieval (CBIR) has been one of the most popular research areas [11–14, 63, 64]. Given an query image, this problem is to retrieval images from a large image database based on the visual similarity between the query and the database images. In this case, representing the visual information of each images as some feature measures are critical [15–20]. Most often, both global and local visual information are important. Cognitive science community has shed some light on this topic, in particular in the ground laying work of Kong [21, 22] on how human visual cognition processes global and local visual information and utilize both type of information to solve complex visual problems. Kong's work on global/local visual information processing has not only made great impact in the field of cognitive science, but also inspired many advances outside cognitive science, such as in the one discussed in this

---

* Corresponding author.

D.-S. Huang et al. (Eds.): ICIC 2014, LNAI 8589, pp. 409–421, 2014.

paper [21, 22]. J. Yang et al. [23] presented a novel approach for foreground objects removal while ensuring structure coherence and texture consistency. The approach used structure as a guidance to complete the remaining scene. The work benefits a wide range of applications especially for the online massive collections of imagery, such as photo localization and scene reconstructions. Moreover, this work is applied to privacy protection by removing people from the scene. Inspired by Kong's work [21, 22], various image representation methods have been proposed. Among them, local visual descriptors has been shown to be outperforming other visual methods [24–28]. This kind of methods are also called bag-of-features since each images is represented as a collection of   local features [29–31].

This work presents a novel content-based   image retrieval system, which is based on the Visual Words (VW) framework. VW framework is a recently introduced framework, and it has been successfully applied to scenery and object image retrieval tasks [32, 33, 15]. The VW model represents each image using a discrete and disjunct visual vocabulary which is composed of   many visual word elements [34–37]. Base on this model, it is possible to split an image into a set of visual features and to represent the image using the statistics of these local visual features. The ideal intrinsic features should be transition, rotation, and scaling invariant [38, 39]. However, in many applications, it is hard to find the ideal features. In our system the visual features are the key points with SIFT (Scale Invariant Feature Transform) descriptors [40–43], or image patches (small sub images) [44–47]. One effective statistical image presentation of VM model is Proximity Distribution (PD) [48, 49]. Moreover, a corresponding kernel function is also proposed to match a pair of proximity distributions. The PD is defined as the distributions of co-occurring local features as they appear at increasing distances from one another. One inherent procedure of the PD model is the discretization of visual words from continuous image local features.

In this study we investigate the effect of ambiguity modeling on ambiguous proximity model. To the best of our knowledge, there are no published studies on the application of visual word ambiguity in constructing proximity distribution. The main contribution of this paper is an investigation of visual word ambiguity leading to explicit *Visual Word Contribution Function* for the *Ambiguous Proximity Distribution* model. Moreover, we try to develop a kernel to match two Ambiguous Proximity Distribution, so that it can be used in image retrieval.

The paper is organized as follows: In section 2, we introduce the proposed Ambiguous Proximity Distribution Kernel. Section 2.1 discusses the original Proximity Distribution Kernel model. Section 2.2 introduces the Visual Word Contribution Function for generalization of hard and soft assignment of a local feature to a word in the dictionary, and by developing three ambiguous contribution functions, we proposed three *Ambiguous Proximity Distributions* in section 2.3. Section 3 presents the experimental results and Section 4 concludes with the final remarks.

2   Ambiguous Proximity Distribution Kernel

Let X be an image and $\{x_l\}$; $l = 1, \cdots\cdots, L$ a collection of L local features extracted from X. Typically, the local features or regions x_l are detected interest regions with SIFT descriptors [40], or densely sampled image patches [50]. In the learning phase, we construct a codebook V using a clustering algorithm. Usually, k-means is used to cluster centers of features which extracted from all images in the database, then, these cluster centers are used as a vocabulary (codebook) $V = \{v1, \cdots, vK\}$ with K visual words for all images to get word vector representation [51–53]. On the set of all available features of query image X, we perform vector-quantization, to arrive at K codebook elements $V = \{v1, \cdots, vK\}$. So, in this first coding stage, the image X is represented by the local feature $\{(x_l, \alpha_l)\}$ $k = 1, \cdots, L$

$$\alpha_l = \arg \min_i (D(v_i, x_l)) \tag{1}$$

where x_l is image region l, and $D(vi, xl)$ is the distance between a codeword v_i and region x_l.

Given a vocabulary of K codewords, and the traditional VW approach describes an image X by a distribution over the visual words. For each word v_i in the vocabulary V, the traditional VW model estimates the distribution of visual words in an image by

$$H_{CB}^X(i) = \frac{1}{L} \sum_{l=1}^{L} I(\alpha_l = i) \tag{2}$$

The indicator $I(x)$ outputs 1 when the Boolean variable x is true and 0 otherwise. By applying $I(\alpha_l = i)$, the region x_l is only assigned to the nearest word v_i in the dictionary. The VW model represents an image by a histogram of word frequencies that describes the probability density over codewords.

2.1   Proximity Distribution Kernel

The *proximity distribution kernel*(PDK) is proposed by Ling and Soatto in [54], which matches distributions of co-occurring local features as they appear at increasing distances from one another in an image X. In this case, each feature point x_l is first mapped to one of K discrete visual words $v_i \in V$, which are the prototypical local features identified via clustering on local features of many training images.

Each local feature set $\{x_l, \alpha_l\}$, $l = 1, \cdots, L$ of X is converted to a $K \times K \times R$-dimensional histogram. Specifically, for a given image input represented as a local feature set X, each histogram element $H^X(i, j, r)$ is defined as the number of times visual word type j occurs within the r spatially nearest neighbors of a visual word of type i. The proximity distribution of X is given as

$$H_{PD}^X(i, j, r) = \sum_{l=1}^{n} \sum_{m=1}^{n} I(\alpha_l = i) I(\alpha_m = j) I(d_{NN}(x_l, x_m) \le r) \tag{3}$$

where $d_{NN}(x_l, x_m) \le r$ indicates that x_m is within the r- th nearest neighbors of x_l. R is the size of the neighborhood. Since $r = 1, \cdots, R$, apparently it is a cumulative distribution of the cooccurring pairs of words. The (unnormalized) proximity distribution kernel (PDK) value between two images with feature sets Y and Z is then

$$K_{PDK}(Y,Z) = \sum_{i=1}^{K}\sum_{j=1}^{K}\sum_{r=1}^{R} \min(H^Y(i,j,r), H^Z(i,j,r)) \qquad (4)$$

where H_Y and H_Z are the associated arrays of histograms computed for the two input feature sets for image Y and Z.

2.2   Visual Word Contribution Function

In this section, we generalize the bag-of-features based image presentation as a visual word contribution form. First of all, for a region x_l in an image X, we define it's contribution in the constructing of an statistical presentation (histogram or proximity distribution) as *Visual Word Contribution Function* as its contribution in accumulating the i-th bin as $F(v_i, x_l)$. Then we consider the Visual Word Contribution for Proximity Distribution H_{PD}. Apparently, (3) can be rewritten as

$$H_{PD}^X(i,j,r) = \sum_{l=1}^{n}\sum_{m=1}^{n} F(v_i, x_1)F(v_j, x_m)I(d_{NN}(x_1, x_m) \le r) \qquad (5)$$

In the above histogram and proximity distribution, one inherent component of the *hard assignment* for codebook model is the assignment of image local feature to visual words in the vocabulary by using *Hard Assignment Contribution Function* as $F_{hard}(v_i, x_l)$. Here, an important assumption is that a discrete visual word is a characteristics representative of a continuous image feature.

2.3   Ambiguous Proximity Distribution

The continuous nature of local visual features indicates selecting a most representative visual word from a visual vocabulary. It is possible that one local feature can have zero, one, or multiple optimal visual words in a visual vocabulary [34]. If there is only one optimal visual word, there is no ambiguity. We propose a robust alternative method of discrete proximity distribution to estimate a probability density function, which is based on kernel density estimation [34].We call it *Ambiguous Proximity Distribution*. By giving three different ambiguous contribution function to replace the hard contribution function F_{hard}, we developed three kinds of Ambiguous Proximity Distributions (APD) as follows:

Kernel Proximity Distribution. In the VW model, the histogram estimation function of the visual words may be replaced by a kernel density estimation function. Moreover, a suitable kernel is used, and it allows kernel density estimation to become a part of the visual word. With this function, the *Kernel Proximity Distribution* is given as,

$$H_{ker}^X(i,j,r) = \sum_{l=1}^{L}\sum_{m=1}^{L}\{K_\sigma(D(v_i, x_1))K_\sigma(D(v_j, x_m))I(d_{NN}(x_1, x_m) \le r)\} \qquad (6)$$

where $F_{ker}(v_i, x_1)$ is the *Kernel Contribution Function*, as follows,

$$F_{ker}(v_i, x_1) = K_\sigma(D(v_i, x_1)) = \frac{1}{\sqrt{2\pi}\sigma}\exp(-\frac{1}{2}\frac{D(v_i, x_1)^2}{\sigma^2}) \qquad (7)$$

In this way, a kernel visual vocabulary replaces the hard mapping of local features in an image region to the visual vocabulary. This soft assignment models is realized by replacing the hard contribution function with the kernel contribution function. This soft assignment models two types of ambiguity between visual words: *codeword uncertainty* and *codeword plausibility*, which are introduced as follows.

Uncertainty Proximity Distribution. Uncertainty Contribution Function indicates that one image local feature x_1 may be assigned to more than one visual words. *Uncertainty Contribution Function* is defined as follows,

$$F_{unc}(v_i, x_1) = \frac{K_\sigma(D(v_i, x_1))}{\sum\limits_{j=1}^{|V|} K_\sigma(D(v_j, x_1))} \qquad (8)$$

F_{unc} normalizes the probabilities to 1 and is distributed over all visual words. By using F_{unc}, we can define the *Uncertainty Proximity Distribution* as

$$H_{unc}^X(i, j, r) = \sum_{l=1}^{L}\sum_{m=1}^{L}\{\frac{K_\sigma(D(v_i, x_1))}{\sum\limits_{k=1}^{|V|} K_\sigma(D(v_k, x_1))}\frac{K_\sigma(D(v_j, x_1))}{\sum\limits_{k=1}^{|V|} K_\sigma(D(v_k, x_1))}I(d_{NN}(x_1, x_m) \leq r)\} \qquad (9)$$

In this way, visual word uncertainty is kept and it has the ability to assign one local feature to multiple visual words. However, it does not take the plausibility of a visual word into account.

Plausibility Proximity Distribution. Visual word Plausibility Contribution Function is proposed by by using ambiguous contribution functions. This function indicates that an image local feature may not be close enough to be represented by all the visual words in the vocabulary. *Plausibility Contribution Function* is defined as

$$F_{pla}(v_i, x_1) = K_\sigma(D(v_i, x_1))I(\alpha_1 = i) = \begin{cases} K_\sigma(D(v_i, x_1)), & \text{if } \alpha_1 = i \\ 0, & \text{otherwise} \end{cases} \qquad (10)$$

F_{pla} selects for an image local feature x_l only the closet visual word v_i and assigns it probability to the kernel value of that visual word. But it should be noted that it cannot select multiple visual word candidates.

In this way, we redefine the Proximity Distribution using Plausibility Contribution Function F_{pla}, resulting the *Plausibility Proximity Distribution* as

$$H_{pla}^{X}(i,j,r) = \sum_{l=1}^{L}\sum_{m=1}^{L}\{F_{pla}(v_i,x_1)F_{pla}(v_j,x_m)I(d_{NN}(x_1,x_m)\leq r)\}$$

$$= \sum_{l=1}^{L}\sum_{m=1}^{L}\{K_\sigma(D(v_i,x_1))I(\alpha_1=i)K_\sigma(D(v_j,x_m))I(\alpha_m=j)I(d_{NN}(x_1,x_m)\leq r)\} \qquad (11)$$

An unified formula is given as follows for these three ambiguous proximity distributions:

$$H^{X}(i,j,r) = \sum_{l=1}^{L}\sum_{m=1}^{L}\{F(w_i,x_1)F_{pla}(w_j,x_m)I(d_{NN}(x_1,x_m)\leq r)\} \qquad (12)$$

where $F(v_i,x_l)$ is to the a version of $K_{ker}(v_i, x_l)$ for Kernel Proximity Distribution,. Moreover, by setting $F(v_i, x_l) = F_{hard}(v_i, x_l) = I(\alpha_l = i)$, the original hard assignment of Proximity Distribution in (4) can also be included in (15).

Then the corresponding (unnormalized) *Ambiguous Proximity Distribution Kernel* (APDK) value between two images with feature sets Y and Z is

$$K_{apd}(Y,Z) = \sum_{i=1}^{K}\sum_{j=1}^{K}\sum_{r=1}^{R}\min(H_{apd}^{Y}(i,j,r),H_{apd}^{Z}(i,j,r)) \qquad (13)$$

where H_{apd}^{Y} and H_{apd}^{Z} are the associated arrays of *Ambiguous Proximity Distribution* computed for the two input feature sets.

3   Experiments

The proposed method is in general applicable for CBIR of large image databases according to learned local visual vocabulary. In this section, we demonstrate our algorithm using image patch exemplars [55] applied to a nearest neighbor classification problem medical image retrieval. We carry the experiments based on ImageClef 2007 medical image classification and ImageClef 2008 large-scale medical image retrieval competitions. In these two groups of experiments, each image is treated as a collection of image local image features.

3.1   Experiment on ImageClef 2007 Dataset

In the ImageClef 2007 medical image classification competition [56], a database of 12,000 categorized radiograph images is used. A set of 11,000 images are used as training set, and the remaining 1000 images are used as test images. There are 116 different classes within this database, based on the differences of either the examined region, the image orientation with respect to the body or the biological system under evaluation. We represent each image as a bag of small patches, which are the representation of local features of an image. We extract a small patch around every pixel,

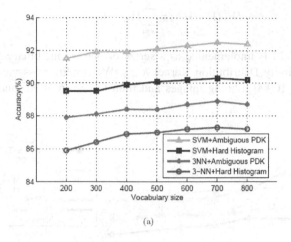

(a)

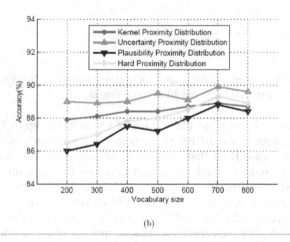

(b)

Fig. 1. (a) Effect of vocabulary size, for K-NN and SVM classifiers. (b) Classification performance results of various types of Ambiguous PDK for the ImageClef 2007 dataset over various vocabulary sizes

using a patch size of 9 × 9 pixels and the level of noise, we have also applied the Principal Component Analysis (PCA) [57, 58] procedure to the data dimensionality of descriptor vector of each local feature from 81 to 7. We use the the k-means algorithm to cluster the patches. Using the generated visual vocabulary, each image X is represented as an Ambiguous Proximity. In this step images are sampled with a dense grid. We firstly use   the k nearest neighbor classifier for the classification problem [59–61]. With our Ambiguous Proximity Distribution, we adopt the Ambiguous Proximity Distribution Kernel K_{APD} to compute the similarity of the Ambiguous Proximity Distribution of two patches collections for images. We use the L_1 norm distance D^{L1} (H^X, H^Y) between the word histograms of the two images X and Y as distance measure for hard histogram,

$$D^{L1}(H^X, H^Y) = \sum_{k=1}^{K} |h_k^X - h_k^Y| \tag{14}$$

The multiclass SVM is implemented as a series of one-vs-one binary SVMs with an Ambiguous Proximity Distribution Kernel K_{APD}. We run 20 cross-validation experiments trained on 10000 training images and then test it on 1000 randomly selected test images.

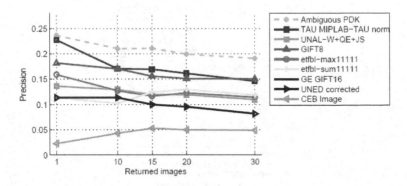

Fig. 2. Precision vs Recall graph of visual retrieval systems; ImageClef 2008 medical database. Predication shown for first 5,10,15,20 and 30 returned images.

As Figure 1 shows, we increase the number of visual words up to 700 words. The performance of vocabulary size of 700 visual words is significantly better than that of the 200 vocabulary for the Hard Histogram baseline. However, with Ambiguous PDK the results are very similar for both sizes, as Ambiguous PDK can utilize more information of provided by vocabulary by Ambiguous Visual Word Contribution Function. Figure 1 (a) also demonstrates that using an SVM classifier provides results that are more than 3% higher than the best K-NN classifier for both Hard Histogram baseline and Ambiguous PDK. We then make a concrete analysis of the results in Fig. 1 (b) with the various types of Ambiguous Proximity Distributions. The results show that Uncertainty Proximity Distributions H_{UNC} consistently outperforms other types of ambiguity for all vocabulary sizes. But we should note that this performance gain is not always that significant. On the other end of the performance scale, there is Plausibility Proximity Distribution H_{PLA}, which always yields the lowest results. We can also see that a Kernel Proximity Distribution H_{KER}, outperforms Hard Proximity Distributions for smaller vocabulary sizes.

3.2   Large Scale Image Retrieval Experiment

In ImageClef 2008 a large-scale medical image retrieval competition was conducted. In this competition, there are 66,000 images in the database, and also 30 query topics. Moreover, each topic is composed of one or more images and also a short textual description. The objective of retrieval is to return a set of 1000 images from the given database. Figure 2 shows the scores of our Ambiguous Proximity Distribution Kernel

(Ambiguous PDK), along with visual retrieval algorithms submitted by additional groups [62]. As the results shown in Fig. 2, The Ambiguous PDK for either original method shows the best possible performance, and again as we decrease the returned images, we can expect more accurate results at the cost of fewer returned images.

4   Conclusions

This paper proposed Ambiguous Proximity Distribution, a novel image presentation combining the advantage of visual word ambiguity and ambiguous proximity distribution. A visual word contribution function framework was used for analyzing its relation with the popular VW model and developing three novel Ambiguous Proximity Distributions. An extensive comparative medical image retrieval and classification experimental analysis with state-of-the art medical image retrieval methods provided empirical evidence of the effectiveness of the proposed technique for enhancing the performance of medical image retrieval system.

References

1. Penatti, O., Silva, F., Valle, E., Gouet-Brunet, V., Torres, R.: Visual word spatial arrangement for image retrieval and classification. Pattern Recognition 47(2), 705–720 (2014)
2. Chen, J., Feng, B., Xu, B.: Spatial similarity measure of visual phrases for image retrieval. In: Gurrin, C., Hopfgartner, F., Hurst, W., Johansen, H., Lee, H., O'Connor, N. (eds.) MMM 2014, Part II. LNCS, vol. 8326, pp. 275–282. Springer, Heidelberg (2014)
3. Furuya, T., Ohbuchi, R.: Visual saliency weighting and cross-domain manifold ranking for sketch-based image retrieval. In: Gurrin, C., Hopfgartner, F., Hurst, W., Johansen, H., Lee, H., O'Connor, N. (eds.) MMM 2014, Part I. LNCS, vol. 8325, pp. 37–49. Springer, Heidelberg (2014)
4. Sun, Q., Hu, F., Hao, Q.: Mobile target scenario recognition via low-cost pyroelectric sensing system: Toward a context-enhanced accurate identification. IEEE Transactions on Systems, Man, and Cybernetics: Systems 44(3), 375–384 (2014)
5. Sun, Q., Hu, F., Hao, Q.: Context awareness emergence for distributed binary pyroelectric sensors. In: 2010 IEEE Conference on Multisensor Fusion and Integration for Intelligent Systems (MFI), pp. 162–167. IEEE (2010)
6. Wang, Y., Jiang, W., Agrawal, G.: Scimate: A novel mapreduce-like framework for multiple scientific data formats. In: 2012 12th IEEE/ACM International Symposium on Cluster, Cloud and Grid Computing (CCGrid), pp. 443–450. IEEE (2012)
7. Wang, Y., Su, Y., Agrawal, G.: Supporting a light-weight data management layer over hdf5. In: 2013 13th IEEE/ACM International Symposium on Cluster, Cloud and Grid Computing (CCGrid), pp. 335–342. IEEE (2013)
8. Su, Y., Wang, Y., Agrawal, G., Kettimuthu, R.: Sdquery dsi: integrating data management support with a wide area data transfer protocol. In: Proceedings of SC13: International Conference for High Performance Computing, Networking, Storage and Analysis, vol. 47. ACM (2013)
9. Xu, L., Zhan, Z., Xu, S., Ye, K.: Cross-layer detection of malicious websites. In: CODASPY, pp. 141–152 (2013)

10. Cui, S., Soh, Y.C.: Linearity indices and linearity improvement of 2-d tetralateral position-sensitive detector. IEEE Transactions on Electron Devices 57(9), 2310–2316 (2010)
11. Xia, Y., Wan, S., Yue, L.: A new texture direction feature descriptor and its application in contentbased image retrieval. In: Proceedings of the 3rd International Conference on Multimedia Technology, ICMT 2013. LNEE, vol. 278, pp. 143–151 (2014)
12. Xia, Y., Wan, S., Yue, L.: Local spatial binary pattern: A new feature descriptor for content-based image retrieval. In: Proceedings of SPIE - The International Society for Optical Engineering, vol. 9069 (2014)
13. Guimaräes Pedronette, D., Almeida, J., Da, S., Torres, R.: A scalable re-ranking method for contentbased image retrieval. Information Sciences 265, 91–104 (2014)
14. Xie, B., Mu, Y., Song, M., Tao, D.: Random projection tree and multiview embedding for large-scale image retrieval. In: Wong, K.W., Mendis, B.S.U., Bouzerdoum, A. (eds.) ICONIP 2010, Part II. LNCS, vol. 6444, pp. 641–649. Springer, Heidelberg (2010)
15. Qian, J., Yang, J., Zhang, N., Yang, Z.: Histogram of visual words based on locally adaptive regression kernels descriptors for image feature extraction. Neurocomputing 129, 516–527 (2014)
16. Zhang, C., Liu, R., Qiu, T., Su, Z.: Robust visual tracking via incremental low-rank features learning. Neurocomputing 131, 237–247 (2014)
17. Wu, F., Pai, H.T., Yan, Y.F., Chuang, J.: Clustering results of image searches by annotations and visual features. Telematics and Informatics 31(3), 477–491 (2014)
18. Zhou, Y., Li, L., Zhao, T., Zhang, H.: Region-based high-level semantics extraction with cedd. In: 2010 2nd IEEE International Conference on Network Infrastructure and Digital Content, pp. 404–408. IEEE (2010)
19. Zhou, Y., Li, L., Zhang, H.: Adaptive learning of region-based plsa model for total scene annotation. arXiv preprint arXiv:1311.5590 (2013)
20. Li, X., Gao, J., Li, H., Yang, L., Srihari, R.K.: A multimodal framework for unsupervised feature fusion. In: Proceedings of the 22nd ACM International Conference on Information & Knowledge Management, pp. 897–902. ACM (2013)
21. Kong, X., Schunn, C.D.: Global vs. local information processing in visual/spatial problem solving: The case of traveling salesman problem. Cognitive Systems Research 8(3), 192–207 (2007)
22. Kong, X., Schunn, C.D., Wallstrom, G.L.: High regularities in eye-movement patterns reveal the dynamics of the visual working memory allocation mechanism. Cognitive Science 34(2), 322–337 (2010)
23. Yang, J., Wang, Y., Wang, H.: K., H., Wang, W., J., S.: Automatic objects removal for scene completion. In: The 33rd Annual IEEE International Conference on Computer Communications (INFOCOM 2014), Workshop on Security and Privacy in Big Data (2014)
24. Cho, W., Seo, S., Na, I., Kang, S.: Automatic images classification using HDP-GMM and local image features. In: Jeong, Y.-S., Park, Y.-H., Hsu, C.-H(R.), Park, J.J(J.H.) (eds.) Ubiquitous Information Technologies and Applications. LNEE, vol. 280, pp. 323–333. Springer, Heidelberg (2014)
25. Javed, U., Riaz, M., Ghafoor, A., Ali, S., Cheema, T.: Mri and pet image fusion using fuzzy logic and image local features. The Scientific World Journal 2014 (2014)
26. Yang, F., Xu, Y.Y., Wang, S.T., Shen, H.B.: Image-based classification of protein subcellular location patterns in human reproductive tissue by ensemble learning global and local features. Neurocomputing 131, 113–123 (2014)

27. Sun, Q., Wu, P., Wu, Y., Guo, M., Lu, J.: Unsupervised multi-level non-negative matrix factorization model: Binary data case. Journal of Information Security 3(4) (2012)

28. Mu, Y., Ding, W., Tao, D., Stepinski, T.F.: Biologically inspired model for crater detection. In: The 2011 International Joint Conference on Neural Networks (IJCNN), pp. 2487–2494. IEEE (2011)

29. Pinto, P., Tome, A., Santos, V.: Visual detection of vehicles using a bag-of-features approach. In: Proceedings of the 2013 13th International Conference on Autonomous Robot Systems, ROBOTICA 2013 (2013)

30. Kranthi Kiran, M., ShyamVamsi, T.: Hand gesture detection and recognition using affine-shift, bag-of-features and extreme learning machine techniques. In: Satapathy, S.C., Udgata, S.K., Biswal, B.N. (eds.) FICTA 2013. AISC, vol. 247, pp. 181–187. Springer, Heidelberg (2014)

31. Yu, J., Jeon, M., Pedrycz, W.: Weighted feature trajectories and concatenated bag-of-features for action recognition. Neurocomputing 131, 200–207 (2014)

32. Zagoris, K., Pratikakis, I., Antonacopoulos, A., Gatos, B., Papamarkos, N.: Distinction between handwritten and machine-printed text based on the bag of visual words model. Pattern Recognition 47(3), 1051–1062 (2014)

33. Fabian, J., Pires, R., Rocha, A.: Visual words dictionaries and fusion techniques for searching people through textual and visual attributes. Pattern Recognition Letters 39(1), 74–84 (2014)

34. van Gemert, Jan C. Veenman, C.J.S.A.W.G.J.M.: Visual word ambiguity. IEEE Transactions on Pattern Analysis and Machine Intelligence 32(7), 1271–1283 (2010)

35. Fiel, S., Sablatnig, R.: Writer identification and writer retrieval using the fisher vector on visual vocabularies. In: Proceedings of the International Conference on Document Analysis and Recognition, ICDAR, pp. 545–549 (2013)

36. Bolovinou, A., Kotsiourou, C., Amditis, A.: Dynamic road scene classification: Combining motion with a visual vocabulary model. In: Proceedings of the 16th International Conference on Information Fusion, FUSION 2013, pp. 1151–1158 (2013)

37. Wang, L., Elyan, E., Song, D.: Rebuilding visual vocabulary via spatial-temporal context similarity for video retrieval. In: Gurrin, C., Hopfgartner, F., Hurst, W., Johansen, H., Lee, H., O'Connor, N. (eds.) MMM 2014, Part I. LNCS, vol. 8325, pp. 74–85. Springer, Heidelberg (2014)

38. Sun, Q., Ma, R., Hao, Q., Hu, F.: Space encoding based human activity modeling and situation perception. In: 2013 IEEE International Multi-Disciplinary Conference on Cognitive Methods in Situation Awareness and Decision Support (CogSIMA), pp. 183–186. IEEE (2013)

39. Sun, Q., Hu, F., Hao, Q.: Human activity modeling and situation perception based on fiber-optic sensing system. IEEE Transactions on Human Machine Systems (2014)

40. Sun, T., Ding, S., Xu, X.: No-reference image quality assessment through sift intensity. Applied Mathematics and Information Sciences 8(4), 1925–1934 (2014)

41. Meng, X., Yin, Y., Yang, G., Xi, X.: Retinal identification based on an improved circular gabor filter and scale invariant feature transform. Sensors 13(7), 9248–9266 (2013)

42. Travieso, C., Del Pozo-Banos, M., Alonso, J.: Fused intra-bimodal face verification approach based on scale-invariant feature transform and a vocabulary tree. Pattern Recognition Letters 36(1), 254–260 (2014)

43. Li, Y., Liu, W., Li, X., Huang, Q., Li, X.: Ga-sift: A new scale invariant feature transform for multispectral image using geometric algebra. Information Sciences (2014)

44. Makar, M., Chang, C.L., Chen, D., Tsai, S., Girod, B.: Compression of image patches for local feature extraction. In: Proceedings of IEEE International Conference on Acoustics, Speech and Signal Processing, ICASSP 2009, pp. 821–824 (2009)

45. Shih, H.C., Chuang, C.Y., Huang, C.L., Lin, C.H.: Gender classification using bayesian classifier with local binary patch features. In: Proceedings of IEEE 4th International Conference on Nonlinear Science and Complexity, NSC 2012, pp. 45–50 (2012)

46. Zhang, Q., Zhang, L., Yang, Y., Tian, Y., Weng, L.: Local patch discriminative metric learning for hyperspectral image feature extraction. IEEE Geoscience and Remote Sensing Letters (2013)

47. Chen, G., Sun, R., Ren, Z., Wang, Z., Sun, L.: Local shape patch based feature matching method. Zhongnan Daxue Xuebao (Ziran Kexue Ban)/Journal of Central South University (Science and Technology) 44(suppl.2), 33–39 (2013)

48. Chandler, D., Field, D.: Estimates of the information content and dimensionality of natural scenes from proximity distributions. Journal of the Optical Society of America A: Optics and Image Science, and Vision 24(4), 922–941 (2007)

49. Ling, H., Soatto, S.: Proximity distribution kernels for geometric context in category recognition (2007)

50. Deselaers, T., Keysers, D.N.H.: Features for image retrieval: an experimental comparison. Information Retrieval 11(2), 77–107 (2008)

51. Ay, M., Kisi, O.: Modelling of chemical oxygen demand by using anns, anfis and k-means clustering techniques. Journal of Hydrology 511, 279–289 (2014)

52. Wang, L., Pan, C.: Robust level set image segmentation via a local correntropy-based k-means clustering. Pattern Recognition 47(5), 1917–1925 (2014)

53. Lin, C.H., Chen, C.C., Lee, H.L., Liao, J.R.: Fast k-means algorithm based on a level histogram for image retrieval. Expert Systems with Applications 41(7), 3276–3283 (2014)

54. Ling, H.B., Soatto, S.: Proximity distribution kernels for geometric context in category recognition. In: 2007 IEEE 11th International Conference on Computer Vision, vol. 1-6, pp. 245–252 (2007)

55. Varma, M., Zisserman, A.: A statistical approach to material classification using image patch exemplars. IEEE Transactions on Pattern Analysis and Machine Intelligence 31(11), 2032–2047 (2009)

56. Deselaers, T., et al.: Overview of the imageCLEF 2007 object retrieval task. In: Peters, C., Jijkoun, V., Mandl, T., Müller, H., Oard, D.W., Peñas, A., Petras, V., Santos, D. (eds.) CLEF 2007. LNCS, vol. 5152, pp. 445–471. Springer, Heidelberg (2008)

57. Li, M., Li, Y., Huang, X., Zhao, G., Tian, W.: Evaluating growth models of pseudomonas spp. in seasoned prepared chicken stored at different temperatures by the principal component analysis (pca). Food Microbiology 40, 41–47 (2014)

58. Lu, Y., Gao, B., Chen, P., Charles, D., Yu, L.: Characterisation of organic and conventional sweet basil leaves using chromatographic and flow-injection mass spectrometric (fims) fingerprints combined with principal component analysis. Food Chemistry 154, 262–268 (2014)

59. Niu, Y., Wang, X.: On the k-nearest neighbor classifier with locally structural consistency. In: Liu, X., Ye, Y. (eds.) Proceedings of the 9th International Symposium on Linear Drives for Industry Applications, Volume 2. LNEE, vol. 271, pp. 269–277. Springer, Heidelberg (2014)

60. Souza, R., Rittner, L., Lotufo, R.: A comparison between k-optimum path forest and k-nearest neighbors supervised classifiers. Pattern Recognition Letters 39(1), 2–10 (2014)

61. Li, C.L., Wang, E., Huang, G.J., Chen, A.: Top-n query processing in spatial databases considering bi-chromatic reverse k-nearest neighbors. Information Systems 42, 123–138 (2014)
62. Müller, H., Kalpathy-Cramer, J., Kahn Jr., C.E., Hatt, W., Bedrick, S., Hersh, W.: Overview of the imageCLEFmed 2008 medical image retrieval task. In: Peters, C., Deselaers, T., Ferro, N., Gonzalo, J., Jones, G.J.F., Kurimo, M., Mandl, T., Peñas, A., Petras, V. (eds.) CLEF 2008. LNCS, vol. 5706, pp. 512–522. Springer, Heidelberg (2009)
63. Shen, J., Su, P.-C., Cheung, S.-C., Zhao, J.: Virtual Mirror Rendering with Stationary RGB-D Cameras and Stored 3D Background. IEEE Transactions on Image Processing 22(9), 1–16 (2013)
64. Shen, J., Cheung, S.C.S.: Layer Depth Denoising and Completion for Structured-Light RGB-D Cameras. In: 2013 IEEE Conference on Computer Vision and Pattern Recognition (CVPR), June 2013, pp. 1187–1194. IEEE (2013)

Indexing SURF Features by SVD Based Basis on GPU with Multi-Query Support

Vibha Patel[1] and Bhavin Patel[2]

[1] Indian Institute of Technology, Kanpur, UttarPradesh 208016
vibhadp@iitk.ac.in
[2] Institute of Technology, Nirma University, Gujarat 382481
12MCEC20@nirmauni.ac.in

Abstract. Space partitioning based indexing technique reduces search space and increases performance of the retrieval system. Geometric hashing is one such space partitioning technique which creates a global model descriptor. Aligning points prior to geometric hashing proves to be better for increasing the retrieval performance. This paper presents a newer technique of aligning points of Speeded Up Robust Features (SURF) using basis direction obtained through Singular Value Decomposition (SVD). Performance gets boosted by 10% on average in online searching as compared to classical approach. Further, the implicit parallelism that exist in proposed technique motivated us to use Graphics Processing Unit (GPU). The GPU based implementation of the proposed indexing technique achieved speed up in the range of 14.3x-82.28x for offline database indexing and 3.54x-4255.88x for online searching. Extension to the single query execution is provided by the newer multi-query approach. It proves to be better than the linear execution of multi-query. For GPU based multi-query implementation, speed up obtained is in the range of 3.74x-3097.55x for 1 to 10 queries to be executed simultaneously.

Keywords: Geometric Hashing, Graphics Processing Unit (GPU), Indexing, Multi-query, Searching, Singular Value Decomposition(SVD).

1 Introduction

In model based image recognition system, there is a need to generate global model descriptor offline from all the available database models. During online identification process, a query is mapped to the given global model descriptor and the closest matching model is considered as an output to the query. Computational cost of matching query with every model descriptor of image in the database is reduced tremendously using global model. Space organizing indexing structures like multidimensional hashing [9], grid-files and kd-trees [5] construct varying global model descriptors for faster recognition. Among all such descriptors, geometric hashing based image recognition is the simplest and widely used technique suitable for both 2D and 3D model points. Instead of applying vector based global descriptor directly on the image points, extracting image features containing selective spatial image points and the

D.-S. Huang et al. (Eds.): ICIC 2014, LNAI 8589, pp. 422–433, 2014.

associated descriptor can improve accuracy and precision. Such extracted features emphasize on the structural analysis of spatial image points and statistical analysis of associated descriptor with each point. Hence, Palm based images [1] are chosen for experimentation, preserving both structural and statistical properties to provide better recognition accuracy. [3], [4] carry out model based recognition on it to achieve speedup in the overall matching process. Scale Invariant Feature Transform (SIFT) [8] and Speeded Up Robust Features (SURF) [15], [14], [12] are two widely used spatial point based feature extraction methods. SIFT provides features invariant to scaling and translation, while SURF also provides invariance against rotation. As palm images may have some rotation we used SURF for our experimentation.

Global model space generated using geometric hashing is accompanied by different transformation techniques for making changes in the spatial location of the points. We hereby propose a new transformation technique to speed up the overall process and achieve invariance against change. There is high level of parallelism in geometric hashing that can be leveraged by the modern GPUs available. The implementation is done on GPUs to speed up overall process of both, the classical and proposed indexing techniques. Global model-based indexing is mostly for single query processing environment. In such systems there is high turnaround and waiting time for the number of queries pipelined for execution. By extending it for the multi-query this execution time can be reduced tremendously.

The salient contribution of this paper is as follows:

- Proposes and evaluates a newer technique to efficiently index SURF based keypoints by basis direction from Singular Value Decomposition (SVD).
- Compares proposed technique with both sequential and GPU based parallel implementation of classical geometric hashing.
- Proposes multi-query support in the proposed technique.
- Validates multi-query technique by implementing both sequential and GPU based parallel approach.

The paper is further arranged as follows: Section II will give information about the related study and work carried out in this domain. Section III will discuss SURF feature extraction technique and presents algorithm for proposed transformation technique for indexing, brief about its parallelization on GPU and provides extension for multi-query environment. Section IV will discuss experimental findings and Section V will provide conclusion.

2   Related Work

Drost et. al [6] presents efficient 3D object recognition using global model descriptor. Searching was done with 2D scene query by matching locally to generated 2D search space. It shows better performance than the traditional matching techniques. Gavrila and Groel [7] did similar model based recognition on 3D images by searching for 2D images.

Badrinath et al. [4] uses palm images for comparison based on geometric hashing with newer score voting strategy. Indexing based on similarity transformation was done to change the locality of the generated feature space. It shows better recognition and invariance to changes in the palm prints. Jayaraman et al. [10], [11] uses both geometric hashing based indexing with novel preprocessing based on mean translation, scaling and alignment of feature points along the principal components. By aligning points along the principal components points, variance of the feature space is considered. This work applied the Principal Component Analysis (PCA) technique to find the direction in which variability is maximized. One of the major drawbacks of PCA is that, it cannot align outliers due to its implicit definition.

Alcantara et al. [2] presents real time two-level hashing based on sub-table creation implemented on GPUs. Verification of this approach was done by 3D painting, collision detection and other methods. We implemented GPU based partitioning for 2D space which we derived from [2] for handling large data with collision resolution. Lefebvre and Hoppe [13] also provide hash table creation on GPUs in real time for image recognition and collision detection.

3   Proposed Indexing Technique

.This section proposes indexing technique which uses geometric properties of the SVD based basis vectors. This technique is based on aligning points along the direction of maximum variance to utilize the spatial properties in the newly transformed space. Figure 1 shows general framework for using such an indexing technique with indexing and searching as its main component. During indexing, all the features extracted from models are mapped into the 2D hash table. While searching, by using hash table, the best match is retrieved without comparing it to the whole database. The techniques presented in this paper provide flexibility in inserting a new model inside the database without reconstructing the whole index structure.

3.1   Speeded Up Robust Features

SURF is a multidimensional feature extraction technique widely used for objects and image recognition. A 64-bit descriptor is associated with each key spatial point selected by this technique. This key spatial point selection is done using the Hessian box operator at different scales and sample rates. It makes this technique scale, rotation and translation invariant. Method to obtain SURF features is discussed in detail in [4].

Suppose there are N model images each containing m average interest point's extracted using SURF method. Each of these interest points is associated with a descriptor of length 64 with it. For an online query containing n interest point's extracted using SURF method, a similar model is identified by comparing all n points, with each of the m points using Euclidean distance measure. Further, indexing technique based on space partitioning is applied to reduce the overall distance.

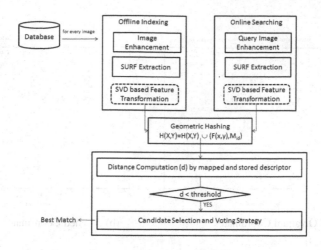

Fig. 1. General Framework for Proposed Indexing Technique

3.2   Indexing

After feature extraction, indexing using SVD based transformation and hash table generation is performed in the feature space. A translated and rotated object in the query is transformed using SVD and compared to the database model. During transformation, SVD is applied directly on the data matrix to generate eigenvectors for each nonzero eigenvalue. Eigenvectors corresponding to two largest eigenvalues will be selected as a basis vector and whole data matrix will be projected with these basis vectors aligned along the coordinate axes. Input to the transformation framework will be the features extracted from each model image having matrix $M_{m \times 2}$. Our objective is to calculate basis vectors B_1 and B_2 corresponding to M using SVD. Unlike SVD, PCA is performed by working with the covariance matrix generated from the mean-invariant data matrix. Variance due to the impact of outliers is larger in SVD than in PCA. Figure 2 shows the transformation of points in the 2D space for a sample feature space generated from the image of a palm. Figure 2(a) shows the original feature space, while 2(b) shows the change in the feature space after transformation by SVD.

The transformed space is then partitioned into a 2D hash space of size $N_1 \times N_2$, where N_1, $N_2 > 0$. For every model image, this common 2D hash space is generated and stored as a global model descriptor. The indexing algorithm is given in Algorithm 1. Indexing using Algorithm 1 over palm databases, figure 3(a) shows an increase in the number of elements per bin by proposed indexing over the classical approach shown in figure 4(a) for a small size of hash bins (say 64*64 in this case). With large hash bins, the scenario gets changed. Figure 3(b) shows lesser elements per bin than 4(b) (Hash bin size is 1024*1024). Hence the number of elements per bin will decrease for larger hash bins, which will also further decrease the overall comparison for the online searching.

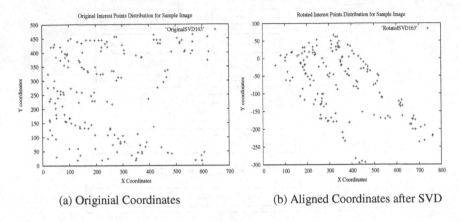

(a) Originial Coordinates   (b) Aligned Coordinates after SVD

Fig. 2. Coordinates Transformation during Indexing

Algorithm 1: Indexing

INPUT: Set of model images
OUTPUT: Global model identity hash table
For each model image M do:

1. Extract SURF based features of an image. Assume created feature Matrix M of size 2*m, where m is extracted features.
2. Determine two basis vectors (each of 2*2) from SVD of matrix M. Basis vectors are eigenvectors with highest eigenvalue.
3. Rotate the feature space with two basis vector aligned to co-ordinate axis. Assume rotated feature Matrix R of size 2*m.
4. Partition the transformed feature space into geometric hash bins. Assume number of hash-bin be N_1 and N_2 in both direction.
5. Index every feature element from R matrix into partitioned hash bins N_1, N_2 to create global model identity.

3.3 Searching

The main advantage of the proposed indexing technique is a reduction in the response time compared to the classical technique with transformation. The searching algorithm is given in Algorithm 2. Suppose $\{I_1, I_2,..., I_q\}$ be q transformed features extracted from a query Q. For finding the match, every query interest point, I_i is mapped to z index of the hash table. Euclidean distance is calculated between I_i and c, where c is an element of $H(z)$. A candidate set is created for every element below threshold and at last all the candidate sets are considered in voting for the best match image.

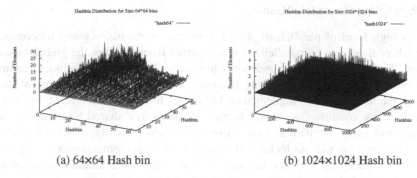

(a) 64×64 Hash bin   (b) 1024×1024 Hash bin

Fig. 3. Hash Distribution for Proposed Indexing Scheme

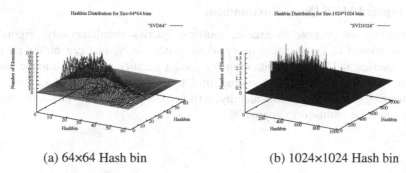

(a) 64×64 Hash bin   (b) 1024×1024 Hash bin

Fig. 4. Hash Distribution for Classical Geometric Hashing

Algorithm 2: Searching

INPUT: Query Image, Global model identity hash table
OUTPUT: Best Match
For Query Image Q do as follows:

1. Extract SURF based features of an image. Assume created feature Matrix Q of size 2*m, for q extracted features.
2. Determine two basis vectors (each of 2*2) from SVD of Matrix Q. Basis vectors are eigenvectors with highest eigenvalue.
3. Rotate the feature space with two basis vectors aligned to co-ordinate axis. Assume rotated feature Matrix QR of size 2×m.
4. Index the transformed feature space QR into hash bins N1, N2 to query identity, N1 and N2 need to be same as indexing.
5. Map every non-zero query identity hash bins into global model identity.
6. Compute Euclidean distance of feature element with every element present in both the indexed bin.
7. Compute Candidate Solution per bin with certain threshold.
8. Calculate total votes for a database model from the candidate solution.
9. Return database model with maximum vote.

3.4   Paralleling Geometric Hashing

There is a high level of parallelism provided by geometric hashing and its components. Each of them can be parallelized easily, apart from creating the global model descriptor. There may be a race condition when multiple threads try to index feature element on the same location. Locking becomes inevitable for incorporating all the changes in the hash bins. We implemented block free based approach for it. For handling such a race condition, we firstly register position in a shared hash-bin for the feature element using compare and swap instruction and secondly, insert it into that registered bin location without locking. It is to be noted that elements need to register a priori before inserting into any bin and this will guarantee mutual-exclusion of changes to be handled properly.

3.5   Proposed Multi-Query Environment

In this section, we propose to execute multiple queries simultaneously. Figure 5 shows that instead of constructing query identity iteratively, one can buffer all the available queries and form a common query model identity with their unique mappings. The overall waiting and turn-around time, by creating query model identity and then comparing with global model identity, will reduce. Algorithm 3 shows the modified algorithm for multi-query approach.

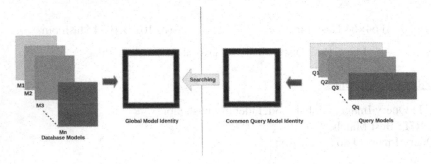

Fig. 5. Constructing Common Query Identity for Multiple Queries

<u>Algorithm 3: Multi-Query Searching</u>

INPUT: Multiple Query Image, Global model Descriptor hash table
OUTPUT: Best Match for all queries
For each query image Q do as follows:

1. Extract SURF based features of an image. Assume created feature Matrix Q of size 2×m, for q extracted features.
2. Determine two basis vectors (each of 2×2) from SVD of Matrix Q. Basis vectors are eigenvectors with highest eigenvalue.
3. Rotate the feature space with two basis vectors aligned to co-ordinate axis. Assume rotated feature Matrix QR of size 2×m.
4. Index the transformed feature space QR into hash bins N1, N2 to create common query identity.

After creating common model Identity do as follows:

1. Map every non-zero query identity bins to global model identity.
2. Compute Euclidean distance of feature element with every element present in both the indexed bin.
3. Compute Candidate Solution per bin with certain threshold for every query.
4. Calculate total votes for a database model from the candidate solution for every query.
5. Return database model with maximum vote for all the queries.

4   Experimental Results

Proposed approach is evaluated with CASIA Palm-print Image Database containing images of 512 subjects [1]. Sequential version of the methods are verified with Intel (R) Core(TM) i3 CPU with 6 GB of RAM and parallel version are verified with Tesla C2070 GPU containing 448 CUDA cores and 5376 MB of global memory for storage. Experimentation is carried out with 1000 total image models due to limited primary memory. Evaluation of the system is done by comparing performance of both classical and proposed indexing technique. This was done using percentage data reduction and multiple distance evaluation mechanisms. We also compared GPU based parallel versions, and looked at the performance of the proposed indexing for multiple queries in both sequential and parallel versions. Performance is calculated for both off-line indexing and online searching. For all experiments we consider different number of hash-bins and different types of queries. For both the methods all parameters are set the same for all the experimentation. The algorithm was implemented in C++ with CUDA 4.0 and OPENCV-2.4.0. All given timings include the overall time take by the system including preprocessing, feature extraction, memory transfer between CPU and GPU, model creation, distance computation, candidate selection and voting. A preprocessing step is required to model images before calculating the model identity. Hence, masking (to blur) and histogram equalization (to enhance images) is performed, resulting in an increase in the number of interest points for searching effectively.

Figure 6 and 7 shows timing of both, offline and online activities in the proposed as well as the classical technique. Online activity, which is the prime target of any retrieval system to optimize, gives good results with our methodology. Figure 8(a) shows speed-up achieved by comparing it with the sequential version of the same method. In comparison with the classical technique, our proposed indexing obtains a better speedup due to the change in the locality and the number of elements per bin. Table 1 and 2 show database reductions for both, proposed indexing and classical geometric hashing. Overall database reduction range for different queries and hash bins for the proposed indexing is 94.3% to 99.8%, while for classical indexing is 91.5% to 99.6%.

Figure 9(a) and 9(b) shows distance evaluation for different queries on both, the proposed and classical techniques respectively. The number of distance computations for a larger bin size is much lesser in the case of the proposed technique, due to the spatial relocation of points. This will reduce the online searching time. Table 3 shows performance of sequential and parallel version of proposed multi-query approach. Experimentation is done with a single query, having an average interest point; and multiple queries to be executed simultaneously for different hash sizes. Figure 8(b) shows corresponding speedup for the experimentation carried out for different hash sizes and number of queries. The proposed multi-query approach for a higher number of queries did not increase linearly as was expected from the naive implementation of the system. Hence our multi-query approach is performing better than expected.

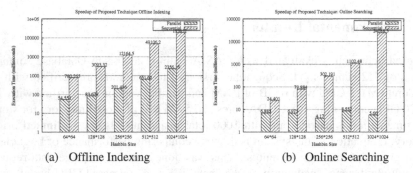

(a)   Offline Indexing   (b)   Online Searching

Fig. 6. Performance of Proposed Indexing Technique (Sequential vs. Parallel)

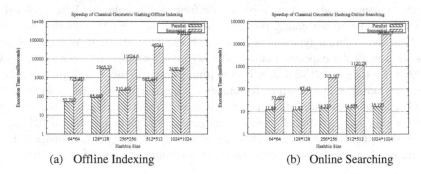

(a)   Offline Indexing   (b)   Online Searching

Fig. 7. Performance of Classical Geometric Hashing (Sequential vs. Parallel)

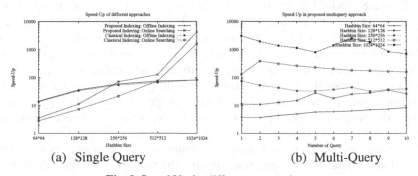

(a)   Single Query   (b)   Multi-Query

Fig. 8. Speed Up for different approaches

Table 1. Database Reduction: Proposed Indexing (Values are in %)

Hash bin Size	Queries			
	Minimum	Average	Median	Maximum
64×64	97.2	94.4	94.3	94.3
128×128	99.2	94.6	94.4	94.3
256×256	99.6	95.9	96.1	95
512×512	99.8	97.9	98.3	96.9
1024×1024	99.8	99	99.4	98.3

Table 2. Database Reduction: Classical Geometric Hashing (Values are in %)

Hash bin Size	Queries			
	Minimum	Average	Median	Maximum
64×64	97.2	94.4	94.3	94.3
128×128	99.2	94.6	94.4	94.3
256×256	99.6	95.9	96.1	95
512×512	99.8	97.9	98.3	96.9
1024×1024	99.8	99	99.4	98.3

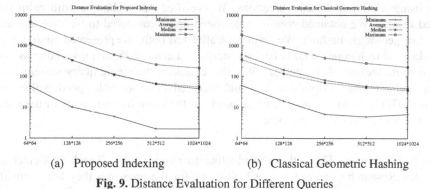

(a)   Proposed Indexing   (b)   Classical Geometric Hashing

Fig. 9. Distance Evaluation for Different Queries

Table 3. Performance of Multi-Query Based Proposed Approach (Online Searching)

No.of Query	64×64		128×128		256×256	
	Sequential	Parallel	Sequential	Parallel	Sequential	Parallel
1.	25.43	6.79	78.84	6.96	301.14	3.96
2.	25.80	6.89	88.44	8.06	344.36	6.43
3.	31.41	7.04	99.41	7.63	397.46	9.06
4.	36.41	7.19	117.54	7.64	428.07	12.54
5.	43.69	7.31	127.57	4.41	476.68	14.09
6.	46.62	7.54	149.21	8.04	659.08	17.34
7.	51.11	7.86	188.79	7.26	884.86	19.31
8.	55.12	7.72	211.41	7.28	926.39	28.18
9.	61.29	8.05	228.13	6.31	999.01	26.29
10.	67.59	7.97	247.09	9.27	1106.76	27.74

Table 3. (*continued*)

No.of Query	512×512		1024×1024	
	Sequential	Parallel	Sequential	Parallel
1.	1073.7	8.02	17302.9	5.58
2.	1297.15	3.32	17434.7	8.98
3.	1197.79	3.76	17248.5	12.45
4.	1242.9	4.67	18786.8	16.22
5.	1312.16	5.59	15974.2	19.7
6.	1403.98	6.79	34275.6	23.96
7.	1464.03	7.93	66159.7	26.15
8.	1524.61	8.54	53968.7	29.24
9.	1591.06	9.39	28533.6	33.4
10.	1641.43	10.18	25144.4	35.23

5   Conclusion and Future Work

In this paper we proposed and evaluated a new indexing technique with transformation using SVD based basis directions. We verified our technique with palm print based images. We obtained considerable speed-up as compared to the classical technique of geometric hashing. We further parallelized both, the presented approach and the classical technique on GPU based systems and achieved a major speedup as compared to its sequential counterpart. We also extended the single query system to execute multiple queries simultaneously and received a considerable speedup after porting it to GPUs. In future we will try to work on the more intrinsic parts of the indexing systems to make it more efficient.

Acknowledgement. The authors would like to thank CUDA Teaching Center and NVIDIA Research Center at Nirma University for the resources they have provided for making the work possible. We would also like to thank Shashwat Chandra, and all the reviewers for their valuable feedback in making this work publishable.

References

1. The casia palmprint database, http://www.cbsr.ia.ac.cn/
2. Dan, A.: Alcantara, Andrei Sharf, Fatemeh Abbasinejad, Shubhabrata Sengupta, Michael Mitzenmacher, John D. Owens, and Nina Amenta. Real-time parallel hashing on the gpu. ACM Trans. Graph. 28(5), 154:1-154:9 (2009)
3. Badrinath, G.S., Gupta, P.: Palmprint verification using sift features. In: First Workshops on Image Processing Theory, Tools and Applications, IPTA 2008, November 2008, pp. 1–8 (2008)

4. Badrinath, G.S., Gupta, P., Mehrotra, H.: Score level fusion of voting strategy of geometric hashing and surf for an efficient palmprint-based identification. Journal of Real-Time Image Processing 8(3), 265–284 (2013)
5. Bohm, C., Berchtold, S., Keim, D.A.: Searching in high dimensional spaces: Index structures for improving the performance of multimedia databases. ACM Comput. Surv. 33(3), 322–373 (2001)
6. Drost, B., Ulrich, M., Navab, N., Ilic, S.: Model globally, match locally: Efficient and robust 3d object recognition. In: 2010 IEEE Conference on Computer Vision and Pattern Recognition (CVPR), June 2010, pp. 998–1005 (2010)
7. Gavrila, D.M., Groen., F.C.A.: 3d object recognition from 2d images using geometric hashing. Pattern Recognition Letters 13(4), 263–278 (1992)
8. Hu, Q., Ai, M.: A scale invariant feature transform based matching approach to unmanned aerial vehicles image geo-reference with large rotation angle. In: 2011 IEEE International Conference on Spatial Data Mining and Geographical Knowledge Services (ICSDM), pp. 393–396 (June 2011)
9. Hutesz, A., Six, H.-W., Widmayer, P.: Globally order preserving multidimensional linear hashing. In: Proceedings of Fourth International Conference on Data Engineering, pp. 572–579 (February1988)
10. Jayaraman, U., Gupta, A.K., Prakash, S., Gupta, P.: An enhanced geometric hashing. In: 2011 IEEE International Conference on Communications (ICC), June 2011, pp. 1–5 (2011)
11. Jayaraman, U., Prakash, S., Gupta, P.: Use of geometric features of principal components for indexing a biometric database. Mathematical and Computer Modeling 58(12), 147–164 (2013); Financial IT & Security and 2010 International Symposium on Computational Electronics
12. Su, J., Xu, Q., Zhu, J.: A scene matching algorithm based on surf feature. In: 2010 International Conference on Image Analysis and Signal Processing (IASP), pp. 434–437 (April 2010)
13. Sylvain, L., Hoppe, H.: Perfect spatial hashing. ACM Trans. Graph. 25(3), 579–588 (2006)
14. Murillo, A.C., Guerrero, J.J., Sagues, C.: Surf features for efficient robot localization with omnidirectional images. In: 2007 IEEE International Conference on Robotics and Automation, pp. 3901–3907 (April 2007)
15. Lei, Y., Jiang, X., Shi, Z., Chen, D., Li, Q.: Face recognition method based on surf feature. In: International Symposium on Computer Network and Multimedia Technology, CNMT 2009, pp. 1–4 (January 2009)

Dynamically Changing Service Level Agreements (SLAs) Management in Cloud Computing

Waleed Halboob[1], Haider Abbas[1,2], Kamel Haouam[3], and Asif Yaseen[4]

[1] Center of Excellence in Information Assurance, King Saud University, Saudi Arabia
{wmohammed.c,hsiddiqui}@ksu.edu.sa
[2] National University of Sciences and Technology, Islamabad, Pakistan
haiderabbas-mcs@nust.edu.pk
[3] Department of Computer Science, College of Computer and Information Sciences,
King Saud University, Saudi Arabia
haouam@ksu.edu.sa
[4] School of Integrative Systems, University of Queensland, Australia
Uqayasee@uq.edu.au

Abstract. Managing Service Level Agreements (SLAs) in cloud computing in a dynamic manner becomes a critical issue for cloud service providers. This is due to the emerging technology and the frequent and continuous change in cloud service requirements and techniques over the time. Current cloud SLAs management methods are rigorous in terms of SLA updation to incorporate new changes. Since updating any SLA, to meet any required change in cloud requirements and techniques, requires reformulation and remapping to all available SLAs (called public SLAs). This paper proposed a mechanism to dynamically manage cloud computing SLAs based on Real Options Analysis (ROA) concept. The proposed model maps the required changes to all public SLAs and sorts out the most related or suitable SLAs (solutions) based on options theory while recording the other solutions for any future change according to emerging circumstances. The technique incorporates any new change dynamically; by mapping it to a limited number of SLAs (recorded solutions) based on various options presented by ROA. The framework presented in this paper would provide a flexible solution in managing cloud SLAs in both cloud provider and the user's perspectives.

Keywords: Cloud computing, Service level agreement (SLA), Real options analysis, User requirements.

1   Introduction

Cloud computing is a modern computation technique in which the data is stored on interconnected servers permanently and cached temporarily through handheld devices, laptops , computer terminals , tablets etc. [1]. Cloud computing differs from grid computing in that the shared pool of virtualized accessible computing resources can be rented by consumers with minimum management and interaction efforts [2][22].

D.-S. Huang et al. (Eds.): ICIC 2014, LNAI 8589, pp. 434–443, 2014.

To establish a service contract between cloud consumer and the service provider, service level agreements (SLAs) reflect the quality of services. The SLA is basically a plain-text document containing information about the service's requirements, goals, etc. To be more specific, it contains: i) service level agreement related parties, such as service provider, consumer and third parties ii) provided services along with their definitions iii) measurement metrics to evaluate and audit the delivered services v) duties of all involved parties and vi) SLA update issues [3]. Cloud SLAs management involves three main steps namely SLA mapping and negotiating, monitoring and profiling, and finally recovering and enforcing. The first step maps the consumer's requirements to service provider's public SLAs (sometimes called SLA templates) and negotiates the subscription issues e.g., price, update, etc. The SLA monitoring and profiling are executed during the subscription period to measure the service provisions and collect runtime information, respectively. The recovery sub-step is essentially used for updating the service SLA in case if any exception occurred. Finally, the SLA enforcement is used to end the SLA lifetime if the service provision is completed [4].

Current cloud computing providers use static and rigorous SLAs. Whereas, the consumer selects the cloud service and agreed on the corresponding public SLA by clicking on a provided checkbox. However, cloud requirements and techniques are continuously changing over the time and, as a result, updating a cloud SLA requires specifying a new SLA or mapping the new change to all public SLAs. Despite the recent progress in SLA understandings, troubling gaps and shortcoming remains. Little is known, for example, about the dynamically updating cloud SLAs with the emerging trends and techniques. Several cloud SLAs management approaches [5,6,7,8] have been proposed to address this issue but these approaches require mapping the new change to all public SLAs, instead of some suitable SLAs only. As such updating the cloud SLAs dynamically still requires more research efforts [9].

This paper proposed a preliminary framework for dynamic cloud SLA management. The proposed model inherits the real options analysis (ROA) concept which helps in providing a dynamic cloud SLAs management. A modest but growing body of research supports employing real option theory to propose more suitable SLA solutions [10]. As such, the design creates option for the most appropriate SLA while prompting to frame the other selected suitable SLAs situational specifics. Whenever a new change is required to a specific SLA; only related and suitable SLAs are analyzed to address the new solution (SLA) for sustainable and efficient computing utility. Thus, one way to address this is dynamic updating of SLA mechanism in real time.

The paper is organized as follows: Section 2 reviews the related works in cloud SLAs management. Then, Section 3 presents the real options analysis (ROA) concept. The proposed model is presented in Section 4. Section 5 presents the SLA specification language for the deployment model used by this research. Finally, Section 6 concludes this work with the future directions of this preliminary framework.

2   Related Work

Several cloud SLAs management approaches have been proposed in the literature. These approaches can be classified into two categories SLA ontologies and SLA mapping approaches [11]. The SLA ontologies approaches [12,13,14,15,16,17] let cloud consumers to specify their requirements in a semantic language for matching mechanism and then the SLA is created based on the consumer requirements. However, these approaches require the consumers to know how to use the semantic language to specify their requirement and not considering the continuously changing mechanisms on both sides (technological and user requirements) [9].

The SLA mapping approaches [5,6,7,8] are proposed to address the issue of adaptability by combining the consumer SLA (called private SLA) with the service provider SLA (called public SLAs) to come out with a new SLA. Using these mapping approaches, SLA templates - popular SLA formats- are used during creation and updating the SLAs provided for consumers. The SLA templates can be either public or private. The public SLA templates are created by service provider while the private SLA templates are created by consumers. In any way, using existing specifications for defining SLA templates, it is difficult to adapt the defined SLAs to meet the new requirements [8].

In addition, using the above mapping approaches, the cost of service subscription process seems to be high as the consumer needs to manually define his private SLA even if the consumer may keep his old private SLA to be used with other services. According to Padhy et al., [9], the consumer must specify only a high level of metrics while the details of SLA must be described by the service provider. In other words, it is clear that there is a need for using only high-level requirements for specifying the new requirements (consumer, service provider, technical requirements and so on) and also updating the cloud SLA without issuing a new SLA or mapping the new requirements to all existing public SLAs. The framework proposed in this paper, would help cloud users to specify and update their SLAs with the advent of new technology and mechanism also for cloud providers (a flexibility to adapt new technology) as well as their customer's requirements.

3   Real Options-Based Dynamically Changing Requirements Management

The real options analysis (ROA) is a "corporate finance method that provides strategic guidelines to model asymmetric response for investment decision making" [10]. The ROA concepts have been transformed and applied in [10,18,19,20] to information systems to address several issues such as dynamically changing security requirements and to incorporate updates in information security systems. The application of this methodology in IT systems (as mentioned in [10]) contains three steps namely identification, formulation and analysis and information recording. As such, such steps can be modeled as real options. During identification phase, the factors that lead to updating the system are identified. These factors can be internal or external i.e. users, system errors, new technologies, etc. Then, the formulation and analysis step is used for analyzing the change request – submitted by users – and find the more appropriate solution from multiple future options (inherited from options theory). The whole solution is divided into sub solutions and adopted after a careful analysis by considering the uncertainty issues that might request a change in future. The anticipation is based on real mechanisms i.e. upcoming technology, previous experience, internal and external factors and the current needs of the customer mapped to the existing technology of the provider. Finally in recording information stage, some information are recorded into cards to be used later on for any further new change request. In this paper, this methodology is applied to address the issue of cloud computing SLAs with dynamically changing requirements management (creation and update).

4   The Proposed Model

The overall architecture of the proposed model is shown in Fig 1. The cloud service provider defines public SLAs for his provided services and uploads them to a database called public SLAs DB. The new consumer, technical or provider's high-level requirements are defined along with their corresponding priorities.

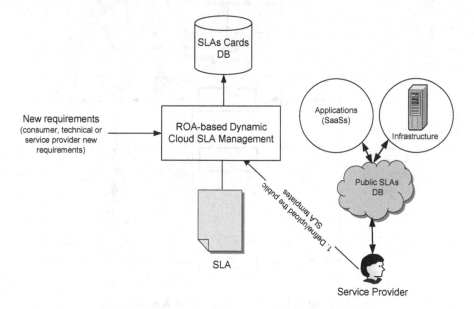

Fig. 1. The overall architecture of the proposed model

As shown in Fig 2, the ROA-based dynamic cloud SLA management has three main sub-components namely change identification, SLA analysis and information recording. The following sub-sections present these components in detail.

4.1   Change Identification

According to Abbas *et al.*, [10], the real option analysis (ROA) process starts with identifying factors/initiators that caused generating new SLA. The change factors can be classified into two categories implicit and explicit change. The implicit change is requested by the system due to any irregular behavior. Whereas, the explicit change is made by any cloud computing party such as consumer, provider, broker, etc. However, a new SLA is required in both cases. In fact, each change involves new requirement(s) measures with different metrics [23]. For example, the memory size can be given several values with different prices. The requirements are specified as high-level requirements, means no need for specifying the new requirements in a new SLA.

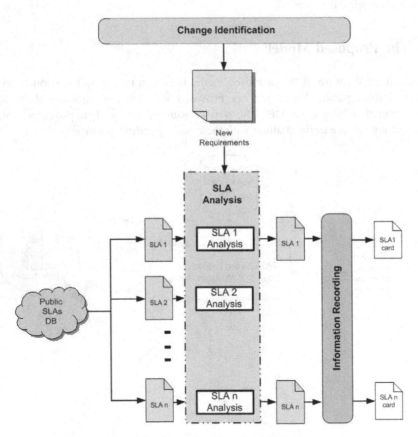

Fig. 2. The ROA-based dynamic cloud SLA management

4.2   SLA Analysis

In this step, new requirements (initiated by any new change) and public SLAs are analyzed to find out the more related SLAs based on the ROA concept. The execution process of the SLA analysis is illustrated in Fig 3 in an algorithm format.

```
01:    INPUT: CR_record; Public_SLAs_DB
02:    OUTPUT: Selected_SLA and SLAs_Cards_List.
03:    Let Selected_SLA=null
04:    Let SLA_card=null;
05:    Let NRSs= required Number of suggested solutions (SLAs)
06:    Enter the NRSs
07:    For each SLA in Public_SLAs_DB
08:    Let i=1; While (i<CRnum)
09:       If CRrecord[i] found inside SLA then rank=rank+1
10:         If ((i- rank>2)
11:            i= CRnum //switching option (jump to another SLA)
12:         If End
13:         Else i=i+1 // Sequential option (test the next step)
14:    While End
15:       If (Rank equal CRnum) then
16:         If (SLA's requirements equal CRnum)
17:            Begin
18:              Selected_SLA= SLA //execute this solution
19:              Add SLA and rank to SLA_card
20:         If End
21:         Else add SLA and rank to SLAs_Cards_List // reformulate
22:    Else
23:    If ((rank +2)=>CRnum){
24:       Add rank to SLA_card
25:       Add additional requirements to SLA_card
26:       Add uncovered requirements to SLA_card
27:    End If
28:    End For
29:    Return Selected_SLA and SLAs_Cards_List
30:    End
```

Fig. 3. The execution process of the SLA analysis

As illustrated in Fig 3, the analysis process receives both the new high-level requirements with their priorities and public SLAs. The new high-level requirements are selected as a metrics and specified in what is called CR_record. For each public SLA, the new metrics (high-level requirements) are compared to the SLA's metrics. The result is calculated based on the number of the matched metrics, additional requirements and uncovered requirements. We assumed that the number of new requirements (change) is 10. As shown in line 19, an SLA is executed if all requirements are satisfied without any extra requirements defined inside the SLA. If there are extra requirements, so the re-formulate option is selected. The output of this step is a list of several related SLA (solutions) where one of them will be executed while the other solutions will be recorded for any further change or update.

4.3 Information Recording

As mentioned earlier, the output of the analysis process will be a list of some suggested SLAs. Whereas, one of this suggested SLA will be executed as the more appropriate solution. However, some information about the suggested SLAs and new requirements is stored in XML-based file called SLAs_Cards_List. This information will help to resolve any required change in the future. Fig 4 shows an example of the

SLAs cards list which contains one or more SLA card. Each card includes four sub-elements namely *SAL_identifier*, *Rank*, *Additional_Requirements*, and *Uncovered_Requirements*.

```
<?xml version="1.0" encoding="UTF-8"?>
<SLAS_Cards_List>
    <SLA_Card>
        <SLA_identifier>PC12</ SLA_identifier>
        <Rank>7</Rank>
        <Additional_Requirement>
            <Name>SCALABILITY</Name>
            <Description>Computing platform scalability
        </Description>
            <Operations> run </Operations>
            <Address>//resources/PC12</Address>
        </Additional_Requirement>
        <Uncovered_Requirement>
            <Name>anyname</Name>
            <Description>asasa</Description>
            <Operations></Operations>
            <XPath>//SigmodRecord/issue[1]</XPath>
        </Uncovered_Requirement>
        <SLA_Level>5</SLA_Level>
    </SLA_Card>
</SLAS_Cards_List>
```

Fig. 4. An example of SLAs cards list

The *SLA_identifier* is a unique identifier of the corresponding SLA. The Rank is a number of the matched requirements found inside the SLA. The *Additional_Requirements* sub-element lists all additional requirements found inside the SLA and not related/listed to/in the new requirements. The *Uncovered_Requirements* contains all new requirements not found inside SLA.

5   SLA Specification

For deploying the proposed model, the SLA specification issue must be considered. In other words, the SLA specification language need to be chosen and modified to meet the proposed model's components. Several SLA specification languages have been proposed such as Web Service Agreement (WS-Agreement) and Web Service Level Agreement language and framework (WSLA). These languages are used for specifying the SLAs contents. In this research, the WSLA will be used as it is the widely accepted language. Using WSLA, each cloud SLA has three main parts, namely parties, service definition and obligation. The parties are service provider, consumer and other supporting parties such as cloud broker. The service definition is the service level parameters which are specified as metrics and used to define how the provided service is measured. Each metric can have several values for the same service. Different metrics are defined for different cloud services.

According to Alhamad *et al.*, [21], the SLA metrics for IaaS are memory size, boot time, storage size, scale up, scale down, scale up time, scale down time, auto scaling, and maximum number of VMs, availability, and response time. For the PaaS, the SLA metrics are integration, scalability, pay-as-you-go billing, servers, browsers, number of developers, and development environments. The SLA metrics for SaaS are defined as reliability, scalability, usability, customizability and availability. Finally, the obligation includes the service level objectives which are actually formal expressions defined as "if …then…" statement.

However, the parameter structure within WSLA is modified in such a way that each SLA parameter will have three elements namely metric, type and priority instead of metric only. The SLA parameter type determines whether the requirement (defined by the parameter) is internal or external. In this research, the cloud services' changes (new requirements) are assumed to be classified into internal and external changes. The internal changes are any change defined/required by the cloud computing parties such as service provider, consumer, broker, etc. It can be technical or financial change. The latter are the external change required by an external party such as government, policy, etc. Finally, the priority is used for ranking the requirement and – later on – deciding which ROA option is suitable for this requirement.

In addition, the new high-level requirements are specified as measurement metrics. The priority of each measurement metric has also to be specified. Regarding the public SLAs, since the cloud computing provides several services (such as IaaS, PaaS, and SaaS) so each service has its own public SLAs. This means that the number of public SLAs categories – inside the public SLAs DB - will be equal to the number of the provided services.

6   Conclusion and Future Work

The paper proposed a preliminary framework for SLA dynamic management model in a component based approach based on the real options analysis (ROA) concept to address the issue of dynamically changing user requirements and the emerging technology. The new requirements need to be mapped to all public SLAs only once. The result will be suggested based on the SLAs list along with some other related solutions (or SLAs) considering current scenario and future expected change. The number of the related solutions can be varied and may have entailed financial and system considerations. As such, cloud administer can rationally factor into their decisions greater volatile solutions. Then, if any new change happens and affects a specific SLA, the new change is mapped to only the related solutions. The future intention of this research is to come up with a comprehensive model for dynamically changing cloud SLA management with deployment details and case studies from cloud providers and cloud user's perspectives. Furthermore, we will also investigate matrices for intelligent choices made by cloud administrator because volatile options hysteresis may increase.

Acknowledgement. This work is supported by the Research Center of College of Computer and Information Sciences (CRC) at King Saud University, Grant Number RC140208. The authors are grateful for this support.

References

1. Hewitt, C.: ORGs for Scalable, Robust, Privacy-Friendly Client Cloud Computing. IEEE Internet Computing 12(5), 96–99 (2008)
2. Halboob, W., Mahmod, R., Alghathbar, K., Mamat, A.: TC-enabled and Distributed Cloud Computing Access Control. Journal of Applied Science 14(7), 620–630 (2014)
3. Patel, P., Ranabahu, A., Sheth, A.: Service Level Agreement in Cloud Computing: Knoesis Center. Wright State University (2009)
4. He, Q., Yan, J., Kowalczyk, R., Jin, H., Yang, Y.: Lifetime service level agreement management with autonomous agents for services provision. Information Sciences 17(2009), 2591–2605 (2009)
5. Brandic, I., Music, D., Dustdar, S., Venugopal, S., Buyya, R.: Advanced QoS Methods for Grid Workflows Based on Meta-Negotiations and SLA-Mappings. In: 3rd Workshop on Workflows in Support of Large-Scale Science (2008)
6. Risch, M., Brandic, I., Altmann, J.: Using SLA mapping to increase market liquidity. In: Dan, A., Gittler, F., Toumani, F. (eds.) ICSOC/ServiceWave 2009. LNCS, vol. 6275, pp. 238–247. Springer, Heidelberg (2010)
7. Brandic, I., Music, D., Dustdar, S.: Service Mediation and Negotiation Bootstrapping as First Achievements Towards Self-adaptable Grid and Cloud Services. Paper Presented at the 6th International Conference Industry Session on Grids Meets Autonomic Computing (GMAC), Barcelona, Spain (2009)
8. Maurera, M., Emeakarohaa, V.C., Brandica, I., Altmannb, J.: Cost–benefit analysis of an SLA mapping approach for defining standardized Cloud computing goods. Future Generation Computer Systems 28(1), 39–47 (2011)
9. Padhy, R.P., Patra, M.R., Satapathy, S.C.: SLAs in Cloud Systems: The Business Perspective. International Journal of Computer Science and Telecommunications 3(1), 481–488 (2012)
10. Abbas, H., Magnusson, C., Yngstrom, L., Hemani, A.: Addressing dynamic issues in information security management. Information Security Management 19(1), 5–24 (2011)
11. Redl, C., Breskovic, I., Brandic, I., & Dustdar, S. Automatic SLA Matching and Provider Selection in Grid and Cloud Computing Markets. Paper Presented at the 13th CM/IEEE International Conference on Grid Computing (GRID 2012), Beijing, China (2012)
12. Ludwig, H., Keller, A., Dan, A., King, R.P., Franck, R.: Web service level agreement (WSLA) language specification. IBM Corporation (2003)
13. Kritikos, K., Plexousakis, D.: Requirements for QoS-Based Web Service Description and Discovery. IEEE Transaction on Service Computing 2(4), 320–337 (2009)
14. Tran, V.X., Tsuji, H.: A Survey and Analysis on Semantics in QoS for Web Services. Paper presented at the IEEE 23rd International Conference on Advanced Information Networking and Applications (AINA), Bradford, United Kingdom (2009)
15. Chudasma, N., Chaudhary, S.: Service Composition Using Service Selection with WS-Agreement. Paper presented at the 2nd Bangalore Annual Compute Conference (Compute, Bangalore, India (2009)

16. Muñoz, H., Kotsiopoulos, I., Micsik, A., Koller, B., Mora, J.: Flexible SLA negotiation using semantic annotations. In: Dan, A., Gittler, F., Toumani, F. (eds.) ICSOC/ServiceWave 2009. LNCS, vol. ICSOC, pp. 165–175. Springer, Heidelberg (2010)
17. Fakhfakh, K., Tazi, S., Drira, K., Chaari, T., Jmaiel, M.: Implementing and Testing a Semantic-Driven Approach towards a Better Comprehension between Service Consumers and Providers. Paper presented at the IEEE 24th International Conference on Advanced Information Networking and Applications Workshops (WAINA 2010), Perth, Australia (2010)
18. Abbas, H., Yngström, L., Hemani, A.: Adaptability infrastructure for bridging IT security evaluation and options theory. Paper presented at the 2nd International Conference on Security of Information and Networks SIN 2009, Famagusta, Cyprus (2009)
19. Abbas, H., Yngström, L., Hemani, A.: Option based evaluation: Security evaluation of IT products based on options theory. Paper presented at the 1st Eastern European Regional Conference on the Engineering of Computer-Based Systems: Setting New ECBS Frontiers, Novi Sad, Serbia
20. Abbas, H., Magnusson, C., Yngström, L., Hemani, A.: A structured approach for internalizing externalities caused by IT security mechanisms. Paper presented at 2nd International Workshop on Education Technology and Computer Science (ETCS 2010), Wuhan, Hubei, China (2010)
21. Alhamad, M., Dillon, T., Chang, E.: Conceptual SLA Framework for Cloud Computing. Paper presented at the 4th IEEE International Conference on Digital Ecosystems and Technologies (IEEE DEST, Dubai, UAE (2010)
22. Latif, R., Abbas, H., Assar, S., Ali, Q.: Cloud Computing Risk Assessment: A Systematic Literature Review. In: Park, J.J(J.H.), Stojmenovic, I., Choi, M., Xhafa, F. (eds.) Future Information Technology. LNEE, vol. 276, pp. 285–295. Springer, Heidelberg (2014)
23. Naveed, R., Abbas, H.: Security Requirements Specification Framework for Cloud Users. In: Park, J.J(J.H.), Stojmenovic, I., Choi, M., Xhafa, F. (eds.) Future Information Technology. Lecture Notes in Electrical Engineering, vol. 276, pp. 297–305. Springer, Heidelberg (2014)

Foundry Material Design with Artificial Intelligence

Jingjing Zhao[1], Xingtong Liu[2], Afeng Yang[2], and Chun Du[2]

[1] College of Humanities and Social Sciences, National University of Defense Technology,
Changsha, Hunan, P.R. China, 410073
[2] College of Electronic Science and Engineering, National University of Defense Technology,
Changsha, Hunan, P.R. China, 410073
zhaojingjing63@gmail.com

Abstract. There are two main development trends in the modeling techniques for the manufacturing procedure of foundry material. One is based on numerical simulation, the other is based on artificial intelligence. The numerical simulation method depend on the strict mathematics model and scientific mechanism, therefore, it is difficult to predict the final structure and properties by computers. The artificial intelligence method can learn from the empirical data, summarize regularity, automatically build models, and predict the future as human brain does. Focusing on the complexity of foundry material design and combining advanced research in the field of artificial intelligence, this paper describes the application and development orientation about artificial intelligence technology in foundry material design, expounds the features of various technologies. In particular, the applications of artificial intelligence in some way of foundry material properties predicting are summarized.

Keywords: foundry material design, artificial intelligence, foundry material performance modeling, quality prediction of sand casting process.

1   Introduction

With the development of technology, the requirement of new foundry materials increases continuously. Study and production of foundry materials become more and more complex, to solve this problem, multi-discipline knowledge (foundry, physics, chemistry, mathematics, artificial intelligence, computer science) and multi-science methods (laboratory analysis, induction and deduction method) should be combined organically. This gives rise to a novel concept of "foundry material design", which focuses on how to predict and optimize the performance of foundry material by modeling the chemical composition, structure and physical property through scientific theory and computation.

Theoretically, the foundry material design can be realized by building a mechanism model. However, in practice, the theory study of the foundry material design processing is far behind the practical application. The researchers could not design the foundry material only in theory. Therefore, it's a viable method to build models that reflect the relationship between the foundry material components, technology, and

D.-S. Huang et al. (Eds.): ICIC 2014, LNAI 8589, pp. 444–455, 2014.
© Springer International Publishing Switzerland 2014

properties of material according to the empirical data, which come from scientific research and productive practice. Optimizing the components and technology by using effective optimal method makes the empirical design of the foundry material to be possible.

In order to design foundry material precisely, we need to create a reasonable and accurate simulation model at first. However, due to the manufacturing procedure of material, whose dynamic process covers many complicated variation such as temperature field, flow field, structure field, stress field, deformation field, melting, solidification and so on, refers to the micro-properties and macro-dimensional accuracy of material structure, the production mechanism of foundry material is still unclear. In recent years, with the rapid development of computer and information science, it is possible to carry out computer simulation for foundry material. Lots of casting numerical simulation technologies have been used in the industry production. Some software (FT-Star, CAE et al), which combine the traditional foundry method and computer simulation technology, as important tools in casting production are playing an increasingly important role [3]. To date, there are two main development trends in the modeling techniques for the manufacturing procedure of foundry material. One is based on numerical simulation, the other is based on artificial intelligence. In general, when the casting method can be described by a strict mathematics model and has exact analytic solution, it can enter the practical stage and have a satisfying application. The most representative casting numerical simulation technologies are: the solidification simulation [3-6] which based on the Fourier's Law, the mold filling simulation [3, 7-9] which based on momentum equation and mass conservation equation, the simulation of mold filling with heat transfer [10-12] which based on momentum equation, mass conservation equation, and energy equation. However, due to the complex of foundry production, there are multiple factors effects the processing, therefore, it is difficult to predict the final structure and properties by computers. Since the mechanisms are unknown, the traditional computer simulation can't build mathematics models exactly to reflect the relations. Numerical simulation is applied under the condition that the mechanism model of foundry material is known. In the circumstances that the mechanism model is unknown, one should rely on artificial experience, i.e. adopting artificial intelligence method.

Artificial intelligence [13] is also called machine intelligence. It is an interdisciplinary subject formed by the computer science, control theory, information theory and neurophysiology. Unlike the traditional computer simulation, artificial intelligence can discover underlying rules and extract useful knowledge from the empirical data when the mechanisms are unknown. The researchers can get a bridge between the foundry theory and experimental data by using the method that combined with induction and deduction. The advent of the artificial intelligence provides an effective method for intelligent foundry material design.

2   Artificial Intelligence Method

It's hoped that the computer has the ability of learning and thinking, since the first computer was developed. For some problems that people knows the mechanism and the mathematics formula, the researcher only need to design the algorithm and data structure for computer, then they will get the solution by the numerical simulation et al. However, there are some problems that we haven't found the complex mechanisms. We do believe that there are some processes that can explain the problems. Though we don't know the details of the underlying process, we know that it is not completely random. Though identifying the complete process may not be possible, we can still detect certain patterns or regularities. We may not be able to identify the problem completely, but we can construct a good and useful approximation. The approximation may not explain everything, however, it can account for some part of the problem. We can get the regularities by observation and learning, and predict the future event by the regularities. Artificial intelligence is programming computers to optimize a performance criterion using example data or past experience. With artificial intelligence, computer can gain knowledge from data automatically, and build a predictive model to make predictions in the future. Predictive ability has always been the key to intelligence.

The foundry material production is one kind of typical complex process, which combines a lot of uncertain factors. In particular, the final structure and performance of the foundry material are affected by lots of uncertain factors. The related structure and parameters are time-variant. What's more, there are a lot of disturbances in the environment, which leads to the fuzzy of the data. It is difficult to obtain the mathematical model of such complex process, thus the traditional numerical simulation methods can't solve the problems effectively. Therefore, we need to use the artificial intelligence technologies to supervise the computer learning, according to the empirical knowledge, and built a model to predict the final properties of the foundry materials. It is subsidiary to the foundry production, providing a new effective way for designing new foundry materials. According to the industrial applications, we introduce some common artificial intelligence technologies that are commonly used in foundry material design: statistical regression, artificial neural network, and support vector machine.

2.1   Statistical Regression

Statistical regression analysis is a common method that deals with the foundry material data. It first studies the relationships among multiple variables by using the experimental data and observation data, then builds the mathematics model based on the correlation, and then uses the model to solve the problems of prediction, optimization and controls. Statistical regression analysis can be used to study the relationship among the material component, organizational structures and physicochemical properties of foundry materials. It also can be used to study the dynamic change rule between the properties and the technology of the foundry processing. In foundry experiment, the statistical regression analysis is performed as follows. It first builds a regression model by using the experimental data, then predicts the properties of the foundry material and optimizes the technology parameters.

The traditional regression methods include least square method, multiple linear regression, multiple nonlinear regression, principal component regression, partial least squares regression, and so on. In the field of foundry industry, the primal data analysis methods are linear regression. As to the nonlinear data, they should be mapped to the linear space firstly. Then the analysis is performed in the linear space. In general, the experimental results are affected by the foundry environment. Since there are measurement errors in the dataset, the traditional statistical regression analysis can only get an approximate expression to show the regularity. Therefore, the modern statistical regression methods often combine with other artificial intelligence technology to have the ability of self-learning and nonlinear fitting. The combination of artificial intelligence technology and computer science application can improve the efficiency of the foundry material data processing, and help the researchers to analyze and predict the performance more exactly.

2.2   Artificial Neural Network

Artificial neural network is one kind of complex network that is constructed by a series of neurons with a certain structure and weight. It can simulate the structure and behavior of the neural network in human brain. Artificial neural network automatically extracts features from the input and output of the objects that need to be processed, gets the empirical knowledge and stores it in the combination of the network. Artificial neural network has some characteristics of the biology neural network, such as self-learning, self-organization, association, and fault-tolerant. Therefore, it is used to solve the linear and nonlinear problems, and is applied widely in the foundry area. It can not only solve the qualitative problem, but also deal with the quantitative problem. Artificial neural network doesn't depend on any empirical knowledge, it automatically summarizes the regularity of the foundry material data. Most of modern regression analysis models are based on artificial neural network, therefore, the regression algorithms have the function of self-learning, and the ability of the nonlinear regression. The typical artificial neural networks are back-propagation neural network, radial basis function neural network, and so on. The combination of the artificial neural network models and the computer science technology improve the efficiency of the regression analysis, and help the researchers to predict the foundry results more accurately.

2.3   Support Vector Machine

Support vector machine is a machine learning method. It is under the principle of structural risk minimization and has best overall solver. Support vector machine gets a good trade-off between complexity and learning ability with limited samples. It performs very well in the problems of classify and regression. Support vector machine can avoid the disadvantage that caused by the traditional machine learning method, such as overfitting, dimension disaster, and local minimization problem. It is easy to generalize under the small-sample size cases, and it is widely used in the foundry material design. On one hand, the foundry material design involves complex process, such as microstructures, physical change, chemical change, electromechanical properties, and so on.

The properties of foundry material are affected by technics parameters, procedures, and machining accuracy. Since there are so many factors affect the foundry processing, the modeling method should be able to solve the problem of dimension disaster. On the other hand, since the working condition of the thermal processing is bad, it is hard to get effective samples of the material design. The traditional statistical analysis methods are robust only if the numbers of samples tends to infinity. However, support vector machine studies the machine learning problem in small-sample size cases, therefore, it is effective to deal with the regression problem when the samples are very limited.

3   Applications of Artificial Intelligence in Foundry Material Design

With the development of foundry production and scientific experiment, lots of foundry material dataset have been built recently. Extracting the regularities from the empirical knowledge and experimental data by using the artificial intelligent technology is a new trend of foundry material design. Applications of artificial intelligence in foundry material design includes building relationship between material structure and properties, predicting chemical change, casting technology modeling and optimization, process control and fault diagnosis et al. From the aspect of application, we review two main technologies: properties predicting and quality prediction of sand casting process in this paper.

3.1   Properties Predicting

In general, the purpose of the properties predicting is to obtain the relationship among material component, material structure and the properties, such as: quantitative structure-activity relationship, the material constitutive relation, and material component-technology-structure-performance relation. On one hand, the properties of the foundry materials are affected by multiple factors. For instance, the mechanical property is the function of material component, organization structure, and foundry technology. There is a complex non-linear relationship between the mechanical property and multiple factors. On the other hand, theoretically, it is need to build an intelligent model to map the complex non-linear relationship. The model can help the researchers to predict the foundry material properties, and reduce the experimental procedure.

As to the application of artificial intelligence in foundry material design, the earlier models are based on the theory of statistical regression. Pickering proposed the method of alloy design [14]. They applied the regression model to design alloy, and obtained the good performance and microstructure parameters by using the control technology. Yu et al [15] measured the mechanical properties and microstructural parameters, and made the comprehensive chemical analysis. They got the regressive formulae of the relationship between mechanical properties and chemical compositions along with microstructural parameters by adopting statistical method. In the paper, they also compared the chemical compositions between steel products of Angang and Baogang Iron and Steel Works. Zhao et al [16] performed the low cycle fatigue testing on Nickel base super alloy at three different temperatures. The strain-life and cyclic stress-strain relationship at various temperatures were given by using the least square method. They discussed the change in fatigue life behavior and fatigue parameters with temperature, and found out that the fatigue life decrease with the increasing of temperature at low and intermediate total strain amplitudes.

Artificial neural network has some special advantages that the traditional regression methods don't possess, such as self-organization, self-adaptation and self-learning. Therefore, artificial neural network become a useful mathematics modeling method and has been extensively applied in the material design. Malinov et al proposed a model for the analysis and prediction of the correlation between processing parameters and mechanical properties in titanium alloys by using artificial neural network [17]. The input parameters of the artificial neural network are alloy composition, heat treatment parameters and work temperature. The outputs of the artificial neural network model are nine most important mechanical properties namely ultimate tensile strength, tensile yield strength, elongation, reduction of area, impact strength, hardness, modulus of elasticity, fatigue strength and fracture toughness. The model can be used for predicting properties of titanium alloys at different temperatures as functions of processing parameters and heat treatment cycle. It can also be used for the optimization of processing and heat treatment parameters. Li et al [18] applied an artificial neural network to acquire the relationships between the mechanical properties and the deformation technological parameters of TC11 titanium alloy. In establishing the relationships, they took the deformation temperature and the true strain as the inputs, and took the ultimate tensile strength, the yield strength, the elongation and the area reduction at fracture as the outputs, respectively. Comparison of the predicted and experimental results shows that the neural network model used to predict the relationships of the mechanical properties of the forged TC11 titanium alloy has good learning precision and good generalization. The neural network method proposed in this paper has been found to have advantage of being able to treat noisy data, or data with strong non-linear relationships. Huang et al [19] applied an artificial neural network to predict the mechanical properties of the ceramic tool. The non-linear mapping relationship between the component, the composition content of raw material, the flexural strength, and the fracture toughness of the composite ceramic tool is investigated. The model for predicting the mechanical properties of the alumina matrix ceramic tool is established by means of an artificial neural network. The mechanical properties of two-phase and three-phase composite ceramic tools such as Al_2O_3–(W, Ti)C and Al_2O_3–TiC–ZrO_2 are used to verify the proposed model. The research results indicate that the model based on artificial neural network is available and effective in simulating the composition content and predicting the mechanical properties of the ceramic tool. Zhou et al [20] developed a model for predicting the correlation between process parameters of second aging treatment and proper ties of 7055 Al alloy by applying the artificial neural networks. The model was based on error back-propagation algorithm and trained by Levenberg-Marquardt training algorithm. The experiment results show that the model has high precision and good generalization performance, and can be successfully used to predict and analyze the influence of secondary aging treatment on the mechanical properties of 7055 Al alloy . Fang et al

[21] qualitatively analyzed the mechanical properties of 7005 Al alloys by partial least squares method and quantitatively calculated the mechanical properties of 7005 Al alloys by using back propagation artificial neural network with the same processing parameters as features. The computation results are in substantial agreement with experimental ones. The partial least squares method can compute the scores for the principal components according to the sorts of components and thus compress the input data for back propagation artificial neural network with linear arithmetic. The proposed model is suitable for forecasting the mechanical properties of the alloys.

Although artificial neural network does very well in the prediction of material performance, it is robust only if the number of samples tends to infinity. As we have discussed above, it is hard to get effective samples of the material sometimes, because the working condition of the thermal processing is bad. Therefore, it is very common case that the material samples are very limited in practice. Fortunately, support vector machine studies the machine learning problems in small–sample size case, it is effective to deal with the regression problem when the samples are very limited. Recently, support vector machine has been widely used in the application of material design. Wu et al [22] used support vector machine to build the mathematical model for predicting the mechanical properties of 42CrMo steel. The experimental results show that the relative maximum predicting error is 4.78% under the condition of using a small quantity of samples, and the predicting precision of the model maintains in a steady extent with the rising of the training precision. They also found out that the generalization of support vector machine model is obviously better than the model of artificial neural network. The support vector machine technology is suitable to be used to build mathematical model for predicting mechanical properties of materials after heat treatment, and it can solve the contradiction better between a small quantity of samples and predicting precision. Cai et al [23] used the support vector regression approach based on the particle swarm optimization, and combined with leave-one-out cross validation to predict the mechanical properties of 7005 Al alloys under different processing parameters including extrusion temperature, extrusion velocity, quenching type and aging time. The results strongly support that the prediction precision of their model is superior to those of partial least squares, back-propagation neural networks and their combination model by applying the identical dataset. According to an experimental data set on the superconducting transition temperature of 21 BiPbSrCaCuOF superconductors under different process parameters including the amount of bismuth, amount of oxygen, sintering time and sintering temperature, a model, which combined support vector regression with particle swarm optimization for its parameter optimization, was proposed to establish a model for prediction of the Tc of BiPbSrCaCuOF superconductors [24]. The performance of the proposed model was compared with that of back-propagation neural network and multivariable linear regression model. The results show that the mean absolute error and mean absolute percentage error of test samples achieved by the proposed model are smaller than those achieved by others. This study suggests that support vector regression as a novel approach has a theoretical significance and potential practical value in development of high-Tc superconductor via guiding experiment. Cai et al [25] applied a support vector machine model to establish the porosities of NiTi alloys synthesized by self-propagation high-temperature synthesis approach under different process parameters, including temperature, particle size and green density. The prediction results indicate that the mean absolute percentage error achieved by the support vector regression model is smaller, and it is more accurate than that of back-propagation neural network for identical training and test samples. The results indicate that the prediction ability of support vector machine is superior to that of back-propagation neural network; the mean absolute percentage error predicted by leave-one-out test of the support vector regression model is also slightly better than that of back-propagation neural network. Therefore, it is demonstrated that support vector regression is a promising and practical technique to estimate the porosity of porous NiTi alloy synthesized under different self-propagation high-temperature synthesis process parameters, and can provide a reasonable guidance for the self-propagation high-temperature synthesis of porous NiTi theoretically. Zhao [26] investigated the influences of the contents of alternative elements in the hydrogen storage alloy, the experimental parameters of copper plating process on the hydrogen storage capacity and electrochemical property by using support vector regression modeling approach combined with particle swarm optimization. The modeling results reveal that support vector regression possesses excellent prediction accuracy and reliability. The comparative analysis between the calculation results of support vector regression model and those of the conventional quadratic polynomial model demonstrate that the accuracy of support vector machine model is superior to that of the traditional polynomial model when dealing with relatively small size of experimental data. They also investigated the influence of the contents of substitution elements (Ce, Pr, Nd) on the maximum discharge capacity and high rate dischargeability. They studied the effect of substitution of Cr with Fe and Mn on the maximum and effective hydrogen storage capacity of Ti-Cr-V hydrogen storage alloys, and optimized the experimental parameters of copper plating on the AB_5 alloy surface. The experimental results show that, in the researches of the hydrogen storage alloy, support vector regression is an effective modeling technique to analyze the experimental data, and play a significant role in the optimization of the foundry material design.

3.2   Quality Prediction of Sand Casting Process

Sand casting is most popular casting process. The molding sand is the most important casting material because of its low cost and high efficiency. The quality of molding sand directly affects the appearance quality of the castings. More than 50% casting defects, such as sand hole, sand inclusion and gas porosity, are caused by sand quality or improper use. Controlling the properties of sand strictly is the best way to deduce casting defects [27]. Meanwhile, the quality of green sand and its casting processing will be benefit of the quality prediction for sand casting. There are so many indicators which determine the qualities of sand castings, most of them are with the characteristics of fuzziness and gray character, therefore, the forecast of sand mold quality is extremely complicated and difficult. What' s more, sand casting processes involve a series of complicated changes, it is hard to obtain accurate mathematical models of sand casting processes by using classical numerical simulation methods.

However, obtaining knowledge models of sand casting processes by using the artificial intelligence technology is helpful for understanding the casting processes and realizing the intelligent control of the foundry processes.

Support vector regression modeling method, which is an artificial intelligence technology that based on structural risk minimization principle, has strong generalization ability. For complex process modeling with unknown mechanism, it has a good applicability. Therefore, Shanghai Jiao Tong University introduced support vector regression modeling method into the sand casting, made necessary formalization of related problems and provided relevant solutions in each step of support vector regression modeling for hot work processes. Eventually, each step was modularized, COSVRKDSHT [28] and MPSVR [29] knowledge modeling software systems were developed. Based on HUNTER sand casting production line for Piston Ring of Huizhong Automotive Axle Works, modeling experiments on obtaining knowledge models of green sand casting process using software systems are carried out. The quality prediction methods not only setup the relation between parameters of sand properties and casting quality, but also predict the quantitative defects. The flow chart of the software MPSVR [29] is shown in Fig. 1, the details are as follows.

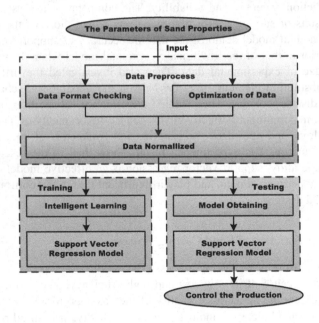

Fig. 1. The flow chart of software MPSVR

The designing of the model is based on the comprehensive experience and the theoretical analysis [29]. The sand compactibility, moisture content, green compression strength, and the ratio of the sand compactibility and moisture content are chosen as the parameters of the sand properties. They are the input variables of the intelligent prediction model. The quality of the casting process is measured by the ratio of the casting defects. It is the output variable of the intelligent prediction model that based on the support vector regression. Five defects are considered here, they are sand in-clusion defects, shrinkage defects, cracking defects, less meat defects and other de-fects. Then the samples are taken from the HUNTER sand casting production line for Piston Ring of Huizhong Automotive Axle Works. They chose 120 groups of data to do the experiment, and each group includes 5000 casting productions. 100 groups of experimental data are randomly chosen to be the training data for intelligent learning, the others are for testing.

To improve the accuracy of the system implementation, the input data need to be pre-processed. The data format is checked first and the noise is removed from the dataset. When the user's input is not belongs to the scope of the technology data, the system will remind the user and ignore the useless data. Then dimensionless method is applied to the data, the data is normalized, thus they can remove the effect of the dimension. At last, they use the empirical data to simulate the sand casting process, then the quality prediction model that based on the support vector regression is built, thus they can predict the quality of sand casting. To obtain the accuracy of the system, the researchers used the real production model to test the performance of the artificial intelligence model. The experimental results indicate that knowledge model of green sand casting process using support vector regression method possesses good generali-zation capability as well as high comprehensibility. The prediction curve coincides with the actual measured curve very well. Support vector regression method is intui-tive, reliable and precise, which makes up for deficiencies in past traditional methods. The artificial intelligence modeling method can effectively obtain the models of com-plex processes with a great number of uncertainties, such as green sand casting pro-cesses.

4   Conclusion

A new type of material is usually developed with the "trial-and-error" method, which wastes a lot of time and effort. With the advancement of artificial intelligence and material science, material design can be carried out based on known knowledge and experience of the materials, with the aid of the computer. The compositions and con-tents of the materials to be developed may be designed and simulated in accordance with the requirement for the mechanical properties. The development process of a new material can be carried out based on the simulated information of the material composition and content.

The study of material design can help the researchers to design the experimental scheme in a cheap and scientific way. By using artificial intelligence, the develop-ment technology processing of the material design will be improved, the period of developing production will be shorten. It provides a new way to design the materials effectively and exactly. Nowadays, designing the materials with artificial intelligence is just in the initial stages. There are some problems need to be further studied, how-ever, artificial intelligence methods have provided useful insights into the develop-ment of material design from the "qualitative" research to the "quantitative" research.

Acknowledgements. This work was supported by the National Natural Science Foundation of China (NO.61302091).

References

1. Saito, T.: Computational Material Design. Springer, Heidelberg (1999)
2. Zhou, J.: Current Status And Development Trend of Casting Numerical Simulation Technology. Foundry 10, 1105–1115 (2012)
3. Zhou, J., Liao, D.: Casting CAD/CAE. Chemical Industry Press (2009)
4. Sun, X., An, G., Su, S., Wang, J.: Recent Development In Numerical Siumulation of Mold Filling And Solidification Processes of Castings. Foundry 49(2), 84–88 (2000)
5. Liu, X., Huang, T., Kang, J.: Analysis On Micro Cracks Defect of Francis Turbine Band Casting Based On Solidification Process Simulation. Foundry 58(10), 1034–1037 (2009)
6. Liu, X., Huang, T., Kang, J., Bian, D., Zhang, L.: Numerical Simulation of The Solidification Process of A Heavy Roll Stand For Temperature Field And Thermal Stress Field. Foundry 55(9), 922–926 (2006)
7. Wang, J., Sun, X., Guan, Y., et al.: Numerical Simulation And Process Optimization For Producing Large-Sized Castings. China Foundry 5(3), 179–185 (2008)
8. Guan, Y., Li, B., Sun, X., Wang, T.: Study On The Pre-Processing And Post-Processing Technology For The Numerical Simulation of Mold Filling And Solidification Processes of Castings. Foundry 53(11), 905–908 (2004)
9. Yan, Q., Tu, Z., Lu, G., Zhang, S., Xiong, B.: Numerical Simulation of Filling Process of Vacuum Counter-Pressure Casting Aluminum Alloy Based On The Fluent Software. Special Casting And Nonferrous Alloys 33(4), 330–333 (2013)
10. Xu, Y., Kang, J., Huang, T.: Numerical Simulation of The Casting Process With Coupled Thermal And Mechanical Effects. Journal of Tsinghua University 48(5), 769–772 (2008)
11. Sun, L., Liao, D., Tu, C., Chen, T., Zhou, J.: Analysis of Thermal Stress During Solidification Process of Casting Based On FDM/FEM. Foundry 61(7), 737–746 (2012)
12. Fu, X., Liao, D., Zhou, J., Chen, T.: Bidirectional Coupled Simulation For Casting Thermal Stress Based On ANSYS. Foundry 60(11), 1103–1110 (2011)
13. Nilsson, N.J.: Artificial Intelligence: A New Synthesis. Morgan Kaufmann Publishers (1998)
14. Pickering, F.B.: Physical Metallurgy And The Design of Steels. Applied Science Publisher (1978)
15. Yu, Z., Yuan, Z., Li, D., Li, S., Wu, J.: Quantitative Relationship Between Mechanical Properties And Chemical Compositions of Steels Made By Anshan Iron And Steel Corporation. Journal of University of Science And Technology Beijing 19(5), 510–515 (1997)
16. Zhao, M., Xu, L., Zhang, K., Ren, P., Yang, B.: Study On High-Temperature Low Cycle Fatigue Behavior Of Nickel Base Superalloy OfGH536. Mechanical Science and Technology 21(2), 279–281 (2002)
17. Malinov, S., Sha, W., Mckeown, J.J.: Modelling The Correlation Between Processing Parameters And Properties In Titanium Alloys Using Artificial Neural Network. Computational Materials Seienee 21, 375–394 (2001)
18. Li, M., Liu, X., Xiong, A.: Prediction OfThe Mechanical Properties Of Forged TC11 Titanium Alloy By ANN. Joumal Of Materials Processing Teehnology 121, 1–4 (2002)
19. Huang, C., Zhang, L., He, L., et al.: A Study On The Prediction of The Mechanical Properties of A Ceramic Tool Based On An Artificial Neural Network. Joumal of Materials Processing Teehnology 129, 399–402 (2002)

20. Zhou, G., Zheng, Z., Li, H.: Predicting Properties For Secondary Aging of 7055 Al Alloy Based On Artificial Neural Networks. The Chinese Journal of Nonferrous Metals 16(9), 1583–1588 (2006)
21. Fang, S., Wang, M., Wang, Z., Qi, W., Li, Z.: Quantitative Study Between Mechanical Properties And Processing Parameters For 7005Al Alloys Predicted Using PLS-BPN Method. The Chinese Journal of Nonferrous Metals 17(12), 1948–1954 (2007)
22. Wu, L., Chen, Z.: Study On SVM-Based Mathematical Model Used To Predict Mechanical Properties of Materials After Heat Treatment. Transactions of Materials And Heat Treatment 28(6), 152–155 (2007)
23. Cai, C., Wen, Y., Zhu, X., et al.: Quantitative Prediction of Mechanical Properties of 7005 Al Alloys From Processing Parameters Via Support Vector Regression. The Chinese Journal of Nonferrous Metals 2, 323–328 (2010)
24. Cai, C., Zhu, X., Wen, Y., Pei, J., et al.: Predicting The Superconducting Transition Temperature Tc of Bipbsrcacu of Superconductors By Using Support Vector Regression. Journal of Superconductivity And Novel Magnetism 23(5), 737–740 (2010)
25. Cai, C., Wen, Y., Pei, J., et al.: Support Vector Regression Prediction of Porosity of Porous Niti Alloy By Self-Propagation High-Temperature Synthesis. Rare Metal Materials And Engineering 39(10), 1719–1722 (2010)
26. Zhao, S.: Study On Electrochemical Properties of Hydrogen Storage Alloys Influenced By Their Compositions Via SVR. Chongqing University (2012)
27. Hu, C.: A Study On Quality Control of Molding Sand. Shanghai Jiao Tong University (2008)
28. Huang, X.: Study On Support Vector Machine-Based Modeling Methods And Their Applications In Material Processing. Shanghai Jiao Tong University (2008)
29. Yang, F.: Research On Modeling of Materials Processing Based On Support Vector Regression Method. Shanghai Jiao Tong University (2010)

An Intelligent Agent Simulation Model to Evaluate Herd Behavior and Sales Effort in a Duopoly Market

Feng Li[1] and Ying Wei[2]

[1] School of Business Administration,
South China University of Technology, Guangzhou 510640, China
[2] Department of Business Administration, Jinan University, Guangzhou 510610, China

Abstract. This paper studies the impact of 'herd behavior' consumers in the market on demand of the products. In the duopoly market, two suppliers locate at two end of the line separately, providing an undifferentiated product. The potential consumers are dispersed along the line randomly. Beside those 'rational' consumers who have willing to buy from one of the two suppliers according to transportation cost from their positions to the two suppliers, other 'herd behavior' consumers may ignore the transportation cost, and change their mind to follow the most popular choice. To represent herd behavior of heterogeneous consumers, multi-agent based modeling and simulation is introduced to model and testify the operations of the market. Our numerical results show that 'herd behavior' consumers have a great influence on demand of two suppliers. Moreover, demand can be obviously increased once the first coming consumers are influenced by the supplier through sales efforts.

Keywords: Herd Behavior, Sales Effort, Duopoly Market, Agent-Based Modeling and Simulation.

1   Introduction

How much does sales effort have on consumers' purchasing decisions? This question is very important in marketing seeing that some consumers in the market are lazy to follow popular decision and promotional activities. Besides those consumers influenced directly by the sales effort, more consumers are influenced by those pioneers as pass-on effects, 'herd effect' named. Therefore, marketers need to understand and measure the impact of sales effort in given market.

Herd behavior, or 'herd effect' called, describes a social phenomenon in human society that a large numbers of people act in the same way at the same time [1]. It is very common in financial market, e.g. stock market bubbles, where individual investors disregard their own private information and mimic actions of the crowd. In research domain of behavior finance and behavior economics, literatures have usually identified the herding behavior of agents in global financial market by means of empirical research. However, seldom works addressed to quantify its social influences on invisible demand of products. Because it is impossible to observe decision process of all the customers, impact of herd behavior is difficult to be directly evaluated.

D.-S. Huang et al. (Eds.): ICIC 2014, LNAI 8589, pp. 456–465, 2014.

To solve this problem, an artificial market was developed, in which each customer was modeled as an intelligent agent. Through agent-based simulation, impact of herd behavior was examined from simulation results.

At the same time, various promotional activities are conventional tools to expand the market [2]. For example, supplier can stimulate demand by customer service before and after sales, celebrity endorsement, or TV advertisement, etc. The adoption of sales effort will significantly increase operational cost, but the increased demand may not benefit the supplier. So, the supplier needs to measure the value of sales effort carefully before it is put into practice. However, the presence of 'herd behavior' consumers complicated evaluation of the effects of sales effort. In this situation, some consumers are influenced by sales effort directly, while more potential consumers are influenced by these actual consumers. Therefore, these two part of consumers need to be identified as the increased demand. It is unrealistic to recognize each influenced consumer in real world market, so that many marketers are unconvinced about the success of sales effort. As a result, this paper evaluated the effects of sales effort through simulation. By modeling and simulating decision-making process of each consumer, the difference of demand is easy to be calculated.

The rest of the paper is organized as follows. We review literature in Section 2 and describe our multi-agent model in Section 3. Then, the numerical results and managerial insights are presented from multi-agent simulation on the given scenarios. Section 5 concludes the whole paper.

2   Related Works

The herd behavior phenomenon is very usual in economics, finance and human behavior, while it is seldom related to operations and supply chain management [3]. In the following, we review the literature in these domains.

Banerjee proposed a simple sequential decision model in which each decision-maker makes her own decision seeing the decisions made by previous decision-makers [1]. Analysis of this decision model shows that the decision rules that are chosen by optimizing individuals can be characterized by herd behavior.

Palley presented a different socio-economic model of managerial herd behavior based on 'safety in numbers' [4]. Assume that managers are individually risk-averse, the model is completely independent of the information signaling model given by Banerjee in [1].

Cipriani and Guarino did an experiment with undergraduate students to study herd behavior [5]. In their laboratory financial market, a phenomenon was observed not accounted for by the theory. In some cases, subjects ignore their private information to trade against the market.

Kim et al analyzed the herd behavior of returns for the won-dollar exchange rate and the Korean stock price index in Korean financial markets by method of empirical study [6]. Similar works include: Chen studied herd behavior of online book purchasing in students from a Taiwanese university [7]; Manahov and Hudson simulated four different artificial stock markets to analyze the behavioral foundations of stylized facts [8]; Chiang and Zheng examined herding behavior in global markets (18 countries) [9].

Generally, empirical study is common methodology to analyze herd behavior in these domains. Seldom research is found using modeling and simulation methods. There is only one paper mentioned ([8]) using agent based modeling and simulation to study herd effect in artificial stock markets. This paper also uses agent technology to model and simulate herd behavior in supply chain and operations system.

Various kinds of sales effort are frequently used to increase sales and profits. Wei and Chen addressed concurrent determination of inventory replenishment and sales effort decisions [2]. Market responses to sales efforts are typically highly uncertain, and demand in each period has its distribution dependent on the sales effort.

Heese and Swaminathan address an optimal strategy problem regarding the timing of promotional effort over a product life cycle [10]. To avoid stock-out occurrences, it is optimal to reduce sales effort at the beginning of a product life cycle. Such a reduction of sales effort allows a high service levels at low cost, and lower uncertain of demand.

Xing and Liu studies a dual-channel supply chain system with one manufacturer and two retailers [11]. One of the two retailers is online retailer who offers a lower price. The other is a brick-and-mortar retailer who has higher operational cost, including the sales effort to promote products. To encourage the brick-and-mortar retailer's sales effort and improve the operations of the supply chain, several contracts are proposed to help the manufacturer to coordinate the two retailers.

Lau et al use simple models (only one retailer and one manufacturer) to identify situations where it is worthwhile for a manufacturer to impose 'resale price maintenance' [12]. Results show that when demand depends on not only price but also sales effort, the classical 'double marginalization' effect may be reversed.

Lau et al investigates the performance of volume discounting schemes in a supply chain where the market demand is sensitive to both retail price and sales effort [13]. Using a simple and parsimonious model, this paper shows that 'continuous' volume discounting schemes lead to perfect channel coordination, while 'regular' volume discounting schemes cannot perfectly coordinate the channel.

These papers show that sales effort is a common variable or parameter to demand function in models of OR/OM. That also means market is sensitive to sales effort. However, the indirectly influence of sales effort to demand by way of 'herd effect' is not studies before. Therefore, this paper tries to assess influence of sales effort on demands by multi-agent based modeling and simulation.

3   A Multi-agent Model of a Duopoly Market

Consider a basic duopoly market model that there are two non-cooperative manufacturers (supplier A and supplier B named) producing a homogeneous product. Each potential consumer will buy and only buy one piece of product from either manufacturer. This market is a classic location model- Hotelling's model that two suppliers located at two end of the line separately. Furthermore, assuming that all consumers are dispersed along the line, and each consumer will buy from the 'closest' supplier because the product from the two suppliers is no difference.

Without loss of generality, suppose that the supplier A is at location 0 of the line, the supplier B at location 1. Heterogeneous consumers located at different position of the line. Assume that $v_i(=v_i-0)$ is the distance value that the customer with index i from the supplier A. Therefore, the customer i is $(1-v_i)$ far from the supplier B. In this paper, the valuation v_i is uniformly distributed within the consumer population from 0 to 1, with a density of 1.

As mentioned above, consumers have willingness to purchase the product from the supplier A instead of from the supplier B only when the distance value (transportation cost), $u_i(A)=v_i$, that consumer from the supplier A is smaller than the distance value from the supplier B, $u_i(B)=(1-v_i)$, and vice versa. If the distance value $u_i(A)$ is equal to $u_i(B)$, the consumer will choose either the supplier randomly. To model that the consumer i likes to buy from the supplier A when the distance value $u_i(A)$ is smaller than $u_i(B)$, a widely accepted discrete choice function (multinomial logit choice model) is adapted [14]:

$$p_i(A) = \exp\left(\frac{1-u_i(A)}{\tau}\right) \Big/ \left[\exp\left(\frac{1-u_i(A)}{\tau}\right) + \exp\left(\frac{1-u_i(B)}{\tau}\right)\right]$$

$$= \exp\left(\frac{1-v_i}{\tau}\right) \Big/ \left[\exp\left(\frac{1-v_i}{\tau}\right) + \exp\left(\frac{v_i}{\tau}\right)\right] \tag{1}$$

$$p_i(B) = \exp\left(\frac{1-u_i(B)}{\tau}\right) \Big/ \left[\exp\left(\frac{1-u_i(A)}{\tau}\right) + \exp\left(\frac{1-u_i(B)}{\tau}\right)\right]$$

$$= \exp\left(\frac{v_i}{\tau}\right) \Big/ \left[\exp\left(\frac{1-v_i}{\tau}\right) + \exp\left(\frac{v_i}{\tau}\right)\right] \tag{2}$$

$$= 1 - p_i(A)$$

In (1), $p_i(A)$ indicates that the customer will chooses to   buy the product from the supplier A with larger probability if the value $u_i(A)$ is smaller. Since the supplier with the smaller transportation cost is not certainly chosen with probability 1.0, it implied that not all the potential customers are 'perfect rationality'. The parameter τ is a constant giving the effects of the distance value v_i and $(1-v_i)$ on probabilities. When the parameter $\tau \to \infty$, $p_i(A)=p_i(B)=0.50$, it means that the consumer disregards for the transportation costs at all and makes choices from two alternatives randomly. On the other hand, when the parameter $\tau \to 0$, $p_i(A)=1.0$ if $v_i<(1-v_i)$, and vice verse. It implies that the consumer is a decision-maker of 'perfect rationality'.

In the market, besides those potential consumers making their decisions themselves (a $\beta(0 \le \beta \le 1)$ percent of all consumers), the rest $(1-\beta)$ are 'herd behavior' consumers. These consumers like to follow the most popular choice. If the most popular choice is different choice to her, she will change to this choice. On the other side, if her choice is also the most popular choice, she will more likely choose the choice. In this situation, it is called that choice of this potential consumer was confirmed by the crowd.

Suppose that p_A and p_B are the ratio of buyers of the supplier A to buyers of the supplier B in the market, the decision-making process of the 'herd behavior' consumer is as followed:

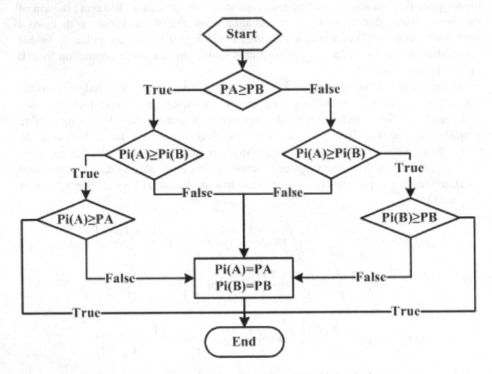

Fig. 1. Decision-making process flowchart of 'herd behavior' consumer

In Fig. 1, 'herd behavior' consumer will choose to buy product from the supplier A more likely than from the supplier B when more consumers in the market prefer to buy from the supplier A $(p_A \geq p_B)$. And, she will more likely prefer to buy from the supplier A if ratio of consumers who buy from the supplier A, p_A, is larger than her expected probability– $p_i(A)$.

Suppose demand of the market has an independent demand arrival rate. Specifically, we assume that consumers arrive at the two suppliers according to a Poisson process with constant intensity λ. Note that the market size N is a constant value, the potential consumers will arrive sequentially till total number of actual consumers reach the value N.

We assume that the first coming consumers are not 'herd behavior' consumers. All these consumers will choose one of the two suppliers considering the transportation costs. Then, the following consumers will be influenced by these first consumers. In this paper, preference bias of these first coming consumers can be affected by sales effort of the suppliers, e.g. advertisement, promotion.

The pseudo code of multi-agent based simulation is shown as followed:

```
//Number of incoming potential consumers is a Poisson
distribution variable with means 'AvgConsumersNum'
IncomingConsumers=PoissonDistribution(AvgConsumersNum);
If(IncomingConsumers>=TotalConsumers-ActualConsumers){
  IncomingConsumers=TotalConsumers-ActualConsumers;
}
//The potential consumers make decision one by one
For(i=0;i<IncomingConsumers;i++){
  //The 'herd behavior' consumer
  If(IsHerdBehaviorConsumer(ConsumersSet[i])){
    //Probability to choose the supplier A is calculated
by decision-making process as Fig. 1 shown
    ProbSupplierA=HerdEffectFunc(ConsumersSet[i], ProbAO-
fActualConsumers,ProbBOfActualConsumers);
  }
  //The 'traditional' consumer
  Else{
    //Probability to choose the supplier A is calculated
by Equation (1)
    ProbSupplierA=DiscreteSelectionFunc(ConsumersSet[i]);
  }
  If(ProbSupplierA<=RandUniformFunc()){
    //The ith consumer does buy from the supplier A
    CustomersOfSupplierA++;
  }
  Else{
    //The ith consumer does buy from the supplier B
    CustomersOfSupplierB++;
  }
  //Ratio of the supplier A' consumers is updated
  ProbAOfActualConsumers=CustomersOfSupplierA/ (Customer-
sOfSupplierA+CustomersOfSupplierB);
  ProbBOfActualConsumers=1-ProbAOfActualConsumers;
}
```

We use a parameter $\theta(0 \leq \theta \leq 1)$ to represent the level of sales effort exerted (by the supplier A) on the first coming consumers. Thus, the utility function for the first coming consumers is adapted to be:

$$u_i(A) = \begin{cases} v_i - \theta, & v_i \geq \theta \\ 0, & v_i < \theta \end{cases} \tag{3}$$

$$u_i(B) = \begin{cases} 1 - v_i + \theta, & v_i \geq \theta \\ 1, & v_i < \theta \end{cases}$$

$$= 1 - u_i(A)$$

(4)

Considering both equation (1) and equation (3), $p_i(A)$ of first coming consumers will increase as $u_i(A)$ becomes smaller. In this paper, change of $p_i(A)$ indicates impact of the sales effort on the individual.

4   Demonstration

This duopoly market model is built on our previous work [14,15]. We set size of the duopoly market N=10,000, the mean number of incoming consumers λ=500, the parameter of discrete choice function (in equation (1)) τ=1.00.

We replicated 100 times for each parameters' settings. Fig. 2 shows the number of incoming consumers each period, and the number of the supplier A's consumers with β=0,20.

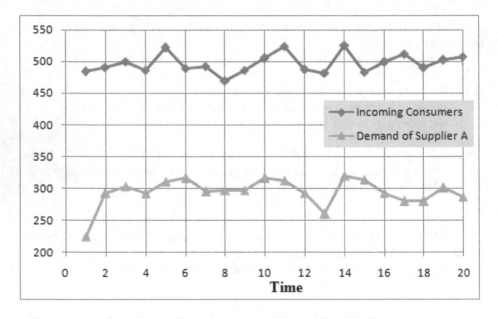

Fig. 2. Number of incoming consumers and consumers of the Supplier A

The results, which are shown in Fig. 3, illustrate the demand of both the supplier A and B with different β (ratio of 'traditional' consumers in the market). Obviously, demand of two suppliers will be more different when 'herd behavior' consumers in the market are more and more.

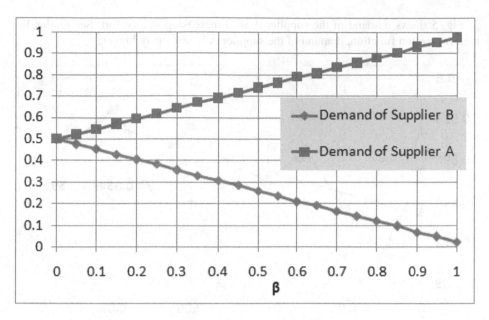

Fig. 3. Demand of both suppliers

The Quantile-Quantile plot in Fig. 4 ($\beta=0{,}20$) shows that the simulation results of 100 replications are normal distributed, which indicates that 100 repeating times of multi-agent simulation is enough.

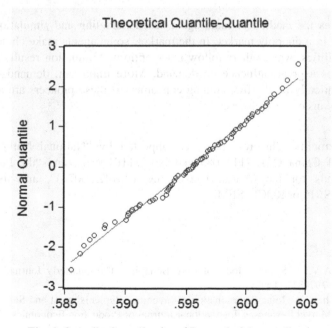

Fig. 4. Quantile-Quantile plot of Demand of the supplier A

Fig. 5 shows demand of the supplier A with increasing sales effort. Seeing that linear regression function, demand of the supplier A is obviously bigger.

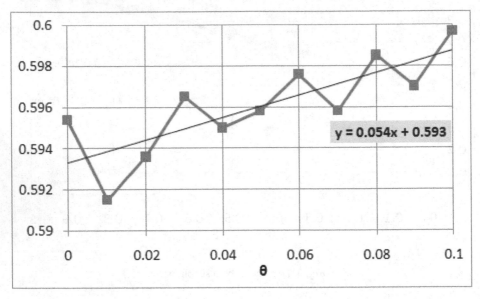

Fig. 5. Demand of supplier A with different θ

5　Conclusion

This paper uses the method of multi-agent based modeling and simulation to study herd behavior in a duopoly market. In the market, some agents make choice to maximize their utilities, while others follow mass options. Simulation results verify that herd effect has a great influence on demand. More important, demand can be increased dramatically by the first coming consumers if these pioneers are affected by supplier by means of sales efforts.

Acknowledgements. This research was supported by "National Natural Science Foundation of China (No. 71171085 and No. 71101063)", and "the Fundamental Research Funds for the Central Universities (2014ZZ0074)", and the Project-sponsored by SRF for ROCS, SEM.

References

1. Banerjee, A.V.: A Simple Model of Herd Behavior. The Quarterly Journal of Economics 107(3), 797–817 (1992)
2. Wei, Y., Chen, Y.: Joint Determination of Inventory Replenishment and Sales Effort with Uncertain Market Responses. International Journal of Production Economics 134, 368–374 (2011)

3. Bikhchandani, S., Sharma, S.: Herd Behavior in Financial Markets. IMF Staff Papers 47(3), 279–310 (2000)
4. Palley, T.I.: Safety in numbers: A Model of Managerial Herd Behavior. Journal of Economic Behavior and Organization 28, 443–450 (1995)
5. Cipriani, M., Guarino, A.: Herd Behavior in a Laboratory Financial Market. The American Economic Review 95(5), 1427–1443 (2005)
6. Kim, K., Yoon, S.M., Kim, Y.: Herd Behaviors in the Stock and Foreign Exchange Markets. Physica A 341, 526–532 (2004)
7. Chen, Y.F.: Herd Behavior in Purchasing Books Online. Computers in Human Behavior 24, 1977–1992 (2008)
8. Manahov, V., Hudson, R.: Herd Behaviour Experimental Testing in Laboratory Artificial Stock Market Settings. Behavioural Foundations of Stylised Facts of Financial Returns. Physica A. 392, 4351–4372 (2013)
9. Chiang, T.C., Zheng, D.: An Empirical Analysis of Herd Behavior in Global Stock Markets. Journal of Banking & Finance 34, 1911–1921 (2010)
10. Heese, H.S., Swaminathan, J.M.: Inventory and Sales Effort Management under Unobservable Lost Sales. European Journal of Operational Research 207, 1263–1268 (2010)
11. Xing, D., Liu, T.: Sales Effort Free Riding and Coordination with Price Match and Channel Rebate. European Journal of Operational Research 219, 264–271 (2012)
12. Lau, A.H.L., Lau, H.S., Wang, J.C.: Usefulness of Resale Price Maintenance under Different Levels of Sales-Effort Cost and System-Parameter Uncertainties. European Journal of Operational Research 203, 513–525 (2010)
13. Lau, H.S., Su, C., Wang, Y.Y., Hua, Z.S.: Volume Discounting Coordinates a Supply Chain Effectively When Demand is Sensitive to Both Price and Sales Effort. Computers & Operations Research 39, 3267–3280 (2012)
14. Li, F., Liu, J., Wei, Y.: Pricing Decision in a Dual-Channel System with Heterogeneous Consumers. In: Zhang, R., Zhang, Z., Liu, K., Zhang, J. (eds.) LISS 2013, pp. 658–665. Springer, Heidelberg (2013)
15. Li, F., Huang, B.: The Impact of Product Category on Pricing Decision in a Two-Echelon Dual-Channel Supply Chain. In: Proceedings of 2012 IEEE International Conference on Service Operations and Logistics, and Informatics, pp. 269–272 (2012)

A "Content-Behavior" Learner Model
for Adaptive Learning System

Qingchun Hu[1], Yong Huang[2], and Yi Li[1]

[1] School of Information Science & Engineering,
East China University of Science and Technology,
Shanghai, China
[2] Shanghai Audio-Video Education Center,
Shanghai Distance Education Group,
Shanghai, China

Abstract. The learner model in adaptive learning system plays an important role. The study aims to explore the main characteristics and components during students' engaging in online learning system, proposing a "content-behavior" learner model based on combining students' learning behavior with the grasp degree towards concepts. Especially, the learner model should be open to students in order to make them look into their learning process. And further, they can review peer's learning portfolio and then reflect their learning. The result of this study should provide some suggestions towards the research in the fields of adaptive eLearning in online learning environment.

Keywords: Learner Model, Learning System, Adaptive.

1 Introduction

The Intelligent Tutor System (ITS) usually consists of three components: learner model, expert model and domain knowledge model. The crucial part in ITS is to characterize, quantify and diagnose the performance of students' learning, giving appropriate feedbacks to students based on the diagnosis. Therefore, the most important element of the ITS is on how to design and model the learners' learning status, so as to the system could provide adaptive learning guidance.

The purpose of this study is to explore how to describe the characteristics of students' learning on current online learning environment. Compared with many years ago, currently, the learning system is usually networked and open to students. The space of students' problem solving is no longer restricted to a closed learning environment. When they encounter problems in the online learning system, students often seek solutions from peers and the internet. The online access behavior of students is an important attribute in the learning process. This study is to explore and propose a method to build a learner model named a "content-behavior" model, which consist of two dimensions. One dimension is the learning contents, which reflect the degree of students' understanding on concepts. Another dimension is learning behavior, which

D.-S. Huang et al. (Eds.): ICIC 2014, LNAI 8589, pp. 466–474, 2014.

comes from the log file. And the "content-behavior" model is open to learners in order to promote students' reflection and learning.

2   Related Work

2.1   ITS and Learner Model

There has been a wide range about the educational applications of artificial intelligence. In the literature, the applications of artificial intelligence in the field of education can be found to follow such a route: from ICAI, then ITS, and then to agent technology. Currently, the artificial intelligence has an effective integration of multimedia technology, network technology and database technology to improve the learning efficiency in online learning system. For example, a study constructed learner model based on students' learning style and graphically presented it to learners and teachers. Another study explored how to make better use of the intelligent tutor system based on the feedback mechanism of the students' learning [1, 2]. And there are also studies explore how to guide students in the learning system based on the interaction with different attributes proxy mechanism, including learning achievements, attitudes, errors, as well as a course holding force [3-7], BDI (Believes, Desires, Intentions) agent mechanism, and according to records related learning paths, and further building a learner model to guide students' learning [8].

2.2   Open Learner Model

In the earlier studies, the learner model, as a part of the ITS, is generally hidden. The data stored in the learner model is automatically updated by a computer system. Students can not browse, but also can not sense the presence of a learner model. The goal is to provide individualized teaching and learning [9-12]. In recent years, there is research that views the learner model from a hidden process (covert process) to an open process (overt process). That is, the learners can check the contents of the learner model, and can even edit the learner model. These studies argue that "to enable students to understand their own, to encourage learners to self-control and self-reflection", which is the idea of open learner model in the early. The students are allowed to browse and access their own learner model, to further reflect on their own learning. It is named as open learner model [13, 14]. It is viewed as benefits of doing so, which encourage and stimulate students' meta-cognition in the theory. There is research which proved that the open learner model helps to improve students' self-awareness for their own learning, self learning difficulties, thereby enhancing learning [15].

In a word, there are few studies which explore the learning behavior of students based on the web log files. Generally speaking, most of the methods can give students feedback mechanism, which mainly are based on the diagnoses of the students' performance on the exercises, the students' domain knowledge, errors in the students' learning process etc. This paper proposes a "content-behavior" open learner model based on integrating the students' learning behavior and learning contents.

3   A "Content-Behavior" Learner Model

A "Content-Behavior" learner model is built with two dimensions. One dimension is to characterize students' understanding on the domain knowledge. Another dimension is the characterization of students' learning behavior.

3.1   The Grasp Degree towards Concepts on Learning Contents

In the literature, the studies of Rafael, Nussbaum and Virvou, constructed a knowledge framework for modeling the students' learning [16-18], which provides a base to our study. Their study showed a theoretical framework. They put forward "concept to grasp degree" which consists of quantitative indicators to describe students' learning. The foremost study described in detail the principles of quantitative indicators, but the "concept to grasp degree" has not been realized in practice. Based on this, we improve, extend the quantitative indicators system and name this mechanism as grasp degree towards concepts. The structure is as follows in Fig.1.

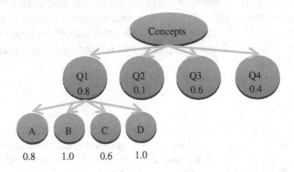

Fig. 1. The structure of grasp degree towards concepts

Firstly, concepts of the course are extracted according to the chapters in the text-book. And it is necessary to make sure that all sections of concepts with non-repetition. Based on the concepts, the appropriate test questions are constructed, such as Q1, Q2, Q3, Q4 in Fig.1. Each test question reflects different proportion of each concept. The different proportion is named as quantitative indicators to describe on what extent the test questions related to one concept, which is described with decimal fraction (a float) between [0,1] (See Fig.1). The average of these quantitative indicators characterizes students' comprehensive grasp of the sets of concepts. In Fig.1, it shows concepts related to four test questions with multiple choices.

Based on grasp degree towards concepts embed in student model, a learning plat-form for C programming contents is established. And the model is put into practice. The following section will detail test contents and explain the mechanism how to be used.

3.2   The Implementation of Test Questions

The open-source MySQL database is adopted as the underlying database system to store test questions. After requirement analysis and abstraction, Question Class is designed as a foundational class in the test questions. The use of Java's inheritance mechanism for different types of specific classes can be extended. So that, it will make codes more economic and easier to be maintained. The property of Question Class is shown in the following Table 1.

Table 1. The Property of Question Class

The Name of Property	Explain
ID	ID identifies items
Chapter	The chapter is that items belongs to
Difficulty	Item difficulty
Content	The details of each item
Explanation	The details of answering the question
Concept_qs	The sets of concepts object
RightCount	Number of correctness
ErrorCount	Number of wrongness
Date	The date of answering the question
Course	Identify different courses

A unified coding system is designed for all test questions. Every test question has the only uniquely identifier in term of ID. For instance, one coding example is as follows:

$$1001 \quad 06 \quad 4 \qquad 0003$$

$$\text{ID+ Chapter + Difficulty + Content}$$

With the property of the ID, the learning system could uniquely identify a particular item. The property of difficulty is described with the numbers from 1 to 4 in ascending order. That is from easy to difficulty. Through analyzing the log file automatically, the learning system records the sum of students' answers and also the number of correctness and wrongness. Based on this, the system could calculate the average correctness rate for further reference when providing the guides. As regards other properties in Question Class, such as Concept_qs, it means that the concept objects. It is a set of objects with an unordered collection. It encompasses all the questions related to the concept. The object encapsulates the concepts' name, key words in the test items. The property of Course is designed for future expansion of the system. At present, the test bank is only used in C programming. In the future, if there are new courses added to the learning platform, then the Question Class could has a special attribute to display different courses.

From a view of database design, ID of the items held is a kind of foreign key in the subclass' mapping table. And this foreign key can achieve foreign key mappings with the table which is initialized from Question Class, and it would connect with two tables. Because the Question Class is designed as a base class to inherit, there will be no specific implementation entities. So, there would be an abstract mapping. There will not be a corresponding entity table in the database, but its all features will be inherited by the Question Class's item. The Selection Class is inherited from Question Class, which extended attributes are showed in the Table 2 below.

Table 2. The Property of Selection Class

The Name of Property	Explain
Options	The options object is encapsulated
Key	The answers' label

Multiple choice questions are the most common style of test exercises in most learning system. And it is easy to quantify. Because that each option should be added a value, which could reflect the comprehensive grasp of the concept after answering the question. So that, the system will also package the select item's property into an entity class named as Option Class, which also includes Key, examples and other content options. It facilitates the design of logic level, and mapped it into physical table, associated with the main table in the database.

Now, it assumes that there is a student who completes a multiple-choice test, submitting the answer. Then the server will receive a test report object, extracting the student's option to analyze and to assess the answer with correctness or not. And further, the system calculated the average value of the grasp towards concepts involved this test item. The student model updates the corresponding data. Thus, a test is completed. Then, in the next stage, the value of grasp degree students owned will provide for adaptive learning programs as a basis for next adaptive guides.

3.3   The Format of Test Questions

To facilitate the extraction of data, the study design multiple-choice format into XML documents, using XSLT (eXtensible Style sheet Language Transformations) technology to convert it into an HTML document and returning it to the client. Such kind of technology has been widely used in various web applications and in the construction of unstructured data management system to store unstructured data [19-20]. To reduce the storage space, the system does not directly stored information in text format which is the document content itself, rather than the address of the document is stored. Each test with a unique ID number for identification, the ID numbering system is relatively simple, as follows:

1001   001

Chapter + ID

In addition to the basic information, the Test Class is designed to save the average score and other information. The information is packaged. And the Test Class will be mapped into physical table. The Test Class is constituted as following Table 3.

The test questions module will be provided to the expert modules for further analysis. Given the complexity of the adaptive learning system, a subsystem is designed for the corresponding entry to guide the teachers to input the test questions expediently in our future work.

Table 3. The Property of Test Class

The Name of Property	Explain
ID	ID identifies items
CourseNo	Which courses questions belongs to
TestAddress	Storage questions' addresses
Time	Average time
AverageScore	The average score
DoneTimes	The times of completions

Base on above design, each option is corresponded to one kind of grasp. As a result, each student participated in a test will be recorded with a numerical value, which show students' grasp towards one concept. And the numerical value of grasp is updated in real time.

Every student, participating in the test, will be assessed with the immediately grasp towards the concept. And then, the system can use data visualization techniques to present students a graph, which could more directly reflect student learning. In addition, the system could provide more intelligent response and supporting according to the data visualization. In our study, a demonstration of understanding on the concept is shown by bar charts in Fig.2.

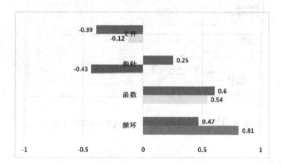

Fig. 2. The bar chart of the grasp degree towards concepts

3.4   The Interface Design of Learning Behavior

In the design of system architecture, students are awards opportunities and encouraged to search and browse the internet during the testing of concepts, as well as peer communication and interaction activities. The students' log files are extracted as a report (See Fig.3).

This report is showed in open student model in the form of feedback to students, including the exchange interaction in the learning process and learning forum. The user-interface design of open learner model is showed in Fig.4, which present a learning process of interaction between the learners in the learning activities. Students could browse this diagram, be aware of their own weaknesses and promote their reflection.

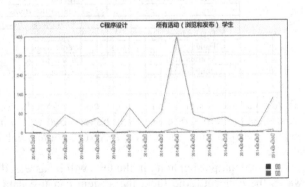

Fig. 3. The learners' behavior from logs

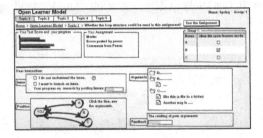

Fig. 4. The user-interface of open learner model

4   Conclusions

Currently, this model is tested by embedding in C programming course website for two semesters in our university. As mentioned above, to facilitate the extraction of data, the study design multiple-choice format into XML documents, using XSLT technology to convert it into an HTML document and returning it to the client. Now, this model is undergoing the usability testing and the reliability testing. For instance, a group of about 100 students' log file during two weeks is showed in Fig.3. During the implementation, it found that there are some tips for us to revise in the next stage study. Firstly, for most lecturers, it is difficult for them to elaborate each test question with the mechanism named the grasp degree towards concepts one by one. It will be better if there is a port with special format for them to upload the test items by a batch. Secondly, the lecturers should take action to encourage students to participate the online learning in a community. One propositional way is to encourage students through compulsory regular works online.

This study suggests that students model are the key component in adaptive learning system, not only record the students' learning performance on learning contents, but also integrate students' learning behavior from the log files and should be presented to students in the learning process. This will help the students to reflect on and self-evaluation of their own learning behavior. After detailed theoretical overview and discussion, we now have a modeling mechanism and implement it which is undergoing the usability testing and reliability testing, so the next step of this study will further thoroughly improve the "content-behavior" student model and refine the model based on the feedback of the test results.

Acknowledgments. This research is supported by Multi-Dimension Assessment Approaches in Programming Course in Online Environment under Shanghai Municipal Education Commission by National Educational Programs Grant.

References

1. Mitrovic, A., Ohlsson, S., Barrow, D.K.: The Effect of Positive Feedback in A Constraint-Based Intelligent Tutoring System. Computers & Education 60(1), 264–272 (2013)
2. Gutierrez, F., Atkinson, J.: Adaptive Feedback Selection For Intelligent Tutoring Systems. Expert Systems With Applications 38(5), 6146–6152 (2011)
3. Yılmaz, R., Kılıç-Çakmak, E.: Educational Interface Agents As Social Models to Influence Learner Achievement, Attitude And Retention of Learning. Computers & Education 59(2), 828–838 (2012)
4. Van Seters, J.R., et al.: The Influence of Student Characteristics on The Use of Adaptive E-Learning Material. Computers & Education 58(3), 942–952 (2012)
5. Clemente, J., RamÍRez, J., De Antonio, A.: A Proposal For Student Modeling Based on Ontologies And Diagnosis Rules. Expert Systems With Applications 38(7), 8066–8078 (2011)
6. Chou, C.-Y., Huang, B.-H., Lin, C.-J.: Complementary Machine Intelligence And Human Intelligence in Virtual Teaching Assistant For Tutoring Program Tracing. Computers & Education 57(4), 2303–2312 (2011)
7. Antal, M., Koncz, S.: Student Modeling For A Web-Based Self-Assessment System. Expert Systems With Applications 38(6), 6492–6497 (2011)
8. Mikic Fonte, F.A., Burguillo, J.C., Nistal, M.L.: An Intelligent Tutoring Module Controlled By BDI Agents For An E-Learning Platform. Expert Systems With Applications 39(8), 7546–7554 (2012)
9. Vandewaetere, M., Desmet, P., Clarebout, G.: The Contribution of Learner Characteristics in The Development of Computer-Based Adaptive Learning Environments. Computers in Human Behavior 27(1), 118–130 (2011)
10. Nkambou, R., Fournier-Viger, P., Nguifo, E.M.: Learning Task Models in Ill-Defined Domain Using An Hybrid Knowledge Discovery Framework. Knowledge-Based Systems 24(1), 176–185 (2011)
11. Murray-Rust, D., Smaill, A.: Towards A Model of Musical Interaction And Communication. Artificial Intelligence 175(9-10), 1697–1721 (2011)
12. Brusilovsky, P.: Knowledgetree, A Distributed Architecture For Adaptive E-Learning. In: Thirteenth International World Wide Web Conference Proceedings, New York (2004)
13. Kay, J.: Learner Know Thyself, Student Models to Give Learner Control And Responsibility. In: International Conference on Computers in Education, ICCE 1997 (1997)
14. Information on Lemore. Learner Modelling For Reflection, http://Www.Eee.Bham.Ac.Uk/Bull/Lemore/Index.html

15. Kay, J., Li, L.: Scrutable learner modelling and learner reflection in student self-assessment. In: Ikeda, M., Ashley, K.D., Chan, T.-W. (eds.) ITS 2006. LNCS, vol. 4053, pp. 763–765. Springer, Heidelberg (2006)
16. Faraco, R.A., Rosatelli, M.C., Gauthier, F.A.O.: An Approach of Student Modelling in a Learning Companion System. In: Lemaître, C., Reyes, C.A., González, J.A. (eds.) IBERAMIA 2004. LNCS (LNAI), vol. 3315, pp. 891–900. Springer, Heidelberg (2004)
17. Nussbaum, M., Rosas, R., Peirano, I., Cardenas, F.: Development of Intelligent Tutoring Systems Using Knowledge Structures. Computers & Education 36(1) (2001)
18. Virvou, M.: Modelling The Knowledge And Reasoning of Users in A Knowledge-Based Authoring Tool. International Journal of Continuing Engineering Education And Life-Long Learning 13(3/4) (2003)
19. Jovanovic, J., Gasevic, D.: Achieving Knowledge Interoperability: An XML/XSLT Approach. Expert Systems With Applications 29, 535–553 (2005)
20. Foetsch, D., Pulvermueller, E.: A Concept And Implementation of Higher-Level Xml Transformation Languages. Knowledge-Based Systems 22(3) (2009)

Aggregate MAC Based Authentication for Secure Data Aggregation in Wireless Sensor Networks

Keyur Parmar and Devesh C. Jinwala

Sardar Vallabhbhai National Institute of Technology, Surat
{keyur.mtech,dcjinwala}@gmail.com

Abstract. Wireless sensor networks perform in-network processing to reduce the energy consumption caused by redundant communication. At the same time, its hostile deployment and unreliable communication raise the security concerns. Thus, there is a need to blend security and data aggregation together to provide secure data aggregation. Secure data aggregation becomes challenging if end-to-end privacy is desired. Privacy homomorphism is used to achieve both en route aggregation and end-to-end privacy of sensor readings. However, privacy homomorphism is inherently malleable. Using privacy homomorphism, one can modify the ciphertext without decrypting it. Thus, it becomes extremely crucial to ensure authentication along with privacy. Symmetric key based Message Authentication Code (MAC) is an efficient solution to provide authentication. In this paper, we use Aggregate Message Authentication Codes (AMAC) to reduce the transmission cost incurred by MAC. However, conflicting requirements of AMAC and data aggregation make its usage limited for certain scenarios. In this paper, we present a cluster based scenario where we can apply AMAC to reduce the number of bits transmitted for authentication.

Keywords: Wireless Sensor Networks, Secure Data Aggregation, Authentication, Aggregate Message Authentication Code.

1 Introduction

The proliferation of Micro-Electro-Mechanical Systems (MEMS) technology facilitates the development of low cost, tiny sensor nodes (motes). These sensor nodes have very limited resources like battery power, memory, processor, bandwidth, etc. [1]. These smart, tiny sensor nodes, with escalating capabilities of sensing, processing and transmission, collaborate to form a network, referred as Wireless Sensor Network (WSN). Wireless Sensor Networks as an emerging technology have envisioned many applications in ubiquitous computing environments, such as battlefield surveillance, environmental & health care monitoring, traffic regulation, etc [1].

Wireless sensor networks have very limited resources and among these resources, scarce energy is the most limiting factor that affects the network lifetime [1]. As communication consumes 1000 times more energy than computation [2], in-network processing (data aggregation) is used to reduce the communication traffic. It improves

D.-S. Huang et al. (Eds.): ICIC 2014, LNAI 8589, pp. 475–483, 2014.

energy efficiency and prolong wireless sensor network's lifetime. The basic idea behind in-network processing is to aggregate the data as close to their sources as possible [3]. In-network processing improves bandwidth and energy utilization but it also negatively affects other performance metrics such as accuracy, data security, latency and fault tolerance.

As wireless sensor networks are deployed in remote and hostile environments, sensor nodes are more prone to node failures and compromise. Thus, it is critical to safeguard sensitive information from malicious attackers. However, there is a strong conflict between in-network processing and data security. Security protocols require encrypted sensor readings prior to any transmission, while in-network processing requires raw sensor readings to perform any computation. Moreover, in-network processing results in alternation of raw sensor readings and therefore the integrity verification of sensor readings becomes formidable. The necessity of providing such conflicting goals, led work in secure data aggregation.

Secure data aggregation schemes are classified according to the way they operate on data. If they operate on raw sensor readings then it is called hop- by-hop secure data aggregation and if they operate on encrypted data then it is called end-to-end secure data aggregation. Though viable, hop-by-hop aggregation becomes problematic, if intermediate sensor nodes (aggregator nodes) are compromised. Thus, to provide end-to-end privacy between sensor nodes and the base station, concealed data aggregation (end-to-end secure data aggregation) has been proposed [4].

To provide authentication, symmetric key based solutions are preferred over asymmetric key based solutions due to their limited communication and computation overhead. One such symmetric key based solution is message authentication code (MAC). The shared secret key used to generate MAC provides source authentication and MAC tag is used for integrity verification.

Our objective in this paper is to reduce the number of bytes transmitted within network through aggregate message authentication codes (AMACs). AMACs [6] [7] can aggregate multiple MACs computed by (possibly) different senders on (possibly) different messages into a same size MAC. However, AMAC still requires all the original readings on which MACs were generated to verify the integrity. Thus, the use of AMACs is formidable in scenarios where the data is aggregated en route. In this paper, we present a cluster based scenario where we can use AMACs to reduce total number of bytes transmitted within the network.

The rest of the paper is organized as follows. Next section summarizes related work to our problem. Section 3 gives the detail description of aggregate message authentication codes. Section 4 shows our proposed cluster based scenario where we can use AMACs. In the last section, we conclude the paper by emphasizing our contributions.

2   Related Work

Any wireless network requires an authentication mechanism. The use of an encryption algorithm without using authentication is proven to be insecure [8]. Moreover, need for authentication becomes obvious due to the encrypted data processing along with data aggregation. An adversary can maliciously add or modify any number of packets to the network and successfully aggregates them with other genuine packets. Thus to prevent such malicious modifications, different authentication mechanisms such as hash functions, MACs and digital signatures have been proposed.

The security strength of a MAC algorithm is dependent on its length and underling encryption algorithm. In traditional networks, 8 or 16 byte MAC is preferred for reasonable amount of security. However, first security architecture for wireless sensor networks, TinySec [8], uses a 4 byte MAC. The reasons to reduce the MAC length are as follows: 1) As the default packet size in TinyOS is 36 bytes, using a 8 or 16 byte MAC is not affordable due to communication cost. 2) Moreover, Mica2 motes with CC1000 radio take 20 months to forge a 4-byte MAC [8]. Motes with a radio CC2420 (MicaZ or TelosB) will forge the same MAC tag in 3 months. However, if the key used to produce MAC is changed within a period of 3 months, such attacks can be mitigated.

The notion of MAC aggregation has been previously utilized by [9]. Moreover, Chan et al. [9] uses a MAC aggregation to verify the legitimate sensor nodes that have submitted their sensor readings. In addition, they do not use aggregate MAC to verify the original sensor readings. Authors in [7] have formally defined aggregate MAC along with the proof of its security. In addition, they show a lower bound on a required length of MAC tag for faster verification. Chan et al. [6] discuss the challenges to provide end-to-end authentication along with end-to-end privacy in secure data aggregation scenarios. In [10], Ozdemir et al. perform MAC aggregation of data aggregated by a cluster head. However, as the number of cluster head per sensor network is negligible compare to the number of sensor nodes, energy improvement due to MAC aggregation at the cluster head is negligible. To the best of our knowledge, the scenario for MAC aggregation presented in this paper has never been utilized in sensor network. Our idea can be effectively utilized to provide authentication and to improve the efficiency of several other hop-by-hop secure data aggregation protocols like [11] or end-to-end protocols like [4, 11].

3   Aggregate MAC Based Authentication

In wireless sensor networks, the quest for authentic communication becomes necessary, as the communication medium is inherently unreliable and unsafe. In addition, any end-to-end secure data aggregation protocol requires authentication mechanism to verify the source and the sensor reading(s) before aggregation. As shown in Figure. 1, a single target region may have multiple sensor networks each detecting different phenomena and/or owned by different owners.

Typically, in order to provide authentication, we use message authentication code. MAC is based on symmetric key cryptography and hence it consumes fewer resources compared to public key cryptography based approaches.

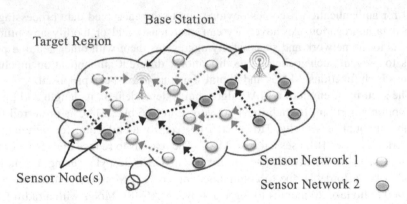

Fig. 1. Wireless Sensor Networks

3.1   Aggregate Message Authentication Code

Aggregate message authentication codes (AMACs) can be used to aggregate multiple MAC tags, computed by (possibly) different senders on multiple (possibly different) messages, into a shorter tag that can be verified by a recipient who shares a distinct key with each sender [6][7]. AMACs can be used to provide authenticated communication in sensor networks where communication is an expensive resource.

Formal Definition [7]: An aggregate message authentication code is a tuple of probabilistic polynomial time algorithms ($\mathcal{M}, \mathcal{A}, \mathcal{V}$) such that

MAC generation $\mathcal{M}$: Given a key $k \in \{0,1\}^n$ and a message $m \in \{0,1\}^*$, algorithm $\mathcal{M}$ outputs a tag $\mathcal{T}$.

$$\mathcal{T} \leftarrow \mathcal{M}_k(m) \tag{1}$$

MAC aggregation $\mathcal{A}$: Given a set of message/identifier pairs (m_1, id_1), (m_2, id_2), ... , (m_l, id_l) and associated tags $\mathcal{T}_1, \mathcal{T}_2, ... , \mathcal{T}_l$ algorithm $\mathcal{A}$ outputs a new tag $\mathcal{T}$.

$$\mathcal{T} \leftarrow \mathcal{T}_1 \oplus \mathcal{T}_2 \oplus ... \oplus \mathcal{T}_l \tag{2}$$

Verification   algorithm $\mathcal{V}$:   Given   a   set   of   key/identifier   pairs $\{(k_1, id_1), (k_2, id_2), ... , (k_l, id_l)\}$,   a   set   of   message/identifier   pairs $\{(m_1, id_1), (m_2, id_2), ... , (m_l, id_l)\}$, and a tag $\mathcal{T}$, where $\mathcal{T} \leftarrow \mathcal{T}_1 \oplus \mathcal{T}_2 \oplus ... \oplus \mathcal{T}_l$, algorithm $\mathcal{V}$ accepts the messages if and only if

$$\mathcal{T} = \oplus_{i=1}^{l} \mathcal{M}_{k_i}(m_i) \tag{3}$$

Without loss of generality, we can say that verification can be done in this way for any deterministic MAC. In addition, most commonly used MACs are deterministic and thus the restriction of using a deterministic MAC is not a serious issue [7].

4   AMAC for Secure Data Aggregation

Message authentication codes require original sensor readings, on which a MACs were computed, in order to be verified. Such requirement is in sharp contrast with the base principle of data aggregation. The data aggregation process aggregates original sensor readings en route and forward the aggregated data while end-to-end authentication requires original non-aggregated sensor readings to be available at the base station. Thus, providing end-to-end authenticity while performing in-network aggregation becomes challenging.

Data aggregation protocols use different topologies like cluster, tree, etc. Depending on topology, they choose aggregator node to perform the aggregation and forward aggregated result towards the base station. In this section, for simplicity, we assume a cluster based data aggregation but the same approach can work on tree based as well as hybrid data aggregation protocols.

In wireless sensor networks, we assume aggregator nodes with higher transmission power compared to the normal sensor nodes. In a cluster based data aggregation protocols, we called them cluster head. Such cluster heads can be chosen according to different criteria, like, selection of cluster heads is based on remaining energy levels [11]. Wireless sensor nodes like micaZ or iris transmit data only up to few meters. Thus, if wireless sensor network is deployed in a large region then data has to be forwarded by intermediate sensor nodes before it reaches the base station.

In a broadcast communication a single packet can reach to bazillion destinations. In such condition, if we choose every other node to perform aggregation then it would be very difficult to keep track of readings that have been aggregated by other sensor nodes. Thus, we assume that aggregation is being done on designated cluster heads. Other intermediate nodes will forward packets towards the cluster head without any aggregation. We would like to use aggregate message authentication codes in such scenarios. Data aggregation process limits the AMAC's use in sensor networks. However, our scenario permits the use of AMAC to reduce the number of bits needed to transfer MAC and supports a data aggregation at each aggregator node.

Figure 2 presents a scenario where sensor nodes are deployed in a region and different cluster heads are chosen to aggregate data. As shown in Figure. 2, cluster heads transmit their data towards the base station through other cluster heads.

Suppose, there are t sensor nodes in a cluster, $S_1, S_2, ..., S_t$. Sensor nodes sense the phenomena and transmit encrypted readings toward the cluster head C. Here, we assume that the cluster head shares a distinct key k_i with each node S_i of that cluster. Any node S_i authenticates the message m_i by computing $T_i = \mathcal{M}_{k_i}(m_i)$. Here, $\mathcal{M}$ is any standard MAC algorithm while T is a tag or checksum produced by MAC algorithm.

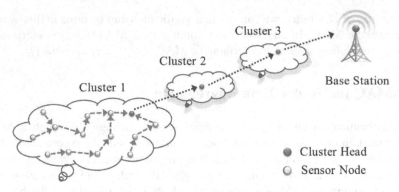

Fig. 2. Cluster based Data Aggregation

4.1   The Proposed Protocol for MAC Aggregation

1. Each node S_i uses MAC Generation algorithm $\mathcal{M}$ to generate tag T_i

$$T_i = \mathcal{M}_{k_i}(m_i), \quad \forall\, i \in (0,1, \ldots ,t) \tag{4}$$

2. S_i transmits a packet m_i and tag T_i towards cluster head C

$$C \leftarrow (m_i \parallel T_i), \quad \forall\, i \in (0,1, \ldots ,t) \tag{5}$$

3. Each intermediate node S_j uses MAC aggregation algorithm $\mathcal{A}$ to aggregate the MAC it has received from S_i and its own MAC $T_j = \mathcal{M}_{k_j}(m_j)$.

$$T = T_i \oplus T_j, \quad \forall\, i,\, j \in (0,1, \ldots ,t)\,,\, i \neq j \tag{6}$$

4. S_j transmits its own packet m_j, packets received from its child node(s) m_i and aggregate MAC tag T towards cluster head C

$$C \leftarrow \left(m_i \parallel m_j \parallel T\right), \quad \forall\, i,\, j \in (0,1, \ldots ,t)\,,\, i \neq j[0,1] \tag{7}$$

5. Cluster head C uses the keys $k_1, k_2, \ldots, k_t$ shared with the nodes $S_1, S_2, \ldots, S_t$ and messages $m_i, m_2, \ldots, m_t$ forwarded by nodes $S_1, S_2, \ldots, S_3$ to generate aggregate MAC.

$$T' = \oplus_{i=1}^{t} \mathcal{M}_{k_i}(m_i) \quad \forall\, i \in (0,1, \ldots ,t) \tag{8}$$

6. Cluster Head accepts sensor readings if and only if

$$T' = T \tag{9}$$

5   Overhead Analysis

In this section, we will consider a cluster based scenario to analyze bandwidth consumption of MAC based schemes 1) No Agg. (Forwarding individual packets without processing) 2) Concat. (Concatenating the payload (sensor readings and MACs) and forwarding) 3) AMAC (Aggregating MACs and forwarding it along with data values)

For ease of calculations, we assume a tree like topologies to forward the data packets towards aggregator node but the same scenario can be seamlessly applied to cluster based topologies or hybrid topologies. We consider a network model similar to [12] to calculate bandwidth consumption.

5.1   Network Model

Let us assume a balanced k-ary tree with a sink node and multitude of sensor nodes. Here, we assume that leaf nodes are sensors and remaining intermediate nodes are forwarders. The difference in network model considered in [12] and our model is the destination of packets. In [12] authors calculate the bandwidth consumptions up to the base station while in our scenario we require it only up to the aggregator node as shown in Figure 3. Here, we assume that aggregator nodes are powerful devices that can directly transmit packets to the base station [13]. In addition, let us assume that the output produced by MAC $\mathcal{M}$ is tag $\mathcal{T}$ of length 128 bits. Suppose the original sensor reading s_i (e.g. Temperature ranges between 0 and 127 Fahrenheit) is 7 bits long. We analyze bandwidth consumption of nodes at various levels in the hierarchy by computing number of bits transmitted by them.

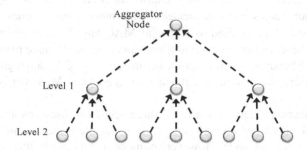

Fig. 3. Three-ary Tree upto Second Level

Here, we consider a standard packet format used in TinyOS operating system, where the packet header is of 56 bits and maximum supported payload is 232 bits. If the payload size exceeds 232 bits, remaining payload data will be transferred to another packet.

5.2   Bandwidth Consumptions

For No Aggregation scenario (No-Agg.), intermediate nodes forward the packet to the next hop without processing. Thus, packet size grows exponentially as it moves toward the aggregator node. As mentioned above, number of bits needed to encode the sensor value is $log_2(128)$ and the MAC tag $\mathcal{T}$ is of 128 bit. Hence, every packet in no aggregation scenario (No Agg.) consists of 135 bits of payload per packet. In a second scenario Concat., payloads of packets are concatenated and transferred to the next hop toward aggregator node. Thus, it utilizes unused payload bits and reduces bandwidth consumption as compared to no aggregation scenario.

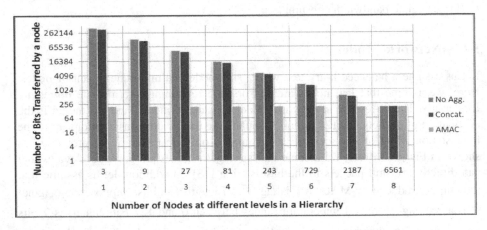

Fig. 4. Number of Bits Sent by the Nodes towards Aggregator Node

As per our aggregate MAC based scenario, the MAC tag $\mathcal{T}$ is aggregated at each hop and sensor values are concatenated with other sensor values. As mentioned above, number of bits required to represent MAC tag $\mathcal{T}$ is an order of magnitude higher than number of bits required to represent an original sensor reading. Hence, we need to reduce bandwidth consumption incurred by MAC by aggregating MAC tags at each hop. Here, we assume that the size of aggregated MAC tag is same as input MAC tags.

Figure 4 presents a comparison of all three scenarios discussed above. At level-8, there are 6581 nodes and they all have to transmit the same amount of information towards the aggregator node. However, number of bits transmitted by intermediate nodes in AMAC is significantly reduced as we move towards higher levels in the hierarchy.

6   Conclusions

The authentication becomes a crucial design parameter for any wireless network. However, the resource constrained sensor network limits the usage of any energy consuming operations. To provide authentication in sensor network, symmetric key based solutions are more suitable due to its less communication and computation overhead. In this paper, we use symmetric key based message authentication code. In addition, we proposed a scenario, where we can use the aggregate message authentication code to reduce transmission cost incurred by MAC. Though we have considered cluster based scenario, it can be applied to other topologies where en route aggregation is not carried out by the immediate next hop.

Acknowledgments. This research is a part of a project "A Secure Data Aggregation System and An Intrusion Detection System for Wireless Sensor Networks". It is supported by the Department of Electronics and Information Technology, Ministry of Communications and Information Technology, Government of India.

References

1. Akyildiz, I.F., Su, W., Sankarasubramaniam, Y., Cayirci, E.: Wireless sensor networks: a survey. Computer Networks: The International Journal of Computer and Telecommunications Networking 38, 393–422 (2002)
2. Hill, J., Szewczyk, R., Woo, A., Hollar, S., Culler, D., Pister, K.: System architecture directions for networked sensors. ACM SIGPLAN Notices 35(11), 93–104 (2000)
3. Fasolo, E., Rossi, M., Widmer, J., Zorzi, M.: In-network aggregation techniques for wireless sensor networks: A survey. IEEE Wireless Communications 14, 70–87 (2007)
4. Girao, J., Westhoff, D., Schneider, M.: CDA: Concealed data aggregation for reverse multicast traffic in wireless sensor networks. In: 40th International Conference on Communications, IEEE ICC 2005 (May 2005)
5. Rivest, R., Adleman, L., Dertouzos, M.: On data banks and privacy homomorphisms, pp. 169–177. Academic Press (1978)
6. Chan, A.C.F., Castelluccia, C.: On the (im)possibility of aggregate message authentication codes. In: ISIT, pp. 235–239. IEEE (2008)
7. Katz, J., Lindell, A.Y.: Aggregate message authentication codes. In: Malkin, T. (ed.) CT-RSA 2008. LNCS, vol. 4964, pp. 155–169. Springer, Heidelberg (2008)
8. Karlof, C., Sastry, N., Wagner, D.: Tinysec: A link layer security architecture for wireless sensor networks. In: Proceedings of the 2nd International Conference on Embedded Networked Sensor Systems. SenSys 2004, pp. 162–175. ACM, New York (2004)
9. Chan, H., Perrig, A., Song, D.: Secure hierarchical in-network aggregation in sensor networks. In: Proceedings of the 13th ACM Conference on Computer and Communications Security. CCS 2006, pp. 278–287. ACM, New York (2006)
10. Ozdemir, S., Xiao, Y.: Integrity protecting hierarchical concealed data aggregation for wireless sensor networks. Computer Networks: The International Journal of Computer and Telecommunications Networking 55(8), 1735–1746 (2011)
11. Ozdemir, S., Xiao, Y.: Secure data aggregation in wireless sensor networks: A comprehensive overview. Computer Networks: The International Journal of Computer and Telecommunications Networking 53, 2022–2037 (2009)
12. Castelluccia, C., Chan, A.C.F., Mykletun, E., Tsudik, G.: Efficient and provably secure aggregation of encrypted data in wireless sensor networks. ACM Transactions on Sensor Networks (TOSN) 5(3), 20:1–20:36 (2009)
13. Ozdemir, S.: Concealed data aggregation in heterogeneous sensor networks using privacy homomorphism. In: IEEE International Conference on Pervasive Services, Istanbul, pp. 165–168 (July 2007)

Active Learning Methods for Classification of Hyperspectral Remote Sensing Image

Sheng Ding[1,2,3], Bo Li[2,3], and Xiaowei Fu[2,3]

[1] School of Electronic Information,Wuhan University,Wuhan, China
[2] College of Computer Science and Technology,
Wuhan University of Science and Technology,Wuhan, China
[3] Hubei Province Key Laboratory of Intelligent Information Processing
and Real-time Industrial System, Wuhan University of Science and
Technology, Wuhan, China
dingwhu@gmail.com, liberol@126.com, 361782849@qq.com

Abstract. Active learning(AL) is an effective method in definition of samples, especially when labeled sample number is small. In this paper, we propose two active learning algorithms, which are Random Sampling (RS) and Margin Sampling(MS) algorithms, the two techniques achieve semiautomatic definition of training samples in remote sensing image classification, starting with a small and representative data set, then according to query criterion, the experts select informative samples to add training data set, the model builds the optimal set of samples which minimizes the classification error. Compared with traditional sample selection methods, the results denote the effectiveness of the proposed AL methods.

Keywords: Active learning(AL), Random Sampling, Margin Sampling(MS), hyperspectral classification.

1   Introduction

Statistical learning models [1] is a common solution for remote sensing data users, statistical learning models such as support vector machines (SVMs), the performances of the supervised algorithms depend on the data which is used to train. Sometimes, the enough sample labels is difficult to acquired, and this lead to low classification accuracy of remote sensing images, in these years, active learning(AL) technique becomes good solution in many fields, such as monitoring network optimization [2], species recognition [3], in computer vision for image retrieval [4]-[8], and in linguistics for text classification. Specially, in recent five year active learning is used in remote sensing[9].In remote sensing classification, sample selection plays an important role for classification accuracy. In the paper, we introduced two samples selection techniques, they can get good training samples by iterative to attain good classification performance The rest of the paper is organized as follows: in Section 2, we provide introduction of the two active learning methods in detail.The experimental results are presented in Section 3. Finally, we provide the concluding remarks and suggestions for future work in Section 4.

D.-S. Huang et al. (Eds.): ICIC 2014, LNAI 8589, pp. 484–491, 2014.
© Springer International Publishing Switzerland 2014

2   Active Learning Methods

The term 'active learning' can be applied to any strategy for sample selection where the algorithm interacts with samples prior to their label with the purpose of choosing only the most informative. This is done by creating some queries that investigate the distribution of samples. At the beginning, the acquired sample labels are lacking, but unlabeled samples are much and they are very easy to acquire. Active learning can make use of a lot of unlabeled samples, the algorithm selects from the pool the candidates that will at the same time maximize the gain in performance and reduce the uncertainty of the model when added to the current training set iteratively, some new samples from unlabeled samples are added to previous training data set. The process repeated many times until stop conditions meet[10-11].In the paper, we propose two active learning methods

2.1   Margin Sampling(MS)

MS is a SVM-specific active learning algorithm taking advantage of SVM geometrical properties, support vectors are labeled examples that lie on the margin with a decision function value Consider the pool of candidates, In a multiclass one-against-all setting, The "margin sampling" (MS) heuristic performs sampling of the candidates by minimizing (7) as follows: we make the assumption that the most interesting candidates are the ones that fall within the margin of the current classifier, as they are the most likely to become new support vectors, this way, the probability of sampling a candidate that will become a support vector is maximized. Consider the decision function of the two-class SVM.

$$f(q_i) = sign \ (\sum_{j=1}^{n} y_i \alpha_j K(x_j, q_i)) \tag{1}$$

where $K(x_j, q_i)$ is the kernel matrix, which defines the similarity between the candidate q_i and the j-th support vector; α are the support vector coefficients; and y_i's are their labels of the form(± 1) . In a multiclass context and using a one-against-all SVM [2], a separate classifier is trained for each class c against all the others, giving a class-specific decision function $f_c(q_i)$ The class attributed to the candidate q_i is the one minimizing $f_c(q_i)$ .Therefore, the candidate included in the training set is the one that respects the condition

$$\hat{x} = \arg\min_{q_i \in Q} |f(q_i)| \tag{2}$$

MS (detailed in Algorithm 1) provides a set of candidates at every iteration.

Algorithm 1 Margin sampling (MS)
Inputs
—Initial training set X.
—Set of candidates Q.
—Number of classes (Nc).
—Pixels to add at every iteration (Np).
for each iteration **do**
  Train current classifier with current training set X.
  Compute test accuracy of the current classifier.
  for each class c **do**
    for each candidate q_i to add **do**
    Compute the distance to the margin for the
      candidate q_i for class c using (1).
    end for
end for
Compute the minimum distance over the Nc classes.
Label the Np pixels associated with minimum distance.
Update X with the Np chosen pixels and remove those from
Q.
end for

2.2   Random Sampling(RS)

Another active learning method is Random Sampling, it also select unlabeled samples actively. At the beginning, the active learning process requires a first step for the initialization of the training set, the initial samples are rare and changeless. The query function is used to select randomly informative samples from unlabeled candidate data set. The supervisor assigns the samples the true class labels. Each time the selected randomly unlabeled samples are included into the training data set by iterative process. The classifier retrains the updated the training data, the closed loop of querying and retraining continues for some predefined iterations or until a stop criterion is satisfied. Because the training data set become bigger and the sampled quality is higher, the classification accuracy becomes higher than former unmodified training data set.

3   Experiment and Discussion

Results are obtained to investigate the performance of our proposed methods. We compared the active learning methods with traditional randomly selecting samples. in this section we use a real hyperspectral data sets collected by the AVIRIS instrument . All experiments were run on an Intel(R)Pentium CPU G20202@90GHz processor. Matlab 2012b is the software workshop in the experiments.

The hyperspectral image in our experiments is the Airborne Visible/ Infrared Imaging Spectrometer (AVIRIS) image Indian Pines [12]. The AVIRIS sensor generates 220 bands across the spectral range from 0.2 to 2.4μm. In the experiments, the number of bands is reduced to 200 by removing 20 water absorption bands. This image has spatial resolution of 20 m per pixel and spatial dimension 145×145. It contains 16 ground truth classes. For each class, we randomly choose around 3% of the labeled samples for training, 50% of whole image data for candidate unlabeled data, and use the whole image data for testing, as shown in Table 1.

Table 1. The training data and testing data in the whole image

| class | | Samples number | | |
No	Name	Train data	Unlabeled	Test data
1	Alfalfa	12	27	54
2	Corn-notil	43	717	1434
3	Corn-min	25	417	834
4	Corn	7	117	234
5	Grass/Pasture	15	249	497
6	Grass/Trees	22	373	747
7	Grass/Pasture-	1	13	26
8	Hay-windrowed	15	245	489
9	CAts	1	10	20
10	Soybean-notill	29	487	978
11	Soybean-min	74	1234	2468
12	Soybean-clean	18	307	614
13	Wheat	6	106	212
14	Woods	39	647	1294
15	Building-grass	8	190	380
16	Stone-steel	3	48	95
	Total number	318	5187	10366

Our experiments on the areas use the training areas, which providing areas with complexity in landcover. The image was taken shortly after the crops were planted, so that each pixel spectral value is not pure. This experiment is designed to test the algorithms proposed to discover classes with strongly overlapping spectra areas.

The classification performance for each of the 16 classes, classification accuracy (CA), on the test set are shown in Table 2. The classification accuracy is the ratio between correctly classified test samples and the total number of test samples.

For the data set, an SVM model trained on all the ground survey pixels has been taken as a reference for the best achievable performance of the classifier[13].

In our experiments, we adopted an SVM classifier with radial basis function (RBF) kernel. Optimal SVM parameters $\{C, \gamma\}$ (RBF kernel is used) are found by grid search on the parameter space. C is a parameter controlling the tradeoff between model complexity and training error. while$\gamma$is the spread of the Gaussian kernel. The values of the RBF kernel parameters C and$\gamma$are found using grid search algorithm ([14]) and both of the values of these parameters are used as 2^n. The values are tried and the one with the best cross-validation accuracy is picked. We found that trying exponentially growing sequences of C andγ is a practical method to identify good parameters (for example, $C = 2^{-5}; 2^{-3}; ..., 2^{15}, \gamma = 2^{-15}; 2^{-13} ... 2^3$).

For all experiments, to assess the effectiveness of the proposed active technique:1) the random sampling (RS); 2) the margin sampling (MS). We compared the two methods with other methods.

First, the SVM classifier is optimized using some pixels for the Indian Pines image from the training sub-area, the tested data set is on whole image area, four experiments are conducted as follows: 1) direct SVM classification of the entire image with the   random 300 training data; 2) classification of the entire image using 3100 pixels randomly selected from the whole image; 3) starting with 300 pixels initial sample data set, 2800 pixels randomly sampling selection by iterative active learning ; and 4) with the same 300 initial set, iterative actively selecting 2800 pixels by MS technique.

For the dataset, the initial training set is 300 labeled samples, each iteration we select 30 samplers to add training set, until the total unlabeled sample number is 3100.For decreasing the training influence of randomly samples, our active learning processes are repeated 20 times, the last classification accuracy is the average value. Performance was evaluated in terms of classification accuracy.

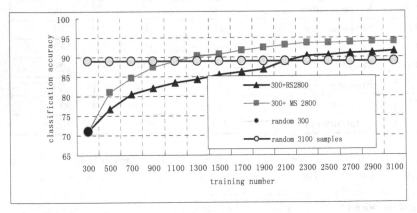

Fig. 1. The classification comparison results between active learning and random samples selection methods

The experiment considers the Indian Pines image. For this complex data set, consisting classes showing it is obvious that with the increase of samples, the classification accuracy of the two active learning methods also enhanced. We can see that the proposed active learning technique resulted is higher than random sampling. classification accuracy than the other techniques.

With continuous iteration, the informative samples are chose to add training set, thus it lead to more training data, so the classification accuracy by the same classifier is improved. When initiate labeled samples are selected randomly and the number is 300,in this case, when the samples dataset is small, the classification accuracy is only 71.05%,but when the random sample number reach 3100,the classification accuracy reach 89%,since the training number is much more than previous training number, the classification accuracy is more than previous training data. It is denoted into a line in Fig.1, the reason is that the labeled 3100 samples number is much more than 300 samples.

Table 2. classification accuracy[%] of two active learning methods with increase of sample number

Sample number	RS	MS
300	71.05	71.05
500	76.71	81.11
700	80.47	84.72
900	82.03	87.49
1100	83.42	88.88
1300	84.27	90.38
1500	85.57	90.73
1700	86.29	91.8
1900	86.96	92.42
2100	88.89	93.1
2300	90.3	93.59
2500	90.6	93.62
2700	91.08	93.72
2900	91.24	93.91
3100	91.54	93.97

At the beginning, the CA of two active learning is low than 3100 random samples. With the increase of the samples. The CA is higher than 3100 randomly. For RS, when the sample number is 2100,the CA is same as CA with random 3100,after this, the CA is higher, when the number is 3100,the CA reach 91.54%.It is worth noting that the proposed two active learning techniques are effective. Fig.1 shows the classification accuracies using active learning and random sample selection methods.

Compared with RS method, MS make use of margin hyperplane of SVM, but RS randomly select samples directly to add unlabeled samples to training data, so the MS classification is better than RS active learning classification accuracy. For MS technique, with increase of samples, the accuracy is improved too, when the sample number is 1100, and the accuracy value is 89%, it is equal to SVM classification accuracy when randomly selecting 3100 samples, after the number,the accuracy become higher, when the number is 3100,the MS classification reach 93.97%.Fig.1 and Fig.2 show the classification accuracies using active learning and random sample selection method. Table 2 shows the results of different sample selection methods.

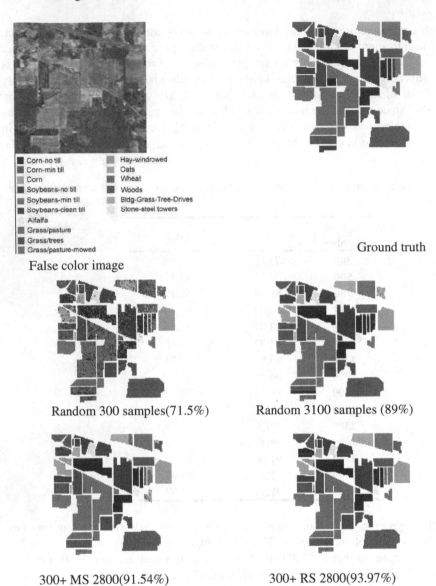

Ground truth

False color image

Random 300 samples(71.5%)   Random 3100 samples (89%)

300+ MS 2800(91.54%)   300+ RS 2800(93.97%)

Fig. 2. The reference map and the classification accuracy result

4   Conclusion

This paper presented two active learning methods. Unlike the traditional sample selection technique that focuses on selecting labeled samples randomly ,By each iteration, the active learning technique allows more unlabeled samples to be selected for manual labeling ,then the selected unlabeled samples are added to training set, As each time select informative samples, the effective can be proved. The experimental results obtained on hyperspectral data sets demonstrate good capability of the proposed method for selecting pixels, which allow good CA to an optimal solution compare traditional method. In the future, we will investigate more active learning technique in remote sensing image classification,we plan to extend the experimental comparison to other methods.

Acknowledgments. This work was supported by National Natural Science Foundation of China (No. 61273303, 61375017, 61201423), the Open Project Program of the National Laboratory of Pattern Recognition (201104212) and Open Project Program of Hubei Province Key Laboratory of Intelligent Information Processing and Real-time Industrial System(znss2013A008)

References

1. Vapnik, V.: Statistical Learning Theory. Wiley, New York (1998)
2. Pozdnoukhov, A., Kanevski, M.: Monitoring Network Optimisation for Spatial Data Classification Using Support Vector Machines. Int. J. Environ. Pollut. 28(3/4), 465–484 (2006)
3. Luo, T., Kramer, K., Goldgof, D.B., Hall, L.O., Samson, S., Remsen, A., Hopkins, T.: Active learning to recognize multiple types of plankton. J. Mach. Learn. Res. 6, 589–613 (2005)
4. Li, X., Wang, L., Sung, E.: Multilabel SVM Active Learning for Image Classification. In: Proc. ICIP, Singapore, pp. 2207–2210 (2004)
5. Jing, F., Li, M., Zhang, H., Zhang, B.: Entropy-based Active Learning with Support Vector Machines for Content-based Image Retrieval. In: Proc. IEEE ICME, Taipei, Taiwan, pp. 85–88 (2004)
6. Cheng, S., Shih, F.Y.: An Improved Incremental Training Algorithm for Support Vector Machines Using Active Query. Pattern Recognition 40(3), 964–971 (2007)
7. Gosselin, P.H., Cord, M.: Precision-oriented Active Selection for Interactive Image Retrieval. In: Proc. IEEE ICIP, Atlanta, GA, pp. 3197–3200 (2006)
8. Ferecatu, M., Boujemaa, N.: Interactive Remote-sensing Image Retrieval Using Active Relevance Feedback. IEEE Trans. Geosci. Remote Sens. 45(4), 818–826 (2007)
9. Liu, A., Jun, G., Ghosh, J.: Active Learning of Hyperspectral Data with Spatially Dependent Label Acquisition Costs. In: 2009 IEEE International Geoscience and Remote Sensing Symposium, IGARSS 2009, pp. 256–259. IEEE (2009)
10. Tuia, D., Volpi, M., Copa, L., et al.: A Survey of Active Learning Algorithms for Supervised Remote Sensing Image Classification. IEEE Journal of Selected Topics in Signal Processing 5(3), 606–617 (2011)
11. Demir, B., Persello, C., Bruzzone, L.: Batch-mode Active-learning Methods for The Interactive Classification of remote sensing images. IEEE Transactions on Geoscience and Remote Sensing 49(3), 1014–1031 (2011)
12. AVIRIS NW Indiana's Indian Pines 1992 data set (1992),
    ftp://ftp.ecn.purdue.edu/biehl/MultiSpec/92AV3C (original files),
    ftp://ftp.ecn.purdue.edu/biehl/PC_MultiSpec/ThyFiles.zip
    (ground truth).
13. Melgani, F., Bruzzone, L.: Classification of Hyperspectral Remote Sensing Images with Support Vector Machines. IEEE Trans. Geosci. Remote Sens. 42(8), 1778–1790 (2004)
14. Chang, C.-C., Lin, C.-J.: LIBSVM: A Library for Support Vector Machines. ACM Transactions on Intelligent Systems and Technology 2, 27:1–27:27 (2011),
    http://www.csie.ntu.edu.tw/~cjlin/libsvm

Time to Fault Minimization for Induction Motors Using Wavelet Transform

Amirhossein Ghods and Hong-Hee Lee

School of Electrical and Computer Engineering,
University of Ulsan, South Korea
{amirhossein,hhlee}@mail.ulsan.ac.kr

Abstract. Time to Fault (TTF) is an important parameter that measures how long it takes that a fault detection algorithm successfully recognizes defect in the motor. If TTF is too long, severe damages can happen to the motor. In this paper, authors try to minimize TTF using Discrete Wavelet Transform (DWT); in other words, the output signals derived from the motor due to an existing fault are analyzed and decomposed in frequency-domain. It will be proved that even though there are n levels for decomposing the signal with 2^n data samples, but after a specific level, the fault characteristics will disappear. This happens because of sporadic form of the signal. Thus, we can finish the analysis in a lower level where all characteristics for fault can be seen. This reduces TTF and consequently possible damages considerably.

Keywords: discrete wavelet transform, fault diagnosis, frequency-domain analysis, induction motors, low energy fault, time to fault.

1   Introduction

In case a fault occurs in induction motor, it can shut down the whole process for possibly a long time and cause severe damages. For preventing this, usually the measured output signals of motor are constantly analyzed.

To address this issue, in [1] using vibration signal of motor, a half-broken-bar detection method is proposed. They have implemented a decision tree in a field-programmable gate array which compares the regions of signal that represents a broken bar or a half-broken bar. By distinguishing between these areas, the exact fault can be recognized. There is a high probability that regions related to half-broken bar will change their location dependent on severity of fault; therefore, locating these areas would be a cumbersome task that makes the proposed algorithm inefficient. Authors in [2-5] derive the stator current signals in all three phases to diagnose the type and severity of the fault happened in induction motor. In [2], mechanical problems cause high magnitude sidebands around major frequency, but in [3-5], rotor bar faults can create repetitive sidebands in specific set of frequency bands. Although these methods can represent the fault characteristics, but usually there is limit for their application.

D.-S. Huang et al. (Eds.): ICIC 2014, LNAI 8589, pp. 492–499, 2014.
© Springer International Publishing Switzerland 2014

By the word "limit", it is meant that some faults have non-stationary and low energy characteristics that cannot be recognized using conventional methods. Wavelets have this feature that can decompose a signal into sub-signals which have lower level of harmonics, and also keep the main fault information. The initial fault signal is passed through a set of high- and low-pass filters and subsampled by 2, where high frequency additional components are removed. The current work is also using advantage of wavelet packets, which will be discussed extensively later on.

In literatures [6-8], faults are detected using Discrete Wavelet Transform (DWT). The structure of signal in healthy and faulty conditions for different decomposition levels are compared to each other. Although, using this method, existence of fault can be verified, but severity of it cannot be measured, because in time-domain DWT analysis, only fault behavior can be seen. We propose frequency-domain DWT analysis which both can verify existence of a defect, and also degree of severity of it.

As mentioned before, a current signal that carries 2^n data samples can be decomposed into n levels by high- and low-pass filters. Typically, for fault detection purposes, signal is passed through n levels of filters, so that the fault characteristics can be observed clearly. The time that it takes this process gets completed is called Time to Fault (TTF) and is a measure of performance of algorithm. TTF should be as less as possible; in other words, a fault should be detected in the earliest time possible.

It will be verified, although the total number of levels which current signal can be decomposed into is n, but there is possibility to decompose until a lower level, while all required information regarding the fault are derived. Therefore, the computations will finish in a shorter time. It means that the fault is recognized earlier.

This paper is structured as follow: In section 2, discrete wavelet transform in frequency-domain will be explained for bearing faults. Also, the modeling of fault and the way fault is represented will be discussed. In section 3, the term Time to Fault will be introduced in the field of fault diagnosis, and its importance for motor performance optimization will be explained. Following in the next section, authors will propose their strategy for minimization of TTF. It will be shown after a certain level of decomposition; there is no fault information in the output. By using this concept, TTF and also possible damages will decrease.

2   DWT Frequency-Domain Analysis for Bearing Faults Detection

Usually in time-domain DWT, the fault characteristics cannot be observed, and just by comparing the signal pattern in healthy and faulty situations, the existence of defect can be proved. As shown in [9], there is no specific fault signature in time-domain spectrum, and authors of this research work just compare the current signal in healthy and faulty cases, and if there are any major differences, then existence of fault can be verified. Also, the type of fault cannot be specified. This is the main deficiency of wavelet transform application in time-domain.

Here we propose frequency-domain DWT analysis. In this method, the frequency responses of high-pass filters are plotted. As it will be shown later, for any specific fault, this spectrum has a particular behavior; therefore, the process of fault diagnosis can be performed in a more precise way.

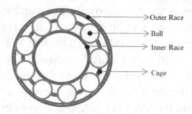

Fig. 1. Ball Bearing Configuration

The DWT analysis strategy is used here to detect bearing faults in induction motors. In Fig. 2, the general configuration of bearings is shown. If a defect happens in bearing section, the fault will be recognized well in frequency spectrum as an abnormal high magnitude component in:

$$f_{bearing} = 0.5 \times n_b \times fr = 0.5 \times 6 \times 50 = 150 \tag{1}$$

In (1), n_b is the number of balls. Usually, the number of balls in a bearing system varies between 6 and 12 [10]. In this simulation 6 balls are considered. With 50 Hz as the major frequency component, the fault frequency band will happen at 150 Hz.

3   DWT Decomposition Levels and Time to Fault Issues

In this section, the number of decomposition levels in process of wavelet transform is determined, and the relation between this number of levels and TTF is explained.

In one decomposition level, a stator current signal is passed through filters and then subsampled. This process should continue up to the time that only two samples are left; in other words, the number of levels is n for a 2^n sample-long signal. The purpose of this action is being able to detect non-stationary faults that have extremely low energy characteristic; such as bearing faults. In this research work, a 512 (2^9)-sample long signal is derived from the sensor that measures the stator current. So, there exist 9 levels of decomposition in the DWT process.

In Table 1, TTFs for a bearing fault are shown in a cumulative way; for example, the time shown for level 4 means time that 4 levels of decomposition takes. Totally, 2 minutes 15 seconds is needed that the fault diagnosis algorithm ends for 9 levels of decomposition.

Table 1. Time to Fault (TTF) for a bearing fault in induction motor

# of levels	Time to Fault (TTF)	
	Minute(s)	Second(s)
Level 1	1	01
Level 2	1	06
Level 3	1	18
Level 4	1	28
Level 5	1	38
Level 6	1	48
Level 7	1	52
Level 8	2	12
Level 9	2	15
Total Time	**2**	**15**

The less TTF, the earlier diagnosis process will be done. It means that dependent on the type of fault, the completion of DWT takes a specific time. If it finishes early, existence of fault can be recognized earlier; this means prevention of severe damages to the motor that can bring the motor into offline mode for an unknown period.

4   Minimization of Time to Fault

As it was stated before, in case that the fault detection process takes too long, possible damages to motor and its failure from online operation will happen. Thus, it seems beneficiary to look for a strategy that will minimize TTF. In this section, a new technique will be presented that is based on fault behavior observation and will reduce TTF considerably.

In the following figures, the output of each level of decomposition is shown in frequency-domain. The initial stator current signal has 512 data-samples, so it can be decomposed into 9 levels. The fault is featured as a high magnitude burst. It can be observed in Fig. 3 that in first level of decomposition, the fault characteristics cannot be recognized clearly; it is because the main fault signal is combined with lots of high magnitude additional components. As the signal is decomposed into higher levels, this problem disappears. In other words, if level 1 and 5, chosen just randomly, are compared together, fault feature in 150 Hz has a more clear view in level 5 rather than level 1.

Considering level 6 of decomposition in Fig. 4, no fault information can be observed. This situation is same for higher levels too; no fault information can be derived for level 7, 8, and 9. There is a signal with n data-samples as $x[n] = \{x_1, x_2, ..., x_n\}$ is passed through a set of low- and high-pass filters and sub-sampled by 2, the output of each filter becomes. The low-pass filter transfer function can be written in time-domain as:

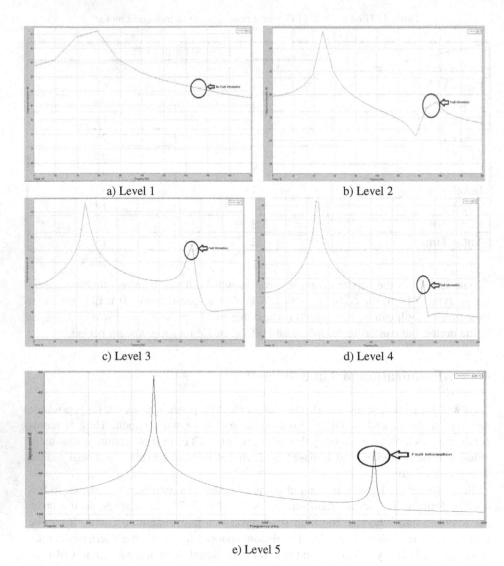

a) Level 1

b) Level 2

c) Level 3

d) Level 4

e) Level 5

Fig. 2. Decomposition levels (1:5); fault information exists

$$\frac{dy}{dt} + f_{cut} y(t) = f_{cut} x(t) \tag{2}$$

Because, in our problem the signal is a discrete signal, the above equation should be discretized:

$$\frac{y_n - y_{n-1}}{T} + f_{cut} y_n = f_{cut} x[n], \tag{3}$$

where y_n is the low-pass filter output of the current time step, y_{n-1} is the filter output of the previous time step, x_n is the current filter input and T is the sampling period. Rearranging (3) to solve for y_n yields the equation for the discrete-time, low-pass filter:

$$y_n - y_{n-1} + f_{cut}Ty_n + f_{cut}Tx[n]$$

$$y_n(1 + f_{cut}T) = y_{n-1} + f_{cut}Tx[n]$$

$$y_{n(low)} = \frac{y_{n-1} + f_{cut}Tx[n]}{(1 + f_{cut}T)} \tag{4}$$

The same reasoning is used for the high-pass filter:

$$y_{n(high)} = \frac{y_{n-1} + Tx[n]}{(1 + f_{cut}T)} \tag{5}$$

The problem that wavelet transform creates is because of filtering process where half of the data samples are eliminated, therefore, x[n]'s length becomes half and signal gets more outspread. Considering (4) and (5), by filtering a signal into several levels, the length of x[n] becomes smaller and smaller, and same for output. It means

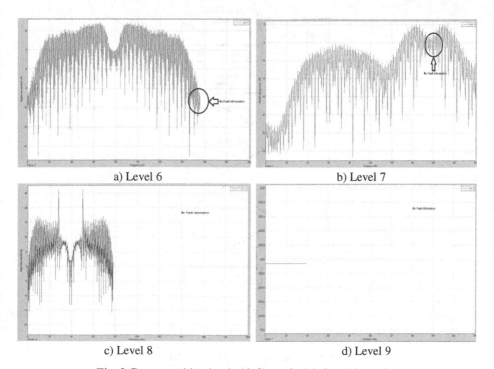

a) Level 6   b) Level 7

c) Level 8   d) Level 9

Fig. 3. Decomposition levels (6~9); no fault information exists

some major fault characteristics are eliminated. In a high energy fault, such as input voltage unbalance, compared to a low energy fault, e.g. bearing defect, the fault signal will lose its features earlier, because the main component of fault has a high frequency; therefore, through low-pass filtering, major component gets eliminated earlier. However, in this paper authors tried to model bearing fault in order to achieve higher levels of decomposition.

The idea behind this work lies in this matter that if we just recognize this level, and do not decompose the signal more than that, because after that level, no fault information exists, the detection process completes earlier and so TTF decreases. By this action, the fault is detected earlier and possible damages to motor are reduced. The flowchart regarding this process can be seen in Fig. 5, which stator current signal gets decomposed until there is no fault characteristics (expect level 1, where the reason that there is no fault characteristic is because of high magnitude noise level, not sporadic signal).

In this paper, the optimum level is 5 steps of decomposition. Considering Table I, if the signal is decomposed into 9 levels, TTF will be 2 minutes 15 seconds, but if it is only decomposed into 5 levels, TTF will be 1 minute 38 seconds that has become less.

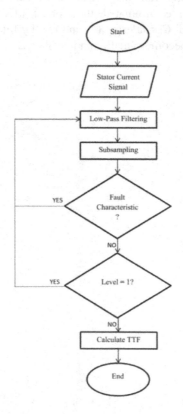

Fig. 4. Flowchart of finding minimized TTF

5   Conclusion

One important parameter in fault diagnosis of induction motor is the time that fault is fully recognized. In case this time becomes long, possible damages to motor can happen and disastrous consequences can happen. In the proposed paper, authors have used DWT in frequency-domain to derive the fault characteristics. For a low-energy fault, such as bearing defect, it is proved that fault will only be visible in the first five levels of decomposition. Filtering a signal several times will reduce the length of data samples, and dependent on the type of fault, fault information will disappear after a certain level, where the signal becomes outspread. Therefore, by decomposing the signal up to only the level that fault information exists- not more than that, the diagnosis time will be reduced significantly; in other words, this technique can decrease possible damages to motor.

Acknowledgment. This work was supported by the National Research Foundation of Korea Grant funded by the Korean Government (No.2013R1A2A2A01016398).

References

1. Rangel-Magdaleno, J., Romero-Troncoso, R., Osornio, R.A., Cabal-Yepez, E.: Novel Methodology for Online Half-Broken-Bar Detection on Induction Motors. IEEE Trans. Instrum. Meas. 58, 1690–1698 (2009)
2. Knight, A.M., Bertani, S.P.: Mechanical Fault Detection in a Medium-Sized Induction Motor Using Stator Current Monitoring. IEEE Trans. Energy Convers. 20, 753–760 (2005)
3. Benbouzid, M.E.H., Kliman, G.B.: What Stator Current Processing-Based Technique to Use for Induction Motor Rotor Faults Diagnosis? IEEE Trans. Energy Convers. 18, 238–244 (2003)
4. Garcia-Perez, A., de Jesus Romero-Troncoso, R., Cabal-Yepez, E., Osornio, R.A.: The Application of High-Resolution Spectral Analysis for Identifying Multiple Combined Faults in Induction Motors. IEEE Trans. Ind. Electron. 58, 2002–2010 (2011)
5. Das, S., Purkait, P., Dey, D., Chakravorti, S.: Monitoring of Inter-Turn Insulation Failure in Induction Motor Using Adnaced Signal and Data Processing Tools. IEEE Trans. Diecltr. Electr. Insul. 18, 1599–1608 (2011)
6. Supangat, R., Ertugrul, N., Soong, W.L., Gray, D.A.: Detection of Broken rotor Bars in Induction Motor Using Starting-Current Analysis and Effects of Loading. IEE Proc. Elec. Power Appl. 153, 848–855 (2006)
7. Mohammed, O.A., Abed, N.Y., Ganu, S.: Modeling and Characterization of Induction Motor Internal Faults Using Finite-Element and Discrete Wavelet Transforms. IEEE Trans. Magn. 42, 3434–3436 (2006)
8. Khan, M., Radwan, T.S., Azizur Rahman, M.: Real-Time Implementation of Wavelet Packet Transform-Based Diagnosis and Protection of Three-Phase Induction Motors. IEEE Trans. Energy Convers. 22, 647–655 (2007)
9. Pineda-Sanchez, M., Riera-Guasp, M., Antonio-Daviu, J.A.: Diagnosis of Induction Motor Faults in the Fractional Fourier Domain 59, 2065–2075 (2010)
10. Zarei, J., Poshtan, J.: Bearing Fault Detection Using Wavelet Packet Transform of Induction Motor Stator Current. Tribology Int. 40, 763–769 (2007)

Legal Reasoning Engine for Civil Court Procedure

Tanapon Tantisripreecha[1], Ken Satoh[2], and Nuanwan Soonthornphisaj[1]

[1] Department of Computer Science, Faculty of Science,
Kasetsart University, Bangkok, Thailand
{g5317400457,fscinws}@ku.ac.th
[2] National Institute of Informatics and Sokendai Tokyo, Japan
ksatoh@nii.ac.jp

Abstract. In this paper, we propose a new inference engine for legal reasoning in civil court (LR engine). Three main modules of the engine consist of the Prove algorithm, the Defense algorithm and the Estoppel algorithm. These modules work in switching style in order to drive the legal reasoning according to the Evidence law and the doctrine of Estoppel. In this study, Thai Civil and Commercial Code Law are applied as a legal knowledge base. Several legal cases are examined to validate LR engine and analyze its performance. The experimental result shows that LR engine can automatically handle the case containing estoppel and can reduce the steps of proof. Moreover, it provides reasonable legal explanation compared to the traditional algorithm.

Keywords: Legal knowledge base, a switch of burden of proof, Estoppel doctrine, legal reasoning.

1 Introduction

Court hearing is conducted in the courtroom where a judge applies a legal procedure (mainly the Evidence law) to drive the legal case. Given a legal argument, the burden of proof lies on the party who claims the fact. In civil cases, the plaintiff is normally charged with the burden of proof, but the defendant can be required to establish certain defenses. However, the duty of the judge includes a legal doctrine called *estoppel* which is typically applied in civil law cases. The concept is that a person wants to claim a fact but the law does not allow him to claim it. Therefore, the fact claimed by the party may be estopped and the truth value of that fact is changed to the opposite value. Estoppel doctrine is important for civil cases because it changes the truth value of the fact that may cause different court conclusion.

Thai Civil Procedural law [2], section no. 84 stated that "*Where a party alleges any fact in support of his plaint or answer, the burden of proof of such fact lies on the party alleging it. However: (1) a party is not required to prove fact which are generally known or are indisputable, or which, in the opinion of the court, are admitted by the opposite party. (2) Where there is a presumption in law favorable to a party, such party shall be required to prove only that the conditions entitling him to avail himself of the presumption have been fulfilled.*" The statement in 84(2) means

D.-S. Huang et al. (Eds.): ICIC 2014, LNAI 8589, pp. 500–512, 2014.

that in case of the absolute presumed fact (estoppel) which comes from legal presumption (e.g. section no. 1008 of civil and commercial code) the court must not allow the opposite party to prove that fact [3].

Satoh, K. et al. [1] formalized a switched of the burden of proof (SBP) using logic programming which provided the realistic court procedure for users. However this logic cannot apply to Thai Law since it cannot handle estoppel principle. The problem is that we cannot apply the Thai procedural law regarding the estoppel doctrine in SBP. Because Thai procedural law does not require the opposite party to further prove the estopped fact. Therefore, we propose a novel inference engine called LR engine to drive the steps of the inference and to obtain the lawful explanation. We will use the algorithm proposed by [1] as a benchmark since Japan is the civil law country and Thailand's substantive law came from Japanese law as well.

2   Related Works

Many researchers developed the legal expert system such as British Nationality [4], TAXMAN [5] and some legal reasoning systems such as PROLEG [6]. In 1986, Sergot, M.J. *et al.,* proposed the formalization of legislation and the development of a computer system to assist with legal problem solving [4]. The system needs an expert to make a set of legal rules combined with logic operators which is the traditional logic. In 2009, Gordon T.F. presented the Legal Knowledge Interchange Format (LKIF) which was a new interchange format developed specifically for application in the legal domain. [7]. Research work done by Beierle, C. *et al.,* employed predefined predicate names to represent the concept of facts and rules [8]. In 2011, Tantisriprecha, T. and Soonthornphisaj, N. proposed a solution to create legal knowledge base with Thai Civil and Commercial Code Law [9]. The system showed the legal explanation of each party by using abductive logic to select the rules in legal knowledge base.

A switch of burden of proof (SBP) was applied for demonstrating a characteristic of argumentation between a plaintiff and a defendant in court hearing. Therefore, several AI and Law researchers studied this principle and formalized the SBP using various techniques. Beierle, C. *et al.,* applied defeasible logic for argumentation system [8]. In 2007, Satoh, K. *et al.,* proposed a formalization of a switch of burden of proof [1] in a different way. They assumed that a plaintiff and a defendant will give all evidences prior to the court hearing process. They proposed the system namely PROLEG for making legal reasoning [6].

However, in Thai law, a fact should be considered before the use of each party in legal inference process. Since some fact cannot be used for arguing because of some general legal doctrines such as good faith and estoppel. Therefore we found that how to control the behavior of facts is important because it affects the final court sentence.

3   The Structure of Legal Knowledge Base

We propose a new structure of legal knowledge base which consists of a set of facts, rules, exceptions and the estoppel sentence (see Figure1).

```
fact(Evidence₁).
…
fact(Evidenceₙ).
RuleName₁ if Conditions.
…
RuleNameₙ if Conditions.
exception(MainRule₁, ExceptionRule₁).
…
exception(MainRuleₙ, ExceptionRuleₙ).
estoppel(Person, Fact, Reason):-
        Reason = cause(Person, Fact, Action),
        fact(Reason).
cause(Person, Fact, Action₁).
…
cause(Person, Fact, Actionₙ).
```

Fig. 1. A structure of legal knowledge base

Fact is an evidence or a situation that supports the party claiming. In this work, we assumed that all facts are ultimate fact. The ultimate fact is a fact that the plaintiff and defendant have accepted, or the fact that has been proved and known the truth value. Thus, every fact in the legal knowledge base can be used for claiming. We use predicate fact with one parameter to represent a set of facts.

Legal rule can be created from the substantive law or legal doctrines. During the legal reasoning process, the question will be asked via the goal rule. To translate legal doctrines or substantive law, a legal engineer has a duty to design a structure of the rule. The simple structure, which we use to create a set of rules in the legal knowledge base, was designed as a top-down model called Horn clauses logic [15]:

Conclusion if Condition1, Condition2, …, Conditionn.

A conclusion is a single predicate consisting of a set of conditions. Condition can be represented by a sub rule and a fact. If all Conditions can be proved as true, the Conclusion will be true, otherwise false.

In the Civil and Commercial code law, a situation of legal case consists of persons and documents. Considering a situation in Civil law, the person does some action on the document with another person. Thus, we propose the rule's structure to represent the situations as follows:

PredicateName({Person}, Document, [Section]).

A PredicateName is a name of each rule consisting of three parameters which are Person, Document and Section. Person is a set of parties. For each rule, the maximum number of persons is two. Thus, if there is one person, it means that the person acts something on the Document without another person. For example, put_signature(a,bill), means that a person a puts a signature on the bill. Moreover, if there are two persons, it means that the first person acts something to another person on the Document (the 1st person is an actor and the 2nd person is an acted). For example, forged_signature(x,c,bill) means that x (the 1st person) forged a signature of c (the 2nd person) on the bill. Document is used to identify evidences in legal case such as bill, bill of exchange or cheque. The last parameter, Section is the optional parameter used to identify the section number in law. For example, a rule name liable(Person,Bill,s900) means that a person is liable for a bill by section no.900 (s900).

Exception rule is a kind of rule that typically found in any written Laws. It is a kind of non-monotonic reasoning and it is used as a counter argument. When the exception rule is proved as true, the result will be changed to the opposite conclusion. In this work, we propose a predicate namely exception in order to identify the exception rule. The exception structure is as follows:

exception(MainRule, ExceptionRule).

The predicate exception consists of two parameters which are MainRule and ExceptionRule. If the parameter ExceptionRule matches the rule in the legal knowledge base, it means that there is an exception of the MainRule which is used to prove by the opposite party. Thus if the exception is true, the truth value of MainRule will be changed to negative. For example:

exception(liable(Person,Bill,s900),

not(liable(Person,Bill,s900))).

If the party proves the MainRule namely liable(Person,Bill,s900) as true, the ExceptionRule namely not(liable(Person,Bill,s900)) will be proved by the opposite party.

To ensure the smooth running of legal inference engine, it is necessary to formulate the rules which would generally work well. However, cases in which some unforeseen state of facts arises, strict rules cannot be used for deciding the cases. Therefore, we need general rule to help a judge for making legal reasoning. Estoppel doctrine is a general principle often found in civil law cases. The theory is used when a person wants to claim a fact but the law does not allow him to claims it. The estoppels' definition is that "*A bar which precludes someone from denying the truth of a fact which has been determined in an official proceeding or by an authoritative body*" [16].

Therefore, we introduce a predicate named estoppel, which consists of three parameters, Person, Fact and Reason. Estoppel is a kind of rule for checking the fact that cannot be used to argue. Thus, we can design the logic of estoppel in Prolog form as follows:

```
estoppel(Person, Fact, Reason) :-

Reason = cause(Person, Fact, Action),

    fact(Reason).
```

LR engine starts from finding a `Reason` that a `Person` is estopped to claim a `Fact` by considering an `Action` that the `Person` did to the `Fact` from predicate `cause`.

The estoppel is handled if the `Reason` matches some fact in legal knowledge base. This means that a `Person` is estopped to claim a `Fact` because the `Reason` that he did to an `Action` which was prohibited.

Let us show the structure of legal knowledge base created from *section no. 1008* in the Thai civil and Commercial Code Law. (See Figure 2 for details)

Section 1008:- *"Subject to the provisions of this Code, where a signature on a bill is forged or placed thereon without the authority of the person whose signature it purports to be, the forged or unauthorized signature is wholly inoperative, and on right to retain the bill or to give a discharge therefor or to enforce payment thereof against any party thereto can be acquired through or under that signature, unless the party against whom it is sought to retain or enforce payment of the bill is precluded from setting up the forgery or want of authority".*

The meaning of section no.1008 is that if there is a forged signature on a bill, a person who is liable can be: 1) The person who forges a signature, 2) The person who endorses the bill of forged a signature, and 3) The person who accepts the bill of the forged signature. Therefore, the legal knowledge base can be represented as shown in figure 2. Note that, there are two causes of estoppel rules. The former is that the person endorses the bill of forged signature and the latter is that the person accepts the bill of forged signature. Thus, if the person makes an action with match the cause, he will be estopped.

```
liable(Person,Bill,s900) if
        put_signature(Person,Bill).
exception(liable(Person,Bill,s900),
        not(liable(Person,Bill,s900))).
not(liable(Person,Bill,s900))if
        not(put_signature(Person,Bill)).
estoppel(Person, Fact, Reason):-
        cause(Person, Fact, Action), fact(Reason).
cause(Person, forged_signature(_,_,_), endorse).
cause(Person, forged_signature(_,_,_), accept).
```

Fig. 2. A sample of legal knowledge base

4   LR Engine

We proposed a legal reasoning engine, called LR engine, which is connected to the legal knowledge base. Given a set of facts obtained from the party, LR engine verifies the facts with the estoppels doctrine before the Prove and the Defense module start their process.

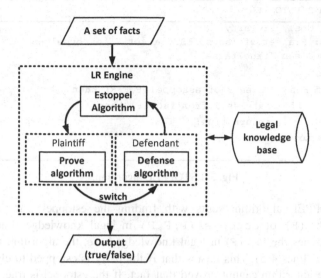

Fig. 3. LR Engine Framework

```
Algorithm 1 LREngine

Input: G: Goal, P: Party
Knowledge: F: a set of facts, R: a set of rules.
Output: Boolean (true/false).
1    begin
2        if ESTOPPEL(P,G,R) == true then
3            return false;
4        if G matches H from knowledge H if B₁,.., Bₙ then
5            for every condition B
6                if  Prove(B, P) == false then
7                    return false;
8        if Defense(H, P) == false then
9            return true;
10       return LREngine(H, P);
11   end
```

Fig. 4. LR engine

From figure 4, the ESTOPPEL algorithm is called to handle the estoppel status. If it is true, it means that if the estoppel occurs, the fact cannot be used. Thus, if there is no estoppel, the algorithm processes the Goal (G) using the Prove algorithm (line 4-7). The Defense algorithm starts if the Prove algorithm returns true. Later, the LREngine(H, P) is processed if the Defense algorithm returns true (line 8-11).

Algorithm 2 ESTOPPEL

Input: G: Goal, P: Party
Knowledge: F:a set of facts, EST: a set of estoppel and R: Reason
Output: Boolean (true/false).

1   **begin**
2   **if** Est matches F of estoppel(P,F,R) **then**
3   **if** R matches F from fact(F) **then**
4   **return** true;
5   **return** false;
6   **end**

Fig. 5. ESTOPPEL algorithm

The ESTOPPEL algorithm starts with finding the estoppel fact(Est) that matches fact (F) of estoppel(P,F,R) in legal knowledge (line 2-3). If a reason (R) matches the fact (F) in legal knowledge base, the algorithm returns true, otherwise false (lines 4-5). This means that if the party is estopped to claim the fact then the Prove algorithm cannot proved that fact. If the estoppel is true, the truth of fact will be the negation. For example, there is the fact that x forged signature of c. Next, a needs to use this fact to defense for himself. If a is estopped from this fact, the truth of forge signature is:

$$\neg(\texttt{forge signature}) \equiv \neg (\texttt{true}) \equiv \texttt{false}.$$

Therefore, the truth value of forged signature is false for a. This means that there is no evidence of forged signature occurs for the person. However, if another party also needs to use this fact and he is not estopped from this fact, he can use the truth value of forged signature as true. The advantage of using estoppel is that the truth value of the fact is different depending on the party who claims that fact whether he is estopped or not.

A switch of burden of proof (SBP) is a court procedure applied in civil law countries includes Thailand. There are two main parties involved which are plaintiff and defendant. The plaintiff has a burden to prove his claims by providing his evidence whereas the defendant must defense himself by proving the exception rule of the goal. This process will switch between both parties until the party cannot prove the goal. However, estoppel must be determined by judges to handle the fact/evidence that may be estopped.

There are two algorithms for making a switch of burden of proof which are the Prove algorithm and the Defense algorithm. If the Prove algorithm returns true, then the Defense algorithm will be activated. In the same way, the Defense algorithm will be switched if the Defense algorithm return true. The switching between both algorithms is stopped when either the Prove algorithm or Defense algorithm cannot successfully proved. However, the ESTOPPEL algorithm will handle the fact that was estopped when the Prove algorithm is processing.

```
Algorithm 3 Prove
────────────────────────────────────────────────────────────
Input: G: Goal, P: Party
Knowledge: F: a set of facts, R: a set of rules
Output: Boolean (true/false).
1    begin
2        if G matches F from knowledge fact(F) then
3            return true;
4        if G matches H from knowledge H if B₁,.., Bₙ then
5            for every condition B
6                if  Prove(B, P) == false then
7                    return  false;
8        if Defense(H, P) == false then
9            return true;
10       else
11           return false;
     end
```

Fig. 6. Prove algorithm

The Prove algorithm starts with finding a fact (F) in legal knowledge base (line 2-3). If the fact is found, it means that there is evidence about the fact (F). Otherwise the algorithm will find the goal (G) that matches the head of rule (H) (line 4-7). For every condition (B), the algorithm checks the condition by calling the Prove algorithm. If the condition is false, it means that the algorithm cannot prove some conditions. Thus the conclusion is false. Later, the Defense algorithm is executed to check the exception of goal (G) (line 8-11). If it is false, then it returns true, otherwise false. This means that if there is no exception or the truth value of the exception is false, the goal (G) is true; otherwise the goal (G) is false.

```
Algorithm 4 Defense
────────────────────────────────────────────────────────────
Input: G: Goal, P: Party
Knowledge: R: a set of rules and E: a set of exceptions.
Output: Boolean (true/false).
1    begin
2        opposite(P, P2);
3        if G matches H from exception(H, E) then
4            return Prove(E,P2);
5        else  return false;
6    end
```

Fig. 7. Defense algorithm

The Defense algorithm (see Fig 7) is called from the Prove algorithm in order to check the exception of goal (G).

The Defense algorithm starts with switching a burden of proof of the parties by calling `opposite(P,P2)` (line no. 2) For instance, if *P* is a plaintiff, it will switch to the defendant (P2). Next (lines 3-7), the algorithm finds the goal (G) that matches H of predicate `exception(H,E)` in legal knowledge base. Then the algorithm proves exception E by calling `prove(E,P2)`. This means that party P2 tries to prove exception E. If there is no exception, the algorithm returns false.

```
opposite(P,D) :- plaintiff(P), defendant(D),!.

opposite(P,D) :- defendant(D), plaintiff(P),!.
```

Consider the opposite predicate, suppose P = plaintiff, D = defendant. After opposite is called, P = defendant, D = plaintiff.

Before using a fact in the step of proof, the fact will be considered whether the estoppel occurs or not. If it occurs the engine notifies the reason of the estoppel to user. Thus we can design logic by considering the estoppel before using the fact as follow:

$$H \text{ :- } \neg Est(Reason), F$$

Where H is conclusion , $\neg Est(Reason)$ is the reason of estoppels, F is a fact

Negative ($\neg$) is placed before the fact (F) which means that an estoppel must be considered before proving the fact. The logic will find the reason of the estoppel from a set of causes;

Reason $\in$ Cause

Cause is a set of estoppel causes that comes from the past action of each party. Thus, if the *Reason* is found, the estoppel will be proved. As we know, condition in any rules can be either a condition or a fact. Thus, the condition is also controlled by estoppel. Accordingly, we can modify the SBP logic with estoppel by checking estoppel before each condition. The logic of Estoppel is shown as follows:

$$H \text{ :- } \neg Est, B_1, B_2, .., B_n,$$
$$exceptions (H, E)$$

Consider the estoppel sentence, $\neg Est$ will be considered before proving the conditions. If no estoppel occurs, the truth of $\neg Est$ is true and the inference engine will check the conditions (B) and exceptions (E) respectively. However, if the estoppel occurs, the truth of $\neg Est$ will be false and the estoppel sentence will be terminated immediately. Compare to the logic used in the traditional SBP[1], there is no Est to control the estoppel situation and every fact can be used to prove in the conditions (B) and exceptions (E). Thus, using only the SBP algorithm leads to the missing of estoppel doctrine.

5   Legal Case Examples

In order to validate the performance of LR engine, we compare the results of legal reasoning between LR engine and SBP[1] algorithm.

Legal Reasoning Engine for Civil Court Procedure

5.1   Case Scenario

In this work we use the situation related to the bill of exchange law. The situation is that a drawer, a, pays a bill to a payee, b, then b endorses the bill to c. However, x who is the son of a endorses the bill to d by forging the signature of c. Next, d endorses the bill to e. Later, a realizes that the signature on the bill is forged but he still accepts that bill. Note that the law states that a party who puts the signature on the bill is liable for the bill and a person who endorses the bill of forged signature is liable as well. (This situation will be used in example 1-5).

5.2   Legal Knowledge Base

Before proving the goal, parties need a set of facts that can support the claiming. Suppose that, there are the evidences about parties who put signature on the bill and some evidences about the bill of forged signature. Moreover, there are the facts and causes that a accepts the bill of forged signature and d endorses that bill.

```
fact(put_signature(a,bill)).
fact(put_signature(b,bill)).
fact(put_signature(c,bill)).
fact(put_signature(d,bill)).
fact(forged_signature(x,c,bill)).
fact(accept(a,forged_signature(x,c,bill))).
fact(endorse(d,e,forged_signature(x,c,bill))).
fact(cause(a,forged_signature(x,c,bill),accept)).
fact(cause(d,forged_signature(x,c,bill),endorse)).
```

Fig. 8. A set of facts obtain from the plaintiff and defendant

```
%rule
liable(A,Bill) if
        put_signature(A,Bill).
liable(A,Bill) if
        forged_signature(A,B,Bill).
%exception
exception(liable(A,Bill),not_liable(A,Bill)).
not_liable(A,Bill) if
          forged_signature(Some1,Some2,Bill).
%estoppel
estoppel(Person,Fact,Reason):-
        Reason = cause(Person,Fact, Action),
        fact(Reason).
cause(Person, forged_signature(Person,_,_), accept).
cause(Person, forged_signature(Person,_,_), endorse).
```

Fig. 9. Legal rules of the ESTOPPEL algorithm

5.3   Legal Explanation

Two categories of legal cases are examined which are the cases with estoppel (example no. 1,2) and without estoppel (example no. 3-5).

Example 1. A plaintiff, e, is a holder and a defendant, a, is a drawer. The holder needs to claim a for the liability of the bill. Thus, we obtain the legal explanation of the question claim(e,a,liable(a,bill)) from LR engine as shown in Figure 10.

```
?- claim(e,a, liable(a,bill)).
     To prove liable(a,bill) e need to prove:
       put_signature(a,bill)
       e need fact: put_signature(a,bill)
       e can find fact: put_signature(a,bill)
     == A switch of burden of proof occur ==
     Exception of liable(a,bill) is not_liable(a,bill)
       To prove not_liable(a,bill) a need to prove:
        forged_signature(x,c,bill)
          a need fact: forged_signature(x,c,bill)
            There is an estoppel
                forged_signature(x,c,bill)
          Because accept(a,forged_signature(x,c,bill))
          a cannot prove fact:
                forged_signature(x,c,bill)
        a cannot prove conclusion: not_liable(a,bill)
      e can prove conclusion: liable(a,bill)
  true
```

Fig. 10. Explanation obtained from LR engine

Example 2. A plaintiff e is a holder and a defendant d is an endorser. e needs to claim d for the liability of the bill. Thus, we obtain the legal explanation of the question claim(e,d,liable(d,bill)) from LR Engine as shown in Fig.11.

```
?- claim(e,d, liable(d,bill)).
     To prove liable(d,bill) e need to prove:
         put_signature(d,bill)
       e need fact: put_signature(d,bill)
       e can find fact: put_signature(d,bill)
     == A switch of burden of proof occur ==
     Exception of liable(d,bill) is not_liable(d,bill)
       To prove not_liable(d,bill) d need to prove:
           forged_signature(x,c,bill)
         d need fact: forged_signature(x,c,bill)
           There is an estoppel
               forged_signature(x,c,bill)
             Because
               endorse(d,e,forged_signature(x,c,bill))
         d cannot prove fact:
            forged_signature(x,c,bill)
        d cannot prove conclusion: not_liable(d,bill)
      e can prove conclusion: liable(d,bill)
  true.
```

Fig. 11. Explanation obtained from SBP

Regarding the performance of LR engine for example 1 and 2, we found that when the estoppel is found, another party has no burden of prove for the exception. For instance, in the example 1, the burden of proof stops when a is estopped and a burden of prove does not switch to d. In example 2, d is also estopped and a burden of prove does not switch to d.

However, SBP allows the party to prove the exception when estoppel is found. For instance, in example 1, a burden of proof is switched to a, and a can prove the fact about bill of forged signature. Later, a switch of burden of proof is switched to d because there is the exception about the forged signature in the legal knowledge base.

We found that ESTOPPEL algorithm needs only 2 steps where as traditional SBP need 3 steps. Hence LR engine algorithm needs fewer steps of proof compared to SBP.

We further examined the performance of LR engine in case of no estoppel occurs in Example 3-5 as follows.

Example 3. A plaintiff e is a holder and a defendant b is an endorser. e needs to claim b for the liability of the bill.

Example 4. A plaintiff *e* is a holder and a defendant *c* is a lawful holder. e needs to claim c for the liability of the bill.

Example 5. A plaintiff, e is a holder and x is a defendant who is an endorser that forged a signature. e needs to claim x for the liability of the bill.

6   Discussion

Regarding the legal explanation, we found that LR Engine can make the proper legal explanation covering all of estoppel cases since the algorithm has a module to handle a set of facts before proving. We found that the knowledge base structure of LR Engine is flexible since there is no effect to other rules. Although we can create the estoppel concept using exception clauses for SBP but the drawback is that we need to manually create exceptions attached to every rule This is a time consuming task and leads to the difficulties for legal engineer. Moreover the proof will be switched recursively and this affects the legal explanation making the misconcept of legal understanding for the court hearing process. Therefore LR Engine is preferred because of its flexibility in designing the set of rules in knowledge base.

We found that SBP algorithm provides different legal explanations compared to the legal reasoning done by judges. SBP has limited applicability since it is suitable only on the case without estoppel. We found that SBP sometimes gives wrong legal explanation if all facts can be used directly to prove. Since in reality, some fact cannot be used to prove due to the estoppels principle.

We know that estoppel doctrine aims to handle the fact that prohibited the party to claim the fact. Thus LR Engine is the answer for dealing with this problem. Moreover, LR Engine can provide the legal explanation of the case without estoppel situation as well. Another advantage of LR Engine is that it can enhance an inference process during the legal reasoning since it reduces the steps of the burden of proof. It terminates when the estoppel event related to that fact is found but SBP does not. We believe that LR Engine has high potential to be applied in the real-world application for civil law countries.

Acknowledgement. This research was supported by The Royal Golden Jubilee (Ph.D. program), Department of Computer Science, Faculty of Science, KU.

References

1. Satoh, K., Tojo, S., Suzuki, Y.: Formalizing a Switch of Burden of Proof by Logic Programming. In: Proceedings of The 1st International Workshop on Juris-Informatics 2007, pp. 76–85 (2007)
2. Tiamchan, B.: The Civil Procedure Code: translate Thai-English (2007)
3. Vichitchonchai, P.: Evidence Law. Chulalongkorn University Press (1999)
4. Sergot, M.J., Sadri, F., Kowalski, R.A., Kriwaczek, E., Hammond, P., Cory, H.T.: The British Nationality Act as A Logic Program. Communications of the ACM, 370–386 (1986)
5. McCarty, L.T.: An Implementation of Eisner v Macomber. In: Proceedings of the Fifth International Conference on AI and Law, College Park, MD, pp. 276–286. ACM Press, New York (1995)
6. Satoh, K., Asai, K., Kogawa, T., Kubota, M., Nakamura, M., Nishigai, Y., Shirakawa, K., Takano, C.: PROLEG: An Implementation of the Presupposed Ultimate Fact Theory of Japanese Civil Code by PROLOG Technology. JURISIN 2010, 153–164 (2012)
7. Gordon, T.F., GovernatoriA., G.R.: Rules and Norms: Requirement for Rule Interchange Languages in the Legal Domain. Rule Representation, Interchange and Reasoning on the Web, RuleML, 282–296 (2009)
8. Beierle, C., Freund, B., Kern-Isberner, G., Thimm, M.: Using Defeasible Logic Programming for Argumentation-Based Decision Support in Private Law. In: Proceedings of the 2010 Conference on Computational Models of Argument: Proceedings of COMMA, pp. 87–98 (2010)
9. Tantisripreecha, T., Soonthornphisaj, N.: Creating Rules Using Abduction for Legal Reasoning by Logic Programming. BIS (Workshops) 2011, 282–293 (2011)
10. Satoh, K., Kubota, M., Nishigai, Y., Takano, C.: Translating the Japanese Presupposed Ultimate Fact Theory into Logic Programming. Information Systems (JURIX 2009), 162–171 (2009)
11. Truszczynski, M.: Logic Programming for Knowledge Representation. In: ICLP, pp. 76–88 (2007)

A Multimodal Fingers Classification for General Interactive Surfaces

Vitoantonio Bevilacqua[1], Donato Barone[1], and Marco Suma[2]

[1] Politecnico di Bari, Dipartimento di Ingegneria Elettrica e dell'Informazione, Bari, Italy
[2] A.M.T. Services s.r.l., Bari, Italy
vitoantonio.bevilacqua@poliba.it
eng.donato.barone@gmail.com
marco.suma@amtservices.it

Abstract. In this paper a multimodal fingers classification to detect touch points over general interactive surfaces is presented. Three different classifiers have been used: artificial neural networks, decision trees and rules learner. The data set has been created extracting statistical parameters from finger ROIs on about 40000 video samples. The accuracy obtained for the three classifiers on the test set is respectively 96,68%, 96,58% and 97,41%. The model classifiers generated work very well in real-time applications, so an innovative software called TouchPAD has been designed and implemented.

Keywords: supervised learning, artificial neural networks, ripper, decision trees, human-computer interaction, ubiquitous computing, finger detection, image processing.

1   Introduction

In the last decades, the impressive technology development has been continuously changing the way humans and computers interact with each other. Recognizing the presence and the pose of hands is an important task in human-computer interaction applications. We can imagine a world in which each object will communicate and interact with humans in a silent way. Indeed the most profound technologies are those that disappear; they weave themselves into the fabric of everyday life until they are indistinguishable from it [1]. We are talking about a new era of computing, the Ubiquitous one, in which augmented reality systems can be used to reach a perfect synergy between humans and computers. For example, interacting with home walls could give us a way to turn on and off lights by simply pressing a virtual button, or to pull up and down blinds by simply sliding fingers over a virtual slider. So hands become a powerful and complete way to interact with projected multimedia objects; in order to do that it is necessary to know exactly hands and fingers positions. The present paper will show how this task can be achieved using an RGB camera. Three different supervised classifiers were used: artificial neural networks, decision trees and rules learner. Moreover, based on the classification results an innovative software called TouchPAD

D.-S. Huang et al. (Eds.): ICIC 2014, LNAI 8589, pp. 513–521, 2014.

(Touchable Projected Augmented Desk) was realized: it is able to transform a generic surface into a touchscreen table using a projector and an RGB-D camera.

Fig. 1. TouchPAD logo

Section 2 overviews TouchPAD; Section 3 presents the hardware and software used; in Section 4 methods and techniques for preparing the training set are explained; Section 5 will tackle the classification problem; Section 6 will show the classification results and finally, in Section 7 conclusions are shown.

2   TouchPAD: Touchable Projected Augmented Desk

TouchPAD is a markerless augmented reality system prototype (see Fig. 2) for advanced desk interaction. This system allows the user to interact with projected multimedia contents by means of bare hands. The Microsoft Kinect and the Picopix 2055 projector are used to achieve this goal.

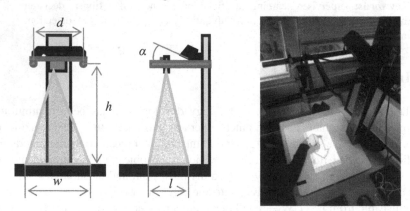

Fig. 2. TouchPAD

TouchPAD's workflow is based on the following tasks:

- Hand detection: hands are recognized by means of skin detection and depth evaluation (only objects in a certain range of distance were considered);
- Fingertips ROI detection and classification: this step will be explained in Section 4 and Section 5;

- Finger centroid tracking during time: an algorithm based on principles of temporal and spatial locality has been used to track finger centroid during time;
- Mapping of contact points in the projector coordinate system using a 3D affine transformation: contact points detected in the Kinect coordinate system are mapped in the projector coordinate system by means of transformation matrices (rotation, translation, shear and scale) obtained by a 3D affine transformation [2]. The calibration process requires only two sets of 4 corresponding 3D points.

The name given to our application is not a coincidence, it tries to recall an "augmented" desk on top of which a user can interact with digital projected images using just bare hands. Furthermore it makes you think of the touchpad, a device present in all laptops that is able to track fingertips movement within its coordinate system and to map them into the screen coordinate system.

The software is composed by two interface modules:

- Processing module;
- Augmented Reality module.

The **processing module** (Fig. 3) shows the results of the algorithms implemented to detect hands, fingers and to track them during time. It is a control panel, therefore it is only visible on the computer's screen.

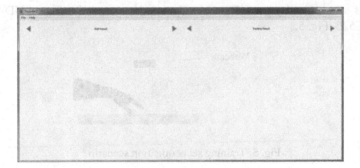

Fig. 3. Processing module GUI

The **augmented reality module** generates and projects the interface shown in Fig. 4 on the desk. It also implements the logic layer of each application implemented.

Fig. 4. Augmented reality panel projected on the desk

The interface is really simple and it is structured in this way to guarantee the maximum room for interaction.

Four applications are available:

- *Optical Character Recognition (OCR)*: the user can draw characters recognized through the Tesseract OCR Engine [3];
- *Qwerty Keyboard*: the user can write text using a simple qwerty keyboard that returns an audio feedback whenever a key is pressed;
- *Paint*: very similar to the Paint software present in all Windows operating systems, the user could pick a color and draw something on the canvas just using his fingertips as pens;
- *Bounce*: the user can track obstacles that change a continuously moving ball trajectory.

3   Materials

An equally divided between finger and not finger dataset of 37982 examples was created. We were able to extract region of interest (ROI) from 11 different videos of hands that differ from each other for color and shape. To ensure a high degree of variability in each video, different actions were performed: rotations, finger extensions, movements along x and y axes. Dataset videos were recorded using an HD Webcam Logitech C525, Fig. 5.

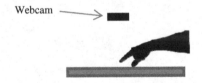

Fig. 5. Training set acquisition scenario

We used **OpenCV** [4] library to extract and process meaningful data from images and we used **Weka** [5] to have easy access to state-of-the-art techniques in machine learning.

For the classification stage, Artificial Neural Networks (ANN) [6], decision trees [7] and a rules learner (Ripper) [8] were used. A brief introduction of these methods will be given below.

4   Methods

In this section methods used to detect hands in the RGB frame and ROIs containing candidate fingers will be explained. In **Fig. 6** the entire workflow is showed.

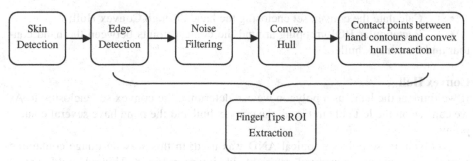

Fig. 6. Extracting feature workflow

4.1 Skin Detection Based on YCbCr

YCbCr is a non-linear transformation of the RGB space used in TV broadcasting studios and for video encoding (MPEGI, MPEGII). The Y in YCbCr denotes the luminance component, and Cb and Cr represent the chrominance factors. If R, G and B are given with 8 bit digital precision, then YCbCr from "digital 8-bit RGB" can be obtained from RGB coordinate using a transformation matrix.

Therefore, unlike RGB, the location of the skin color in the chrominance channels will not be affected by changing the intensity of the illumination. In the chrominance channels the skin color is typically located as a compact cluster with an elliptical shape. The dataset used to determine the position and shape of the ellipse was created from different people from different races: Asian, African and Caucasian. We created a method that converts the RGB image in the YCbCr and at the same time takes into account just skin pixels using the ellipse defined in [9].

In Fig. 7 is illustrated the result of the skin detection before and after filtering operations.

Fig. 7. Skin detection and filtering results

To filter out noise in the binary image obtained from the skin detection, we used a median filter with kernel size 5 to remove salt and pepper noise and several cycles of dilation and erosion.

4.2 Finger Tips ROI Extraction

The algorithm used to detect the fingertips ROI in the image follow these steps:

- Edge detection with Canny;
- Noise filtering based on objects area;

- Calculate the convex set enclosing the hand through Convex Hull;
- Rectangular ROI cropping around the contact points between the hand contour and the convex hull.

Convex Hull

If we think at the hand as a polygon we can determine the convex set enclosing it. As we can see on the left side in Fig. 8, the convex hull and the hand have several contact points.

To obtain these points a logical AND was used; in this way an image containing contact segments was generated, see right side image in Fig. 8. This set could contain points belonging to the fingertips, but also points belonging to the wrist or the hand side.

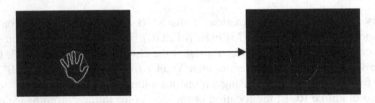

Fig. 8. Logical AND between hand contour and hand convex hull

5   Classification

For each segment detected with the previous method a centroid was computed. Then a ROI proportional to the finger size was cropped from the image containing the hand contour, see Fig. 9.

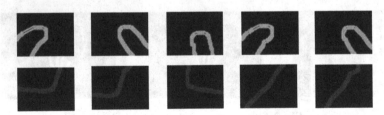

Fig. 9. Fingertips ROI examples. Green contours is related to fingertips, blue contours is related to other hand part contours

The terminal part of a finger has a certain shape, completely different from the one of other segments detected. Based on this assumption we decided to classify these segments through supervised classification based on statistical moments and Hu moments [10, 11].

Region moment representations interpret a normalized gray-level image function as a probability density of a 2D random variable. Properties of this random variable can be described using statistical characteristics: **moments** [11, 12].

In addition to the ones stated before seven rotation, translation and scale invariant moment characteristics were calculated: the Hu Moments [11, 13].Moments, Hu moments, length and area contour parameters were calculated for each ROI and saved on the file system.

Since we decided to use supervised classification, each ROI, previously saved on the file system, has been checked and moved on a particular folder named 0 (Not Finger) or 1 (Finger). Then an automatic procedure to add the correspondent class was created. For each entry, the FName (filename) field was substituted with 0 (Not Finger) if the related ROI was in the "0" directory, 1 otherwise.

Using Weka we trained different classifiers:

- Decision Tree;
- Neural Network;
- Rules learner.

6   Results

This section will illustrate the classification results achieved. The dataset structure used was:

- Training set: 25623 examples (66%);
- Validation set: 12359 examples (34%).

An accuracy of 96.68% was obtained with an 18 leaves - decision tree. The training was performed with the J48 algorithm [7].

Table 1. Decision Tree Confusion matrix

	NOT FINGER	FINGER
NOT FINGER	6279 (TN)	179 (FP)
FINGER	231 (FN)	5670 (TP)

An accuracy of 96.56% was reached with a 3-layer neural network with 33 inputs, 2 outputs and 4 neurons in the hidden layer.

Table 2. Artificial Neural Network Confusion Matrix

	NOT FINGER	FINGER
NOT FINGER	6127 (TN)	331 (FP)
FINGER	94 (FN)	5807 (TP)

At last, an accuracy of 97.41% was obtained using a propositional rule learner, the Ripper, that returns 6 rules.

Table 3. Ripper Confusion Matrix

	NOT FINGER	FINGER
NOT FINGER	6173 (TN)	285 (FP)
FINGER	35 (FN)	5866 (TP)

$$Sensitivity = \frac{TP}{TP + FN} \qquad\qquad Precision = \frac{TP}{TP + FP}$$

$$Specificity = \frac{TN}{TN + FP} \qquad\qquad NPV = \frac{TN}{TN + FN}$$

Table 4. Comparison results of the three classifiers

	Decision Tree	Neural Network	Ripper
Sensitivity	0,9608	0,984	0,994
Precision	0,9693	0,946	0,953
Specificity	0,9722	0,948	0,955
NPV	0,964	0,984	0,994

The software has been tested using each single classifier mentioned. Despite the logic necessary to interpret touches and gestures, in each case the frame rate achieved was 30 FPS average.

7   Conclusions

In this paper, we described and evaluated an innovative technique for finger detection by means of a supervised approach. Moreover, we demonstrated how it can be used in real-time applications, like TouchPAD. Accuracy results are very good: in all cases we reached an accuracy of about 96%, so we tried to implement each model in TouchPAD and checked performances: we chose the Ripper model since it returns the best performance, as it is more robust against skin detection noise. If we assume that skin detection performs well, the model classifier will be able to detect fingers and then touch points with high precision.

Being able to detect fingers and touches allow interactions traditionally reserved for dedicated touch surfaces, to be portable into the environment, according to the needs of ubiquitous computing requirements: so, if touch inputs can be achieved on everyday surfaces, there are many applications that can be developed in home automation, entertainment, rehabilitation. This could really change the way we think about human-computer interaction: no more desktop monitor, no more keyboard, no more mouse.

References

[1] Weiser, M.: The computer for the 21st century. SIGMOBILE Mob. Comput. Commun. Rev. 3, 3–11 (1999)
[2] Fischler, M., Bolles, R.: Random Sample Consensus: A Paradigm for Model Fitting with Applications to Image Analysis and Automated Cartography. Commun. ACM 24, 381–395 (1981)
[3] Smith, R.: An Overview of the Tesseract OCR Engine. In: Proceedings of the Ninth International Conference on Document Analysis and Recognition (ICDAR 2007), vol. 2, pp. 629–633 (2007)
[4] Bradski, G.: The OpenCV Library. Dr. Dobb's Journal of Software Tools (2000)
[5] Hall, M., Frank, E., Holmes, G., Pfahringer, B., Reutemann, P., Witten, I.H.: The WEKA data mining software: an update. SIGKDD Explor. Newsl. 11, 10–18 (2009)
[6] Bevilacqua, V.: Three-dimensional virtual colonoscopy for automatic polyps detection by artificial neural network approach: New tests on an enlarged cohort of polyps. Neurocomputing 116, 62–75 (2013)
[7] Salzberg, S.: C4.5: Programs for Machine Learning. In: Ross Quinlan, J. (ed.) Machine Learning, vol. 16, pp. 235–240. Morgan Kaufmann Publishers, Inc. (1993, September 1,1994)
[8] Cohen, W.: Fast Effective Rule Induction. In: Proceedings of the Twelfth International Conference on Machine Learning, pp. 115–123 (1995)
[9] Elgammal, A., Muang, C., Hu, D.: Skin Detection - a Short Tutorial. Enciclopedia of Biometrics (2009)
[10] Sonka, M., Hlavac, V., Boyle, R.: Moments. In: Image Processing, Analysis, and Machine Vision, pp. 357–359 (2008)
[11] D. T. OpenCV. Structural Analysis and Shape Descriptors - OpenCV 2.4.8.0 Documentation (2013) http://docs.opencv.org/modules/imgproc/doc/structural_analysis_and_shape_descriptors.html?#findcontours
[12] Green, G.: An Essay on the Application of mathematical Analysis to the theories of Electricity and Magnetism (1828)
[13] Hu, M.-K.: Visual pattern recognition by moment invariants. IRE Transactions on Information Theory 8, 179–187 (1962)

A Signal Modulation Type Recognition Method Based on Kernel PCA and Random Forest in Cognitive Network

Xin Wang[1], Zhijun Gao[1], Yanhui Fang[2], Shuai Yuan[1], Haoxuan Zhao[1], Wei Gong[1], Minghao Qiu[1], and Qiang Liu[3]

[1] School of Information & Control Engineering, Shenyang Jianzhu University,
Shenyang, Liaoning, 110168, China
[2] Hebei College of Science and Technology, Baoding, Hebei, 071000, China
[3] Jilin Teachers Institute of Engineering and Technology, Changchun, Jilin, 130052, China
wangxin7988@163.com

Abstract. This paper develops a solution for the problem of the low accuracy on signal modulation type recognition of the weak primary users in low signal-to-noise ratio, a novel signal recognition method based on dimensionality reduction and random forest (RF) is proposed. Firstly, the kernel principal component analysis (KPCA) is applied to extract the most discriminate feature vector. Secondly, the detecting signal is classified by the trained random forest. Performance evaluation is conducted through simulations, and the results reveal the benefits of adopting the proposed algorithm comparing with support vector machine (SVM) and PCA-SVM algorithms.

Keywords: cognitive network, random forest, kernel principal component analysis, spectral correlation characteristics.

1   Introduction

Spectrum sensing by far is the most important component for the establishment of cognitive radio, which detects the availability of frequency bands for secondary users (SUs) in CN[1,2,3]. As one of the key technologies of cognitive network, spectrum sensing can quickly, accurately and effectively detect the situation of spectrum utilization. Due to the fact of the radio channel complex environment, the spectrum sensing technology is focused on spectrum detection problem under the circumstances of low signal-to-noise ratio. Some traditional techniques proposed for spectrum sensing are energy detection, matched filter detection and cyclostationary feature detection[4,5,6]. In [7], the authors proposed the idea of projecting the spectral coherence function (SCF) of the signal to cyclic domain profile and picking the peak values to determine the modulation format. Intelligence and learning are key factors to the classification, there is a trend of applying machine learning algorithms to spectrum sensing in CN recently[8]. The SVM developed by Vapnik and others [9] is used for many machine learning tasks such as pattern recognition and object

D.-S. Huang et al. (Eds.): ICIC 2014, LNAI 8589, pp. 522–528, 2014.
© Springer International Publishing Switzerland 2014

classifications. A SVM-based spectrum sensing algorithms proposed in [10] was applied to the field of cognitive radio.

To improve the performance of the signal recognition, we proposed an algorithm based on the kernel principal component analysis (KPCA) and random forest (RF) algorithm. The detecting signals are classified by utilizing the proposed algorithm. Comparing with the SVM and PCA-SVM, the simulations indicate that the proposed algorithm can effectively enhance the performance of the classification.

The rest of this paper is organized as follows: Section 2 describes the system model of the spectrum sensing. The procedure of combining KPCA dimensionality reduction with RF is introduced in Section 3. It is well validated by using the simulation methods in Section 4. Finally, several conclusions are given in Section 5.

2   Problem Statement

We assume that a cognitive network with N primary users and W secondary users, each primary user signal can be transmitted in multipath channels with a certain delay respectively and is independent from each other. For each time, only one path signal of multipath channels can be received by the secondary users independently. For any one of the secondary users, the recognition of the primary user can be summarized as a hypothesis test model with two elements

$$\begin{cases} H_0 : y(t) = n(t) \\ H_1 : y(t) = s(t) + n(t) \end{cases} \tag{1}$$

where H_0 denotes that the primary user is absent, H_1 denotes that the primary user is present, $0 \le t \le T$, T represents the sampling time of the received signal. A process $s(t)$ (the primary user signal) is cyclostationary in wide sense when its mean and variance are periodic with a period T; $n(t)$ represents the additive white Gauss noise with the mean zero and variance σ_n^2.

3   The Algorithm Based on KPCA Dimensionality Reduction and Random Forest

3.1   Characteristic Parameters Extraction

We assume that each secondary user receiving signal is $y(t)$, The SCF is calculated by utilizing the spectrally smoothed cyclic periodogram method. $S_y^\alpha(K)$ indicates the SCF which is given by:

$$S_y^\alpha(k) = \frac{1}{TL} \sum_{l=-(L-1)/2}^{(L-1)/2} Y(k+l+\frac{\alpha}{2})Y^*(k+l-\frac{\alpha}{2}), \quad k = 1, 2, \cdots, K \tag{2}$$

where $Y(k)$ is the discrete Fourier transform of $y(t)$, L indicates the sample length participated in the frequency domain smoothing, $Y^*(k)$ is the conjugate of $Y(k)$.

$$S(k) = S_y^\alpha(k)\Big|_{\alpha = 1/T_0} \tag{3}$$

The spectrum energy $\mathcal{E}$ is defined as,

$$\mathcal{E} = \frac{1}{K} \sum_{k=0}^{K-1} |S(k)|^2 \tag{4}$$

According to the central limit theorem, the $S(k)$ obeys the Gauss distribution under $L \ggg 1$.

3.2   The Procedure of Kernel Principal Component Analysis for Dimensionality Reduction

KPCA is the best-known nonlinear dimensionality reduction method [11]. We assume that the $h_{l1}, h_{l2}, \cdots, h_{lM}$ express the M values of cyclic spectrum $S_y^\alpha(K)$ with $\alpha \neq 0$ respectively. Then we make up a matrix of the original data set $H = (h_{lm})_{M \times L}$, where l is the number of samples, $l = 1, 2, \cdots, L$, $m = 1, 2, \cdots, M$. The procedure of kernel principal component analysis for dimensionality reduction can be summarized into the following steps.

Step1.   Choose a kernel function k.
Step2.   Compute the covariance matrix of $\varphi(x_l)$.
Assuming the mean of feature space data $\varphi(x_l)$ is zero,

$$\sum_{l=1}^{L} \varphi(x_l) = 0 \tag{5}$$

The covariance matrix in F is

$$C_F = \frac{1}{L} \sum_{l=1}^{L} \varphi(x_l)\varphi(x_l)^T \tag{6}$$

Step3. Calculate eigenvalues and the corresponding eigenvectors of the covariance matrix C_F.

$$\lambda V = C_F V \tag{7}$$

The eigenvectors V of C_F must lie in the span of $\varphi(x_l)$, $l = 1, 2, \cdots, L$.

$$V = \sum_{l=1}^{L} \alpha_l \varphi(x_l) \tag{8}$$

where α_l express the coefficient of feature space data $\varphi(x_l)$.

Step4. Calculate eigenvalues and the corresponding eigenvectors of the kernel matrix K.

By combining, we get the formula (9).

$$\lambda \sum_{k=1}^{L} \alpha_k (\varphi(x_j)^T \varphi(x_k)) = \frac{1}{L} \sum_{l=1}^{L} \sum_{k=1}^{L} \alpha_k (\varphi(x_j)^T \varphi(x_l))(\varphi(x_l)^T \varphi(x_k)) \tag{9}$$

The $L \times L$ dimensional kernel matrix K can be given by

$$K_{ij} = (\varphi(x_i)^T \varphi(x_j)) \tag{10}$$

By combining the formula (10) and the formula (10), we get the formula (11).

$$L\lambda\alpha = K\alpha \tag{11}$$

The eigenvalues and the corresponding eigenvectors of the kernel matrix K can be obtained by the above formula.

Step5. Obtain the kernel principal components $s_k(x)$ of the original sample by

$$s_k(x) = V\varphi(x) = \sum_{l=1}^{L} \alpha_l \varphi(x_l)\varphi(x) = \sum_{l=1}^{L} \alpha_l K(x_l, x) \tag{12}$$

Actually, this centering can be done on the Gram matrix without ever explicitly computing the feature map $\varphi(x_l)$.

The kernel matrix for this centering or zero mean data can be derived by

$$\tilde{K} = (I - H^T / L)K(I - H^T / L) \tag{13}$$

Where H^T expresses the identity matrix of $L \times L$.

Gaussian RBF kernel which is used in our algorithm is

$$k(x_i, x_j) = \exp(-\frac{\|x_i - x_j\|^2}{2\sigma^2}) \tag{14}$$

3.3   The Procedure of Random Forest for Spectrum Sensing

Random forest (RF) is a combination of tree predictors. Each tree is depended on the values of a random vector sampled independently and with the same distribution for all trees in the random forest[12].

According to the above procedure of dimensionality reduction, we assume that $y_{l1}^0, y_{l2}^0, \cdots, y_{lM}^0, \varepsilon^0$ are eigenvalues under the condition of H_0, and the characteristic vector expresses ; $y_{l1}^n, y_{l2}^n, \cdots, y_{lM}^n, \varepsilon^n$ are eigenvalues under the condition of H_1, $Y_l^0 = [y_{l1}^0, y_{l2}^0, \cdots, y_{lM}^0, \varepsilon^0]^T$ where n expresses the $n-th$ primary user, $Y_l^n = [y_{l1}^n, y_{l2}^n, \cdots, y_{lM}^n, \varepsilon^n]^T$ indicates the characteristic vector. The details of the random forest algorithm are summarized as the following steps:

Step1. Assuming the a training sample subset is formed by M_f of characteristic vectors under the condition of H_0, and each subset of different primary user is composed by the M_t characteristic vectors under the condition of H_1, respectively.

Step2. The samples are randomly selected in the above sample subsets as the training samples for random forest under the condition of H_1 and H_0, respectively.

Step3. Every decision tree of random forest is not growing until the percentage of the classification purity in each node achieves the desired growth or a given layer. After that, one decision tree in the random forest is generated.

Step4. Repeat the above steps, the random forest with K trees is eventually established.

Step5. Finally, all the signals are classified by the trained random forest.

4   Simulation Results

In this section, we evaluate the performance of the proposed algorithm with various features. By using BPSK, MSK and OFDM signals as input, the resolution of the carrier frequency is 1 MHZ, and input signals appear to uniform distribution between 3.1 ~ 4.8 GHz. The SVM, PCA-SVM and the proposed algorithm (KPCA-RF) are compared under the signal-to-noise ratio (dB) in different circumstances, such as -15, -10, -5 , 0. The DFT length of cyclic spectrum is 256 and 512. We randomly select 600 and 200 samples from each set composed by 1000 samples as the training samples and the testing samples, respectively. The random forest in the proposed algorithm is established by 100 trees. The channel is assumed to have a multipath (3-path) delay profile with 150ns delay spread to transmit the three modulation signals. Simulations are carried out 10^4 orders of magnitude times.

The overall identification performance of the proposed algorithm comparing with the SVM and PCA-SVM algorithms for the three signals with 512 and 256 data length are shown in fig.1 and fig.2, respectively. The results indicate that the overall identification performance of the three different algorithms increases with the increasing of SNR. The correct identification rate is closer to the ideal value (100%) with the increase of data length. The advantage of random forest as a strong classifier consisted of multiple weak classifiers and the KPCA algorithm's accurate extraction for the nonlinear structure of data. In addition, the KPCA algorithm can preserve most

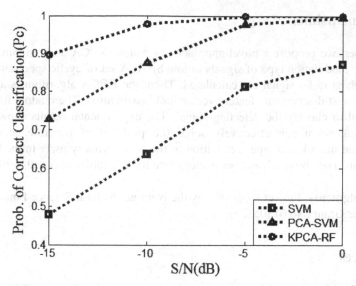

Fig. 1. The detection performance of the proposed algorithm versus SVM and PCA-SVM algorithms with 512 data length

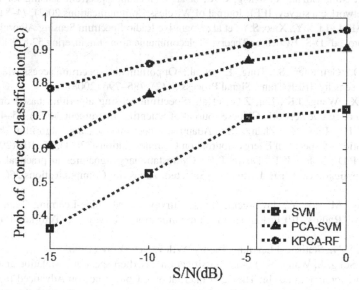

Fig. 2. The detection rates of the proposed algorithm versus SVM and PCA-SVM algorithms with 256 data length

discriminate information and effectively decrease the required feature dimensions in classification. The defects of fails to extract the nonlinear structure of data in PCA and cause overfitting in SVM that impact signal recognition in the comparison methods are effectively overcome in the proposed algorithm.

5   Conclusions

In this paper, we propose a novel approach combining KPCA and random forest to identify the modulation type of signals in low SNR. A set of cyclic spectrum features of the received radio signal are calculated. Then, the KPCA algorithm is applied to extract the most discriminate feature vector for classification. The random forest (RF) is generated to classify the detecting signal. The experimental results show that the proposed algorithm can effectively solve the problem of the low accuracy on detection and modulation type recognition of the weak primary users in the low SNR environment. The above advantages demonstrate the feasibility of our algorithm.

Acknowledgments. Project supported by the National Natural Science Foundation of China under Grant (No.61305125).

References

1. Wang, B., Liu, K.J.R.: Advances in cognitive radio networks: A survey. IEEE Journal of Selected Topics in Signal Processing 5(1), 5–23 (2011)
2. Sun, H., Nallanathan, A., Wang, C.X., et al.: Wideband spectrum sensing for cognitive radio networks: a survey. IEEE Journal of Wireless Communications 20(2), 74–81 (2013)
3. Wang, M.M., Li, S.Y., Xiao, S.Y., et al.: Cognitive Radio Spectrum Sensing Algorithm Based on Improved D-S Evidence Theory. Telecommunication Engineering 52(8), 1303–1307 (2012)
4. Zhao, Q., Geirhofer, S., Tong, L., et al.: Opportunistic spectrum access via periodic channel sensing. IEEE Trans. Signal Process 56(2), 785–796 (2008)
5. Wang, X., Wang, J.K., Liu, Z.G., et al.: Spectrum sensing algorithm based on random forest in cognitive network. Chinese Journal of Scientific Instrument 34(11), 35–41 (2013)
6. Zhang, H., Bao, Z., Zhang, S.: Adaptive spectrum sensing algorithm based on cyclostationary spectrum Energy. Journal on Communications 32(11), 95–102 (2011)
7. Sutton, P.D., Nolan, K.E., Doyle, L.E.: Cyclostationary signatures in practical cognitive radio applications. IEEE Journal of Selected Areas in Communications 26(1), 13–24 (2008)
8. Bkassiny, M., Li, Y., Jayaweera, S.: A Survey on Machine-Learning Techniques in Cognitive Radios. IEEE Journal of Communications Surveys & Tutorials 45(5), 1–24 (2012)
9. Vapnik, V.N.: Statistical Learning Theory. Wiley, New York (1998)
10. Hu, H., Song. J., Wang, Y.: Signal classification based on spectral correlation analysis and SVM in cognitive radio. In: The 22nd International Conference on Advanced Information Networking and Applications (AINA 2008), Okinawa, Japan, pp. 883–887 (March 2008)
11. Scholkopf, B., Smola, A., Muller, K.: Nonlinear component analysis as a kernel eigenvalue problem. Neural Computation 10(5), 1299–1319 (1998)
12. Breiman, L.: Random forests. Machine Learning 45, 5–32 (2001)

Privacy-Preserving Data Mining Algorithm Based on Modified Particle Swarm Optimization

Lei Yang[1], Jue Wu[1,2], Lingxi Peng[3], and Feng Liu[4]

[1] College of Computer Science and Technology,
Southwest University of Science and Technology, Mianyang, Sichuan, China
[2] College of Information Science and Technology,
Southwest Jiaotong University, Chengdu, China
[3] College of Computer Science and Education Software,
Guangzhou University, Guangzhou, Guangdong, China
[4] International School of Software, Wuhan University, China

Abstract. The privacy preserving data mining is a research hotspot. Most of the privacy preserving algorithms are focused on the centralized database. The algorithms on the distributed database are very vulnerable to collusion attack. The Privacy-Preserving data mining algorithm based on particle swarm optimization is proposed in this paper. The algorithm is based on centralized database, and it can be used on the distributed database. The algorithm is divided into two steps in the distributed database. In the first step, the modified particle swarm optimization algorithm is used to get the local Bayesian network structure. The purpose of the second step is getting the global Bayesian network structure by using local ones. In order to protect the data privacy, the secure sum is used in the algorithm. The algorithm is proved to be convergent on theory. Some experiments have been done on the algorithm, and the results prove that the algorithm is feasible.

Keywords: Particle Swarm Optimization, Bayesian network, data mining, privacy preserving.

1 Introduction

The goal of data mining is to do some analysis such as query, search and classification on one or more databases. With the development of the information technology, the data mining are used widely in many fields. This has led to the concerns that the personal data may be misused for a variety of purposes. In order to alleviate these concerns, a number of techniques have been proposed recently in order to perform the data mining tasks in a privacy-preserving way. The privacy preserving data mining is becoming a research hotspot.

The traditional data mining is performed on centralized database, and the privacy preserving algorithms also concentrate on the centralized one. Privacy preserving data mining on the distributed database need higher security level than on the centralized one. It is more difficult and more complex to fulfill the privacy preserve on the

D.-S. Huang et al. (Eds.): ICIC 2014, LNAI 8589, pp. 529–541, 2014.

distributed database. Most of the privacy preserving algorithms on centralized database are not suitable for the distributed ones. To satisfy the privacy-preserving need of the data mining on the distributed database, some new algorithms have been proposed.

Classification is one of the most common applications in the real world. The goal of the classification is to build a model which can predict the value of a variable on the base of the values of the other variables. The classification can help us understand the data more efficiently. There are many classification algorithms, such as decision tree classifier, Bayes classifier, rule based classifier, support vector Machine (SVM) and so on.

Bayesian network is a graphic model which presents the relationship of variables. It provides a method to present the causality of the information. This method can find the relationship in the data. Bayesian theory provides belief function calculation method on mathematics, so Bayesian network has firm mathematical basement. Bayesian network can process incomplete data and data with noisy in data mining. The probability measurement weight is used in Bayesian network to present the relativity among the data, and this method can solve the inconsistency even dependency problem of the data, even dependency problem. The graphic method is used in the method to present the relationship among the data. It is clear and comprehensible for the forecast analysis. Bayesian network has been used extensively in data mining [1-5].

Privacy-preserving data mining algorithm based on modified particle swarm optimization (PPDM-MPSO) is proposed in this paper. This algorithm is focused on the centralized database, and it can also be used in the distributed database. In the distributed database, the modified particle swarm optimization is used to solve the problem of Bayesian network structure learning on each data node. The privacy-preserving problem is focused on the communicate process on the distributed database. Only the local Bayesian network structure is need to be preserved in the algorithm. The secure sum scheme is used to preserve the local data in the algorithm. The Bayesian network learning based on modified particle swarm optimization algorithm is proved to be convergent on theory, and the experiment result is also shows that the algorithm is convergent. As to the privacy-preserving, the secure sum scheme is proved to be feasible according to the experiment result.

2   Algorithm Model

2.1   Bayesian Network

The method of the graph theory is used to express the joint distribution of the variables set according to the relationship of the variables. The Bayesian network can be defined as follows.

Definition 1. The Bayesian network can be described as a triple $G = \langle V, \lambda, P \rangle$. Where V means a nodes set, $V = \{X_1, X_2, ..., X_n\}$, every node represents an attribution. λ represents a set of edges with direction, and $\lambda = \{< X_i, X_j > | X_i \neq X_j, X_i, X_j \in V\}$. $< X_i, X_j >$ represents that there is a dependency relationship between X_i and X_j.

The purpose of Bayesian network structure learning is to get an entire Bayesian network which includes the network topology and the conditional possibility table. The purpose of network learning is to find a Bayesian network which is matched well to the data samples. There is a fitness function which can describe the accuracy of the Bayesian network. The BIC measure function is used as the fitness function in the algorithm. The problem can be described as formula (1).

$$f = BIC(\xi \mid D) = \sum_{i=1}^{n}\sum_{j=1}^{q_i}\sum_{k=1}^{r_i} m_{ijk} \lg \frac{m_{ijk}}{m_{ij}} - \sum_{i=1}^{n} \frac{q_i(r_i-1)}{2} \lg m \qquad (1)$$

Where the ξ means the a Bayesian network structure which is composed by n variables $X=\{x_1,x_2,...x_n\}$. q_i presents the number of x_i parent node's value combination. If there is no parent node, $q_i=1$. r_i presents the number of X_i value. m_{ijk} presents the value of sample when X_i parent node is j and x_i equals to k. $m_{ij} = \sum_{k=1}^{r_i} m_{ijk}$,

$\theta_{ij} = \frac{m_{ijk}}{m_{ij}}$ presents the likelihood conditional probability. Where, $0 \le \theta_{ijk} \le 1$,

$\sum_{k} \theta_{ijk} = 1$.

2.2   Modified Particle Swarm Optimization Algorithm

The position of a particle represents a possible solution in search space in the PSO algorithm. Each particle has its own speed and position. It also has a fitness function values which is decided by the fitness function. Each particle remembers and traces the optimal particle in search space. Each iterative is not random, it is determined by the optimal solution of the last iterative.

Supposed that the possible solutions are initialized as random swarm particles, the particles update themselves according to two optimal particles Pbest$_i$ and Gbest. The update method is described as formula (2) and (3).

$$v_{k+1} = \omega * v_k + c_1 r_1 * (Pbest_i - x_k) + c_2 r_2 * (Gbest - x_k) \qquad (2)$$

$$x_{k+1} = x_k + v_{k+1} \qquad (3)$$

Where k represents the iterations; V_k represents the speed of iteration k; X_i presents the position in of the particle i in the iteration k; r_1 and r_2 are random numbers uniformly distributed within the range [0,1]; c_1 and c_2 are two positive constants, called cognitive parameter and social parameter respectively ; ω is the inertia weight factor; Pbest$_i$ is the best previous position of particle i during the iterations; Gbest is the best position of all particles during the iterations.

2.2.1   Encoding Method

Bayesian network structure is a Directed Acycline Graph (DAG). In the algorithm the adjacency matrix is used to describe the Bayesian network structure. The structure matrix is used to encode the Bayesian network. Supposed that the adjacency matrix of the Bayesian network structure is described as follow, and the code of the Bayesian network structure is 010001100.

$$
\begin{bmatrix} 0 & 1 & 0 \\ 0 & 0 & 1 \\ 1 & 0 & 0 \end{bmatrix} \tag{4}
$$

Where, <i,j>=1 means that there is a directed edge $i \rightarrow j$ in the network. If <i,j>=0, there does not exist the directed edge $i \rightarrow j$. After the initialization, the matrix can be encoded according to the attributes of the directed acyclic graph.

2.2.2   Particle Update

A sequence of 0 and 1 is used to describe the position of a particle in the algorithm. The speed and the motion trace of the particle are defined from the standpoint of the probability. The trace represents the coordinate changing of a particle. Speed represents the probability of the coordinate changes from one status to another in the status space. Each particle x_{id}(the dimension d of particle i) only have two value to choose, 0 or 1 in the algorithm. v_{id} represents the probability of x_{id} equals to 1. The speed iteration is the same to the continuous problem. The position iteration method is described as formular (5).

$$
x_{id}(t+1) = \begin{cases} 0 & sign(v_{id}(t+1)) \leq r_3 \\ 1 & sign(v_{id}(t+1)) > r_3 \end{cases} \tag{5}
$$

Where, r_3 is a random number. Sign (x) is the ambiguous function.

$$
sign(x) = 1/(1 - e^x) \tag{6}
$$

After the updating operation, some topology structures may be illegal, because the Bayesian network structure is a directed acyclic graph without circle, the modify operation must be done on the illegal topology structure. And so the illegal topology must be modifed into the legal one. The process of modification is shown as follow.

1) Calculate the transitive closure according to the matrix of the graph.
2) If the data on the diagonal of the transitive closure are all equal to 0, then the topology structure is legal, otherwise it is illegal. If the topology structure is illegal, the nodes of the graph corresponding to the data on the transitive closure have to be preserved (this nodes are all on the circle in the graph.
3) Find all the parents nodes of the nodes in the circle.
4) Delete or reverse the edges between the nodes and its parents nodes, and then there will no circle in the graph.

2.2.3   Mutation

The mutation operation is used in the algorithm to reduce the probability of getting the local optimal solution. The operation of mutation is a reverse operation in fact. Select a particle according to the probability p, and select a bit of the particle. If the original value of the bit is 1, then change it to 0. If the original value of the bit is 0, then change it to 1.

2.2.4   Algorithm MPSO Procession

1) Initialize the structure graph, and get the adjacency matrix. Initialize the size of the swarm, the mutation probability, the maximum iteration time and so on.

2) Calculate the fitness function, and take it as the optimal solution of this particle; calculate the global optimal solution and get the global optimal particle.

3) Loop

While ($t < t_{max}$)

{

   Iteration numbers +1

   For $i=1:n$

   {

   4) Updating the local optimal particle and particle i, and then execute the mutation operation.

   5) If the new matrix is illegal, then modify it. If it is legal, keep it as the new particle.

   }

   6) Calculate the fitness value of the new particle, and compare it to its original value. If the fitness value is larger than the original one, change the particle and the fitness value, otherwise keep the original one.

   7) Updating the global optimal particle and the fitness value according to the new local optimal particle.

   8) If the new global fitness value is larger than the original one, update the fitness value and the global particle. Otherwise keep the original one.

}

9) Output the global particle, and get the Bayesian network structure.

2.3   Secure Sum

The main idea of the algorithm is to generate a matrix which has the same size with the local data matrix, and the random matrix will be added to the local data matrix, sends the merged matrix to the following party. Each party receives the perturbed matrix, adds it with its local matrix, and passes it to the following party (the last party sends the matrix back to the first party). The first party subtracts the random matrix from the received matrix, which results in the matrix that adds the matrices of all parties, without disclosing their local matrices to each other. The method preserves the data privacy, sine only the original party gets to exactly see the data. The local model is directly computed from the local data. But this algorithm is very fragile to the collusion attack. The parties preceding and following a party can collude to recover its local matrix. This paper uses a modified secure sum algorithm, and the method is described as follow.

Let x_i represents the data held by the party i, k is the number of the data slice. All the data split into k slices. And the x_{ij} satisfies the equations (7). The x_{ij} and x_i are in the same interval.

$$x_i = \sum_{j=1}^{k} x_{ij} \tag{7}$$

1. Every party sends the $x_{i2},...,x_{im}$ to the other m-1 parties, and keep the x_{i1} as private.
2. Every party receives k-1 numbers which come from the other k-1 parties.
3. Every party adds the k-1 numbers to its own x_{i1}, and sends the sum to the data center.
4. What the data center gets is as follow.

$$\sum_{i=1}^{n} \sum_{j=1}^{k} x_{ij} = \sum_{i=1}^{n} x_i \tag{8}$$

What the data center gets is the global gram matrix of the distributed database. During the procedure of the algorithm, every party keeps the x_{i1} as private. And so it is very difficult to get the data of other parties. If the data center don't take part in the collusion attack, what the other m-1 parties can get is $\sum_{i=2}^{k-1} x_i$, and they can't obtain the data x_{i1}. And so they can't get the exact data of the party. This has nothing to do with other coefficients. If the data center takes part in the collusion attack, the collusion parties can get the exact data of one party only when they receive the k-1 data.

The data needed to be protected in the algorithm is the local Bayesian network structure. Supposed that the local Bayesian network is G=<V_i, E_i>, E_i represents the edge set, V_i represents the point set. The Bayesian network is described by matrix. The matrix element is -1 or 1. The secure sum scheme is used in the algorithm. Supposed the number of the node is S. If the matrix element is equal to S, the oriented edge saved in edges set E_a, otherwise the edges saved in edges set E_b. E_a is the edges in the global Bayesian network. E_b is the undetermined edges in the global Bayesian network. The mutual information of the edges in the set E_b is calculated in the next step. If the value of the mutual information is larger than threshold value, than add the corresponding edge to edges set E_a. The edges in set E_a is the global Bayesian network structure.

3   Convergence Analysis

The speed and the motion trace of the particle are defined from the standpoint of the probability in the modified PSO algorithm. The motion trace represents the probability of the value equals 0 or some other value. The speed represents the status changing probability of a certain coordinate of a particle.

Gbest $_d$ represents the global optimal value of the dimension d. If the probability of the particle values in dimension d equals to 1 is larger than 50%, then the value of the Gbest $_d$ equals 1. If the probability of the Pbest $_d$ equals to 0 is larger than 50%, then the values of the Gbest $_d$ equals 0. If the probability of the pbest $_d$ =0 is 50%, then the G_{best_d} equals 0 or 1 randomly.

The updating of the particle on each dimension has nothing to do with each other, and so we can simply the proof process of one dimension. The note of dimension d can be ignored, and so only the particle i is take into account, the other particles are thought to be still. The note i is also can be ignored.

The formula (2) and (5) are used to update the particle in the iteration. The formula (5) is the kernel function sigmoid function. The sigmoid function is very similar to formula (9), so the x(t) can be replaced by the v(t) during the calculation. The x(t) can be got by the round calculation according to the r_3.

$$f(x) = \begin{cases} 1 & x > r_3 \\ 0 & x \le r_3 \end{cases} \tag{9}$$

Supposed that p_i represents the optimal position of particle i, p_g is the optimal position of the all particle. $\varphi_1 = r_1 c_1$, $\varphi_1 = r_1 c_1$, $\varphi = \varphi_1 + \varphi_2$, $p = \dfrac{\varphi_1 p_i + \varphi_2 p_g}{\varphi_1 + \varphi_2}$.

And then the formula (2) can be changed as follow.

$$v(t+1) = \omega v(t) + \varphi(p - x(t)) = \omega v(t) + \varphi(p - v(t)) = (\omega - \varphi)v(t) + \varphi p \tag{10}$$

The formula (5) can be modified as formula (11).

$$x'(t+1) = v(t+1) \tag{11}$$

After getting the optimal solution of the problem x'(t), the binary solution can be gotten according to the formula (9).

According to the formula (10) and (11), we can get the formula (12)

$$x'(t+1) = v(t+1) = (\omega - \varphi)^t v(0) + \frac{1 - (\omega - \varphi)^t}{\omega - \varphi} p \tag{12}$$

The limitation of the formula (12) is described as follow.

$$\lim_{t \to \infty} v(t) = \begin{cases} \infty & \omega - \varphi > 1 \\ \dfrac{p}{\omega - \varphi} & \omega - \varphi \le 1 \end{cases} \tag{13}$$

Form the formula (13), we can draw a conclusion that the particle status will converge to a certain value if it satisfy the condition $0 < \omega - \varphi < 1$, and it can come to a conclusion that the algorithm is converge in the condition $0 < \omega - \varphi < 1$.

4   Simulation Experiment

The Asia network is used to be an example to do some experiment in the paper. It is a very small belief network for a fictitious medical example about whether a patient has tuberculosis, lung cancer or bronchitis, related to their X-ray, dyspnea, visit-to-Asia and smoking status. It is also called "Chest Clinic". There are eight variables in the Asia network. Each variable only has two possible values. The data in the experiment are created by the software Netica according to the probability of the Asia network. There are there data sets with different scale.

Algorithm K2 is an important algorithm on Bayesian network structure learning. It is used to compare with the algorithm supposed in this paper. K2 algorithm and the modified ant colony optimization algorithm are used to get Bayesian network structure on different data sets separately. The computer with the main frequency P4.6G is used in the experiment. And the matlabR2007a is used to simulate the algorithm. The comparison is focused on two aspects, the run time of the algorithm, and the error rate of the result. In the simulation process, the population number equals to 50, the iteration number equals to 500, the inertia factor $\omega = 0.5$, the cognitive parameter $c_1=2$, the social parameter $c_2=2$, the mutation probability $p_m=0.05$. The two algorithms were run ten times separately and the average values are used to do the comparison in the experiment. The average values include the redundant edges, the reverse edges, the missing edges and the run time.

The Figure1 shows that the Bayesian network structure learnt by the MPSO has less redundant edges than the Bayesian network structure which learnt by the K2. The Figure 2 and figure 3 shows that the structure learnt by the MPSO has less missing edges and reverse edges than which learnt by the algorithm K2. From the Figure4, it can be seen that the MPSO is also more efficiency than K2 algorithm. And with the increasing data number, the rate of the run time increase of the MPSO is less than the K2. The best result of the MPSO is show in Figure5. There is only one reverse edge compared with the standard network as shown in Figure6. From the simulation we can come to a conclusion that the MPSO algorithm can get a better Bayesian network structure than K2.

As to the privacy preserving, what we want is the sum the gram matrix of all the distributed parties. And the most important thing is to keep the data from disclosing to the other parties in the procedure of data exchanging. In the experiment, we supposed that there are 200 parties in the distributed database. In the experiment we set m as 3,5,10,20 and 30 separately. Each party selects other m-1 parties to send the data slices initiatively, and the data center don't take part in the data sending. This method makes the low possibility of the collusion attack. The collusion attack possibility of each algorithm is shown in the Figure7. From the Figure7 we can see that the PPDM-MPSO algorithm has low possibility of disclosing the private data. And the larger the number m is, the smaller the disclosure possibility is. When the number of the collusion attack is small the disclosure possibility is very small. With the development of the collusion number, the disclosure possibility will become large, but we can solve this problem by using a larger m. From the experiment we can see that the PPDM-MPSO on distributed database is feasible.

The PPDM-MPSO algorithm has the same performance of the MPSO which used in the centralized database. And so the stabalization of the algorithm on the distributed databse, we are mainly focus on the stabilization of the algorithm with the increasing number of the parties. In the experiment we use a 80×80 matrix, and suppose they are on 3,10,20,30,40,50,60,70,80 parties separately. The data center gets the data from all the parties, and then they do the classification on the global data. The run time of the algorithm with the different party number is described on the Figure8. From the Figure8, we can draw a conclusion that the run time does not enhance too more with the enhancement of the party number. That is to say that the algorithm has the run time stabilization with the increasing party number.

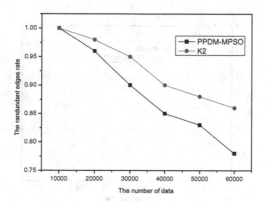

Fig. 1. The comparison of redundant edges

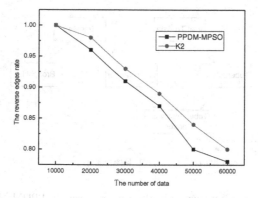

Fig. 2. The comparison of reverse edges

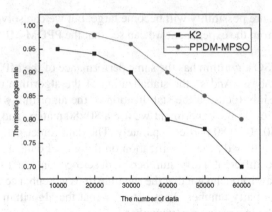

Fig. 3. The comparison of the missing edges

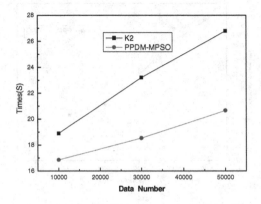

Fig. 4. The comparison of the run time

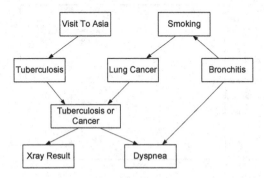

Fig. 5. The Bayesian network structure based the PPDM-MPSO

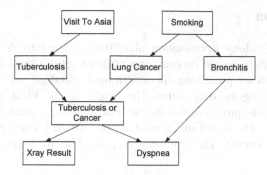

Fig. 6. Standard Asia Bayesian network structure

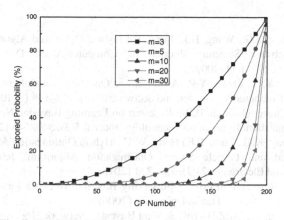

Fig. 7. The PPDM-MPSO performance in collusion attack

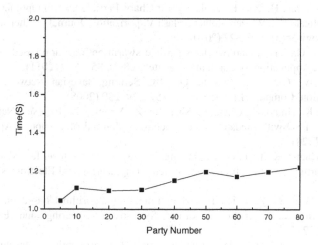

Fig. 8. Run time with different distributed node number

5   Conclusion

The modified ant colony optimization algorithm is used to do data mining on distributed database. It is used on each data node to get the local Bayesian network structure. The privacy-preserving in distributed database is focused on the communication among the data nodes. The results of the local Bayesian network learning is needed to be protect during the process of private preserving data mining on distributed database. The secure sum is used to preserve the privacy in the algorithm to do the privacy-preserving. The algorithm is proved to be feasible on theory and experiment.

References

[1] Xu, L.J., Huang, J.G., Wang, H.J., Long, B.: Hybrid Optimized Algorithm for Learning Bayesian Network Structure. Journal of Computer-Aided Design & Computer Graphics 21(5), 633–639 (2009)

[2] Zhao, J.S., Li, Y.G., Zhang, Y.P.: An Improved Quantum Ant Colony Algorithm and Its Application. Computer Application and Software 27(7), 133–138 (2010)

[3] Xing, C.H., Zhang, Q., Xian, H.W.: Research on Learning Bayesian Networks by Particle Swarm Optimization. Information Technology Journal 5(3), 540–545 (2006)

[4] Li, M.W., Kang, K.G., Zhou, P.F., Hong, W.C.: Hybrid Optimization Algorithm Based on Chaos, Cloud and Particle Swarm Optimization Algorithm. Journal of Systems Engineering and Electronics 24(2), 324–334 (2013)

[5] Wang, H.M.: Research on Privacy-Preserving Bayesian Network Learning, Dissertation for Philosophy Degree in Tianjin University (2006)

[6] Wang, H.S., Liu, G., Qi, Z.H.: BIC Scoring Bayesian Network Model and Its Application. Computer Engineering 34(15), 229–230 (2008)

[7] Sheng, Y., Pan, H., Xia, L., et al.: Hybrid Chaos Particle Swarm Optimization Algorithm and Application in Benzene-toluene Flash Vaporization. Journal of Zhejiang University of Technology 38(3), 319–322 (2010)

[8] Dong, Y., Guo, H.: Adaptive chaos particle swarm optimization based on colony fitness variance. Application Research of Computers 28(3), 855–859 (2011)

[9] Haishu, W., Gang, L., Zhaohui, Q.: BIC Scoring Bayesian Network Model and Its Application. Computer Engineering 34(15), 229–230 (2008)

[10] Yuyuan, K., Jintao, Y., Qiang, L., Shenglin, Z., Minghu, Z.: Bayesian Network Classifier based on PSO with predatory escape behavior. Journal of Computer Application 31(2), 454–457 (2011)

[11] Lu, X., Huang, K., Lian, G.: Optimizing strategy of system level fault diagnosis based on chaos particle swarm optimization. Systems Engineering and Electronics 32(1), 217–220 (2010)

[12] Yu, S., Wei, Y., Zhu, K.: Hybrid optimization algorithms based on particle swarm optimization and genetic algorithm. Systems Engineering and Electronics 33(7), 1648–1652 (2011)

[13] Zhang, Y., Shao, S.: Cloud hyper mutation particle swarm optimization algorithm based on cloud model. Pattern Recognition and Artificial Intelligence 24(1), 91–95 (2010)

[14] Liu, H., Zhou, Y.: A cloud adaptive particle swarm optimization algorithm based on mean. Computer Engineering & Science 33(5), 97–100 (2011)

[15] Hongmei, W., Yuan, Z., Zheng, Z., Chengshan, W.: Privacy-preserving Bayesian network Learning on distributed heterogeneous data. Journal of Tianjin University 40(9) (2007)

[16] Weiping, G., Wei, W., Haofeng, Z., Baile, S.: Privacy Preserving classification Mining. Journal of Computer Research and Development 43(1), 39–45 (2006)

[17] Wang, J.: Matrix Decomposition for Data Disclosure Control and Data Mining Applications, Dissertation for Philosophy Degree in University of Kentucky (2008)

[18] Han, X., Min, J., Ma, H.-X.: Soving Fuzzy Chance-chonstrained Programming Using Chaos Quantum ACO Algorithm and Its Convergence. Journal of System Simulation 21(20), 6462–6468 (2009)

[19] Huang, J.-W., Qin, C.-Y.: Real-coded quantum ant colony optimization algorithm for global numerical optimization. Application Research of Computer 26(10), 3660–3662 (2009)

[20] Ashrafi, M.Z., Taniar, D., Smith, K.: Towards Privacy Preserving Distributed Association Rule Mining. In: Proceedings of Distributed Computing-IWDC, pp. 835–835 (2003)

Topic Extraction Based on Knowledge Cluster in the Field of Micro-blog

Ming Li, Chunhong Zhang, Li Sun, and Xianlei Shao

Mobile Life and New Media Laboratory,
Beijing University of Posts and Telecommunications, Beijing, China
{liming00bupt,zhangch.bupt.001,buptsunli}@gmail.com,
shaoxianlei@163.com

Abstract. Views in the field of Micro-blog can always express what people concerns. Topic extraction in this domain is necessary to help focus on events that are popular and important. At present LDA, a probabilistic topic model based on word co-occurrence is widely used. However, when it comes to deal with microbloggings that are all short texts with a lot of noise, the result can be not satisfactory. In this paper, we propose a new index called M-membership that is proved to be more suitable to indicate the importance of each term in this scenario, and present a topic extraction method based on Fuzzy C-Means algorithm, then use fuzzy set to represent clusters and topics, which can lead to a more reasonable and thematic result. Comparative experiments with K-Means and LDA illustrate our ability to cluster and extract topics in the field of Micro-blog.

Keywords: Topic Extraction, Fuzzy C-Means, Fuzzy Set, Micro-blog.

1 Introduction

With the social media revolution, Micro-blog has become a significant and popular media that draws wide attention. People express their thoughts, show their daily lives, and communicate with each other by sending posts within 140 characters, pasting images, or embedding videos. In these massive and heterogeneous data resources, which contain knowledge with great potential value, it is urgent to be able to discover the useful information efficiently.

Topic extraction is a common technique for information discovery,which can be implemented by topic models, based upon the assumption that documents are a mixture of topics [1]. Some well-known topic models have been widely used such as Latent Dirichlet Allocation (LDA), which is a kind of unigram model based on word co-occurrence. Anothor is clustrering or classification based on natural language grammar structure unit, as mentioned in paper [2]. At the same time clustering is also a popular process of topic model. A clustering approach distinguishes topics based upon the similarity of words used in documents and groups documents into topic clusters[1]. As is known to all, microbloggings are conveyed by natural languages, which have multiple and different meanings and are inherently fuzzy, that makes it

D.-S. Huang et al. (Eds.): ICIC 2014, LNAI 8589, pp. 542–550, 2014.

difficult to define the specific categories. Moreover, the typical topic model such as LDA is not very suitable to short texts like microbloggings [3]. Furthermore those methods based on natural language grammar structure increases the computational overhead by bring more side information. Therefore, in this paper, we analyze the Micro-blog content based on fuzzy clustering algorithm, and study how to find the main topics merely from the Micro-blog content, then extract keywords to represent each topic in the form of fuzzy set. Here we focus on the content and ignore other social graph information in the social media, for Marco Pennacchiotti and Ana-Maria Popescu from Yahoo! Labs have shown that a machine learning algorithm using user-centric features achieves good performance [4].

Our contributions are as follows: Firstly, we propose an index based on the membership from FCM, defined as *M-membership,* to measure the importance of one term in one cluster. Secondly, we introduce an unsupervised method called TEBF (Topic Extraction Based on FCM) to extract salient topics from microbloggings effectively. Finally, we provide a general method to extract topics from short text, for TEBF can be extended to any kinds of ducuments not only short text.

The rest of this paper is organized as follows. In Section 2, we introduce the related research status. Section 3 describes in detail the topic extraction method Topic Extraction Based on FCM (TEBF) we proposed here in detail. Then the experiment result and discussions are presented in Section 4. Finally Section 5 gives some concluding remarks and discusses future extensions of this work.

2   Related Work

Micro-blog as a form of social media, is fast emerging in recent years. There have been many researches based on Twitter, the original Micro-blog in USA. For example, in 2009, Grinev M [5] had studied how to extract user Interest from Micro-blogging stream. Ishikawa S [6] proposed a novel detection scheme for hot topics on Twitter in 2012, which focused on a real-time system that detected hot topic in a local area during a particular period.

With the population booming of Sina-Microblog in China, researches based on it is developing in recent years. Liu, Zitao [7] proposed a novel feature selection method based on part-of-speech and HowNet in 2009. Tu, Hao [8] proposed a mean based top k category incomplete clustering algorithm, finally developed a system that can detect the hot event online. He, Huang [9] put forward three methods of Micro-Blog orientation classification to resolve the problem of Micro-Blog sentiment analysis, and compare the accuracy and performance of each classification method.

All above researches didn't discussed in depth about the clustering method used in topic extraction. In this paper, we incorporate fuzzy set theory with FCM algorithm and introduce a new unsupervised method for topic extraction. Fuzzy sets are sets whose elements have degrees of membership, which were introduced by Lotfi A. Zadeh and Dieter Klaua in 1965 as an extension of the classical notion of set. It permits the gradual assessment of the membership of elements in a set, which is described with the aid of a membership function valued in the real unit interval [0, 1] [10]. Membership is a specific attribute for fuzzy and is the basis for achieving a soft topic classification, since it solves the problem that word is polysemous or multi-word with one meaning. Based on the membership matrix generated from FCM, which is a method of clustering that allows one piece of data to belong to two or more clusters, and developed by Dunn in 1973 and improved by Bezdek in 1981 [11], we propose an index defined as M-membership to measure the importance of one term in one fuzzy set. The following is a detailed description of our method.

3   Topic Extraction From Micro-blog Content

3.1   Framework Design

In this section, we present a detailed description of our proposed topic extraction method. Firstly, the original microbloggings from Sina Micro-blogs is used as our data set. Since the observation that many microbloggings are meaningless that should be filtered out. Then after filtering, all the remaining microbloggings are divided into tokens using the popular freeICTCLAS system [12], which has a high accuracy in Chinese word segmentation. The collection of tokens is a high dimension space, so we use a weighted TF-IDF algorithm to extract tokens as features of each microblogging.

Secondly, the microbloggings are represented by the selected features in the form of Vector Space Model (VSM). And the basic Fuzzy C-Means (FCM) algorithm is applied to reach a preliminary cluster result, which include a membership matrix to show the degree of each microbloggings belongs to each cluster. With the assumption that each individual cluster mainly reflect a distinct semantic topic which is discriminate with other clusters, we select the most representative features as keywords in each cluster for representation of topics based on the index we defined as M-membership that comes from the membership matrix. Then each topic can be expressed as a fuzzy set that consists of TOP-n keywords and their M-membership. With the proposed fuzzy representation of each cluster, we are in a position to formulate the semantic similarity measures between any two fuzzy sets. When the similarity is greater than a certain threshold, we consider merging those fuzzy sets as one. At the same time, the corresponding clusters would be achieved.

3.2   Feature Extraction

Microbloggings are short text messages consisting of Chinese sentences, as a result, the first thing that should be done is to divide these sentences into tokens (that is Chinese word) before clustering, which is called Chinese word segmentation. Then we use TF-IDF [13] to find the important tokens as the features of each microblogging, since it is simple and can retain most of the original token information.

3.3   Topic Extraction Based on FCM (TEBF)

Preliminary Clustering. Here we utilize the basic FCM algorithm to achieve a preliminary clustering result with a membership matrix. The element u_{rj} of the membership matrix $U_{c \times N}$ is calculated as

$$u_{rj} = \left\{ \sum_{r=1}^{c} \left[\left(\frac{d_{ij}}{d_{rj}} \right)^{\frac{2}{m-1}} \right] \right\}^{-1} \tag{1}$$

Where u_{rj} is the value of membership of the *j-th* microblog in the *r-th* cluster acquired by FCM algorithm. The parameter d_{rj} is the distance defined in vector space between microblog j and the center of cluster r. Similarly, d_{ij} is the distance between microblog j and any other microblog i. The parameter c is the number of cluster, m is the ambiguity that always be 2 [14]. The membership u_{rj} shows the degree of one microblog belonging to one cluster. And the higher the u_{rj} of a microblog is, the more sematic correlation the microblog to the topic of the cluster. This characters of membership matrix $U_{c \times N}$ makes FCM quiet feasible as primary cluster approach to deal with the prevalent fact of topic mixture nature observed in microblog. The process of basic FCM algorithm is described in detail in [15].

Key-words Extraction. The result of preliminary clustering provides us the membership matrix $U_{c \times N}$. With the membership matrix, we can know the degree that one micro-blog belongs to each cluster. Then we use the membership-MSE (the Mean Squared Error of one's membership) of each Micro-blog to indicate the discriminate capability of a single microblog to different clusters, and filter out those whose membership-MSE is smaller. This is based on the observation that microbloggings with small membership-MSE are not distinguishing, most of them should belong to none cluster on earth. After filtering them we can get the candidate microbloggings that are discriminative for topic extraction. For *j-th* microblogging the membership-MSE can be denoted as

$$(U_\delta)_j = \sqrt{\frac{\sum_{r=1}^{c} (u_{rj} - \frac{1}{c})^2}{N}}, j \in (1, N) \tag{2}$$

While the average membership-MSE of all microbloggings is denoted as $\overline{U_\delta}$. When $(U_\delta)_j > \overline{U_\delta}$, the *j-th* microblogging is considered to be the candidate from whom to extract topics.

It should be again noted that not all words in the candidate microblogs have the same significance in the sense of topic representation even though they have high U_δ. Simultaneously, the large amount of words used by primary sentences of microblog will confuse the understanding of human about the principle semantic opinions contained in large corpus. Therefore as mentioned before, we will then try to find a way to acquire a small subset of keywords with the most representative capability for the topic of each cluster from candidate microblogs.

In order to extract useful keywords from each cluster, we define the M-membership ϕ_{tj} of each feature t in j-th Micro-blog. The M-membership ϕ_{tj} consists of two factors: one is the weight of the feature, the other is the membership of the microblogging. The higher the feature weights, the more important it will be to the microblog than other words in the same microblog. So the weight of a word in a single microblog could be viewed as its local relative significance indicator to the microblog. On the other side, the membership of a microblogging can also represent the importance of its features to the cluster. When two words with equal weight but appear in two different microbloggings belonging to the same cluster, their value of membership could be used as global regulators on the local weights of them to further rank their representative significance in the particular cluster. The one with a higher membership must have a greater chance to be the cluster-keywords than the other. So here we combine the weight and the membership to define the M-membership of one feature to one cluster. The definition is as follows.

Definition. For the j-th Micro-blog vector with dimension m, the t-th feature has a weight $w_{j,t}$, the Micro-blog vector is denoted as $\langle w_{j,1}, w_{j,2}, ..., w_{j,m} \rangle$, and u_{rj} is the membership value to represent the degree that the j-th microblogging vector belonging to the r-th cluster. If the number of microbloggings in cluster r is q, the M-membership ϕ_{rt} of the feature t in cluster r is defined as

$$\phi_{rt} = \frac{1}{q}\sum_{j=1}^{q}(u_{rj} \times w_{j,t}) \tag{3}$$

ϕ_{rt} reflects the degree that the t-th feature belonging to the r-th cluster, which can make the important feature more important. Features in the distinguishable microbloggings can be considered as candidate description of the cluster they belong to. For the microbloggings in cluster r, the M-membership ϕ_{rt} of the t-th feature can be calculated. Then we can rank features according to their M-membership and select the TOP-n features as keywords to compose a fuzzy set S_r to represent the r-th cluster as

$$S_r = \{\phi_{rt} / feature_{rt}, t \in [1, n]\} \tag{4}$$

Here $feature_{rt}$ is the t-th feature in the r-th cluster, and ϕ_{rt} is the M-membership of $feature_{rt}$ to the r-th cluster. At the same time, features in the clustering center can also be the candidate description of the current cluster as well. Similarly, we can get a fuzzy set denoted as S_r^c. Then with the Union operation of fuzzy set, the candidate topic fuzzy set S_r^T of the r-th cluster can be expressed as

$$S_r^T = S_r \cup S_r^c = \{\phi_{rt} + \phi_{rt}^c - \phi_{rt} \times \phi_{rt}^c\} \tag{5}$$

Similarity-Based Merger. Through the processes above, the Top-n keywords and its corresponding M-membership of each cluster have been extracted and each cluster is represented in a form of fuzzy set. Given the proposed fuzzy representation of the candidate topic, we are in a position to formulate the semantic document similarity measures between two sets S_1 and S_2 as $sim(S_1, S_2)$. We approach the similarity measure based on Tversky's Ratio Model, which is a set-theoretic model that represents objects as set of features as follows [16].

$$sim(S_i, S_j) = \frac{|S_i \cap S_j|}{|S_i \cap S_j| + |S_i \setminus S_j| + |S_j \setminus S_i|} \tag{6}$$

Where S_i and S_j are two fuzzy sets, $|\cdot|$ is the scalar cardinality of a fuzzy set, $\cap$ is the set intersection, and $\setminus$ is the set difference [17]. According to the formulas above, we can get the Similarity matrix with element of pairwise similarity between two arbitrary fuzzy sets, which is symmetric obviously. A reasonable threshold should be selected in order to merge two similar small topic fuzzy sets into a big one. At the same time, the corresponding clusters should be merged as well. After merging, the new M-membership is the average of the old ones.

4   Experiment and Evaluation

4.1   Data Set

The data set comes from the Project that the Analysis of Social Index of the 2014 Spring Festival Gala of CCTV [18]. The principle goal of the project is to find the dominant opinions and comments to the Gala from huge amount audiences with large diverse social backgrounds according to their microblogs. The topics acquired by TEBF are applied to fulfill the requirements. During the project experts tagged 15,000 microbloggings from Sina-Microblog to a certain topic, which are used as the ground truth of our approach as well as other two comparative algorithms. Later we will show our experiment to verify the effect of K-Means, TEBF and LDA respectively.

4.2   Experiment

For our data set is microbloggings talked about the 2014 Spring Festival Gala of CCTV, the name of actors and programs are accordingly meaningful keywords found by both TEBF and LDA. Following we give There topics as example, and show top-10 keywords and their weight in each topic in order to give an intuitive understanding of TEBF.

Table 1. Topic Fuzzy Sets Based on TEBF

Topic1	Topic2	Topic3
0.854/Xiao Caiqi (小彩旗)	0.919/Spring Festival Gala(春晚)	0.942/Last resort (情非得已)
0.564/circling(转)	0.552/director(导演)	0.349/Lee Minho (李敏镐)
0.439/think(想)	0.288/Director Feng(冯导)	0.114/Oba(欧巴)
0.139/stop(停)	0.082/Feng Xiaogang(冯小刚)	0.074/listen(听)
0.098/hard work(辛苦)	0.032/the Spring Festival(春节)	0.053/really(真)
0.095/clock(时钟)	0.026/thing(事)	0.039/music(音乐)
0.063/spinning top (陀螺)	0.025/half(半)	0.024/Minho(敏镐)
0.063/tired(累)	0.022/dream(梦想)	0.021/handsome (帅)
0.049/child(孩子)	0.014/expect(期待)	0.016/happy(快乐)
0.045/praise (点赞)	0.009/program(节目)	0.015/long legs(长腿)

Table 2. Topic Result achieved by LDA

Topic1	Topic2	Topic3
0.224/distressed (心疼)	0.319/Feng Xiaogang(冯小刚)	0.286/Lee Minho(李敏镐)
0.114/Xiao Caiqi(小彩旗)	0.149/Director Feng(冯导)	0.167/Oba(欧巴)
0.045/expect(期待)	0.149/director(导演)	0.117/very(太)
0.029/circling(转圈)	0.137/Spring Festival Gala (春晚)	0.113/Not(没)
0.028/song(歌)	0.123/spring(春)	0.112/Last resort (情非得已)
0.025/scene(镜头)	0.084/program(节目)	0.056/blank(呆)
0.023/return(回)	0.078/hour(小时)	0.049/long leg(长腿)
0.023/Tucao(吐槽)	0.057/get out(滚出)	0.036/handsome (帅)
0.017/memory(记忆)	0.054/thing(事)	0.027/Korea (韩国)
0.017/god(神)	0.029/Feng(冯氏)	0.027/foreign(外)

Compared Table 2 to Table 3, we can see that the topics achieved through TEBF are easy to be understood by human and closer to the labeled benchmark than those generated through LDA. For example, Topic1 is talks about Xiao Caiqi and her program that she circled like a clock for 4 hours. After analyzing by experts who participate in the project, most of the sampling microbloggings talk about how hard for such a young child to perform this program, make praise to her performance, and show people's distressed feeling to Xiao Caiqi. In Table 1, Topic1 is mainly consists of "Xiao Caiqi", "circling", "hard work", "clock", "praise" and "child", "tired", "spinning top", which is obviously more similar to the content of microbloggings, rather than Topic1 in Table 2 that includes a lot of less related words such as "memory", "song", "return", "scene", "god" and so on. This also proves that with the higher cluster precision, the topics are closer to the sematic meanings of microbloggings.

When we apply the TEBF and LDA as well as K-Means to the total 15,000 labeled microbloggings which contains noise microbloggings. With the tagged data set as benchmark, we can use evaluation such as Rand Index and Precision [19]. Then we also consider the typical measure of performance called Perplexity that is widely used for LDA. For a fair comparison, we use this parameter to measure the effect of our method, too. The model with low Perplexity value is better in performance [20]. The evaluation parameters are showed in Fig1.

As can be seen in Fig.1 (a), while the Rand Index of TEBF is a little higher than LDA, both of their performance is better than K-Means in terms of Rand Index as shown. At the same time, the clustering result of TEBF can reach a relative high precision as 43.87% while K-Means and LDA only can reach 17.60.11% and 33.48% respectively as shown in Fig.1 (b). This is because the use of membership-MSE in the part of Key-words Extraction in TEBF can filter out microbloggings that do not belong to any cluster on earth. While K-Means as a hard clustering method that puts some meaningless microbloggings that had not belong to any cluster to a certain topic mandatorily, which cause a disturbance on topic extraction and affect the extraction effect. Given its topic extraction is so poor, we only discuss the topic extraction results of LDA and TEBF above. We can see from Fig.1 (c) that TEBF can reach a significant lower perplexity than LDA, which indicates that TEBF is effective in topic extraction. This is also a verification that LDA is useful in long texts, but is not very suitable to short texts like microbloggings [3]. While TEBF itself is no limit on the length of the text, and the parameters M-membership we proposed in this paper can effectively extract keywords to represent topics.

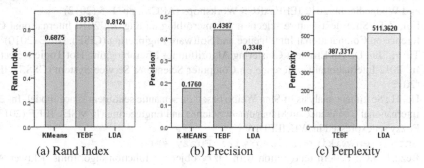

(a) Rand Index   (b) Precision   (c) Perplexity

Fig. 1. The evaluation index for the total labeled microbloggings

5   Conclusions

In this paper, we proposed an unsupervised method to extract main topics from Micro-blog content based on FCM algorithm. Considering the inherently fuzzy of the Micro-blog content, we proposed a new parameter defined as M-member to quantify the semantic representation capability of a word to fuzzy clusters and the corresponding topics, and extracted keywords from clusters according to their M-membership. The most advantage of TEBF is that the found topics are more aligned to the meaningful interpretation of human being, which makes it a potential feasible approach of topic analysis especially for short social media corpus.

For further work, we wish to explore in detail running our system in parallel computing platform to improve the clustering performance. In addition, we consider to introduce the word2vec tool that is developed by Google in the process of VSM.

Acknowledgment. This work is supported by the National Science and Technology Support Program that TV content cloud services technology integration and demonstration based on social network (No.2012BAH41F03).

References

1. Lu, Y.J., Zhang, P.Z., Deng, S.S.: Exploring Health-Related Topics in Online Health Community Using Cluster Analysis. In: 2013 46th Hawaii International Conference on System Sciences (HICSS). IEEE (2013)
2. Kim, H., Ren, X., Sun, Y.: Semantic Frame-Based Document Representation for Comparable Corpora
3. Quercia, D., Askham, H., Crowcroft, J.: TweetLDA: supervised topic classification and link prediction in Twitter. In: Proceedings of the 3rd Annual ACM Web Science Conference, pp. 247–250. ACM (2012)
4. Marco, P., Popescu, A.M.: Democrats, republicans and Starbucks aficionados: user classification in twitter. In: Proceedings of the 17th ACM SIGKDD International Conference on Knowledge Discovery and Data Mining. ACM (2011)
5. Grinev, M., Grineva, M., Boldakov, A.: Sifting micro-blogging stream for events of user interest. In: Proceedings of the 32nd International ACM SIGIR Conference on Research and Development in Information Retrieval, pp. 837–837. ACM (2009)
6. Ishikawa, S., Arakawa, Y., Tagashira, S.: Hot topic detection in local areas using Twitter and Wikipedia. In: 2012 IEEE ARCS Workshops (ARCS), pp. 1–5 (2012)
7. Liu, Z.T.: Short text feature selection for micro-blog mining. In: 2010 International Conference on Computational Intelligence and Software Engineering (CiSE). IEEE (2010)
8. Tu, H., Jin, D.: An Efficient Clustering Algorithm for Microblogging Hot Topic Detection. In: 2012 International Conference on Computer Science & Service System (CSSS). IEEE (2012)
9. He, H.: Sentiment analysis of Sina Weibo based on semantic sentiment space model. In: 2013 International Conference on Management Science and Engineering (ICMSE). IEEE (2013)
10. Fuzzy set expressed in WIKIPEDIA:
http://en.wikipedia.org/wiki/Fuzzy_set
11. Bezdek, J.C.: Pattern recognition with fuzzy objective function algorithms. Kluwer Academic Publishers (1981)
12. ICTCLAS Chinese word segmentation system: http://www.ictclas.org/
13. Zhang, S.Q., Shi, B.F., Ma, F.: An Improved Text Feature Weighting Algorithm Based on TFIDF. Computer Applications and Software 2, 007 (2011)
14. He, Y.: The optimal cluster number of FCM in complex economic systems. In: IEEE International Conference on Fuzzy Systems, FUZZ-IEEE 2008 (IEEE World Congress on Computational Intelligence). IEEE (2008)
15. Liao, Q., He, Z.F., Chen, Z.H.: Data mining and mathematical modeling. National Defense Industry Press, Beijing (2010)
16. Tversky, A.: Features of similarity. Psychological Review 84(4), 327 (1977)
17. Huang, H.H., Kuo, Y.H.: Cross-lingual document representation and semantic similarity measure: a fuzzy set and rough set based approach. IEEE Transactions on Fuzzy Systems 18(6), 1098–1111 (2010)
18. The Project that the Analysis of Social Index of the, Spring Festival Gala of CCTV is led by the Institute of High Energy Physics (IHEP), and the analysis result is published as a report in the Central Propaganda Department briefing (2014)
19. Rand, W.M.: Objective criteria for the evaluation of clustering methods. Journal of the American Statistical Association (American Statistical Association) 66(336), 846–850 (1971)
20. Perplexity is expressed in WIKIPIDIA:
http://en.wikipedia.org/wiki/Perplexity

Multi-strategy Based Sina Microblog Data Acquisition for Opinion Mining

Xiao Sun[1], Jia-qi Ye[1], and Fu-ji Ren[1,2]

[1] School of Computer and Information, Hefei University of Technology, Hefei, China
sunx@hfut.edu.cn, lane_3000@163.com, ren@is.tokushima-u.ac.jp
[2] Department of Information Science and Intelligent Systems,
Faculty of Engineering, The University of Tokushima, Tokushima, Japan

Abstract. As an important media for social interactions and information dissemination through the internet, Sina microblog contains emotional state and important opinion of participants. Dealing with microblog data belongs to big data areas, the premise of which is to obtain a large amount of microblog data for further analysis and data mining. For commercial interests as well as security considerations, the access to the data is becoming increasingly difficult and the API Sina microblog officially provided doesn't support large amount of data mining. In this paper, we try to design a platform that is mainly based on the access mechanism of multi-strategy and existing resources to collect data stably from Sina microblog. The results demonstrate that a combination of API and web crawler allows efficient data mining. In such way, sentiment analysis and opinion mining are performed on the data obtained by the multi-strategy method, which proved that the proposed solutions will be allowed to build straightforward application of hot words searching, opinion mining and sentiment analysis.

Keywords: Sina Microblog, Big Data, Data Mining, Web Crawler, Multi Strategy, Opinion Mining, Sentiment Analysis.

1   Introduction

"Behind every microblog user is a living consumer."[3] From testing the water in 2009, to the outbreak of microblog in 2010, and then to triumph in 2011, until Sina microblog apprehension and confusion in 2012, Sina microblog is not prevented from becoming a synonym for web culture. On December 19th, Sina microblog released 2012's annual inventory of web hot words. "London 2012 Olympic Games," "Yancan" and "PSY" ranked first in three different series, which were totally mentioned in microblog for more than 2.2 billion times [5]. It can be found that Sina microblog with it's over 400 million users have had the power to lead Chinese netizens' opinion. According to Sina fourth quarter and annual financial statements on February 20th, 2013, by the end of December 2012, Sina microblog has attracted more than 500 million registered users, representing a substantial increase of 74%. Daily active users (DAU) reached 46.2 million, accounting for 9.24%. Sina published data, in 2011 and 2012, the development trend of microblog stability, every year the new growth of 200 million users, and the active degree does not reduce [6].

D.-S. Huang et al. (Eds.): ICIC 2014, LNAI 8589, pp. 551–560, 2014.

Recently researching on Sina microblog has gradually become a hot topic. Zhang [4] conducts the research to the network of public opinion based on Sina Microblog, Zhang [2] experiments on Chen Yao's microblog case for example to study microblog propagation mechanism. Wang [10] studies the characteristic and relationship between microblog users. Lian[1] introduces early Sina Microblog scheme of data mining. The acquisition of Sina microblog data mainly two methods: the first one is the use of Sina microblog API access to data; the second method is to use the web crawler microblog webpage access to data, and the two methods are tested. But the original scheme is no longer applicable after the API upgraded, and Sina microblog for its data acquisition have adopted a more stringent restrictions. Comparison of Chen S [9] twitter and Chinese local microblog service focuses on the research on the value of microblog service, Sina microblog as also not full field of study. Researchers deserve attention. Amparo[8] studies on twitter topic sensitivity based on the forwarding path, which has an impact on the user, and the theme content related influence. Fang [11] studies the large number twitter data acquisition scheme based on Twitter List API and Lookup API. LIU [7] tries to dig Chinese microblog interest generated by keyword. More and more researchers begin to concentrate on the emotional content of Microblog distribution, Microblog user's emotional state analysis and other topics.

This paper tries to put forward the adoption of Sina microblog API (OAuth2.0) and the use of web crawler multi-strategy method to parse the weibo.com and weibo.cn on a collaborative program. We try to achieve efficient way to capture Sina microblog data, and so as to obtain valid data to the largest extent in the limited authority. The feasibility of web crawler is tested and the crawling strategy seems perfect. In addition, we use revised SVM Model to training micro-blog emotion analysis model, for trend analysis and passionate tendency for a hot topic in microblog.

2   Data Acquisition and User Database Establishment

2.1   Web Crawler Based Method

The Web2.0 era is arrived, making open platform as the current trend. Sina API is no exception. Developers can submit arguments to the network server and the server issues the requested data to the developer's application. Compared to the way to get data from web crawler, Sina microblog open API interface is more simple and convenient to use. If API calls too frequently, it will lead to temporarily unable to obtain data from the API. All requests will be no return. In order to prevent this, on the one hand by Sina real-time query interfaces under the current access token remaining visits, on the other hand, a program thread of control should be established, with an average frequency of visits.

Microblog data can be requested by calling the API interface and we can analyze them conveniently. But with the coming of the area of big data, the amount of data has become a new standard to measure. So to avoid the data releases too much, the Sina Company updates a series of rules to limit developer catch data. On this occasion, we turned our eyes to the multi-strategy web crawler with web analytic techniques to obtain more information.

The web Sina microblog was divided into PC browser and mobile phone version. Before using the program crawling specified microblog page we need to analyze the whole login process in detail. Based on the login page source code analysis, we find out that two kinds of algorithms are completely different. On the mobile microblog webpage (http://weibo.cn), the user name password is plaintext transmission when logining. The main parameters of the form are named of password and vk of value. After submit the correct form you will get the cookie to continually catch data. Meanwhile, a long string named gsid is contained in the cookie. If it is put in the request URL, the effect is the same as user being logined to access Sina microblog.

On Sina microblog webpage (http://weibo.com) for PC platform, the process simulation is basically the same. Because of constantly update of Sina microblog, the operation is more complex than microblog for mobile platform, so a detailed algorithm process should be attached. The first step, we need to request the pre-login_URL link and get the parameters such as extract server time, from nonce, pubkey and rsakv values. The username in the submit form should be coded by Base64, the password encryption as RSA, Sina microblog RSA encryption [12] as follows:

① First create rsa public key with two parameters which Sina microblog have given a fixed value when request. These two parameters, the first one is in the first step named pubkey, the second is in the encrypted JavaScript file, it is10001.

②After the encryption process, all the parameters are flexible enough to login. Then request the URL: http://login.Sina.com.cn/sso/login.php?client=ssologin.js (v1.4.4). When the value of the response of the record is 0 indicates successful login. The cookie will be easily got to access other information.

③ In the web page, the Chinese characters are encoded by UTF-8, same as the URL with Chinese characters.

For the above two different ways of logining Sina microblog, acquired data with different features and limitations, which will be compared within the next part.

2.2   Establishment of Microblog User Database

In the experiments and analysis of microblog information, researchers need a large number of complete data package, and for these data, does not require the timeliness is very strong. To the timeliness, whether from the API or through the web crawler, on the premise of all kinds of restrictions, cannot provide timely data very well.

Table 1. User information fields and   Microblog content fields

User information fields		Microblog content fields	
uid	username	Meeage_id	Audio_url
screen_name	Sex	Message_text	Video_url
Address	Description	Uid	Repost_count
image_url	attention_num	User_name	Comment_count
fans_num	message_num	Screen_name	Post_time
is_verified	is_daren	image_url	Name_count
verify_info	insert_time	Repost_id	Picture_url
Tag	education_info	Message_url	Platform
career_info	base_info		
create_time	follower_userid		

To a single user, the unique identifier is microblog user's uid, through which we can obtain user's personal information, such as fan list, attention list and even the mood vote in the new version. But for large-scale microblog user's information retrieval, you first need to collect a large number of uid as the entrance to the crawler program. To this question, we artificially screen some microblog uid in good quality, such as hot microblog users. We set the number is 1000 as database. Then we obtain a list of their fans uid through the API to deal with the uid again, expanding the uid library. After looking for a few times, until it reaches the number of threshold setting. From the uid library, we can make use of the crawler to crawl the data. While the efficiency is relatively low, but can get a lot of effective data continually. Compared to the API web crawler has higher flexibility after experiment. In Table 1 above is the user's personal information field and microblog user information fields. Based on these fields we can design data forms to describe and store user's information and microblog data.

Fig.1 below describes the flow chart of user database establishment. The main program can be divided into two parts. One part on the left side is the program which won't stop scanning uid in the database and chose the uid which haven't been used before to grab Sina microblog user's followers to enlarge uid database in order to make sure the part of right side always have work to do. The other part on the right side is a program designed to get user's detail information and the user's whole microblog content data to fill in the two tables above. All the data will gather together to store in the database in case of further research.

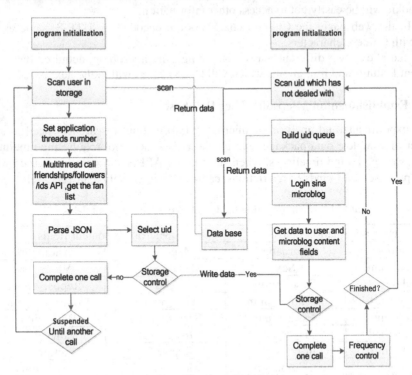

Fig. 1. User database build scheme

2.3   Establishment of Microblog Sentiment Analysis Model

Sentiment analysis of Chinese microblog is mainly divided into text preprocessing, emotional information extraction and emotion classification. Emotional information extraction is divided into emotional words, themes and relationship extraction. The subjective emotional text classification method of microblog is mainly based on semantic dictionary and machine learning [14]. At present, the method based on SVM (support vector machine) of machine learning in Chinese microblog emotion classification is widely used.

Support vector machine is a supervised learning model with associated learning algorithms that analyze data and recognize patterns, employed to classification and regression analysis. In this paper, we choose libSVM package developed by Chih-Jen Lin and revise it as our analysis tool. After observing a lot of microblog with Chinese characters, we find the length of Chinese microblog text is limited to 140 Chinese characters. The text can contain more than one sentence and the emotion in different sentences may be quite discrete. If the whole sentence gives a polarity may affect training with bad effect; more vocabulary and new words on the network appear endlessly. Some of them will become the buzzwords over a period time and get a deeper meaning. Such as "打酱油", "弃疗" etc. There are various forms of expression as well as casual expression. In the process of Chinese microbial emotion analysis researchers need careful consideration on these factors.

Xie [15] and other researchers experiment on microblog data forms for rules of emoticons method and emotional dictionary method. Moreover they applied the hierarchical structure of the SVM multi-strategy method. The experimental results show that compared with the former two methods of rules, the method based on SVM is best. They further analyzed the characteristics in detail. In this paper, we will use the method of multiple SVM classifiers training emotional classification model to judge the microblog emotional tendency. The emotional dictionary is HowNet and emotional words library from Dalian University of Technology. Training the SVM classifier process is as follows:

(1) Depending on the method of subjective and objective classification feature description [15]. Extract microblog content classification of subjective and objective characteristics from the data forms. Then start training objective and subjective SVM classifier.

(2) Extract polarity classification features from training corpus clauses, then using chi-square statistics to choice feature words. The chi-square formula is presented in formula (1):

$$\delta^2(w,e) = \frac{(AD-BC)^2}{(A+B)(C+D)} \tag{1}$$

W is candidate words; e means the emotional tendency where w is it; A means the number of microblog which contains word w and belong tendency e; B means the number of microblog which contains word w but doesn't belong tendency e; C means the number of microblog which doesn't contain word w but belongs tendency e; D means the number of microblog which neither contains word w nor belongs tendency e. After the feature words have been generated, we use the TF-IDF to calculate the weight of feature words. Then according to the positive and negative tag we can train SVM classifier.

3   Experiment and Results

The experiment platform is SONY CS36 notebook, Intel core P8700 dual-core, ECLIPSE EE program platform, 4 GB of memory, WINDOWS 7 platform, the program is developed by Java, with access networks of 10 MB sharing education network.

3.1   Comparison of File Size of Query Results

In this paper, based on the Sina microblog API and web crawler, we design total of three kinds of Sina microblog data acquisition scheme. In the test, we first manually choose 100 Sina microblog users whose fans is more than 5000 people, in order to make sure all of the user's data are available through the API, and web crawler. Sina microblog API can only return 5000 fans uid information, and at weibo.com, the uid information can query to only 1000 fans. In http://weibo.cn and can query to the 5000 fans as well as API.

Table 2. The comparison of the file size of query results in user fans

Request	total fans	total files	file size
JSON	500000	100	4.95MB
weibo.cn	500000	500	558MB
weibo.com	100000	100	442MB

Table 3 shows the size of queried uid whose fans is over 100 users. During the test, JSON is built on the API query results. Compared through the API to get the user fans, we can find web crawler gets the data more vast. On the premise of getting the same amount the number of fans, the file size is almost one hundred times more. This is because the web crawler retained a large number of useless information in the text, including the HTML source, JavaScript tags. So, the efficiency through the way of web crawler to parse user state is less than the approach to get user state by using the API method.

3.2   Access Request Number Limitation Test

The request limitation of the API can be found directly from the open platform of Sina microblog and also can find that from the API documentation. Access restrictions on weibo.com and weibo.cn, we cannot have a clear data on the Internet. Therefore, we design crawlers to maximum to the limit of single account's request number. The experimental results are shown in Table 4 below. We can find that, for web microblog (weibo.com), the user profiles and user access is not restricted, so it can be used as building microblog user database. And the mobile phone microblog (http://weibo.cn) can also be used as a data access or a secondary data source. For the application of crawler on keywords search, Sina microblog have the corresponding limit. Looked at the API it is not provided, it just has been able to get part of the data. For this limit, we need to establish an account pool so that we can gain a large number of data.

Table 3. Request times on crawler limit

crawler way	weibo.com	weibo.cn
Data access	unlimited	unlimited
Keyword	40times/h	500times/h
State of the user access	unlimited	500 times/h

3.3   Search and Analysis of Hot Keywords from Microblog

Researchers can perform the corresponding research after having the corresponding source of data. The following is keywords search and analysis, for example. In the first half of 2013, the food safety issue often appears on microblog. Our experiment was fixed on the dead pig events in the Shanghai area. We selected a list of keywords and search the keywords by province. Acquisition time is taken from March 3, 2013 to April 5, 2013. Because Sina microblog API does not provide the corresponding interface, therefore the experiment first registered a series of microblog account, then get microblog content of 34 provinces and regions through the crawler. Finally we have collected a total of more than 68000 microblog content. The microblog provinces distribution of the data is shown in Fig.2 and Fig.3 below.

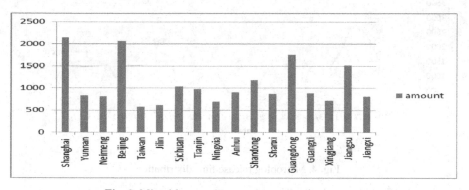

Fig. 2. Microblog number provinces distributio part 1

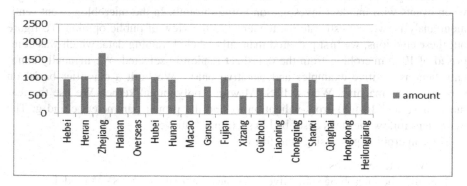

Fig. 3. Microblog number provinces distributio part 2

From Fig.2 and Fig.3, of the dead pig event, the number of microblog in Shanghai, Beijing, Guangdong, Jiangsu, and Zhejiang is more than other province. Shanghai, Jiangsu, Zhejiang, Guangdong are the areas where dead pig event affected. And Beijing area's number is large because the government department is located in the Beijing area, which mainly releases news about dead pig incident investigation process and results by means of the official microblog. The number of microblog on other provinces is in a low degree. This visible indicates that the whole country is highly concerned about food safety.

In addition to microblog province distribution, microblog released time are also used to statistics. As showed in Fig.4. It can be found in the line chart that the overall trend of microblog release quantity takes on the form of two peaks. Around March 10, microblog release quantity began to increase, to March 14, began to decline and maintain in a certain number, and then around March 28 microblog release quantities began to increase again, then to April 1 began to decline. Microblog quantity changes around an event topic have close relationship with the event processing of the incident.

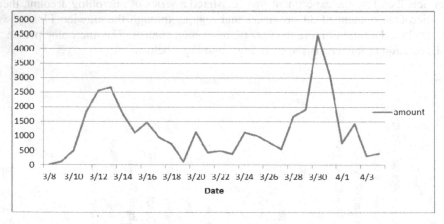

Fig. 4. Microblog release time distribution

Microblog is the expression of public opinion platform. Public emotional changes when events with incident investigation going through. In the microblog event sentiment analysis, we can estimate the tendency to the view of public opinion. To figure out these emotions, we first prepared manually labeled training data, which was comprised of 1000 microblog from the datasheet randomly selected. We manually annotate them as positive examples and negative ones. We built a classifier based on support vector machine. We use libSVM with a polynomial kernel. We use the features above and 10-fold cross-validation to find the proper parameters c and g. The main steps follow the steps described in section 2.4 above.

Topic microblog processing is as follows:

1. Divide the sentences.
2. Predict the microblog subjective and objective tendency by SVM model

3. To the subjective microblog use polarity SVM model to predict their polarity tendency, and the according to the number of positive and negative numbers to classify microblog according to the formula (2):

$$\begin{cases} \text{positive (positive sentence number > negative sentence number)} \\ \text{negative (positive sentence number < negative sentence number)} \\ \text{neutral (positive sentence number = negative sentence number)} \end{cases} \quad (2)$$

The experimental results are shown in Fig.5. From the results, microblog negative emotion change trend similar to microblog release time of the trend. When microblog release quantity increases the negative emotion increases; positive microblog emotional change trend was smoothly from the beginning to fluctuations.

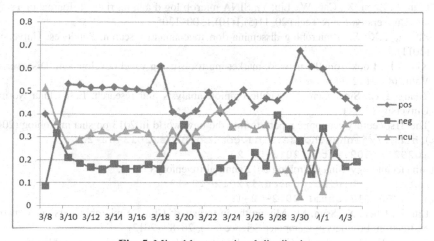

Fig. 5. Microblog emotional distribution

4   Conclusions

With the popularity of microblog in social interaction, microblog text analysis research has also become a research hotspot nowadays. In order to facilitate the microblog text analysis experiment, this paper introduces a design of data mining to microblog: a small number of user's uid for the entrance, smoothly get plenty of microblog data.

This paper first analyzes the Sina official API, the usage and limitations are described, and then analysis of two kinds of webpage microblog login process. After that, we designed the storage of users' data and completed the process of data mining. Finally the experiments show that, through the combination of API and multi-strategy web crawler can efficiently obtain microblog data. At present, the method also has a few shortcomings; we will continually develop it from the following aspects, for further research and exploration.

First of all, in Sina microblog data mining scheme, there has many customized mining scheme, in order to adapt to different data requirements. We need to solve the efficiency and low cost problem to get the needed data for research.

Secondly, to microblog data analysis, large-scale data is needed. Data with increased time mean the storage scheme should redesign and how to better organize these microblog data need further consideration. In addition, relations of users in microblog data are very valuable, on how to effectively deal with it also needs further study.

Acknowledgment. This research is supported by National Natural Science Funds for Distinguished Young Scholar(No.61203315) and 863 National Advanced Technology Research Program of China (NO. 2012AA011103).

References

1. Lian, J., Zhou, X., Cao, W., Liu, Y.: SINA microb log d a ta ret ri ev al. Journal of Tsinghua Un iversity(Sci & Tech) 20, 11(5),1(10),1300–1305
2. Zhang, Y.X.: Sina microblog dissemina-tion mechanism research. Southwest University (2011)
3. Xie, A.P.: Look from the network market-ing tools using vancl development. Discovering Value (4), 21 (2011)
4. Zhang, L.L.: Sina microblog public opinion analysis and research. East China Normal University (2011)
5. The 21st century economic report, to find the new world in 2013 on sina microblog (2013), http://jinhua.house.sina.com.cn/news/2013-02-23/ 10292163302.shtml,2013-02-23
6. Sina technology, Sina microblog 2012 annu-al inventory (2012), http://tech.sina.com.cn/i/2012-12-19/ 13447902817.shtml,2012-12-19
7. Liu, Z., Chen, X., Ssun, M.: Mining the in-terests of Chinese microbloggers via keyword ex-traction. Frontiers of Computer Science 6(1), 76–87 (2012)
8. Cano, A.E., Mazumdar, S., Ciravegna, F.: Social influence analysis in microblogging platforms–A topic-sensitive based approach. Semantic Web.
9. Chen, S., Zhang, H., Lin, M., et al.: Com-parision of microblogging service between Sina Weibo and Twitter. In: 2011 International Conference on Computer Science and Network Technology (ICCSNT), vol. 4, pp. 2259–2263. IEEE (2011)
10. Wang, X.G.: Empirical analysis on behavior characteristics and relation characteristics of microblog users – take "sina microblog " for ex-ample. Library and Information Service 54(14), 66–70 (2010) (in chinese)
11. Fang, W.W., Li, J.Y., Liu, Y.: Re-search on twitter data collection. Journal of Shandong University(Natural Science) 47(5) (2012)
12. Microblog login with program (Python and RSA encode algorithm) (May 5, 2013), http://www.2cto.com/kf/201303/192970.html (in chinese)
13. Microblog open plate-form, http://open.weibo.com/,2013-3-15
14. Zhou, S.C., Zhai, W.T., Shi, Y.: Overview on sentiment analysis of chinese microblog. Computer Application and Software 30(3), 161–164 (2013)
15. Xie, L.X., Zhou, M., Sun, M.S.: Hierarchical structure based hybrid approach to sentiment analysis of chinese microblog and its feature extraction. Journal of Chinese Information Processing 26(1), 73–83 (2012)
16. Sina news report about Shanghai dead pig event (April 2013), http://roll.news.sina.com.cn/s_sizhu_all/index.shtml

Mining Longest Frequent Patterns with Low-Support Items

Qinhua Huang and Weimin Ouyang

Modern Education Technology Center
Shanghai University of Political Science and Law
201701 Shanghai, China
{hqh,oywm}@shupl.edu.cn

Abstract. Finding the longest sequential pattern is a basic task in data mining. Current algorithm often depends on the cascading support counting, including the famous apriori algorithm, FP-growth, and some likewise derived algorithms. One thing should be pointed out that in these sorts of algorithms the items with very high support may lead to a poor time performance and very huge useless search space, especially when the items in fact are not the member of the result longest pattern. We reconsider the role of connection between the items and carefully analysis the hidden chains connecting items. Thus a new method of mining the longest pattern in transaction database is proposed in this paper. Our algorithm can have better performance when overcoming the bad side-effect of big-support items, especially in case of the items being not members of the result longest pattern.

Keywords: data mining, long frequent pattern, low-support item, HMM.

1 Introduction

Data mining and knowledge discovery has well developed and applied during several past decades. Among plenty branches of mining techniques and algorithms, frequent sequential pattern mining is a very interesting application. Many applications based on sequential pattern have been come to the world, such as DNA analysis, financial data analysis, varied kinds of human behavior analysis, etc.. In this paper we are going to discuss problem of mining sequential patterns while reduce the bad impact bought by non-result but with high occurrence items

First we present the problem description of sequential pattern mining. We are given a data base $\mathcal{D}$ of customer transactions. Each transaction may have contain some items, which are subset of item dictionary, $I = \{i_1, i_2, ..., i_n\}$. Denote the transaction with TId. These transactions are appeared in time order. And within a transaction TId, the items are arranged in lexicographic order. If a item occurs in many transactions, and it's occurrences is accumulated to be great than a user-specified frequency $\theta * |\mathcal{D}|$, we call this item frequent item and name the user-specified frequency, θ, as support. For simplicity, sometimes we define support to be $\theta * |\mathcal{D}|$. If a set of items occurred frequently in transactions, we call the itemset frequent pattern. A vertical format of transaction often used in application, described in Table 1.

D.-S. Huang et al. (Eds.): ICIC 2014, LNAI 8589, pp. 561–570, 2014.

Table 1. A demo transaction database $\mathcal{D}$

TID	Itemset
1	2, 3, 4, 5
2	1, 3, 4, 5
3	1, 2, 4, 5
4	1, 2, 3, 5
5	1, 2, 3, 4
6	6, 7, 8, 9
7	6, 7, 8, 9
8	6, 7, 8, 9
9	6, 7, 8, 9

From Table 1., the item dictionary is $\{1,2,3,4,5,6,7,8,9\}$. The supports of all of the single items are same, 4. Let's specify the threshold support to be 2. Only one 4-item set, $\{6,7,8,9\}$, is frequent. This is the longest frequent pattern in the database.

For the simplicity of calculation, the transactions database can be presented in binary format. As Table 2 shows.

Table 2. Binary format of transaction database

TID	1 2 3 4 5 6 7 8 9
1	0 1 1 1 1 0 0 0 0
2	1 0 1 1 1 0 0 0 0
3	1 1 0 1 1 0 0 0 0
4	1 1 1 0 1 0 0 0 0
5	1 1 1 1 0 0 0 0 0
6	0 0 0 0 0 1 1 1 1
7	0 0 0 0 0 1 1 1 1
8	0 0 0 0 0 1 1 1 1
9	0 0 0 0 0 1 1 1 1

The above mentioned term itemset can be taken as sequence of items. In the sequence, items occurred must follow the lexicographic order. For example, sequence $\langle 6,7,8,9 \rangle$ and sequence $\langle 9,8,7,6 \rangle$ are same. This simplification from itemset to sequence will benefit our mining method in this context.

When coming to describe relations between items in a sequence, the term association is often used. An association rule takes the form of $\alpha \Rightarrow \beta$, where $\alpha \subset I$, $\beta \subset I$, and $\alpha \cap \beta = \phi$. Association can be expressed in probability representation: $\Pr\{\beta|\alpha\}$. And support and confidence are two measures of rule interestingness. An association rule is considered interesting if it satisfies both a user-specified support threshold and a minimum conditional probability threshold.

With associations between item-pairs, we can get a whole chain connect all of the items. HMM is a widely used model to deal with connections within sequence. Now we briefly introduce the Hidden Markov Model.

A discrete HMM, denoted by a triplet $\lambda = (A, B, \pi)$, is defined by the following elements: [10,11]

- A including set $S = \{S_1, S_2, \ldots, S_N\}$ of hidden states; transition matrix $A = \{a_{ij}\}$, representing the probability to go from state S_i, to state S_j. $a_{ij} = P[q_{t+1} = S_j | q_t = S_i]$, $1 \leq i, j \leq N$, $a_{ij} \geq 0$, $\Sigma_{j=1}^{N} a_{ij} = 1$; set $V = \{v_1, v_2, \ldots, v_T\}$ of observation symbols.
- B is emission matrix, $B = \{b_j(k)\}$, indicating the probability of emission of symbol v_k when system state is S_j. $b_{j(k)} = P[v_k \text{ at time } t | q_t = S_j]$, $1 \leq j \leq N, 1 \leq k \leq T$, with $b_i(k) \geq 0$ and $\Sigma_{j=1}^{T} b_j(k) = 1$.
- π is the initial state probability distribution, $\pi = \{\pi_i\}$, representing probabilities of initial states. $\pi_i = P[q_1 = S_i]$, $1 \leq i \leq N, \pi_i \geq 0, \Sigma_{i=1}^{N} \pi_i = 1$

Our context of using HMM is that given a sequential pattern, say $\{1,1,1,1\}$, which items in table 2 emitted the bit 1. Note that under the representation of binary itemset, we do not need to care about bit 0, because there is no 0 symbol in our pattern.

Data mining employ HMM is far from scarce. But in the following problem, it will meet big challenge.

Consider a 40×40 square table with each row being the integers from 1 to 40 in increasing order. Remove the integers on the diagonal, and this gives a 40×39 table, which we call Diag$_{40}$. Add to Diag$_{40}$ 20 identical rows, each being the integers 41 to 79 in increasing order, to get a 60×39 table. Take each row as a transaction and set the support threshold at 20. Obviously, it has an exponential number (i.e., $\binom{40}{20}$) of midsized closed/maximal frequent patterns of size 20, but only one that is the longest: $\alpha = (41, 42, \ldots, 79)$ of size 39. [9]

Literature [9] has analyzed the challenge when mining using apriori-based method. The exponential increasing in searching space makes mining process task unaffordable in calculation. But if we consider HMM, another problem appeared. The items that with high support, but are not the member of the longest result pattern, will take very bad effect on deciding pattern member items.

2   Related Work

The problem of sequential patterns mining was originally proposed by Agrawal and Srikan [1]. A lot of algorithms have been proposed during the past decade, such as Apriori[1], GSP[2], PSP[3], PrefixSPAN [4], FPtree [5][6], Spade [7], Spam [8]. Apriori and GSP both need to generate candidate set, while PrefixS-pan and Spam do not. In comparison with PrefixSpan, Spam prevails in running speed, but with lower efficiency in storage context, relatively. In mining progress, the problem of search space increases in a exponential speed greatly worse performance of algorithms. Pattern Fusion [9] try to overcome this difficulty by fusing the core patterns in the database, thus find some long frequent patterns with a certain probability. If mining using a random method, the result pattern could be constructed in a probability. That also means if an item have very high frequency, but not the member of the result long pattern, it might be kept with high probability to construct searching space.

HMM is a very popular framework in processing sequence. In our sequential pattern mining context, HMM could be possibly employed. But, our experiment showed, using HMM directly could not overcome the bad effect of high frequency as well. We studied the Viterbi algorithm and find that, if we want to employ HMM, the emission probability part should be modified.

3   The Proposed Method

In this section we describe our method in details. First, we present the framework of our algorithm.

3.1   Algorithm Overview

The aim of our algorithm is to discover the longest pattern, while overcome the bad effect from high frequency of items other than the result items. Here we refer result item to items in the discovered long frequent patterns. For simplicity, we define the term non-result item to be item that is not in the destination frequent pattern.

Our proposed algorithm can be divided into 3 steps. First, the raw data should be preprocessed so as to simplify the calculation of support counting and connection building. After preprocessing, the single items are kept and the connections between any two items are constructed for future processing. Second, to evaluate the connections between more items, a modified HMM calculation mode is presented. The basic HMM model have some problems in our mining context. Third, we backtrack the chain to get the most probably item of long frequent pattern.

3.2   Preprocessing

According to apriori principle, if a itemset is frequent, every sub itemset of the itemset must be frequent. This is the most basic way to discover frequent pattern. But we can't rely on apriori principle all the way in discovering process, or else it will bring unaffordable calculation cost. To make the best of the principle, we employ it to get all of the frequent 1-item set, thus remove items do not reach the specified threshold of support.

After weeding out the unqualified items, next we will try to build connections between items. The connection between 2 items is within the simplest case. Naturally the conditional probability can well describe this relation. Let $\Pr\{i|j\}$ be the probability of item i in existence of item j.

Now we are considering the simplification of counting. In transactions databases, transactions or sequences often have unequal length. This adds difficulty to our counting job. We proposed the following method to get this done easier.

Algorithm 1. Building binary transaction data matrix.

```
1 For each sequence in dataset
2 Generate an empty binary sequence with size of item
    dictionary.
3 For each item in sequence
4   Set the relative bit to be 1
5 End for
6 End for
```

Algorithm 2. Counting the support of each item and the conditional probability of between each two-items

```
1  For each column in binary transaction data matrix
2 Sum all of the bits to get support of each item.
3 If sum < specified threshold support
4   Delete the item from matrix
5 End if
6 End for
//we get a binary matrix and a support list, both have
// same column size, each column represents an element.
7  let L = the columns of the binary matrix
8  build a empty matrix (Trans) of size L×L to store the
    2-item connections, name this matrix CMAT
9  for each 2 column pair (column_i, column_j) in the bi-
    nary matrix, where i< j
10     Set product to be 0
11     For each row
12        product = product + bit(i) ×bit(j)
13     End for
14     set CMAT(I,j) to be procuct/support(i)
15  End for
```

When finishing the processing, we get a upper-triangle matrix CMAT, which stores the transition probability between i to j, (i<j). Beware that the CMAT is a conditional probability matrix and items in pattern are in lexicographic order.

3.3   Constructing the Hidden Connection

The purpose of this step is to get a sequential pattern of length n as long as possible. When dealing with sequence processing problem, a very popular way is to consider HMM. To calculate in HMM, the conditional probability of any two items is needed as well as the emission probability and the initial states. From the definition of support, support in fact is the probability emission. We take the initial states to be equal distributed.

First, initialization step:

$$\begin{cases} \delta_1(i) = \pi_i b_i(o_1) & 1 \le i \le N \\ \psi_1(i) = 0 & (no\ previous\ states) \end{cases}' \tag{1}$$

where N is the is the size of item dictionary after preprocessing, π is equal distributed.

Second, Recursion step:

$$\left.\begin{array}{l} \delta_t(j) = \max_{1 \le i \le N}\left[\delta_{t-1}(i)a_{ij}\right]b_j(o_t) \\ \psi_t(j) = \arg\max_{1 \le i \le N}\left[\delta_{t-1}(i)a_{ij}\right] \end{array}\right\} 2 \le t \le T; 1 \le j \le N, \tag{2}$$

where a_{ij} is the transition probability in matrix CMAT.

The aim of our mining algorithm is to find the sequential pattern with max length, while this pattern should be existed in a curtain number (threshold) of transactions. But when checking the definition of emission rate b_j, we find that it can't be the taken as the frequency of item j. The reason can be seen intuitively from Eq.2. As we know, as j increasing, which means sequential pattern increasing in size, item with higher support will get more weight for staying in the hidden chain. If the item is a non-result item, it will most probably have higher P^* by multiply a higher support. To minimize the negative effects of b_j and simplify the calculation, we set b_j to be constant during recursion. The experiment result proved it benefits us a lot.

Third, Termination step: $\begin{cases} P^* = \max_{1 \le i \le N}[\delta_T(i)] \\ q_T^* = \arg\max_{1 \le i \le N}[\delta_T(i)] \end{cases}$

In termination step, a problem will arise. When to terminate? Actually we do have strategy options when mining. Option 1, we can choose a size T, which is enough big, by user experience. When T is set too big, it is easy to find the P^* get values of 0 in last several stages. Let the length of mined longest pattern be nFP, N be size of item dictionary. Obviously we have nFP≤n. Table 3 shows the calculation map in a matrix. This condition can easily be loosed to nFP≪n, because rarely we have most of the items in a frequent pattern. The new calculation map is described in Table 4. Clearly a great of calculation cost can be reduced. Option 2, keep calculating along the route on Table 3, terminate till we get all zeros in one column.

Table 3. Calculation map matrix when itemset size equal pattern length

Item	pattern	1	2	3	...	nFP
1		→				
2		→	→			
3		→	→	→		
⋮				...		
n		→	→	→	→	→

Table 4. Calculation map matrix when itemset size great greater than pattern length

Item	pattern	1	2	3	...	nFP	
1		→					
2		→	→				
3		→	→	→			
⋮			...		→		
n-nFP		→	→	→		→	
			→	→		→	
				→		→	
					→	...	
n						→	

3.4   Back Tracking

And the optimal state sequence can be retrieved by backtracking.

$$q_t^* = \psi_{t+1}(q_{t+1}^*) \ t = T - 1, T - 2 \ldots 1$$

By checking q_t^* we can decide which item is the most probable item in each bit of a pattern like $\{1,1,1,1\}$

4   Experiments

In this section, we tested our algorithms on synthetic and real datasets. First we run our algorithm on a small demo dataset to illustrate the accuracy of our algorithm. Second, we adjust the support of items with high frequency in a larger synthetic data set, which are not member items of result long patterns, to demonstrate the performance of our strategy. All experiments were done on a PC with a 2.6G dual-core CPU and 6GB main memory running Linux of Ubuntu 64-bit 12.04.4 LTS. All algorithms were implemented in Matlab.

4.1   Experimental Datasets

We use synthetic datasets to test our proposed algorithm. They are described in the following 2 subsections.

4.2   Small Synthetic Demo

In order to demonstrate the algorithm process in an intuitive way, we first run our algorithm on a synthetic small demo data set. This data set has 8 transaction sequence, contains 9 items. All the items in the first 5 transactions have same support of 4, while all the items in the last 3 transactions have same support of 3. Let the support threshold to be 3. It can be easily find that the longest frequent pattern is ⟨6,7,8,9⟩, that is same with the last 3 transaction sequence.

Table 5. Small data set with 7 transactions

TID	Sequence
1	2, 3, 4, 5
2	1, 3, 4, 5
3	1, 2, 4, 5
4	1, 2, 3, 5
5	1, 2, 3, 4
6	6, 7, 8, 9
7	6, 7, 8, 9
8	6, 7, 8, 9

Table 6 shows the result probability matrix of algorithm before backtracking. The cells in the table head row label the member order of discovered pattern, the cells in left column are the items of dataset.

Table 6. The probability matrix of pattern members

Pattern item	1	2	3	4
1	0.5000	0	0	0
2	0.5000	0.3750	0	0
3	0.5000	0.3750	0.2813	0
4	0.5000	0.3750	0.2813	0.2109
5	0.5000	0.3750	0.2813	0.2109
6	*0.3750*	0	0	0
7	0	*0.3750*	0	0
8	0	0	*0.3750*	0
9	0	0	0	*0.3750*

From column 5 in Table 6, the biggest probability is 0.3750, thus item 9 is the start point of backtracking. When the backtracking process is being called, sequential pattern ⟨6,7,8,9⟩ will be discovered. The result pattern is represented in italic, bold and underlined font.

4.3   Attenuating the Importance of Non-result Items with High Support

If an item is not a member of the discovered result frequent pattern, we denote is non-result item.

To analysis the effect of non-result items but with high support, we generate a series of datasets based on dataset described in Introduction. Let us call the base dataset dataset_1. We will run our proposed algorithm on these datasets to check the impacts of high supports of non-result items on our proposed algorithm. Remember the last sequence, ⟨41, ... ,79⟩, has a support of 20. And it is easy to find the 39th sequence is ⟨1, ... ,38,40⟩. The support of item 40 is 39. We repeat the 39th sequence one more time in order get item 40's support of 40. Name this new dataset to be dataset_2. Be aware that the support of item 40 is twice of support of sequential pattern $\langle 41, \dots, 79 \rangle$.

Now let's continue to generate the rest synthetic datasets. To generate synthetic dataset_3, dataset_4, dataset_5, dataset_6, we repeat the 39th sequence in dataset_1 20, 40, 60 and 80 times respectively. Thus the frequencies of item 40 in the 39th sequence in dataset_2, dataset_3, dataset_4, dataset_5 and dataset_6 are 2, 3, 4, 5, 6 multiples of sequential pattern $\langle 41, \dots, 79 \rangle$'s frequency, respectively. Table 7 shows the relative multiples in value.

Table 7. The multiples of frequency of item 40 compared to frequency of sequential pattern $\langle 41, \dots, 79 \rangle$ in each dataset

dataset	Item 40' frequency represented in multiples of $\langle 41, \dots, 79 \rangle$'s frequecy
Dataset_2	2×
Dataset_3	3×
Dataset_4	4×
Dataset_5	5×
Dataset_6	6×

Table 8 presents comparison of probability being the 39th result pattern element. When the frequency of item 40 increases, from 2 multiples to 6 multiples, item 79 remains greater probability of being the 39th element in the result pattern.

Table 8. The Probability comparison of item 79 and item 40, where item 79 is the last element of result long pattern, while 40 is not but with several mutiple higher support.

Times	Probability of item 79	Probability of item 40
2	0.3279	0.0957
3	0.2469	0.1090
4	0.1980	0.1171
5	0.1653	0.1224
6	0.1418	0.1263

5   Conclusion and Future Work

In this paper we introduce a data mining problem that takes into account of the side-effect of frequent items which are not the members of result long discovered patterns. The problem of non-result items with high frequency is widely existed, which could make most algorithms focus on useless branch search space. We purpose a method, which emphasize the connections between items rather than items with high support, to cope with this problem. Our algorithm was tested on both synthetic datasets. From the experimental results, our purposed algorithm can be well suit for the mining environment when discovering long patterns.

As future work we will go further to discuss how to employ the hidden connection to mining big data, in which item may have millions to billions multiples of gap in frequency.

References

1. Agrawal, R., Srikant, R.: Mining Sequential Patterns. In: Proceedings of the Eleventh International Conference on Data Engineering, pp. 3–14. IEEE Computer Society Press (1995)
2. Srikant, R., Agrawal, R.: Mining Sequential Patterns: Generalizations and Performance Improvements. In: Apers, P.M.G., Bouzeghoub, M., Gardarin, G. (eds.) EDBT 1996. LNCS, vol. 1057, pp. 3–17. Springer, Heidelberg (1996)
3. Masseglia, F., Cathala, F., Poncelet, P.: The PSP Approach for Mining Sequential Patterns. In: Żytkow, J.M. (ed.) PKDD 1998. LNCS, vol. 1510, pp. 176–184. Springer, Heidelberg (1998)
4. Pei, J., Han, J., Mortazavi-Asl, B., Pinto, H.: Prefixspan:Mining Sequential Patterns Efficiently by Prefix-projected Pattern Growth. In: Proc. of the 17th International Conf. on Data Engineering, ICDE 2001, pp. 215–226 (2001)
5. Han, J., Pei, J., Yin, Y., Mao, R.: Mining Frequent Patterns without Candidate Generation. In: Proc. 2000 ACM-SiGMOD Int'l Conf. Management of Data, SIGMOD 2000, pp. 1–12 (2000)
6. Han, J., Pei, J., Mortazavi-Asl, B., Chen, Q.: FreeSpan: FrequentPattern-Projected Sequential Pattern Mining. In: Proc. 2000 Int. Conf. Knowledge Discovery and Data Mining, KDD 2000, pp. 355–359. Boston, MA (2000)
7. Zaki, M.J.: Spade: An Efficient Algorithm for Mining Frequents Sequences. Machine Learning 42, 31–60 (2001)
8. Ayres, J., Gehrke, J., Yiu, T., Flannick, J.: Sequential PAttern Mining using A Bitmap Representation. In: SIGKDD 2001, Edmonton, Alberta, Canada (2001)
9. Zhu, F., Yan, X., Han, J., Yu, P.S.: Mining colossal frequent patterns by core pattern fusion. In: Proc. 2007 Int. Conf. Data Engineering, ICDE 2007 (2007)
10. Rabiner, L.R.: Proceedings of the IEEE, vol. 77(2) (February 1989)
11. Panuccio, A., Bicego, M., Murino, V.: A hidden markov model-based approach to sequential data clustering. In: Caelli, T.M., Amin, A., Duin, R.P.W., Kamel, M.S., de Ridder, D. (eds.) SPR 2002 and SSPR 2002. LNCS, vol. 2396, p. 734. Springer, Heidelberg (2002)

Integrating Time Stamps into Discovering the Places of Interest

Jun Zhou, Qinpei Zhao*, and Hongyu Li

School of Software Engineering, Tongji University, Shanghai, China
{jackie8756,qinpeizhao,hongyuli.tj}@gmail.com

Abstract. With the employment of GPS embedded device, large numbers of data has been collected from location aware applications. It is interesting and challenging to discover meaningful information behind the data. Since the GPS data contains the time information, we take use of the time stamps of the GPS data in this paper for better discovering the places of interest. The collection usually contains large amounts of trajectories, where not every point has information. Therefore, a time stamp clustering algorithm is firstly proposed to reduce the size of raw data and also extract the points with more information. Different clustering algorithms are then conducted on the pre-processed data for extracting the places of interest. Finally, we compare the clustering algorithms on the GPS data by several external validity indexes.

Keywords: Time Stamp Clustering, Trajectories, Data Mining, Clustering Algorithm, External Validity.

1 Introduction

With the emergence of GPS embedded devices, such as smart phones, navigators etc., life has been changed in recent years. It is much faster for a person to locate his/her position by GPS devices than by maps, which provide more information about surroundings. With the development of social networks, users are willing to share their geo-tagged photos or trajectories on travelling. Hence, it is much easier for them to plan their routes of trips or find the most popular restaurants immediately. Such applications could be available when the features of people's activities are extracted from tremendous GPS data which could be easily acquired nowadays. And the places of interest are those with features of people's activities. They could be famous shopping malls, highly popular restaurants or scenic spots.

The advantage of using GPS data to extract important places is that it is a standardized, globally available location system. And one of the GPS attributes is that the signal could be lost indoors, and be recovered after certain duration of time when users get out. Therefore, the attribute can be used to detect facilities or places of interest. With a large collection of GPS data from different users, there are always patterns

---

* Corresponding author.

D.-S. Huang et al. (Eds.): ICIC 2014, LNAI 8589, pp. 571–580, 2014.

behind the data. Therefore, the frequently visited locations are also considered to find places of interest.

Lots of research has been carried on GPS data. Some focused on extracting trajectory pattern from GPS data. It is introduced as concise descriptions of frequent behaviors in [1]. They tried to explain the mobility phenomena behind the GPS data by developing extension of sequential pattern mining paradigm. A framework was proposed in [2] that analyses, manages, and queries object movements following hidden maximal periodic patterns with an effective and fast mining algorithm. S. Hwang et al. focused on mobile group patterns in [3] which were determined by a distance threshold and a time threshold. They proposed ATGP algorithm and TVG-growth to discover group patterns. In [4], researchers aimed to predict locations with a certain level of accuracy based on trajectory mining. They built a T-pattern tree which is learned from trajectory patterns to be used as a predictor of the next location by finding the best matching route. Besides that, some studied on assigning the importance of locations. In [5], a list of interesting locations were mined and ranked by the analyses of an aggregation on GPS information from multiple users. Then, a ranked list of interesting locations was obtained.

Clustering methods are commonly used to discover semantic places from GPS data. In [6], the algorithm kmeans, a classical clustering method, was used to extract important place. In their work, they considered points where the interval time of the GPS signal disappeared and recovered more than 10 minutes later as a place, then clustered the points as a location by using a variant of kmeans. DJ-cluster algorithm [7] is a density-based clustering method similar to DBSCAN that was used to process sets of trajectories for detecting places. Stops and Moves model was used to extract important places by integrating trajectories with geographic information in [8]. An incremental clustering was proposed in [9], they extracted semantic places by means of eliminating the intermediate location. In this work, the clusters of trace represented an important place which determined by the distance and time thresholds. In [10], a similar idea of [9] was proposed. Important places were extracted by using time stamp information. Firstly, they extracted the instances of users by stays derived from the location history. A stay indicates a location where the users spent longer time. Then the method clustered the stays of lots of users to obtain the interesting places.

For a single trajectory or a set of GPS data, not every point contains important information. When the data size is too large, the traditional clustering algorithms are usually restricted by the data size. Therefore, we introduce a time stamp clustering method for pre-processing of the raw data, since time stamps are a part of GPS data, which contains very important information. The time stamp clustering provides an adaptive way of processing the GPS data. On the extracted data from the time stamp clustering algorithm, three clustering algorithms such as kmeans, DBSCAN and Greedy EM are compared. According to the experimental results, the kmeans and DBSCAN have similar performance, but the Greedy EM is very different from the other two algorithms surprisedly.

This paper is organized as follows. We introduce the time stamp clustering in Section 2. Preliminary introduction on different clustering algorithms such as kmeans,

DBSCAN and Greedy EM is also given in Section 2. Experimental results are shown in Section 3 while a conclusion is given in Section 4.

2 Methodology

The collection of GPS data is always in large size. The large size of data might bring problems such as "out of memory" to the traditional clustering algorithms. On the other hand, there are quite many points in the data have little information. Removing these points from the raw data has two benefits, one of which is to make the clustering algorithms more applicable and the other benefit is to extract the points with more information. Motivated by this, we firstly introduce a time stamp clustering algorithm for removing the points. Based on the pre-processed data from the time stamp clustering algorithm, we then give a brief introduction on the clustering algorithms used in this paper for extracting places of interest.

We give the definitions of GPS data, GPS trajectories and place of interest in the following:

Definition 1. *GPS data* : Basically, GPS data is a collection of points with location information. GPS data p with attributes of latitude, longitude, and a time stamp constitutes of a GPS collection which is defined as $P = \{p_1, p_2 \dots p_n\}$, where n is the size of the collection.

Definition 2. *GPS trajectory*: It is a GPS sequence connecting GPS points on a two dimensional plane. The point $p_{i,t} \in T$ is the i_{th} point with time stamp t of the trajectory T.

Definition 3. *Event* : An event is happened when the consecutive points have significant difference on their time stamps.

Definition 4. *Place of interest*: We define that frequently visited locations with events are the places of interest, which is notated with $C = \{c_1, c_2, \dots, c_k\}$, where c_i represents a place of interest and k is the number of places extracted.

2.1 Event Detection by Threshold Method

An event can be detected when a user stayed at one place in a longer duration than a normal case, i.e., the consecutive points $p_{i,t}$, $p_{i+1,t}$ and $p_{i+2,t}$ are recorded in a longer duration than other points. A simple way to detect the events is to set thresholds. If $(p_{i+1,t} - p_{i,t}) > \Delta t$ and $(p_{i+2,t} - p_{i+1,t}) > \Delta t$, points A and B are considered place with events. Of course, the setting of threshold Δt is an open question.

2.2 Time Stamp Clustering Algorithm

Setting thresholds is quite easy; however, select a proper threshold is difficult. We propose a time stamp clustering algorithm to extract the points with events adaptively. The procedure of the proposed algorithm is shown in Fig. 1.

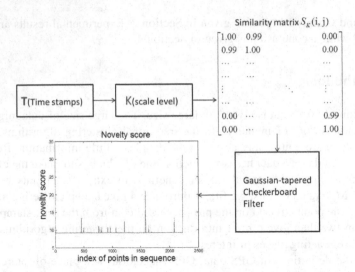

Fig. 1. The procedures on the proposed time stamp clustering algorithm

The algorithm (see Algorithm 1) firstly constructs a similarity matrix from the time stamp of points, where the similarity matrix can be calculated by

$$S_K(i, j) = \exp\left(-\frac{|t_i - t_j|}{K}\right) \tag{1}$$

Here, in Eq. 1, the similarity matrix S_K of t_i and t_j is based on a Gaussian kernel with different scale level of K. The value of S indicates the difference between two time stamps. The scale level K controls the sensitivity of exponential similarity measurement. When K is smaller, the generated matrix will be rougher. A similarity matrix is shown in an image in Fig. 1, where the higher the S value is, the lighter the pixel is in the image and the difference between the time stamps is higher. Therefore, the event can be visualized from the checkboard pattern along the main diagonal of the matrix.

We use an adaptive media segmentation algorithm [11] to detect the events by calculating a score of an event which is defined in Eq. 2. An event is defined as observations that deviate from other observations as to arouse suspicions that they were generated by a different mechanism. Therefore, the score is defined as follows:

$$v_K(i) = \sum_{l,m=-e}^{e} S_K(i+l, i+m) g(l, m) \tag{2}$$

A Gaussian-tapered checkerboard kernel $g(l,m)$ is used as a filter on the similarity matrix. The l and m are the size of the filter. The scores of each time stamp obtained can be plotted as shown in Fig. 1. The peaks in the plot indicate the events. In order to obtain the points with events, we perform a peak detection step on the scores. A simple way for peak detection is to get first difference on the scores. Thereby, all of the peaks can be obtained.

To get a more accurate result, these steps can be repeated at different scale level K. A criterion can be defined to choose the proper K. In clustering, the compactness within a group and separation between groups can be employed for validity of clustering results. In order to obtain a proper K, a criterion is defined as:

$$O(C_K) = \sum_{l=1}^{|C_K|-1} \sum_{i,j=c_l}^{c_{l+1}} \frac{S_K(i,j)}{(c_{l+1}-c_l)^2}$$

$$-\sum_{l=1}^{|C_K|-2} \sum_{i=c_l}^{c_{l+1}} \sum_{j=c_{l+1}}^{c_{l+2}} \frac{S_K(i,j)}{(c_{l+1}-c_l)(c_{l+2}-c_{l+1})}$$

(3)

where C_K represents the clustering result at scale level K and c_l is the l_{th} group. The first part in the equation indicates the compactness within a group and the latter one is the separation between groups. We select the proper K through getting the maximum $O(C_K)$.

Input: GPS data $P = \{p_{1.t}, p_{2.t} \dots p_{n.t}\}$
Output: a list of points with event p_o
1 Take out the time stamp of data;
2 for scale level K do
3   Compute the similarity matrix S_K;
4   Compute the score of an event v_K;
5   Peak detection: detect peaks of first difference of v_K ;
6   List the points with events p_o at K;
7   Calculate the confidence score $O(C_K)$;
8 end
9 Select the p_o at scale level with maximum value of $O(C_K)$;
10 return p_o;

Algorithm 1. Time stamp clustering algorithm

2.3 Places of Interest by Clustering Algorithms

With the points that contain event information from the raw data, the clustering algorithms can be performed to extract places of interest. Classical algorithms such as kmeans, DBSCAN and Greedy EM (see Fig. 2) are commonly used in this purpose.

The kmeans clustering is a typical partitioning-based clustering, which is very popular in different areas. It has been widely used because the algorithm is simple and efficient. However, the algorithm is suffered from initialization problem, which makes the algorithm produces different results at different runs. Since Euclidean distance is commonly used in the algorithm, it is not working well on data in non-Gaussian structure or with imbalanced class distribution. Another issue of kmeans is that the number of clusters has to be decided by users.

DBSCAN is a density based clustering algorithm, which is very suitable for GPS data. The algorithm has two parameters that are ε (eps, a given distance determine neighborhood retrieved) and *MinPts* (the minimum number of points to form a cluster). The algorithm is not restricted to the shape of clusters. It can also detect outliers automatically. However, the setting of the parameters is quite problematic.

Greedy EM algorithm is an improved variant of the conventional EM algorithm. As an iterative algorithm, the GEM takes use of the mixture of Gaussian to model the clusters. The algorithm has the same problems as kmeans has.

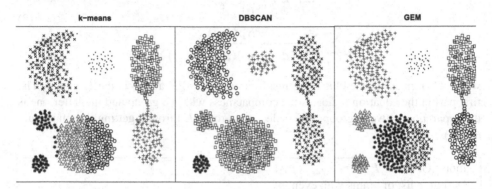

Fig. 2. Clustering results (partitions) from different clustering algorithms

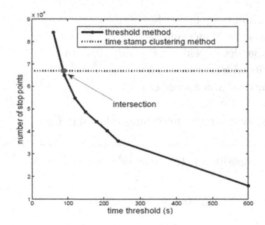

Fig. 3. The number of events detected by threshold method and the time stamp clustering algorithm. The result from the threshold method varies on different settings of Δt; however, the clustering method is deterministic on the result. The intersection of the threshold method and the time stamp clustering method is 90 seconds with the number of stop points as about $6.8*10^4$.

3   Experiment

We tested the methods on a geo-life data in [12] that contains 182 users with large amounts of trajectories. For example, there are 171 trajectories for user000. Fig. 1 is a data collection with more than 2000 points of user000's one trajectory. The time stamps are taken to test the time stamp clustering algorithm. An artificially generated data Aggregation is employed to show three different clustering algorithms, which are kmeans, DBSCAN and GEM. All of the algorithms are implemented in Matlab, which can be reached at *http://sse.tongji.edu. cn/zhaoqinpei/Software/*. For the time stamp clustering algorithm, we set $K = 100, 1000, 10000, 100000$ to compute the similar matrix and set the kernel $g(l,m)$ as $8 * 8$, which is mentioned in Eq. 2. We set K from small to large enough to cover the range widely, since if K is smaller, the generated matrix will be rougher.

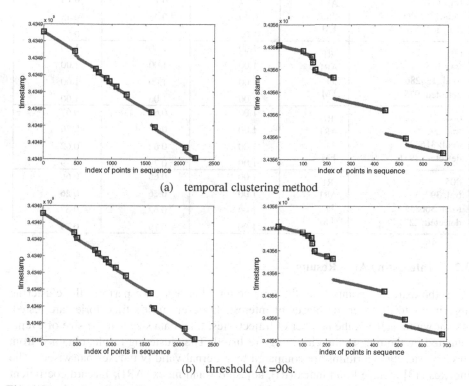

(a)   temporal clustering method

(b)   threshold $\Delta t = 90s$.

Fig. 4. Events detected by the temporal clustering algorithm and the threshold method with Δt =90s on two trajectories (left and right)

3.1   Event Detection

The events can be extracted by setting thresholds. However, it is intuitive to select the threshold. In order to compare the results from the threshold method and the proposed time stamp clustering algorithm fairly, we firstly determine the threshold value through Fig. 3. When the Δt is set as 90, both the threshold method and time stamp clustering algorithm produce the same number of events. Thereby, a comparison of results from two methods is shown in Fig. 4, where the trajectories of user001 are tested. As shown in Fig. 4, the events detected are emphasized by black squares.

The result from the time stamp clustering is quite similar to that from the threshold method. However, the benefit of the proposed method needs no empirical parameter as the threshold method does.

Table 1. Comparing the results from three clustering algorithms by external validity index

User Number	index	DBSCAN vs kmeans	DBSCAN vs GEM	kmeans vs GEM
001-010 traj. 1218 total:1384783 detected:7968	RI	1.00	0.50	0.50
	ARI	0.97	0.14	0.13
	Jac	0.99	0.45	0.45
	FM	1.00	0.67	0.67
001 traj. 165 total: 139286 detected: 975	RI	1.00	1.00	1.00
	ARI	1.00	1.00	1.00
	Jac	1.00	1.00	1.00
	FM	1.00	1.00	1.00
004 traj. 292 total: 316404 detected: 1931	RI	1.00	0.67	0.67
	ARI	1.00	0.30	0.30
	Jac	1.00	0.62	0.62
	FM	1.00	0.79	0.79
005 traj. 397 total: 366852 detected: 2371	RI	1.00	0.66	0.66
	ARI	1.00	0.26	0.26
	Jac	1.00	0.62	0.62
	FM	1.00	0.79	0.79

3.2   Interesting Area Results

With the extracted points from the time stamp clustering, we perform the clustering algorithms to discover the places of interest. Different user's trajectories are tested. As shown in Table 1, the number of trajectories, total data size and the size of events detected by the time stamp clustering are listed. The clustering results obtained from three clustering algorithms are compared by external validity indexes pairwisely. The indexes [13] include Rand index (RI), adjusted Rand index (ARI), Jaccard coefficient (jac) and Fowlkes and Mallows (FM), which evaluate the clustering quality by the similarity of the pairs of data objects in different partitions. Higher value of the index indicates higher similarity. Therefore, according to Table 1, the clustering results from DBSCAN and kmeans are very close in general; however, the GEM produces different results from DBSCAN and kmeans.The clustering results on multiple users' GPS trajectories from user 000 to 010 are shown in Fig. 5 and Fig. 6. The clusters are displayed in different colors on a google map. The number of clusters is set as six for all the clustering algorithms. The figures verify the results from Table 1. The groups from DBSCAN and kmeans (the result on DBSCAN therefore is skipped) are mostly the same. However, as we can see from the enlarged part in Fig. 6, three clusters from the GEM have overlaps between each other. Only one cluster is obtained from the GEM at the bottom right. However, there are two clusters obtained from the

DBSCAN and kmeans. Based on the results, it is obvious that the GEM is suitable for the cases that there exist subclasses. The kmeans is a good choice to conduct similar work because it is more efficient than the DBSCAN.

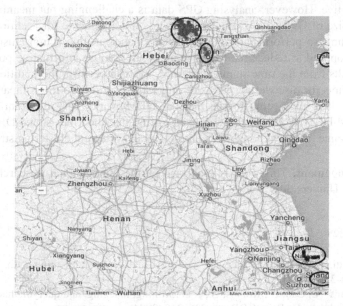

Fig. 5. Places of interest detected by kmeans

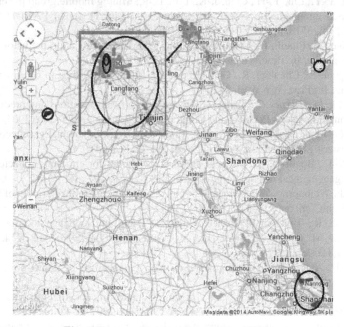

Fig. 6. Places of interest detected by GEM

4   Conclusions

With the explosion of the GPS data collected, the hidden information in the data is quite attractive. However, analysing GPS data is a challenging but meaningful work. Since the size of GPS data is usually large, it is not applicable for traditional clustering algorithms to perform the analysis. In this paper, a time stamp clustering algorithm is proposed to reduce the data size and extract more informative points. Based on the results from the time stamp clustering, we then perform the traditional clustering algorithms for discovering places of interest. According to the comparison on the algorithms, it is interesting to see that the GEM works very differently from the kmeans and DBSCAN. It is also surprised to find that the kmeans and DBSCAN obtains very same results although they employ totally different clustering strategies.

Acknowledgement. This work is supported by the Fundamental Research Funds for the Central Universities.

References

1. Giannotti, F., Nanni, M., Pinelli, F., et al.: Trajectory pattern mining. In: Proc. KDD, pp. 330–339 (2007)
2. Mamoulis, N., Cao, H., Kollios, G., et al.: Mining, indexing, and querying historical spatiotemporal data. In: Proc. KDD, pp. 236–245 (2004)
3. Hwang, S.-Y., Liu, Y.-H., Chiu, J.-K., Lim, E.-p.: Mining mobile group patterns: A trajectory-based approach. In: Ho, T.-B., Cheung, D., Liu, H. (eds.) PAKDD 2005. LNCS (LNAI), vol. 3518, pp. 713–718. Springer, Heidelberg (2005)
4. Monreale, A., Pinelli, F., Trasarti, R., et al.: Wherenext: a location predictor on trajectory pattern mining. In: Proc. KDD, pp. 637–646 (2009)
5. Khetarpaul, S., Chauhan, R., Gupta, S., et al.: Mining GPS data to determine interesting locations. In: Proc. WWW, p. 8 (2011)
6. Ashbrook, D., Starner, T.: Using GPS to learn significant locations and predict movement across multiple users. Personal and Ubiquitous Computing 7, 275–286 (2003)
7. Zhou, C., Frankowski, D., Ludford, P., et al.: Discovering personal gazetteers: an interactive clustering approach. In: Proc. GIS, pp. 266–273 (2004)
8. Alvares, L., Bogorny, V., Kuijpers, B., et al.: A model for enriching trajectories with semantic geographical information. In: Proc. GIS (2007)
9. Kang, J., Welbourne, W., Stewart, B., et al.: Extracting places from traces of locations. In: Proc. MobiCom, pp. 110–118 (2004)
10. Hariharan, R., Toyama, K.: Project lachesis: Parsing and modeling location histories. In: Egenhofer, M., Freksa, C., Miller, H.J. (eds.) GIScience 2004. LNCS, vol. 3234, pp. 106–124. Springer, Heidelberg (2004)
11. Foote, J.: Automatic audio segmentation using a measure of audio novelty. In: Proc. ICME, pp. 452–455 (2000)
12. Zheng, Y., Zhang, L., Xie, X., et al.: Mining interesting locations and travel sequences from GPS trajectories. In: Proc. WWW, pp. 791–800 (2009)
13. Zhao, Q., Xu, M., et al.: Expanding external validity measures for determining the number of clusters. In: Proc. Intelligent Systems Design and Applications, pp. 931–936 (2012)

A Hybrid Solution of Mining Frequent Itemsets from Uncertain Database

Xiaomei Yu, Hong Wang[*], and Xiangwei Zheng

School of Information Science and Engineering,
Shandong Provincial Key Laboratory for Distributed Computer software Novel Technology,
Shandong Normal University, Jinan, Shandong, China
yxm0708@126.com

Abstract. With the emergence of new applications, the traditional way of mining frequent itemsets is not available in uncertain environment. In the past few years, researchers presented different solutions in extending conventional algorithms into uncertainty environment. In this paper, we review previous algorithms and proposed a hybrid solution to mine frequent itemsets from uncertain databases. The new scheme bases on traditional Eclat algorithm and mines frequent itemsets under the definition of frequent probability. Furthermore, the hybrid solution exerts fuzzy mining and precise mining adaptively according to the characters of the candidate databases, which addresses the problem of tradeoff between computation and accuracy. We tested our solution on a number of uncertain data sets, and compared it with the well known uncertain frequent itemsets mining algorithms. The experimental results show that our solution is efficient and accuracy.

Keywords: frequent itemsets mining, uncertain database; algorithm, data mining.

1 Introduction

As an important data mining problem, frequent itemsets mining plays an essential role in real applications. In the traditional databases, the present or absence of an item is known with certainty and an itemset is frequent if and only if its support is not less than a specified minimum support. In the past twenty years, a great effort has been dedicated to this field and researchers have made tremendous achievements in developing efficient and scalable algorithms for frequent itemsets mining in precise and certain databases. Three of the most classic algorithms are Apriori [1], FP-growth [2] and Eclat [3].

However, in many emerging applications, the underlying databases may contain errors or even incomplete, thus the existence of an itemset in a transaction is best captured by probability, which put forward a new topic of exploiting efficient algorithms for mining frequent itemsets in uncertain environment.

Obviously, extending traditional algorithms into uncertain cases is preferred in the past several years. For example, UApriori [4], UFP-growth [5] and UH-mine [5] algorithms are three well known adaptations of traditional algorithms for mining frequent

---

[*] Corresponding author.

D.-S. Huang et al. (Eds.): ICIC 2014, LNAI 8589, pp. 581–590, 2014.

itemsets in uncertain databases. Besides, in order to handle the exponentially large possible worlds induced under the Possible World Semantics in uncertain environment, some approximant improvements are also developed, such as PDUApriori [6], NDUApriori [7] and NDUH-mine [8] algorithms.

Besides frequent itemsets mining algorithms using horizontal representation outlined above, there are also vertical mining algorithms, such as UEclat [9], UV-Eclat [10], and U-Eclat algorithms [11]. In 2008, Toon Calders [11] presented a sample-based approach in uncertain database, named U-Eclat algorithm. This technique works by sampling the database several times to find the frequent itemsets with a very accurate approximation of support. By comparing the existing probability with random number produced for each item in a transaction, an item is identified in the currently sampled transaction once the number is greater than the probability of the item. In 2010, Laila A extended the Eclat algorithm and developed UEclat algorithm. The only difference with certain Eclat algorithm is that each node in Utidset structure also stores the existential probability of each item besides the item itself. In 2011, Carson K. L. proposed UV-Eclat algorithm with further improvements than UEclat algorithm, which corrected the method in calculating $esup(X)$ of itemset X and associated each item with an augmented tidset. Thanks to its excellent properties in sparse databases, UV-Eclat algorithm achieved better performance than most horizontal mining algorithms in uncertain environments.

Although vertical mining is regarded as a promising approach that is experimentally proved to be more efficient than the horizontal approach [12], it has not received much attention and there is not much efficient vertical mining algorithm in uncertain environment. Moreover, summarizing the achievement and the shortage of previous research, we find that all the three Eclat-based algorithms focus on mining expected support-based frequent itemsets in uncertain databases and there is not any Eclat-based algorithm at present that can mine probabilistic frequent itemsets efficiently in uncertain environment, which is a problem worthy of further research.

In this paper, based on the fundamental status of frequent itemsets mining in uncertain database communities, we put forward a new hybrid solution with Eclat framework in uncertain databases under the definitions of probabilistic frequent itemset. The new solution exerts a fuzzy mining and a precise mining adaptively according to the characters of the candidate databases, which addresses the problem of tradeoff between computation and accuracy.

The rest of the paper is organized as follows. In section 2, we introduce some basic definitions on frequent itemsets mining in uncertain databases. Then we present our hybrid solution with Eclat framework in section 3. The experimental results are shown in section 4 and we draw our conclusions in section 5.

2   Problem Definitions

There are two different data models to represent uncertain databases of transactions with existential uncertain items, in which the attribute uncertainty model is usually adopted for frequent itemsets mining. In the attribute uncertainty model, each transaction consists of attributes (items) with values carrying uncertain information. Given an uncertain transaction database UDB which contains N transactions, each transaction is denoted as a tuple $<tid, X>$ where tid is the transaction ID, and $X=\{x_1(p_1), x_2(p_2), ..., x_m(p_m)\}$ is the item sets occurs in a transaction. Each unit $x_i(p_i)$ in X consists of an item x_i and its existential probability p_i, denoting the probability of this item appearing in the tid tuple.

Since the items in transactions are probabilistic in nature, it is impossible to count the frequency of the itemset deterministically. An approach is proposed by Chui et al [4] to compute the expected support of itemsets in their U-Apriori algorithm.

Definition 1. (Expected Support) Given an uncertain transaction database UDB which includes N transactions. The expected support of an itemset X is defined as:

$$exsup(X) = \sum_{i=1}^{N} p_i(X) \tag{1}$$

Based on the definition of expected support, an itemset X is expected support-based frequent itemset if and only if $exsup(X) \geq N \times min\_sup$, where min_sup is a minimum expected support ratio specified by user.

However, recent studies declared that determining a frequent itemset only by its expected support may cause loss of important information. Any uncertain transaction database naturally involves uncertainty on the support of an itemset, which is important when evaluating whether an itemset is frequent or not. Therefore, the confidence with which an itemset is frequent is introduced as another measurement to evaluate the uncertain data.

Definition 2. (Frequent Probability) Given an uncertain transaction database UDB which includes N transactions, and a minimum support ratio min_sup. Frequent probability of an itemset X, is defined as follows:

$$Pr(X)=Pr\{sup(X) \geq N \times min\_sup\} \tag{2}$$

Based on the definition of support probability, an itemset X is a probabilistic frequent itemset if X's frequent probability is larger than the probabilistic frequent threshold, namely:

$$Pr(X)=Pr\{sup(X) \geq N \times min\_sup\}>min\_prob \tag{3}$$

Here, min_prob is a minimum probabilistic support threshold specified by user.

By experimental study and theoretical analysis, Yongxin Tong et al [13] find that the two definitions have a close relationship. Considering the support of an itemset as a random variable following Poisson Binomial distribution, the expected support of an itemset equals to the expectation of the random variable. Because a Poisson Binomial distribution can be approximated as a Poisson distribution or a Normal distribution under high confidence in a large enough uncertain dataset, the frequent probability of an itemset is equal to the expected support of the itemset if the latter also considers the variance of the support at the same time.

Therefore, the efficiency of mining probabilistic frequent itemsets can be greatly improved by employing efficient expected support-based frequent itemset mining algorithms. Provided that the variance of support of each itemset is also calculated during the calculation of expected support on each itemset, we can use the expected support-based FIM algorithms to mine probabilistic frequent itemsets as well.

3   The Hybrid Solution with Eclat Framework

Compared with the definition of expected support, the definition of frequent probability treats the set of support holistically. Whether an item is supported in a particular possible world does not only depend on its own frequency, but also other item's combined frequency. Unlike the definition of expected support-based frequent itemset, which simply relies on single item's expected support, the definition of probabilistic frequent itemset captures the intricate interplay between the uncertain items in a randomly generated possible world.

On the other hand, provided that the variance of support of each itemset is also calculated as well as the expected support, we can use the expected support-based FIM algorithms to mine probabilistic frequent itemsets as long as the database under consideration is large enough. Taken together the actual situation in uncertain databases, we present an Eclat-based framework to solve the probabilistic frequent itemset mining problem in uncertain environment, with consideration of the characteristics of uncertain databases being analyzed.

a)   A Data-related Issue

As we know, most of uncertain databases are usually sparse ones and the frequent items in which have low relative support. Although a large enough uncertain dataset is given, the relative support of frequent items will be far below 1% in the sparse cases [14]. With a low *min_sup* given, although the frequent items generated are large at the first scan of database, the total number of candidate frequent itemsets will decrease sharply after several recursions. Furthermore, if a little big *min_sup* is given, the number of frequent items generated will also reduce quickly, which aggravates the whole trend of declining in sparse cases. Moreover, thanks to the various effective pruning techniques, after mining of several rounds, the number of k frequent itemsets left will drop dramatically, so that it will be impossible to produce a large enough candidate frequent database for mining (k+1) frequent itemsets (Shown in Fig.1 [7]).

Even in the case of dense uncertain databases, the same problem maybe occurs in the mining of (k+1) frequent itemsets, although the k is much larger than that in sparse databases. Then, the condition of Lyapunov Central Limit Theory no longer holds, and those approximate probabilistic frequent algorithms would have difficulty to accomplish the effective mining.

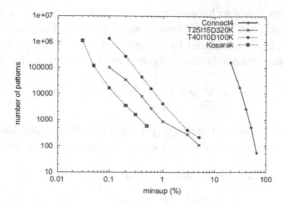

Fig. 1. Varying Support & The Number of Frequent Itemsets [7]

Under the semantics of Possible Worlds, there are $2^{/T/*/N/}$ possible worlds, where $/T/$ is the total number of transactions and $/N/$ the total number of items in the dataset. Along with the decrease in the number of N or T, the number of possible world drops steeply, and the amount of calculation declines, too. In fact, the matter of computation in 2-frequent itemsets gains more attention in discussion so far. There is a good reason to do so: the most main memory is usually required for determining the 2-frequent itemsets in practice.

Therefore, considering the whole process of frequent itemsets mining, we propose a hybrid solution to mine frequent itemsets in uncertain databases, which is divided into the approximate mining phase and the exact mining phase.

b) Eclat-based Approximate Probabilistic Frequent Mining Phase

As we have seen in previous section, computing probabilistic frequent itemsets is much more difficult than computing the expected support-based ones. To tackle the problem of costly computation, T. Calders et al [7] presents a Normal distribution-based approximate algorithm. However, it is impractical to apply this algorithm on very large sparse uncertain databases since it employs the Apriori framework. With the consideration that uncertain databases are often considered as sparse ones and Eclat algorithm is efficient especially in sparse databases, we propose a novel NDUEclat algorithm which integrates the framework of traditional Eclat algorithm and Normal distribution-based approximation in order to achieve a win-win partnership in spare uncertain databases. The main steps are outlined as follows:

Step 1: Scan the uncertain database and transform it into vertical format

In this step, the transformed DB is represented as a collection of items, and each item x is associated with a tidlist including 1) the transaction identification t_i 2) the existential probability of item x in the transaction t_i, i.e., $tidlist(x)=\{(t_1, p_x(t_1)), (t_2, p_x(t_2)), ..., (t_m, p_x(t_m))\}$.

Step 2: Prune all itemsets and calculate the frequent probability of each itemset to find probabilistic frequent itemsets

In this step, some other technologies can be applied to prune the itemsets besides Monotonicity Criteria and Chernoff Bound based pruning. Then the expected support of each item *exsup(x)* is calculated by summing the existential probability up in the same *tidlist*. And the variance is also computed by the formula as follows:

$$Sn^2 = \sum_{k=1}^{n} Pr(x_k = 1)(1 - Pr(x_k = 1))$$

Thus, the approximation of frequent probability will be calculated based on the cumulative distribution function Φ, as long as the database is large enough:

$$freqprob(x) = \Phi(\frac{min\_sup - 0.5 - expsup(x)}{Sn})$$

If *freqprob(x)* $\geq$ *min_prob*, then the itemset X is considered frequent under the definition of frequent probability in uncertain database.

Step 3: Construct (k+1) candidate itemsets and mine all frequent itemsets

With the k-frequent itemsets mined in step 2, we generate (k+1) candidate itemsets by intersecting the *tidlists* of two frequent k-itemsets, then exert multiply operation to get the expected support of (k+1) itemsets.

According to the Equation (1), we get the expected support of (k+1)-itemset *Axy* similarly by multiplying *exsup(Ax,t_i)* and $p_y(t_i)$ or *exsup(Ay,t_i)* and $p_x(t_i)$:

exsup(Axy)={*t_i*:*exsup(Ax,t_i)*×*p_y(t_i)*| *Ax,Ay* $\in D_k$ *and t_i*:*exsup(Ax,t_i)* $\in$*tidlist(Ax) and t_i*:*p_y(t_i)* $\in$*tidset(y)}* or
exsup(Axy)={*t_i*:*exsup(Ay,t_i)*×*p_x(t_i)*| *Ax,Ay* $\in D_k$ *and t_i*:*exsup(Ay,t_i)* $\in$*tidlist(Ay) and t_i*:*p_x(t_i)* $\in$*tidset(x)}.*

Here, A is a k-frequent itemset and $k>2$.

In this way, our NDUEclat algorithm recursively mines all approximate probabilistic frequent itemsets in the uncertain database of D.

c) Eclat-based Exact Probabilistic Frequent Mining Phase

Approximate probabilistic frequent algorithms are proved to have the same efficiency as expected support-based algorithms and also guarantee to return frequent probabilities of all probabilistic frequent itemsets with high confidence, as long as the databases handled is large enough.

However, as described in subsection 3.1, after several rounds of mining operation, the size of the candidate database will drop sharply and conditions of exerting Lyapunov Central Limit Theory no longer holds. At this time, we have to apply exact probabilistic frequent approach based on Eclat framework, named UDPEclat phase.

The UDPEclat phase is exerted on the projected database D_k, which consists of k-frequent itemsets in vertical representation. Each itemset is associated with a *tidlist* including 1) the tid of t_i 2) the expected support in t_i, i.e., *tidlist(X)={(t_1, p_X(t_1)), (t_2, p_X(t_2)), ..., (t_m, p_X(t_m))}*. All itemsets in D_k are sorted in the descending order of expected support.

Step 1: Prune the itemsets and calculate frequent probability of each itemset to find probabilistic frequent itemsets

Thanks to the Monotonicity Criteria in uncertain environment, some infrequent itemsets can be identified quickly. Then under the definition of probabilistic frequent itemset, we calculate the frequent probability of each itemset by dynamic programming. The recursive relationship is defined as follows:

$$Pr_{\geq i,j}(X) = Pr_{\geq i-1,j-1}(X) \times Pr(X \subseteq T_j) + Pr_{\geq i,j-1}(X) \times (1 - Pr(X \subseteq T_j))$$

Here $Pr_{\geq i,j}(X)$ denotes the probability that at least i of j transactions contain itemset X. As far as we know, the probability $Pr_{\geq i,j}(X)$ can be computed using the previous result of $Pr_{\geq i,j-1}(X)$, and also be useful in the following calculation of $Pr_{\geq i+1,j}(X)$, thus the frequent probability $Pr_{\geq minsup,|T|}(X)$ is achieved by iterative procedure until i reaches min_sup. With consideration of certain itemsets in some transactions, the whole computation maybe further reduces. After comparing the probability with min_prob given by specific user, we find all probabilistic (k+1) frequent itemsets exactly.

Step 2: construct (k+2) candidate itemsets and mine all frequent itemsets in the remaining candidate database

Continually, we generate (k+2) candidate itemsets based on original Eclat framework, i.e., by intersecting the *tidlists* of two (k+1)-frequent itemsets with same k-prefix, we get the (k+2) candidate itemset.

Using this dynamic programming scheme, we find the rest probabilistic frequent itemsets in an iterative process. It is noted that information about probability $Pr_{i-1,j-1}(X)$, $Pr_{i,j-1}(X)$ and $Pr(X \subseteq t_j)$ should be preserved for next iteration, so the probabilities have to be stored too.

Additionally, during the whole mining process, we need to weigh the number of frequent itemsets based on conditions of the special form of the Central Limit Theorem as well as the memory usage provided, and switch to the exact probabilistic frequent phase adaptively if necessary. Otherwise, we continue the approximate probabilistic frequent phase, and test again after several rounds of frequent itemset mining, until the number of frequent itemsets is not large enough to satisfy the Central Limit Theorem. Thus, using a hybrid frequent itemsets mining algorithm consists of approximate and exact mining phases, we implement a more efficient and accurate mining algorithms with consideration of the characteristics of uncertain databases being analyzed.

In the following section, we can observe that the hybrid algorithm has an efficient performance in large sparse uncertain datasets, which confirms our goal.

4   Experiments

In this section, we introduce the results of preliminary experiments conducted. The experiments run on a PC with a 2.5 GHz Intel(R) core(TM) i5-2520M processor and 4 Gbytes of RAM. The operating system was Window 7. All algorithms were implemented and compiled using Microsoft's Visual C++ 2010.

All of the experiments were conducted on several databases with different characteristics, including both dense databases and sparse ones. We introduce uncertainty
into the datasets to generate an existential probability for each item. The relatively
high or low probability is assigned to each item in the dataset following a Gaussian
distribution with different mean and variance. For example, the dataset of Kosarak,
which comes from the Frequent Itemset Mining Database Repository provided by
Ferenc Bodon, contains click-stream data of a Hungarian on-line news portal. It consists of 990002 transactions, the average length of a frequent itemset is 8.1, and the
number of item is 41935. The existential probability of each item is a variance, drawn
from a Gaussian distribution with mean 0.5 and variance 0.5. In this spare dataset of
Kosarak with high mean and low variance, we run our hybrid algorithm, NDUApriori,
and NDUH-mine algorithm respectively.

Table 1. Accuracy in Dataset of Gazelle

Min_sup	NDUH-Mine	Hybrid
0.002	0.92	0.96
0.005	0.95	0.97
0.01	0.96	0.98
0.02	0.98	1
0.05	1	1

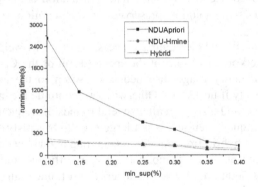

Fig. 2. Results in Dataset of Kosarak

We also conducted the experiment on a sparse dataset of Gazelle. It is a clickstream
data set from Gazelle.com. It contains 59602 transactions, up to 498 items per
transaction, and the average length of a frequent itemset is 2.5. With high mean and
low variance, namely with the mean (0.95) and the variance (0.05), we run our hybrid
algorithm with NDUEclat phase shifting to UDPEclat phase as well as NDUH-Mine
algorithm. The preliminary results show that our hybrid algorithm is efficient and
more accuracy (Shown in Fig.2), especially when the cadidate itemsets dataset is
sparse with low *min_sup* relatively (Shown in Table 1).

5 Conclusions and Future Study

Recently, frequent itemsets mining in uncertain databases become an important research topic and attracted much attention. Hence, many efficient frequent itemsets mining algorithms have been proposed. While these algorithms work well for databases with precise values, it is not clear whether they can be used to mine probabilistic data efficiently. In this paper, we compared the present algorithms for extracting frequent item sets from uncertain databases and proposed our hybrid algorithm with Eclat framework. Although our algorithm is developed based on the traditional Eclat framework, it can be conducted for supporting other improved algorithms in handling uncertain data.

In the future, we will continue our research on frequent itemsets mining in uncertain environment. With further experiments, we will explore the characters of specific uncertain databases and improve frequent itemset mining in imperfect database, thus utilize and develop the previous studies into emerging applications.

Acknowledgement. We are grateful for the support of the Natural Science Foundation of China (61373149) and the Shandong Provincial Project for Science and Technology Development (2012GGB01058).

References

1. Agrawal, R., Srikant, R.: Fast Algorithms for Mining Association Rules. In: 20th International Conference on Very Large Data Bases, September 12-15, pp. 487–499 (1994)
2. Han, J., Pei, J., Yin, Y.: Mining Frequent Patterns without Candidate Generation. In: ACM SIGMOD International Conference on Management of Data, pp. 1–12. ACM Press (2000)
3. Zaki, M.J.: Scalable Algorithms for Association Mining. IEEE Transactions on Knowledge and Data Engineering 3, 372–390 (2000)
4. Chui, C.-K., Kao, B., Hung, E.: Mining frequent itemsets from uncertain data. In: Zhou, Z.-H., Li, H., Yang, Q. (eds.) PAKDD 2007. LNCS (LNAI), vol. 4426, pp. 47–58. Springer, Heidelberg (2007)
5. Aggarwal, C.C., Li, Y.: Frequent Pattern Mining with Uncertain Data. In: KDD 2009, Paris, France, June 28–July 1, pp. 29–37 (2009)
6. Wang, L., Cheng, R., Lee, S.D.: Accelerating Probabilistic Frequent Itemset Mining: a Model-based Approach. In: CIKM 2010, Toronto, Ontario, Canada, pp. 429–438 (2010)
7. Calders, T., Garbini, C., Goethals, B.: Approximation of Frequentness Probability of Itemsets in Uncertain Data. In: ICDM, pp. 749–754 (2010)
8. Tong, Y., Chen, L., Cheng, Y., Yu, P.S.: Mining Frequent Itemsets over Uncertain Databases. In: VLDB 2012, Istanbul, Turkey, August 27, vol. 5(11), pp. 1650–1661 (2012)
9. Abd-Elmegid, L.A.: Vertical Mining of Frequent Patterns from Uncertain Data. Computer and Information Science 3(2), 171–179 (2010)
10. Leung, C.K.-S., Sun, L.: Equivalence Class Transformation Based Mining of Frequent Itemsets from Uncertain Data. In: SAC 2011, pp. 983–984 (2011)

11. Calders, T., Garboni, C., Goethals, B.: Toon Calders, Calin Garboni, and Bart Goethals: Efficient Pattern Mining of Uncertain Data with Sampling. In: PAKDD, vol. (1), pp. 480–487 (2010)
12. Song, M., Rajasekaran, S.: A Transaction Mapping Algorithm for Frequent Itemsets Mining. IEEE Transactions on Knowledge and Data Engineering 18(4), 472–481 (2006)
13. Tong, Y., Chen, L., Philip, S.: UFIMT: An Uncertain Frequent Itemset Mining Toolbox. In: KDD 2012, Beijing, China, August 12–16, pp. 1508–1511 (2012)
14. Pei, J., Han, J., Lu, H.: H-Mine: Hyper-Structure Mining of Frequent Patters in Large Databases. In: IEEE International Conference on Data Mining, ICDM 2001, San Jose, CA, November 29–December 2, pp. 441–448 (2001)

DVT-PKM: An Improved GPU Based Parallel K-Means Algorithm

Bo Yan[1,*], Ye Zhang[1], Zijiang Yang[2], Hongyi Su[1], and Hong Zheng[1]

[1] Beijing Institute of technology, Beijing, CN
{yanbo,zhangxiaoye,henrysu,hongzheng}@ bit.edu.cn
[2] School of information technology, York University, Toronto, Canada
zyang@yorku.ca

Abstract. K-Means clustering algorithm is a typical partition-based clustering algorithm. Its two major disadvantages lie in the facts that the algorithm is sensitive to initial cluster centers and the outliers exert significant influence on the clustering results. In addition, K-Means algorithm traverses and computes all the data multiple times. Thus, the algorithm is not efficient when dealing with large data sets. In order to overcome the above limitations, this paper proposes to exclude the outliers using the minimum number of points in the d-dimensional hypersphere area. Then k cluster centers can be obtained by adjusting the threshold making use of density idea. Finally, K-Means algorithm will be integrated with Compute Unified Device Architecture (CUDA). The time efficiency is improved considerably through taking advantage of computing power of Graphic Processing Unit (GPU). We use the ratio of distance between classes to distance within classes and speedup as the evaluation criteria. The experiments indicate that the proposed algorithm significantly improves the stability and running efficiency of K-Means algorithm.

Keywords: K-Means, density, Graphic Processing Unit.

1   Introduction

Cluster analysis known as group analysis [1], is a statistical analysis method. It is a process that divides abstract set into several clusters which consist of similar objects. K-Means algorithm is one of the most popular algorithms and has been extensively examined. According to the principle that similarity within clusters is large while similarity among clusters is small, the clustering results can be optimized. This algorithm is proposed in 1967 by MacQeen [2]. He summed up the research results of Cox [3], Fisher [4], Sebestyen [5] and others and put forward detailed steps of K-Means algorithm which had been proved by mathematical methods. The advantage of this algorithm is fast, simple and easy to implement. Since 1970s, the algorithm has been applied to many areas, including mathematical science and chemistry [6], image analysis [7], soil [8] and so on. With in-depth study of the K-Means algorithm, some

---

* Corresponding author.

D.-S. Huang et al. (Eds.): ICIC 2014, LNAI 8589, pp. 591–601, 2014.

limitations are found including the need for a pre-determined k value, interference of the initial cluster centers, converging to local optima easily and low efficacy when dealing with big data. Since Selim [9] found that K-means algorithm was liable to converge to local optima, the selection of the initial cluster centers has been a research direction in the field of K-means algorithm. To solve this problem, some researchers have proposed the improved heuristic algorithms. Bagirov [10] improved the reported global k-means algorithm. A starting point for the k-th cluster center in this algorithm is computed by minimizing an auxiliary cluster function. Results demonstrate the superiority of the new algorithm. However, it requires more computational time than the global k-means algorithm. Lee et al. [11] use a method which maximizes the distance among the initial cluster centers and distributed the centers evenly. It is time-consuming, but can reduce the total clustering time by minimizing allocation and recalculation. Khan [12] proposes a simple seed selection algorithm along one attribute that draws initial cluster boundaries along the "deepest valleys" or greatest gaps in dataset. Unlike existing initialization methods, no additional parameters or degrees of freedom are introduced to the algorithm. This improves the replicability of cluster assignment by as much as 100% over k-means.

Nowadays, Graphics Processing Unit (GPU), previously used in computer graphics, has become common components in many organizations [13]. Many scholars have been using CUDA to speed up a large number of applications [14] [15]. With the explosive growth of data, the running time of k-Means algorithm grows with the increase of the size and dimension of data set. Hence clustering large-scale data sets is usually a time consuming task. It is necessary to accelerate the K-Means algorithm with CUDA. Bai et al. [16] proposed GPUs based k-means algorithm. Both data objects assignment and k centroids recalculation were offloaded to the GPU. But there is a data transmission between these two steps in each iteration. Li et al. [17] used two different strategies for low-dimensional and high-dimensional data sets respectively to make the best use of GPU computing horsepower. The results show that their algorithms are three to eight times faster than the best reported GPU-based algorithm. Kijsipongse [18] presented the design and implementation of an efficient large-scale parallel K-Means on GPU clusters. They employed the dynamic load balancing to distribute workload equally on different GPUs installed in the clusters and improve the performance of the parallel K-Means at the inter-node level.

To meet the challenges in K-Means algorithm and the explosive growth of data, we propose a density based variable threshold parallel K-Means algorithm. The outlier will be removed to improve the robustness of clustering. k relatively dispersed samples are set as the initial cluster centers to improve the clustering accuracy. Combining GPU architecture platform we improve the efficiency of the algorithm.

The paper is organized as follows. Section 2 presents the related works on CUDA and K-Means algorithm. Section 3 discusses the improved K-Means algorithm. The evaluation functions for clustering algorithm, experimental results and comparison to the other methods are provided in Section 4. Finally, Section 5 concludes the paper.

2   Related Works and Backgrounds

Clustering is a method to divide n data objects into k clusters where each cluster has maximal similarity as defined. K-Means is a commonly used clustering algorithm in data mining fields. CUDA is a minimal extension of C and C++ programming languages. This section provides the K-means algorithm basics and CUDA basics.

2.1   Partition-Based Clustering Algorithm

There are many clustering algorithms in literature. In general, clustering algorithm can be divided into several categories: partitioning method, hierarchical method and etc. Partition-based method includes K-mean algorithm, K-center algorithm and so on. The basic principles of K-Means algorithm are provided as below [3]:

- Choose the number of clusters, k.
- Randomly generate k clusters and determine the cluster centers, or directly generate k random points as cluster centers.
- Assign each object to the nearest cluster center.
- Re-compute the new cluster centers.
- Repeat steps 3 and 4 until some convergence criterion is met.

2.2   CUDA Model

NVIDIA Corporation (NVIDIA) officially released CUDA in 2007. CUDA model [15] is a new hardware and software architecture to conduct computations on GPU as a data-parallel computing device without mapping them to a graphics API. CUDA programming is a heterogeneous system programming, i.e. the hybrid of CPU and GPU. The key point of CUDA model is to take CPU as the host (Host) to process logic and transactional program, while the GPU works as a coprocessor (co-processor) to process highly parallel procedures. GPU is not suitable to compute complex data structure while it can do a large amount of same pattern calculation.

On microcosmic level, there are a lot of thread blocks operating on the GPU. Each thread block contains many threads, and each thread is responsible to execute the same kernel function. Generally, CUDA is collectively composed of two processes: the parallel process executing kernel function by GPU and the serial process executing control function by CPU. In CUDA model, several threads constitute of a thread block (block) and several blocks constitute of a grid (Grid). Fig 1 provides the details.

CUDA computing core is composed of Stream Multiprocessors (SMs) with eight Stream Processors (SPs) in each SM. Five types of GPU memories are supplied to take advantage of the flexibility to develop efficient coding. Among the five types of memories, registers are fastest, and the time overhead can be negligible when accessing them. However, they are limited in numbers. Shared memory is as fast as registers, but it may have bank conflicts when the threads are assigned poorly. Threads in the same block can communicate through shared memory. Global memory is off-chip memory. Both CPU and GPU can access global memory.

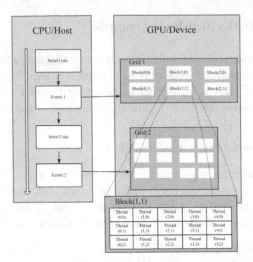

Fig. 1. CUDA parallel architecture

3   Proposed Methodology

In order to solve the problem of local optimal solution and low efficiency when dealing with big data, a density based variable threshold parallel K-Means algorithm (DVT-PKM) is put forward. The algorithm includes two parts: the selection of initial cluster centers and GPU based parallel improvement for K-Means method. The details of the algorithm are specified as follows.

3.1   The Selection of Initial Center Points

The traditional K-Means algorithm is liable to converge to a locally optimal solution. Meanwhile, the sensitivity to the outliers leads to inaccurate clustering results. This paper proposes the DVT-PKM algorithm to solve these limitations. It takes a group of relatively dispersed samples as the initial cluster centers to increase the possibility to find the global optimal solution. We use the Euclidean distance as the distance function. First of all, several concepts based on the density are introduced as follows:

1. σ neighborhood $N_\sigma(x_i): M = \{x_1, w_2, \ldots, x_n\}, M \subseteq R^d$ is a dataset in d dimensions. For any object $x_i \in M$, $N_\sigma(x_i)$ is defined as a d-dimensional hypersphere region which takes x_i as the center point and σ as the radius, i.e.

$$N_\sigma(x_i) = \{x_j \in M \mid 0 \le dist(x_i, x_j) \le \sigma, j = 1,2,\ldots,n\} \quad (1)$$

$dist(x_i, x_j)$ is the Euclidean distance between object x_i and x_j in M.

2. Density: the density of x_i in dataset M is defined as the number of objects within the distance of σ neighborhood.

VTD-KM procedures are provided as below.

Data description: k, number of clusters; δ, threshold, i.e. the selected radius; ω, the density within the δ neighborhood; M, the set of all points; S, the set of initial cluster centers; $S1$, temporary set of initial cluster centers. ε, a small value.

1. Calculate the distances for any pair of points in M to get the distance matrix. Choose the distance between the first two points as σ.

2. Choose the point $x_i (x_i \in M)$ which has the maximum density as the initial center V_1 and add it into set S. Calculate the density ψ_i within the δ neighborhood of all points in M, the mean density $\omega = \sum_{i=1}^{n} \psi_i / n$ is taken as a reference density. Finally, all points in M meet the conditions of $N_\delta(v_1)$ are deleted.

3. The points in M are put into set $S1$ if their distances from V_1 are $d2 = 2\delta + \varepsilon$ and their density are greater than ω. The two points with the largest distance in $S1$ are chosen as the initial center points and move them to S.

4. If $S1$ is not empty, point x_i in $S1$ will be moved to S if the distance from x_i to V_i in S is beyond the $dist(x_i, v_i) > \delta$. Repeat this step until all the points in $S1$ do not meet this condition. Then empty $S1$. All points within $d1(d1 = 2\delta)$ neighborhood of V_1 are deleted from M.

5. Set $d1 = 2 \cdot \delta \cdot i$ and $d2 = 2 \cdot \delta \cdot i + \varepsilon$ where i is the current number of loops. Repeat steps (3) and (4). i is incremented by one in each iteration until M is empty. Then go to step (6).

6. When M is empty, we can decide whether the number of center points in S is k. If yes, k center points are found. If it is larger than k, δ is replaced by a bigger number and back to step (2). If it is smaller than k, δ is replaced by a smaller number, then go back to step (2).

3.2   GPU Based K-Means Method

When dealing with large data sets, K-Means algorithm has high scalability and good performance. The complexity of this algorithm is O (nkt) where n is the number of objects, k is the number of clusters and t is the number of iterations. When n is too large, the distance calculation process is time consuming. In this case, the computing power of GPU can be applied to parallelize K-Means algorithm.

The main idea of the GPU-based K-Means algorithm is to improve the performance by separating data and transferring the intensive calculation from host to device. It can be observed that the main computational tasks are carried by steps (3) and (4). Although these two steps are performed repetitively, the data for each calculation are independent from each other. Thus, they can be separated into two kernel functions and parallel computed by the GPU core. In this way, we manage to optimize the GPU-based K-Means algorithm in steps (3) and (4).

In step (3), since currently a GPU has more than 100 processing cores, data points can be operated in threads within these cores. In each thread, distances from each data point to all the cluster centers are calculated. According to the distance, data point is assigned to a label i between 1 and k, i is determined by the center c_i it is nearest to. The label array is saved in global memory. Access to the global memory multiple times is required when doing the distance calculation between a certain point from the data set and k centers. Considering the high latency of global memory we can put the cluster centers in shared memory to save time. The detail is presented as below.

```
Program Distance calculation
  var Data[], k, Centers[], D_dim, D_label[], temp, min;
  begin
    int tidX = blockIdx.x*blockDim.x+threadIdx.x;
    __shared__ float s[][];
    s[k][D_ dim] = Centers[k][D_ dim];
    if(tidX<n)
      for j=1 to k
        for i=1 to D_men
          temp is the Euclidean distance between
              Data[tidX*pitch/sizeof(float)+i] and s[j][i]
        end loop
        if(temp is smaller than min)
          d_out[tidX*pitch/sizeof(int)] = j;
          min ← temp
        endif
      end loop
    endif
end
```

In step (d), after calculation, we can then move the output array to shared memory to do the next step. For k clusters, each calculation is operated by a thread in CUDA. Each thread decides whether the data point in the output array belongs to its cluster. If yes, accumulate the data point. To avoid the latency that caused by bank conflict, each thread accesses the corresponding location element in the array, and moves on sequentially. In other words, the first thread accesses the first element in array, meanwhile the second thread accesses the second one and it goes on sequentially. After the k thread accesses the last element in array, it returns to the first one until each element in the array is accessed. Fig 2 describes the detail.

```
Program Updating cluster centers
  var Data[],D_label[],D_dimension,num,Centers[],n;
  begin
    Int tidX ← blockIdx.x * blockDim.x + threadIdx.x
    __shared__ float temp_cluster[n];
    temp_cluster[n] = c_cluster[n];
    If(tidX < k)
```

```
    for j = 0 to n
        tid = blockIdx.x*blockDim.x+(threadIdx.x+j)%n;
        if(the data belong to this cluster)
            for i=0 to D_men
                temp_data[i] +=D[j*pitch/sizeof(float)];
            end loop
            num++
        endif
    end loop
    for i=0 to D_dimension
    Cent[[tidX*pitch/sizeof(int)+i] = temp_data[i] /num
    end loop
  if it is converged
    then break;
end
```

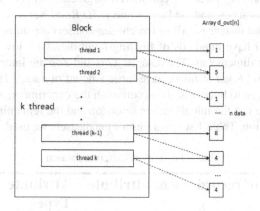

Fig. 2. The process of cluster centers recalculation

4   Experimental Results and Discusses

4.1   Evaluation Function for Clustering Algorithm

Generally, there are two principles to evaluate clustering algorithm: Tightness, i.e., the members in a cluster must be close to each other as much as possible; Separation, i.e., the distance between clusters should be as far as possible. Therefore, we adopt the ratio of within class distance to between class distance as the criterion.

The within class distance (*within(k)*) is the average value of distance between mean point and other points in the same cluster. The between class distance (*Between(k)*) is determined by the average value of distance between any two mean points of clusters. The smaller the J (k) is, the higher the quality of clustering algorithm is.

$$within(k) = 1/k \cdot \sum_{i=1}^{k} (1/|c_i| \cdot \sum_{x_i \in c_i, i \neq u}^{|c_i|} \delta(x_i - x_u)) \qquad (2)$$

$$between(k) = 2\sum\nolimits_{i \neq j, i > j}^{k} \delta(y_i - y_j)\Big/k(k - 1) \qquad (3)$$

$$J(k) = within(k) / between(k) \qquad (4)$$

where c_i is the cluster, $|c_i|$ is the number of points which belongs to c_i, x_i is the point in c_i, x_u, y_i, y_j are the mean points, k is the number of clusters, and $\delta(x_i - x_j)$ is the distance functions which represent the distance between two points.

4.2   Experimental Datasets

In order to validate the accuracy and efficiency of our proposed algorithm, eight data sets (*Glass, Zoo, Iris*, etc.) from UCI database are selected. Five of them are relatively small, while the remaining three are considered large datasets. The smallest dataset is *Zoo* with only 101 instances, and the largest one is *3D Road Network* dataset with about four hundred thousand instances. All of the chosen datasets are numerical. The datasets *Iris, Glass* and *Yeast* have hundreds of floating-point data with low dimensions while *Libras* has very high dimensions. The datasets *Gas* and *Zoo* are integer. The remaining datasets *Letter* and *3D Road Network* have large amount of data.   The number of clusters is range from three to hundreds. To accomplish the experiment, we use the top five datasets in Table I to test the initial centers selection and the remaining three to test the parallel implementation. Table 1 summarizes all the datasets we used.

Table 1. Features of different datasets

DataSet	# of records	# of attribute	Attribute Type	Data Type
Iris	150	4	Numerical	float
Glass	214	9	Numerical	float
Zoo	101	16	Numerical	Integer
Yeast	1480	8	Numerical	float
Libras	360	90	Numerical	float
Gas	10756	128	Numerical	Integer
Letter	20000	16	Numerical	float
Road Network	40000	3	Numerical	float
Road Network	70000	3	Numerical	float
Road Network	434874	3	Numerical	float

4.3   Experimental Setting and Evaluation Criteria

All experiments are implemented on MATLAB and Visual Studio. The test system is a PC with an Intel Core 6 Exon 2620 CPU. The GPU is an NVIDIA Tesla k20 with 2496 stream processors and CUDA is the 5.5 version.

To verify the accuracy of the improved algorithm, we test the traditional K-Means and VTD-KM algorithm 15 times. Then the two average ratios $\sum_{i=1}^{15} J_i(k) / 15$ are compared. The smaller the ratio is, the better the clustering result is.

To verify the effectiveness of the GPU based algorithm, we test ten times on the CPU based K-Means and GPU based K-Means algorithm. Then the average run time and speedup are compared. The larger the speedup is, the better the efficiency is.

4.4   Analysis of Results

This section tests the DTV-PKM algorithm on eight datasets. Table 2 and Fig 4 show the experimental results for the DTV-PKM algorithm based on five datasets. It can be seen that the ratio of the proposed algorithm is less than the traditional K-means algorithm. Clustering results of our algorithm are stable and overcome the limitations including selecting center points randomly in the K-Means algorithm. Moreover, our proposed algorithm has good performances on a variety of data sets and is not affected by the number of samples and properties. However, when we compared our improved algorithm and the traditional K-Means algorithm on different datasets, the growth of accuracy is slightly different. As is shown in Fig 3, the growth of accuracy is large on dataset *Glass* while relatively small on dataset *Yeast*. The reason is that our algorithm is based on density, the growth of accuracy on initial centers selection depends on the distribution of data to some extent. If the data are evenly distributed, the improvements are considerably slight.

Table 2. The ratio of within class distance and between class distance

Dataset	K-Means	DVT-PKM
Iris	0.2214	0.1550
Glass	0.8584	0.5153
Zoo	0.4648	0.3286
Yeast	0.1171	0.0896
Libras	0.4763	0.4052

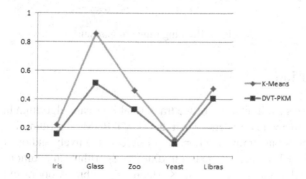

Fig. 3. The ratio of within class distance and between class distance

Table 3 and Fig 4 show that GPU can accelerate K - Means algorithm. However, if the number of cluster centers k is changed, it affects the number of times that thread accesses memory and the number of threads, as well as the running time. Therefore, we take the same value k on each dataset. The result shows that the speedup ratios are different based on the size of data sets, data dimensions and presented characteristics. When the number of data points reaches four hundred thousand, the speedup is 43 times. The acceleration is obvious. In the case of small size of data set, the speedup is not obvious due to the delay of GPU memory copies, starting up of device and data-completion. For the same size of dataset, as we make full use of the shared memory and register and avoid bank conflicts, our algorithm is slightly faster than other existing algorithms.

Table 3. Running time comparison

DataSet	K-Means	DVT-PKM	Speedup
Gas	2.570s	0.335s	7.672
Letter	99.737s	5.445s	18.317
Road Network	40.055s	1.648s	24.305
Road Network	81.677s	2.663s	30.671
Road Network	837.157s	19.571s	42.775

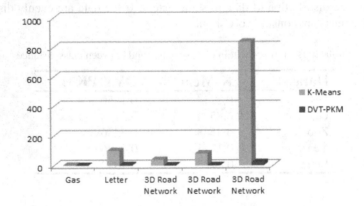

Fig. 4. Running time comparison

5   Conclusion

This paper analyzes the limitations of the traditional K-Means algorithm based on numerous studies by many researchers. Then the DVT-PKM algorithm is proposed. The improved algorithm can remove the impact of outliers effectively and overcome the sensitivity issues of initial centers selected in traditional algorithm. In addition, with the introduction of parallel implementation, the K-Means algorithm is optimized on GPU. The results show that the speedup can be improved considerably. It is proved that the proposed method can improve the stability and operation efficiency effectively on a single machine.

References

1. Wang, C.F., Tang, Y.Z.: Research of K-means Algorithm Combined with Neighbors And Density. Computer Engineering and Applications 47(19), 147–149 (2011)
2. MacQueen, J.: Some Methods for Classification And Analysis of Multivariate Observations. In: Proceedings of the Fifth Berkeley Symposium on Mathematics, Statistics and Science, pp. 281–297 (1967)
3. Cox, D.R.: Note on Grouping. Journal of the American Statistical Association, 543–547 (1957)
4. Fisher, W.D.: On Grouping for Maximum Homogeneity. Journal of the American Statistical Association, 789–798 (1958)
5. Sebestyen, G.S.: Decision Making Process in Pattern Recognition, p. 162. Macmillan, New York (1962)
6. Cheng, M.Y., Huang, K.Y., Chen, H.M.: K-Means Particle Swarm Optimization with Embedded Chaotic Search for Solving Multidimensional Problems. Applied Mathematics and Computation 219(6), 3091–3099 (2012)
7. Rajab, M.: Segmentation of Dermatoscopic Image by Frequency Domain Filtering And K-means Clustering Algorithms. Skin Research and Technology 17(4), 469–478 (2011)
8. Nunes, J., Madeira, M., Gazarini, L., et al.: A Data Mining Approach to Improve Multiple Regression Models of Soil Nitrate Concentration Predictions in Quercus Rotundifolia Montados(Portugal). Agroforestry Systems 84(1), 89–100 (2012)
9. Selim, S.Z., Ismail, M.A.: K-Means Type Algorithms: A Generalized Convergence Theorem and Characterization of Local Optimality. IEEE Transactions on Pattern Analysis and Machine Intelligence 6(1), 81–87 (1984)
10. Bagirov, A.M.: Modified Global K-means Algorithm for Minimum Sum-of-squares Clustering Problems. Pattern Recognition 41(10), 3192–3199 (2008)
11. Lee, W., Lee, S.S., An, D.-U.: Study of a Reasonable Initial Center Selection Method Applied to a K-Means Clustering. IEICE Transactions on Information and Systems 96(8), 1727–1733 (2013)
12. Khan, F.: An Initial Seed Selection Algorithm for K-means Clustering of Georeferenced Data to Improve Replicability of Cluster Assignments for Mapping Application. Applied Soft Computing 12(11), 3698–3700 (2012)
13. Zhang, S., Chu, Y.: High-performance Computing of GPU CUDA (2009)
14. Ryoo, S., Rodrigues, C.I., et al.: Optimization Principles and Application Performance Evaluation of a Multithreaded GPU Using CUDA. In: Proceedings of the 13th ACM SIGPLAN Symposium on Principles and practice of parallel programming, pp. 73–82 (2008)
15. Wu, J., Hong, B.: An Efficient k-means Algorithm on CUDA. In: IEEE International Symposium on Parallel and Distributed Processing Workshops and Phd Forum (IPDPSW), pp. 1740–1749 (2011)
16. Bai, H.T., et al.: K-means on Commodity GPUs with CUDA. In: 2009 WRI World Congress on Computer Science and Information Engineering, pp. 651–655 (2009)
17. Li, Y., Zhao, K., Chu, X., Liu, J.: Speeding up K-Means Algorithm by GPUs. Journal of Computer and System Sciences 79(2), 216–229 (2013)
18. Kijsipongse, E.: Dynamic Load Balancing on GPU Clusters for Large-scale K-Means Clustering. In: 2012 International Joint Conference on. Computer Science and Software Engineering (JCSSE), pp. 346–350. IEEE (2012)

A Multi-Intelligent Agent for Knowledge Discovery in Database (MIAKDD): Cooperative Approach with Domain Expert for Rules Extraction

Mohammed Abbas Kadhim[1,2], M. Afshar Alam[1], and Harleen Kaur[1]

[1] Department of Computer Science, Hamdard University, New Delhi, India
[2] College of Computer Science and Mathematics, University of Al-Qadisiyah, Iraq
moh_abbas74@yahoo.com, {mailtoafshar,harleen_k1}@rediffmail.com

Abstract. In last decade, autonomous intelligent agents or multi-intelligent agents and knowledge discovery in database are combined to produce a new research area in intelligent information technology. In this paper, we aim to produce a knowledge discovery approach to extract a set of rules from a dataset for automatic knowledge base construction using cooperative approach between a multi-intelligent agent system and a domain expert in a particular domain. The proposed system consists of several intelligent agents, each one has a specific task. The main task is assign to associative classification mining intelligent agent to deal with a database directly for rules extraction using Classification Based on Associations (CBA) rule generation and classification algorithm, and send them to a domain expert for a modification process. Then, the modified rules will be saved in a knowledge base which is used later by other systems (e.g. knowledge-based system). In other words, the aim of this work is to introduce a tool for extracting knowledge from database, more precisely this work has focused on produce the knowledge base automatically that used rules approach for knowledge representation. The MIAKDD is developed and implemented using visual Prolog programming language ver. 7.1 and the approach is tested for a UCI heart diseases dataset.

Keywords: KDD, data mining, intelligent agent, multi-intelligent agent, CBA algorithm, Multi-Intelligent Agent Knowledge Discovery in Database (MIAKDD).

1   Introduction

One of the essential challenges of Artificial Intelligence (AI) is automatic acquisition of knowledge from a large volume of data to support reasoning in intelligent information system. In construction knowledge-based systems the traditional knowledge extraction from its resources is the bottleneck of the approach. Database is one resources of knowledge for constructing knowledge base in decision support systems using automatic knowledge acquisition process. The concept of finding useful patterns in large data has been given different names including information discovery, knowledge extraction, data mining, information harvesting, data archeology, and data

D.-S. Huang et al. (Eds.): ICIC 2014, LNAI 8589, pp. 602–614, 2014.

pattern processing [1], [2]. Data mining is relatively a new concept of tools used to extract useful information from large databases. It is also can be used to discover a new healthcare knowledge from clinical databases or generate scientific hypotheses from experimental data [3], [4]. Fayyad, Shapiro, and Smyth view Knowledge Discovery in Database (KDD) as all processes of discovering useful knowledge (patterns) from data while data mining is a particular step in these processes, it includes applying specific algorithms for extracting useful patterns from volume of data. They defined KDD as "The nontrivial process of identifying valid, novel, potentially useful, and ultimately understandable patterns in data" [1], [2], while the data mining is defined as the extraction of useful patterns or models from observed data [5], or it is the process of extracting knowledge from dataset and transforms it into human understandable representation for further use [6]. This term apply on all process that associates with discovering useful patterns from data using a combination of techniques including neural network, statistical analysis, fuzzy logic, data visualization, and intelligent agent [7]. There are several kinds of data mining techniques, which are: clustering, classification, association, evolution, pattern matching, and meta-rule guided mining. These techniques can be used for extracting knowledge from different kinds of databases which involves transactional, relational, spatial, and active databases [8]. Association rule mining is one of the most popular data mining techniques used to discover frequent patterns in large database. It is suitable for discovering predictive rules that related with the set of attributes in dataset [9], [10]. It has been effectively used to build classification model to produce rules for describing relationships between attributes value and class attribute [11].

Intelligent agents or software agents are programmed software entities that carry out a series of operations on behalf of users or other programs with some degree of autonomy [12]. A Multi-Intelligent Agent (MIA) is a collection of autonomous agents which interact with each other or with their environments to achieve one or more objectives. Agents can achieve and support the knowledge discovery process. For example, agents can contribute in data selection, extraction, preprocessing, and acting as interfaces between humans and software systems [13], [14]. Data mining and multi-intelligent agent have been used for building a complex system, they are combined to produce automatic data mining system which is used to extract and discover a set of useful patterns from dataset using autonomous intelligent agents. In this paper, we proposed the development of a multi-intelligent agent system that is capable to convert raw data in a dataset to a set of rules and save it in a knowledge base for further use (e.g. knowledge-based systems or rule-based expert systems). The proposed MIAKDD consists of several autonomous intelligent agents: user interface agent, associative classification agent, control agent, and visualization agent. Each one has specific sub-task, the combination of the whole sub-tasks will produce the main objective of proposed multi-intelligent agent system.

The rest structure of this paper is organized as follows. In section 2 we present an overview of related works on using intelligent agent or multi-intelligent agents in data mining or knowledge discovery approaches. Section 3 discusses the architecture of the proposed MIAKDD in details. In section 4 and 5, we will discuss the experimental results and system evaluation on UCI heart diseases dataset respectively. Finally, section 6 describes the conclusions of this work.

2   Related Works

In this section, we will briefly review of the previously proposed studies in autonomous intelligent agent systems or multi-agent systems for data mining and knowledge discovery in database. The techniques used in these studies include association rule mining, associative classification mining, computational intelligence, rule generation algorithms, and greedy algorithm. The proposed MIAKDD is mainly related to two areas of research, knowledge extraction from large dataset and knowledge modeling using multi-intelligent agent system.

Popa, Pop, Negru, and Zaharie proposed a multi-agent based intelligent recommendation system called AgentDiscover. The aim of this system is to deal with the complexity of KDD processes and produce a tool to support researchers and non-expert users for exploring KDD method and looking for quick results in this field [15]. Warkentin, Sugumaran, and Sainsbury produce a study which is discuss the role of intelligent agents and data mining in electronic partnership management. The procedures of data mining used in this process can be enhanced by using intelligent agents [7]. Tong, Sharma, and Shadabi proposed a real time Data Mining and Multi-Agent System called DMMAS, it is heterogeneous data mining techniques and distributing agent based processing for modeling and improve efficiency by combining results from all the agents [16].

Ordonez, Ezquerra, and Santana used association rules in the medical domain, where datasets are generally small and high dimensional. They aim to discover only most important association rules and optimal search process. A greedy algorithm is used to compute rule covers in order to summarize rules having the same results. The evaluation of the most important association rules is processed by using three metrics: support, confidence and lift. The experiment results focus on discovering association rules on real dataset to predict existence or absence of heart disease [9]. Nahar, Imam, Tickle, and Chen discussed a paper in which they used association rule mining and a computational intelligence to identify the factors which contributes to heart diseases for males and females. This research presents rule extraction experiments on heart disease data using three rule generation algorithms Apriori, Predictive Apriori and Tertius [17]. F. A. Thabtah and P.I. Cowling proposed a new associative classification technique, Ranked Multilabel Rule (RMR) algorithm which generates rules with multiple labels. Rules derived by current associative classification algorithms overlap in their training objects, resulting in many redundant and useless rules. The proposed algorithm solves the overlapping problem between rules in the classifier by generating rules that does not share training objects during the training phase, resulting in a more accurate classifier [18]. Kaur, Chauhan, Alam, Aljunid, and Salleh produced a spatial data mining techniques which is applied to extract implicit knowledge from spatial attributes. These techniques are applied in different fields such as healthcare, marketing, environmental and remote sensing databases to improve planning and decision making process [19].

3   Proposed MIAKDD System Architecture

In this section, we present the overall proposed system architecture as shown in fig. 1 which consists of four autonomous intelligent agents, a dataset file, and a knowledge base. The user interface agent interact with a human (user or domain expert) in order to allow him/her to determine user specifications. The control agent is responsible for managing and controlling for data transmission among the other components of the proposed system. The main function of associative classification mining agent is extracting or discovering knowledge from dataset using CBA algorithm for rules generation and classification. The visualization agent is responsible for displaying all the constructed rules which are extracted by associative classification mining agent. The dataset is saved in the database file and used by rule generation algorithm for extracting rules and save them in the knowledge base for later use by diagnostic systems (e.g. rule-based expert system for back pain diagnosis [20]). The following subsections describe in details the components of the proposed MIAKDD architecture.

3.1   User Interface Agent

The user interface intelligent agent task is interact with users or domain experts to determine a set of specifications such as name and number of attributes, number of records, class attribute for some cases, minimum support threshold value, and other detail information of the processed dataset. It is responsible for receiving the set of specifications from environment and send to the control agent for another process. The interface agent is user friendly and hide the complexity of other components of the multi-intelligent agent system. The multi-intelligent agent system success may be determined by the nature of its interface agent. It interacts with its environment using questions and answers and menu driven techniques to achieve the main goal of that agent. The user or expert should enter all the attributes of the dataset and should separate between attributes and a class attribute as well as the number of records he/she want to process in that dataset.

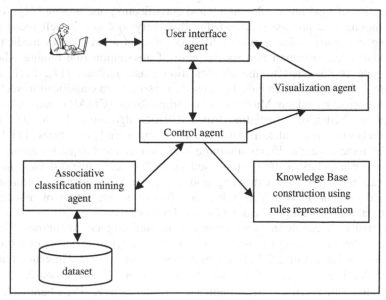

Fig. 1. Architecture of proposed MIAKDD system

3.2   Control Agent

The main function of control intelligent agent is managing the communication for all components of proposed MIAKDD and control the data transmission among them. It is responsible for transferring the user specifications from user interface agent as output and sending it to associative classification mining agent as input for mining process. Then, send back all mined rules from associative classification mining agent to user interface agent for modification process (if necessary) by domain expert and save them in knowledge base, at the same time it sends a copy to visualization agent to display final form of these rules. The control agent transfers attributes and a class attribute as input to a rule generation algorithm in the associative classification mining agent and send a set of rules as antecedence-consequence pairs to the interface agent for testing and modifying by the domain expert to ensure those rules are correct and we can depend on contains of the knowledge base for another usage. That means, the proposed MIAKDD will be producing automatically a knowledge base with high dependency and reliability.

3.3   Associative Classification Mining Agent

Most exist works have considered the dataset as a classification problem. We view in this paper, the dataset is not only a classification problem but also a knowledge extraction problem. The most important step in MIAKDD system is rules extraction from dataset. This task is a responsibility of associative classification mining agent as well, carrying out the data mining operation. Data mining is a very important activity of the KDD process which consists of four different steps; selection, preprocessing, data mining, and interpretation [2], [21]. This agent contains a specific algorithm for rules generation and classification processes. Associative classification is the integration of associative rule mining and classification rule mining [22]. Association rule mining is a promising knowledge discovery tool and a well recognized data mining procedure which is widely used in different areas such as medical domain. Associative classification is a special case of association rule mining whose right-hand-side is restricted to the classification class attribute [11], [22]. There are different algorithms which have been used for association classification such as CBA, Classification Based on Multiple Association Rules (CMAR), and Classification Based on Predictive Association Rules (CPAR) algorithms. Each one use several approaches to extract rules, discover rules, rank rules, and prune rules [11], [18].

The associative classification mining is chosen as a technique for association rules generation and classification in proposed agent. This technique is domain independent and can be applied for extracting and modeling knowledge in different area of medical domain dataset. This subsection focus on integration of association rule mining and classification to extract knowledge from dataset.

Formally we can define association rule mining problem as follows [9]; let $I=\{i_1, i_2,..., i_m\}$ be a finite set of m items, let $T=\{t_1, t_2, ..., t_n\}$ be a dataset of n transactions, where each transaction $t_i \subseteq I$ ($1 \leq i \leq n$) represent a set of items. In general, a subset of I is called itemset. Let X,Y be two itemsets, where $X,Y \subset I$ and $X \cap Y = \emptyset$. An association rule is defined as an implication of the form $X \rightarrow Y$, where X is called the antecedent (or body) and Y is called the consequent (or head) of the rule. The support (sup(X)) of itemset X is defined the number of transaction that contain X, P(X) can be estimated as P(X)=sup(X). The support of a rule X→Y is defined as sup(X ∪ Y)=P(X ∪ Y). The measure of reliability of a rule X→Y is called confidence, which is defined as confidence (X→Y)= sup(X ∪ Y)/sup(X). The problem of association rule mining is to discover all association rules in database having support more than a user defined threshold minimum support and a confidence more than a user defined threshold minimum confidence, these values are two measures of rule usefulness and certainty [6], [23], [24].

The associative classification mining agent received user specifications from control agent and it performs the following four tasks; the first task includes selecting a dataset or data samples in which the discovery will be performed. The second task is preparing the dataset for processing by matching the user specification with a database file and remove all unwanted records from that file and open a new file contains all wanted records to extract association rule from these records in the next step. The next task is attributed the data mining (whether clustering, classification, association rule mining, etc.) and data modeling is completed by discovering patterns in large dataset using data mining algorithm. In the proposed MIAKDD we used a CBA algorithm to extract rules from dataset. It consists of two parts: a Rule Generator (CBA-RG) and a Classifier Builder (CBA-CB). In the rule generator part, the CBA algorithm generates all the association rules using Apriori algorithm which are known as candidate rules. In the classifier builder part, the CBA algorithm build a classifier which focuses on a special subset of candidate rules whose right hand side is constrained to the classification class attribute. This subset of association rules is called Class Association Rules (CARs). Finally, the last task includes interpreting the discovered patterns and representing them using rules representation method. Although the above four tasks are the function of KDD process, the main intelligent agent is called associative classification mining agent because of associative classification mining is the most important activity in proposed MIAKDD system. The most important activity in associative classification mining agent consists of the following steps:

1- Suppose we have a database which consists of N transactions and specify the user-definition threshold values (minimum support and minimum confidence).
2- Find all frequent itemsets with their supports more than minimum support threshold value using Apriori algorithm.
3- Using frequent itemsets result from step 2 to generate association rules (candidate rules) with their confidence more than minimum confidence threshold value.
4- Building a classifier based on the generated association rules (candidate rules) which results from step 3

The Apriori finds frequent itemsets in a database transactions and having at least a user specified threshold. For each iteration in the algorithm attempts   specify the set of frequent patterns with k-itemsets from previous frequent (k-1)-itemsets, and then tests the support threshold value of candidates to find frequent k-itemsets. This algorithm scans several passes on the database. The correctness and effectiveness of frequent itemsets depends on Apriori property which means any subset of frequent itemsets must be frequent [25]. In the input of Apriori algorithm we refereed to dataset consists of more or less items. The output of the algorithm is a set of association rules describe the relationships that these items have in their sets [6].

3.4   Visualization Agent

The visualization agent is responsible to display all extracted and modified rules by proposed intelligent agents. It receives extracted and modified rules from control agent and display them as a model of antecedence-consequence pairs which are mined from dataset by associative classification mining agent and modified by the domain expert. This agent contains predefined template which is consists of three parts: antecedence, consequence and treatment. The antecedence or condition part is represented the attributes (symptoms) which is extracted from dataset. The consequence or action part represents class attribute (disease name) which is extracted from dataset or added by domain expert during modification process if it doesn't exist in dataset. The treatment of that disease is added by domain expert because it almost doesn't exist in dataset.

3.5   Dataset

A dataset is a collection of data usually presented as a table form, which consists of rows and columns. The column represents a particular variable and the row corresponds to a member of the dataset. There are many datasets already available in websites on the Internet which are used for scientific researches especially for data mining field such as UCI machine learning repository [26], KEEL dataset repository, and SEER dataset. We can also construct a dataset by collecting data from different organizations such as companies, hospitals, statistical data centers and so on and use it for scientific researches. In our case study of MIAKDD we used UCI heart disease dataset for knowledge discovery.

3.6   Knowledge Base Construction

The proposed MIAKDD system will produce automatically the knowledge base which is contain up to date rules in a particular domain because it is constructed by cooperation between association rule mining agent and domain expert. Once the rules have been discovered by associative classification mining agent, the control agent send them to interface agent for modification process by the domain expert. The modification process includes display all extracted rules to the domain expert for adding, deleting, and modifying the components of these rules. For modification process, we are preparing a preformatted template which consists of three parts: the consequence or action part which represents disease name, the antecedence or condition part which represents the symptoms of that disease, and the treatment part which represents the treatment of that disease.

4   Experimental Results

In this section, we will apply the proposed system on a real dataset as a case study to illustrate how the proposed MIAKDD system can be extracted a set of rules from real dataset to construct knowledge base automatically. We use the publicly available UCI heart disease dataset which consists of 76 attributes but most studies use a maximum of 14 attributes as follows [24]:

1. Age: age in years; numeric.
2. Sex: sex (male=1; female=0).

3. CP: chest pain type; 4 values (typical angina=1; atypical angina=2; non-anginal pain=3; asymptomatic=4).
4. Trestbps: resting blood pressure (in mm Hg on admission to the hospital); numeric.
5. Chol: serum cholesterol in mg/dl; numeric.
6. Fbs: 2 values; fasting blood sugar > 120 mg/dl?; (true=1; false=0).
7. Restecg: resting electrocardiographic results; 3 values (normal=0; ST-T wave abnormality=1; probable or definite left ventricular hypertrophy by Estes' criteria=2).
8. Thalach: maximum heart rate achieved; numeric.
9. Exang: exercise induced angina; 2 values (yes=1; no=0).
10. Oldpeak: ST depression induced by exercise relative to rest; numeric.
11. Slope: the slope of the peak exercise ST segment; 3 values (upsloping=1; flat=2; downsloping=3).
12. Ca: number of major vessels (0-3) colored by fluoroscopy; numeric.
13. Thal: the heart status; 3 values (normal=3; fixed defect=6; reversible defect=7).
14. Num: diagnosis of heart disease (angiographic disease status) or the class attribute value is either healthy class or existence of heart disease (sick class) (0;1).

The user or domain expert enter all the attributes for heart disease UCI dataset such as Age, Sex, CP, Trestbps, ..., etc. and enter number of records, in our case study we select 100 records for process. The control agent transfers those 13 attributes (Age, Sex, CP, Trestbps, ...,etc) and a class attribute (Num) as input to CBA algorithm in associative classification mining agent which consists of two phases:

I-CBA-RG: In rules generation phase, we applied the Apriori algorithm on UCI heart disease dataset, we obtained the following sample of association rules in Prolog programming language:

```
disease_name("class sick"):-
chest_pain_type(4),cholesterol(X),X>=225,X=<310,resulting_ecg(Y),(Y=0;Y=
1),exercise_induced_angina(1),
slope(2).
disease_name("class sick"):-
chest_pain_type(4),resulting_ecg(Y),(Y=0;Y=1),exercise_induced_angina(1)
,slope(2),heart_status(7).
disease_name("class healthy"):-
chest_pain_type(3),chol(X),X>=225,X=<310,resulting_ecg(2),oldpeak(1.24),
slope(1).
disease_name("class sick"):- chest_pain_type(4),fasting_blood_suger(0),
   exercise_induced_angina(1),heart_status(7).
   disease_name("class healthy"):-
   ca(0),thal(3)slope(1).
```

The above rules consists of two parts, antecedents or conditions part (set of symptoms) on Right Hand Side (RHS) and consequence or action part which has two classes (sick or healthy) on Left Hand Side (LHS). The confidence threshold values of above rules are between (0.9-1). These rules are association rules (candidate rules) with certain support and confidence threshold values.

II-CBA-CB: In this phase, we constructed a classifier to select an appropriate subset of association rules which is satisfy our objective of proposed agent. That means, the rules with class sick on LHS for the above candidate rules have been selected as a class, observe the follow rules in Prolog code:

```
disease_name("class sick"):-
chest_pain_type(4),cholesterol(X),X>=225,X=<310,resulting_ecg(Y),(Y=0;Y=
1),exercise_induced_angina(1),
slope(2).
disease_name("class sick"):-
chest_pain_type(4),resulting_ecg(Y),(Y=0;Y=1),exercise_induced_angina(1)
,slope(2),heart_status(7).
disease_name("class sick"):- chest_pain_type(4),fasting_blood_suger(0),
   exercise_induced_angina(1),heart_status(7).
```

But these rules are before the modification process which discussed In section III-F to capture correct and complete knowledge base. We will take two rules as an example for modification process, Table 1 illustrated the components of that template before and after modification process.

Table 1. Components of a preformated template
A: Before modification process

Consequence	Antecedence	Treatment
class sick	chest_pain_type=asymptomatic and cholesterol=225-310 and resulting_ecg=normal or abnormal and exercise_induced_angina=true and slope=flat	-
class sick	chest_pain_type= asymptomatic and resulting_ecg= normal or abnormal and exercise_induced_angina=true and slope=flat and heart_status=reversible defect	-

The modification process in our case study include change each consequence has "class sick" with real heart disease name that satisfy antecedence part (symptoms) of that rule. The domain expert can add, delete, and modify the contents of antecedence part (if necessary). Finally, he/she will add a treatment for each disease name in these rules.

B: After modification process

Consequence	Antecedence	Treatment
cardiovascular disease	chest_pain_type=asymptomatic and cholesterol=225-310 and resulting_ecg=normal or abnormal and exercise_induced_angina=true and slope=flat	Cholesterol-modifying medications. By reducing the cholesterol in the blood, you can choose from a range of medications, such as statins, niacin, fibrates and bile acid sequestrants. or aspirin taking a daily or other blood thinner.
cardiac arrhythmia	chest_pain_type= asymptomatic and resulting_ecg= normal or abnormal and exercise_induced_angina=true and slope=flat and heart_status=reversible defect	Medicine control arrhythmias and produce treatment related with symptoms such as high blood pressure or a pacemaker that uses batteries to help heart beat more regularly or Surgery

The rules yielded after modification process will be saved in knowledge base by control agent and it will send one copy of rules to visualization agent for displaying. The constructed knowledge base can be divided into two parts, the condition-action part and the treatment part. The condition-action part contains all rules that are discovered and modified as condition-action pairs. The second part of the knowledge base is disease's treatment and it is related with the condition-action part by the disease name. The following rules illustrated Prolog language codes after modification process and they will be saved in knowledge base files:

```
disease_name("cardiovascular disease"):-
chest_pain_type(4),cholesterol(X),X>=225,X=<310,resulting_ecg(Y),(Y=0;Y=
1),exercise_induced_angina(1),slope(2).
disease_name("cardiac arrhythmia"):-
chest_pain_type(4),resulting_ecg(Y),(Y=0;Y=1),exercise_induced_angina(1)
,slope(2),heart_status(7).
treatment("cardiovascular disease", "Cholesterol-modifying medications.
By reducing the cholesterol in the blood, you can choose from a range of
medications,   such   as   statins,   niacin,   fibrates   and   bile   acid
sequestrants. Or aspirin taking a daily or other blood thinner.").
treatment("cardiac arrhythmia","Medicine control arrhythmias and produce
treatment  related  with  symptoms  such  as  high  blood  pressure  or  a
pacemaker  that  uses  batteries  to  help  heart  beat  more  regularly  or
Surgery").
```

Now we capture a complete knowledge base which results from cooperation between associative classification mining agent and domain expert using rules method for knowledge representation. The knowledge base which is constructed automatically can be used in expert system tools to construct expert system in that domain (domain of knowledge base) [27].

5   System Evaluation

In this section, we will focus on the associative classification mining agent's performance because it has the main function of the proposed MIAKDD system. To evaluate the performance of that agent, we requested from a knowledge engineer to extract a set of rules from same UCI heart disease dataset file. The purpose of this experiment was to analyze the ability of the proposed associative classification mining agent to extract correct knowledge and compare the results with the knowledge engineer's performance. After that, four domain experts were asked to evaluate the performance of the associative classification agent and the knowledge engineer. Table 2 illustrates the accuracy scores shown by four domain experts compared with the associative classification agent and knowledge engineer performance across the UCI heart disease dataset file.

Table 2. The accuracy shown by four domain experts compared with the associative classification agent and knowledge engineer performance

Domain expert	Associative Classification Agent (ACA)	Knowledge Engineer (KE)
Domain expert 1	80%	90%
Domain expert 2	75%	85%
Domain expert 3	70%	83%
Domain expert 4	68%	75%
average	73.25%	83.25%

From Table 2, we can calculate average accuracy of performance produced by four domain experts. Based on the results in table 2, we can conclude the performance of associative classification agent is near to the knowledge engineer's performance. That means, the knowledge base which was produced by two of them is similar.

6   Conclusions

In this paper, we propose a multi-intelligent agent system has presented the rules extraction from medical dataset using CBA algorithm for association rule generation and classification. We also produce a cooperative approach between associative classification mining agent and domain expert to capture complete and consistence knowledge base automatically. The modification process is very important step because of the domain expert review, check, and modify all discovered rules to produce effective and reliable knowledge base. The proposed multi-agent system can speed up the construction of knowledge base by decreasing the required time for domain expert to explain his expertise to a knowledge engineer for constructing knowledge in traditional method. In other words, the automatic knowledge acquisition approaches can reduce the required time for constructing of the knowledge base. Based on experimental results and system evaluation, we proved that the knowledge base which was produced automatically by the proposed agent is similar to the knowledge base which was produced manually by the knowledge engineer.

Acknowledgement. The first author wish to thank Ministry of Higher Education and Scientific Research, and University of Al-Qadisiyah, Iraq for supporting me during the research study. The authors wish to acknowledge to the four domain experts (heart disease doctors) in Max hospital for their roles in the system evaluation.

References

[1] Fayyad, U., Shapiro, G.P., Smyth, P.: Knowledge Discovery and Data Mining: Towards a Unifying Framework. In: Proceeding of the Second International Conference on Knowledge Discovery and Data Mining (KDD 1996), pp. 82–88. AAAI Press, Portland (1996)

[2] Fayyad, U., Shapiro, G.P., Smyth, P.: The KDD process for Extracting Useful Knowledge from Volumes of Data. Communication of the ACM 39(11), 27–34 (1996)

[3] Herawan, T., Deris, M.M.: A soft set approach for association rules mining. Knowledge-Based Systems 24(1), 186–195 (2011)

[4] Yoo, I., Alafaireet, P., Marinov, M., Hernandez, K., Gopidi, R., Chang, J., Hua, L.: Data Mining in Healthcare and Biomedicine: A survey of the literature. J. Med. Syst. 36, 2431–2448 (2012)

[5] Marcano-Cedeno, A., Chausa, P., Garcia, A., Caceres, C., Tormos, J., Gomez, E.: Data mining applied to the cognitive rehabilitation of patients with acquired brain injury. Expert Systems with Applications 40(4), 1054–1060 (2013)

[6] Foguem, B.K., Rigal, F., Mauget, F.: Mining association rules for the quality improvement of the production process. Expert Systems with Applications 40(4), 1034–1045 (2013)

[7]   Warkentin, M., Sugumaran, V., Sainsbury, R.: The role of intelligent agents and data mining in electronic partnership management. Expert Systems with Applications 39(18), 13277–13288 (2012)

[8]   Liao, S., Chu, P., Hsiao, P.: Data mining techniques and applications – A decade review from 2000 to 2011. Expert Systems with Applications 3(12), 11303–11311 (2012)

[9]   Ordonez, C., Ezquerra, N., Santana, C.: Constraining and summarizing association rules in medical data. Knowl. Inf. Syst. 9(3), 259–283 (2006)

[10] Ordonez, C., Omiecinski, E., Braal, L., Santana, C., Ezquerra, N., Taboada, J., Cooke, D., Krawczynska, E., Garcia, E.: Mining Constrained Association Rules to Predict Heart Disease. In: Proceedings of the IEEE International Conference on Data Mining (ICDM 2001), California, pp. 433–440 (2001)

[11] Thabtah, F.: Rules pruning in associative classification mining. In: Proceedings of the IBIMA Conference, Cairo, Egypt, pp. 7–15 (2005)

[12] Duan, Y., Ong, V.K., Xu, M., Mathews, B.: Supporting decision making process with "ideal" software agents – What do business executives want? Expert Systems with Applications 39(5), 5534–5547 (2012)

[13] Cao, L., Gorodetsky, V., Mitkas, P.: Agent Mining: The Synergy of Agents and Data Mining. IEEE Intelligent Systems, 64–72 (May/June 2009)

[14] Kadhim, M.A., Alam, M.A., Kaur, H.: A Multi-intelligent Agent Architecture for Knowledge Extraction: Novel Approaches for Automatic Production Rules Extraction. International Journal of Multimedia and Ubiquitous Engineering 9(2), 95–114 (2014)

[15] Popa, H., Pop, D., Negru, V., Zaharie, D.: AgentDiscover: A Multi-Agent System for Knowledge Discovery from Databases. In: Ninth International Symposium on Symbolic and Numeric Algorithms for Scientific Computing, pp. 275–281. IEEE, Timisoara (2008)

[16] Tong, C., Sharma, D., Shadabi, F.: A Multi-Agents Approach to Knowledge Discovery. In: IEEE/WIC/ACM International Conference on Web Intelligence and Intelligent Agent Technology, Sydney, pp. 571–574 (2008)

[17] Nahar, J., Imam, T., Tickle, K., Chen, Y.: Association rule mining to detect factors which contribute to heart disease in males and females. Expert Systems with Applications 40(4), 1086–1093 (2013)

[18] Thabtah, F.A., Cowling, P.I.: A greedy classification algorithm based on association rule. Applied Soft Computing 7(3), 1102–1111 (2007)

[19] Kaur, H., Chauhan, R., Alam, M. A., Aljunid, S., Salleh, M.: SpaGRID: A Spatial Grid Framework for High Dimensional Medical Databases. In: Corchado, E., Snášel, V., Abraham, A., Woźniak, M., Graña, M., Cho, S.-B. (eds.) HAIS 2012, Part I. LNCS, vol. 7208, pp. 690–704. Springer, Heidelberg (2012)

[20] Kadhim, M.A., Alam, M.A., Kaur, H.: Design and Implementation of Fuzzy Expert System for Back pain Diagnosis. International Journal of Innovative Technology & Creative Engineering 1(9), 16–22 (2011)

[21] Kaur, H., Chauhan, R., Aljunid, S.: Data Mining Cluster analysis on the influence of health factors in Casemix data. BMC Journal of Health Services Research 12(suppl. 1), O3 (2012)

[22] Das, K., Vyas, O.P.: A Comparative Study of Four Feature Selection Methods for Associative Classifiers. In: International Conference on Computer & Communication Technology (ICCCT 2010), pp. 431–435 (September 2010)

[23] Agrawal, R., Srikant, R.: Fast Algorithms for Mining Association Rules. In: Proceeding of the 20th VLDB Conference, Santiago, pp. 487–499 (1994)

[24] Kaur, H., Chauhan, R., Alam, M.: Spatial Clustering Algorithm using R-tree. Journal of Computing 3(2), 85–90 (2011)

[25] Kuo, R.J., Lin, S.Y., Shih, C.W.: Mining association rules through integration of clustering analysis and ant colony system for health insurance database in Taiwan. Expert Systems with Applications 33(3), 794–808 (2007)

[26] UCI, Heart disease dataset (2010),
http://archive.ics.uci.edu/ml/datasets/Heart+Disease
(accessed October 03 2013)

[27] Kadhim, M.A., Alam, M.A.: To Developed Tool, an Intelligent Agent for Automatic Knowledge Acquisition In Rule-based Expert System. International Journal of Computer Applications 42(9), 46–50 (2012)

The Role of Pre-processing in Twitter Sentiment Analysis

Yanwei Bao[1], Changqin Quan[1], Lijuan Wang[1], and Fuji Ren[1,2]

[1] AnHui Province Key Laboratory of Affective Computing and Advanced Intelligent Machine,
School of Computer and Information, HeFei University of Technology, Hefei, 230009, China
baoyanwei007@sina.com, quanchqin@gmail.com,
wlj1034900987@163.com
[2] Faculty of Engineering, University of Tokushima, 2-1 Minami-Josanjima,
Tokushima 770-8506, Japan
ren@is.tokushima-u.ac.jp

Abstract. Recently, increasing attention has been attracted to Social Networking Sentiment Analysis. Twitter as one of the most fashional social networking platforms has been researched as a hot topic in this domain. Normally, sentiment analysis is regarded as a classification problem. Training a classifier with tweets data, there is a large amount of noise due to tweets' shortness, marks, irregular words etc. In this work we explore the impact pre-processing methods make on twitter sentiment classification. We evaluate the effects of URLs, negation, repeated letters, stemming and lemmatization. Experimental results on the Stanford Twitter Sentiment Dataset show that sentiment classification accuracy rises when URLs features reservation, negation transformation and repeated letters normalization are employed while descends when stemming and lemmatization are applied. Moreover, we get a better result by augmenting the original feature space with bigram and emotions features. Comprehensive application of these measures makes us achieve classification accuracy of 85.5%.

Keywords: Twitter, Sentiment Analysis, Pre-processing.

1 Introduction

Twitter is one of the most fashional micro-blog service platforms where registered users can update their messages and follow others' statuses expediently. Twitter's contents (named tweets) include users' behaviors, states of mind, comments on certain topics etc, and a lot of these contents express the users' sentiments unavoidably. Twitter sentiment analysis is a significative topic because the research findings can feedback much important information we have never thought of (e.g. the trend of the stock market [3, 4]).

Under routine circumstances, sentiment analysis is researched as a problem of classification. This technique is also known as opinion mining aiming to identify the kind of emotions, mainly as positive, neutral or negative at present. There have been a large amount of researches in the area of sentiment classification. Applications utilizing sentiment classification technique might be classified into two main categories: trends prediction [3, 4, 6, 7, 8, 10] and products recommendation [11, 12, 13, 18, 19, 20, 21].

D.-S. Huang et al. (Eds.): ICIC 2014, LNAI 8589, pp. 615–624, 2014.
© Springer International Publishing Switzerland 2014

The works mentioned above are all based on sentiment analysis. However, these works didn't focus on pre-processing methods like [5] did, who discussed the role of pre-processing for reviews. Although pre-processing for Twitter data was used for dimensionality reduction in [1, 2, 16], there is no work specialising in researching the role of pre-processing in Twitter sentiment analysis. In this paper, we concentrate on exploring pre-processing methods to elevate the performance of sentiment analysis besides dimensionality reduction. In our work, we show the role of pre-processing in Twitter sentiment classification, including the effects of URLs, negation, repeated letters, stemming and lemmatization. Experimental results on the Stanford Twitter Sentiment Dataset show that sentiment classification accuracy rises when URLs features reservation, negation transformation and repeated letters normalization are employed while descends when stemming and lemmatization are applied. We also augment the original feature space with bigram and emotions features. All positive-impact methods are applied on the Stanford Twitter Sentiment Dataset comprehensively and we achieve the sentiment classification accuracy of 85.5%.

The rest of the paper is organized as follows. Section 2 outlines existing works on Twitter sentiment analysis. Section 3 describes the methodology we use in this paper. Experiments and results are discussed in Section 4. Finally, we conclude our work and outline future work in Section 5.

2   Related Work

A great deal of researches have been done in the field of sentiment analysis. Several works have pursued a study in this dimension with a particular emphasis on products reviews[2, 11, 12, 13].

The above works are focused on products reviews, which are always well-organized sentences and relate to a particular domain. In contrast, tweets, whose length is limitted to 140 characters, are casual in the language style and multifarious in fields and themes. Moreover tweets contain a large amount of noises, such as hashtags, URLs, and emotions. These characters make Twitter sentiment analysis a challenging assignment.

Researches [1, 3, 4, 6, 7, 8, 14, 15, 16] are working on Twitter sentiment classification. Researches [3, 4, 6, 7, 8] are good examples of Twitter sentiment classification based application. In these works, the sentiment orientations of online information like tweets and reviews were analyzed to predict the trends of the stock market and election or outcomes of the box office.

Go et al. [1] used emotions as noisy labels to label tweets as positive and negative. They set a corpus containing 1.6 million tweets (the Stanford Twitter Sentiment Dataset[1]), which can be obtained by public. Based on this dataset, they continued to train classifiers as Pang et la. [2] did. The best result was 83% reported by MaxEnt Classifier using a combination of unigrams and bigrams as features.

---

[1] http://twittersentiment.appspot.com/

Under normal circumstances, the first step of sentiment analysis is text pre-processing [1, 2, 5, 16], especially for tweets. Haddi et al. [5] explored the role of text pre-processing in online text sentiment analysis. Agarwal et al. [16] applied some novel methods to pre-process tweets. The result of their experiment illustrated that appropriate text pre-processing methods can significantly increase the classifier's performance. There have also been some researches like [14], who excavated the original features to improve the performance of sentiment classification.

3   Methodology

We follow the following procedures for Twitter sentiment classification.

3.1   Denoising

For visual differentiation, we call pre-processing of this step as denoising. We take the following measures to denoise:

Username There are usernames like "@Tonny", that starts with symbol "@". The usernames indicate who is the information pointing to, i.e. the target. We replace all usernames with a space.

Hashtags Hashtags, marked by symbol "#", mean that the tweets are associated with the particular topics. We replace all hashtags with a space as well.

Emotions Users are used to express sentiment with emotions, e.g. ":)" means happy or other positive affections. Emotions with obvious emotional colouring ar e persisted while others are replaced with a space also.

Others We delete all digital symbols, single letters, punctuation except "'" in "won't", "can't" and "n't" and other non-alphabetic symbols.

3.2   Pre-processing

In this step, five steps are proposed to process tweets, named as URLs features reservation, negation transformation, repeated letters normalization, stemming and lemmatization. We verify the role of URLs in Twitter sentiment analysis by observing the change of classification accuracy when URLs features are reserved or not. Considering the significant impact of negation on sentiment classification, we put forward four kinds of transformations to propose negation in tweets and explore which is the most effective way to improve Twitter sentiment classification accuracy.

Words with repeated letters, e.g. "cooooold", are common in tweets, and people tend to use this way to express their sentiments. Thus it is necessary to deal with these words to make them more formal. Here, repeated letter means a letter emerges more than 3 times consecutively in a word.

In this work, we propose a method based on lexicon to cope with these words. We first build a lexicon using words provided by WordNet[2] and other frequent terms.

---

[2] http://wordnet.princeton.edu/wordnet/

And then, we perform the following conditional statement. If a word contains repeated letters, we replace repeated letters with two occurrences and output the processed word if the lexicon contains the processed word, else we replace repeated letters with one occurrence and output the result.

Both of stemming and lemmatization are used to achieve feature reduction. Stemming is the process for reducing inflected words to their stem, base or root form, for example, "stemmer", "stemming", and "stemmed" all can be reduced into "stem". Lemmatization is the process for converting inflected words to their root form, for example, "drove" can be converted into "drive".

3.3   Features Selection

All of the unigram features and part of bigram features are used to structure the feature space. TF, IG, and χ2 Statistics are three standards we apply to select bigram features.

TF means the frequency of a feature appears in the corpus, and more frequent more important.

IG, i.e. information gain, is used to evaluate the importance of features on the classification system. The standard of IG is the amount of information feature contributes, more information, more important. We use the formula (1) to calculate the information gain of features:

$$IG(T)=H(C)-H(C/T)=$$

$$-\sum_{i=1}^{n}P(C_i)\log_2^{P(C_i)} + P(t)\sum_{i=1}^{n}P(C_i/t)\log_2^{P(C_i/t)} + P(\bar{t})\sum_{i=1}^{n}P(C_i/\bar{t})\log_2^{P(C_i/\bar{t})} \qquad (1)$$

Where IG(T) is the information gain of feature t, H(C) is the entropy of class C, H(C/T) is the conditional entropy of C in the condition of feature T. $P(C_i)$ is the probability of class C_i occurrence in the corpus, P(t) is the probability of t, $P(C_i/t)$ is the probability of C_i in the condition of t occurrence, and $P(C_i/\bar{t})$ is the probability of C_i in the condition of t absence. What needs to be pointed out is that we use Laplace-like smoothing method when we calculate the probability mentioned above to avoid zero probability and zero denominator.

χ^2 Statistics is the measurement of interdependency between feature and class. Interdependency and χ^2 Statistics are proportional relations. The value of χ^2 Statistics between feature t and class C_i is calculated by the formula (2):

$$\chi^2(t, C_i) = \frac{(AD-BC)}{(A+B)(C+D)} \qquad (2)$$

Where N is the total number of tweets in the corpus, A represents the number of tweets belong to class C_i and contains feature t, B is the number of tweets don't belong to class C_i but contain feature t, C is the number of tweets belong to C_i but don't contain feature t, and D means the number of tweets when both are negative.

3.4   Classification

In this work, we employ Liblinear[3] to train a linear classifier. Liblinear is able to handle large-scaled dataset, especially for text classification, whose features space is extremely sparse. And it runs really faster than Libsvm because it doesn't have to compute the kernel for any two points. For optimization, trust region Newton method is applied, which is the combination of trust region method and truncated Newton method. We can see the mathematics foundations of Liblinear in [17].

4   Experiments and Results

In this section, we present the results obtained when several kinds of pre-processing methods are applied individually and jointly. After that, we compare the results with the existing approaches.

4.1   Corpus

The experiments are carried on the Stanford Twitter Sentiment Dataset. The dataset was collected by using emotions as noisy labels to label tweets as positive and negative. For the training set, there are 800,000 positive tweets and 800,000 negative tweets. A set of 177 negative tweets and 182 positive tweets were manually marked, for a total of 359 test tweets.

4.2   Experiments

In the experimental stage, we take the safe-launch mode to carry out the experiment. Alleged safe-launch mode means that we implement the experiment step by step, and we choose the better result in each step. Then the next step will be conducted based on the previous better result. If the effect of a certain pre-processing method is not so good, we will abandon this method.

We use WEKA[4] to perform Liblinear classification. We set the parameter S to be 0 meaning L2-regularized logistic regression, which is appropriate for binary classification. Other parameters are set to be the default values.

We start the experiments with URLs features reservation and negation transformation. In order to verify the effectiveness of processing, we establish four datasets using unigram features named as: **UN0** for the dataset after denoising. **UN1** when dataset contains URLs features and negation is processed, **N** when dataset contains URLs features while negation is not processed, **U** when all URLs features are deleted and negation is not processed.We use the method in **UN0** as our baseline. Here, the method used to process negation is transforming "won't","can't" and "n't" into "will not", "can not" and " not", respectively.

---

[3] http://www.csie.ntu.edu.tw/~cjlin/liblinear/
[4] http://weka.wikispaces.com/

Table 1. The role of URLs and negation

	U	UN1	N	UN0
Accuracy	82.73%	83.01%	81.62%	81.62%

Table 1 indicates the effect of URLs and negation. The result does be better when URLs and negation are processed. The accuracy grows by 0.28% when URLs features are utilized. That is because URLs may contain some helpful features. The effect of negation is more powerful for the accuracy drops down to 81.62% when negation is not processed. Note that, the result of **UN1** is similar to Go [1] while we use 308306 unigram features only.

In consideration of the effect of negation, we propose another three methods to deal with negation:

UN2 transforming all verb negatived by "not" or its abbreviations into "verb not", e.g. "didnt" to "did not".

UN3 transforming "won't","can't" and "n't" into "willnot", "cannot" and "not", respectively.

UN4 transforming "won't","can't" and "n't" into "wont", "cant" and "nt", respectively.

Table 2 shows the result of four datasets when unigram features are used only. **UN3** and **UN4** get the best result, followed by **UN1**, and **UN2** gets the worst result. This result may be due to the fact that **UN3** and **UN4** transform one negation word into another negation word while **UN1** and **UN2** transform one negation word into two words. We conjecture that **UN3** and **UN4** may enhance the unigram features while **UN1** and **UN2** weaken the unigram features, or enhance the bigram features in another way. We name this conjecture as **conjecture1**.

Table 2. Compare of four negation datasets

	UN1	UN2	UN3	UN4
# of features	308306	308284	308452	308332
Accuracy	83.01%	82.73%	83.57%	83.57%

In order to verify **conjecture1**, we augment the original unigram feature space with bigram features. Here we select part of bigram features to be used. So we carry out the subsequent experiment based on **UN1** to ascertain the standard of selecting bigram features and the best number of bigram features. We select features with TF, IG, and χ^2 Statistics. For each method, we perform four experiments to contrast the impact established by the number of bigram features.

Table 3. Accuracy of different feature sets based on **UN1**

Accuracy	# of bigram features			
	200	300	500	1000
Unigram	83.01%			
Unigram+Bigram(TF)	84.40%	83.57%	83.01%	82.73%
Unigram+Bigram(IG)	84.40%	**84.68%**	84.12%	83.29%
Unigram+Bigram(χ^2)	**84.68%**	**84.68%**	84.12%	82.73%

Table 3 presents the results of different feature sets based on **UN1**. We get better results after bigram features are affiliated with feature space. Also IG and χ^2 are more effective than TF when used to select features. We experiment with different number of bigram features and get the best accuracy of 84.68% at 300.

Table 4. Results of using unigram and bigram features

Accuracy	UN1	UN2	UN3	UN4
Unigram	83.01%	82.73%	83.57%	83.57%
Unigram+Bigram(IG)	**84.68%**	84.40%	84.40%	84.40%
Unigram+Bigram(χ^2)	**84.68%**	**84.68%**	84.40%	84.40%

To go on verifying **conjecture1**, we carry on the next experiments on **UN2**, **UN3** and **UN4** where 300 bigram features, selected by IG and χ^2 respectively, are used to augment unigram features set. Results are shown inTable 4.

As shown from Table 4, the accuracy increases more on **UN1** and **UN2** than **UN3** and **UN4**. That means bigram features are more effective when be used in **UN1** and **UN2**, and **conjecture1** is verified commendably.

Table 5 shows the results after repeated letters normalization. We get a better result while the number of features is even less (282087). This shows that normalization of repeated letters is effective. We achieve 84.96% sentiment classification accuracy.

Table 5. Results of repeated letters normalization

Accuracy	UN1	UN2
# of unigram features	281787	281837
Unigram	83.29%	83.01%
Unigram+Bigram(IG)	**84.96%**	84.68%
Unigram+Bigram(χ^2)	**84.96%**	84.40%

To continue reducing features from the original feature space, we introduce stemming and lemmatization to process the corpus. Table 6 shows the effect of stemming and lemmatization. There is a sharp decline in classification accuracy though the feature space is reduced. The results illustrate that stemming and lemmatization are not helpful for Twitter sentiment classification.

Table 6. Effect of Stemming and Lemmatization

Accuracy	UN1	Stemming	Lemmatization
# of unigram features	281787	234446	257364
Unigram	83.29%	82.45%	81.62%
Unigram+Bigram(IG)	**84.96%**	81.89%	81.62%
Unigram+Bigram(χ^2)	**84.96%**	81.62%	80.78%

Continually, we augment the Unigram-Bigram feature space with emotions. Here we define emotions such as ":)", ":-)", ":)" etc. as positive emotions, and define emotions ":(", ":-(", ": (" etc. as negative emotions. Once an instance contains one of the positive emotions, the "Positive" feature would be "1" or else it would be "0". "Negative" feature meets the parallel situation. Another 641 instances (321 positive instances and 320 negative instances) selected randomly from training set are removed from the training set and appended to test set.

Table 7 reveals the effect of emotions features. The result based on **UN1** doesn't get better when emotions features are put in. Analysising the wrongly classified instances, we can see that all these instances do not contain emotions in their content. However, the rise of accuracy based on **newdata** illustrates the effectiveness of emotions features and we achieve 85.5% sentiment classification accuracy, which outperforms the baseline by 3.9%. Moreover, our result is better than [1], who reported the best accuracy of 83%, while their number of features is more than 364464.

Table 7. Results of using emotions features

Accuracy	UN1	newdata
# of unigram features	281789	281780
Unigram+Bigram(IG)	84.96%	85.1%
Unigram+Bigram(χ^2)	84.96%	85.2%
Unigram+Bigram(χ^2)+Emotions	84.96%	**85.5%**
UN0 baseline	81.62%	
Go et al.,2009	83%	

5   Conclusions and Future Work

Twitter has been one of the most popular social networking platforms and it is a really meaningful topic to research Twitter sentiment analysis. However, length limitation, multi-topics, casual language, and rich in symbols, all of these characteristics of tweets make Twitter sentiment analysis a challenging assignment. Pre-processing can improve the classification accuracy besides reducing the original feature space.

In this paper, we conduct a series of experiments to verify the effectiveness of several pre-processing methods. Our experiments results on Stanford Twitter Sentiment Dataset show that URLs features reservation, negation transformation and repeated letters normalization have a positive impact on classification accuracy while stemming and lemmatization have a negative impact.

We also augmented the original feature space with bigram features and emotions simultaneously to improve Twitter sentiment classification appearance. Compared to the existing researches, we get a better result by using fewer features.

There are some possible directions we could try in the future. First, for pre-processing, spelling correction can be used to make tweets more regular. Second, different sets of features should be exploited to enhance the classification performance, and hashtags might be useful features in topic-based sentiment analysis. Finally, semi-supervised classification algorithm should be the focus of our research.

Acknowledgments. This research has been partially supported by National Natural Science Foundation of China under Grant No. 61203312 and the National High-Tech Research & Development Program of China 863 Program under Grant No. 2012AA011103 and the Scientific Research Foundation for the Returned Overseas Chinese Scholars, State Education Ministry, and Key Science and Technology Program of Anhui Province, under Grant No. 1206c0805039.

References

1. Go, A., Bhayani, R., Huang, L.: Twitter sentiment classification using distant supervision. CS224N Project Report, Stanford, pp. 1–12 (2009)
2. Pang, B., Lee, L., Vaithyanathan, S.: Thumbs up?: sentiment classification using machine learning techniques. In: Proceedings of the ACL 2002 Conference on Empirical Methods in Natural Language Processing, vol. 10, pp. 79–86. Association for Computational Linguistics (2002)
3. Zhang, X., Fuehres, H., Gloor, P.: Predicting Stock Market Indicators Through Twitter "I hope it is not as bad as I fear". Procedia - Social and Behavioral Sciences 26, 55–62 (2011)
4. Bollen, J., Mao, H., Zeng, X.: Twitter mood predicts the stock market. Journal of Computational Science 2(1), 1–8 (2011)
5. Haddi, E., Liu, X., Shi, Y.: The Role of Text Pre-processing in Sentiment Analysis. Procedia Computer Science 17, 26–32 (2013)
6. Asur, S., Huberman, B.A.: Predicting the future with social media. In: 2010 IEEE/WIC/ACM International Conference on Web Intelligence and Intelligent Agent Technology (WI-IAT), vol. 1, pp. 492–499. IEEE (2010)
7. Stieglitz, S., Dang-Xuan, L.: Political Communication and Influence through Microblogging-An Empirical Analysis of Sentiment in Twitter Messages and Retweet Behavior. In: 2012 45th Hawaii International Conference on System Science (HICSS), pp. 3500–3509. IEEE (2012)
8. Tumasjan, A., Sprenger, T.O., Sandner, P.G., et al.: Predicting Elections with Twitter: What 140 Characters Reveal about Political Sentiment. ICWSM 10, 178–185 (2010)
9. Williams, C., Gulati, G.: What is a social network worth? Facebook and vote share in the 2008 presidential primaries. American Political Science Association (2008)
10. Mishne, G., Glance, N.S.: Predicting Movie Sales from Blogger Sentiment. In: AAAI Spring Symposium: Computational Approaches to Analyzing Weblogs, pp. 155–158 (2006)
11. Aciar, S., Zhang, D., Simoff, S., et al.: Informed recommender: Basing recommendations on consumer product reviews. IEEE Intelligent Systems 22(3), 39–47 (2007)
12. Aguwa, C.C., Monplaisir, L., Turgut, O.: Voice of the customer: Customer satisfaction ratio based analysis. Expert Systems with Applications 39(11), 10112–10119 (2012)
13. Kang, H., Yoo, S.J., Han, D.: Senti-lexicon and improved Naïve Bayes algorithms for sentiment analysis of restaurant reviews. Expert Systems with Applications 39(5), 6000–6010 (2012)
14. Saif, H., He, Y., Alani, H.: Alleviating data sparsity for twitter sentiment analysis. In: The 2nd Workshop on Making Sense of Microposts (2012)
15. Speriosu, M., Sudan, N., Upadhyay, S., et al.: Twitter polarity classification with label propagation over lexical links and the follower graph. In: Proceedings of the First workshop on Unsupervised Learning in NLP, pp. 53–63. Association for Computational Linguistics (2011)

16. Agarwal, A., Xie, B., Vovsha, I., et al.: Sentiment analysis of twitter data. In: Proceedings of the Workshop on Languages in Social Media, pp. 30–38. Association for Computational Linguistics (2011)
17. Lin, C.J., Weng, R.C., Keerthi, S.S.: Trust region newton method for logistic regression. The Journal of Machine Learning Research 9, 627–650 (2008)
18. Quan, C.Q., Ren, F.J.: Target Based Review Classification for Fine-grained Sentiment Analysis. International Journal of Innovative Computing, Information and Control 10(1) (2014)
19. Quan, C.Q., Ren, F.J.: Unsupervised Product Feature Extraction for Feature-oriented Opinion Determination. Information Sciences (2014), doi: http://dx.doi.org/10.1016/j.ins.2014.02.063
20. Quan, C.Q., Wei, X.Q., Ren, F.J.: Combine Sentiment Lexicon and Dependency Parsing for Sentiment Classification. In: SII 2013 (December 2013)
21. Quan, C.Q., Ren, F.J., He, T.T.: Sentimental Classification Based on Kernel Methods. International Journal of Innovative Computing, Information and Control 6(6) (2010)

Construction of a Chinese Emotion Lexicon
from Ren-CECps

Lijuan Wang[1], Changqin Quan[1], Yanwei Bao[1], and Fuji Ren[1,2]

[1] AnHui Province Key Laboratory of Affective Computing and Advanced Intelligent Machine,
School of Computer and Information, HeFei University of Technology,
Hefei, 230009, China
[2] Faculty of Engineering, University of Tokushima,
2-1 Minami-Josanjima, Tokushima 770-8506, Japan
wlj1034900987@163.com, quanchqin@gmail.com,
baoyanwei007@sina.com, ren@is.tokushima-u.ac.jp

Abstract. This paper presents an automatic method to build a Chinese emotion lexicon based on the emotion corpus Ren-CECps. The method includes word extraction and emotion classification. Firstly, sentences are parsed to extract candidate emotional words. By making use of the words co-occurrence in the corpus, we get the similarity between words. And then Support Vector Machine (SVM) is adopted to classify the candidate emotional words. Experiment on the manual labeled words has shown that our classification method achieved high precision. Finally we apply our method on unlabeled corpus to get emotional words.

Keywords: emotion lexicon, corpus, sentence parse, SVM.

1 Introduction

Currently, the number of blogs, network comments and other texts on Internet is increasing every day, and these texts contain huge amount of information. How to quickly recognize the emotional information on characters, events, products, etc., becomes a hot research topic in the field of natural language processing and affective computing [1, 2, 3, 4].

Emotion lexicon plays a very important role in textual emotion classification and recognition. The General Inquirer was developed at Harvard and is a lexical database that uses tags to carry out its tasks. It contains about 3500 entries and each entry consists of a term and a number of tags. The two tags Positive and Negative express valence [5]. MPQA Subjectivity Cues Lexicon is created from the Multi-perspective Question Answering (MPQA) Opinion Corpus, and the lexicon contains 2718 positive words and 4912 negative words [6]. SentiWordNet is developed from WordNet. It is formed by conducting emotion classification and labeling the positive weight and negative weight of the terms in WordNet [7].

D.-S. Huang et al. (Eds.): ICIC 2014, LNAI 8589, pp. 625–634, 2014.

As to Chinese emotion lexicons, HowNet has published a set of emotion words, such as the degree words, positive emotion words and proposal words[1]. But there are no relatively complete Chinese emotion lexicons with precise emotion categories. The manual assignment of emotion categories and intensities needs a great amount of labor and time cost. In order to support the task of emotion recognition in text, this paper proposes an automatic approach to build a Chinese emotion lexicon with precise emotion category. We present a new algorithm of building a Chinese emotion lexicon; the process of our method includes word extraction and emotion classification by making use of the words co-occurrence in the corpus.

The rest of this paper is organized as follows: Section 2 presents a review of methods for building emotion lexicon. Section 3 introduces the emotion corpus Ren-CECps 1.0. Section 4 describes the main algorithm for building emotion lexicon. Section 5 is the experiments. And in Section 6, we come to a conclusion and present some future work.

2   Related Work

There are mainly two methods to build emotion lexicon: lexicon-based method and corpus-based method. Lexicon-based method is to extend the set of emotional labeled words by synonyms and antonyms relations in dictionaries. Corpus-based method is to calculate the similarity between words according to words' co-occurrence and use words semantic similarity computation tendencies.

Li J and Ren F proposed a method to automatically build Chinese emotion lexicon with emotional intensity marked by the use of Tongyici Cilin and HowNet [8]. The advantage of this method is that the initial mass of the existing dictionary words, thus avoiding dependence on the basis of the word.

One approach to creating Chinese emotion lexicon is to use a semantic lexicon, such as WeiPing Liu et al. [9]. They use Chinese emotion words to create a basic emotion dictionary for the different areas. Based on the Chinese word similarity calculation method, proposes a method of calculation the emotional weight of a Chinese word.

Xu Ge et al. [10] described an algorithm based on graph theory and multiple resources to build emotion lexicon. The advantage of the method is used four methods when constructing the similarity matrix; the final result is a weighted sum of the four methods, which improves the accuracy of the similarity matrix. The disadvantage is dependencies of the reference corpus.

Rohwer R et al. [11] proposed an automatic approach of creating lexicon by analyzing statistical data obtained from the corpus. According to the obtained statistical data, words are clustered. Calculating mutual information between the given words and words in the corpus, the most common words in the corpus is clustered by using information theory joint cluster method.

---

[1] http://www.keenage.com/html/c_index.html

The above methods are to construct dictionary for given words, the method proposed in this paper is automatically extracting words from the corpus, and to determine the emotional category of this extracted words.

3   The Emotion Corpus Ren-CECps

The corpus we used in this experience is Ren-CECps1.0. It is a Chinese blog emotion corpus with manual annotations for linguistic expressions of emotion. The frame of emotion annotation includes four levels (document, paragraph, sentence and word). For each level, Eight basic emotion classes (surprise, sorrow, love, joy, hate, expect, anxiety, and anger) are used in the emotional expression space model. Emotion of each level is represented by an 8-dimensional vector: where is a basic emotion class contained in the level, the values of range from 0.0 to 1.0 (discrete) [12]. The annotated files are organized into XML documents. And we can extract emotional keywords from them. The corpus now contains 1487 articles.

4   The Method

In this section, we present our method of automatically extracting words from Ren-CECps1.0 corpus, and classifying the emotional category of the extracted words. The method has two basic principles; one is that the emotion words mostly exist in some specific syntactic dependencies, the other is that the emotion words co-occurred in one sentence always has the same emotion category if the sentence doesn't contain any negative modifiers [13].

Based on these principles, we first divide the corpus into two parts, the first 1000 articles are used to extract the seed emotion words, the remaining articles are used to extract the candidate emotion words, and we select feature words from the seed words. Then we get the co-occurrence of the candidate words and the seed words from the corpus and compute the similarity based on the co-occurrence. All seed words and candidate words have characteristic value of each feature.

4.1   Extracting Candidate Emotional Words

To extract candidate emotional words, Stanford Parser is used to parse sentences. After parsing sentences, we observe the dependencies of words in sentences and find that the emotional words exist mostly in some specific syntactic dependencies. The dependencies include adverbial modifier (advmod), adjectival modifier (amod) and relative clause modifier (rcmod).

Stanford Parser is a statistical parser that works out the grammatical structure of sentences. The parser provides Stanford Dependencies output as well as phrase structure trees. Typed dependencies are otherwise known grammatical relations. The grammatical relations outputted by Stanford parser are arranged in a hierarchy, rooted with the most generic relation, dependencies. The hierarchy contains 48 grammatical relations and includes grammatical relations for NPs (amod – adjective modifier, rcmod - relative clause modifier, det - determiner, partmod - participial modifier, infmod - infinitival modifier, prep - prepositional modifier)[14].

After parsing, we choose the selected grammatical relations and extract the words that may be emotion words as the candidate emotional words in our method. For example, the Chinese segmentation sentence: "我 很 开心 ，你 喜欢 这 份 礼物 。 (I am very happy that you like this gift.)" The dependencies parsed of this sentence are that "nsubj(开心-3, 我-1), advmod(开心-3, 很-2), conj(喜欢-5, 开心-3), dobj(开心-3,你-4), dep(开心-3, 喜欢-5), det(礼物-8, 这-6), clf(这-6, 份-7), dobj(喜欢-5, 礼物-8)". We extra "开心 (happy)" from advmod dependence and "礼物 (gift)" from det dependence as candidate emotional words.

4.2 Calculating the Similarity between Words

Our method of building an emotion lexicon is based on corpus. And the approach of computing two words' similarity is relied on the co-occurrence number of words.In the computing co-occurrence number procedure, we set each sentence as an independent window and find if there is co-occurrence of words in the window [15]. Finally we get the number of word pairs and the number of words appeared alone in the whole corpus.

The similarity calculation algorithm used in our method is Jaccard. The Jaccard similarity is a common index for binary variables and is computed as following formula.

$$Jacsim=num(A*B)/(num(A)+num(B)-num(A*B)) \qquad (1)$$

In the formula, and is respectively the presence number of word A and B in the corpus, is the co-occurrence number of word A and word B in the corpus [16].

5   Experiments

In this section, we mainly introduce experiment preparation, evaluate our algorithm using the manual labeled words from the bottom 487 articles and summarize the experimental results.

5.1 Preprocessing

In order to be consistent with the emotional corpus' category, the lexicon we constructed includes eight basic emotion categories. The seed words are extracted from the top 1000 articles of the corpus. To select the feature words, we rank the seed words according to the words' number of occurrences and keep the top 15 words of each category as feature words. To extract candidate words, we parse the sentences in the segmentation articles using Stanford Parser. If the sentence doesn't contain the negative dependency, we extract the words in advmod, amod and rcmod dependencies as candidate words.

The process to obtain high quality emotional seed words is as follow:

Step1: We extract the emotional key words and their emotion vector.
Step2: We compute the average emotion vector of each word and get 6752 emotion words with emotion vector.
Step3: We select a threshold β. If vector of a word has a value more than β, we put the word to the category that the value more than β indicates and this category is the domain category of the word.

In order to select an appropriate threshold, we calculate the number of words that have one, two or three domain categories on different threshold. Experimental data is the 6752 words obtained on the previous step. The results are as follows:

Table 1. The number of emotion words on different threshold

β	N_1	N_2	N_3
0.4	4260	268	8
0.5	3153	91	0
0.6	1697	24	0
0.7	671	6	0
0.8	127	2	0

In the table, we find 0.6 and 0.7 can be selected as an appropriate threshold. Observing the emotion words and their vectors, we find some words have emotion intensity value just big than 0.6, and emotion category of the word is obvious. Such as "美满(happy) (0,0,0,0,0.67,0.31,0,0)", "贼(thief) (0,0,0,0.6,0,0,0,0)". So we choose 0.6 as the threshold, this can provide more data for subsequent experiments.
Step4: According to the intensity value, we rank words in each category and save about the preceding 50 words in each rank list as seed emotion words.

5.2   Experimental Settings

Our method uses SVM model [17] to predict the emotion category of candidate words. We calculate the similarity between seed words and feature words as feature values and constitute the training file. At the same time, we calculate the similarity between candidate words and feature words as feature values and form the testing file.

In order to examine the effectiveness of SVM classification method, we classify tagged words in the last 487 articles of the corpus and obtain a high precision. After examining, we predict the candidate words' category using our method. Then we delete the words that don't have co-occurrence words that the words' category is the same with the classification result.

5.3   Experimental Results

The Seed Emotion Words and the Feature Words
The seed emotion words and the feature words respectively contain 493 words and 120 words. Table 2 lists the seed emotion words; table 3 lists the feature words.

Table 2. The seed emotion words

Anger	气愤(indignant) 生气(angry) 争吵(quarrel) … (60 words)
Surprise	大吃一惊(surprise)难以置信(incredible)惊讶(confound)…(55 words)
Sorrow	惨剧(disaster)悲痛(grieved)悲伤(sad)绝望(despair)…(67 words)
Love	尽善尽美(perfect)热爱(love) 珍爱(cherish)… (66 words)
Joy	欢天喜地(joy)欢呼(cheer)高高兴兴(happy)…(67 words)
Hate	深恶痛绝(hate) 谩骂(diatribe) 厌恶(disgust) … (58 words)
Expect	期待(expect)希望(hope) 祝愿(wish) 追寻(pursue)… (57 words)
Anxiety	恐惧(fear)忐忑不安(uneasy) 烦躁(irritable)… (63 words)

Table 3. The feature words

Anger	争吵(quarrel)批评(criticism) 抱怨(complain)…(15 words)
Surprise	莫名其妙(surprise) 不可思议(incredible) … (15 words)
Sorrow	忧伤(sad)伤感(sorrow)绝望(despair)…(15 words)
Love	爱人(lovers)珍贵(precious)偏爱(preference)… (15 words)
Joy	快乐(happy) 高兴(joy) 幸运(lucky) … (15 words)
Hate	恨(hate) 贪婪(greedy) 无耻(shameless) … (15 words)
Expect	希望(hope) 期待(expect) 渴望(desire) … (15 words)
Anxiety	恐惧(fear) 担心(worry) 惊慌(panic) … (15 words)

The candidate words that we parsed from the sentences of the rest 487 segmentation articles using Stanford Parser. We extract the nouns, verbs and adjectives words in advmod, amod and rcmod dependencies from the sentences that don't contain the negative dependency as candidate words. After deleting the stop words and the words that the seed words list contained, we get 5,131 words as candidate emotion words.

Using the Manual Labeled Words from the Bottom 487 Articles to Test Our Classification Method

Table 4. The experimental results of labeled words in the last 487 articles

category	number of manual labeled words	number of correctly classified words	Accuracy
Anger	6	2	33.33%
Surprise	0	0	/
Sorrow	34	16	47.06%
Love	168	105	62.50%
Joy	25	16	64.00%
Hate	34	13	38.24%
Expect	11	4	36.36%
Anxiety	103	53	51.46%
total	381	209	54.86%

In this section, we use the manual labeled last 487 articles of the Ren-CECps corpus. The number of labeled words that have emotional intensity more than 0.6 in these articles is 664. We use the seed words' characteristic value file as the test file, and we get the similarity of the manual labeled words and the feature words from the remaining articles, organized the testing file. We find 381 words that have co-occurrence with the seed words. After classifying, we find 209 words have been classified correctly. Table 4 lists the number of each category, the number of correctly classified words and the accuracy rate of each category.

Results of Extracting the Candidate Emotional Words and Classification
Our method extracts candidate words from the unlabeled last 487 articles of the Ren-CECps corpus. Using the method described in 4.1, we get 5131 candidate emotional words. After calculating the words co-occurrence with feature words and their feature value, we finally get 762 words that have co-occurrence with feature words and classify the 762 words.

For the emotion words co-occurred in one sentence always have the same emotional category if the sentence doesn't contain any negative modifiers, we delete some candidate emotional words from the 762 words. If a candidate word's co-occurred words don't contain any word that the emotional category of the word is the same with the word's classification results. Finally we get 229 emotional words. Table 5 lists the emotional words got from the unlabeled last 487 articles of the Ren-CECps corpus.

Table 5. The emotional words got from the unlabeled last 487 articles

Category	words
Anger	错误(error) 生活(life) 折磨(torture) ... (18 words)
Surprise	感动(moving) 大叫(shouted) 眼睛(eye) ... (11 words)
Sorrow	错过(miss) 悲哀(sorrow) 痛(pain) ... (36 words)
Love	宝贵(precious) 微笑(smiling) 旅途(journey) ... (23 words)
Joy	享受(enjoy)浪漫(romantic)烦(bother) 自然(nature) ... (48 words)
Hate	/
Expect	重生(rebirth)花朵(flower)梦想(dream)伤害(hurt)...(64 words)
Anxiety	后遗症(sequela) 灾难(disaster) 急(emergency) ... (29 words)

5.4   Analysis of Experimental Results

In this section, we analyze the accuracy of results in chapter 5.3.3 and cause of words misclassification.

The Accuracy Analysis
We manual pick out the right words for each category. The accuracy is listed in table 6.

Table 6 shows the accuracy for each category. Correctly classified words are the words that all annotators mark the words are correctly classified. The accuracies vary from 36.36% to 50.00%, and the average value is more than 40%, this value is considered as good performance for an eight classification experiments. According to table 4, we find that the classification method can get a better performance than the result showed in table 6. The reason maybe we extract some neutral words or some words only have emotions in some context.

Table 6. The accuracy of our method

category	number of clas- sified words	number of correctly classified words	Accuracy
Anger	18	9	50.00%
Surprise	11	4	36.36%
Sorrow	36	16	44.44%
Love	23	9	39.13%
Joy	48	20	41.67%
Hate	0	0	/
Expect	64	24	37.50%
Anxiety	29	11	37.93%
total	229	93	40.61%

Problem Analysis

The main reason is that we extract some neutral words as candidate words and these words have been classified, such as "生活 (life)", "眼睛 (eye)" and "自然 (nature)". This kind of words always simultaneously occurs with some modifying words and is extracted according to the grammatical relation.

Another important reason for words misclassification is that emotion category of some words is determined by the context. For example, "花朵 (flower)" is a neutral word. But in sentence "青少年是祖国的花朵 (Teenagers are flowers of the mother-land) 。" The word "花朵 (flower)" is an emotion word that means expectation. In another sentence "花园里盛开了五颜六色的花朵 (Colorful flowers are in full bloom in the garden) 。" The word "花朵 (flower)" is an emotion word that express someone's love of flowers.

Besides, some emotion words are misclassified. For example, "伤害 (hurt)" is an emotion word in sorrow category, but we classified it into expect category; "烦 (bother)" is an emotion word in anger category, but we classified it into joy category.

6   Conclusion and Future Work

In this paper, we present a method of building a Chinese emotion lexicon; the process of our method includes word extraction and emotion classification by making use of the words co-occurrence in the corpus. The lexicon we built contains words and their emotional category. A major disadvantage of our method is that error classified word cannot be automatically removed. In future, we will do our efforts on this respect. In addition, we will continue to explore more information such as emotional intensity and one word with multiple categories to improve our lexicon.

Acknowledgements. This research has been partially supported by National Natural Science Foundation of China under Grant No.61203312 and the National High-Tech Research & Development Program of China 863 Program under Grant No. 2012AA011103 and the Scientific Research Foundation for the Returned Overseas Chinese Scholars, State Education Ministry, and Key Science and Technology Program of Anhui Province, under Grant No. 1206c0805039.

References

1. Ren, F.J., Quan, C.Q., Matsumoto, K.: Enriching Mental Engineering. International Journal of Innovative Computing, Information and Control 9(8), 1–12 (2013)
2. Quan, C.Q., Ren, F.J.: Unsupervised Product Feature Extraction for Feature-oriented Opinion Determination. Information Sciences (2014),
   doi: http://dx.doi.org/10.1016/j.ins.2014.02.063
3. Ren, F.J., Quan, C.Q.: Linguistic-based emotion analysis and recognition for measuring consumer satisfaction: an application of affective computing. Information Technology and Management 13(4), 321–332 (2012)
4. Quan, C.Q., Ren, F.J., He, T.T.: Recognition of Word Emotion State in Sentences. IEE J. Transactions on Electrical and Electronic Engineering 6(1), 35–41 (2011)
5. Stone, P.J., Hunt, E.B.: A computer approach to content analysis: studies using the general inquirer system. In: Proceedings of the Spring Joint Computer Conference, May 21-23, pp. 241–256. ACM (1963)
6. Wilson, T., Wiebe, J., Hoffmann, P.: Recognizing contextual polarity in phrase-level sentiment analysis. In: Proceedings of the Conference on Human Language Technology and Empirical Methods in Natural Language Processing, pp. 347–354. Association for Computational Linguistics (2005)
7. Baccianella, S., Esuli, A., Sebastiani, F.: SentiWordNet 3.0: An Enhanced Lexical Resource for Sentiment Analysis and Opinion Mining. In: LREC 2010, vol. 10, pp. 2200–2204 (2010)
8. Li, J., Ren, F.: Creating a Chinese emotion lexicon based on corpus Ren-CECps. In: 2011 IEEE International Conference on Cloud Computing and Intelligence Systems (CCIS), pp. 80–84. IEEE (2011)
9. Liu, W., Zhu, Y., Li, C.: Research on building Chinese Basic Semantic Lexicon. Journal of Computer Applications 29(11), 2882–2884 (2009)
10. Ge, X., Meng, X.F., Wang, H.F.: Build Chinese emotion lexicons using a graph-based algorithm and multiple resources. In: Proceedings of the 23rd International Conference on Computational Linguistics. Association for Computational Linguistics (2010)
11. Rohwer, R., Freitag, D.: Towards full automation of lexicon construction. In: Proceedings of the HLT-NAACL Workshop on Computational Lexical Semantics, pp. 9–16. Association for Computational Linguistics (2004)
12. Quan, C., Ren, F.: A blog emotion corpus for emotional expression analysis in Chinese. Computer Speech & Language 24(4), 726–749 (2010)

13. Zhu, Y.L., Min, J., Zhou, Y.: Semantic orientation computing based on HowNet. Journal of Chinese Information Processing 20(1), 14–20 (2006)
14. De Marneffe, M.C., MacCartney, B., Manning, C.D.: Generating typed dependency parses from phrase structure parses. In: Proceedings of LREC, vol. 6, pp. 449–454 (2006)
15. Lund, K., Burgess, C.: Producing high-dimensional semantic spaces from lexical co-occurrence. Behavior Research Methods, Instruments, & Computers 28(2), 203–208 (1996)
16. Cheng, Y., Huang, H., Qiu, L.: Similarity-Based Dynamic Multi-Dimension Concept Mapping Algorithm. Minimicro Systems 27(6), 975 (2006)
17. Drucker, H., Wu, D., Vapnik, V.N.: Support vector machines for spam categorization. IEEE Transactions on Neural Networks 10(5), 1048–1054 (1999)

Word Frequency Statistics Model for Chinese Base Noun Phrase Identification

Lu Kong[1], Fuji Ren[1,2], Xiao Sun[1], and Changqin Quan[1]

[1] Anhui Province Key Laboratory of Affective Computing and Advanced Intelligent Machine,
School of Computer and Information, Hefei University of Technology,
Hefei, 230009, China
[2] Faculty of Engineering, University of Tokushima,
2-1 Minami-Josanjima, Tokushima 770-8506, Japan
konglu0812@126.com, ren@is.tokushima-u.ac.jp,
sxdlut@qq.com, quanchqin@gmail.com

Abstract. The Chinese base phrase identification plays an important role in the field of natural language processing. It needs to be improved in the recognition scope and methods currently. This paper presents a method based on word frequency statistics model for Chinese base noun phrase identification: Building the noun phrase dictionary by training corpus, calculating the co-occurrence frequency and threshold of the noun phrase, and constructing word table according to the different roles of the words in the noun phrase. Unknown word processing and rule templates are added. Improve the results with error correction processing at last. Experiments on the test corpus show that the average precision and average recall rate of the base noun phrases identification in different areas are 91.28% and 93.22%.

Keywords: word frequency statistics model, co-occurrence frequency, rule templates, Chinese base noun phrase.

1 Introduction

In the field of natural language processing, the correct identification and analysis of the noun phrase have an important role in machine translation, text classification and syntactic analysis. However, due to the characteristics of Chinese language and other factors, such as: the accuracy of speech tagging, the multi-speech of the same word, difficulties in learning the methods in English because of the difference between English and Chinese. It is difficult in Chinese base noun phrase identification.

Currently, there are many ways to identify the noun phrase at home and abroad. These methods are generally divided into three types: statistics-based, rules-based, combining statistical and rules [1, 2, 3]. For example, in English terms, Church used the boundaries statistical methods to identify the noun phrase [4]. Brill proposed transformation -based error-driven learning method [5] .In Chinese terms, Zhao proposed a model based on transformation [6], combining the structure templates with transformation rules, which precision and recall rate are 89.3% and 92.8% in open

D.-S. Huang et al. (Eds.): ICIC 2014, LNAI 8589, pp. 635–644, 2014.

test. Zhou used the maximum entropy method to identify the base noun phrase in both English and Chinese [7], the precision and recall rate in the open test corpus were 88.09 % and 87.43%. Tan proposed a mixed statistical model that combines the CRF, TBL and SVM [8], the experimental results of precision and recall rate were 88.95% and 88.40%. The above typical methods of noun phrase identification are a combination of statistics or rules, the division of the meaning of a word that constituting the noun phrase is not obvious. But it is very important for the Chinese that dividing the phrase effective and making full use of the connection between word and the word's semantic information.

Based on the analysis, this paper uses the word as a unit to divide noun phrase according to the word's different roles in it, we propose a method for Chinese base noun phrase identification based on the frequency statistics model, while adding to unknown word processing and rule templates, increasing the ability to identify new words, combining the word's meaning and statistics and rules, and adding error correction processing to optimize the results in the end. According to research, there are currently no reports of using the method for the base noun phrase identification. The results show that this method has good results in the recognition of Chinese base noun phrase.

The rest of this paper is organized as follows: Section 2 gives Chinese base noun phrase definition of this paper. Section 3 introduces the word frequency statistics model for Chinese base noun phrase identification. Section 4 is the experiments. And in Section 5, we come to a conclusion and present some future work.

2   The Definition of Chinese Base Noun Phrase

Base noun phrase is a phrase which center word is a noun [9], such as: "自然语言处理 (Natural Language Processing)", "国际政治经济秩序 (International political and economic order)"and so on. Reference [6] presents a formal description of base noun phrase from restrictive attributive:

> Base noun phrase (baseNP for short)
> baseNP → baseNP + baseNP
> baseNP → baseNP + 名词(noun)|名动词(noun-verb)
> baseNP → 限定性定语(limiting attribute) + baseNP
> baseNP → 限定性定语+ 名词|名动词
> 限定性定语→ 形容词(adjective)|区别词(distinguishing words)|动词(verb)|名词|
> 处所词(locative words)|西文字串(western string)|(数词(numeral)+ 量词(quantifier))

This paper references the above formal definition of the baseNP, defining baseNP as a single semantic core, non-nested noun phrase (not including a single noun). Including a noun phrase without any modification of components, a bunch of noun that difficult to determine the modified relationship between them, listed noun and so on, such as: "美好时刻(beautiful moment)", "时代主题(era theme)".

3   Automatic Identification of Chinese Base Noun Phrase Based on Word Frequency Statistics Model

3.1   The Structure of Chinese Base Noun Phrase Automatic Identification

It includes two parts: the frequency statistics model and baseNP identification module, the structure is shown in figure 1.

3.2   The Word Frequency Statistics Model

This part includes preparation work of baseNP identification: building noun phrase dictionary and constructing word table which containing the frequency of each word.

Building the Noun Phrase Dictionary
The marker of Noun phrase (NP) in Chinese Treebank is conforming to the requirements of the experiments because of its specification, artificial-modification and high accuracy [10, 11]. This paper select Chinese Treebank as the training corpus, extracting after-marked noun phrases (NP) and modifying them artificial according to section 2. The NP dictionary is established with each noun phrase as separate line (only contains the meaning without speech information).

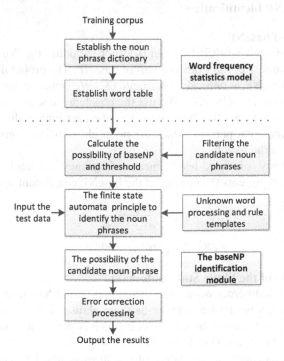

Fig. 1. The structure of Chinese base noun phrase automatic identification

Constructing the Word Table

We split the baseNP with word as a unit according to the position of each word in it: word-head, word-middle and word-tail. Word-head is the first word of NP, the middle words of noun phrase are word-middle and the last word of noun phrase is word-tail (the times of word-middle appears in a noun phrase greater than or equal to zero).

Then we construct each word table. Word-head table: all the word-heads in the same column and all the co-occurrence words of each word-head in the same line with it. Word-middle table and word-tail table are similar to word-head table. Calculate the frequency or co-occurrence frequency of each word, as shown in formula (1)-(2).

Calculate the frequency of each word with formula (1). means the total occurrence number of the word as word-head or word-middle or word-tail in the training corpus, represents the total number of the correspondent word, means the total number of the co-occurrence word in the training corpus and means the total number of the co-occurrence word in word table [12,13,14].

$$F\_word = \frac{w_i}{a_i} \tag{1}$$

$$F\_co = \frac{c_i}{k_i} \tag{2}$$

3.3 The baseNP Identification

The Possibility of baseNP

The frequency of each word is between 0-1 (not including 0). We can use them to indicate the probability of the word as a part of baseNP. The probability of each NP in dictionary as a baseNP is defined as formula (3). We improve formula (3) and get which means possibility of baseNP. We use the absolute value to ensure that the value of is positive. Then we get the threshold value (the maximum value of) and use it to filter the candidate noun phrases. Mark the noun phrase if, else give up the marker and continue down the search.

means frequency of word-head. means frequency of each subsequent co-occurrence word, represents the number of the baseNP constituent words.

$$p_{NP} = \frac{\prod_{i=1}^{n} p_i}{n} \tag{3}$$

$$p_w = |\left(\sum_{i=1}^{n} \log p_i\right) - \log n| \tag{4}$$

States Transition of the Finite State Automata

We extract all the left border words (the words that before NP) from the training corpus and form the left border table. Describe noun phrase identification as conversion between the finite state automata in each state which is shown in figure 2 [15].

Among them, the state 000 is the initial state, which means starting to read test data. The state 101 indicates the processing of unknown words. The state 110 represents adding the rule templates processing. The state 111 is the ending state, which represents the end of a baseNP marker and begins to mark the next.

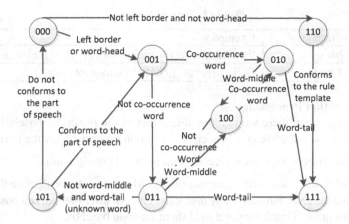

Fig. 2. States transition of the finite state automata

Unknown Word Processing and Rule Templates

Unknown word processing

Unknown word means the one which is not in word table. This condition reaches state 101 in figure 2. By section 2, the baseNP constitutes compliance with certain rules. Therefore, to judge the position of the unknown word in the baseNP and the part of speech may appear on the location when meeting the unknown word, according to the judgment result of speech to choice next state. At this point, we use the average frequency of word-middle or word-tail to instead of in calculating . We get the table 1 after analyzing which contains the position of word in the baseNP and the part of speech may appear on the position (using the North-speech tagging set).

Rule templates

The baseNP comply with the rules according to the definition in section 2. In figure 2, without rule templates, the algorithm cannot identify the baseNP which is not beginning with left border or word-head. Analysis of such baseNP, the rule templates are summarized as shown in table 2.Judge the current word and the one behind it whether conform to the rule templates when the current word reaching the state 110, if they meet rule templates in table 2, the noun phrase will be marked directly.

Table 1. The position of word in the baseNP and the part of speech may appear on the position

The position in baseNP	The part of speech may appear on the position	Examples
The first	m|a|b|n|vn	丰富(rich)/a 经验(experience)/n
The second	q|a|b|n|vn	一(a)/m 桌(table)/q 强人(strongman)/n
Third (not word-tail)	a|b|n|vn	不少(many)/m 常用(commonly used)/a 服务(service)/vn 命令(orders)/n
The fourth or more	n|vn	一 (a)/m 项 /q 最新 (new)/a 研究 (study)/vn 结果(result)/n

Table 2. The rule templates

The rule templates	Examples
m + q + n|vn	六(sixth)/m 周年(anniversary)/q 庆典(celebration)/n
a|b|s|n|vn + n|vn	街头(street)/s 摄影师(photographer)/n

Error Correction Processing

After the processing, there are two possible errors in the results of baseNP identification. Then we use the corresponding error correction processing for the two errors.

Two or more baseNP appear continuous with no word among them

The mark of baseNP will end when meeting word-tail and then calculate threshold of the candidate baseNP, but there may also exist baseNP after the marked baseNP.

When this error occurs, we combined them into one baseNP.

The baseNP do not reach the right border

This occurs for two reasons: First, the word-tail of marked baseNP may also be word-middle. There is a real word-tail after the marked baseNP. Second, there may be other rule templates such as m + q + n\|vn + n\|vn or a\|b\|s\|n\|vn + n\|vn + n\|vn.

Therefore, moving the right boundary of the marked baseNP to the right one, getting the subsequent n \| vn into the marked baseNP and outputting the results.

4   Experiment

4.1   The Corpus

Experimental training corpus is the Chinese Treebank that previously described. It has a total of 3.32MB. We extract 25,000 noun phrases from the training corpus and get more than 12,000 baseNPs that meeting the requirements after pretreatment and manual processing. The longest noun phrase composed of 10 words, the shortest composed of two words, the average length composed of four words. These noun phrases constitute a dictionary as the experimental training corpus.

Test data are divided into three categories: People's Daily corpus in January 1998, microblog corpus, Sogou corpus. We randomly select two different data from each class in the average size of 13KB, each containing an average of 127 baseNPs. People's Daily corpus in January 1998 is through speech tagging text. Microblog corpus is grabbed from Sina microblog. Sogou corpus comes from Sogou laboratory. The latter two types of data are not part of speech tagging, in this paper, using NLPIR Chinese word segmentation system (also known as ICTCLAS2013) for part-of-speech tagging.

4.2   The Experimental Results

After calculated，the. In order to quantify the effects of unknown word processing, rule templates and error correction processing, the comparative experiments as following, and the experimental results are shown in table 3 and table 4:

Experiment I: Identify baseNP with noun phrase dictionary and the threshold .

Read test data, start tagging when meeting the left border or word-head, namely the states except 110 and 101 of the finite state automata in figure 2. Give up the marker if the current word is unknown word or not left border or not word-head.

Experiment II: Add into unknown word processing on the basis of Experiment I

Judge the part of speech if the current word is unknown word. This approach is detailed in unknown word processing section.

Experiment III: Add into rule templates on the basis of Experiment II

Determine whether the current word conform to the rule templates if it is not left border or word-head, jumping to the state 110 of the finite state automata in figure 2.

Experiment IV: Add into error correction processing on the basis of Experiment III

This paper uses error correction processing for the marked results after completion of the threshold calculation and filtering of candidate baseNP and outputs the final result.

Table 3. The experimental results

	Experiment I		Experiment II	
	Precision	Recall	Precision	Recall
People's Daily in January 1998_1	69.57%	60.38%	73.15% ↑	74.53% ↑
People's Daily in January 1998_2	64.06%	59.42%	64.90% ↑	71.01% ↑
Microblog corpus_3	68.29%	41.18%	68.10% ↓	58.09% ↑
Microblog corpus_4	59.68%	28.24%	66.37% ↑	57.25% ↑
Sogou corpus_5	58.59%	36.94%	55.48% ↓	51.59% ↑
Sogou corpus_6	47.06%	25.53%	43.42% ↓	35.11% ↑
Average value	61.21%	41.99%	61.90% ↑	57.93% ↑

Table 4. The experimental results

	Experiment III		Experiment IV	
	Precision	Recall	Precision	Recall
People's Daily in January 1998_1	70.18% ↓	75.47% ↑	94.39% ↑	95.28% ↑
People's Daily in January 1998_2	61.64% ↓	71.74% ↑	92.25% ↑	93.57% ↑
Microblog corpus_3	66.45% ↓	74.26% ↑	91.49% ↑	93.48% ↑
Microblog corpus_4	66.89% ↑	75.57% ↑	88.41% ↑	93.13% ↑
Sogou corpus_5	57.75% ↑	68.79% ↑	90.63% ↑	92.36% ↑
Sogou corpus_6	52.78% ↑	60.64% ↑	90.53% ↑	91.49% ↑
Average value	62.62% ↑	70.96% ↑	91.28% ↑	93.22% ↑

4.3   Experimental Analysis

We can see from table 3 and table 4:

In experiment II, the precision is improved and also decreased compared with the experiment I, but the recall rate all raise. The average precision and recall rate are improved. The most increased of precision and recall rate are 6.69% and 29.01% of "microblog corpus_4". In the corpus above, "microblog corpus_4" is living language from microblog. We can see from that, the precision and recall rate of this type of text are up sharply after adding unknown words.

In experiment III, the precision is improved and also decreased, but the recall rate all raise compared with the experiment II. The average precision and recall rate are improved. The most increased of precision is 9.36% (Sogou corpus_6). The most increased of recall rate is 18.32% (microblog corpus_4). It can be seen that the precision and recall rate all raise after adding rule templates.

In experiment IV, the precision, recall rate and the average of them are largely improved compared with the experiment III. The precision and recall rate are above 90% except "microblog corpus_4". We can see from that, either for formal People's Daily corpus or informal microblog corpus, we obtain good recognition results after adding into error correction processing.

The Comparison Result with Other Traditional Models
Through the experiments in this paper, we get better results in open test. Besides, test corpuses are different from the training corpus, they are from different fields: People's Daily corpus in January 1998, microblog corpus, Sogou corpus. They contain the traditional news corpus, also include informal corpus. It is better than the traditional methods because it is not limited in test corpus. Compared the word frequency statistics model with other traditional models, the results are shown in table 5 (open test).

Table 5. The comparison result with other traditional models

The paper	The train corpus	The test corpus	Precision	Recall
References [6]	Corpus in five different areas	Corpus in the five areas above	89.3%	92.8%
References [7]	Chinese Treebank	Chinese Treebank	88.09%	87.43%
References [8]	ACE2005 Chinese corpus	ACE2005 Chinese corpus	87.43%	88.09%
The paper	Chinese Treebank	Open test in different fields	91.28%	93.22%

5   Conclusion

In this paper, an approach based on word frequency statistics model is proposed to solve the problem of the baseNP recognition in unknown text. The noun phrase dictionary is established based on the Chinese Treebank, adding to unknown word processing and rule templates and error correction processing to build a baseNP recognition system. Contrast experiments on test corpus from different platforms and different fields have been conducted, especially in the special language phenomenon of irregular and informal language, such as the microblog corpus that achieves good results.

But according to the above analysis, there are still improvements:

Training corpus selection still has some influence on the results, how to expand the training corpus while more reducing dependence on training corpus is a direction for future improvements.

Improve the accuracy of speech tagging. Reduce recognition errors due to speech tagging.

For multi-word speech phenomena, how to better take into account the elements of speech.

The speech tagging of noun phrases may meet the requirements of baseNP, but not the actual baseNP in microblog informal language phenomenon.

Acknowledgements. This research has been partially supported by National Natural Science Foundation of China under Grant No. 61203312 and the National High-Tech Research & Development Program of China 863 Program under Grant No. 2012AA011103 and the Scientific Research Foundation for the Returned Overseas Chinese Scholars, State Education Ministry, and Key Science and Technology Program of Anhui Province, under Grant No. 1206c0805039.

References

1. Xu, F., Zong, C.Q., Wang, X.: Chinese baseNP chunking by Error-driven Combination Classifiers. Journal of Chinese Information Processing 21(1), 115–119 (2007)
2. Xu, Y.H.: Corpus-based studies of base noun phrase. Language Application (1), 120–125 (2008)
3. Hu, N.Q., Zhu, Q.M., Zhou, G.D.: Hybrid Method to Chinese Base Noun Phrase Recognition. Computer Engineering 35(20), 199–201 (2009)
4. Church, K.W.: A Stochastic Parts Program and Noun Phrase Parser for Unrestricted Text, pp. 136–143 (1988)
5. Eric, B.: Transformation-Based Error-Driven Learning and Natural Language Processing: A Case Study in Part-of-Speech Tagging. Computational Linguistics 21(4) (1995)
6. Zhao, J., Huang, C.N.: A Chinese base noun phrase identification model based on the transformation. Journal of Chinese Information Processing 13(2), 1–7 (1998)
7. Zhou, Y.Q., Guo, Y., Huang, X.J.: Chinese and English BaseNP Recognition Based on a Maximum Entropy Model. Journal of Computer Research and Development 40(3), 440–446 (2003)
8. Tan, W., Kong, F., Ni, J.: A mixed statistical model-based method for identifying Chinese base noun phrase. Computer Applications and Software 28(8), 254–256 (2011)
9. Zhang, Y.Q., Zhou, Q.: Automatic identification of Chinese Base Phrases. Journal of Chinese Information Processing 16(6), 1–8 (2002)
10. Liu, S., Li, Y., Zhang, L.: Chinese Text Chunking Using Co-training Method. Journal of Chinese Information Processing 19(3), 73–79 (2005)

11. Huang, C.N., Jin, G.J.: To observe three Chinese grammar problems from Chinese TreeBank. The language of science 12(2), 178–192 (2013)
12. Yin, B.Y., Fang, S.Z.: New concepts and methods of word frequency statistics. In: Proceedings of Language Application (1995)
13. Hu, W.T., Yang, Y., Yin, H.F.: Organization name recognition based on word frequency statics. Application Research of Computers 30(7), 2014–2016 (2013)
14. Wu, X.Q., Lv, N.: An analysis method based on keyword co-occurrence frequency. The Intelligence Theory and Practice 35(8), 115–119 (2012)
15. Mao, T., Yang, J.D., Wang, W.G.: State transition method of natural language based on finite automata machine. Journal of Liaoning Technical University (Natural Science) 31(6), 885–888 (2012)
16. Fu, G.H., Wang, P., Wang, X.L.: Research on the approach of integrating Chinese word segmentation with Part-of-speech Tagging. Application Research of Computers (7), 24–26 (2001)

Structure Constrained Discriminative Non-negative Matrix Factorization for Feature Extraction

Yan Jin[1], Lisi Wei[2], Yugen Yi[2,3], and Jianzhong Wang[2,*]

[1] Changchun Medical College, Changchun, Jilin Province, China
[2] College of Computer Science and Information Technology,
Northeast Normal University, Changchun, China
[3] School of Mathematics and Statistics, Northeast Normal University, Changchun, China
wangjz019@nenu.edu.cn

Abstract. In this paper, we propose a novel algorithm called Structure Constrained Discriminative Non-negative Matrix Factorization (SCDNMF) for feature extraction. In our proposed algorithm, a pixel dispersion penalty (PDP) constraint is employed to preserve spatial locality structured information of the basis obtained by NMF. At the same time, in order to improve the classification performance, intra-class graph and inter-class graph are also constructed to exploit discriminative information as well as geometric structure of the high-dimensional data. Therefore, the low-dimensional features obtained by our algorithm are structured sparse and discriminative. Moreover, an iterative updating optimization scheme is developed to solve the objective function of the proposed SCDNMF. The proposed method is applied to the problem of image recognition using the well-known ORL, Yale and COIL20 databases. The experimental results demonstrate that the performance of our proposed SCDNMF outperforms the state-of-the-art methods.

Keywords: NMF, PDP, Label Information, SCDNMF, face recognition, objection recognition.

1 Introduction

Feature extraction plays a fundamental role in many image processing and pattern recognition tasks. A good feature extraction method typically reveals the latent structure of a dataset, and also can reduce the dimensionality of the data. In real world applications, many data are non-negative, and the non-negativity makes a non-subtractive combination of "parts" to form a whole, and makes the encoding of data easier to be interpreted [1]. Recently, non-negative matrix factorization (NMF) [2] has been proposed to incorporate the non-negativity constraints and obtain a parts-based representation. In particular, it represents data as a linear combination of a set of basis vectors, in which both the combination coefficients and the basis vectors are nonnegative. Though NMF has been successfully used in diverse fields of science, such as

---

[*] Corresponding author.

D.-S. Huang et al. (Eds.): ICIC 2014, LNAI 8589, pp. 645–657, 2014.
© Springer International Publishing Switzerland 2014

biometrics [6], image processing [3], computer vision [2], and pattern recognition [4], it still suffers from the following limitations for practical applications.

Firstly, NMF can learn a part-based representation, but the bases learned by it are not always sparse enough [5]. In order to overcome this limitation, several NMF variants have been proposed by introducing additional constraints to the original NMF in recent years. Li et al. proposed a local NMF (LNMF) [6], which modified the objective function of the NMF algorithm by imposing a spatially localized constraint on basis. Hoyer [7] proposed a new algorithm called Non-negative Sparse Coding (NNSC) which combined the sparse coding and non-negative matrix factorization to minimize the sparseness and the residual of the reconstruction. Like LNMF and NNSC methods, non-smooth NMF (nsNMF) [8] was proposed by Montano, which allowed a controlled sparseness degree in both factors. However, all the methods described above only concerned about the sparsity of the extracted basis. Thus, they did not directly or explicitly exploit the structured information of data during learning process, and the extracted basis may not completely describe the "parts" in its matrix form very well. As a result, a structured sparse NMF (SSNMF) [9] has been proposed to learn structured basis. However, the structured patterns in SSNMF need to be predefined manually, which would hinder its applications in some applications. Consequently, Zheng et al. proposed another extension of NMF called spatial non-negative matrix factorization (SNMF) [10]. In SNMF, a pixel dispersion penalty (PDP) describing the spatial dispersion of pixels in a basis matrix was introduced into NMF to learn a structured sparse basis. Compared with SSNMF, SNMF is more flexible since it does not need to predefine the structured patterns.

Secondly, the original NMF and its sparse extensions are unsupervised, i.e., the discriminative information of different classes is neglected in them, so they cannot perform well for classification or recognition tasks. In order to improve the classification performance, a supervised NMF method called Discriminant NMF (DNMF), which explored the label information during the feature extraction process, was proposed in [11]. By taking the class labels of the data into account, DNMF can learn more discriminative low-dimensional feature than original NMF.

Thirdly, the unsupervised and supervised NMF algorithms described above are all linear models that can only learned a parts-based representation in Euclidean space. Therefore, they may fail to discover the intrinsic geometrical structure of the data space, which is essential for data analysis. Cai et al. proposed a Graph regularized NMF approach [12] to handle this limitation by constructing the k-nearest-neighbor (kNN) graph to model the local geometrical structure information of high-dimensional data. GNMF can achieve better performance on clustering task, but, it cannot perform well for classification task due to the label information is not considered during the feature extraction.

In this paper, we propose a new non-negative matrix factorization method called Structure Constrained Discriminative Non-negative Matrix Factorization (SCDNMF) to improve the performance of NMF algorithm for recognition and classification tasks. Firstly, in order to make the basis learned by our algorithm to be structured sparse, a pixel dispersion penalty (PDP) is employed. Secondly, two k-NN graphs (intra-class graph G_w and inter-class graph G_b) are also constructed in our algorithm to exploit discriminative information as well as geometric structure of the high-dimensional data. Moreover, an iterative updating optimization scheme is developed to solve the objective function of the proposed SCDNMF. The proposed method is applied to the problem of image recognition using the well-known ORL, Yale and COIL20 databases. The experimental results demonstrate that the performance of our proposed outperforms the state-of-the-art methods.

The rest of the paper is organized as follows. In Section 2,the proposed SCDNMF algorithm is presented. Section 3 shows the experimental results obtained by the proposed SCDNMF algorithm. Finally, the conclusions are given in Section 4.

2   Structure Constrained Discriminative Non-negative Matrix Factorization

In this section, we first present our Structure Constrained Discriminative Non-negative Matrix Factorization (SCDNMF) algorithm, which integrates the structure information of the learned basis and discriminative information of high-dimensional data into non-negative matrix factorization framework. Then, an optimization scheme based on iterative updating rules is proposed to solve the objective function of SCDNMF.

2.1   SCDNMF Algorithm

The first aim of the proposed SCDNMF is to explore structure information of the basis learned by NMF without using any manually predefined. The matrix form of a basis w_i is spatially localized if the non-zero elements of the basis matrix are spatially and locally non-dispersive, i.e. those non-zero elements should be clustered and close to its center. Suppose the size of basis matrix w_i is $b{\times}a$, that is, the matrix has b rows in height and a columns in width. The pixel dispersion penalty (PDP) can be defined as following

$$D(w_i) = \sum_{x=1}^{a}\sum_{y=1}^{b}\sum_{x'=1}^{a}\sum_{y'=1}^{b} \mid w_i^{2D}(y,x) \parallel w_i^{2D}(y',x') \mid \times l([y,x],[y',x']) \tag{1}$$

where $w_i^{2D}(y,x)$ is the corresponding matrix form of the basis vector w_i, and l is an association function to measure the distance between two coordinate vectors $[y,x]$ and $[y',x']$ in w_i^{2D}. It can be computed by

$$l([y,x],[y',x']) = \mid y-y' \mid - \mid x-x' \mid \tag{2}$$

From (1), it is found that minimizing this function will suppress the scenario that two disjoint elements at $[y, x]$ and $[y', x']$, which are far away from each other, both have large values.

In order to optimizing (1), an indicator vector $e_{y,x} \in R^d$ is used as

$$e_{y,x}(j) = \begin{cases} 1, & \text{if } j = (x-1){\times}b+y \\ 0, & \text{otherwise} \end{cases} \tag{3}$$

We can see clearly that $w_i^{2D}(y,x) = w_i^T e_{y,x}$. By simple algebra formulation, (1) is deduced to

$$D(w_i) = w_i^T \{\sum_{x=1}^a \sum_{y=1}^b \sum_{x'=1}^a \sum_{y'=1}^b l([y,x],[y',x'])e_{y,x}e_{y',x'}^T\}w_i = w_i^T E w_i \qquad (4)$$

where $E = \sum_{x=1}^a \sum_{y=1}^b \sum_{x'=1}^a \sum_{y'=1}^b l([y,x],[y',x'])e_{y,x}e_{y',x'}^T$.

The second aim of the proposed SCDNMF is to exploit discriminative information as well as geometric structure of the high-dimensional data. Therefore, two k-NN graphs are constructed. One is the within-class graph (G_w) which encodes the local similarity of the neighboring data points sharing the same class labels. The other is the between-class graph (G_b) which encodes the local diversity of the neighboring data points from different classes. The weights in the two graphs are defined as:

$$S_{ij}^W = \begin{cases} \exp(-\|x-x\|^2/t^2), & \text{if } x_i \in N_{k_1}(x_j) \text{ or } x_j \in N_{k_1}(x_i) \text{ and } La(x_i) = La(x_j) \\ 0, & \text{otherwise} \end{cases} \qquad (5)$$

$$S_{ij}^B = \begin{cases} \exp(-\|x-x\|^2/t^2), & \text{if } x_i \in N_{k_2}(x_j) \text{ or } x_j \in N_{k_2}(x_i) \text{ and } La(x_i) \neq La(x_j) \\ 0, & \text{otherwise} \end{cases} \qquad (6)$$

where $La(x_i)$ is the class label of x_i, $N_{k_j}(x_i)$ $(j=1, 2)$ is the k_j neighbors of x_i and t is a heat kernel parameter which determines the decay rate of similarity function. In order to extract the discriminative information, we expand margins between inter-class neighborhoods, while shrink margins among intra-class neighborhoods. Therefore, we can formulate the objective function as

$$\frac{1}{2}\sum_{i,j=1}^N (\|h_i - h_j\|^2 S_{ij}^W - \|h_i - h_j\|^2 S_{ij}^B)$$

$$= (\sum_{i=1}^N h_i^T h_i D_{ii}^W - \sum_{i,j=1}^N h_i^T h_j S_{ij}^W) - (\sum_{i=1}^N h_i^T h_i D_{ii}^B - \sum_{i,j=1}^N h_i^T h_j S_{ij}^B) \qquad (7)$$

$$= tr(H^T(D^W - S^W)V) - tr(H^T(D^B - S^B)H) = tr[H^T(L^W - L^B)H]$$

where $L^W = D^W - S^W$ and $L^B = D^B - S^B$ are the Laplacian matrices of G_w and G_b. D^W and D^B are the diagonal matrices.

Now, we can incorporate (4) and (7) into the original NMF to obtain the objective function of our SCDNMF:

$$\min \| X - WH^T \| + \alpha tr(W^T E W) + \beta tr(H^T(L^W - L^B)H)$$

$$s.t \ W \geq 0, H \geq 0 \qquad (8)$$

where $X \in R_+^{d \times N}$ is the input data matrix including N non-negative training samples in R^d, $W \in R_+^{d \times r}$ is a non-negative basis matrix containing r basis vectors and

$H \in R_+^{r \times N}$ is the non-negative coefficient matrix. From (8), we can clearly find that if the two parameters are set as $\alpha=0$ and $\beta=0$, the proposed SCDNMF reduces to the original NMF [2]. In addition, if $\beta=0$, the proposed SCDNMF reduces to the SNMF [10] and if $\alpha=0$, the proposed SCDNMF reduces to the NMF-kNN [13]. Thus, NMF, SNMF and NMF-kNN can all be regarded as special cases of our proposed algorithm.

2.2   Optimization Algorithm

The objective function of SCDNMF in (8) is not convex in both W and H together. Therefore it is unrealistic to expect an algorithm to find the global minima. In the following, we introduce iterative algorithms to solve (8) which can achieve local minima.

The objective function of SCDNMF can be rewritten as

$$
\begin{aligned}
J &= \| X - WH^T \| + \alpha tr(W^T EW) + \beta tr(H^T (L^W - L^B)H) \\
&= tr(XX^T) - 2tr(XHW^T) + tr(WH^T HW^T) + \alpha tr(W^T EW) + \beta tr(H^T (L^W - L^B)H) \\
&s.t. W \geq 0, H \geq 0
\end{aligned}
\tag{9}
$$

Let ψ_{ij} and φ_{ij} be the Lagrange multiplier for constraints $W_{ij} \geq 0$ and $H_{ij} \geq 0$, respectively. Then the Lagrange function of (16) can be written as

$$
\begin{aligned}
J &= tr(XX^T) - 2tr(XHW^T) + tr(WH^T HW^T) + \alpha tr(W^T EW) + \beta tr(H^T (L^W - L^B)H) \\
&= tr(XX^T) - 2tr(XHW^T) + tr(WH^T HW^T) + \alpha tr(W^T EW) + \beta tr(H^T (L^W - L^B)H) \\
&+ tr(\psi W) + tr(\varphi(H))
\end{aligned}
\tag{10}
$$

The partial derivatives of J with respect to W and H are

$$
\frac{\partial J}{\partial W} = -XH + WH^T H + \alpha EW + \psi
\tag{11}
$$

$$
\frac{\partial J}{\partial H} = -X^T W + H^T W^T W + \beta (L^W - L^B)H + \varphi
\tag{12}
$$

Since $L^W = D^W - S^W$ and $L^B = D^B - S^B$, (12) can be written as

$$
\frac{\partial J}{\partial H} = -X^T W + H^T W^T W + \beta (D^W - S^W)H - \beta (D^B - S^B)H + \varphi
\tag{13}
$$

Using the KKT condition $\psi_{ij} W_{ij} = 0$ and $\varphi_{ij} H_{ij} = 0$, we can get

$$
[-XH + WH^T H + \alpha EW + \psi]_{ij} W_{ij} = 0
\tag{14}
$$

$$
[-X^T W + H^T W^T W + \beta (D^W - S^W)H - \beta (D^B - S^B)H + \varphi]_{ij} H_{ij} = 0
\tag{15}
$$

According to (14) and (15), the formula to update the W and H can be obtained as

$$W_{ij} = W_{ij} \frac{[XH]_{ij}}{[WH^T H + \alpha EW]_{ij}} \tag{16}$$

$$H_{ij} = H_{ij} \frac{[X^T W + \beta(D^B + S^W)H]_{ij}}{[HW^T W + \beta(D^W + S^B)H]_{ij}} \tag{17}$$

3 Experiment and Results

In this section, we conducted a set of experiments to verify the effectiveness of the proposed SCDNMF method and compare it with NMF [2], GNMF [12], SNMF [10] and NMF-kNN [13]. Two face image databases and one object image database including Yale [14], ORL [15] and COIL20 [16] are used in this study. In each experiment, the images set are partitioned into a training set and a testing set with different numbers. For ease of representation, the experiments are named as p-train, which means that p images per individual are selected for training. To fairly evaluate the performance of different algorithms in different training and testing conditions, we selected images randomly and repeated the experiment 10 times.

3.1 Experiments on Yale Face Image Database

The Yale face database was constructed at the Yale Center for Computation Vision and Control. There are 165 images of 15 individuals (each person providing 11 different images).The images demonstrate variations in lighting condition, facial expression, and with or without glasses. In our experiment, all images are cropped and resized to the resolution of 32×32 pixels. Some images from Yale database shown in Fig. 1.

Fig. 1. Some images from Yale face database

In the first experiment, p (p=5) images for each person are randomly selected for training and the rest were used for testing. For GNMF, the neighborhood size k is set to 4. For our proposed SCDNMF, the intra-class neighborhood size k_1 and the inter-class neighborhood size k_2 are set to 4 and 8 respectively, the parameters α and β are all set as 1, and the heat kernel parameter t is set to 100. The performances of different algorithms are shown in Fig. 2. From this figure, we can see that the recognition results of NMF, GNMF, SNMF and NMF-kNN are very low when the feature dimension is equal to 20 at the beginning. However, the performance of our proposed

SCDNMF is obviously better than other algorithms. Then, with the increase in feature dimension, the recognition results are also improved. But this trend is not maintained for all the dimensions. The best average recognition accuracy rates on Yale face image database is listed in Table 1. From this result, we can observe that GNMF outperforms NMF. This implies that preserving geometrical structure of data is important in matrix factorization process. Secondly, SNMF also outperforms the NMF, which indicates that the structure information of basis images provides benefits to recognition tasks. Thirdly, we can see that NMF-kNN outperforms NMF, GNMF and SNMF, which is due to that the label information is incorporated into non-negative matrix factorization algorithm. At last, it can be found that our proposed SCDNMF performs better than all other algorithms. This is because SCDNMF combines both the structure information and label information into the NMF framework.

Table 1. The maximal recognition rates (%) and standard deviations (%) of different algorithms on the Yale face database

NMF	GNMF	SNMF	NMF-kNN	SCDNMF
79.56±3.32(140)	80.22±2.27(200)	80.44±2.35(160)	80.67±2.04(180)	82.11±3.03(200)

Note that the number in parentheses is the corresponding feature dimensions with the best result.

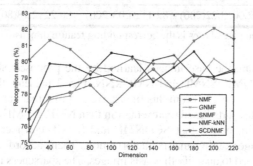

Fig. 2. Recognition rates vs. dimensionality on the Yale face database

In the second experiment, the performances of the proposed SCDNMF algorithm under different α and β values are evaluated. Here, we set the values of α and β from 0.01 to 1. The maximal average recognition rates of the proposed SCDNMF are given in Table 2. From this table, we can find that the performance of our SCDNMF depends on the choice of α and β. With the increase of α and β values, the recognition results are also improved in most cases. SCDNMF achieves its best recognition rate under $\alpha=1$ and $\beta=1$, respectively. This may due to that a larger α value can make the basis image more structure sparse and a larger β value can bring more discriminative information to improve the discriminative ability of the low-dimensional features.

Table 2. The average recognition accuracy rate of proposed SCDNMF with different α and β on the Yale database

	$\beta=0.01$	$\beta=0.10$	$\beta=0.50$	$\beta=1.00$
$\alpha=0.01$	79.89±4.07(180)	80.33±2.87(200)	80.44±2.88(100)	80.33±2.67(80)
$\alpha=0.10$	80.22±3.62(80)	79.89±3.30(220)	80.56±3.02(180)	80.44±2.83(80)
$\alpha=0.50$	81.00±2.79(100)	80.22±3.18(200)	81.00±3.16(220)	81.67±2.78(140)
$\alpha=1.00$	80.22±3.39(100)	80.33±3.06(180)	81.56±3.32(80)	82.11±3.03(200)

Note that the number in parentheses is the corresponding feature dimensions with the best result.

In the third experiment the impact of the inter-class neighborhood parameters (k_2) on the performance of SCDNMF algorithm is tested. In this experiment, because the number of training samples per person is small, we fix the parameter k_1 to 4, and then set the parameter k_2 to 4, 6, 8, 10 and 12 respectively. The corresponding best average recognition rates achieved by SCDNMF are demonstrated in Table 3. From this table, we can see that the proposed SCDNMF performs best when the parameter k_2 is neither too small nor too large ($k_2=8$).

Table 3. The average recognition rate (%) and standard deviation (%) of SCDNMF with varied k_2 on the Yale database

$k_2=4$	$k_2=6$	$k_2=8$	$k_2=10$	$k_2=12$
81.67±1.73(200)	82.00±3.22(200)	82.11±3.03(200)	80.22±2.18(80)	80.56±2.18(160)

Note that the number in parentheses is the corresponding feature dimensions with the best result.

At last, we compare the base images learned by the proposed SCDNMF with those learned by NMF, GNMF, SNMF and NMF-kNN. Fig. 3 presents these base images in subspace of dimensionality 36. From this figure, we can find that our proposed SCDNMF and SNMF learn better parts-based representation than NMF, GNMF and NMF-kNN. The reason to this phenomenon is that SCDNMF and SNMF consider the structure information of basis images during matrix factorization. Then, according to Hoyer [7], the sparseness index is used to quantify the basis images. The sparseness index is defined as

$$\text{sparseness}(W) = (\sqrt{W} - (\textstyle\sum_{i=1}^{r}|W_i|) / \sqrt{\textstyle\sum_{i=1}^{r}W_i^2}) / (\sqrt{r} - 1) \tag{18}$$

where r is the dimensionality of basis vector W. The average sparseness indices of the basis vector learned by different methods are listed in Table 4. From this table, we can see that both SCDNMF and SNMF obtain sparser basis than those of NMF, GNMF and NMF-kNN.

(a) NMF (b) GNMF (c) NMF-kNN (d) SNMF (e) SCDNMF

Fig. 3. Basis images learned by different algorithms on the Yale face dataset. Each subfigure represents a basis image and the higher intensities are assigned to the pixel with larger values.

Table 4. The average sparseness indices of the basis vector learned by different methods on the Yale face database

NMF	GNMF	NMF-kNN	SNMF	SCDNMF
53.56%	50.82%	52.00%	85.00%	83.81%

3.2 Experiments on ORL Face Image Database

The ORL face database contains 400 images of 40 persons, 10 images per person. The images are taken at different times, varying lighting conditions, facial expressions and facial details. All the images are taken against a dark homogeneous background. The faces are in up-right position of frontal view, with slight left-right out-of-plane rotation. In our experiments, we use the cropped images with the resolution of 32×32. Some images of one object selected from ORL database shown in Fig. 4.

Fig. 4. Images of one persons from ORL face database

Firstly, like the experiment on Yale database, p ($p=5$) images of each person are randomly selected for training and the rest were used as testing samples. For our proposed SCDNMF, the parameters α and β are all set as 0.1. And other parameters are set as same as those in experiment on Yale database. The performance of different algorithms is shown in Fig. 5. From this figure, we can see that the recognition results of all methods improve with the increase of feature dimension at the beginning. However, when they achieve their best results, their performances begin to decrease with the increase in dimension. The reason to this performance loss may be that too many features will bring some noises into the subspace, which leads to the decrease of the recognition rates. The best average recognition accuracy rates on ORL face image database is listed in Table 5. From this result, we can find that the proposed SCDNMF outperforms other methods, which is consistent with the experimental results on Yale face database.

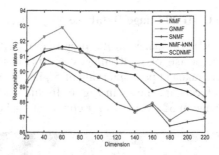

Fig. 5. Recognition rates vs. dimensionality on ORL face database

Table 5. The maximal average recognition rates of different algorithms on ORL database

NMF	GNMF	SNMF	NMF-kNN	SCDNMF
90.40±1.81(40)	91.50±2.16(40)	90.85±1.93(40)	91.65±1.89(60)	92.90±1.74(60)

Note that the number in parentheses is the corresponding feature dimensions with the best result.

Then, the performances of the proposed SCDNMF algorithm under different α and β values are also evaluated. Like the experiment on Yale face database, we also set α and β value from 0.01 to 1. The maximal average recognition rates of the proposed SCDNMF are given in Table 6. From this table, we can find SCDNMF achieves its best recognition rate under α=0.1 and β=0.1, respectively.

Table 6. The average recognition accuracy rate of proposed SCDNMF varies with α and β on the ORL database

	β=0.01	β=0.10	β=0.50	β=1.00
α=0.01	91.80±1.40(80)	91.10±2.57(40)	91.65±1.55(40)	92.45±1.50(40)
α=0.10	91.60±1.22(40)	92.90±1.74(60)	91.60±1.52(40)	91.40±1.73(40)
α=0.50	92.10±2.41(60)	91.80±1.53(60)	91.50±1.83(60)	92.05±1.82(60)
α=1.00	91.30±2.16(40)	92.20±2.18(60)	91.25±1.80(40)	92.15±1.36(60)

Note that the number in parentheses is the corresponding feature dimensions with the best result.

At last, the impact of the inter-class neighborhood parameters (k_2) on the performance of SCDNMF algorithm is also tested. Like the experiment on Yale face database, we also fix the parameter k_1 to 4, and then set the parameter k_2 to 4, 6, 8, 10 and 12 respectively. The best average recognition rates of SCDNMF are listed in Table 7. From this table, we can see that the proposed SCDNMF performs best when k_2=8, which is consistent with the experimental results on Yale face database.

Table 7. The average recognition rate (%) and standard deviation (%) of SCDNMF with varied k_2 on the ORL database

k_2=4	k_2=6	k_2=8	k_2=10	k_2=12
91.10±1.64(40)	91.35±1.58(60)	92.90±1.74(60)	91.55±2.03(40)	91.65±1.93(40)

Note that the number in parentheses is the corresponding feature dimensions with the best result.

3.3 Experiments on COIL20 Image Database

In the last experiment, the performance of the proposed algorithm is tested using the COIL20 database which consists of 1,440 images of 20 objects. In this database, the images of the object were taken at pose intervals of 5°, i.e. 72 poses per object. Some sample images of four objects are shown in Fig. 6. In this study, all images are converted to a gray-scale image of 32×32 pixels for computational efficiency.

Fig. 6. Some images of four objects from COIL20 database

Similar to experiments on Yale and ORL face database, p ($p=12$) images of each object are randomly chosen to form the training set and other images are regarded as testing set. For our proposed SCDNMF, the parameters α and β are set as 0.1 and 1, respectively. The intra-class neighborhood size k_1 and the inter-class neighborhood size k_2 are all set as 8. The other parameters are set the same as those in face recognition experiments. The maximal average recognition rate of different algorithms on COIL20 database is shown by Fig. 7. From this figure, we can see that that the proposed SCDNMF outperforms other methods, which is consistent with the experimental results on Yale and ORL face database. Then, we test the impact of parameters α and β on the performance of our proposed SCDNMF. Here, the values of α and β are set as 0.01, 0.10, 0.50 and 1.00. The maximal average recognition rates of the proposed SCDNMF are recorded in Table 8. From this table, we can find SCDNMF achieves its best recognition rate under $\alpha=1$ and $\beta=1$, which justifies the choice of parameter values in this experiment. At last, we also compare the basis images learned by different algorithms. In this experiment, the subspace of dimensionality is set as 25. Fig. 8 shows the basis images obtained by different algorithms. The average sparseness indices of the bases learned by different methods are listed in Table IX. We can see that the sparseness of our algorithm is a little bit lower than SNMF but much higher than other algorithms.

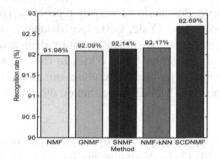

Fig. 7. The maximal average recognition rate of different algorithms on the COIL20 database

Table 8. The average recognition accuracy rate of proposed SCDNMF varies with α and β on COIL20 database

	$\beta=0.01$	$\beta=0.10$	$\beta=0.50$	$\beta=1.00$
$\alpha=0.01$	92.10±0.93(20)	92.11±1.32(20)	92.33±1.06(20)	92.64±1.88(20)
$\alpha=0.10$	92.17±1.52(20)	92.01±0.97(20)	92.32±1.21(20)	92.63±1.12(20)
$\alpha=0.50$	92.36±1.17(20)	92.18±0.98(20)	92.50±1.39(20)	92.53±0.89(20)
$\alpha=1.00$	92.58±1.22(40)	92.69±1.60(40)	92.66±1.17(40)	92.63±1.12(20)

Note that the number in parentheses is the corresponding feature dimensions with the best result.

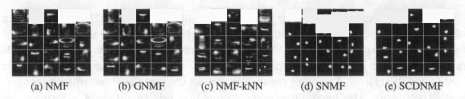

(a) NMF   (b) GNMF   (c) NMF-kNN   (d) SNMF   (e) SCDNMF

Fig. 8. Basis images learned by different algorithms on the COIL20 database. Each subfigure represents a basis image and the higher intensities are assigned to the pixel with larger values.

Table 9. The average sparseness indices of the basis vector learned by different methods on the COIL 20 database

NMF	GNMF	NMF-kNN	SNMF	SCDNMF
62.08%	61.88%	60.26%	83.15%	80.92%

4   Conclusions and Future Work

In this paper, we developed a new matrix factorization method called Structure Constrained Discriminant Non-negative Matrix Factorization (SCDNMF) for feature extraction. Our proposed method combines discriminative information and structure information together during the matrix factorization process. The proposed SCDNMF are applied for face recognition (Yale, ORL face database) and object recognition (COIL20 database) tasks. These experimental results demonstrate that SCDNMF is more effective than other algorithms based NMF. In this work, we only concentrate on face and object recognition problem. Therefore, our future work will evaluate the performance of the proposed SCDNMF algorithm on other tasks such as iris and palm print recognition.

Acknowledgment. This work is supported by the Science Foundation for Post-doctor of Jilin Province (No. 2011274), Young scientific research fund of Jilin province science and technology development project (No. 20130522115JH).

References

1. Mel, B.W.: Think positive to find parts. Nature 401, 759–760 (1999)
2. Lee, D.D., Seung, H.S.: Learning the parts of objects by non-negative matrix factorization. Nature 401, 788–791 (1999)
3. Zafeiriou, S., Petrou, M.: Nonlinear non-negative component analysis algorithms. IEEE Transactions on Image Processing 19(4), 1050–1066 (2009)
4. Buciu, I., Pitas, I.: Subspace image representation for facial expression analysis and face recognition and its relation to the human visual system. In: Organic Computing, pp. 303–320. Springer (2008)
5. Chen, X., Gu, L., Li, S.Z., Zhang, H.J.: Learning representative local features for face detection. In: CVPR, pp. 1126–1131 (2001)

6. Li, S.Z., Hou, X.W., Zhang, H.J., Cheng, Q.S.: Learning spatially localized, parts-based representation. In: CVPR, pp. 207–212 (2001)
7. Hoyer, P.O.: Non-negative sparse coding. In: IEEE Workshop on Neural Networks for Signal Processing, pp. 557–565 (2002)
8. Pascual-Montano, A., Carazo, J.M., Kochi, K., Lehmean, D., Pascual-Marqui, R.: Nonsmooth nonnegative matrix factorization (nsNMF). IEEE Transactions on Pattern Analysis and Machine Intelligence 28(3), 403–415 (2006)
9. Jenatton, R., Obozinski, G., Bach, F.: Structured sparse principal component analysis. In: International Conference on Artificial Intelligence and Statistics (2010)
10. Zheng, W.S., Lai, J.H., Liao, S.C., He, R.: Extracting Non-negative Basis Images Using Pixel Dispersion Penalty. Pattern Recognition 45(8), 2912–2926 (2012)
11. Zafeiriou, S., Tefas, A., Buciu, I., Pitas, I.: Exploiting discriminant information in nonnegative matrix factorization with application to frontal face verification. IEEE Transactions on Neural Networks 17(3), 683–695 (2006)
12. Cai, D., He, X., Han, J., Huang, T.: Graph regularized nonnegative matrix factori-zation for data representation. IEEE Transactions on Pattern Analysis and Machine Intelligence 33(8), 1548–1560 (2011)
13. An, S., Yoo, J., Choi, S.: Manifold-Respecting Discriminant Nonnegative Matrix Factorization. Pattern Recognition Letters 32(6), 832–837 (2011)
14. Yale University Face Database (2002),
    http://cvc.yale.edu/projects/yalefaces/yalefaces.html
15. Samaria, F.S., Harter, A.C.: Parameterisation of a stochastic model for human face identification. In: IEEE Workshop on Applications of Computer Vision, pp. 138–142 (1994)
16. Nene, S.A., Nayar, S.K., Murase, J.: Columbia Object Image Libraray (COIL20). Technical Report, CUCS-005-96, Columbia, university (1996)

Simulated Annealing Based Algorithm for Mutated Driver Pathways Detecting

Chao Yan[1], Hai-Tao Li[2], Ai-Xin Guo[1], Wen Sha[1], and Chun-Hou Zheng[1,2,*]

[1] College of Electrical Engineering and Automation, Anhui University, Hefei, China
zhengch99@126.com
[2] College of Information and Communication Technology, Qufu Normal University,
Rizhao, China

Abstract. With the development of Next-generation DNA sequencing technologies, one of the challenges is to distinguish functional mutations vital for cancer development, and filter out the unfunctional and random "passenger mutations." In this study, we introduce a modified method to solve the so-called maximum weight submatrix problem which is based on two combinatorial properties, i.e., coverage and exclusivity. This problem can be used to find driver mutations. Particularly, we enhance an integrative model which combines mutation and expression data. We apply our method to simulated data, the experiment shows that our method is efficiency. Then we apply our proposed method onto real biological datasets, the results also show that it is applicable in real applications.

Keywords: Next-generation DNA sequencing data, driver pathways identification, maximum weight submatrix problem, SAGA method.

1 Introduction

Cancer is very difficult to be treated. Researchers have found that cancer should be arisen by single-nucleotide mutations, larger copy-number aberrations or structural aberrations [1]. One of the challenge is to distinguish functional mutations vital for cancer development, which is so-called "driver mutations", and filter out the unfunctional and random "passenger mutations" [2]. In the past years, in gene level, even cancer genomes from the same tumor type, no two genomes exhibit exactly the same complement of somatic aberrations. In other words, the approaches cannot capture the heterogeneity of genome mutations [3,4].

As well known, same pathway may result from different genome aberrations [5,6]. Hence, it is significant to study gene in pathway level, rather than in gene level. Until now, most of studies analyze known pathway for enrichment of somatic mutations [3,4,7,8,9,10]. Unfortunately, knowledge of pathway remains incomplete, and many pathway databases contain overlap and unavailable data. Therefore, it is indispensable to develop Do novo discovery of mutated driver pathways without relying on prior knowledge.

---

[*] Corresponding author.

D.-S. Huang et al. (Eds.): ICIC 2014, LNAI 8589, pp. 658–663, 2014.

In these studies, the researchers find that there are two constraints on combinatorial patterns of mutations in cancer. First, generally, a driver mutation is rare. Particularly, researchers find that a single mutation is frequently enough to perturb one way. In other words, there is a phenomenon of mutual exclusivity between driver mutations. Second, a significant cancer pathway should cover a great majority of patients. Thus, the mutations will be had by most patients in the pathway. This is the property called high coverage. Lately, based on these two constraints, Vandin et al.[11] proposed a new and effective method, They defined the maximization of this method as the maximum weight submatrix problem. However, this problem is computationally difficult to solve. In order to solve this problem, in this paper, we proposed a new solution method, which is based on the simulated annealing (SA) method.

2   Materials and Methods

To identify driver pathways are extremely difficult. Considering this point, some researchers transformed this problem into maximum weight submatrix problem using two criteria [11], i.e., 'high coverage' and 'high exclusivity'. However, this problem is NP-hard. Many researchers use stochastic search methods to solve this problem. Particularly, Vandin et al. [11] proposes a method using two properties. The first one is 'high coverage', which means the majority of samples have at least one mutation in driver pathway; the second one is 'high exclusivity', which means that lots of samples have no more than one mutation in one driver pathway. The maximum weight submatrix problem is defined as selecting a submatrix M of size $m \times k$ in the mutation matrix A by calculating maximizing the scoring function:

$$W(M) = |\Gamma(M)| - \sum_{g \in M} |\Gamma(g)| \tag{1}$$

Where $\Gamma(g) = \{i : A_{ig} = 1\}$ denotes that gene g in i-th row (sample) is mutated. $\Gamma(M) = \bigcup_{g \in M} \Gamma(g)$ represents the set of patients, in which at least one of the genes in M is aberrations. So, $|\Gamma(M)|$ indicates the coverage of M.

$$\omega(M) = \sum_{g \in M} |\Gamma(g)| - |\Gamma(M)| \tag{2}$$

Eqn.(2) denotes the coverage overlap weight. In order to solve this problem, Vandin et al. [11] propose the Markov chain Monte Carlo (MCMC) method. After that, Zhao et al. [13] used the genetic algorithm (GA) to solve this problem, and achieve good experimental results.

However, local search method proposed by them is not good enough to solve this problem. The result may be trapped in a local solution. Taking into account this situation, in this paper, we propose to use simulated annealing hybrid genetic algorithm (SAGA) to solve this problem. Simulated annealing (SA) as an optimization and heuristic algorithm mimics certain thermodynamic principles of producing an ideal crystal, which can be used to solve large-scale optimization problems in order to achieve a global optimal solution [12]. SA has been widely used in operational research problems. The details of our implementation, named simulated annealing hybrid genetic algorithm (SAGA), for the maximum weight submatrix problem based on SA are described as follows:

Step 1.Initialize the temperature S0.

Step 2.Using GA method to generate initial solution submatrix M, and generate the scoring function W(M).

Step 3.Using GA method to generate a new solution submatrix M', in the neighborhood of current solution X, revaluate the scoring function W(M').

Step 4.If the generated solution submatrix scoring W(M') is larger than former W(M), putting M=M'. Update the existing optimal solution and go to step 6.

Step 5.Else accepts M' with probability:

$$p = e^{\frac{\Delta S}{T}} \text{ , where } \Delta S = W(M') - W(M)$$

If the solution is accepted, then M=M'. Update the existing optimal solution.

Step 6.Decrease the temperature periodically.

Step 7.Repeat step 2 through 6 until stopping criterion is met.

Figure 1 shows the process of SA. It can be seen clearly that we can solute the global maximum solution by using SA.

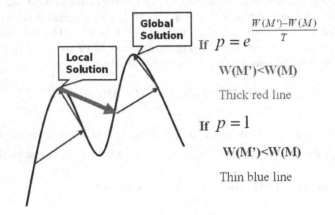

Fig. 1. Simulated annealing: escape from local maximum solution

3   Results

3.1   Simulation Study

The simulated mutation data has with five gene sets M1, M2, ... , Mk. Each set has k members (k=5 has been used in this study). For each row, we set the number to 1(chosen uniformly at random) in Mi (i=1, 2, ... , 5) with probability pi (pi=1-i·Δ, Δ=0.05 has been used in this study), and if gene is 1 already, after that, with probability p0 we set the others to 1 in Mi (p0=0.04 has been used in this study). We can see that pi indicates the coverage of Mi and p0 indicates exclusivity of Mi. The others in Mi are mutated using a random model based on the observed characteristics of the glioblastoma data. This is the background mutation rate in M.

We compared the time complexity of MCMC, GA and SAGA on selecting the submatrix of maximum weight (Fig. 2). From figure 2 we can see clearly that the GA is faster than MCMC when n less than 5000. Particularly, SAGA is always faster than GA from $n=1000$ to $n=10000$. In the research, it is well know that for almost all of real applications, the n is smaller than 5000. On the other hand, the results of SAGA are the same as GA method.

We use an exact approach to test the accuracy of these methods, which is called binary linear programming (BLP) model [13]. We run the BLP method to compare MCMC, GA performance with SAGA. The accuracy of GA and SAGA are THE same, which is 95%, higher than that of MCMC, which is 44%.

In summary, our SAGA method has competitive efficiency with GA and MCMC.

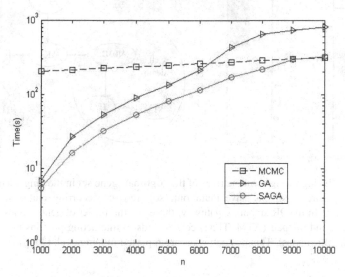

Fig. 2. Compared of computational time of SAGA, GA and MCMC in terms of gene number n from 1000 to 10000. In this polt, we use semilog coordinate (the y axis) to show the computational time in seconds. All the makers correspond to the results of an average over 10 times.

3.2   Lung Adenocarcinoma

We then apply our SAGA method for analyzing a dataset of 1013 somatic mutations identified in 188lung adenocarcinoma patients' 623 sequenced genes from the Tumor Sequencing Project [3]. According to statistics, there are 365 genes mutated in at least one patient. We run the SAGA for sets of size $2 \leq k \leq 10$. After running this algorithm, when $k=2$, the pair (EGFR, KBAS) has the maximum weight submatrix. When $k=3$, the most significant triplet is (EGFR, KRAS, STK11). The frequency is 8.4% for all the samples. When $k \geq 4$, the frequency of all the sets are less than 0.3%. After that we performed a permutation test [11], the p-value obtained is 0.018, which is larger than that of the triplet (EGFR, KRAS, STK11). In other word, the triplet (EGFR, KRAS, STK11) is at least as significant as the pair (EGFR, KRAS). In biological, we find that EGFR, KRAS and STK11 are all involved in the pathway of mTOR (Fig. 3), and the mTOR pathway is very important in lung adenocarcinoma [3]. Hence, our method can seek out driver pathway.

We remove the above three genes and apply the methods to detect the additional gene sets. On the remaining genes, when $k=2$, we identify the gene set (ATM, TP53) that is mutated with frequency 56%, and find that the weight of the pair is significant ($p<0.01$). Previous studies have shown that both of ATM and TP53 are in the cell cycle checkpoint control and direct interaction [14,15](Fig. 3).

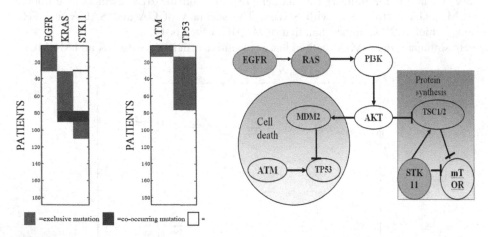

Fig. 3. (Left) The high weight submatrices of the 'optimal' gene set in the lung adenocarcinoma data. In this picture: (red) exclusive mutation; (soft red) co-occurring mutation; (white) no mutation; (Right) In mTOR signaling pathway, there are the triplet of genes codes for proteins (orange nodes), and the pair (ATM, TP53) corresponds to interacting proteins in the cell cycle pathway (light blue nodes). This two pathways are reported in Ding et al. (2008).

4   Discussion and Conclusion

In this paper, we introduce an algorithm for discovering mutated driver patterns de novo using somatic mutation data from biological datasets, which is based on recently exploring by Vandin et al. [11] and Zhao et al.[13]. We proposed an optimization and heuristic algorithm simulate annealing hybrid genetic algorithm, which is named SAGA. Compared published methods, our methods can achieve satisfactory results. In future, we will study the biological data, such as DNA methylation and copy-number variant (CNV), exploring the regular pattern of co-occurring and the others mutated driver pathways.

Acknowledgments. This work was supported by the National Science Foundation of China under Grant nos. 61272339 and 61271098, and the Key Project of Anhui Educational Committee, under Grant no. KJ2012A005.

References

1. Hanahan, D., Weinberg, R.A.: The hallmarks of cancer. Cell 100, 57–70 (2000)
2. Greenman, C., et al.: Patterns of somatic mutation in human cancer genomes. Nature 446, 153–158 (2007)
3. Ding, L., et al.: Somatic mutations affect key pathways in lung adenocarcinoma. Nature 455, 1069–1075 (2008)
4. Jones, S., et al.: Core signaling pathways in human pancreatic cancers revealed by global genomic analyses. Science 321, 1801–1806 (2008)
5. Hahn, W.C., et al.: Modelling the molecular circuitry of cancer. Nat. Rev. Cancer 2, 331–341 (2002)
6. Vogelstein, B., et al.: Cancer genes and the pathways they control. Nat. Med. 10, 789–799 (2004)
7. The Cancer Genome Atlas Research Network (TCGA): Comprehensive genomic characterization defines human glioblastoma genes and core pathways. Nature 455, 1061–1068 (2008)
8. Kanehisa, M., Goto, S.: KEGG: Kyoto Encyclopedia of Genes and Genomes. Nucleic Acids Res. 28, 27–30 (2000)
9. Jensen, L.J., Kuhn, M., Stark, M., Chaffron, S., et al.: STRING 8—a global view on proteins and their functional interactions in 630 organisms. Nucleic Acids Res. 37, D412–D416 (2009)
10. Keshava Prasad, T.S., Goel, R., Kandasamy, K., Keerthikumar, S.A., et al.: Human Protein Reference Database—2009 update. Nucleic Acids Res.37, D767–D772 (2009)
11. Vandin, F., et al.: De novo discovery of mutated driver pathways in cancer. Genome Res. 22, 375–385 (2012)
12. Sharda, R., Vob, S., Woodruff, D.L., Fink, A.: Optimization software class libraries, pp. 91–94. Springer, Berlin (2003)
13. Zhao, J., et al.: Efficient methods for identifying mutated driver pathways in cancer. Bioinformatics 28(22), 2940–2947 (2012)
14. Khanna, K.K., Keating, K.E., Kozlov, S., Scott, S., Gatei, M., Hobson, K., Taya, Y., Gabrielli, B., Chan, D., Lees-Miller, S.P., et al.: ATM associates with and phosphorylates p53: Mapping the region of interaction. Nat. Genet. 20, 398–400 (1998)
15. Chehab, N.H., Malikzay, A., Appel, M., Halazonetis, T.D.: Chk2/hCds1 functions as a DNA damage checkpoint in G1 by stabilizing p53. Genes. Dev. 14, 278–288 (2000)

A Grouped Fruit-Fly Optimization Algorithm for the No-Wait Lot Streaming Flow Shop Scheduling

Peng Zhang and Ling Wang

Tsinghua National Laboratory for Information Science and Technology (TNList),
Department of Automation, Tsinghua University, Beijing, 100084, P.R. China
zhangpenghit@aliyun.com, wangling@mail.tsinghua.edu.cn

Abstract. Lot streaming involves splitting products into several sublots to enhance the processing flexibility. In this paper the no-wait lot streaming flow shop scheduling problem is discussed with the constraint of unintermingling consistent-size sublots. The number of sublots, sublots sizes and the sequence of products are optimized by a proposed grouped fruit fly optimization algorithm (GFOA). We design three basic neighborhood structures and a structure based on the no-wait property to generate new solutions. A speed-up evaluation method is presented to improve the efficiency. Numerical test results and comparisons are provided, which demonstrate the effectiveness of the proposed GFOA.

Keywords: fruit fly optimization, lot streaming, no wait flow shop, job splitting, sequencing.

1 Introduction

Lot streaming (LS) is to split jobs into several sublots so that the work-in-process can be reduced and the cycle time can be shortened. The LS has been widely applied in many manufacturing industries [1-4]. One typical application environment of LS is flow shop scheduling problem (LSFSP). Earlier work managed to provide mathematical models and exact methods for small-scale problems. A comprehensive review can be found in [5]. However, exact methods are not suitable to large-scale LSFSP with multi-product, multi-machine since they might incur an unacceptable long time to obtain optimal solutions. Recently, meta-heuristics have gained increasing attention. Most existing work is about the FSP with unintermingling equal-size sublots, like genetic algorithm (GA) [6], ant colony optimization (ACO) [7], threshold accepting (TA) [7], discrete differential evolution (DDE) [8], discrete particle swam optimization (DPSO) [9], shuffled fog-leaping algorithm (SFLA) [10], discrete artificial bee colony algorithm (DABC) [11], harmony search (HA)[12], and estimation of distribution algorithm (EDA) [13]. For the no-wait FSP with unintermingling consistent-size sublots, a GA mixed with a linear programming [14] and a generalized traveling salesman problem (GTSP) based heuristic [15] have been presented. For the FSP with intermingling consistent-size sublots, a GA combining

D.-S. Huang et al. (Eds.): ICIC 2014, LNAI 8589, pp. 664–674, 2014.
© Springer International Publishing Switzerland 2014

with mathematical programming based heuristic [16] has been applied. Please refer to [2] for more details about recent studies.

This paper deals with the no-wait lot streaming FSP (nwLSFSP) with unintermingling consistent-size sublots. The no-wait constraint demands that every sublot should be processed continuously from its start on the first machine to its completion on the last machine without any interruption on machines and any waiting time between machines [15]. Such requirement widely exists in steel-making and pharmaceutical industries [16]. Moreover, for unintermingling consistent-size sublots it means that the sublots of the same product are consistent over all machines and they must be processed continuously. So, it needs to find the optimal number of sublots and sublots sizes for all the products as well as the sequence of products on machines. Obviously, the problem is very difficult to solve.

The FOA is inspired by the knowledge from the food finding behavior of fruit flies [17]. Recently, it has been applied to financial distress [17], power load forecasting [18], and web auction logistics service [19]. Due to the simplicity of the FOA, we will propose a novel grouped fruit-fly optimization algorithm (GFOA) to solve the nwLSFSP in this paper. The organization of the remainder of the paper is as follows. Section 2 introduces the problem description. Section 3 presents the neighborhood structures. In Section 4, the GFOA is presented in details. Test results and comparisons are provided in Section 5. Finally, we end the paper in Section 6.

2   Problem Description

The nwLSFSP is described as follow: Each job $j \in \{1,2,...N\}$ has L_j identical items which can be split into s_j sublots. Let X_{jk} denote the size of the kth sublot of job j. Let p_{ij} denote the unit processing time of job j on machine $i \in \{1,2,...M\}$. Let p_{ijk} denote the processing time of the kth sublot of job j on machine i. So, $p_{ijk} = p_{ij} \times X_{jk}$. All the sublots of any job should be processed continuously on any machine, and each sublot should be processed continuously from its start on the first machine to its completion on the last one without any waiting time between machines. Machine i costs a setup time R_{ij} when processing the first sublot of job j. Suppose that release time of each job is zero, buffer size is infinite, and transportation time of each sublot is included in the processing time. The objective is to minimize the makespan by determining the sublot quantity s_j and size X_{jk} for each job j and the sequence of all the jobs.

Given a solution, i.e., s_j, X_{jk} and the sequence, it needs to calculate the makespan, which is usually called decoding. Next, we first analyze the structure of single product schedule, and then show that the computational complexity of the decoding can be reduced, and finally present the decoding process for the multi-product case.

For a single product, it illustrates the structure of a schedule with sublot size vector $[X_1, X_2,..., X_s]$ in Fig. 1. The second subscript is omitted since only a single product is considered. The setup time R_i is attached for the first sublot on machine i. Let d_l denote the idle time on the first machine after sublot l, and d_0 be the sum of idle time and setup time before the first sublot. It can be seen that makespan C_{max} is the sum of the idle time and the processing time on the first machine with the following formula.

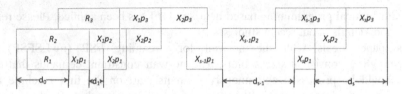

Fig. 1. The structure of single product schedule

$$C_{max} = d_0 + \sum_{l=1}^{s-1} d_l + d_s + p_1 \times L; \tag{1}$$

$$d_0 = \max\{R_1, \max_{2 \leq i \leq M}\{R_i - X_1 \sum_{j=1}^{i-1} p_j\}; \tag{2}$$

$$d_l = \max\{0, \max_{2 \leq i \leq M}\{X_l \sum_{j=2}^{i} p_j - X_{l+1} \sum_{j=1}^{i-1} p_j\}, 2 \leq l \leq s-1; \tag{3}$$

$$d_s = X_s \sum_{i=2}^{M} p_i; \tag{4}$$

The calculation of d_l ($1 \leq l \leq s$-1) contains M times of sums with the time complexity $O(M^2)$, which includes some unnecessary repeated sums. Since $\sum_{j=2}^{i+1} p_j$ equals $\sum_{j=2}^{i} p_j + p_{i+1}$, the time complexity can be reduced to $O(M)$, which can be further reduced to $O(\ln M)$ [20].

The structure of multi-product case is shown in Fig. 2. The idle time appears when processing either sublots of a product, i.e., d_{jl}, or sublots of two products, i.e. $D_{j,j+1}$. The makespan can be calculated by (5), where $D_{j,j+1}$ is calculated by (6) and d_{jl} is calculated by (2). Suppose that each of the N jobs has s sublots. Then, the time complexity is $O(NM+Ns\ln M)$.

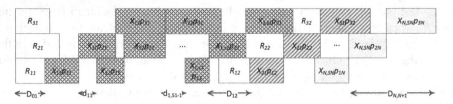

Fig. 2. The structure pattern of multi-product schedule

$$C_{max} = \sum_{j=0}^{N} D_{j,j+1} + \sum_{j=0}^{N} \sum_{l=1}^{S_j-1} d_{jl} + \sum_{j=0}^{N} P_{1j} \times L_j \tag{5}$$

$$D_{j,\,j+1} = \max_{1 \leq i \leq M-1} \{R_{0,j+1}, \ x_{j,s_j} \sum_{m=2}^{i+1} P_{mj} + R_{i+1,j+1} - x_{j+1,0} \sum_{m=1}^{i} P_{m,j+1}\}, 1 \leq j \leq N-1; \tag{6}$$

3   Neighborhood Structures and Evaluation

3.1   Basic Neighborhood Structures

Meta-heuristics move solutions with certain neighborhoods [21]. For the nwLSFSP, three basic structures can be designed.

The first one is called job-sequence move, where a job of a sequence is randomly selected and then inserted into another site. An example is illustrated in Fig. 3, where job 3 is selected and moved to site 2.

The second one is called sublot-size move, where two sublots (or their sizes) of a randomly selected job are exchanged or adjusted. An example is illustrated in Fig. 4. A sublot is removed if its size is reduced to 0.

The third one is called sublot-quantity move. The quantity of sublots of a randomly selected job will be increased or decreased by one, if the new quantity does not exceed the upper or lower bound. Then, the sizes of sublots will be adjusted by the equal-size heuristic as Eq. (7). An example is illustrated in Fig. 5.

$$q = \lfloor L_j / s_j \rfloor; n = L_j - q \times s_j;$$
$$X_{jk} = q, 1 \le k \le s_j - n; \tag{7}$$
$$X_{jk} = q + 1, s_j - n < j \le s_j;$$

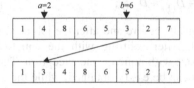

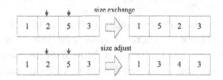

Fig. 3. An example of job-sequence move   **Fig. 4.** An example of sublot-size move

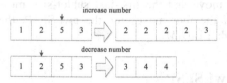

Fig. 5. An example of sublot-quantity move

3.2   No-Wait Property Based Neighborhood Structure

For the above moves, each sublot has an equal chance to be adjusted in size. It has been shown that the first and last sublot-sizes of jobs affect the result, and then it can be decomposed to a job sequencing problem and N sublot sizing problems if the first and last sublot-sizes are fixed [15]. So, we design a new neighborhood structure by directly changing the first and last sublots, which is called first-last sublot-size move or the no-wait property based neighborhood with the following steps.

Step1: Random select job j;

Step2: If the sublot quantity of job j is 1, split job j into two sublots with random
sizes. If the quantity is larger than 1, go to Step 3;

Step3: Select either the first or the last sublot of job j;

Step4: For the selected sublot, increase or decrease its size by one, and decrease
or increase the size of another randomly selected sublot of job j;

Step5: If the size is 0, remove the sublot and update the sublot number of job j.

These four moves can be adopted with different probabilities. In this paper, the
job-sequence, sublot-size, sublot-quantity and first-last sublot-size moves are used
with probabilities 0.4, 0.2, 0.2 and 0.2.

3.3   Evaluating Neighbors

Idle time depends on two sublots processed consecutively. Since a move only changes
a few sublots, makespan is only changed by a few idle times influenced by the
adjusted sublots.

For the job-sequence move, consider a move from $V_1=[J_1,\ldots,J_i,\ldots,J_j,\ldots J_N]$ to
$V_2=[J_1,\ldots,J_{i-1},J_{i+1},\ldots,J_j, J_i,\ldots J_N]$. The makespans of V_1 and V_2 are Q_1 and Q_2, and the
idle times between jobs are $D_{i,i+1}$ and $D_{i,i+1}'$, respectively. Then Q_2 can be calculated
as follows with the time complexity $O(M)$.

$$Q_2 = Q_1 - D_{i-1,i} - D_{i,i+1} - D_{j,j+1} + D_{i-1,i}' + D_{j-1,j}' + D_{j,j+1}'; \qquad (8)$$

For the sublot-quantity move, the idle time inside, before and after the job need re-
calculating as there is only one job changed. After splitting with the equal-size
heuristic, most adjacent sublots have the same size with none or one exception. Then,
the idle time between the same kinds of sublots can be easily obtained, while the idle
time between the different kinds of sublots should be re-calculated with the time
complexity $O(M)$.

For the sublot-size move and the first-last sublot-size move, the analysis is the
same. So, for each of the four moves for the nwLSFSP, the time complexity of
evaluating a new neighbor is $O(M)$.

4   GFOA for nwLSFSP

The traditional FOA is developed to solve continuous problems [17]. For the
nwLSFSP, the FOA should be designed in a specific way. In this paper, a three-level
encoding scheme is designed as Fig. 6 with 5 products. It implies that the processing
sequence of the jobs is 2-1-4-5-3; job 1 has 3 sublots with sizes 3, 5, 4; job 2 has 5
sublots with sizes 1, 2, 4, 3, 1; and so on. Once a solution with the encoding is
generated, its makespan can be calculated.

The FOA uses the smell-based and vision-based operators to perform search
process. To balance the exploration and exploitation, we design a special grouped
FOA (GFOA) to solve the nwLSFSP with two main search procedures, individual
neighborhood-based search and global cooperation-based search.

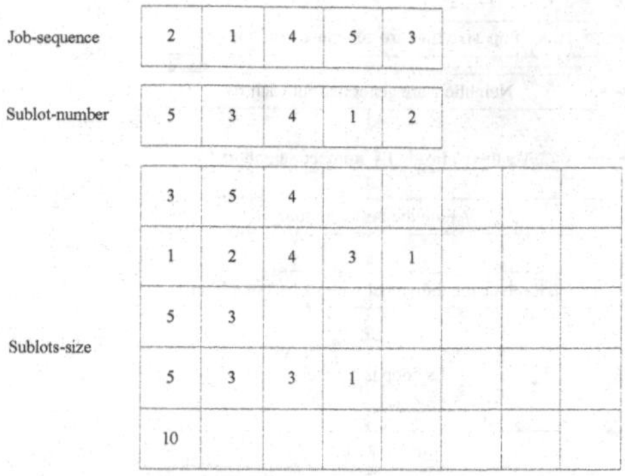

Job-sequence	2	1	4	5	3
Sublot-number	5	3	4	1	2

Sublots-size				
3	5	4		
1	2	4	3	1
5	3			
5	3	3	1	
10				

Fig. 6. An example of encoding scheme

For the first one, each fruit fly performs search around itself to generate several neighbors. Then, the best neighbor will replace the old one. The GFOA contains a group of fruit flies. Thus, the parallel exploration can be enhanced. To perform communication among fruit flies, a cooperation-based search is applied after performing the neighborhood-based search several times. To be specific, the better half of all the fruit flies with smaller makespan will guide the move of the other worse half pair by pair in the following way: the worst with the best, the second worst with the second best, and so on. Fig. 7 shows an example. For the worse one, the better one is used to perform the two-position based crossover to generate a new one, which inherits the partial job-sequence between two positions from the worse one. And then the corresponding sublot-quantity and sublot-size are inherited too. Such a cooperation can help the worse fruit flies jump out of bad regions with the guide of the better ones. The flowchart of the GFOA is illustrated in Fig. 8.

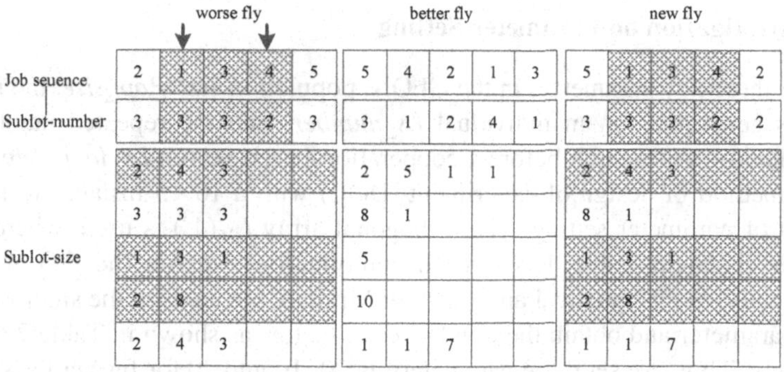

	worse fly					better fly					new fly				
Job sequence	2	1	3	4	5	5	4	2	1	3	5	1	3	4	2
Sublot-number	3	3	3	2	3	3	1	2	4	1	3	3	3	2	2
Sublot-size	2	4	3			2	5	1	1		2	4	3		
	3	3	3			8	1				8	1			
	1	3	1			5					1	3	1		
	2	8				10					2	8			
	2	4	3			1	1	7			1	1	7		

Fig. 7. An example of cooperation procedure

5   Simulation and Comparison

Same as the literature [14]-[15], random instances with different scales are produced for tests. The related data are random discrete integers with uniform distributions. To be specific, $p_{ij} \in U[1,5]$, the maximum sublot number $S_i \in U[2,10]$, $L_j \in U[2s_i, 7s_i]$, and

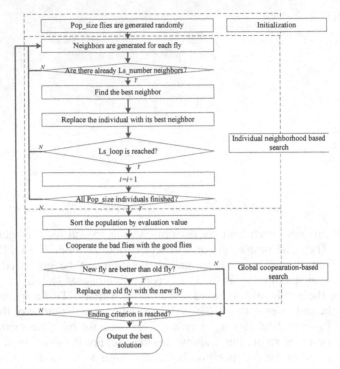

Fig. 8. The flowchart of GFOA

$R_{ij} \in U[1,5]$. For an instance, 30×20 means 30 jobs and 20 machines. The algorithm is coded in C++ and run on a PC of Core2(TM) P8400, 1.81GHz.

5.1   Investigation on Parameter Setting

There are three key parameters in the GFOA, population size *Pop_size*, the number of neighbors generated for an individual *Ls_number*, and the repeated number of the neighborhood-based search before a cooperation-based search *Ls_loop*. We apply the Taguchi method of design of experiment (DOE) with a 10×5 instance to investigate the effect of parameter setting. The orthogonal array $L_{16}(4^3)$ is used, where different values are listed in Table 1. For each combination, we run the GFOA 20 times independently. With statistical analysis tool Minitab, we analyze the significance rank of each parameter and obtain the trend of each factor, as shown in Table 2 and Fig. 8, respectively. Thus, we set these parameters as 20, 10 and 20 for further tests.

Table 1. Combination of Parameter Values

parameter	level			
	1	2	3	4
Pop_size	5	10	15	20
Ls_number	5	10	15	20
Ls_loop	5	10	15	20

Table 2. Response Value

Level	*Ls loop*	*Ls number*	*Pop size*
1	1440.21	1450.81	1427.52
2	1422.57	1413.30	1430.27
3	1422.41	1418.75	1424.84
4	1420.62	1422.96	1423.19
Delta	19.59	37.51	7.09
Rank	2	1	3

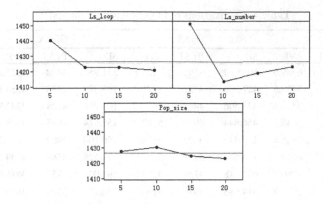

Fig. 9. Factor level trend of DFOA

5.2   Test on Designs

Next we compare the GFOA to two others without cooperation-based search, where FOA1 only uses three basic neighborhoods, and FOA2 uses three basic moves and the first-last sublot-size move. For each scale, 10 instances are generated. For each instance, we run every algorithm 10 times and then list the obtained best results (BEST), the average results (AVE) and the standard deviation (SD) in Table 3 and Table 4. Besides, RPI denotes the relative percentage increase from the best result of three algorithms for an instance to measure the search quality, and ARPI is the average values of all the instances. And RSD= SD/AVE×100 is used to measure the robustness, and ARSD implies the average RSD value for all the instances.

Table 3. Comparisons of FOAs (10×5 instances)

Instance no.	FOA1			FOA2			GFOA		
	BEST	AVE	SD	BEST	AVE	SD	BEST	AVE	SD
1	1372	1476.9	127.461	**1325**	1376.8	23.202	1341	**1357.8**	**10.768**
2	1065	1131.5	45.516	1058	1081.2	26.869	**1050**	**1066.8**	**9.053**
3	990	1116	127.635	990	1024.8	46.074	**986**	**993.5**	**4.528**
4	1182	1268.3	83.743	**1160**	1196.3	24.166	1162	**1168.5**	**5.233**
5	976	1089.6	105.631	986	1019.2	24.827	**974**	**983.3**	**7.528**
6	943	1099	195.964	937	960.4	38.842	**932**	**936.7**	**7.528**
7	929	972.2	70.783	907	956.8	93.326	**900**	**905.4**	**3.658**
8	**1379**	1570	181.126	1392	1479.3	57.077	1394	**1403**	**6.848**
9	**724**	810.1	100.375	**724**	740.8	15.223	726	**731**	**4.000**
10	1638	1807.5	254.116	**1631**	1701.3	71.887	1640	**1645.1**	**4.433**
ARPI	11.6121			4.1848			**1.12**		
ARSD	10.3485			3.7059			**0.5823**		

Table 4. Comparisons of FOAs (30×20 instances)

Instance no.	FOA1			FOA2			GFOA		
	BEST	AVE	SD	BEST	AVE	SD	BEST	AVE	SD
1	5316	6589.6	1035.699	4542	5224.8	351.957	**4408**	**4523.7**	**65.437**
2	4761	5459.5	427.506	4322	4725.7	196.322	**4193**	**4350.5**	**147.704**
3	4716	5459.5	427.506	4401	4657.2	104.756	**4128**	**4242.9**	**72.035**
4	5311	6046.2	456.844	4970	5187	157.474	**4645**	**4808.4**	**101.394**
5	6062	6505.8	249.84	5323	5560.1	156.104	**5089**	**5211.7**	**71.819**
6	5509	6569.3	755.57	4957	5246.7	271.528	**4569**	**4749.2**	**98.372**
7	4029	4980.7	659.178	4158	4283.6	99.502	**3753**	**3816.8**	**46.901**
8	4176	5087.8	514.985	3849	4166.2	150.131	**3779**	**3870.3**	**92.335**
9	4906	5898.4	536.785	4922	5183.1	202.838	**4694**	**4798.8**	**66.077**
10	6018	6760.8	560.892	5420	5904.7	371.977	**4906**	**5000.3**	**92.904**
ARPI	34.4549			13.4973			**2.7308**		
ARSD	9.5029			4.0298			**1.8947**		

By comparing FOA1 and FOA2, the significance of the first-last sublot-size move is proved. By comparing FOA2 and GFOA, the effectiveness of the cooperation-based search is further demonstrated.

5.3   Comparison with GA

To further test the GFOA, we compare it with the hybrid GA [14]. Both use four millions evaluations in each run. For each scale, 10 instances are generated and 10 runs are performed for each instance. The results are summarized in Table 5. And we carry out analysis of variance (ANOVA) for a 10×5 instance and a 30×20 instance. The results are shown in Table 6 and Table 7, and the 95% confidence intervals are shown in Fig. 9 and Fig. 10. Clearly, the GFOA dominates the GA for all the cases.

Table 5. Comparisons between GA and GFOA

Instance scale	GFOA			GA		
	ARPI	ARSD	ATIME	ARPI	ARSD	ATIME
10×5	**0.6735**	**0.4665**	**15.78**	6.2492	1.6140	49.90
10×20	**0.7874**	**0.9058**	**25.60**	7.7394	1.5357	99.82
20×5	**0.6203**	**0.4264**	**29.88**	7.1402	0.8602	95.865
20×20	**1.7269**	**1.3779**	**42.19**	9.1994	1.4729	197.12
30×5	**0.7672**	**0.4986**	**42.81**	7.2244	1.0149	142.74
30×20	**2.1491**	**1.5107**	**59.64**	7.8882	1.2555	302.04
40×5	**0.7648**	**0.4780**	**59.80**	7.1357	0.8398	210.28
40×20	**2.0632**	1.3225	**67.51**	6.8182	**1.2937**	450.95
50×5	**0.8562**	**0.4632**	**54.14**	6.6849	0.6528	223.80
50×20	**2.28226**	1.5228	**82.68**	6.6121	**1.0633**	566.56

Table 6. GA-GFOA ANOVA (10×5 instance)

Source	Dof	SS	MS	F	P
Between groups	1	27380	27380	177.93	0.000
Within group	18	2770	154		
Total	19	30150			

Table 7. GA-GFOA ANOVA (30×2 instance)

Source	Dof	SS	MS	F	P
Between groups	1	485161	485161	34.42	0.000
Within group	18	253701	14095		
Total	19	738863			

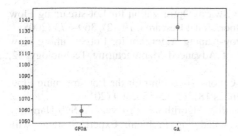

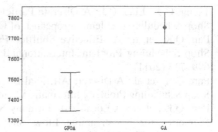

Fig. 10. Confidence interval (10×5 instance)   **Fig. 11.** Confidence interval (30×20 instance)

6   Conclusion

This was the first research paper to solve the no-wait flow shop scheduling problem with unintermingling consistent-size sublots (nwLSFSP) by using the fruit fly optimization algorithm (FOA). To optimize the job-sequence, sublot-quantity and sublot-size simultaneously with a special encoding scheme, we presented three basic neighborhood structures and a no-wait property-based neighborhood. To enhance parallel search ability, grouped mechanism was used for the FOA. A cooperation-based search was designed for sharing information among population, and a speed-up evaluation method was adopted. Testing results verified the effectiveness of the designs, and comparative results demonstrated the superiority of the GFOA to the existing GA in solving nwLSFSP. Future work could consider the FSP with intermingling case or other constraints like limited buffers.

Acknowledgments. This work is supported by National Key Basic Research and Development Program of China (2013CB329503), NSFC (61174189) and Doctoral Program Foundation of Institutions of Higher Education of China (20130002110057).

References

1. Potts, C.N., Baker, K.R.: Flow Shop Scheduling with Lot Streaming. Operations Research Letters 8(6), 297–303 (1989)
2. Cheng, M., Mukherjee, N., Sarin, S.: A Review of Lot Streaming. International Journal of Production Research 51(23-24), 7023–7046 (2013)
3. Kalir, A., Sarin, S.: A Near-optimal Heuristic for the Sequencing Problem in Multiple-batch Flow-shops with Small Equal Sublots. Omega 29(6), 577–584 (2001)
4. Chang, J., Chiu, H.: A Comprehensive Review of Lot Streaming. International Journal of Production Research 43(8), 1515–1536 (2005)
5. Sarin, S., Jaiprakash, P.: Flow Shop Lot Streaming Problems. Springer (2007)
6. Marimuthu, S., Ponnambalam, S.G., Jawahar, N.: Evolutionary Algorithms for Scheduling m-Machine Flow Shop with Lot Streaming. Robotics and Computer-Integrated Manufacturing 24(1), 125–139 (2008)
7. Marimuthu, S., Ponnambalam, S.G., Jawahar, N.: Threshold Accepting and Ant-colony Optimization Algorithm for Scheduling m-Machine Flow Shop with Lot-streaming. Journal of Materials Processing Technology 209(2), 1026–1041 (2009)
8. Sang, H.Y., et al.: A Discrete Differential Evolution Algorithm for Lot-Streaming Flow shop Scheduling Problems. In: International Conference on Natural Computation 10-13 (2010)
9. Tseng, C.T., Liao, C.J.: A Discrete Particle Swarm Optimization for Lot-streaming Flow Shop Scheduling Problem. European J. of Operational Research. 191(2), 360–373 (2008)
10. Pan, Q.K., et al.: An Effective Shuffled Frog-leaping Algorithm for Lot-streaming Flow Shop Scheduling Problem. International J. of Advanced Manufacturing Technology. 52, 699–713 (2011)
11. Pan, Q.K., et al.: A Discrete Artificial Bee Colony Algorithm for the Lot-streaming Flow Shop Scheduling Problem. Information Sciences 181(12), 2455–2469 (2011)
12. Pan, Q.K., et al.: A Local-best Harmony Search Algorithm with Dynamic Sub-Harmony Memories for Lot-streaming Flow Shop Scheduling Problem. Expert Systems with Applications 38, 3252–3259 (2011)
13. Pan, Q.K., Ruiz, R.: An Estimation of Distribution Algorithm for Lot-streaming Flow Shop Problems with Setup Times. Omega 40, 166–180 (2012)
14. Kumar, S., Bagchi, T.O., Srikandarajah, C.: Lot Streaming and Scheduling Heuristics for m-Machine No-wait Flowshops. Computers & Industrial Engineering 38(1), 149–172 (2000)
15. Hall, G., Sriskandarajab, C.: A Survey of Machine Scheduling Problems with Blocking and No-wait in Process. Operations Research 44(2), 510–525 (1996)
16. Marin, C.H.: A Hybrid Genetic Algorithm/Mathematical Programming Approach to the Multi-family Flow Shop Scheduling Problem with Lot Streaming. Omega 37(1), 126–137 (2009)
17. Pan, W.T.: A New Fruit Fly Optimization Algorithm: Taking the Financial Distress Model as an Example. Knowledge-Based Systems 26(2), 69–74 (2012)
18. Li, H., et al.: A Hybrid Annual Power Load Forecasting Model Based on Generalized Regression Neural Network with Fruit Fly Optimization Algorithm. Knowledge-Based Systems 37, 378–387 (2013)
19. Lin, S.M.: Analysis of Service Satisfaction in Web Auction Logistics Service using a Combination of Fruit Fly Optimization Algorithm and General Regression Neural Network. Neural Computing & Applications 22(3-4), 783–791 (2013)
20. Zhang, P.: Research on Lot Streaming Flow Shop Scheduling Problem Based on Fruit Fly Optimization Algorithm. Master Dissertation. Tsinghua University, Beijing (2013)
21. Eugeniusz, N., Czeslaw, S.: A Fast Tabu Search Algorithm for the Permutation Flow-shop Problem. European J. of Operational Research. 91, 160–175 (1996)

An Enhanced Estimation of Distribution Algorithm for No-Wait Job Shop Scheduling Problem with Makespan Criterion

Shao-Feng Chen[1,2], Bin Qian[1,2,*], Rong Hu[1,2], and Zuo-Cheng Li[1,2]

[1] Department of Automation, Kunming University of Science and Technology,
Kunming 650500, China
[2] Key Laboratory of Computer Technologies Application of Yunnan Province,
Kunming 650500, China
bin.qian@vip.163.com

Abstract. In this paper, an enhanced estimation of distribution algorithm (EEDA) is proposed for the no-wait job shop scheduling problem (NWJSSP) with the makespan criterion, which has been proved to be strongly NP-hard. The NWJSSP can be decomposed into the sequencing and the timetabling problems. The proposed EEDA and a shift timetabling method are used to address the sequencing problem and the timetabling problem, respectively. In EEDA, the EDA-based search is applied to guiding the search to some promising sequences or regions, and an Interchange-based local search is presented to perform the search from these promising regions. Moreover, each individual or sequence of EEDA is decoded by applying a shift timetabling method to solving the corresponding timetabling problem. The experimental results show that the combination of the EEDA and the shift timetabling method can accelerate the convergence speed and is helpful in achieving more competitive results.

Keywords: No-wait job shop scheduling problem, sequencing, timetabling, enhanced estimation of distribution algorithm.

1 Introduction

This paper considers the no-wait job shop scheduling problem (NWJSSP) with the makespan criterion. The no-wait job shop is a job shop with the constraint that no waiting time is allowed between operations within any job. The no-wait constraints usually arise from requirements for processing environments or characteristics of jobs, such as steel production [1], computer systems [2], food processing [3], chemical industry [4], pharmaceutical industry, production of concrete wares [5] and semiconductor testing facilities [6].

The classical job shop scheduling problem is well-known to be strongly NP-hard [7], and it has been subject to intensive research during the past decades. In contrast, there are considerably fewer contributions dealing with the NWJSSP, which is also

---

* Corresponding author.

D.-S. Huang et al. (Eds.): ICIC 2014, LNAI 8589, pp. 675–685, 2014.
© Springer International Publishing Switzerland 2014

known to be NP-hard in the strong sense. The review of NWJSSP given by Hall and Sriskandarajah [8] showed that the considered problem is very difficult, especially for the large-size instances. The NWJSSP was proven to be NP-hard in the strong sense even for two-machine cases [9]. Thus, it is meaningful and practical to develop an effective algorithm for the considered problem.

Recently, a complete local search with limited memory (CLLM) algorithm [10] has been proposed for the NWJSSP to minimize the makespan criterion. CLLM is a recent metaheuristic that makes use of limited memory in a explicit manner. That is, all explored solutions are recorded to avoid the exploration of already visited solutions. The concept is similar to the idea of "tabu search with structured moves" in [11]. CLLM can achieve better results than variable neighborhood search (VNS), hybrid simulated annealing/genetic algorithm (GASA), and complete local search with memory (CLM) algorithm [12] when solving the NWJSSP [10]. CLLM is the best algorithm so far for the considered problem.

As a novel probabilistic-model based evolutionary algorithm, an estimation of distribution algorithm (EDA) was introduced by Baluja [13] for solving traveling the salesman problem (TSP) and job shop scheduling problem. EDA uses a variable-independence probability model to generate new population and guide the search direction. The evolution process of EDA is regarded as a process of competitive learning, whose probability model is updated by the current best solution at each generation to accumulate the information of excellent individuals. Due to its simple framework and outstanding global exploration ability, EDA has attracted much attention and has been used to solve some production scheduling problems. Salhi et al. [14] proposed an EDA to deal with the hybrid flow shop scheduling problem. Wang et al. [15] developed a hybrid EDA for the flexible job shop scheduling problem, whose local search scheme is designed based on the critical path. Zhang et al. [16] proposed an EDA for permutation flow shops to minimize total flowtime. Bai et al. [17] used an improved hybrid differential evolution-estimation of distribution algorithm for the nonlinear programming and the mixed integer nonlinear programming problems in engineering optimization fields. However, to the best of our knowledge, there has no published word addressing the NWJSSP by using the EDA-based algorithm.

In the current paper, an enhanced estimation of distribution algorithm (EEDA) is presented to deal with the NWJSSP. In our EEDA, the EDA-based search is adopted to perform global exploration in the sequence or solution space and guide the whole search to the promising regions/sequences, while an Interchange-based local search is developed to emphasize exploitation from those regions. Moreover, a so-called shift timetabling method is utilized to decode each individual or sequence of EEDA. Test results and comparisons with CLLM demonstrate the effectiveness of the presented EEDA.

The remainder of this paper is organized as follows. Section 2 briefly depicts the NWJSSP. Section 3 introduces the shift timetabling method. Section 4 presents EEDA in details. Section 5 provides simulation results and comparisons. Finally, Section 6 gives some conclusion and future work.

2   NWJSSP Description

2.1   NWJSSP

The NWJSSP can be depicted as follows. There are a set of m machines and a set of n jobs. The processing time of each job on each machine is deterministic. At any time, preemption and interruption are forbidden and each machine can process at most one job. To satisfy the no-wait restriction, each job must be processed without interruptions between consecutive machines. A sequence or schedule of jobs to be processed can be denoted as $\pi = [\pi_1, \pi_2, \cdots, \pi_n]$. A processing of each job in π on each machine is called an operation. Each job must be processed in order of its predefined operations.

Let $C_{\max}(\pi)$ denote the makespan calculated function. The aim of NWJSSP is to find a sequence π^* in the set of all permutations Π such that

$$C_{\max}(\pi^*) = \min_{\pi \in \Pi} C_{\max}(\pi). \tag{1}$$

2.2   Sequence and Timetabling Problems of NWJSSP

Obviously, the NWJSSP can be decomposed into the lower-level and the upper-level problems. The lower-level problem is called the sequencing problem, in which a processing sequence π of an optimal schedule is obtained for a given NWJSSP. The upper-level problem is called the timetabling problem, in which a set of the feasible start processing times of jobs in π on each machine (i.e., the timetable of the sequence obtained from the sequencing problem) are found to minimize $C_{\max}(\pi)$.

3   Shift Timetabling Method

It is a common belief and intuitively quite clear that the major difficulty in solving the NWJSSP lies in the sequencing problem. Existing algorithms spend most computation time on the sequencing problem while employing simple strategies for the timetabling problem [18]. In fact, even though a sequence for the problem is given, different timetabling methods' decoding results for this sequence is quite different. Effectiveness of an algorithm for the NWJSSP also depends greatly on the timetabling methods. In this paper, a shift timetabling method presented in [10] is utilized, which is briefly introduced in this section.

There some notations to be used are given as follows.

n the number of jobs

m the number of machines

$k \in \left[2, \lceil n/2 \rceil\right]$ the number of sub-sequences

$J = \{J_1, J_2, ..., J_n\}$ the set of jobs

$\boldsymbol{\pi} = [\pi_1, \pi_2, \cdots, \pi_n]$ the sequence or schedule of jobs to be processed

$\pi_i \in \{J_1, J_2, ..., J_n\}$ the ith job in $\boldsymbol{\pi}$ and $i = 1, 2, \cdots, n$

$\boldsymbol{\pi^j} = [\pi_{(j-1)*\lfloor n/k \rfloor+1}, \pi_{(j-1)*\lfloor n/k \rfloor+2}, \cdots \pi_{j*\lfloor n/k \rfloor}]$ the jth subsequence of $\boldsymbol{\pi}$ and $j = 1, 2, \cdots, k-1$

$\boldsymbol{\pi^k} = [\pi_{(j-1)*\lfloor n/k \rfloor+1}, \pi_{(j-1)*\lfloor n/k \rfloor+2}, \cdots \pi_n]$ the last subsequence of $\boldsymbol{\pi}$ and $j = k$

$len(\boldsymbol{\pi^j})$ the length of $\boldsymbol{\pi^j}$ and $j = 1, 2, \cdots, k$

$\boldsymbol{\pi} = [\boldsymbol{\pi^1}, \boldsymbol{\pi^2}, \cdots, \boldsymbol{\pi^k}]$ the sequence or schedule of jobs to be processed

$\pi_i^j \in \{J_1, J_2, ..., J_{\lfloor n/k \rfloor}\}$ the ith job in $\boldsymbol{\pi^j}$

$t_{\pi_i^j}$ the start processing time of ith job in $\boldsymbol{\pi^j}$

$L_{\pi_i^j}$ the process time of the ith job in $\boldsymbol{\pi^j}$

Based on the non-delay timetabling and enhanced timetabling strategies [18], the shift timetabling method is selected for solving the timetabling problem. Next, a brief description of the shift timetabling method is provided. A more extensive discussion can be found in [10]. Prior to the description of the shift timetabling, two concepts about shift are reviewed as follows.

1. *left shift*: Set the job as early as possible, if there is no conflicting.
2. *right shift*: 1) The start time of the setting job is greater than the time of prior job. 2) Under the premise of satisfying the condition 1), set the job as early as possible, if there is no conflicting.

Then, the shift timetabling method is described as bellows.

Step 1: $t_{\pi_1^1} \leftarrow 0, i \leftarrow 1, C_{\max} \leftarrow 0 . //^* Initialization^* //$

Step 2: Divide sequence $\boldsymbol{\pi}$ into k sub-sequences $\boldsymbol{\pi^1}, ..., \boldsymbol{\pi^k}$.

Step 3: Repeat

     Step 3.1: Perform *left shift* on all jobs in $\boldsymbol{\pi^i}$, and then calculate start times $t'_{\pi_1^i}, \cdots, t'_{\pi_{len(\boldsymbol{\pi^i})}^i}$. Set the generated schedule as $[\boldsymbol{\pi^1}, \cdots, \boldsymbol{\pi^i}]$ and calculate the makespan of $[\boldsymbol{\pi^1}, \cdots, \boldsymbol{\pi^i}]$ by $C' \leftarrow \max \{ \max_{j=1,...,len(\boldsymbol{\pi^i})} \{ t'_{\pi_j^i} + L_{\pi_j^i} \} , C_{\max} \}$.

Step 3.2: Re-schedule π^i by applying *right shift* to the head job in π^i while *left shift* to the other jobs in π^i sequentially. Calculate the start times $t''_{\pi_1^i}, \ldots, t''_{\pi_{len(\pi^i)}^i}$ and then calculate the corresponding makespan of

$$[\pi^1, \cdots, \pi^i] \text{ by } C'' \leftarrow \max \{ \max_{j=1,\ldots,len(\pi^i)} \{ t''_{\pi_j^i} + L_{\pi_j^i} \}, C_{\max} \}. \text{ If } C' > C'',$$

then $C_{\max} \leftarrow C''$ and $t_{\pi_1^i}, \ldots, t_{\pi_{len(\pi^i)}^i} \leftarrow t''_{\pi_1^i}, \ldots, t''_{\pi_{len(\pi^i)}^i}$, otherwise

$$C_{\max} \leftarrow C' \text{ and } t_{\pi_1^i}, \ldots, t_{\pi_{len(\pi^i)}^i} \leftarrow t'_{\pi_1^i}, \ldots, t'_{\pi_{len(\pi^i)}^i}.$$

Step 3.4: $i \leftarrow i+1$.

Step 4: Until $i > k$.

Step 5: Set $C_{\max}(\pi) = C_{\max}$.

Step 6: Output $C_{\max}(\pi)$.

4 EEDA

In this section, EEDA presented after explaining the solution representation, probability model and updating strategy, new population generation method, *Insert*-based mutation, and *Interchange*-based neighborhood search with first move strategy.

4.1 Solution Representation

Based on the properties of the NWJSSP, we adopt the job-based solution representation, that is, every individual of the population is a feasible solution of the NWJSSP, for example, $[\pi_1, \pi_2, \ldots, \pi_4] = [1, 2, 4, 3]$ is an individual when the problem's scale n is set to 4.

4.2 Probability Model and Updating Mechanism

Different from other evolutionary algorithms, EDA generates new population by sampling a probability model (i.e., probability matrix). Hence, the probability model has a key effect on the performances of the EDA. In this study, the probability matrix is defined as follows:

$$P_{matrix}(gen) = \begin{pmatrix} P_{11}(gen) & \cdots & P_{1n}(gen) \\ \vdots & \ddots & \vdots \\ P_{n1}(gen) & \cdots & P_{nn}(gen) \end{pmatrix}_{n \times n}, \tag{2}$$

where $\sum_{w=1}^{n} P_{wj}(gen) = 1$, and $P_{wj}(gen)$ is the probability of job w appearing in the jth position of π at generation gen.

Let $\pi_i(gen) = [\ \pi_{i1}(gen), \pi_{i2}(gen), ..., \pi_{in}(gen)\ ]$ denote the ith individual in EEDA's population at generation gen, π_{gbest} denote the global best individual and α the learning rate. The matrix $P_{matrix}(gen+1)$ is updated according to π_{gbest} by the following two steps:

Step 1: Set $x = \pi_{ij}(gen)$ and $p_{xj}(gen+1) = p_{xj}(gen) + \alpha(gen)$ for $j = 1,..., n$.

Step 2: Set $p_{wj}(gen+1) = p_{wj}(gen+1) / \sum_{y=1}^{n} P_{yj}(gen+1)$ for $w = 1,..., n$ and $j = 1,..., n$.

4.3 New Population Generation Method

In each generation of the EDA, the new individuals are generated by sampling the probability matrix mentioned in 4.2. Denote PS the size of the population and $SelectJob(\pi_i(gen+1), j)$ the function of selecting a job in the jth position of $\pi_i(gen+1)$ by using the matrix $P_{matrix}(gen)$. The procedure of $SelectJob(\pi_i(gen+1), j)$ is described as follows:

Step 1: Randomly create a probability r where $r \sim [0,1)$.
Step 2: Get a candidate job CJ by the roulette-wheel selection scheme.
  Step 2.1: If $r \sim [0, p_{1j}(gen))$, then set $CJ = 1$ and go to Step 3.
  Step 2.2: If $r \sim [\sum_{y=1}^{w} P_{yj}(gen), \sum_{y=1}^{w+1} P_{yj}(gen))$ and $w \in \{1, \cdots, n-1\}$, then set $CJ = w+1$ and go to Step 3.
Step 3: Return CJ.

Let $l(CJ, \pi_i(gen+1))$ denote the repeat times of CJ in $\pi_i(gen+1)$. Then, the new population generation method is given as the following steps:

Step 1: Set $i = 1$.
Step 2: Generate a new individual $\pi_i(gen+1)$.
  Step 2.1: Set $\pi_{ij}(gen+1) = 0$ for $j = 1, \cdots, n \times m$.
  Step 2.2: Set $j = 1$.
  Step 2.3: $CJ = SelectJob(\pi_i(gen+1), j)$.
  Step 2.4: If $l(CJ, \pi_i(gen+1)) = 1$, then go to Step 2.3.
  Step 2.5: Set $\pi_{ij}(gen+1) = CJ$.
  Step 2.6: Set $j = j+1$.
  Step 2.7: If $j \leq n \times m$, then go to Step 2.3.
Step 3: Set $i = i+1$.
Step 4: If $i \leq PS$, then go to Step 2.

4.4   Insert-Based Mutation

Mutation is an important element of evolutionary algorithm, which is usually used to enhance the diversity of population and prevent the search from falling into local optima. Thus, in our EEDA, the *Insert*-based mutation is utilized to guide the search to more different regions. Denote P_m the mutation probability of each individual and $Insert(\pi_i(gen), u, v)$ the insertion of $\pi_{iu}(gen)$ in the vth dimension of $\pi_i(gen)$, $u \neq v$. The *Insert*-based mutation is described as the following steps:

Step 1: Set $i = 1$.
Step 2: Randomly create a probability r where $r \sim [0,1)$.
Step 3: If $r \leq P_m$, then $\pi_i(gen) = Insert(\pi_i(gen), u, v)$.
Step 4: Set $i = i + 1$.
Step 5: If $i \leq PS$, then go to Step 2.

4.5   Interchange-Based Neighborhood Search with First Move Strategy

Because *Interchange* is a simple neighborhood in the existing literatures, we use it to execute exploitation in local search. The $Interchange(\pi_i(gen), u, v)$ operation denotes the interchange of the job at the uth dimension and the job at the vth dimension. The *Interchange*-based neighborhood of $\pi_i(gen)$ can be expressed as

$$N_{interchange}(\pi_i(gen)) =$$

$$\{ \pi_i(gen)^{n,\mu,v} = Interchange(\pi_i(gen), u, v) \mid n-1 \text{ and } v = u+1,...,n \} \tag{3}$$

Let $FindBestN_{interchange}(\pi_i(gen))$ denote the procedure of obtaining the best $\pi_i(gen)^{n,\mu,v}$ (denoted by $\pi_i(gen)^{n,\mu,v}_{best}$) in $N_{interchange}(\pi_i(gen))$. For searching more regions, a first move strategy is applied in $FindBestN_{interchange}(\pi_i(gen))$. That is, when the first $\pi_i(gen)^{n,\mu,v}$ (denoted by $\pi_i(gen)^{n,\mu,v}_{firstl}$) in $N_{interchange}(\pi_i(gen))$ that improves $\pi_i(gen)$ is obtained, $FindBestN_{interchange}(\pi_i(gen))$ terminates and outputs $\pi_i(gen)^{n,\mu,v}_{firstl}$.

Based on the previous tests, we find that applying $FindBestN_{interchange}(\pi_i(gen))$ with the first move strategy to the top 10%-15% individuals of the population can perform a wider and deeper local search and obtain good results. Therefore, the top 10% individuals are chosen to apply the proposed neighborhood search. If $C_{max}(\pi_i(gen)^{n,\mu,v}_{best})$ or $C_{max}(\pi_i(gen)^{n,\mu,v}_{firstl})$ is less than $C_{max}(\pi_i(gen))$, $\pi_i(gen)$ will be updated with $\pi_i(gen)^{n,\mu,v}_{best}$ or $\pi_i(gen)^{n,\mu,v}_{firstl}$.

4.6   Procedure of EEDA

According to the above subsections, the procedure of EEDA is proposed as follows:

Step 0: Let *genMax* the maximum generation, *nonICount* the consecutive non-improved times of π_{gbest}, and *CountMax* the maximum number of *nonICount*.

Step 1: Initialization.

Step 1.1: Set $gen = 1$ and $nonICount = 0$.

Step 1.2: Randomly generate individual $\pi_i(1)$ for $i = 1, ..., PS$, and calculate makespan of each individual by using the ***Shift Timetabling Method***, and update π_{gbest}.

Step 1.4: Set $p_{wj}(gen) = 1/n$ for $w = 1, ..., n$ and $j = 1, ..., n$.

Step 1.5: Update the matrix *Pmatrix(gen)* by π_{gbest}.

Step 2: Set $gen = gen + 1$.

Step 3: Generate individual $\pi_i(gen)$ for $i = 1, ..., PS$ by new population generation method, and calculate makespan of each individual by using the ***Shift Timetabling Method***, and update π_{gbest}.

Step 4: If π_{gbest} can be updated, then

$nonICount = 0$

Else

$nonICount = nonICount + 1$.

Step 5: If $nonICount < CountMax$, then update the matrix *Pmatrix(gen)* by π_{gbest} and go to Step 2.

Step 6: Set $nonICount = 0$.

Step 7: Apply *Insert*-based mutation to the current population and calculate makespan of each individual by using the ***Shift Timetabling Method***.

Step 8: Apply *Interchange*-based neighborhood search with first move strategy to the top 10% individuals of the current population, and calculate makespan of each neighbor by using the ***Shift Timetabling Method***.

Step 9: Update π_{gbest}.

Step 10: Update the matrix *Pmatrix(gen)* by π_{gbest}.

Step 11: If $gen < genMax$ then go to Step 2.

Step 12: Output π_{gbest}.

It can be seen from the above procedure that Step 10 uses π_{gbest} to update the probability matrix, which means new generated individuals can share the information of the global best individual and then guide the search to more promising regions, and Step 7 is the perturbation operator, which can restrain the search from dropping into local optima and drive the search to quite different regions. Moreover, Step 8 performs exploitation from the regions obtained by Steps 3 and 7. Due to well-balanced between exploration and exploitation, EEDA is hopeful to achieve good results.

5   Simulation Result and Comparisons

5.1   Experimental Design

In order to test the performance of the proposed EEDA, the well known job shop benchmarks are tested. The $n \times m$ combinations include 10×5 and 10×10. All algorithms are coded in Delphi 2010 and are executed on Mobile Intel Core 2 Duo 2.2 GHz processor with 8GB memory.

In EEDA, we use the following parameters: the population size $PS = 50$, the learning rate $\alpha = 0.02$, the maximum consecutive non-improved times $CountMax = 20$, the maximum generation $genMax = 200$, the mutation probability $P_m = 0.3$, and the length of π^j (i.e., len) is set to 3 and $n - 3 \times (k-1)$ for $j = 1, 2, \cdots, k-1$ and $j = k$, respectively.

5.2   Comparisons of EEDA and CLLM

For the purpose of manifesting the effectiveness of EEDA, we compare EEDA with CLLM [10]. The CLLM method is the best algorithm so far for the considered problem. EEDA and CLLM are compared on some classic benchmarks (i.e., La01- La05, Orb01-Orb10, La16-La20). We set the maximum generation of EEDA=300 and let CLLM run at the same amount of computer time as EEDA. Each benchmark is independently run 20 times for comparison. The simulation results are listed in Table 1, where $BEST$ denotes the best makespan, AVG denotes the average makespan. $WORST$ denotes the worst makespan, and SD denotes the standard derivation.

Table 1. Comparisons of *BEST*, *WORS*, *AVG* and *SD* of CLLM and EEDA

Instances	n	m	CLLM[10]				EEDA			
			BEST	WORST	AVG	SD	BEST	WORST	AVG	SD
La01	10	5	975	1128	1057.65	37.58	**971**	1031	991.15	24.50
La02	10	5	988	1041	1009.35	14.80	**961**	1005	978.70	11.53
La03	10	5	862	919	889.55	19.62	**820**	867	860.25	10.41
La04	10	5	917	983	940.15	12.37	**887**	923	896.80	12.81
La05	10	5	784	886	841.67	27.94	**781**	829	791.15	13.38
Orb01	10	10	**1615**	1771	1680.32	40.74	**1615**	1663	1631.60	21.30
Orb02	10	10	1572	1694	1633.73	51.28	**1485**	1518	1516.35	7.19
Orb03	10	10	1603	1745	1678.49	35.18	**1599**	1685	1633.10	22.53
Orb04	10	10	1750	1874	1812.39	30.34	**1653**	1732	1679.25	18.17
Orb05	10	10	1424	1574	1508.80	34.42	**1370**	1451	1402.85	20.75
Orb06	10	10	1616	1716	1665.55	43.75	**1555**	1557	1556.60	0.80
Orb07	10	10	711	797	753.15	33.25	**706**	725	711.45	7.74
Orb08	10	10	**1319**	1452	1353.95	35.90	**1319**	1350	1320.55	6.70
Orb09	10	10	1581	1710	1646.25	41.83	**1445**	1560	1520.80	34.24
Orb10	10	10	1691	1749	1724.95	20.33	**1557**	1601	1582.60	9.55
La16	10	10	1650	1804	1729.77	25.34	**1575**	1628	1591.85	10.14
La17	10	10	1484	1592	1536.84	25.36	**1371**	1417	1387.80	11.43
La18	10	10	1597	1687	1649.34	27.82	**1555**	1630	1589.35	16.37
La19	10	10	1610	1764	1710.91	39.17	**1482**	1533	1508.90	7.83
La20	10	10	1655	1846	1768.10	42.67	**1526**	1683	1616.25	23.58
Average			1370.2	1486.6	1429.5	31.98	**1311.65**	**1369.4**	**1338.36**	**14.54**

From Table 1, it is shown that the EEDA is better than CLLM with respect to solution quality. The values of *BEST*, *AVG* and *SD* obtained by EEDA are much better than those obtained by CLLM. Moreover, the *WORST* values of EEDA are smaller than the *AVG* values of CLLM for all benchmarks except Orb03. Thus, EEDA is an effective algorithm for the NWJSSP.

6   Conclusion and Future Work

To the best of the current authors' knowledge, this is the first report on the application of estimation of distribution algorithm (EDA) for the no-wait job shop scheduling problem (NWJSSP) with the makespan criterion. In our presented enhanced EDA (EEDA), EDA-based global search was used to execute exploration for promising regions within the whole sequence or solution space, while an *Interchange*-based local search was designed to stress exploitation from these promising regions. In addition, a shift timetabling method is selected to decode each individual or sequence of EEDA. Simulations and comparisons based on benchmarks demonstrate the effectiveness of the presented EEDA. Our future work is to develop some EDA-based algorithms to deal with the re-entrant NWJSSP.

Acknowledgments. This research was partially supported by National Science Foundation of China (No. 60904081), 2012 Academic and Technical Leader Candidate Project for Young and Middle-Aged Persons of Yunnan Province (No. 2012HB011), and Discipline Construction Team Project of Kunming University of Science and Technology (No. 14078212).

References

1. Wismer, D.A.: Solution of the Flowshop Scheduling-problem with No Intermediate Queues. Operations Research 20(6), 89–97 (1972)
2. Reddi, S., Ramamoorthy, C.: A Scheduling-problem. Operational Research Quarterly 24(44), 1–6 (1973)
3. Raaymakers, W., Hoogeveen, J.: Scheduling Multipurpose Batch Process Industries with No-wait Restrictions by Simulated Annealing. European Journal of Operational Research 126, 131–151 (2000)
4. Rajendran, C.: A No-wait Flowshop Scheduling Heuristic to Minimize Makespan. Journal of The Operational Research Society 45(4), 472–478 (1994)
5. Grabowski, J., Pempera, J.: Sequencing of Jobs in Some Production System. European Journal of Operational Research 125(5), 35–50 (2000)
6. Ovacik, I., Uzsoy, R.: Decomposition Methods for Complex Factory Scheduling-Problems. Kluwer Academic Publishing, Dordecht (1997)
7. Lenstra, J.K., Rinnooy Kan, A.H.G.: Computational Complexity of Discrete Optimization Problems. Annals of Discrete Mathematics 4(1), 21–40 (1979)
8. Hall, N.G., Sriskandarajah, C.: A survey of machine scheduling problems with blocking and no-wait in process. Operations Research 44, 510–525 (1996)

9. Sahni, S., Cho, Y.: Complexity of Scheduling Shops with No-wait in Process. Mathematics of Operations Research 4, 448–457 (1979)
10. Zhu, J., Li, X.P., Wang, Q.: Complete Local Search with Limited Memory Algorithm for No-wait Job Shops to Minimize Makespan. Eur. J. Oper. Res. 198(2), 378–386 (2009)
11. Glover, F.: Tabu search – Part II. ORSA Journal on Computing 2, 4–32 (1990)
12. Framinan, J.M., Schuster, C.: An Enhanced Timetabling Procedure for The No-wait Job Shop Problem: A complete local search approach. Computers & Operations Research 331, 1200–1213 (2006)
13. Baluja, S.: Population-based Incremental Learning: A Method for Integrating Genetic Search based Function Optimization and Competitive Learning. Technical Report CMU-CS-94-193. Carnegie Mellon University, Pittsburgh (1994)
14. Salhi, A., Rodriguez, J.A.V., Zhang, Q.F.: An Estimation of Distribution Algorithm with Guided Mutation for a Complex Flow Shop Scheduling Problem. In: Proceedings of the 9th Annual Conference on Genetic and Evolutionary Computation, London, UK, pp. 570–576 (2007)
15. Wang, S.Y., Wang, L., Zhou, G., Xu, Y.: An Estimation of Distribution Algorithm for the Flexible Job-shop Scheduling Problem. In: Huang, D.-S., Gan, Y., Gupta, P., Gromiha, M.M. (eds.) ICIC 2011. LNCS (LNAI), vol. 6839, pp. 9–16. Springer, Heidelberg (2012)
16. Zhang, Y., Li, X.P.: Estimation of Distribution Algorithm for Permutation Flow Shops with Total Flowtime Minimization. Computers & Industrial Engineering 60, 706–718 (2011)
17. Bai, L., Wang, J., Jiang, Y.H., Huang, D.X.: Improved Hybrid Differential Evolution-Estimation of Distribution Algorithm with Feasibility Rules for NLP/MINLP Engineering Optimization Problems. Chinese Journal of Chemical Engineering 20(6), 1074–1080 (2012)
18. Pan, J.C.H., Huang, H.C.: A Hybrid Genetic Algorithm for No-wait Job Shop Scheduling Problems. Expert Systems with Applications 36, 5800–5806 (2009)

A Bayesian Statistical Inference-Based Estimation of Distribution Algorithm for the Re-entrant Job-Shop Scheduling Problem with Sequence-Dependent Setup Times

Shao-Feng Chen[1,2], Bin Qian[1,2,*], Bo Liu[3], Rong Hu[1,2], and Chang-Sheng Zhang[1]

[1] Department of Automation, Kunming University of Science and Technology,
Kunming 650500, China
[2] Key Laboratory of Computer Technologies Application of Yunnan Province,
Kunming 650500, China
[3] Academy of Mathematics and Systems Science Chinese Academy of Sciences,
Beijing 100190, China
bin.qian@vip.163.com

Abstract. In this paper, a bayesian statistical inference-based estimation of distribution algorithm (BEDA) is proposed for the re-entrant job-shop scheduling problem with sequence-dependent setup times (RJSSPST) to minimize the maximum completion time (i.e., makespan), which is a typical NP hard combinatorial problem with strong engineering background. Bayesian statistical inference (BSI) is utilized to extract sub-sequence information from high quality individuals of the current population and determine the parameters of BEDA's probabilistic model (BEDA_PM). In the proposed BEDA, BEDA_PM is used to generate new population and guide the search to find promising sequences or regions in the solution space. Simulation experiments and comparisons demonstrate the effectiveness of the proposed BEDA.

Keywords: bayesian statistical inference, estimation of distribution algorithm, re-entrant job-shop scheduling problem, setup times.

1 Introduction

This paper presents a meta-heuristic for the re-entrant job-shop scheduling problem with sequence-dependent setup times (RJSSPST). The criterion is to minimize the maximum completion time (i.e., makespan). The RJSSPST is a general job-shop scheduling problem characterized by re-entrant work flows and sequence-dependent setup times. This problem extends the classical job-shop scheduling problem (JSSP) by allowing for sequence-dependent setup times between adjacent operations on any machine and relaxing the restriction that each job visits a machine only once. The JSSP is well-known to be NP-complete [1]. From the point of view of computational complexity, the JSSP obviously reduces to the RJSSPST, which means the RJSSPST is also NP-complete. Moreover, the RJSSPST is commonly encountered in today's

---

* Corresponding author.

D.-S. Huang et al. (Eds.): ICIC 2014, LNAI 8589, pp. 686–696, 2014.

manufacturing environment. Thus, it is meaningful and practical to make the work of developing effective scheduling algorithms for the considered problem.

Although the RJSSPST appears frequently in commercial printing, plastic manufacturing, metal processing, cylinder-head manufacturing, side frame press shop, and semiconductor manufacturing, but mainly because of its complexity, it has received little attention from researchers. Recently, Sun [2] proposed a genetic algorithm (GA), a modified version of the nearest-setup heuristic [3], and a modified version of the shifting-bottleneck heuristic [4] to solve the RJSSPST. The test results show that GA performs better than the other algorithms.

In recent years, a novel probabilistic-model based evolutionary algorithm, i.e., the estimation of distribution algorithm (EDA), was first introduced by Baluja [5] for addressing the traveling salesman problem and the JSSP. EDA adopts a bran-new evolution scheme to drive the evolution process, which has no traditional crossover and mutation operations in the algorithm. The evolution process of EDA can be regarded as a process of competitive learning, and its probabilistic model is updated by the better solutions at each generation to accumulate the information of excellent individuals. Due to its simple framework and outstanding global exploration ability, EDA has attracted much attention and has been used to solve some production scheduling problems. Salhi et al. [6] presented an EDA for the complex flow shop scheduling problem. Wang et al. [7] designed an EDA to deal with the flexible JSSP. Liu et al. [8] used a combination of particle swarm optimization with EDA for the permutation flowshop scheduling problem. Wang et al. [9] developed a bi-population based EDA for the flexible JSSP. Zhou et al. [10] applied a decomposition-based EDA for the multiobjective traveling salesman problems.

However, there are some shortcomings in the current EDA's model. As shown in Ruiz et al. [11], there are many similar blocks of jobs within the individuals' sequences in the latter stages of evolutionary methods. These similar blocks may occupy the same positions or different positions. If these blocks are disrupted, the algorithm has a high probabilistic to produce offspring with worse makespan values. In this paper, a bayesian statistical inference-based estimation of distribution algorithm (BEDA) is proposed. The purpose of bayesian statistical inference (BSI) is to describe the condition probabilistic of each decision variable [12]. And the new model of conditional probabilistic is built based on the frequencies of sub-sequences or neighbor operations in the high quality individuals. By combining the BSI with the EDA, the proposed BEDA can overcome the shortcomings in the current EDA's model to some extent.

The rest of this paper is arranged as follows. In section 2, the RJSSPST is briefly introduced. In section 3, the BEDA is proposed and described in details. In section 4, computational results and comparisons are provided. Finally, we end the paper with some conclusions and future work in section 5.

2   RJSSPST

2.1   Mathematical Model of RJSSPST

The following describes the mathematical model for the RJSSPST. For the purpose of simplification, in the following model, each operation of each job is denoted by a unique notation.

i, j Notations used for operations

B a large enough positive number

D The set of operations in the shop floor

M The set of machines available in the shop floor

L The repeat or reentrant times

M_j The specific machine that operation j requires according to its process route

t_j Standard processing time of operation j

C_j Completion time of operation j

F_j the start time of operation j

R_k The dummy operation that describes the first activity in machine k

S_{ij} The setup time between operation i and operation j if operation i performs just before operation j

x_{ijk} The indicator variable

With the above notations, the RJSSPST can be formulated as a mixed integer programming problem as follows. Let

$$x_{ijk} = \begin{cases} 1, & \text{if operation } j \text{ is just after operation } i \text{ in the machine } k \\ 0, & \text{otherwise} \end{cases}$$

Then,

$$\text{Min makespan} = \max(C_i = F_i + t_i) \quad i \in D \tag{1}$$

s.t.

$$\sum_{i \in D} x_{ijk} = L \tag{2}$$

$$\sum_{j \in D} x_{ijk} = L \tag{3}$$

$$F_i + t_i + S_{ij} \leq F_j + B(1 - X_{ijk}) \tag{4}$$

$$F_i + t_i \leq F_j \qquad\qquad i, j \in S_{ij} \tag{5}$$

$$t_{R_k} = 0 \qquad\qquad k \in M \tag{6}$$

$$S_{R_k i} = 0, S_{iR_k} = 0, \qquad\qquad i \in D, (i, k \mid k = M_i) \tag{7}$$

Equation 1 presents the objective function, which aims to reduce the cycle time of all jobs called makespan. Constraint 2 and constraint 3 force the job scheduling to have an available sequence in a schedule sequence of each machine. Constraint 4 requires that the start time of each operation in one machine should be larger than the completion time of the operation that performs just before this operation considering the setup times between the pervious operation and current operation. Constraint 5 ensures that an operation could not start until its preceding operation is done. In constraint 6 and constraint 7, processing time and the relationship of setup times between dummy operations and other operations, which require the same machine, are set to zero. The existence of variable F_i in the mathematical model eliminates the subtours in the solutions. It should be noted that the operation, which starts in the next location after dummy operation of a machine, is the starting point of sequence and consequently no setup is required to perform this operation.

2.2   Statement of RJSSPST

The above mathematical model is only used to depict the characteristics of RJSSPST. The following statement is utilized to code and decode each sequence or individual in the proposed BEDA. The RJSSPST has the following usual assumptions. There are m machines, n jobs and L times re-entrant in the production system. The problem size can be denoted by $n \times m \times L$. Each job consists of a set of operations that have to be processed in a specified sequence. Each operation has to be processed on a definite machine and has a processing time and a setup time which are deterministically known. Moreover, each job should be processed on each machine L times and the setup times of operations on each machine are sequence dependent. Once an operation is started, it must be completed without any interruption. At any time, each machine can process at most one job.

Denote $\pi = [\pi_1, \pi_2, ..., \pi_{n \times m \times L}]$ the sequence or schedule of operations to be processed. The RJSSPST with the makespan criterion is to find a schedule π^* in the set of all schedules Π such that

$$\pi^* = \arg(C_{max}) \to \min , \ \forall \pi \in \Pi . \tag{8}$$

3   BEDA for RJSSPST

In this section, we will present BEDA after explaining the solution representation, decoding scheme, population initialization, probabilistic model and updating mechanism, new population generation method and new probabilistic model construction strategy.

3.1   Solution Representation

Based on the properties of RJSSPST, we adopt the operation-based solution representation, that is, every individual of the population is a feasible solution of the

RJSSPST, for example, $[\pi_1,\pi_2,...,\pi_{3\times2\times2}]=[1,2,1,3,2,1,3,3,1,2,3,2]$ is an individual when the problem's scale $n\times m\times L$ is set to $3\times2\times2$.

3.2   Decoding Scheme

Due to the constraints of RJSSPST, we can improve the solution to minimize the makespan criterion by converting the semi-active schedule to the active one. Thus, when decoding we check the possible idle interval before appending an operation at the last position, and fill the first idle interval before the last operation (called active decoding). The details of active decoding scheme used in this paper are very similar to that used in [13]. The difference between the two schemes only lies in the treatment of the setup times. In [13], the setup times of each operation are not considered.

3.3   Population Initialization

In order to search different regions in the huge solution space, we need to ensure that the initial individuals have widely distributed in the space. Therefore, BEDA adopts stochastic method to generate the initial population.

3.4   Probabilistic Model and Its Updating Mechanism

3.4.1   Probabilistic Model

A bayesian network in [14] is used as the probabilistic model of BEDA, whose structure (including nodes and directed arcs in the network) is partially predetermined. This probabilistic model can describe the probabilistic distribution of favorable values based on the high quality individuals in population. The operations in each high quality individual can be represented by a directed acyclic network as shown in Fig. 1.

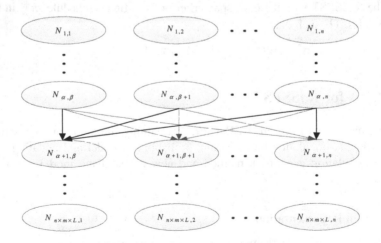

Fig. 1. The structure of bayesian network

The directed arc from node $N_{\alpha,\beta}$ to node $N_{\alpha+1,\beta'}$ ($\beta' \in 1,...,n$) represents the dependent relationship between the two nodes, which records the sub-sequence information of the high quality individuals.

3.4.2 Updating Mechanism

The probabilistic model is a most important part for the BEDA, because it is built for guiding generation new population. Building or updating the probabilistic model is equivalent to determining the bayesian network's parameters (i.e., the weights of directed arcs) according to the high quality individuals. After that, each individual of the next generation can be produced by iteratively sampling the probabilistic model from the first layer (i.e., $N_{1,\beta}$) to the last layer (i.e., $N_{n \times m \times L,\beta}$), which is useful for obtaining promising offspring.

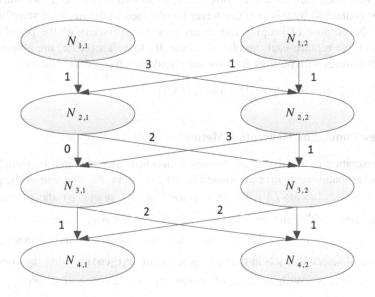

Fig. 2. A example network

The weights of directed arcs can be obtained simply by calculating the frequencies of sub-sequences or neighbor operations in the high quality individuals. Here we provide a concrete example to illustrate the calculation process in Fig. 2. In this example, $n \times m \times L$ is set to $2 \times 2 \times 1$, the high quality individuals are set as $\pi_1^{\text{high\_quality}} = [1,2,1,2]$, $\pi_2^{\text{high\_quality}} = [1,2,1,2]$, $\pi_3^{\text{high\_quality}} = [1,2,2,1]$, $\pi_4^{\text{high\_quality}} = [2,1,2,1]$, $\pi_5^{\text{high\_quality}} = [1,1,2,2]$, $\pi_6^{\text{high\_quality}} = [2,2,1,1]$. Then, the weight of each directed arc (i.e., the number placed on the arc) can be updated.

By using the weights of directed arcs as approximation probabilities, the conditional probabilities of all the near nodes can be calculated as follows:

$P(N_{1,1}) = (1+3)/6$, $P(N_{1,2}) = (1+1)/6$,

$P(N_{2,1}|N_{1,1}) = 1/(1+3)$, $P(N_{2,2}|N_{1,1}) = 3/(1+3)$,

$P(N_{2,1}|N_{1,2}) = 1/(1+1)$, $P(N_{2,2}|N_{1,2}) = 1/(1+1)$,

$P(N_{3,1}|N_{2,1}) = 0/(0+2)$, $P(N_{3,2}|N_{2,1}) = 2/(0+2)$,

$P(N_{3,1}|N_{2,2}) = 3/(3+1)$, $P(N_{3,2}|N_{2,2}) = 1/(3+1)$,

$P(N_{4,1}|N_{3,1}) = 1/(1+2)$, $P(N_{4,2}|N_{3,1}) = 2/(1+2)$,

$P(N_{4,1}|N_{3,2}) = 2/(2+1)$, $P(N_{4,2}|N_{3,2}) = 1/(2+1)$.

Obviously, if a weight of directed arc is equal to 0, it should be deleted from the bayesian network. This means a corresponding sub-sequence (i.e., two successive operations connected by this arc) can never be obtained in the iterative sampling process. For the purpose of keeping the diversity and dispersivity of the population to some extent, we replace each number 0 on the arc to 1. Therefore, the above conditional probabilities with number 0 on the arc should be changed as follows:

$$P(N_{3,1}|N_{2,1}) = 1/(1+2) , \quad P(N_{3,2}|N_{2,1}) = 2/(1+2) .$$

3.5   New Population Generation Method

In each generation of the algorithm, the new individuals are generated by sampling the conditional probabilistic matrix mentioned in 3.4.2. Denote PS the size of the population, π_j ($j \in \{1,...,n \times m \times L\}$) the jth operation in π , $pop(gen)$ the population at generation gen , $\pi^{\text{best}}(\textbf{gen})$ the best individual of $pop(gen)$, $\pi_i(\textbf{gen}) = \{\pi_{i1}(gen)$, $\pi_{i2}(gen)$,..., $\pi_{i(n \times m \times L)}(gen)\}$ the ith individual in $pop(gen)$, and $Job(\pi_i(\textbf{gen}), j)$ the function of selecting a job in the jth position of $\pi_i(\textbf{gen})$ by using the conditional probabilistic matrix. Without loss of generality, we assume that $P(N_{1,\beta}|N_{0,\beta'})$ $= P(N_{1,\beta})$ ($\beta, \beta' \in 1,...,n$). The procedure of $Job(\pi_i(\textbf{gen}), j)$ is described as follows:

Step 1: Set $\alpha = j$, $\beta' = \pi_{i(j-1)}(gen)$.

Step 2: Randomly create a probabilistic r where $r \sim [0,1)$.

Step 3: Get a candidate job CJ by the roulette-wheel selection scheme.

Step 3.1: If $r \sim [0, P(N_{\alpha,1}|N_{\alpha-1,\beta'})$, then set $CJ = 1$ and go to Step 4.

Step 3.2: If $r \sim [\sum_{\beta=1}^{w-1} P(N_{\alpha,\beta}|N_{\alpha-1,\beta'}), \sum_{\beta=1}^{w} P(N_{\alpha,\beta}|N_{\alpha-1,\beta'}))$ and $w \in \{2,$ $\cdots, n\}$, then set $CJ = w$ and go to Step 4.

Step 4: Return CJ .

Let $l(CJ, \pi_i(\text{gen}+1))$ denote the repeat times of CJ in $\pi_i(\text{gen}+1)$. Then, the new population generation method is given as the following steps:

Step 1: Set $i = 1$.

Step 2: Generate a new individual $\pi_i(\text{gen}+1)$.

    Step 2.1: Set $\pi_{ij}(gen+1) = 0$ for $j = 1, \cdots, n \times m \times L$.

    Step 2.2: Set $j = 1$.

    Step 2.3: $CJ = Job(\pi_i(\text{gen}+1), j)$.

    Step 2.4: If $l(CJ, \pi_i(\text{gen}+1)) = m \times L$, then go to Step 2.3.

    Step 2.5: Set $\pi_{ij}(gen+1) = CJ$.

    Step 2.6: Set $j = j+1$.

    Step 2.7: If $j \leq n \times m \times L$, then go to Step 2.3.

Step 3: Set $i = i+1$.

Step 4: If $i \leq PS$, then go to Step 2.

3.6 New Probabilistic Model Construction Strategy

Based on Schiavinotto and Stützle [15], the diameter of *Insert* is $n \times m \times L - 1$, which is one of the shortest diameters. That is, using *Insert* at most $n \times m \times L - 1$ times, one solution π can transit to any other solution π'. This means *Insertion*-based neighborhood can perform a more efficient and thorough search than the other kinds of neighborhoods with the same running time. So, we utilize the *Insertion*-based neighborhood to generate $(n \times m \times L - 1)^2$ neighbors of $\pi^{best}(\text{gen})$. Then, the best e% of the generated neighbors and $pop(gen)$ (i.e., the high quality individuals) are selected to construct the probabilistic model of BEDA.

3.7 Procedure of BEDA

Based on the contents in the above subsections, we propose the procedure of BEDA as follows:

Step 0: Denote π^{g_best} the global best individual and *genMax* the maximum
      generation.

Step 1: Initialization.

    Step 1.1: Set $gen = 0$.

    Step 1.2: Generate the initial population $pop(1)$ by using the method in subsec-
          tion 3.3.

Step 2: Set $gen = gen + 1$.

Step 3: Calculate the makespan of each individual in $pop(gen)$.

Step 4: Construct new probabilistic model by using the strategy in subsection 3.6,
      and update π^{g_best}.

Step 5: Generate $pop(gen+1)$ by using the new population generation method in
subsection 3.5.

Step 6: If $gen < genMax$, then go to Step2.

Step 7: Output π^{g_best} .

It can be seen from Steps 4 and 5 that the new generated individuals can aptly absorb the sub-sequences information of the high quality individuals during the evolution process and then guide the search to more promising regions. Thus, the algorithm is hopeful to obtain good results.

4   Simulation Result and Comparisons

4.1   Experimental Design

In order to test the performance of the proposed BEDA, a set of instances under different scales is randomly generated. The $n \times m \times L$ combinations include $10 \times 10 \times 2$, $10 \times 10 \times 3$, $20 \times 10 \times 3$, and $30 \times 10 \times 3$. For each scale, we generate 5 groups of data. The processing time t_j is generated from a uniform distribution [1,100] and the set-up time S_{ij} is generated from [1,20]. All algorithms are coded in Delphi 2010 and are executed on Mobile Intel Core 4 Duo 2.2 GHz processor with 8GB memory.

For each instance, each algorithm is run 20 times independently. Based on our previous experiments, the parameters of BEDA are set as follows: the population size $PS = 100$, and the proportion of the high quality individuals e%=20%. The parameters of GA are set the same as those in [2].

4.2   Comparisons of GA, BEDA_V1 and BEDA

For the purpose of showing the effectiveness of BEDA, we compare BEDA with GA and BEDA_V1 (i.e., a variant of BEDA). The difference between BEDA and BEDA_V1 lies in the probabilistic model and updating mechanism. The BEDA_V1 uses the probabilistic model and updating mechanism proposed by Baluja [5]. The learning rate α of BEDA_V1 is set to 0.1. The maximum generations of algorithms are set to 500. The running time of the algorithms are decided only by the scale of problems. The simulation results are listed in Table 1, where *BEST* denotes the best makespan, *AVG* denotes the average makespan and *WORST* denotes the worst makespan.

From Table 1, it can be seen that BEDA performs much better than GA and BEDA_V1 with respecting to solution quality. The values of *BEST*, *AVG* and *WORST* obtained by BEDA are much better than those obtained by GA and BEDA_V1. Specifically speaking, for instances 1 to 20, BEDA is better than GA and BEDA_V1 on all statistical indicators (i.e., *BEST* , *AVG* and *WORST*) except the *WORST* indicator of instances 09 and 12, and the *BEST* indicator of instance 18. So, it can be concluded that BEDA is an effective algorithm for the RJSSPST. Moreover, the test results also manifest a bayesian statistical inference-based probabilistic model is more suitable for guiding the search to the promising regions in the solution space of RJSSPST.

Table 1. Comparisons of *BEST*, *WORS* and *AVG* of GA, BEDA_V1 and BEDA

instances	n	m	L	GA[2]			BEDA_V1			BEDA		
				BEST	WORST	AVG	BEST	WORST	AVG	BEST	WORST	AVG
01	10	10	2	2771	3543	2971.95	2328	3307	2580.95	**1886**	2493	**2131.95**
02	10	10	2	2625	3864	3450.05	2135	2929	2500.35	**1612**	2499	**2129.70**
03	10	10	2	2815	3258	2949.40	1956	2852	2332.90	**1605**	2089	**1896.15**
04	10	10	2	2583	3387	2871.75	2176	2955	2477.70	**1867**	2398	**2119.05**
05	10	10	2	2911	4048	3771.25	2469	3376	2916.65	**1827**	2905	**2420.50**
06	10	10	3	5137	8789	6481.75	4720	9926	7473.25	**4613**	8763	**5986.85**
07	10	10	3	5098	11086	8974.95	4827	8662	6547.00	**3727**	7088	**5079.25**
08	10	10	3	5524	9112	6812.40	4644	7739	6045.90	**3424**	6209	**4810.05**
09	10	10	3	4992	**6713**	5658.35	4837	9019	6866.45	**4080**	7473	**5537.65**
10	10	10	3	5713	8771	7423.65	5179	12207	7592.95	**4012**	8438	**5860.35**
11	20	10	3	6757	7606	7155.10	6280	6592	6430.45	**5962**	6541	**6412.60**
12	20	10	3	6943	7816	7336.45	6442	**6685**	6663.40	**6256**	6691	**6490.60**
13	20	10	3	6901	7396	7208.65	6175	6724	6430.45	**6076**	6667	**6424.00**
14	20	10	3	6523	7342	7056.55	6130	6589	6335.50	**6022**	6562	**6323.80**
15	20	10	3	7030	7696	7357.90	6493	6832	6697.30	**6442**	6685	**6663.40**
16	30	10	3	8767	10348	9748.75	7993	8665	8489.05	**7882**	8566	**8388.90**
17	30	10	3	9550	10747	10294.00	8668	9211	8924.30	**8473**	9208	**8914.80**
18	30	10	3	8860	10015	9524.95	**7885**	8572	8263.00	7966	8320	**8224.00**
19	30	10	3	9193	10270	9812.05	8095	8839	8486.35	**7981**	8614	**7485.00**
20	30	10	3	9325	10561	10162.75	8182	9193	8850.25	**8089**	9103	**8827.30**

5   Conclusion and Future Work

This paper proposed a bayesian statistical inference-based estimation of distribution algorithm (BEDA) to solve the re-entrant job-shop scheduling problem with sequence-dependent setup times (RJSSPST). In BEDA, the initial population was generated by using stochastic method, and the search in the solution space was executed through sampling and updating the probabilistic model of BEDA. Simulation results and comparisons based on a set of instances showed the effectiveness of the proposed BEDA. To the best of our knowledge, this is the first paper on the application of estimation of distribution algorithm (EDA) for the RJSSPST. Our future work is to develop some BEDA-based algorithms to deal with re-entrant no-wait job-shop scheduling problem.

Acknowledgments. This research was partially supported by National Science Foundation of China (No. 60904081, 71101139), 2012 Academic and Technical Leader Candidate Project for Young and Middle-Aged Persons of Yunnan Province (No. 2012HB011), and Discipline Construction Team Project of Kunming University of Science and Technology (No. 14078212).

References

1. Garey, M.R., Johnson, D.S., Sethi, R.: The Complexity of The Flowshop And Jobshop Scheduling. Computer Science Department 1(2), 117–129 (1976)
2. Sun, J.U.: A Genetic Algorithm for A Reentrant Job-shop Scheduling Problem with Sequence-dependent Setup Times. Engineering Optimization 41(6), 505–520 (2009)
3. Karg, L.L., Thompson, G.L.: A Heuristic Approach to The Traveling Salesman. Management Science 10(2), 225–248 (1964)
4. Adams, J., Balas, E., Zawack, D.: The Shifting Bottleneck Procedure for Job Shop Scheduling. Management Science 34(3), 391–401 (1988)
5. Baluja, S.: Population-based Incremental Learning: A Method for Integrating Genetic Search Based Function Optimization And Competitive Learning. Technical Report CMU-CS-94-193. Carnegie Mellon University, Pittsburgh (1994)
6. Salhi, A., Rodriguez, J.A.V., Zhang, Q.F.: An Estimation of Distribution Algorithm with Guided Mutation for A Complex Flow Shop Scheduling Problem. In: Proceedings of the 9th Annual Conference on Genetic and Evolutionary Computation, London, UK, pp. 570–576 (2007)
7. Wang, S.Y., Wang, L., Zhou, G., Xu, Y.: An Estimation of Distribution Algorithm for The Flexible Job-shop Scheduling Problem. In: Huang, D.-S., Gan, Y., Gupta, P., Gromiha, M.M. (eds.) ICIC 2011. LNCS (LNAI), vol. 6839, pp. 9–16. Springer, Heidelberg (2012)
8. Liu, H., Gao, L., Pan, Q.K.: A Hybrid Particle Swarm Optimization with Estimation of Distribution Algorithm for Solving Permutation Flowshop Scheduling Problem. Expert Systems with Applications 38(4), 4348–4360 (2011)
9. Wang, L., Wang, S.Y., Xu, Y., Zhou, G., Liu, M.: A Bi-population Based Estimation of Distribution Algorithm for The Flexible Job-shop Scheduling Problem. Computers & Industrial Engineering 62(4), 917–926 (2012)
10. Zhou, A., Gao, F., Zhang, G.: A Decomposition Based Estimation of Distribution Algorithm for Multiobjective Traveling Salesman Problems. Computers and Mathematics with Applications 66(10), 1857–1868 (2013)
11. Ruiz, R., Maroto, C., Alcaraz, J.: Two New Robust Genetic Algorithms for The Flowshop Scheduling Problem. Omega 34(5), 461–476 (2006)
12. Huelsenbeck, J.P., Ronquist, F.: MrBayes: Bayesian Inference of Phylogenetic Trees. Bioinformatics 17(8), 754–755 (2001)
13. Qian, B., Li, Z.H., Hu, R., Zhang, C.S.: A Hybrid Differential Evolution Algorithm for The Multi-objective Reentrant Job-shop Scheduling Problem. In: The 10th IEEE International Conference on Control & Automation, pp. 485–489 (2013)
14. Zhang, R., Wu, C.: A Hybrid Local Search Algorithm for Scheduling Real-world Job Shops with Batch-wise Pending Due Dates. Engineering Applications of Artificial Intelligence 25(2), 209–221 (2012)
15. Schiavinotto, T., Stützle, T.: A Review of Metrics on Permutations for Search Landscape Analysis. Computers & Operations Research 34(10), 3143–3153 (2007)

An Effective Estimation of Distribution Algorithm for Multi-track Train Scheduling Problem

Shengyao Wang and Ling Wang

Tsinghua National Laboratory for Information Science and Technology (TNList),
Department of Automation, Tsinghua University, Beijing, 100084, P.R. China
wangshengyao@tsinghua.org.cn, wangling@mail.tsinghua.edu.cn

Abstract. In this paper, an effective estimation of distribution algorithm (EDA) is presented for solving the multi-track train scheduling problem (MTTSP). The individual of the EDA is represented as the permutation of train priority. With a proper track assignment rule, the individual is decoded into feasible schedule. In addition, the EDA builds a probability model for describing the distribution of the solution space. In every generation, it samples the promising region for generating new individuals and updates the probability model with the superior population. Moreover, the influence of parameter setting is investigated based on design-of-experiment method and a set of suitable parameter values is suggested. Simulation results based on some instances and comparisons with the existing algorithm demonstrate the effectiveness and efficiency of the EDA.

Keywords: train scheduling, assignment rule, estimation of distribution algorithm, probability model.

1 Introduction

Due to the increasing traffic of the railway, the scheduling for trains becomes more and more important to the transportation efficiency. As a complicated combinational optimization problem, the train scheduling problem is difficult to solve [1]. Ghoseiri et al. [2] developed a multi-objective optimization model for the passenger train-scheduling problem on a railroad network including single and multiple tracks as well as multiple platforms with different train capacities. Dorfman and Medanic [3] developed a local feedback-based travel advance strategy based on a discrete-event model of the traffic on the railway network. Zhou and Zhong [4] developed a branch-and-bound (B&B) algorithm with effective dominance rules and a beam search algorithm with utility evaluation rules for multi-objective planning applications of a double-track train scheduling problem. D'ariano et al. [5] developed a B&B algorithm by using the estimation of time separation among trains and modeling the scheduling problem with an alternative graph formulation.

Besides, the train scheduling problems have some similar features compared with the shop scheduling problems [6-9]. Burdett and Kozan [6] addressed the representation and construction of accurate train schedules by a solution algorithm for the hybrid

D.-S. Huang et al. (Eds.): ICIC 2014, LNAI 8589, pp. 697–708, 2014.

job shop problems with capacitated buffers. Liu and Kozan [7] modeled the train scheduling problem as a blocking parallel-machine job-shop scheduling problem and proposed a constructive heuristic algorithm to obtain the feasible schedule. Recently, Zhang and Chen [8-9] modeled the problem as the blocking hybrid flow shop scheduling problem and proposed a hybrid particle swarm optimization (HPSO) algorithm for solving it.

As a relatively new population-based evolutionary algorithm, the estimation of distribution algorithm (EDA) has gained increasing study during recent years [10]. Moreover, the EDA has already been successfully applied to solve a variety of engineering optimization problems, especially the scheduling problems [11-15]. However, to the best of our knowledge, there is no research work about the EDA for solving the train scheduling problem. Inspired by the successful applications of the EDA, this paper attempts to present an effective EDA for solving the multi-track train scheduling problem (MTTSP). To be specific, the individual of the population is represented as the permutation of the train priority and is decoded into feasible schedule by a track assignment rule. A probability model is built for describing the distribution of the solution space. It samples the promising region for generating new individuals and updates the probability model with the superior population in every generation. The influence of parameter setting is investigated based on design-of-experiment method and a set of suitable parameter values is suggested consequently. Simulation results based on some instances and comparisons with the existing algorithms demonstrate the effectiveness and efficiency of the EDA.

The remainder of the paper is organized as follows. The MTTSP is described in Section 2. In Section 3 the basic EDA is introduced, and the EDA for solving the MTTSP is presented in Section 4. Simulation results and comparisons are provided in Section 5. Finally, the paper ends with some conclusions and future work in Section 6.

2   Problem Description

2.1   Notation

n: the number of the trains;

S: the number of the railway segments;

$m(k)$: the number of the tracks in segment k;

X_{ijk}: a binary variable that is equal to 1 if train i is assigned to track j in segment k and is equal to 0 otherwise;

Y_{ijk}: a binary variable that is equal to 1 if train i precedes train j on the same track in segment k and is equal to 0 otherwise;

S_{ik}: the starting time of train i in segment k;

T_{ijk}: the travel time of train i on track j in segment k;

C_{ik}: the completion time of train i in segment k;

L: a large constant.

2.2   The MTTSP

The MTTSP is described as follows: A set of n trains are going to pass S segments, where each segment has at least one track and some segments have multiple tracks. Tracks in each same segment are unrelated. One track can be used for only one train at a time. When a train completes a segment, if the planning track in the next segment is occupied by another train, it waits on the current track until the planning track is available. Due to these features, the MTTSP can be modeled as the blocking hybrid flow shop scheduling problem (BHFSP) [8-9]. The corresponding relationship between the MTTSP and the BHFSP is listed in Table 1.

Table 1. Corresponding relationship between the MTTSP and the BHFSP

Problem	Properties				
MTTSP	Train	Track	Segment	Travel time	Waiting on the track
BHFSP	Job	Machine	Stage	Processing time	Blocking on the machine

The MTTSP is supposed that: All the n trains are independent and ready at the initial time; All the tracks are available at time 0; For all the n trains, the travel times on each track are deterministic and known in advance; The transportation times between different segments are negligible. An example of MTTSP with 5 segments is illustrated in Fig. 1. The configuration of track numbers is $\{1, 3, 1, 2, 1\}$.

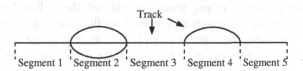

Track

Segment 1   Segment 2   Segment 3   Segment 4   Segment 5

Fig. 1. An example of the MTTSP

The MTTSP can be formulated as follows:

$$\text{Minimize: } C_{\max} \tag{1}$$

$$\text{Subject to: } C_{\max} = \max C_{iS}, i = 1, 2, \cdots, n \tag{2}$$

$$\sum_{j=1}^{m(k)} X_{ijk} = 1, i = 1, 2, \cdots, n; k = 1, 2, \cdots, S \tag{3}$$

$$C_{ik} = S_{ik} + \sum_{j=1}^{m(k)} X_{ijk} T_{ijk}, i = 1, 2, \cdots, n; k = 1, 2, \cdots, S \tag{4}$$

$$S_{ik} \geq C_{i,k-1}, i = 1, 2, \cdots, n; k = 2, 3, \ldots, S \tag{5}$$

$$S_{jk} \geq C_{ik} - L(1 - Y_{ijk}), \text{ for all the pairs } (i, j) \tag{6}$$

$$S_{jk} \geq S_{i,k+1} - L(1 - Y_{ijk}), \text{ for all the pairs } (i, j) \tag{7}$$

As for the above formulation: Eq. (1) denotes the objective function to minimize the completion time of all the trains as shown in (2); Eq. (3) ensures that a train can use exactly one track in each segment; Eq. (4) describes the computation of C_{ik}; Constraint (5) means that the trains pass all the segments from the first one to the last one in order; Constraint (6) means that a train can use the track until all its preceding train passes the track; Constraint (7) means that the trains will wait on the current track if the next segment is busy.

3   Estimation of Distribution Algorithm

The EDA [18] is a relatively new member of the evolutionary algorithms. The EDA employs explicit probability distributions to perform optimization procedure. It establishes a probability model of the most promising area by statistical information. Based on the probability model, it samples the search space to generate new solutions. To trace the more promising search area, it adjusts the probability model in each generation by using the information of some superior individuals of the new population.

The key content of the EDA is to estimate the probability distribution. Since the probability model is use to describe the distribution of the solution among the space, it is crucial to build a proper probability model. In addition, the probability model should be well adjusted to make the search procedure tract the promising search area. As a consequence, a reasonable mechanism should be design to adjust the model. Since different problems have different properties, both the probability model and the updating mechanism should be well adopted for a specific problem. Nevertheless, the EDA pays more attention to global exploration while its exploitation capability is relatively limited. Therefore, an effective EDA should balance the exploration and the exploitation abilities [13].

4   EDA for MTTSP

In this section, an effective EDA is presented for solving the MTTSP. First, the encoding, decoding, probability model, updating mechanism and local search scheme are introduced. Then, the flowchart of the EDA is illustrated.

4.1   Encoding and Decoding

Every individual of the population denotes a solution of the MTTSP. A solution is expressed by a permutation of the train priority for scheduling. In other words, an integer number sequence with the length of n determines the scheduling order of the trains. For example, a solution $\{2, 5, 3, 4, 1\}$ represents that train 2 is scheduled first, and next are train 5, train 3 and train 4 in sequence. Train 1 is the last one to be scheduled.

In the decoding procedure, it assigns tracks in each segment for a certain train. After all the segments are assigned, the scheduling of a certain train completes and the next train in the solution sequence is considered. To be specific, for a given train i, the track of each segment is selected according to the track assignment rule as follows.

Step 1: $j = 1$. Choose track s if $T_{i1s} + a_s$ has the smallest value, where a_s denotes the available time of track s. Record s as the selected track for segment 1 and denote it as s^1. Update the completion time of train i (i.e., $C_{i,1}$) and a_s.

Step 2: $j = j + 1$. Decide the earliest starting time according to the completion time at the previous segment ($C_{i,j-1}$) and the earliest available time of each track (a_k). Obtain $\max\{a_k, C_{i,j-1}\}$.

Step 3: Choose track s if $T_{ijs} + \max\{a_s, C_{i,j-1}\}$ has the smallest value. Record s as the selected track for segment j, which is denoted as s^j.

Step 4: Update the completion time of train i (i.e., $C_{i,j}$) and the available time of track s (i.e., a_s). Update the available time of the track s^{j-1}, i.e., $a_{s^{j-1}} = S_{ij}$.

Step 5: Go to Step 2 until all the segments are assigned.

From step 4, it can be seen that the available time of the previous track is set as the starting time of the current track. The reason is that a train has to wait on the current track until the next track is available. With the above assignment rule, a feasible schedule is formed and the completion time of the schedule is obtained.

4.2   Probability Model and Updating Mechanism

The EDA produces the new individuals by sampling a probability model. In this paper, the probability model is designed as a probability matrix P. The element $p_{ij}(l)$ of the probability matrix P represents the probability that train j appears before or in position i of the solution sequence at generation l. The value of p_{ij} implies the priority of a train when deciding the scheduling order. For all i and j, p_{ij} is initialized as $p_{ij}(0) = 1/n$, which ensures that the whole solution space can be sampled uniformly.

In each generation of the EDA, the new individuals are generated via sampling the probability matrix P. For every position i, train j is selected with a probability p_{ij}. If train j has already appeared, it means the scheduling procedure of train j is completed. Then, the whole j-th column of probability matrix P will be set as zero and all the elements of P will be normalized to maintain that each row sums up to 1. In such a way, an individual is constructed until all the trains appear in the sequence. Finally, P_Size individuals are generated.

Next, it determines the superior sub-population that consists of the best SP_Size solutions, where $SP\_Size = \eta\% \cdot P\_Size$. The probability matrix P is updated according to the following equation:

$$p_{ij}(l+1) = (1-\alpha)p_{ij}(l) + \frac{\alpha}{i \times SP\_Size} \sum_{k=1}^{SP\_Size} I_{ij}^k, \forall i, j. \tag{8}$$

where $\alpha \in (0,1)$ is the learning rate of P and I_{ij}^k is the following indicator function of the k-th individual in the superior sub-population.

$$I_{ij}^k = \begin{cases} 1, \text{if train } j \text{ appears before or in position } i \\ 0, \text{else} \end{cases} \tag{9}$$

4.3   Local Search Scheme

To enhance the exploitation of the EDA, several different local search operators for the shop scheduling problems [17] are employed, which are described as follows:

Swap: Randomly select two different elements from a sequence and then swap them.

Insert: Randomly choose two different elements from a sequence and then insert the back one before the front one.

Inverse: Invert the subsequence between two different random positions of a sequence.

In each local search step, the above three operators are performed in sequence and the new individual will replace the old one if it has a smaller completion time. The above procedure is applied on the best individual of the current population ls times in every generation, where ls represents the intensity of the local search.

4.4   Procedure of EDA for MTTSP

With the design above, the procedure of EDA for solving the MTTSP is illustrated in Fig. 2.

From the above procedure, it can be seen that the EDA samples the probability model to generate the new individuals and learns the information of the superior ones to update the probability model. Meanwhile, the local search strategy is performed to enhance local exploitation capability. The algorithm stops when the maximum decoding time MAX_D is satisfied.

4.5   Computation Complexity Analysis

For each generation of the EDA, its computational complexity can be roughly analyzed as follows.

For the updating process, first it is with the computational complexity $O(P\_Size \log P\_Size)$ by using the quick sorting method to select the best SP_Size individuals from population; Then, it is with the complexity $O[n(SP\_Size+n)]$ to update

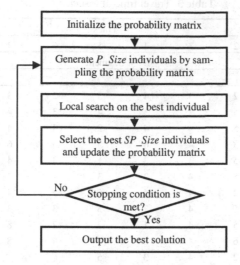

Fig. 2. Procedure of the EDA for solving the MTTSP

all the $n{\times}n$ elements of P. Besides, local search is performed on the best individual of the population ls times with the complexity $O(ls{\times}n)$. For the sampling process, the element for every position is generated with the roulette strategy by sampling the probability matrix P to obtain a new individual. To generate a train priority sequence, it is with the complexity $O(n^2)$. Thus, the computational complexity for generating P_Size individuals is $O[n^2P\_Size]$. It can be seen that the complexity of the EDA is not large and the algorithm could be efficient.

5   Simulation and Comparisons

5.1   Benchmark Instances

In this section, three instances with different scales taken from literatures [8-9] are used to test the performance of the EDA and for comparison. The parameters of the instances are provided in Table 2. Besides, the travel times of each train on every track are listed in Tables 3-5 for each instance.

Table 2. Parameters of the instances

Instance	Train	Track	Segment	Track configuration
1	12	9	3	{3, 2, 4}
2	10	14	9	{1, 2, 1, 2, 1, 3, 1, 2, 1}
3	12	10	4	{3, 3, 2, 2}

Table 3. Travel time of instance 1

Train	Track								
	1	2	3	4	5	6	7	8	9
		Segment							
		1			2			3	
1	2	2	3	4	5	2	3	2	3
2	4	5	4	3	4	3	4	5	4
3	6	5	4	4	2	3	4	2	5
4	4	3	4	6	5	3	6	5	8
5	4	5	3	3	1	3	4	6	5
6	6	5	4	2	3	4	3	9	5
7	5	2	4	4	6	3	4	3	5
8	3	5	4	7	5	3	3	6	4
9	2	5	4	1	2	7	8	6	5
10	3	6	4	3	4	4	8	6	7
11	5	2	4	3	5	6	7	6	5
12	6	5	4	5	4	3	4	7	5

Table 4. Travel time of instance 2

Train	Track													
	1	2	3	4	5	6	7	8	9	10	11	12	13	14
						Segment								
	1		2		3		4		5		6		7		8		9
1	3	1	2	2	1	3	4	2	3	5	4	6	9	5
2	3	2	4	4	2	3	3	3	4	2	5	4	6	4
3	4	2	3	2	2	1	2	3	6	9	2	4	5	4
4	2	2	5	5	4	2	3	1	3	6	2	3	7	7
5	3	1	2	3	1	4	5	4	3	4	5	6	4	5
6	2	3	6	1	3	5	2	6	2	3	2	5	6	3
7	1	3	3	1	3	2	5	3	4	6	3	4	8	7
8	1	4	2	3	4	3	6	3	2	7	5	2	4	4
9	4	1	2	3	2	3	4	2	3	5	4	6	7	5
10	2	2	3	4	2	3	3	2	4	4	3	5	6	6

5.2   Parameter Setting

The EDA has four parameters: P_Size (the population size), η (the parameter associated with the superior sub-population), α (the learning rate of P), and ls (the intensity of local search). To investigate the influence of these parameters on the performance of the EDA, the Taguchi method of design-of-experiment (DOE) [18] is implemented base on instance 1. The combinations of different values of these parameters are listed in Table 6.

Table 5. Travel time of instance 3

Train	Track									
	1	2	3	4	5	6	7	8	9	10
	Segment									
	1			2			3		4	
1	45	48	50	35	35	30	30	35	25	26
2	45	50	45	35	36	35	35	34	25	30
3	50	45	46	35	36	36	31	34	30	31
4	50	48	48	34	38	35	32	33	27	31
5	45	46	48	30	35	50	34	32	28	31
6	45	45	45	30	35	50	33	32	30	26
7	47	50	47	31	30	35	35	31	29	25
8	50	45	48	32	30	34	34	30	24	27
9	48	46	46	33	34	30	34	30	25	25
10	45	47	47	33	33	30	35	34	32	26
11	46	50	45	34	30	50	30	35	31	25
12	48	50	47	35	31	35	32	30	25	30

Table 6. Combinations of parameter values

Parameters	Factor level			
	1	2	3	4
P_Size	20	30	40	50
η	10	20	30	40
α	0.1	0.2	0.3	0.4
ls	5	10	15	20

In all the experiments, it sets $Max\_D = 10000$ as the stopping condition. For each parameter combination, the EDA is run 20 times independently and the average result obtained is calculated as the response variable (RV) value. Besides, the number of factor levels is set as 4. Accordingly, the orthogonal array $L_{16}(4^4)$ is used. The orthogonal array and the obtained RV values are listed in Table 7.

According to the orthogonal table, the trend of each factor level is illustrated in Fig. 3. Then, the average response value of each parameter is figured out in Table 8 to analyze the significance rank.

From Table 8, it can be seen that the intensity of local search ls is the most significant one among the four parameters. A large value of ls may help the EDA enhance the searching capability. In addition, the significance of the population size P_Size ranks the second. With a fixed maximum number of evaluation times, a small population size allows more generations so that it may provide a deeper search. According to above analysis, a better choice of the parameter combination is suggested as P_Size=20, η=30, α=0.3, ls=20 and MAX_D=10000, which is also used for the following experiments.

Table 7. Orthogonal array and ARV values

Experiment Number	Factor				RV
	P Size	η	α	ls	
1	1	1	1	1	24.1
2	1	2	2	2	24.0
3	1	3	3	3	23.6
4	1	4	4	4	23.9
5	2	1	2	3	24.0
6	2	2	1	4	23.8
7	2	3	4	1	24.1
8	2	4	3	2	24.0
9	3	1	3	4	23.9
10	3	2	4	3	23.9
11	3	3	1	2	24.0
12	3	4	2	1	24.1
13	4	1	4	2	24.0
14	4	2	3	1	24.2
15	4	3	2	4	23.9
16	4	4	1	3	24.0

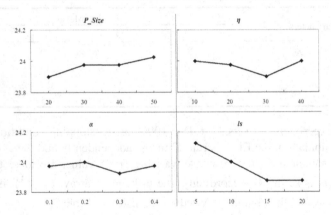

Fig. 3. Factor level trends of the parameters

Table 8. Average response values

Level	P Size	η	α	ls
1	**23.9**	24	23.975	24.125
2	23.975	23.975	24	24
3	23.975	**23.9**	**23.925**	**23.875**
4	24.025	24	23.975	**23.875**
Delta	0.125	0.1	0.075	0.25
Rank	2	3	4	1

5.3   Simulation Results and Comparisons

The EDA is compared with the HPSO [8-9] for solving the MTTSP. For each instance, the algorithm runs 10 times as in literatures [8-9]. The best and average completion times as well as the CPU time are used as the performance measure. The results are listed in Table 9, where the results of the comparative algorithm are directly taken from the literatures [8-9].

Table 9. Simulation results

Instance	HPSO			EDA		
	Best	Average	CPU(s)[a]	Best	Average	CPU(s)[b]
1	25	26.2	95	23	23.5	0.047
2	72	72	<1	70	70	0.094
3	301	N/A	190	297	297	0.062

a. Pentium 1.6 GHz CPU.
b. Core i5 2.3 GHz CPU.

From Table 9, it can be seen that the EDA outperforms the HPSO in solving all the three instances. The best and average results obtained by the EDA are better than that by the HPSO, which means that the EDA is more effective for solving the MTTSP. Besides, the CPU time employed by the EDA is less than 0.1 second for every instance. It demonstrates that the EDA is much more efficient than the HPSO.

6   Conclusions

In this paper, an effective estimation of distribution algorithm is presented for solving the multi-track train scheduling problem. The permutation of train priority is employed to represent the individual of the EDA. A track assignment rule is designed for decoding the individual into a feasible schedule. A probability model with its updating mechanism is presented. Based on the design-of-experiment method, a set of suitable parameter values is suggested. Simulation results based on some instances and comparisons with the existing algorithms demonstrate the effectiveness and efficiency of the EDA. As for the future work, it is worth to design EDA-based algorithm for other scheduling problems, such as the semi-conductor manufacturing scheduling.

Acknowledgments. This work was supported by the National Key Basic Research and Development Program of China (2013CB329503), the National Science Foundation of China (61174189) and the Doctoral Program Foundation of Institutions of Higher Education of China (20130002110057).

References

1. Gholami, O., Sotskov, Y.N., Werner, F.: Job-Shop Problems with Objectives Appropriate to Train Scheduling in a Single-Track Railway. In: Proceedings of 2nd International Conference on Simulation and Modeling Methodologies, Technologies and Applications, Rome, pp. 425–430 (2012)
2. Ghoseiri, K., Szidarovszky, F., Asgharpour, M.J.: A Multi-Objective Train Scheduling Model and Solution. Transportation Research Part B: Methodological 38(10), 927–952 (2004)
3. Dorfman, M.J., Medanic, J.: Scheduling Trains on a Railway Network Using a Discrete Event Model of Railway Traffic. Transportation Research Part B: Methodological 38(1), 81–98 (2004)
4. Zhou, X., Zhong, M.: Bicriteria Train Scheduling for High-Speed Passenger Railroad Planning Applications. European Journal of Operational Research 167(3), 752–771 (2005)
5. D'ariano, A., Pacciarelli, D., Pranzo, M.: A Branch and Bound Algorithm for Scheduling Trains in a Railway Network. European Journal of Operational Research 183(2), 643–657 (2007)
6. Burdett, R.L., Kozan, E.: A Disjunctive Graph Model and Framework for Constructing New Train Schedules. European Journal of Operational Research 200(1), 85–98 (2010)
7. Liu, S.Q., Kozan, E.: Scheduling Trains as a Blocking Parallel-Machine Job Shop Scheduling Problem. Computers & Operations Research 36(10), 2840–2852 (2009)
8. Zhang, Q.L., Chen, Y.S.: Train Scheduling Problem and Its Solution Based on Hybrid Particle Swarm Optimization Algorithm. China Mechanical Engineering 24(14), 1916–1922 (2013)
9. Zhang, Q.L., Chen, Y.S.: Hybrid Particle Swarm Optimization Algorithm for Hybrid Flow Shop Scheduling Problem with Blocking. Information and Control 42(2), 252–257 (2013)
10. Wang, S.Y., Wang, L., Fang, C., Xu, Y.: Advances in Estimation of Distribution Algorithms. Control and Decision 27(7), 961–966 (2012)
11. Wang, L., Fang, C.: An Effective Estimation of Distribution Algorithm for the Multi-Mode Resource-Constrained Project Scheduling Problem. Computer & Operations Research 39(2), 449–460 (2012)
12. Wang, L., Wang, S.Y., Liu, M.: A Pareto-Based Estimation of Distribution Algorithm for the Multi-Objective Flexible Job-Shop Scheduling Problem. International Journal of Production Research 51(12), 3574–3592 (2013)
13. Wang, L., Wang, S.Y., Xu, Y., Zhou, G., Liu, M.: A Bi-Population Based Estimation of Distribution Algorithm for the Flexible Job-Shop Scheduling Problem. Computers & Industrial Engineering 62(4), 917–926 (2012)
14. Wang, S.Y., Wang, L., Liu, M., Xu, Y.: An Effective Estimation of Distribution Algorithm for the Flexible Job-Shop Scheduling Problem with Fuzzy Processing Time. International Journal of Production Research 51(12), 3778–3793 (2013)
15. Wang, S.Y., Wang, L., Liu, M., Xu, Y.: An Effective Estimation of Distribution Algorithm for Solving the Distributed Permutation Flow-Shop Scheduling Problem. International Journal of Production Economics 145(1), 387–396 (2013)
16. Larranaga, P., Lozano, J.A.: Estimation of distribution algorithms: A new tool for evolutionary computation. Springer, Netherlands (2002)
17. Wang, L.: Shop scheduling with genetic algorithms. Tsinghua University & Springer Press, Beijing (2003)
18. Montgomery, D.C.: Design and analysis of experiments. John Wiley & Sons, Arizona (2005)

Construction of Basic Education Cloud Computing Platform Based on Virtualization Technology

Minghui Zhang[1] and Jinchen Zhou[2]

[1] College of Information Science & Technology, Zhengzhou Normal University, Zhengzhou
[2] Department of Physics and Electronic Science, Zhengzhou Normal University, Zhengzhou
450044, China

Abstract. The information technology of basic education is constantly popularized, basic education curriculum and textbook reform are carried forward step by step, high-quality educational resources are seriously deficient, and these factors directly affect sustainable development of basic education. 'Education cloud' concept is introduced in the paper by transferring cloud computing technology in education field. Service frames and functions in three types of education clouds are described with virtual technique as core, in addition, concepts and methods for constructing basic education cloud platform are provided on the basis.

Keywords: education cloud, virtualization technology, cloud services.

1 Introduction

The state, educational institutions at all levels; schools and enterprises invest manpower, material and financial resources greatly in promoting implementation of basic education informational project and construction of education resources with prevalence of network and promotion of education information technology in recent years. Although preliminary results have been achieved, there are still some problems in construction of basic education resources, mainly including the follows:

(1) Unbalanced proportion in educational resource investment. The investment is mainly based on hardware facilities in the aspect of primary and secondary education resource, which accounts for more than 80% of total fund investment. However, software investment proportion is relatively low.

(2) Educational resources are mismatched. Existing resources have more prominent problems in the aspect of systematization and match with text materials, etc. Schools and teachers also pay the most attention to these problems. In addition, high-quality education and teaching resources which are in line with the concept of curriculum reform, suitable for IT environment characteristics, and adapted to moral education and quality education demand are deficient .

(3) Shared educational resources do not have uniform technical standards. Educational institutions and a number of software resource enterprises in different regions adopt different technical standards in production of education and teaching resources due to difference in different versions of textbooks, users of educational resource, teaching methods and ideas, thereby forming teaching resource database platforms in separate ways and with own systems.

D.-S. Huang et al. (Eds.): ICIC 2014, LNAI 8589, pp. 709–716, 2014.

(4) The regions of education resource construction are imbalanced. IT infrastructure is constructed and invested weakly in economically underdeveloped provinces and rural areas. Teachers are apparently lack of material environment for applying digital teaching resources, and concepts of digitally constructing and sharing curriculum resources are deficient.

(5) Resource software is provided with uneven technical contents. Teachers in primary and secondary schools bear heavy teaching work currently, and they are lack of centralized time for creating teaching resources. Meanwhile, technology application level is not high as a whole; therefore interactive software fully reflecting advantages of computer and network technology are seriously deficient. [1]

Therefore, it is extremely critical to construct authorized, systematic, scientific, selective, open and high-quality education resource resources convergence platform which is adapted to characteristics of modern education, combines information technology and network technology development, realizes education teaching as main body, and achieves personal development, teaching tool resource base based on the platform and resource mode based on interconnection and communication should be established.

2   Concept

2.1   Cloud Computing

Large-scale massive data should be processed with the rapid growth of basic education information and data in Internet era. However, desktop computers, which handle these data alone at present, can not meet current demand for large data processing. System hardware performances and hardware amount can be generally increased to meet the growing requirements in system scalability. Cloud computing concept is proposed due to limitation of traditional parallel programming model application. New parallel programming frame which can be easily learnt, used and deployed is demanded, which not only can save cost, but also can realize system scalability. Cloud computing is further developed from distributed computing, parallel processing and grid computing. It belongs to a system which is based on Internet-based computing and provides various Internet applications with hardware services, infrastructure services, platform services, software services and storage services [1]. Cloud computer is produced by combining various traditional computer techniques and network technology development, computer techniques include virtualization, distributed computing, grid computing, etc. Computing resources (such as storage, hardware, platform and software) are shared under cloud environment, and the concept was firstly proposed by Google in 2007, which was rapidly developed since then. Cloud computing resources can be provided for users in three service modes: Infrastructure as a Service (IaaS), Platform as a Service (PaaS) and Software as a service (SaaS). [2]

2.2   Education Cloud

Education cloud refers to transfer of cloud computing in education field, which provides cloud services for education field, including Cloud Computing Assisted Instructions (CCAI), Clouds Computing Based Education (CCBE) and other forms. Wherein, cloud computing assistant teaching refers that schools and teachers utilizes education 'cloud services' to provide convenient conditions of teaching resource sharing and unlimited storage space through cloud services of cloud computing. Cloud computing assistant teaching is also known as 'education based on cloud computing', which assists education teaching activities by services provided by cloud computing in education teaching. Services provided by cloud computing is mutually supported and combined with basic classroom teaching, thereby providing corresponding education and other teaching resources as well as design and management in the teaching process.

Education cloud will be applied in infrastructure of education information technology in the future, including all hardware computing resource necessary for education information technology. These resources can provide education institutions, education practitioners and students with a good platform after being virtualized.

2.3   Cloud Learning

'Cloud learning' refers that cloud knowledge, cloud tasks, cloud resources, cloud components, cloud website, leaner cognitive structure and other key models are constructed around learning services in cloud computing environment with psychology, education, knowledge engineering and system engineering theories as guidance. [3] Software architecture and Web interactive technology are utilized for developing open, sustainably-developed, personalized and distributed learning system with interactive inquiry features, and it is technical code for developing, trading, operating and evolving interaction inquiry learning resources. Cloud learning system includes cloud learning platform and cloud terminal learning machine. Cloud learning platform, as 'cloud' resource network, is open all over the world, which is constructed, cooperated and shared by the whole society. 'Knowledge cloud' is formed by knowledge engineering, learning theory, semantic network, information technology and other components. Currently, cloud learning platform mainly includes cloud learning master station, IVIDEO interactive video website and cloud learning open platform. Cloud terminal learning machine can be adopted for interactively exploring and learning knowledge clouds anywhere and anytime. Cloud terminal learning machine belongs to a lightweight client, such as learning computer, Ipad, smart phone, browser, etc.

2.4   Virtualization Technology

Virtualization technology was derived in IBM mainframe system in 1960s, which became popular gradually in System 370 series in 1970s. Virtual Machine Monitor (VMM) program forms the core of the system. Various Virtual Machine examples with independent operation are virtualized in physical hardware. Virtualization technology is mainly divided into three categories, namely: Platform Virtualization, Resource Virtualization and Application Virtualization. [4] Platform virtualization technology refers to virtualization of computer and operation system. Resource virtualization technology refers to virtualization of specific system resources, such as memory, network resources, network storage, etc. Application virtualization technology includes explanation technology, phantom, simulation, etc. Virtualization technology refers to platform virtualization usually, which is also known as virtual machine technology. It usually operates control program (also known as Virtual Machine Monitor or Hypervisor), and hides physical characteristics of actual work platform, thereby providing users with abstract and simulated computing environment.

Currently, basic education resources are distributed dispersedly. Education administrators such as cantonal educational institutions and third party participants of education such as educational software development enterprises do not have uniform contents, technical standards and even file format in education and teaching resources, thereby developing teaching resource database platforms in separate ways and with own systems, primary and secondary schools have to accept various education resources in different technical standards. Hardware resource utilization rate statistics is adopted as an example:

Table 1. Education Platform Hardware Resources and Resource Utilization Rate in Primary and Secondary Schools

Education Platform	Hardware and Network Equipment	Storage	CPU Utilization	Memory Utilization
1	Six servers, two switchboards and two firewalls	2T	18.3%	24.6%
2	Five servers, one switchboard and one firewall	1T	9.4%	11.3%
3	Eight servers, two switchboards and two firewalls	2T	22.8%	29.2%
4	Six servers, two switchboards and two firewalls	2T	13.2%	18.8%
5	Four servers, one switchboards and one firewall	2T	14.7%	20.5%

3   Implementation Mode of Education Cloud Platform of Virtualization Technology

A series of operations will be implemented on education cloud platform, such as resource joint construction and sharing, teacher community construction (teaching and scientific research), home-school community construction, teacher- student community construction, school management, assessment modernization, etc., thereby optimizing course content, transforming teaching and learning mode, sufficiently sharing high-quality education resources, meeting individual educational needs of students, implementing quality education and innovative talent training strategies, improving quality of education teaching, promoting balanced development of education, and realizing the principle of fair education.

3.1   Construction of Education Information Public Support Environment (Basic Cloud IaaS)

IaaS basic cloud business platform is mainly composed of network, servers, storage systems, peripheral interfaces, etc. as shown in Figure 1. Mature virtualization technology is adopted on business platform to create multiple functional resource pools which can be managed, thereby sharing network, computing power (servers), storage resources, etc. Resource management can be standardized, resource regulation and distributed can be finer, coordination can be more flexible, convenient and rapid in the data center by realization of virtualization technology, thereby improving resource utilization rate, and reducing total cost.

IaaS service mode is adopted. Education or government departments can use existing equipment for purchasing and deploying multiple servers in a centralized mode for composing 'cloud' infrastructure. Memory, I/O equipment, storage and computing power are integrated into a virtual network infrastructure resource pool, which is provided for cloud primary and secondary schools, especially schools in remote areas or private schools without procurement ability in measurement service mode, thereby providing these schools with demanded storage resources, virtualized servers and other services, saving school expense in basic hardware, and effectively realizing balanced development of education. [7]

For example, primary and secondary schools should invest in foundational network hardware and operation maintenance during construction of digital campus. Through cloud IaaS service mode, the education department can purchase scale storage array by education department and virtualization server. Combination of PaaS cloud platform and comprehensive show deployment model of SaaS cloud, which can realize zero input of the construction of campus network. Avoid the school network hardware problems of decentralized investment and repeated construction, reduce the operational cost of the network hardware, improve the efficiency of the use of the network hardware.

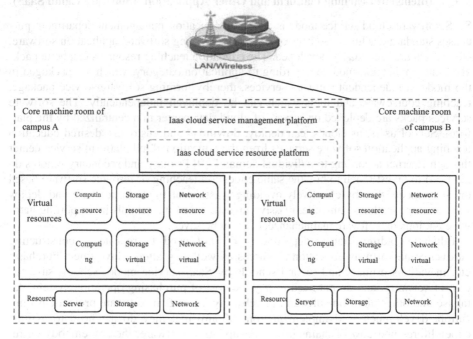

Fig. 1. Public Facility Environment Drawing of Education Cloud

3.2   Construction of Educational Public Service Platform (Platform Cloud PaaS)

Currently, primary and secondary schools independently uses a part of application software basically, thereby leading to high purchase and operation maintenance costs, low efficiency, poor reuse and other phenomena with investment waste. Small and medium-sized schools can not afford some outstanding application software due to the high cost. Common application service of school can be realized by cloud service PaaS mode, thereby improving school digital education, digital scientific research and digital management level by services with low cost and high quality. [8] PaaS cloud service provides platform to develop environment services and also provides development template. Various school users at all levels can customize own application programs under the cloud platform development environment, such as: digital campus portal website, teaching management system, and school letter passes services. These services are opened to cloud users through servers in cloud service center. Personalized, self-developed and customized middleware platforms can be provided for various school units of basic education through PaaS services.

Most primary and secondary schools generally have difficult technical problems such as digital campus website construction or complex code application development of various teaching affair management systems. Manpower and financial resources are demanded. These problems can be solved by program package services and customization services provided by PaaS cloud platform. Data centralization and management can be achieved.

3.3   Intelligent Learning Platform and Other Applications (Software Cloud SaaS)

SaaS software cloud service mode is adopted. Education management department purchases standard teaching resources of education teaching software, application software, etc., which are secondarily developed. The education teaching resources can be unpackaged into application module according to application category, which are packaged in the mode of independent software services, thereby forming software service package, teaching courseware and tools which can be deployed independently. The software services packs are deployed on servers of cloud data center in a centralized by the platform. School users, as clouds, can deploy, download and customize desired education teaching application software services from the education cloud platform service center through Internet according to individual demand. [9] Primary and secondary schools do not need to invest much to buy the same education teaching or resource software based on SaaS mode, thereby effectively avoiding wastage of funds to purchase and deploy software repeatedly. Meanwhile, massive investment of the schools in supporting hardware equipment, software maintenance, etc. can be saved.

Education administrative departments do not need to repeatedly constructing hardware and software education resources as well as scattered investment purchase, etc. in various primary and second schools by SaaS service mode. Schools save the cost of software, and do not need to pay the cost of purchasing operation system, database and other platform software, thereby saving costs in software project customization, development and implementation. Meanwhile, they do not need to assume expenditures and cost of maintaining and updating software, thereby embodying resource data comprehensive construction and comprehensive sharing more effectively. Investment in scattered purchase of software can be avoided on one hand, and high-quality resource sharing and services also can be improved at the same time.

4   Overall Program of Education Platform Based on Cloud Computing

Information infrastructure of various education institutions at all levels should be sufficiently integrated and utilized during construction of regional basic education cloud platform. Open basic education cloud environment characterized by comprehensive coverage and rational distribution should be constructed for supporting the formation of hierarchical structure, including cloud basic platform, cloud resource platform and cloud education management service platform. The education cloud basic platform should be able to support effective deployment and application of education cloud resource platform and management service platform. Education basic cloud services can be provided for IPv4 and IPv6 users at the same time. [10] IDC machine room (Internet Data Center) can be transformed and upgraded for realizing IDC services supporting cloud computing in the specific implementation process. Education users are served in SaaS form through main network. Meanwhile, service and education resource sharing services aiming at individual education should be introduced at the same time. Users in the network can be served through integrating, customizing and constructing education cloud platform.

Cloud-based service framework model is shown as follows based on the functions provided by education cloud platform, as shown in Figure 2.:

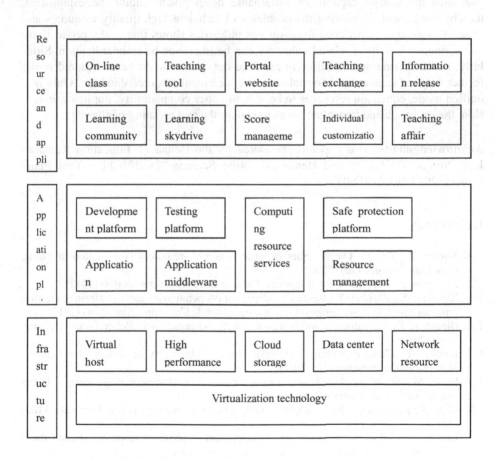

Fig. 2. Education Cloud Platform Service Framework

5   Conclusion

Advantage compensation, risk mutual assumption and resource sharing of digital education teaching resource construction can be comprehensively realized by establishing education cloud. Since resource construction is related to various fields of education, politics, economy, culture, business, scientific research, etc., forces from various fields such as government departments, universities, primary and secondary schools, various associations, foundations, companies, etc. should be integrated by relying on government organizations, thereby establishing exchange mechanisms in various forms, and constructing dynamic development trend of self-improvement. The education cloud platform, education resources, rules, protocols and related certification standards should be planed as a whole and coordinated from the national level. Development, synchronous integration, sharing mechanism and effective investment guarantee mechanism capable of sustainable development should be established, thereby fundamentally solving the problems of deficient high-quality resources and structure shortage. Experience from various industries shows that scale, productivity and economic benefit of related industries can be improved by standardization. Similarly, better sharing among different resource databases should be considered in the former stage of standardized formulation of educational resource database. Only in a unified mode, education resources to be able to better compatibility, duplicate, extension, the exchange and transplant, so as to realize the real meaning of sharing.

Acknowledgments. This paper is sponsored by the Henan IT Education Research Key Project (ITE12039) and Henan Education Science "Twelfth Five Year Plan" Project (2012-JKGHAD-0324).

References

[1] Jingtao, Z., Xin, L.: On strategies of basic education resources construction in China. Education Research (10) (2006)
[2] Zhaoqing, Z.: Benefits of cloud services. Chinese Education Network (08) (2012)
[3] Xiaofeng, X., Zhaohui, Y.: Research on operator cooperation business platform realization program based on cloud computing technology. Mobile Communications (04) (2012)
[4] Bingchao, T.: Virtualization technology research and exploration. Police Force Academy (04) (2010)
[5] Lanjing, T.: Cloud computing knowledge database: Inventory of strange cloud names. Network and Information (01) (2012)
[6] Lijun, M.: Study on class I information virtualization sharing strategy design. Hubei Normal University Thesis (2011)
[7] Zhen, T.: Research on cloud computing virtualization technology. IT and Information (02) (2012)
[8] Based on social major education, and development of localized resources (Henan Basic Education Resource Network), http://www.hner.net/

A New Compact Teaching-Learning-Based Optimization Method

Zhile Yang, Kang Li, and Yuanjun Guo

School of Electronics, Electrical Engineering and Computer Science,
Queen's University Belfast, Belfast, BT9 5AH
{zyang07,k.li,yguo01}@qub.ac.uk

Abstract. Population based heuristic optimization techniques, though powerful, are often limited by the memory size of hardware context when implemented on micro-controllers, embedded systems and commercial robots platforms etc. On the other hand, the teaching-learning based optimization algorithm (TLBO) is a recently proposed algorithm of high performance on both constrained and unconstrained optimization problems. In this paper, a new compact teaching-learning based optimization algorithm (cTLBO) is proposed to combine the strength of the original TLBO and reduce the memory requirement through a compact structure that utilizes an adaptive statistic description to replace the process of a population of solutions. Numerical results on test benchmark functions show that the new algorithm does not sacrifice the efficiency within the limited hardware resources.

Keywords: Compact Genetic Algorithms, Teaching-Learning Based Optimization, Estimation Distribution Algorithms, Normal Distribution.

1 Introduction

Meta-heuristic optimization techniques see a tremendous boost in recent several decades and have been widely used to tackle both simulation and real-world problems. Among this family, most nature-inspired approaches, such as genetic algorithm (GA) [1], particle swarm optimization (PSO) [2] and harmony search (HS) [3], maintain a population of solutions and mutate them through the process of mimicking the behaviours of specific natural phenomenon. Such operation converges fast under a proper mutation scheme but might easily fall into local minima of the function. On the other hand, model based approaches such as estimation of distribution algorithm (EDA) update solutions through an adaptive distributed model rather than directly modifying the value. Such scheme might be helpful to avoid the solution being trapped within a local optima but have low exploitation ability due to the deficiency of model adaptation. Few researches have focused on the combination of these two types of algorithms, mainly due to the lack of proper algorithm structure.

Fortunately, a compact real-code heuristic algorithm structure proposed by [4] is able to take advantage of both aforementioned algorithm families. However, the idea of the compact structure is derived from the real-world optimizations. The

D.-S. Huang et al. (Eds.): ICIC 2014, LNAI 8589, pp. 717–726, 2014.
© Springer International Publishing Switzerland 2014

implementations of the optimization to tackle the real-world problems are realized on hardware platforms or portable devices. For instance, the portable Global Positioning System (GPS) plans optimal routes by micro-controllers, and space-bound bots, such as Chinese 'Jade Rabbit' on the moon, control the behaviour of their actuators based on the strategies determined by the internal embedded system. One of the bottlenecks of such realizations is that the micro-controller might not be able to provide enough memory size for the optimization process due to the multi-task operation and limited computational resources. To address this problem, the concept of compact algorithms is first proposed in [5]. The compact structure only need to maintain several (generally no more than 5) solutions rather than a population of them during the whole process, thus can significantly reduce the necessary memory size. The structure has been applied to the optimization of both binary [5,6] and real-value optimization problems [7-10]. However, only a few population based algorithms have been integrated in this efficient structure, and there are many other powerful algorithms remain to be developed.

Teaching-learning based optimization is a recently proposed nature-inspired algorithm [11,12] and has gained wide attention [13-17] due to its fast convergence speed and good ability of both exploration and exploitation. Moreover, a profound characteristic of TLBO is its freedom from parameter tuning. Sensitive tuneable parameters, such as crossover rate and mutation rate in GA, social and cognitive parameters in PSO as well as harmony memory consideration rate and pitch adjusting rate in HS, sometimes have significant impact on the algorithm performance. TLBO has completely freed the evolution process from tuning the parameters, and only the initialization parameters including population and generation numbers need to be set.

In this paper, a compact TLBO is proposed by integrating the original TLBO within a compact structure to keep the evolution logic of TLBO and generate a single solution according to an adaptive normal distribution. Thus, the advantage of the desirable convergence speed from TLBO has been preserved and the required memory has been reduced considerably. The rest of the paper is organized as follows. An extensive study on compact heuristic algorithm is given in Section 2, followed by the discussion of the TLBO algorithm in Section 3. The new cTLBO algorithm is presented in detail in Section 4 while the numerical results carried out on diverse benchmarks are analysed and compared in Section 5. Finally, Section 6 concludes the paper.

2   Compact Algorithm

The compact algorithm is first realized with the combination of GA. Harik et.al [5,18] proposed a compact GA (cGA), analysing the positive and negative impact of changes in each gene bit in turns all along the whole length of a single chromosome in one generation by the way of maintaining a probability vector (**PV**) for each gene position and updating it in every generation. The cGA has logically maintained the basic process of original GA, but the crossover and the mutation steps have been replaced by their counterparts, a **PV** updating step and a new solution generating step. The brief process of cGA could be illustrated as follows,

1. Initialize probability vector
   for $i = 1$ to l do $\psi[i] = 0.5$;

2. Generate two solutions from the vector
   $\alpha = generate(\psi)$
   $\beta = generate(\psi)$

3. Compare the fitness of the two solutions
   $[winner, loser] =$ compare (α, β)

4. Update the probability vector towards the
   better one for $i = 1$ to l do
   if $winner[i] \neq loser[i]$ then
   if $winner[i] = 1$ then $\psi[i] = \psi[i] + 1/n$
   else $\psi[i] = \psi[i] - 1/n$;

5. Check if the array has converged
   for $i = 1$ to l do
   if $\psi[i] > 0$ and $\psi[i] < 1$ then
   return to step 2;

6. ψ represents the final solution

   n:population size.
   l:chromosome length.

Fig. 1. Pseudo code of the compact GA

As illustrated in Fig. 1, the cGA first initializes a vector $\psi[i]$ with the length l of a single gene. The value in the vector represents the probability value of the certain gene position and is initialized as 0.5 respectively. In Step 2, two new solutions α and β will be generated based on the **PV**, and would be compared by fitness value in Step 3, leaving a winner and a loser. A traversal of the gene position would be implemented and the specific value of the winner will increase or decrease the **PV** by adding or subtracting a weight number $1/n$ where n is a virtual population size in Step 4. The evolution will stop when the all the value in the vector $\psi[i]$ have achieved the stable status (e.g. either 0 or 1), and the final vector is the optimal solution.

It should be noted that cGA created a transparent process that enable the observability of the impact of each gene bit on the fitness value through tracing the change of **PV**. Meanwhile, only one vector and two solutions need to be maintained and updated, leading to a significant reduction of the necessary memory size compared with the situation of original n solutions. Therefore, cGA and some variants [6,19,20] have successfully implemented on platforms and hardware. However, they are only able to optimize the binary coding based problem due to the **PV** design and the update scheme. For real valued problems, a real-valued compact GA (rcGA) has been proposed by Mininno et.al [4]. On the basis of compact structure, the rcGA utilizes a normal distribution to collect evolutionary statistical information rather than a **PV**, generating a single generation from the distribution and updating the expectation and standard deviation in every generation. Inspired by the similar idea, compact differential evolution (DE) [8,9] and compact PSO [10] have been proposed and successfully implemented on the robot control platform [7]. It is worth noticing that parameter tuning based on the empirical knowledge is inevitable both in PSO and DE process, therefore, the design of more efficient and robust algorithms remains a key issue.

3   Teaching-Learning-Based Optimization

Teaching-learning based optimization is a novel nature-inspired algorithm proposed by Rao et.al [11,12]. It mimics a social process that a teacher (e.g. the best solution in a population) shares his/her knowledge to the class and students (other solutions) gains knowledge from the teacher as well as the interactions of each other. Therefore, it possesses similar algorithm structure with its counterpart such as PSO, DE and HS, in the ways of maintaining a group of solutions and updating their values by a specific scheme. In the meanwhile, the elegance of TLBO lies on its non-parameter evolutionary principle and the process of TLBO could be divided into two phases: the teaching phase and the learning phase.

3.1   Teaching Phase

In the teaching phase, a teacher will firstly be selected from a population of students, and the means of the students in each dimension are calculated. The difference between the teacher and the mean value of students are calculated and adjusted by a random weight, which will be further taken as the step length of the group evolution. Such process is illustrated as follow,

$$DMean_i = rand_1 \times (T_i - T_F Mean_i) \qquad (1)$$

where $DMean_i$ is the difference of values and $Mean_i$ is the mean value, both dimensions are in the solutions of i-th iteration. T_i denotes the selected teacher, and T_F is the teaching factor evaluating the impact of the teacher's knowledge on the students which can be either 1 or 2 denoted as:

$$T_F = round(1 + rand_2(0,1)) \qquad (2)$$

Then students gain the knowledge from the difference as:

$$St_{ij}^{new} = St_{ij}^{old} + DMean_i \qquad (3)$$

where St_{ij}^{new} and St_{ij}^{old} denote the j-th solution of a particle in i-th iteration imitating learners before and after gaining knowledge, respectively. The new learners after gaining knowledge will compete with their predecessors, and replace them if the performance is better.

3.2   Learning Phase

Learning phase is the next stage of learning for students. After gaining knowledge from the teacher, each of the students will randomly select a classmate and share knowledge with him/her. Their self-knowledge storage could be updated through the interaction. The learning phase could also be called student interactive learning stage due to this process. The detailed process is as follows,

$$St_{ij}^{new} = \begin{cases} St_{ij}^{old} + rand_3(St_{ik} - St_{ij}) & if \quad f(St_{ik}) < f(St_{ij}) \\ St_{ij}^{old} + rand_3(St_{ij} - St_{ik}) & if \quad f(St_{ij}) < f(St_{ik}) \end{cases} \tag{4}$$

where the j-th learner St_{ij} and k-th learner St_{ik} are randomly selected from the population in the i-th iteration. The fitness values of the two learners are compared and St_{ij} benefits from the deviation of the two learners with the better performer remaining in the population.

The whole process of these two phases will repeat until it arrives the maximum number of iterations or a certain criteria is met. No parameters need to be tuned except for the number of the population and iterations, and the particles evolve by the statistics of previous iteration. The remarkable performance of TLBO has been discussed and analysed in [11, 12].

4   Compact Teaching-Learning-Based Optimization

For the purpose of taking advantages of both the compact structure for decreasing the required memory size and the high evolutionary logic performance of TLBO for increasing the convergence speed, a compact TLBO is proposed in this section. The new process only keeps memory space of three solutions consisting of a teacher as the global optimum and two students for evolution process. Inspired by compact PSO and compact DE [9,10], a normal distribution is introduced into the algorithm for tracing and recording the evolution process. In each iteration, the distribution would be updated twice and the new solutions would be generated randomly based on the latest normal distributed statistics. Within a single iteration, two new solutions would be generated successively from the two phases while a virtual population number would also be introduced to represent the weight of the specific solution to the whole population. The detailed implementation is illustrated as Fig. 2.

In the optimization process as in Fig. 2, an initialization will be implemented in the beginning, where the expectation $\mu_t[i]$ and standard deviation $\sigma_t[i]$ of the normal distribution for the i-th dimension is empirically pre-set as 0 and λ, respectively [4], which compose a $m \times 2$ matrix **PV** similar functions as in cGA. A teacher Tr_t is generated by means of **PV** to finish the initialization stage.

counter $t = 0$;
for $i = 1 : n$ do
  // **PV** *initialization* //;
  initialize $\mu_t[i] = 0$;
  initialize $\sigma_t[i] = \lambda$;
end for
generate $\mathbf{Tr}_t$ by means of **PV**;
while counter t is not arrived its maximum value **do**
  // **Teaching Phase** //
  $\mathbf{St}_t = generate(\mathbf{PV})$;
  $DMean_t = rand_1 \times (\mathbf{Tr}_t - round(1 + rand_2(0,1)) \times \mu_t)$;
  $\mathbf{St}_t^{new} = \mathbf{St}_t + DMean_t$;
  $[winner, loser] = compete(\mathbf{St}_t^{new}, \mathbf{Tr}_t)$;
  for $i = 1 : n$ **do**
    $\mu_{t+1}[i] = \mu_t[i] + \dfrac{1}{Np}(winner[i] - loser[i])$;

    $\sigma_{t+1}[i] = \sqrt{(\sigma_t[i])^2 + (\mu_t[i])^2 - (\mu_{t+1}[i])^2 + \dfrac{1}{Np}(winner^2[i] - loser^2[i])}$;
  end for
  // **Learning Phase** //
  $\mathbf{St}_t^{new2} = generate(\mathbf{PV})$;

$$\mathbf{St}_t^{new} = \begin{cases} \mathbf{St}_t^{new} + rand_3(\mathbf{St}_t^{new2} - \mathbf{St}_t^{new}) & if \ \ f(\mathbf{St}_t^{new2}) < f(\mathbf{St}_t^{new}); \\ \mathbf{St}_t^{new} + rand_3(\mathbf{St}_t^{new} - \mathbf{St}_t^{new2}) & if \ \ f(\mathbf{St}_t^{new}) < f(\mathbf{St}_t^{new2}); \end{cases}$$

  $[winner, loser] = compete(\mathbf{St}_t^{new}, \mathbf{Tr}_t)$;
  for $i = 1 : n$ **do**
    $\mu_{t+1}[i] = \mu_t[i] + \dfrac{1}{Np}(winner[i] - loser[i])$;

    $\sigma_{t+1}[i] = \sqrt{(\sigma_t[i])^2 + (\mu_t[i])^2 - (\mu_{t+1}[i])^2 + \dfrac{1}{Np}(winner^2[i] - loser^2[i])}$;
  end for
  $\mathbf{Tr}_{t+1} = winner$;
  $t = t + 1$;
end while
$St_{opt} = \mathbf{Tr}_{tmax}$

Fig. 2. Pseudo code of the compact TLBO

In the teacher phase, a new student solution $\mathbf{St}_t$ which is firstly generated from **PV** gains knowledge from the teacher. It is worth mentioning that the $Mean_i$ in (1) of TLBO has been replaced by the expectation $\mu_t[i]$ of **PV**. The new enhanced student would compete against the teacher, leading to a winner and a loser representing them respectively. The original teacher will be replaced by the new student if he/she loses. The parameters of **PV**, both the expectation $\mu_t[i]$ and the standard deviation $\sigma_t[i]$, would be mathematically updated by an elite strategy that the loser will be replaced by the winner [4]. The N_P in the calculation of the standard deviation represent a virtual particle number, which determine the impact weight of a single solution on the whole distribution of population.

The learning phase also performs in the similar scheme. Another new student $St_{t,i}^{new2}$ generated from **PV** would be competed with the student $St_{t,i}^{new}$ in previous phase. The interactive between students would be implemented and the student $St_{t,i}^{new}$ would benefit again from the knowledge transfer. The newly enhanced student would compete with the teacher and form a new winner and a loser. The winner of the updated **PV** would be the teacher of the next iteration. The evolutionary process would not stop until the counter   achieves the maximum value.

5   Numerical Results

To compare the algorithm performance, some well-known benchmarks listed below have been used to compare the results.

— Sphere function: from [21] with $dimension = 30$
— Rosenbrock function: from [21] with $dimension = 30$
— Rastrigin function: from [21] with $dimension = 30$
— Quartic function: from [21] with $dimension = 30$

In terms of the parameters initialization, the expectation $\mu_i[i]$ are all set as 0 while the value λ of standard deviation $\sigma_i[i]$ is chosen as 10 in each dimension according to [4] for cTLBO. To equally compare the performance of the algorithms, the calling times of the fitness function is taken equally. The original DE and PSO are also compared. The social and cognitive factors of PSO are set as 2, and the learning factor and mutation rate are set as 0.8 and 0.5 respectively. The maximum generation is set as 200 for DE and PSO and 100 for TLBO due to the natural dual functional evaluation of TLBO. The particle number is 20 for all the approaches. The dimension has been illustrated in the above benchmarks. The average performance trends of PSO, DE, original TLBO and cTLBO are illustrated in Fig. 3. It is evident from the graph that the new algorithm converge much faster than DE, PSO as well as his ancestors. The exact result is listed in Table. 1 with 30 runs implementing respectively for both algorithms.

According to the test on benchmark $f1$ and $f2$, cTLBO outperforms the original TLBO within a limited number of generations. The results are highly unstable for benchmark $f3$ due to the significant impact of the initialization value on both optimizing processes. The cTLBO is outperformed by his ancestor on $f4$. More benchmark functions should be further tested, and comparative study should be conducted to analyse the performance and suitable applications of the new algorithm.

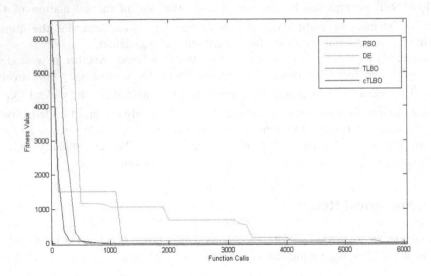

Fig. 3. Average performance trends of cTLBO and other methods

Table 1. The Comparison of cTLBO against other methods

TP	PSO	DE	TLBO	cTLBO
$f1$	1.224e-07±1.586e-06	0.0044±0.0609	3.192e-27±1.725e-26	**8.127e-57±1.459e-55**
$f2$	2.094±8.081	6.396±3.096e01	2.888 ±4.508	**2.227±5.001**
$f3$	2.242e02±1.302e02	1.160e02±5.001e02	6.387 ± 1.884e02	**7.775 ± 1.281e2**
$f4$	2.736e03±6.512e03	2.526e04±1.094e05	**9.776 ± 3.241**	1.283e2 ± 1.563e3

6   Conclusion

In this paper, a novel optimization algorithm, compact TLBO (cTLBO) has been proposed. The search logic of TLBO has been employed and a probabilistic representation has been used, guaranteeing the fast converging speed of the algorithm. The compact structure provides a better choice for embedded implementation by dramatically reducing the required memory size. Simulation studies have demomstrated the performance of cTLBO. More numerical experiments need to be implemented to analyse the strength and weakness of the proposed method. Further, the proposed algorithm will be tested on more industrial applications and complicated problems . The family of compact algorithms which combine the distribution of population statistics and the search logic of population based algoirthms could be a future direction.

Acknowledgments. The authors would like to thank Chinese Scholarship Council (CSC) and UK Engineering and Physical Sciences Research Council (EPSRC) International Doctoral Scholarship for sponsoring their research. This work is also partially supported by UK RCUK and EPSRC under grant EP/G042594/1 and EP/L001063/1 and Science and Technology Commission of Shanghai Municipality under grant 11ZR1413100.

References

1. Goldberg, D.E., Holland, J.H.: Genetic Algorithms And Machine Learning. Machine Learning 3(2), 95–99 (1988)
2. Shi, Y., Eberhart, R.: A Modified Particle Swarm Optimizer. In: The 1998 IEEE International Conference on Evolutionary Computation Proceedings of the IEEE World Congress on Computational in Telligence, pp. 69–73. IEEE (1998)
3. Geem, Z.W., Kim, J.H., Loganathan, G.V.: A New Heuristic Optimization Algorithm: Harmony Search. Simulation 76(2), 60–68 (2001)
4. Mininno, E., Cupertino, F., Naso, D.: Real-Valued Compact Genetic Algorithms for Embedded Microcontroller Optimization. IEEE Transactions on Evolutionary Computation 12(2), 203–219 (2008)
5. Harik, G.R., Lobo, F.G., Goldberg, D.E.: The Compact Genetic Algorithm. IEEE Transactions on Evolutionary Computation 3(4), 287–297 (1999)
6. Gallagher, J.C., Vigraham, S., Kramer, G.: A Family Of Compact Genetic Algorithms for in trinsic Evolvable Hardware. IEEE Transactions on Evolutionary Computation 8(2), 111–126 (2004)
7. Neri, F., Mininno, E.: Memetic Compact Differential Evolution for Cartesian Robot Control. IEEE Computational intelligence Magazine 5(2), 54–65 (2010)
8. Neri, F., Iacca, G., Mininno, E.: Disturbed Exploitation Compact Differential Evolution for Limited Memory Optimization Problems. Information Sciences 181(12), 2469–2487 (2011)
9. Mininno, E., Neri, F., Cupertino, F., Naso, D.: Compact Differential Evolution. IEEE Transactions on Evolutionary Computation 15(1), 32–54 (2011)
10. Neri, F., Mininno, E., Iacca, G.: Compact Particle Swarm Optimization. Information Sciences 239, 96–121 (2013)
11. Rao, R.V., Savsani, V.J., Vakharia, D.P.: Teaching-Learning-Based Optimization: A Novel Method for Constrained Mechanical Design Optimization Problems. Computer Aided Design 43(3), 303–315 (2011)
12. Rao, R.V., Patel, V.: An Improved Teaching-Learning-Based Optimization Algorithm for Solving Unconstrained Optimization Problems. Scientia Iranica (2012)
13. Crepinsek, M., Liu, S.-H., Mernik, L.: A Note on Teaching-Learning based Optimization Algorithm. Information Sciences 212, 79–93 (2012)
14. Waghmare, G.: Comments on A Note on Teaching-Learning-Based Optimization Algorithm. Information Sciences 229, 159–169 (2013)
15. Niknam, T., Azizipanah-Abarghooee, R., Aghaei, J.: A New Modified Teaching-Learning Algorithm for Reserve Constrained Dynamic Economic Dispatch. IEEE Transactions on Power Systems 28(2), 749–763 (2013)
16. Yang, Z., Li, K., Foley, A., Zhang, A.C.: A New Self-Learning Tlbo Algorithm for Rbf Neural Modelling Of Batteries in Electric Vehicles. In: 2014 IEEE World Congress on Computational intelligence. IEEE (2014)
17. Niknam, T., Fard, A.K., Baziar, A.: Multi-Objective Stochastic Distribution Feeder Reconfiguration Problem Considering Hydrogen And Thermal Energy Production by Fuel Cell Power Plants. Energy 42(1), 563–573 (2012)

18. Harik, G., Cantu-Paz, E., Goldberg, D.E., Miller, B.L.: The Gambler's Ruin Problem, Genetic Algorithms, And The Sizing Of Populations. Evolutionary Computation 7(3), 231–253 (1999)
19. Ahn, C.W., Ramakrishna, R.S.: Elitism-Based Compact Genetic Algorithms. IEEE Transactions on Evolutionary Computation 7(4), 367–385 (2003)
20. Kramer, G.R., Gallagher, J.C., Raymer, M.: On The Relative Efficacies Of Cga Variants for in trinsic Evolvable Hardware; Population, Mutation, And Random Immigrants. In: Proceedings of the 2004 NASA/Dod Conference on Evolvable Hardware, 2004, pp. 225–230. IEEE (2004)
21. Yao, X., Liu, Y., Lin, G.: Evolutionary Programming Made Faster. IEEE Transactions on Evolutionary Computation 3(2), 82–102 (1999)

ANFIS Modeling of PMV
Based on Hierarchical Fuzzy System

Yifan Luo, Ning Li, and Shaoyuan Li

Department of Automation, Shanghai Jiao Tong University,
Key Laboratory of System Control and Information Processing,
Ministry of Education of China,
Shanghai, China
ning_li@sjtu.edu.cn

Abstract. The calculation of predicted mean vote (PMV) index is complex in real time when estimates indoor thermal comfort. As a result, some suitable model had been built to tackle this problem. In this paper, sensitivity analysis is used to sort the importance of each potential input variable on PMV. According to the results of ranking, the dimensional reduction and distribution of input space will be available. Then a T-S type hierarchical fuzzy system will be utilized to reflect PMV index by combining expert knowledge and the association analysis methods. After that the ANFIS is used to train and adjust the parameters of each subsystem through existing dataset. Simulation results show that it not only improves the accuracy but also reduce the total number of fuzzy rules.

Keywords: PMV, hierarchical fuzzy system, sensitivity analysis, ANFIS.

1   Introduction

As we all know, the most important function of HVAC system is to adjust the indoor thermal environment in order to meet the requirement by occupants. Therefore, the research of thermal comfort in HVAC system is of great significance. Under this circumstance, quite a lot of criteria that evaluate indoor thermal comfort have been proposed [1-3], especially the PMV (predicted mean vote) thermal comfort model proposed by Fanger [2]. Considering computational complexity of PMV index, many effective researches using fuzzy method on PMV modeling can be found in previous articles [4-7], for the advantage that fuzzy model can only use input and output data as well as combine with some existing expert knowledge during fuzzy structure identification. However, in real condition some input values may obviously have greater importance than others on thermal comfort sensation. Despite of using fuzzy method, the importance of those factors which impact on the PMV output were mostly considered to be same. Meanwhile common fuzzy structure may lead to 'fuzzy rules explosion' [8]. For this reason, some appropriate structure should be chosen to replace the conventional one making PMV model less complex.

D.-S. Huang et al. (Eds.): ICIC 2014, LNAI 8589, pp. 727–738, 2014.
© Springer International Publishing Switzerland 2014

As is mentioned before, sensitivity analysis method is given in this study to rank the importance of each potential input value on PMV. Then a T-S type hierarchical fuzzy structure which has been proved that it can dramatically reduce the fuzzy rules [9-10] will be used to build the PMV model. Under this structure, the factors associated with thermal comfort can be considered in different levels according to their contribution degree. The expert knowledge will also be fully considered during input space division and fuzzy rules extraction. Lastly, the hybrid learning algorithm [11] given by ANFIS (Adaptive Neural-Network-Based Fuzzy Interference System) will be used to optimize the parameters of PMV model.

This paper is organized as follows. Section 1 describes the purpose and background of the research. Section 2 briefly presents the theoretical knowledge about TS type hierarchical fuzzy system and PMV index. Section 3 gives the explanation of the proposed approach about PMV modeling in detail. In Section 4, the simulation results will show its performance. The conclusion is remarked in Section 5.

2   Background Knowledge

2.1   T-S Type Hierarchical Fuzzy System

Suppose that in a fuzzy system there are n input variables and each variable has m membership functions. It needs m^n fuzzy rules to constitute this fuzzy system. It is obvious that the number of rules increases exponentially and the rule base quickly overload with n increasing in a standard fuzzy system. The hierarchical fuzzy system [8] was proposed to deal with this problem. A typical hierarchical fuzzy system structure is shown in Fig.1. Particularly, in this fuzzy structure, the output y_j of jth layer becomes an input value to the $(j+1)$th layer. The intermediate outputs y_j ($j = 1, \cdots n-1$) have the same domain which is equal to that of the real output y. Similarly, we define m fuzzy sets for n input variables respectively. Then, under this structure, each low-dimensional fuzzy system consists of two inputs with m^2 rules. Therefore, the total number of rules is $(n-1)m^2$, which increases only linearly with the number of input variables n.

As mentioned before, many factors can influence occupant's thermal comfort. With the number of influence factors increases, the rules of thermal comfort model will grow exponential. To tackle this problem, in this paper, a hierarchical fuzzy system (HFS) with n inputs is proposed on account of thermal comfort index PMV. It has been proved that [8] the total number of rules is minimal among all HFS with n inputs when each low-dimensional fuzzy system consists of only two inputs. Therefore, this kind of hierarchical structure is chosen to reduce the complexity of thermal comfort model. Each sub-fuzzy-system in Fig.1 is the T-S fuzzy system. The lth rule in jth subsystem can be expressed as follow

$$R_j^l: \text{ IF } x^{j+1} \text{ is } A_{j1}^l \text{ and } y^{j-1} \text{ is } A_{j2}^l, \text{ THEN } y_j^l = f_{j1}^l x^{j+1} + f_{j2}^l y^{j-1} + f_{j3}^l \tag{1}$$

the output of j th is

$$F_j(x^{j+1}, y^{j-1}) = y_j$$

$$= \frac{\sum_{l=1}^{L_j} w_j^l a_{j1}^l}{\sum_{l=1}^{L_j} w_j^l} x^{j+1} + \frac{\sum_{l=1}^{L_j} w_j^l a_{j2}^l}{\sum_{l=1}^{L_j} w_j^l} y^{j-1} + \frac{\sum_{l=1}^{L_j} w_j^l a_{j3}^l}{\sum_{l=1}^{L_j} w_j^l} \tag{2}$$

$$w_j^l = \mu_{A_{j1}}^l(x^{j+1}) \mu_{A_{j2}}^l(y^{j-1}) \tag{3}$$

where L_j is the number of rules in j th subsystem and $A_{j,i}^l$ is fuzzy sets of i th input in i th rule of j th subsystem. The membership function of $A_{j,i}^l$ is Gauss function as follow.

$$\mu_{A_{j1}}^l(x_{j,i}) = \exp\left[-\frac{1}{2}(\frac{x_{j,i} - c_{j,i}^l}{\sigma_{j,i}^l})^2\right] \tag{4}$$

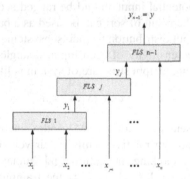

Fig. 1. Hierarchical fuzzy system

This special structure makes T-S type hierarchical fuzzy system quite suitable for PMV modeling. On the one hand, hierarchical modeling approach reduces the number of rule base caused by large potential input variables on thermal comfort. On the other hand, T-S fuzzy systems are universal approximators which can quantify people's thermal sensation well and give good approximation accuracy.

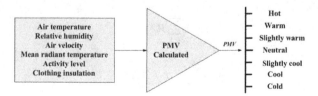

Fig. 2. PMV and thermal sensation

2.2   Thermal Comfort Index PMV

Based on the thermal sensation indicator given by ASHRE [16], Fanger proposed predicted mean vote (PMV) which is the most widely used thermal comfort index. It divides the thermal comfort of human body into 7 different levels which is able to quantify the mean thermal feeling for a group of persons in a given indoor environment. The values of the index have a range from -3 to +3, which corresponds to the occupant's feeling from cold to hot. The PMV index considers four environmental variables i.e. air temperature, air relative humidity, air velocity and mean radiant and two human variables i.e. clothing insulation and human activity. Fig. 2 shows a combination of each thermal variable affecting the PMV level.

3   Modeling of Thermal Comfort

3.1   Input Selection Based on Sensitivity Analysis

As mentioned before, the factors which impact on thermal comfort may be various in different area and condition. Even same factors may cause thermal sensation to varying degree. And different thermal sensation will lead to different PMV value. Under this circumstance, every potential input should be ranged according to their contribution to thermal comfort. The result of sort can be used as a basis for input selection of hierarchical system and input distribution to each subsystem.

In order to analysis the sensitivity of each input, a single-stage T-S fuzzy system should be built. The input and output structure of system is like this

$$y = f(x_1, x_2, x_3, x_4, x_5, x_6) \tag{5}$$

where the six inputs represent six factors of PMV index, including four environmental variables i.e. air temperature, air relative humidity, air velocity and mean radiant and two human variables i.e. clothing insulation and human activity. Suppose that $TD = [x_1(k), x_2(k), \cdots, x_6(k); y(k)], k = 1, \cdots, N$ is the training data set and N is the number of I/O data pairs. If ith input has M_i membership function, the total rules of this single-stage T-S model is $L = \prod_{i=1\cdots6} M_i$. For this model is just built for sensitivity analysis, high model accuracy is not necessity. So it can take the number of M_i few. As the consequent parameters of T-S model is $y^l = a_1^l x_1 + a_2^l x_2 + \cdots + a_6^l x_6 + a_7^l$. The output of this model can be shown as follow.

$$y = \sum_{i=1}^{7} p_i x_i \tag{6}$$

$$p_i = \sum_{l=1}^{L} w^l a_i^l / \sum_{l=1}^{L} w^l \tag{7}$$

where $x_7 = 1$ and $w^l = \prod_{i=1}^{6} \mu_{A_i^l}(x_i)$. $\mu_{A_i^l}(x_i)$ is the membership function value of i th input under A_i^l.

$$\mu_{A_i^l}(x_i) = \exp\left[-\frac{1}{2}\left(\frac{x_i - c_i^l}{\sigma_i^l}\right)^2 \right] \tag{8}$$

where the membership function is Gauss function ($i = 1, \cdots, 6$)

Then, this paper draws on existing research results [12], defining the sensitivity of a T-S model output y with respect to its input x_i as their partial derivative,

$$D_i = \sqrt{\frac{1}{N}\sum_{k=1}^{N}\left(\frac{\partial y(k)}{\partial x_i(k)}\right)^2} \tag{9}$$

where

$$\frac{\partial y(k)}{\partial x_i(k)} = p_i(k) + \sum_{j=1}^{7} x_j(k)\frac{\partial p_j(k)}{\partial x_i(k)} \quad i \neq j \tag{10}$$

$$\frac{\partial p_j(k)}{\partial x_i(k)} = \frac{\sum_{l=1}^{L} w^l (a_j^l - p_j^l)(c_i^l - x_i)}{(\sum_{l=1}^{L} w^l)\cdot(\sigma_i^l)^2} \tag{11}$$

It takes their square values due to the possibility of cancelations of negative and positive values. The final sensitivity can be expressed as follow

$$S_i = \frac{D_i}{\max(D_i)} \tag{12}$$

where S_i is the relative sensitivity of x_i that the ratio form can make us understand it more clearly. Therefore, in this paper the large the S_i is, the more important the input value x_i to the output PMV value y will be. According to the result of sensitivity analysis, we can sort the input values and choose appropriate inputs.

3.2   Input Division and Rule Base Extraction

After sorting the importance of input values, the input distribution to each subsystem can be determined with the results of [8]. Then the input space division and rules extraction of each subsystem should be discussed.

Different from those clustering algorithms which simply utilize data-driven method without any physical significance, Input variables are divided in their region respectively only rely on the prior knowledge and their physical significance. In [13], an approach of input division based on expert knowledge about thermal sensation in Shanghai area was given. Fig.3&4 respectively show the initial fuzzy division of relative humidity and metabolic rate. Similarly, intermediate outputs y_j are also divided equally in their domain.

Then, the first T-S subsystem of hierarchical structure is taken as example to explain the rule extraction method. The extraction process can be divided into six steps,

Step1: Suppose that $X_1 = [x_1, x_2] = [T_a, Met]$ (T_a is air temperature and *Met* is metabolic rate) are the input values in first subsystem and y_1 is its intermediate output.

Step2: x_1 is divided into five fuzzy sets among its domain according to the initial input division above while x_2 is three. y_1 will be divided into five sets with equal size.

Step3: Number each set of input and output values which is briefly shown in Fig.4. After that every I/O data in the training data set can be mapped into a specific number pattern rule on the basis of their membership function value. For example, if in a certain data $X_1^N = [x_1^N, x_2^N]$, $x_2^N = 95$, then x_2^N is located in the 3rd fuzzy set which means x_2^N is mapped into number three according to Fig.4&5. On the basis of this way, each data in data set can be written as a number pattern rule.

Step4: Build an initial rule base R^1 using number pattern rule given above. In the rule base, some rules may have same antecedent part but different consequent part. For instance, $r_1^1 = [1, 2; 4]$ and $r_2^1 = [1, 2; 3]$ ($r_1^1, r_2^1 \in R^1$). However, same origin can absolutely not lead to different result which means the conflict of consequence among these rules. So an extracting method based on consequent part sort order is proposed through traversing the rule base. In this method, $R_{spt}^1 (r_i^1)$ is the frequency of r_i^1. If $R_{spt}^1 (r_1^1) > R_{spt}^1 (r_2^1)$, r_1^1 will be reserved for its consequent part owns most frequency of occurrence while removing rest rules from rule base. It ensures that there is no conflict of consequent part in the rule base.

Step5: Compute the confidence coefficient R_{conf}^1 of each rules in the new rule base as follows

$$R_{conf}^1 (A_i \Rightarrow B_i) = \frac{R_{spt}^1 (A_i B_i)}{R_{spt}^1 (A_i)} \tag{13}$$

where A_i is antecedent of rule while B_i represents consequent part. If $R_{conf}^1 (A_i \Rightarrow B_i) > \theta_1$ (θ_1 is a given threshold on confidence) and $R_{spt}^1 (A_i) > \theta_2$ (θ_2 is a given threshold on frequency), the rule $r_i^1 = A_i \Rightarrow B_i$ will be finally extracted.

Step6: Add some rules which satisfy the prior knowledge but the frequency is less than θ_2 into final rule base. The number of rules in this subsystem is f_{num}^1.

Similarly, this method can be used in other level of subsystems

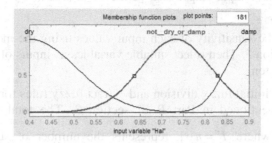

Fig. 3. Initial fuzzy division of relative humidity

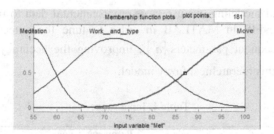

Fig. 4. Initial fuzzy division of metabolic rate

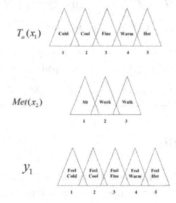

Fig. 5. Linguistic Rule Mapping to Number Rule

3.3   Parameters Identification

After finishing building each low-dimensional hierarchical fuzzy system, we utilize ANFIS to train the parameters of antecedent and consequent part. Then the hierarchical fuzzy model on PMV will be established.

3.4   PMV Modeling Process Based on Hierarchical Fuzzy System

As mentioned before, the modeling process can be concluded into four steps,

Step1: Give a set of input-output data pairs $TD = [x_1(k), x_2(k), \cdots, x_6(k); y(k)]$ under various conditions.

Step2: Compute the sensitivity of each input values using the method mentioned in **Part A** of section 3. Then select suitable variables as inputs of each subsystem in hierarchical system.

Step3: Obtain the initial fuzzy division and extract fuzzy rules for each subsystem by the method mentioned in Part B of section 3. The total number of rule is $f_{num} = \sum_{j=1}^{K} f_{num}^j$, where $K = n-1$ represents the number of subsystem in the whole hierarchical fuzzy system.

Step4: After initial modeling, we use large environmental data to train each subsystem model by ANFIS in MATLAB in order to tune the antecedent parameters $\{c_i^l, \sigma_i^l\}$ and consequent parameters $\{a_i^l\}$, improving the accuracy of final model. And finally build this hierarchical fuzzy model.

Table 1. Input range

Parameters	Parameter range
Air temperature T_a ($^\circ C$)	20 ~ 28
Radiant temperature T_r ($^\circ C$)	20 ~ 28
Relative humidity H_{ai} ($\%$)	40 ~ 90
Air velocity V (m/s)	0 ~ 0.5
Metabolic rate Met (W/m^2)	55 ~ 100
Clothing insulation I_{cl} (clo)	0.411 ~ 0.699

4   Simulation Results

The training data set for the hierarchical model are obtained from Fanger's model with a feasible range for input parameters as shown in Table.1. $TD = [x_1(k), x_2(k), \cdots, x_6(k); y(k)] = [T_a, T_r, H_{ai}, v, I_{cl}, Met; PMV]$. $k = 1, \cdots, N$. The PMV can be computed by means of [17] a heat-balance equation. The training data points and evaluating data points covering the above range are $N = 3200$ and $N_1 = 600$ respectively.

We make $M_i = 3$, $i = 1, \cdots 6$ and divide all inputs equally in order to build an initial single-stage T-S fuzzy model. This model is just built for sensitivity analysis, which we only need an input and output formula. So we train this model with only one step by using ANFIS. After that the method mentioned in step2 is used to analyze the sensitivity of all inputs. The result is $S_1 = 1$, $S_2 = 0.0612$, $S_3 = 0.4843$, $S_4 = 0.3293$, $S_5 = 0.0197$, $S_6 = 0.8174$. Because of low sensitivity, x_2 and x_5 are excluded from input sets. The input data set is reduced to $TD = [T_a, H_{ai}, v, Met; PMV]$ and the input values of each subsystem can be affirmed according to sensitivity size. Two variables of maximum sensitivity become the two inputs of first subsystem. The rest can be done in the same manner. So that $x_1 = T_a$, $x_2 = Met$, $x_3 = H_{ai}$, $x_4 = v$. As we can see in Fig.6&7, the fuzzy sets number of each input are 5, 3, 3 and 3, respectively and the intermediate output is 5. The maximum number of rules that can be generated is $45(= 15 + 15 + 15)$. As is mentioned in step3, we extract and form the final rule base. Finally, the hierarchical model is tuned by ANFIS and its output is compared with the actual output by Fanger's model. The evaluation of the proposed method is based on root mean-squared error (RMSE). The number of rules in each layer and their respective RMSEs are shown in Table.2. We can find that the RMSE dropped sharply in the third subsystem which means the modeling can be finished in this layer. In this case, the final output $RMSE_3$ is $\varepsilon = 0.0451$ and the total number of rules is only 23. Fig. 8&9 show the performance of modeling. We can find that most of the errors are within the limits of $[-0.1\ 0.1]$. The model accuracy and number of fuzzy rules are both considered. And the performance comparison with some other models [5][14-15] is shown in Table. 3. It is clear that the propose method has higher model accuracy than previous method and less number of rule than common T-S model structure [13] which includes 38 rules and its RMSE is 0.0462.

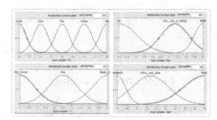

Fig. 6. Initial Division of Input Values

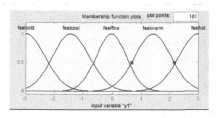

Fig. 7. Initial Division of Intermediate Output

Table 2. Simulation Results

layer j	f_{num}^{j}	RMSE$_j$	f_{num}
1	9	0.2381	
2	8	0.2203	23
3	6	0.0451	

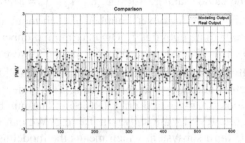

Fig. 8. Comparison between the output of proposed hierarchical model and actual output by Fanger's model

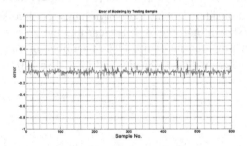

Fig. 9. Error of modeling

Table 3. RMSE comparison table

Modeling Method	RMSE	Rule Number
Interval type-2 T-S fuzzy model	0.0508	
Interval type-2 fuzzy neural network	0.100	
Neural network	0.3316	
common ANFIS model	0.0462	38
Proposed model by this paper	0.0451	23

5   Conclusion

In this paper, we utilize sensitivity analysis to select and distribute the input from all the six variables associated with PMV. We also take a full account of expert knowledge about thermal comfort under specific area when dividing fuzzy division. Some statistics and association analysis method are applied in rule ex-traction. What's more, quite different from other fuzzy PMV model, the hierarchical fuzzy systems based on ANFIS is used to building this model which dramatically reduce the number of rules. The simulation results validate that the proposed modeling method achieves a higher accuracy and generates less fuzzy rules than the existing methods.

Ackowledgments. This work was supported by the National Nature Science Foundation of China (61233004, 61221003, 61374109, 61104091, 61304078), the National Basic Research Program of China (973 Program-2013CB035500), and partly sponsored by the International Cooperation Program of Shanghai Science and Technology Commission (12230709600), the Higher Education Research Fund for the Doctoral Program of China (20120073130006 & 20110073110018), and the China Postdoctoral Science Foundation (2013M540364).

References

1. Fanger, P.: Thermal Comfort Analysis and Applications in Environmental Engineering. McGraw-Hill, New York (1972)
2. EN ISO 7730: Moderate Thermal Environments-determination of the PMV and PPD Indices and Specification of the Conditions for Thermal Comfort (1994)
3. Ye, G., Yang, C., Chen, Y., Li, Y.: A New Approach for Measuring Predicted Mean Vote (PMV) And Standard Effective Temperature (SET*). Building and Environment 38(1), 30–40 (2003)
4. Raad, H., Sahari, S.M.: RLF and TS Fuzzy Model Identification of Indoor Thermal Comfort Based on PMV/PPD. Building and Environment 49, 141–153 (2012)
5. Chen, M., Li, N.: An Interval Type-2 TS Fuzzy Model of Thermal Comfort Index PMV. Transactions on Intelligent Systems 6(3), 219–224 (2011)
6. Tahir, A., Ertugrul, C.: Adaptive Neuro-fuzzy Inference Systems (ANFIS) Application to Investigate Potential Use of Natural Ventilation in New Building Designs in Turkey. Energy Conversion and Management 48, 1472–1479 (2007)
7. Jassar, S., Liao, Z., Zhao, L.: Adaptive Neuro-fuzzy Based Inferential Sensor Model for Estimating the Average Air Temperature in Space Heating Systems. Building and Environment 44(8), 1609–1616 (2009)
8. Raju, G.V.S., Zhou, J., Kisner, R.A.: Hierarchical Fuzzy Control. International Journal of Control 54(5), 1201–1216 (1991)
9. Wang, L.X.: Analysis And Design of Hierarchical Fuzzy Systems. IEEE Transactions on Fuzzy Systems 7(5), 617–624 (1999)
10. Lee, M.L., Chung, H.Y., Yu, F.M.: Modeling of Hierarchical Fuzzy Systems. Fuzzy Sets and Systems 138(2), 343–361 (2003)
11. Jang, J.S.R.: ANFIS: Adaptive-network-based Fuzzy Inference System. IEEE Transactions on Systems, Man, and Cybernetics 23(3), 665–685 (1993)

12. Zurada, J.M., Malinowski, A.: Sensitivity Analysis for Minimization of Input Data Dimension for Feed Forward Neural Network. Circuits and Systems 6, 447–450 (1994)

13. Luo, Y., Li, N., Li, S.: ANFIS Modeling of the PMV Thermal Comfort Index Based on Prior Knowledge. In: 2014 9th IEEE Confer. on Industrial Electronics and Applications, ICIEA (2014)

14. Li, C., Yi, J., Wang, M., Zhang, G.: Prediction of Thermal Comfort Index Using Type-2 Fuzzy Neural Network. In: Zhang, H., Hussain, A., Liu, D., Wang, Z. (eds.) BICS 2012. LNCS (LNAI), vol. 7366, pp. 351–360. Springer, Heidelberg (2012)

15. Atthajariyakul, S., Leephakpreeda, T.: Neural Computing Thermal Comfort Index for HVAC Systems. Energy Conversion & Management 46(15), 420–426 (2005)

16. ANSI, ASHRAE, Thermal Environmental Conditions for Human Occupancy (2004)

17. ISO 7730: Moderate Thermal Environment

Static Security Risk Assessment with Branch Outage Considering the Dependencies among Input Variables

Xue Li, Xiong Zhang, and Dajun Du

Shanghai Key Laboratory of Power Station Automation Technology,
School of Mechatronic Engineering and Automation,
Shanghai University, Shanghai 200072 China
{lixue,zhangxiong,ddj}@shu.edu.cn

Abstract. This paper presents a method of static security risk assessment for wind-integrated power system with the consideration of network configuration uncertainties and correlated parameters. A probabilistic load flow (PLF) model is firstly constructed for grid-connected induction wind power system. Modeling correlated parameters and network configuration uncertainties is then taken with Cholesky decomposition, Nataf transformation, compensation method and total probability theorem. Finally, the transmission line overload risk index and the over-limit voltage index of static security are quantified, which can be used as an indicator for power system security. The proposed method is tested on the modified IEEE 30-bus system.

Keywords: static security, risk assessment, probabilistic load flow, correlated parameters, network configuration uncertainties.

1   Introduction

With the integration of various renewable energy sources, power systems are more often operating under highly stressed and unpredictable conditions [1]. Some accidental contingencies may break down system running and render overload, low voltage, voltage collapse, and even cascading fault [2]-[4]. Hence, static security risk assessment needs to be performed in order to assist system operators in maintaining system security level within an acceptable range.

Since the risk assessment of power system was proposed, power flow calculation is regarded as the premise and foundation to conduct risk assessment. Traditional static security risk assessment using deterministic power flow with N-1 contingency can't effectively take the uncertainty of generation and load into account, only with limited samples, from which the results are wrong to a certain extent [5]. However, PLF can fully consider the uncertainties in power system, it has been commonly used as an effective tool to assess the performance of a power network over most of its working conditions and report the operation state of system [6]-[7].

To reduce the computational effort, most PLF algorithms have considered the following two assumptions, independent input parameters and constant network configuration.

D.-S. Huang et al. (Eds.): ICIC 2014, LNAI 8589, pp. 739–750, 2014.

The assumptions means the PLF solution becomes a sum of independent random variables weighed by sensitivity coefficients. However, correlation between different sources of generation, between generation and loads, and between the loads themselves may exist [8]. Assumption (1) can generally meet the requirements when the random variables have week correlation. But when the random variables have strong correlation, the correlation can have a significant impact on the results if ignoring the correlation [9]. Reference [10]-[14] have done valuable research in this field. Therefore, it is necessary to quantify the risk indices of the wind-integrated power system with correlated parameters.

The network configuration is assumed to be a fixed parameter in the great majority of PLF formulations. Consequently, the probability of the basic configuration is assumed to be unity, and therefore the probability of losing any network element, such as transmission lines, transformers etc., is neglected. However, in power system operation, the network element often malfunctions, resulting in several possible network configurations, will have significant influence on the power flow results. Therefore, assumption (2) may be considered unrealistic and a more complex PLF method needs to be considered when the network configurations are uncertain.

Unlike the case when only considering random power injections, the network structure is also "uncertain" after incorporating network outages [15]. Therefore, the conventional power flow equations and the method obtains the probability density function (PDF) of desired variables can't be used. The computational burden could be very heavy if finding other solutions. So far, only a few works [15]-[17] have considered network configuration uncertainties in the PLF formulations. In [15], random branch outages are simulated by fictitious power injections with 0-1 distributions at the corresponding nodes. Reference [16] uses Distribution Factor Method to model the impact of n-1 fault to the distributions of line flows. Reference [17] considers the influence of line outage on the distribution of state variables by conditional probability.

Motivated by the above observations, this paper aims at considering several key issues for evaluating the transmission line overload risk and voltage beyond limits risk in wind-integrated power systems. The main contributions of the paper include:

The improved PLF model is proposed in this paper to report the performance of a wind-integrated power system after some accidental contingencies, with the consideration of network configuration uncertainties and correlation among random variables.

The proposed model using Cholesky Factorization method to model normal correlated parameters, using Nataf transformation to model non-normal correlated parameters, using compensation method to model branch outage and using total probability theorem to model random branch outage.

The static security risk assessment, both transmission line overload risk assessment and voltage beyond limits risk assessment, are carried out.

The rest of the paper is organized as follows. Section 2 presents the probabilistic models for wind- integrated power systems. Improve PLF models for static security risk assessment is described in section 3. Section 4 proposes the static security risk assessment. The simulation results on the modified IEEE 30-bus system are described in section 5, followed by the conclusions in section 6.

2   PLF Models for Grid-Connected Induction Wind Power System

Assuming that a wind farm is connected at bus i ($i \in N_W$), the corresponding PLF equations for a wind-integrated power system are given as (2):

$$\begin{cases} P_{ei}(e_i, f_i, s_i) - P_{Li} - \sum_{j \in i}(e_i a_i + f_i b_i) = 0 \\ Q_{ei}(e_i, f_i, s_i) - Q_{Li} - \sum_{j \in i}(f_i a_i - e_i b_i) = 0 \\ P_{mi} - P_{ei}(e_i, f_i, s_i) = 0 \end{cases} \tag{1}$$

where, P_{Li} and Q_{Li} are the load real and reactive powers in ith bus, respectively, $a_i = G_{ij}e_j - B_{ij}f_j$ and $b_i = G_{ij}f_j + B_{ij}e_j$. P_{ei} and Q_{ei} the real power injected into grid and the reactive power absorbed from the grid generated by induction wind generator, respectively [18]. e_i and f_i are the real part and imaginary part of voltage at the ith bus, respectively. s_i is the slip of induction machine at the ith bus. P_{mi} is the mechanical power of wind turbine at the ith bus [18] and is shown as

$$P_m = \frac{1}{2}\rho A v^3 C_p \tag{2}$$

where, ρ is the density of air (Kg/m^3), A is the area swept out by the turbine blades (m^2), v is the wind speed (m/s) and C_p is the dimensionless power coefficient. C_p can be expressed as a function of the blade tip speed ratio and be obtained by interpolation method. Weibull distribution is used for simulating wind speed uncertainty in this paper.

The Newton-Raphson algorithm is used to solve PLF equations (1). Note that when uncertainties of wind speed and load are considered, the mechanical power and the real/reactive power from wind power generator are described by probabilistic equations.

3   Modeling Correlated Parameters and Network Configuration Uncertainties

How to incorporate correlated parameters and network configuration uncertainties in PLF model is a key issue and is explained as follows.

3.1   Models of Normal and Non-normal Correlated Parameters

For the normal correlated parameters, Cholesky Factorization method [19] is used to generate multivariate dependent random variables. Meanwhile, Nataf transformation is used to model the non-normal correlated parameters and is explained as follows.

Define standard normal variables $Z = (Z_1, \cdots, Z_n)$, which can be obtained by marginal transformations of $X = (X_1, \cdots, X_n)$ as follows [20].

$$Z_i = \Phi^{-1}\left[F_{X_i}(X_i)\right], \quad i = 1, \cdots, n \tag{3}$$

where $\Phi(\cdot)$ is the standard cumulative normal probability distribution. The distribution for X can be obtained by assuming that Z is jointly normal. Using the rules of probability transformation, the joint PDF of X is

$$f_X(X) = f_{X_1}(x_1) f_{X_2}(x_2) \cdots f_{X_n}(x_n) \frac{\varphi_n(z, R')}{\varphi(z_1)\varphi(z_2)\cdots\varphi(z_n)} \tag{4}$$

where $z_i = \Phi^{-1}\left[F_{X_i}(x_i)\right]$, $\varphi(\cdot)$ is the PDF of standard normal variables, and $\varphi_n(z, R')$ is the n-dimensional normal PDF with zero means, unit standard deviations and correlation matrix R'. The elements ρ'_{ij} of R' are defined in terms of the correlation coefficients ρ_{ij} through the integral relation

$$\begin{aligned}
\rho_{ij} &= \int_{-\infty}^{\infty}\int_{-\infty}^{\infty} \left(\frac{x_i - \mu_i}{\sigma_i}\right)\left(\frac{x_j - \mu_j}{\sigma_j}\right) f_{X_i}(x_i) f_{X_j}(x_j) \times \frac{\varphi_2(z_i, z_j, \rho'_{ij})}{\varphi(z_i)\varphi(z_j)} dx_i dx_j \\
&= \int_{-\infty}^{\infty}\int_{-\infty}^{\infty} \left(\frac{x_i - \mu_i}{\sigma_i}\right)\left(\frac{x_j - \mu_j}{\sigma_j}\right) \varphi_2(z_i, z_j, \rho'_{ij}) dz_i dz_j
\end{aligned} \tag{5}$$

3.2   Model of Branch Outage with Compensation Method

The configuration of a power network change with branch outage. In order to make the state of the system be equivalent to that after the line outage, branch outages are simulated by suitable power injections at corresponding nodes. Each line outage is simulated by injecting fictitious powers at both ends of the line. When the injected powers are equal to powers leaving from each end of the line, the state of the system is equivalent [21].

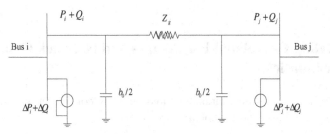

Fig. 1. Branch outage simulated by equivalent power injections

As shown in Fig.1, suppose $P_{ij} + jQ_{ij}$ and $P_{ji} + jQ_{ji}$ are powers leaving from the two ends before the line outage, respectively. The fictitious power injections $\Delta P_i + j\Delta Q_i$ and $\Delta P_j + j\Delta Q_j$ make them change $\Delta P_{ij} + j\Delta Q_{ij}$ and $\Delta P_{ji} + j\Delta Q_{ji}$, respectively. The post-outage system state should satisfy the following equations:

$$\Delta P_i + j\Delta Q_i = \left(P_{ij} + \Delta P_{ij}\right) + j\left(Q_{ij} + \Delta Q_{ij}\right) \tag{6}$$

$$\Delta P_j + j\Delta Q_j = \left(P_{ji} + \Delta P_{ji}\right) + j\left(Q_{ji} + \Delta Q_{ji}\right) \tag{7}$$

They can be rewritten in matrix form as

$$\begin{bmatrix} \Delta P_i \\ \Delta Q_i \\ \Delta P_j \\ \Delta Q_j \end{bmatrix} = H^{-1} \begin{bmatrix} P_{ij} \\ Q_{ij} \\ P_{ji} \\ Q_{ji} \end{bmatrix} \tag{8}$$

$$H = I_{4\times4} + \begin{bmatrix} -H_{ij} & 2P_{ij}-N_{ij} & H_{ij} & N_{ij} \\ -J_{ij} & 2Q_{ij}-L_{ij} & J_{ij} & L_{ij} \\ H_{ji} & N_{ji} & -H_{ji} & 2P_{ji}-N_{ji} \\ J_{ji} & L_{ji} & -J_{ij} & 2Q_{ji}-L_{ji} \end{bmatrix} \times \begin{bmatrix} S_{ii}^{(1)} & S_{ii}^{(2)} & S_{ij}^{(1)} & S_{ij}^{(2)} \\ S_{ii}^{(3)} & S_{ii}^{(4)} & S_{ij}^{(3)} & S_{ij}^{(4)} \\ S_{ji}^{(1)} & S_{ji}^{(2)} & S_{jj}^{(1)} & S_{jj}^{(2)} \\ S_{ji}^{(3)} & S_{ji}^{(4)} & S_{jj}^{(3)} & S_{jj}^{(4)} \end{bmatrix} \tag{9}$$

where, $I_{4\times4}$ is a identity matrix, $H_{ij}, N_{ij}, J_{ij}, L_{ij} \cdots$ are the corresponding elements of Jacobian matrix, $S_{ii}^{(1)}, S_{ii}^{(2)}, S_{ii}^{(3)}, S_{ii}^{(4)} \cdots$ are the elements of sensitivity matrix corresponding to the outage line. When the fictitious power injections $\Delta P_i + j\Delta Q_i$ and $\Delta P_j + j\Delta Q_j$ are obtained by (8). Thus, the PLF equations for wind-integrated power system can be used.

3.3   Models of Random Branch Outage with Total Probability Theorem

There is no need to consider all the network configurations, so it is available to select certain configurations that of great interesting to form sample space, such as random single outage, several lines outage simultaneously or network configurations under certain standard. Then total probability theorem is employed to obtain the probabilistic distributions of power flows with consideration of network configuration uncertainties [22]:

$$P(B) = P(B / A_1)P(A_1) + \cdots + P(B / A_k)P(A_k) \tag{10}$$

where A_k represents the kth network configuration. $P(A_k)$ is the probability of the kth network configuration occurrence. B represents the power flow results of the output variables and $P(B)$ is the probabilistic distribution of it. $P(B / A_k)$ is the probabilistic distribution of power flow results under the kth network configuration.

$P(A_k)$ can be acquired by failure rate of lines, that is

$$P(A_k) = \prod_{i=1}^{n} u_i \prod_{j=1}^{m} (1 - u_j)$$

(11)

where u_i is failure rate of the ith component; n is the number of components that are out of service; m is the number of components that are in service.

Because the selected sample space is only a subset of the complete set of network configurations, so $\sum_k P(A_k) < 1$. Than the following transmission is needed:

$$P'(A_k) = \frac{P(A_k)}{\sum_k P(A_k)}$$

(12)

Obviously, $\sum_k P'(A_k) = 1$. Thus in the selected sample space, the probability of the kth network configuration A_k occurrence is $P'(A_k)$.

4   Model of Static Security Risk Assessment

4.1   Risk Assessment Model of Transmission Line Overload

The risk of transmission line overload, which indicates the probability and severity of the line real power violation under fault occurrence, may be effectively used to make decisions regarding the line loading and system operation. The assessment, hence, consists of two issues: probability evaluation and impact evaluation. In a given operating condition, the risk is equal to the sum of line overload risk of individual's contingency and it is given by

$$R(F/A) = \sum_k \sum_j P(A_k) \times P(F_{kj}/A_k) \times S(F_{kj})$$

(13)

where, A is the selected sample space of the network configurations; F_{kj} is the real power of the jth transmission line under the kth network configuration; $P(F_{kj}/A_k)$ is the probabilistic distribution of F_{kj} under the kth network configuration A_k; $S(F_{kj})$ is the severity function of F_{kj} occurrence.

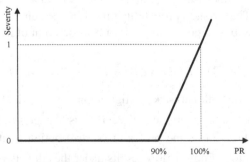

Fig. 2. Severity function of overload

Severity function for line overload is related to real power flow of a transmission line and is specified for each transmission line. The real power flow as a percentage of power rating (PR) of each line determines the overload severity of that line [23]. The severity function for line overload is illustrated in Fig.2. For each line, when the real power flow is no larger than 90% of the rated value, the risk severity value is zero. When the line real power flow is larger than 90% of the rated value, the risk severity value is a linear function of the real power flow, and is equal to one when the line real power flow is equal to the rated value.

4.2   Risk Assessment Model of Over-Limit Voltage

The risk of voltage beyond limits, which indicates the probability and severity of the bus voltage violation under fault occurrence, may be effectively used to make decisions regarding the system operation. The assessment also consists of two issues: probability evaluation and impact evaluation. In a given operating condition, the risk is equal to the sum of voltage beyond limits risk of individual's contingency and it is given by,

$$R(V/A) = \sum_k \sum_j P(A_k) \times P(V_{kj}/A_k) \times S(V_{kj}) \qquad (14)$$

where A is the selected sample space of the network configurations; V_{kj} is the real power of the jth bus voltage under the kth network configuration; $P(V_{kj}/A_k)$ is the probabilistic distribution of V_{kj} under the kth network configuration A_k; $S(V_{kj})$ is the severity function of V_{kj} occurrence.

Severity function for voltage beyond limits is related to the magnitude of a bus voltage and is specified for each bus nodes. The magnitude of each bus voltage determines the severity of that bus. The severity function for voltage beyond limits is illustrated in Fig.3. For each bus, when the magnitude of bus voltage is 0.95 p.u. or 1.05 p.u., the risk severity value is one. When the magnitude of bus voltage is larger or smaller than 1.0 p.u., the risk severity value is a linear function of the bus voltage, and is equal to zero when the magnitude of bus voltage is equal to 1.0 p.u..

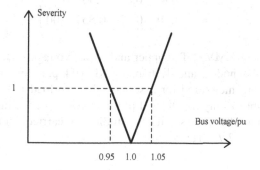

Fig. 2. Severity function of voltage beyond limits risk

5   Test Results and Discussions

The static security risk assessment considering network configuration uncertainties and dependencies among input non-normal random variables was tested on a modified IEEE 30-bus test system as shown in Fig.4, which connects four wind farms to 3, 19, 26, 30 via step-up transformers, double circuit transmission lines and extra buses. The wind farm is composed of 20 identical wind turbines, each of which has the nominal capacity of 600KW and other characteristic parameters are given in [18].

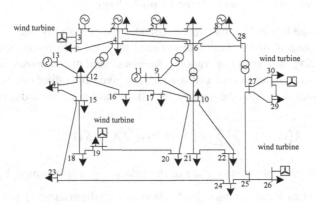

Fig. 3. Wind farm connected to 30-bus system

Loads and generators are correlated with a correlation coefficient of 0.9 and their uncertainties are modeled as normal functions, of which the means were taken as per-unit value and the standard deviations were taken as 5% of their means. The equivalent wind farms are also considered correlated, the correlated matrix is given as follow:

$$R = \begin{bmatrix} 1.00 & 0.88 & 0.87 & 0.91 \\ 0.88 & 1.00 & 0.85 & 0.87 \\ 0.87 & 0.85 & 1.00 & 0.85 \\ 0.91 & 0.87 & 0.85 & 1.00 \end{bmatrix}$$

The base value is 100MVA. The upper and lower voltage bounds are 1.1 p.u. and 0.95 p.u. for the PV nodes, and 1.06 p.u. and 0.94 p.u. for all other nodes. All transmission lines are monitored for assessing the overload risk. Given the limited space available, Table 1 only list the real power capacity of ten lines, which and will be focused throughout the studies. All case studies are carried out using MATLAB in an AMD 64 Dual Core 2.71GHz PC.

Table 1. Bounds of Line Real Powers

Line	Real power capacity (p.u.)
3-4	0.4226
3-11	0.3222
9-12	0.0662
9-17	0.0527
10-11	0.0398
13-14	0.0742
16-18	0.0582
21-23	0.0590
23-24	0.0318
25-26	0.0215

The selected sample space A of the network configurations is the branch outages of transmission line 3-4, 1-28 and 20-26. Their failure rate u_i is 0.3%, 0.2% and 0.6%, respectively.

Assuming that A_1, A_2 and A_3 donate the branch outages of transmission line 3-4, 1-28 and 20-26, the corresponding probability of the network configuration occurrence can be obtained from (20):

$$P(A_1) = 1 - e^{-\lambda_1} = 1 - e^{-0.003} = 0.003$$

$$P(A_2) = 1 - e^{-\lambda_2} = 1 - e^{-0.002} = 0.002$$

$$P(A_3) = 1 - e^{-\lambda_3} = 1 - e^{-0.006} = 0.006$$

Thus in the selected sample space, according to (21), the probability of the network configuration is shown:

$$P'(A_1) = \frac{P(A_1)}{P(A_1) + P(A_2) + P(A_3)} = \frac{3}{11}$$

$$P'(A_2) = \frac{P(A_2)}{P(A_1) + P(A_2) + P(A_3)} = \frac{2}{11}$$

$$P'(A_3) = \frac{P(A_3)}{P(A_1) + P(A_2) + P(A_3)} = \frac{6}{11}$$

5.1   Risk Assessment of Transmission Line Overload

When the network configuration A_1 occurrence, the selected line real power before and after A_1 are shown in Table 2.

Table 2. PLF Results with Correlated Parameters Before and After the Outage of Transmission Line 3-4

PLF results	Before A_1 occurrence		After A_1 occurrence	
	μ	σ	μ	σ
P_{2-3}	0.70577	0.03865	0.79163	0.03910
P_{3-35}	0.29036	0.01495	0.29099	0.01504
P_{6-15}	0.15131	0.00795	0.15120	0.00802
P_{8-9}	0.01320	0.00126	0.01396	0.00129
P_{12-13}	0.01971	0.00337	0.02160	0.00432
P_{19-20}	0.01556	0.00788	0.01573	0.00760
P_{33-34}	0.81304	0.04143	0.78645	0.04140

It can be seen from Table 2 that the means and standard deviations of the line real power are affected to an extent after the network configuration A_1 occurrence. For some line with great changes, it indicates the network configuration A_1 has a big influence on these lines. For some line with slight changes, it indicates the network configuration A_1 has a small influence on these lines. This can further lead to different overload risk index of each line and ultimately result in a large difference in overload risk indices of the whole system.

Therefore, based on the total probability theorem, the overload risk index of the whole system under the selected network configurations can be calculated as

$$Risk = \sum_{i=1}^{3} Risk(A_i) = 12.52451$$

5.2   Risk Assessment of Over-Limit Voltage

Similarly, when the network configuration A_2 occurrence, the selected bus voltage before and after A_2 are shown in Table 3.

Table 3. PLF Results Before and After the Fault Occurrence

PLF results	Before A_2 occurrence		After A_2 occurrence	
	μ	σ	μ	σ
V_1	1.0222	0.00107	1.0313	0.00409
V_5	1.0517	0.00123	1.0525	0.00132
V_{11}	1.0408	0.00220	1.0418	0.00238
V_{20}	1.0098	0.00439	1.0108	0.00444
V_{28}	1.0249	0.00147	1.0339	0.00421
V_{30}	1.0335	0.00506	1.0345	0.00519
V_{32}	1.0461	0.00142	1.0554	0.00427

It can be seen from Table 3 that the means and standard deviations of bus voltage are affected to an extent after the network configuration A_2 occurrence. This can further lead to different voltage beyond limits risk index of each bus and ultimately result in a large difference in voltage beyond limits risk indices of the whole system.

Based on the total probability theorem, the voltage beyond limits risk of the whole system under the selected network configurations also can be calculated as

$$Risk = \sum_{i=1}^{3} Risk\left(A_i\right) = 0.07837$$

6   Conclusions

This paper presents a method to carry out static security risk assessment in wind-integrated power system with the consideration of network configuration uncertainties and correlated parameters. Using Cholesky Factorization method to model normal correlated parameters, Nataf transformation to model non-normal correlated parameters, compensation method to model branch outage and total probability theorem to model random branch outage, the performance of a wind-integrated power system after some accidental contingencies is reported. Through probability evaluation and impact evaluation under contingency, the transmission line overload risk index and the voltage beyond limits index of static security can be obtained. The effectiveness of the proposed method is verified by the modified 30-bus system.

Acknowledgment. This work was supported in part by the National Science Foundation of China under Grant No. 51007052 and 71201097. Science and Technology Commission of Shanghai Municipality under grant No. 14ZR1415300.

References

1. Billinton, R., Salvaderi, L., McCalley, J.D., et al.: Reliability Issues in Today's Electric Power Utility Environment. IEEE Trans. on Power Systems 12(4), 1708–1714 (1997)
2. Chen, W., Jiang, Q., Cao, Y., Han, Z.: Risk-based Vulnerability Assessment in Complex Power Systems. Power Systems Technology 29(4), 12–17 (2005)
3. Karatekin, C.Z., Ucak, C.: Sensitivity Analysis Based on Transmission Line Susceptances for Congestion Management. Electric Power Systems Research 78(9), 1485–1493 (2008)
4. Leonardi, B., Ajjarapu, V.: Investigation of Various Generator Reactive Power Reserve (GRPR) Definitions for Online Voltage Stability/Security Assessment. In: IEEE Power and Energy Society General Meeting, pp. 1–7. IEEE Press, Pittsburgh (2008)
5. Wehenkel, L., Lebrevelec, C., Trotignon, M., Batut, J.: Probabilistic Design of Power-system Special Stability Controls. Control Engineering Practice 7(2), 183–194 (1999)
6. Borkowska, B.: Probabilistic Load Flow. IEEE Trans. on Power Apparatus and Systems PAS 93(3), 752–759 (1974)
7. Dopazo, J.F., Klitin, O.A., Sasson, A.M.: Stochastic Load Flows. IEEE Trans. on Power Apparatus and Systems PAS 94(2), 299–309 (1975)

8. da Leite Silva, A.M., Arienti, V.L., Allan, R.N.: Probabilistic Load Flow Considering Dependence Between Input Nodal Powers. IEEE Trans. on Power Apparatus and Systems PAS 103(6), 1524–1530 (1984)
9. Li, X., Zhang, X., Du, D., Cao, J.: Overload Risk Assessment in Grid-connected Induction Wind Power System. In: Xiao, T., Zhang, L., Ma, S. (eds.) ICSC 2012, Part II. CCIS, vol. 327, pp. 44–51. Springer, Heidelberg (2012)
10. Allan, R.N., Al-Shakarchi, M.R.G.: Linear Dependence between Nodal Powers in Probabilistic a.c. load flow. Proceeding of the Institution of Electrical Engineers 124(6), 529–534 (1977)
11. Li, X., Li, Y.Z., Li, H.Y.: Comparison and Analysis of Several Probabilistic Power Flow Algorithms. Proceedings of the CSU-EPSA 21(3), 12–17 (2009)
12. Morales, J.M., Baringo, L., Conejo, A.J., Mínguez, R.: Probabilistic Power Flow with Correlated Wind Sources. IET Generation, Transmission & Distribution 4(5), 641–651 (2010)
13. Usaola, J.: Probabilistic Load Flow With Correlated Wind Power Injections. Electric Power Systems Research 80(5), 528–536 (2010)
14. Ertel, R.B., Reed, J.H.: Generation of Two Equal Power Correlated Rayleigh Fading Envelopes. IEEE Communications Letters 2(10), 276–278 (1998)
15. Hu, Z., Wang, X.: A probabilistic load flow method considering branch outages. IEEE Trans. on Power Systems 21(2), 507–514 (2006)
16. Min, L., Zhang, P.: A Probabilistic Load Flow with Consideration of Network Topology Uncertainties. In: International Conference on Intelligent Systems Applications to Power Systems, pp. 5–8. IEEE Press, TokiMesse (2007)
17. Lu, M., Dong, Z., Saha, T.K.: A Probabilistic Load Flow Method Considering Transmission Network Contingency. In: Power Engineering Society General Meeting, pp. 24–28. IEEE Press, Tampa (2007)
18. Li, X., Pei, J.X., Zhang, S.H.: A Probabilistic Wind Farm Model for Probabilistic Load Flow calculation. In: Asia-Pacific Power and Energy Engineering Conference, pp. 1–4. IEEE Press, Chengdu (2010)
19. Li, X., Cao, J., Du, D.: Two-Point Estimate Method for Probabilistic Optimal Power Flow computation including wind farms with correlated parameters. In: Li, K., Li, S., Li, D., Niu, Q., et al. (eds.) ICSEE 2012. CCIS, vol. 355, pp. 417–423. Springer, Heidelberg (2013)
20. Kiureghian, A.D., Liu, P.L.: Structural reliability under Incomplete Probability Information. Journal of Engineering Mechanics 112(1), 85–104 (1986)
21. Mamandur, K.R.C., Berg, G.J.: Efficient simulation of line and Transformer Outages in power systems. IEEE Trans. on Power Apparatus and Systems PAS 101(10), 3733–3741 (1982)
22. Dong, L., Yang, Y., Zhang, C., et al.: Probabilistic Load Flow Considering Network Configuration Uncertainties. Trans. of China Electrotechnical Society 27(1), 210–216 (2012)
23. Ni, M., McCalley, J.D., Vittal, V., Tayyib, T.: Online Risk-based Security Assessment. IEEE Trans. Power Syst. 18(1), 258–265 (2003)

On Spacecraft Relative Orbital Motion Based on Main-Flying Direction Method[*]

Yanchao Sun, Huixiang Ling, Chuanjiang Li, Guangfu Ma, and Wenrui Zhao

Harbin Institute of Technology, Department of Control Science and Engineering, Harbin, China
{sunyanchao,linghx,lichuan,magf}@hit.edu.cn

Abstract. This paper introduces an analysis of a relative orbit control problem that how to let the tracking spacecraft be in the range of a small angle which is outward from the non-cooperative target spacecraft's field of view, and at the same time satisfy the requirement of flight time and distance. This analysis is called "main-flying direction" analysis method. Based on this method, we provide a design method of the relative orbit in the fluttering mode that the tracking spacecraft flies in its own orbit after it enters into the spatial range relevant to the target spacecraft. This design method of solving the relative orbit control problem is demonstrated by simulations. The analysis thought and the calculation of this method are simple, and it avoids some disadvantages in other methods, such as the direction limitation of the field of view, attitude and orbit control coupling, etc.

Keywords: spacecraft, relative orbit control, main-flying direction, fluttering mode.

1 Introduction

Nowadays, spacecraft relative orbital motion control in the short distance is one of important issues in the field of space technology. Spacecraft relative orbital motion is the research of continuous motion law between the tracking spacecraft and the target spacecraft, which are in short distance [1-2]. It is often used on some occasions, such as formation flying, on-orbit maintenance, rendezvous and docking, tracking, monitoring and other space missions [2]. The problem of tracking and monitoring the non-cooperative target spacecraft more reflects the significance of relative orbit control. This paper mainly studies the problem that how to let the tracking spacecraft in the range of a small angle which is outward from the non-cooperative target spacccraft's field of view, and at the same time satisfy the requirement of flight time and distance.

The most common relative orbital motion forms include hovering (two spacecraft maintain the relative position unchanged), accompanying flying (the tracking spacecraft flies with a closed orbit which is around a point near the target spacecraft),

[*] This work is supported by National Natural Science Foundation (NNSF) of China under Grant (No. 61304005, 61174200); Research Fund for the Doctoral Program of Higher Education (RFDP) of China (NO. 20102302110031).

D.-S. Huang et al. (Eds.): ICIC 2014, LNAI 8589, pp. 751–762, 2014.

fly-around (a special form of accompanying flying, the point is the target spacecraft) and so on [3-19]. Without considering the effect of perturbations and orbit control, references [3-8] provide some relative motion forms based on analytical solutions of hill equation, and they include non-closed natural spiral orbit, accompanying flying, fly-around and hovering. However, if we use them to solve the problem which this paper studies, without orbit control, non-closed natural spiral orbit will face some difficulties of designing the initial values and the parameters, complex computation; the period of accompanying flying and fly-around equal the orbital period of the target spacecraft, so the flight time is difficult to guarantee, and it cannot satisfy the requirement that the direction of the target spacecraft's field of view can be arbitrary; hovering can just maintain the tracking spacecraft in the flying direction of the target spacecraft, so it also cannot satisfy the direction requirement. References [1], [2], [8-11] refer to a class of fast fly-around orbit with the velocity pulse control. The fast fly-around orbits are global, however the problem this paper studies is the range of the specified sight space near the target spacecraft. References [1], [8], [12], [13] provide other flight orbits, so it can maintain the tracking spacecraft in the specified range and satisfy the flight time requirement by periodically giving the velocity pulse in the largest point of x coordinate for maintaining "drop" shape in the projection of xy-plane. However, it should periodically give the velocity pulse, so it will necessarily face the problem of attitude and orbit control coupling. It not only affects the accuracy of orbit control, but also affects the accuracy of some space missions, for example, high-precision attitude pointing. References [14-19] focus on spacecraft hovering. The methods can maintain the tracking spacecraft hovering in any position relative to the target spacecraft with continuous control. However, if using them to solve the problem this paper studies, it will also necessarily face the problem of attitude and orbit control coupling. Moreover, to think of the non-cooperative target spacecraft, hovering is very easy to be perceived, tracking and monitoring mission will fail.

This paper aiming to the relative orbital motion problem that the tracking space-craft is in the range of the non-cooperative target spacecraft's field of view, and at the same time, satisfies the flight time requirement, provide "main-flying direction" analysis method, and based on this analysis method, we provide a design method of the relative orbit in the fluttering mode (the tracking spacecraft travels in its own orbit after it enters into the spatial range relevant to the target spacecraft). The main design steps are as follows. In hill coordinate system, we can analyze the specified spatial range by dividing it into xy-plane and z-axis, respectively [4]. Make linear approximation of the orbit. Select the main-flying direction in x or y directions. Calculate the initial velocity, the beginning and the end points of the main-flying direction. Calculate the initial velocities of the non-main-flying directions. In the fluttering mode, the tracking spacecraft does not need the orbit control, so it can avoid the problem of attitude and orbit control coupling. After finding the beginning and the end points of the main-flying direction, this analysis first transforms the three-dimensional motion problem into a problem of determining the three-dimensional initial velocities by a series of simple principles. The analysis thought and calculation of this method is simple, and it is easy to satisfy the index requirement.

By simulation validation, using the method this paper provided can perfectly solve the problem. If we leave some margin when designing, then the flight time of the tracking spacecraft in the specified spatial range can satisfy the requirement. As the result, the calculation of this method is simple, the tracking spacecraft can fly in its own orbit in the specified spatial range, and the flight time is enough.

2   Theoretical Basis

2.1   Two-Body Orbital Model Basis

Spacecraft orbit is the position of spacecraft in the inertial coordinate system. It is used for describing the translational motion of spacecraft. To two-body problem, two celestial bodies are treated as particles. Think of the motion equation between two celestial bodies whose mass are m and M, respectively:

$$\ddot{r} = -\frac{G(M+m)}{r^3}r = -\frac{\mu}{r^3}r \tag{1}$$

Through solving two-body problem, the motion of celestial body (spacecraft) can be described by six classical orbital elements: semi-major axis a, orbital eccentricity e, longitude ascending node Ω, orbital inclination i, argument of perigee ω, true anomaly f.

2.2   Relative Orbital Model Basis

Relative Orbital Dynamics Equation. The target spacecraft is denoted as s, and the tracking spacecraft is denoted as c. We take the orbit coordinate system s-xyz of the target spacecraft as the relative motion coordinate system. The origin is the center of

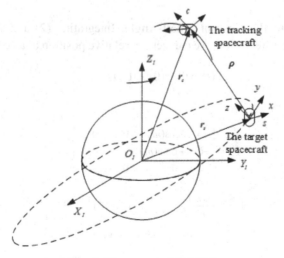

Fig. 1. The s-xyz and O_I-$X_IY_IZ_I$

mass of the target spacecraft, and move in orbit with it. The x-axis coincides with geocentric vector r_s of the target spacecraft, and points from geocenter to s. The y-axis is normal to x-axis in the orbit plane, and points to the direction of motion. The z-axis determined by right-hand rule, coincides with the angular momentum vector of the target spacecraft. The relative motion coordinate system s-xyz and geocentric inertial coordinate system O_I-$X_IY_IZ_I$ is shown in Fig. 1

The dynamics equations of the tracking spacecraft and the target spacecraft in geocentric inertial coordinate system are substituted into the relative motion formula. To the situation that the target spacecraft is in a circular orbit ($e=0$) and the distance between two spacecraft is not large, with linear approximation, the relative motion dynamics equation is simplified to a linear differential form with constant coefficients:

$$\begin{cases} \ddot{x} - 2n\dot{y} - 3n^2 x = a_x \\ \ddot{y} + 2n\dot{x} = a_y \\ \ddot{z} + n^2 z = a_z \end{cases} \tag{2}$$

In (2), a_x, a_y, a_z are the projection of difference between acceleration vector a_s and a_c of resultant force which act on two spacecraft except the gravity of the Earth in relative motion coordinate system s-xyz; the force includes control thrust and perturbations which caused by the shape of the Earth, atmospheric drag force, solar radiation pressure, etc. $n = \sqrt{\mu / a^3}$ is the average angular velocity of the target spacecraft.

Equation (2) is called the hill equation, and also called Clohessey-Whiltshire equation (C-W equation).

Independent Analysis of xy-Plane and z-Axis. In (2), if a_x, a_y, a_z are constant, relative motion along z-axis is independent, but relative motion along x-axis and y-axis in the orbit plane are coupled with each other, so we can separate (2) into xy-plane and z-axis to analyze.

Approximation in the Case of Small-Angle. Integrating (2) and solving it, if the initial values $(r_0, \dot{r}_0)$ are known, we can get the relative position at a certain time,

$$r(t) = \Phi_{11} r(0) + \Phi_{12} \dot{r}(0) \tag{3}$$

where

$$\Phi_{11} = \begin{bmatrix} 4 - 3\cos(nt) & 0 & 0 \\ -6nt + 6\sin(nt) & 1 & 0 \\ 0 & 0 & \cos(nt) \end{bmatrix} \tag{4}$$

$$\Phi_{12} = \begin{bmatrix} \dfrac{\sin(nt)}{n} & \dfrac{2}{n}(1 - \cos(nt)) & 0 \\ -\dfrac{2}{n}(1 - \cos(nt)) & -3t + \dfrac{4}{n}\sin(nt) & 0 \\ 0 & 0 & \dfrac{\sin(nt)}{n} \end{bmatrix} \tag{5}$$

If nt is the small-angle, we can simplify the trigonometric function approximately

$$\sin(nt) \approx nt, \cos(nt) \approx 1 \qquad (6)$$

Substituting (6) in (4) and (5):

$$\Phi_{11} \approx \begin{bmatrix} 1 & 0 & 0 \\ 0 & 1 & 0 \\ 0 & 0 & 1 \end{bmatrix}, \Phi_{12} = \begin{bmatrix} t & 0 & 0 \\ 0 & t & 0 \\ 0 & 0 & t \end{bmatrix} \qquad (7)$$

By the small-angle approximation, we can get the linear-variation conclusion of relative coordinate r in the hill equation. Assuming $r = [x;y;z]$, $r_0 = [x_0;y_0;z_0]$, $\dot{r}_0 = [\dot{x}_o; \dot{y}_0; \dot{z}_0]$, it satisfies that

$$r = r_0 + \dot{r}_0 t \qquad (8)$$

3 Analysis and Design of the Fluttering Orbit Based on Main-Flying Direction Method

3.1 Problem Analysis

This paper mainly studies the problem how to let the tracking spacecraft be in the range of a small angle which is outward from the non-cooperative target spacecraft's field of view, and at the same time satisfy the requirement of flight time and distance.

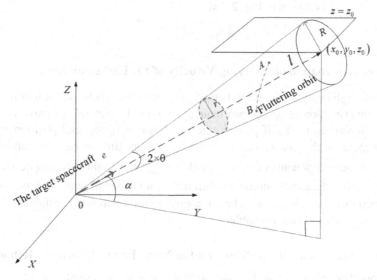

Fig. 2. The fluttering area

For example, the target spacecraft is located in the geostationary earth orbit (GEO). The range of sight angle is $\pm\,\theta$. There are upper and lower boundaries of the relative distance. After the tracking spacecraft entering into the specified spatial range, it gets into the fluttering mode that it flies in its own orbit. The fluttering area is shown in Fig. 2

In Fig. 2, $e = [e_1\ e_2\ e_3]^T$ is the unit vector of the sight line vector l of the non-cooperative target spacecraft. e_i (i = 1,2,3) are the components in hill coordinate system. It can be seen that the range of the allowing fluttering area of the tracking spacecraft is determined by the direction of the sight line and the relative distance, and it's a cone-frustum in the hill coordinate system. The design goal is that properly choose the beginning point A ($[r_0,v_0]$) and the end point B ($[r_f,v_f]$) based on hill coordinate system; Let the tracking spacecraft freely flutter in the cone-frustum without any active orbit control; At the same time, satisfy the flight time requirement.

3.2   Choosing z_0

Comparing the orbital period of GEO, the time of keeping in the specified spatial range is very short, and r varies periodically with the orbital period of GEO. So we can consider that nt is a small-angle. Thus if we choose a proper z_0 in the case of the velocity of z direction $\dot{z}_0$ being very small at initial moment, we can consider it will not vary obviously. Then we can deal with the three-dimensional fluttering problem as a two-dimensional problem approximately, namely just to consider the relative fluttering orbit in the xy-plane.

The principle of choosing z_0 is keeping the xy-plane which is intercepted by the plane $z = z_0$ and the cone-frustum as large as possible. In this paper, we choose z coordinate of circle center of the bottom of the cone-frustum z_0. According to the definition of e, z_0 is shown in Fig. 2 and

$$z_0 = l\sin\alpha = l\cdot e_3 \qquad (9)$$

3.3   The Explanation on Entering Velocity of the Fluttering Area

GEO is the high orbit, so the perturbations are small relatively, and it will hardly impact the fluttering process (as shown in analysis with the concrete example in the next section). According to (8), if proper entering position r_0 and end position r_f is provided, proper $\dot{r}_0$ is chosen, we can guarantee the flight time to be long enough. Note that if only the components of $\dot{r}_0$ are small enough, the flight time can be very long. But if it is too long, the assuming condition that nt is a small-angle will be not established, then this analysis based on linear approximation will not be established. So the choosing of $\dot{r}_0$ should be reasonable.

3.4   The Analysis about xy-Plane and the Main-Flying Direction Method

For the problem this paper studies, the sight angle is usually small, namely the vertex angle of the cone is small, only $2\times\theta$, so the projection of the figure which is bounded by the cone genertrix is an isosceles triangle, and its bottom corner $(90^\circ\text{-}\theta)$ is approximately 90°. Actually in a relatively short range, the fluttering area of the tracking spacecraft is a cylinder approximately. The area which is intercepted by $z = z_0$ plane and the cylinder is a part of ellipse, as shown in Fig. 3. Because of the difference in the direction of the projection of the axis of symmetry in xy-plane, there is a shorter component of the orbit in x or y direction, called the main-flying direction, and it is sufficient to guarantee satisfying the mission requirement in this direction.

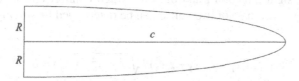

Fig. 3. A part of approximate ellipse which is intercepted by $z = z_0$ plane and the cylinder

3.5   Choosing the Main-Flying Direction

We present the example that the included angle between the symmetry axis and the x-axis is smaller than 45°, as shown in Fig. 4. The components among y-axis are all smaller than the components among x-axis in every orbit, namely y-axis is more sensitive, so we choose y-axis direction as the main-flying direction, it is sufficient to guarantee that the initial velocities of x-axis and z-axis are small enough.

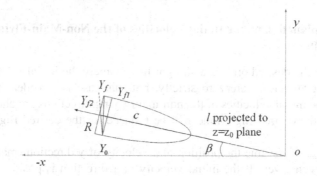

Fig. 4. Analysis figure of main-flying direction

3.6   The Design of the Desired Orbit in Main-Flying Direction

As shown in Fig. 4, the desired main-flying direction is from point Y_0 to point Y_f. The calculation of both points in hill coordinate system is given as follows.

Firstly, the x coordinate of Y_0 and Y_f are equal:

$$x_f = X_0 - R\sin\beta \tag{10}$$

where X_0 is the x coordinate of the bottom of the cone. R is the radius of the bottom, and β is calculated as

$$\beta = \arctan\left(\frac{e_2}{e_1}\right) \tag{11}$$

Y_0 and Y_f are on the surface of cone frustum, so the included angle between the target spacecraft's sight direction l and the line which connect either of them and the origin is θ. The y coordinate of two points can be determined by solving the following equation

$$\arccos((x_f e_1 + y_{o,f} e_2 + z_0 e_3) / \sqrt{x_f^2 + y_{0,f}^2 + z_0^2}) = \theta \tag{12}$$

where y_{0f} are the y coordinates of two points. There are two solutions. The z coordinate has been determined by (9).

After providing the beginning and end points of the desired orbit, we provide the method of calculating the desired initial velocity of the main-flying direction. Assume that t_{lve} is the desired flight time, the desired initial velocity of the main-flying direction can be calculated as

$$\dot{y}_0 = \frac{|Y_f - Y_0|}{t_{lve}} \tag{13}$$

3.7 The Explanation on the Initial Velocities of the Non-Main-Flying Directions

It is hoped that the desired orbit is a straight line, namely the initial velocities of the non-main-flying directions are zero strictly, but this case is too ideal to establish. Moreover, if the initial velocities of the non-main-flying directions are chosen properly, the loss of the desired flight time will be reduced, or the desired flight time will lengthen.

First, see Fig. 4. Assume the position and velocity of y direction (main-flying direction) have been given. If the initial velocity of x direction is positive and a little large, the fluttering orbit may be along $Y_0 Y_{f1}$, so the orbit projection of y direction will be shorter than the length of the desired fluttering orbit; If the initial velocity of x direction is negative, the tracking spacecraft may fly out the specified spatial range ahead of schedule along $Y_0 Y_{f2}$. We choose that the initial velocity of x direction is positive and small.

The analysis of z direction is shown in Fig. 5. If the initial velocity of z direction is negative but not large, the actual orbit may be along $Y_0 Y_{f3}$, and the orbit projection on the xy-plane will be longer than the length of the desired fluttering orbit. So the orbit projection of the y direction will lengthen. So if designing the initial velocity of y direction properly, the flight time will lengthen. Similarly, if the initial velocity of z direction is positive, the flight time will be reduced.

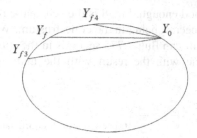

Fig. 5. The rear view of the range of allowing fluttering area

4 Simulation

4.1 Simulation parameters

The target spacecraft is in the GEO, and the initial six orbital elements are that $a_1 = 4.225 \times 10^7$m, $e_1 = 0$, $i_1 = 5°$, $\Omega_1 = 31°$, $\omega_1 = 0°$, $M0_1 = 0°$.

The parameters of perturbation coefficients are that aerodynamic force coefficient $C_D = 1$, drag force coefficient $C_d = 2.2$, reflection coefficient $C_r = 0.8$.

The sight direction vector of the target spacecraft are $e_{-hill} = [-49647;-5194;2863]$, the minimum distance along the sight direction is $R_{min} = 50$km, maximum distance is $R_{max} = 99.2$km, the sight angle $\theta = 0.35°$. The desired flight time is 800s.

4.2 The Fluttering Orbit Design

Firstly, to design as section 3.5, we choose the y direction as the main-flying direction. The initial position and velocity are determined by the principles in sections 3.6 and 3.7. After some calculation, the position coordinate of the beginning point Y_0 is [-9.8437;-1.0908;0.5680]$\times 10^4$m, and the initial velocity is [0.200;1.5222;-0.100]m/s. Assume that the tracking spacecraft has reached the beginning point with the desired initial velocity by orbit maneuver control, we begin to simulate and analyze its fluttering orbit.

4.3 The Simulation Based on Kepler Equation and Hill Equation

We provide the simulations based on kepler equation and hill equation. Considering the effect of multiple orbit perturbations based on two-body orbit model, we transform the position vector of the tracking spacecraft relative to the target spacecraft back to hill coordinate system, and compare it to the case of hill dynamical equation without any orbit perturbations.

The partial enlarged schematic drawing of the three-dimensional curves is shown in Fig. 6. It indicates that the curves based on two equations can be coincident basically. Due to the perturbations, there are a little differences between the fluttering orbits based on the two equations. As shown in Fig. 7, the y and z components of the fluttering orbit maintain the linear-degree well in the fluttering process, the linear-degree of the *x* component is not good enough. Fig. 8 is the schematic drawing of time intervals of which the tracking spacecraft being in the cone-frustum. With these initial parameters, the tracking spacecraft can flutter 775 seconds in the specified spatial range, and it is basically anastomotic with the result with the main-flying direction analysis method.

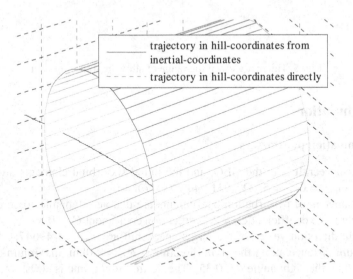

Fig. 6. The enlarged schematic drawing of the fluttering orbit

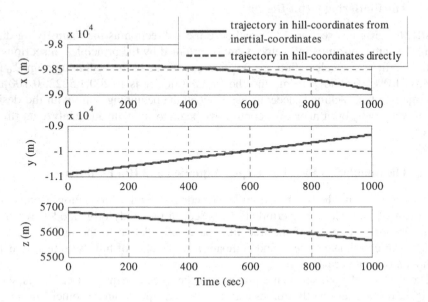

Fig. 7. The variation curves of three axis components along with time in hill coordinate system

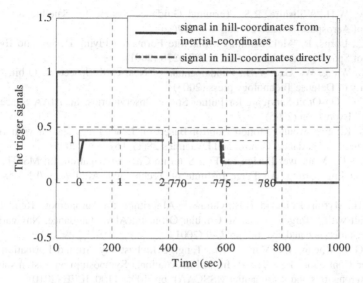

Fig. 8. The schematic drawing of triggering signals of the flight time

5   Conclusions

This paper investigated the relative orbital motion problem. The tracking spacecraft should fly in the range of field of view of the non-cooperative target spacecraft and satisfy the flight time requirement. We provided the "main-flying direction" analysis method. Based on this method, we provided a design method of the relative orbit in the fluttering mode which did not use the orbit control. The relative orbit was designed with the steps which included choosing main-flying direction, calculating the beginning and end points of the fluttering orbit, calculating the initial velocity of the main-flying direction, and designing the initial velocities of the non-main-flying directions. The analysis thought and calculation of this method was simple, and it avoided some disadvantages in other methods, such as the direction limitation of the field of view, attitude and orbit control coupling, etc. The effectiveness of this design method on solving the relative orbit control problem was proved by simulations. To the need of simplified calculation, approximation in the case of small angle is used in the design process. So the approximation conditions limit the mission time. In the future, we will study the result of this method affected by the flight time which the mission requires.

References

1. Zhu, Y.W., Yang, L.P.: Mission Trajectory Design and Control for Spacecraft Proximity Relative Motion. Journal of Astronautics 30(5), 1834–1841 (2009)
2. Luo, J.J., Yang, Y.H., Yuan, J.P.: Trajectory Design and Control of In-Plane Fast Controlled Fly-around. Journal of Astronautics 27(6), 1389–1392 (2006)

3. Clohessy, W.H., Wiltshire, R.S.: Terminal Guidance System for Satellite Rendezvous. Journal of Aerospace Science 27, 653–674 (1960)
4. Sabol, C., Burns, R., McLaughlin, C.: Satellite Formation Flying Design and Evolution. Journal of Spacecraft and Rockets 38(2), 270–278 (2001)
5. Xi, X.N., Wang, W., Gao, Y.D.: Fundamentals of Near-earth Spacecraft Orbit. National University of Defense Technology press (2003)
6. Lillie, C.F.: On-Orbit Servicing for Futher Space Observatories. In: AIAA Space Conference and Exposition (2005)
7. Hirohisa, K.: Fly-Around Motion Control Based on Exact Linearization with Adaptive Law. Journal of Guidance, Control and Dynamics 28(1), 167–169 (2005)
8. Straight, S.D.: Maneuver Design for Fast Satellite Circumnavigation. In: MS Thesis, Astronautical Engineering. Air Force Institute of Technology, AFIT /GA /ENY/04 - M05 (2004)
9. Hari, B.H., Myron, T., David, D.B.: Guidance Algorithms for Autonomous Rendezvous of Spacecraft with a Target Vehicle in Circular Orbit. In: AIAA Guidance, Navigation, and Control Conference and Exhibit, pp. 6–9 (2001)
10. Wang, G.B., Zheng, W., Meng, Y.H., Tang, G.J.: Fast Fly Around Formation Design Based on Continuous Low Thrust. In: 3rd International Symposium on Systems and Control in Aeronautics and Astronautics (ISSCAA), pp. 1095–1100. IEEE (2010)
11. Qi, Y.Q., Jia, Y.M.: Forming and Keeping Fast Fly-around under Constant Thrust. Advances in Space Research 48(8), 1421–1431 (2011)
12. Woffinden, D.: On-orbit Satellite Inspection Navigation and $\triangle v$ Analysis. In: MS Thesis. Massachusetts Institute of technology (2004)
13. Irvin Jr., D.J.: Optimal Control Strategies for Constrained Relative Orbits. In: PhD Dissertation. Air Force Institute of Technology, AFIT/DS/ENY/07-03 (2007)
14. Yan, Y.: Study of Hovering Method at Any Selected Position to Space Target. Chinese Space Science and Technology 1, 1–5 (2009)
15. Leonard, C., Bergmann, E.: A Survey of Rendezvous and Docking Issues and Development. Orbital Mechanics and Mission Design, 85–101 (1989)
16. Broschart, S.B., Scheeres, D.J.: Control of Hovering Spacecraft near Small Bodies: Application to Asteroid 25143 Itokawa. Journal of Guidance, Control and Dynamics 28(2), 343–354 (2005)
17. Sawaim, S., Scheeres, D.J., Broschart, S.B.: Control of Hovering Spacecraft Using Altimetry. Journal of Guidance, Control, and Dynamics 25(4), 786–795 (2002)
18. Zhang, J.R., Zhao, S.G., Yang, Y.Z.: Characteristic Analysis for Elliptical Orbit Hovering Based on Relative Dynamics. IEEE Transactions on Aerospace and Electronic Systems 49(4), 2742–2750 (2013)
19. Li, Y.K., Jing, Z.L., Hu, S.Q.: Optimal Sliding-mode Control for Finite-thrust Spacecraft Hovering around Elliptical Orbital Target. International Journal of Innovative Computing, Information and Control 7(5) (2011)

Community Detection Method of Complex Network
Based on ACO Pheromone of TSP

Si Liu[1], Cong Feng[1], Ming-Sheng Hu[1,2,*], and Zhi-Juan Jia[1]

[1] College of Information Science and Technology,
Zhengzhou Normal University, Zhengzhou, China
[2] Institute of Systems Engineering,
Huazhong University of Science & Technology, Wuhan, China
ukyosama@163.com

Abstract . community detection method of complex network with a combination of TSP model and ant colony optimization is proposed in this paper. The topology relationship of network node is transformed into distance, thus the community detection problem is transformed into a path optimization problem (TSP) and solved by using ant colony algorithm, and then the pheromone matrix is used to achieve the community clustering by the convergence of algorithm. Experimental results show that, the use of TSP path length as fitness is feasible, and compared with some representative algorithms, TSPP algorithm can cluster out the number of real communities in network effectively, which has a higher clustering accuracy.

Keywords :omplex network, Community detection, Ant colony optimization, Traveling salesman problem, Pheromone.

1   Introduction

With the in-depth study of the physical significance and mathematical properties of the complex network, it was discovered that many real networks such as social networks, biological networks, Web network have a common nature, namely community structure [1]. That is, the whole network can be considered as constituted by a plurality of communities. The node connections in the same community are relatively very intense, and the node connections in different communities are relatively sparse.

Community detection methods of complex network proposed in recent years can be broadly divided into optimization-based methods and heuristic methods. Optimization-based methods mainly include: FN (Fast Newman) algorithm [2], GA (Guimera-Amaral) algorithm [3], KL algorithm (Kernighan Lin) algorithm [4], EO (Extremal Optimization) algorithm [5], FUA (Fast Unfolding Algorithm) algorithm [6] and ITS (Iterated Tabu Search) algorithm [7]. Heuristic methods mainly include: GN (Girvan-Newman) algorithm [1], FEC (Finding and Extracting Communities) algorithm [8],

---

* Corresponding author.

D.-S. Huang et al. (Eds.): ICIC 2014, LNAI 8589, pp. 763–770, 2014.

CPM (Clique Percolation Method) algorithm [9] and LPA (Label Propagation Algorithm) algorithm [10].

Each of the above community detection method exists some individual defects, such as the accuracy of FN algorithm is not high when facing large-scale communities, the time complexity of GA algorithm is too high and a relatively long time required to calculate, KL algorithm needs priori knowledge (such as the number of communities), GN algorithm based on the maximum value of Q function to achieve the division of communities, but the results are sometimes have a big difference with the real community. In short, how to further improve the accuracy of community detection algorithms [11], especially in the case of the number of network community is unknown in advance [12], which still is a not well solved problem.

In response to this problem, this paper proposed a heuristic community detection method based on Traveling Salesman Problem (TSP) [13] and ant colony optimization algorithm (ACO). The relationship between the network nodes is processed as distance, thereby changed the community detection problem into a TSP model, and then the path pheromone matrix of ACO is used to achieve community detection.

2   Community Detection and TSP

This paper introduced TSP (Traveling Salesman Problem) to community detection. TSP is one of the famous mathematical problems which suppose a salesman had a travel business needing to visit N cities, he must choose which path to go but limit the path is to visit each city only once and at last return to the original departure city. The solution of TSP is to obtain the shortest path from all feasible paths, which is an NP-hard problem. Also note that for the TSP solution, even the nearest city to city A is city B, but the path AB may not be selected for the sub-path of the TSP shortest path, because solving the TSP requires global topology considerations. And same in community detection, two nodes have close relationship may not necessarily belong to the same community, which is a very important meeting point between TSP and community detection.

Of course, the above discussion is based on the ideal case. In practice, there are the following three key problems need to be solved:

1 The determine problem of node distance

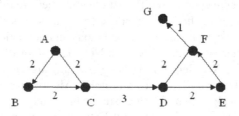

Fig. 1. Determine the distance between nodes

The distance between nodes is the basic to solve TSP. In principle that the shorter the distance between nodes, the greater the chance of becoming TSP sub-path, so the provisions of node distance should meet the needs for community detection. If node j is the neighbor node of node i and the degree of node j is 1, then node i is actually the only neighbor of node j so that node j can only follow the node i belongs to the same community in general, which indicates that the tightness between a node and its neighbors also depends on the degree of its neighbors, so as shown in Figure 1, we set the distance between the two nodes is the smaller degree of the two nodes, such as nodes B, C, D with the degrees of 2, 3, 3 respectively, then the distance between node B and C is 2, and the distance between node C and D is 3.

2 The problem of path feasibility

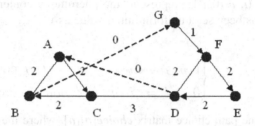

Fig. 2. Air-pass between nodes

In Figure 1, we can find that as a solution to TSP that the path from A to G is not complete, because it must go back to node A and then it is a complete path of TSP, but there is no link between node A and G, so it is impossible to form a complete path of TSP. And in fact, this disconnection is often the case. In general TSP model where the graph is fully connected, but the actual complex network is basically a sparse graph, so we introduce a mechanism (called "air pass") to resolve this path feasibility problem. As shown in Figure 2, there is no way to go after from node A to C to B, then we can randomly give node B a "air pass" path to visit a node it haven't visited before, for example air pass to node G as shown in Figure 2, then after reached node D and finally returned node A by air pass, thus completed a full path of TSP. The distance of "air pass" path is 0 such as the paths B-G and D-A.

For convenience, the proposed algorithm is referred to as TSPP algorithm (TSP Pheromone algorithm). Specific algorithm introduced in Section 3.

3   TSPP Algorithm

In order to quickly solve the TSP, this paper used the ant colony system (ACS) algorithm with some simplifications and made corresponding adjustments and optimizations according to the features of community detection. TSPP algorithm is described below, set the number of nodes is n; the number of ants is m.

1 Data initialization

Step1 Import the topology of complex network and get the distance between nodes by the above distance determination method, then save the information in the distance matrix $dist[n][n]$, where the value of array element $dist[i][j]$ ($i, j\in[0,n-1]$) represents the path length from node i to j, as shown in formula (1).

$$dist[i][j] = \begin{cases} \min\{degree_i, degree_j\} & i, j \text{ connected} \\ 0 & i = j \text{ or } i, j \text{ non} - \text{connected} \end{cases} \quad (1)$$

Where $degree_i$ and $degree_j$ represent the degree of node i and j respectively.

Step2 Initialize the pheromone matrix $phero[n][n]$, where the value of array element $phero[i][j]$ ($i, j\in[0,n-1]$) represents the pheromone content of the path from node i to j which has been set a uniform initial value as $v_{init\text{-}pheromone}$ or 0, as shown in formula (2).

$$phero[i][j] = \begin{cases} Vinit\_pheromone & i, j \text{ connected} \\ 0 & i = j \text{ or } i, j \text{ non} - \text{connected} \end{cases} \quad (2)$$

Step3 Initialize the path choice matrix $choice[n][n]$, where the value of array element $choice[i][j]$ ($i, j\in[0,n-1]$) is shown in formula (3), it can be seen, the value of $choice[i][j]$ is proportional to the pheromone content in its path, and inversely proportional to the path length.

$$choice[i][j] = \begin{cases} \dfrac{phero[i][j]}{dist[i][j]} & i, j \text{ connected} \\ 0 & i = j \text{ or } i, j \text{ non} - \text{connected} \end{cases} \quad (3)$$

Step4 Initialize the parameters of ants and so on in algorithm, such as the maximum number of iteration set as $epoch\_max$, the length of global optimal path set as $dist\_best$, the lower limit of pheromone content set as $pheromone\_min$ and so on.

2 Iteration

When the iteration number is equal to $epoch\_max$, then the iteration is end; otherwise circulating the following acts:

Step1 Clear the way-finding memory of all ants on the previous iteration;

Step2 All ants were randomly assigned to an initial node;

Step3 Every ant builds a full path: in each step of the path construction, ant k ($k\in[1,m]$) in accordance with the roulette wheel selection method [13] to decide the next node to go. The characteristic of this selection method is that the value of $choice[i][j]$ corresponding to a path is greater, the probability of the path chosen by ant k is greater. So on the one hand ants tend to choose the paths which is considered better currently, on the other hand the other paths are also have the chance of being selected which can avoid algorithm falling into local convergence prematurely. The probability of an ant in node i will choose the node j as its next visit node is shown in formula (4), where N represents the set of nodes has not yet visited by the ant. If this set is empty, then the node conducts air pass.

$$p_{ij} = \frac{choice[i][j]}{\sum\limits_{l \in N} choice[i][l]}, \qquad j \in N \qquad (4)$$

Step4 Ants return to the starting node, then calculate the length of the paths they constructed and reduce the pheromone content of each path they have visited (set the reduction as V_{sub}), so that the probability of other paths chosen by ants in next iteration is increased which could prevent the algorithm falling into a local optimum. If the pheromone content of a path is less than *pheromone_min*, then make it equal to *pheromone_min*, which can avoid the pheromone content of a path is too low to be selected.

Step5 Select the ant which constructed the shortest path in this iteration and increase the pheromone content (set the increment as V_{add}) of this path, which is to reward the best path in this iteration so the probability of this path chosen by ants in next iteration is increased, and V_{add} is significantly greater than V_{sub}. If the length of this path is less than *dist_best*, then use it to replace *dist_best* and record the node order of this path;

Step6 Go to the next iteration.

3 Community Detection

After the iteration we can analyze the converged pheromone matrix ***phero[n][n]*** to obtain the results of network community detection. As the convergence characteristic of ACO algorithm, the community structure indicated by pheromone matrix is very obvious, and in the paper a splitting method is used to achieve community detection from the pheromone matrix ***phero[n][n]***.

Step1 Set the threshold value as *threshold*, this value can be obtained by an empirical formula, as shown in formula (5).

$$threshold = \log(epoch\_max * v_{sub}) \qquad (5)$$

Step2 Take the first row of pheromone matrix ***phero[n][n]***, extract the elements whose value is greater than the *threshold* (namely extract the neighbor nodes of the 1st node whose community is same as the first node in network), and add the 1st node to constitute community 1.

Step3 Like the operations in step 2, then extract the rest nodes in other rows belong to commuity1. If found new node then to add it to the commuity1, repeat this until no new node joining, and then delete all rows corresponding to the nodes in community 1. If the pheromone matrix is not empty, return to step 2 and to discover new communities.

Step4 Return the community detection results.

4   Case Study

In order to better analyze the performance of TSPP algorithm, we selected three real-world networks to test it, which are Zachary's karate club [14], dolphin social network [15] and American college football [1] respectively. Experimental platform is Matlab 7.0. In the parameter setting of TSPP algorithm, the number of ants is 10, the initial pheromone content of connected path (vinit-pheromone) is 1, the pheromone award of best path (Vadd) is 0.1, the amount of pheromone volatilization (Vsub) is 0.01, the pheromone limit of paths (pheromone_min) is 0.5.

Table 1 gives the experimental results compared with the algorithms GN, FEC, FUA and RWACO algorithm proposed in [12]. In parentheses which at the right of network name are the real numbers of communities in each network. *Q-Value* is a reference value to evaluate the community detection results of each algorithm which calculated by *Q* modularity function. It can be seen that the *Q-Value* of FUA whose objective function also depends on *Q-Value* is significantly better than the heuristic algorithms TSPP, FEC and RWACO. The value *Division* of is the number of communities divided by each algorithm. It can be seen that TSPP algorithm and RWACO algorithm correctly divided the number of communities in all three networks, and other algorithms showed poor performance in automatically determine the number of communities.

Table 1. Comparison of community detection algorithms

Algorithm	Karate(2)		Dolphin(2)		Football(12)	
	Q-Value	Division	Q-Value	Division	Q-Value	Division
TSPP	0.386	2	0.483	2	0.598	12
GN	0.401	5	0.515	5	0.600	10
RWACO	0.371	2	0.377	2	0.601	12
FEC	0.374	3	0.522	4	0.569	9
FUA	0.418	4	0.531	4	0.605	10

Limited to picture clarity problem, here we only specifically analyze the TSP path graph of the smaller karate network. As shown in Figure 3, karate network showed the social relationship between each karate club members in an American university, which consisting of 34 nodes and a total of 78 edges, divided into two communities (using the mark circle and triangle to distinguish these two communities in Figure 3).

In the experiment of karate network, the iteration number is set to 1000 and TSPP algorithm will reach convergence after 800 iterations on average, the shortest path length after TSP converged is generally about 44. As shown in Figure 4 there is a converged TSP shortest path with a length of 43, where the solid line represents the feasible sub-path and the dashed line represents the air pass path.

It can be seen From Figure 4 that in the feasible sub-paths (solid lines) only the 3rd node (ambiguity node) not connected to the nodes with same community. In fact the TSPP algorithm also only divided the 3rd node to the wrong community. And note

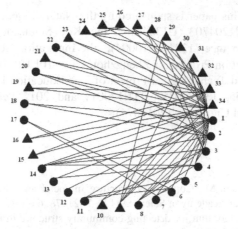

Fig. 3. Topology of the karate network

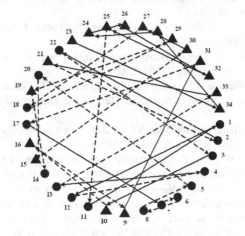

Fig. 4. The length of shortest TSP path is 43

that the different lengths of converged TSP paths do not affect the community detection results of karate network (the result of larger networks will change slightly), so long as the algorithm reached convergence and correspondingly the pheromone matrix is mature, the community detection of network can be achieved effectively.

5   Conclusion

In this paper, the length of TSP path is used as the fitness function of ACO algorithm; thereby the converged pheromone matrix is used to achieve community detection. Compared to similar algorithms, the computing cost of using the length of TSP path as fitness function is lower, and experiments proved that the idea of TSPP algorithm is feasible, and showed good performance in automatically determine the number of communities.

Acknowledgment. This paper is sponsored by the NaturalScienceFoundation of China (NSFC, Grant U1204703, U1304614), the Key Scientific and Technological Project of Henan Province (122102310004), the Innovation Scientists and Technicians Troop Construction Projects of Zhengzhou City (10LJRC190, 131PCXTD597), the Key Scientific and Technological Project of The Education Department of Henan Province (13A413355, 13A790352, ITE12001) and 2012 year university subject of Zhengzhou Normal University (2012074).

References

1. Girvan, M., Newman, M.: Community structure in social and biological networks. Proceedings of National Academy of Science 9(12), 7821–7826 (2002)
2. Newman, J.: Fast algorithm for detecting community structure in networks. Physical Review E 69(6), 66133 (2004)
3. Guimerà, R., Amaral, L.: Functional Cartography of Complex Metabolic Networks. Nature 433(7028), 895–900 (2005)
4. Newman, J.: Detecting Community Structure in Networks. European Physical Journal B 38(2), 321–330 (2004)
5. Duch, J., Arenas, A.: Community Detection in Complex Networks Using Extremal optimization. Physical Review E 72(2), 27104 (2005)
6. Blondel, V.D., Guillaume, J.L., Lambiotte, R., Lefebvre, E.: Fast unfolding of communities in large networks. Journal of Statistical Mechanics: Theory and Experiment 10, 10008 (2010)
7. Lü, Z., Huang, W.: Iterated Tabu Search for Identifying Community Structure in Complex networks. Physical Review E 80(2), 26130 (2009)
8. Yang, B., Cheung, W., Liu, J.: Community mining from signed social networks. IEEE Trans. on Knowledge and Data Engineering 19(10), 1333–1348 (2007)
9. Palla, G., Derényi, I., Farkas, I., Vicsek, T.: Uncovering the overlapping community structure of complex networks in nature and society. Nature 435(7043), 814–818 (2005)
10. Raghavan, U., Albert, R., Kumara, S.: Near Linear-time Algorithm to Detect Community structures in largescale networks. Physical Review E 76(3), 36106 (2007)
11. Rosvall, M., Bergstrom, C.T.: An Information-theoretic Framework for Resolving community structure in complex networks. Proc. Natl. Acad. Sci. USA 104(18), 7327–7331 (2007)
12. Jin, D., Yang, B., Liu, J., Liu, D.: Ant Colony Optimization Based on Random Walk for Community Detection in Complex Networks. Journal of Software 23(3), 451–464 (2012)
13. He, Z., Wang, J., Liu, S.: TSP-Chord: An Improved Chord Model with Physical Topology Awareness. In: 2012 International Conference on Information and Computer Networks, vol. 27, pp. 176–180 (2012)
14. Zachary, W.: An Information Flow Model for Conflict and Fission in Small Groups. Journal of Anthropological Research 33(4), 452–473 (1977)
15. Lusseau, D.: The Emergent Properties of a Dolphin Social Network. Proceedings of the Royal Society B: Biological Sciences 270, 186–188 (2003)

Complex Network Construction Method
of Disaster Regional Association
Based on Optimized Compressive Sensing

Si Liu[1], Cong Feng[1], Zhi-Juan Jia[1,*], and Ming-Sheng Hu[1,2]

[1] College of Information Science and Technology,
Zhengzhou Normal University, Zhengzhou, China
[2] Institute of Systems Engineering, Huazhong University of Science & Technology,
Wuhan, China
zyyyb1@163.com

Abstract. iming at the disaster regional association issues, a complex network construction method of disaster regional association based on compressive sensing is proposed in this paper. The disaster system dynamic equations of network node are obtained through the use of power series expansion and the correlation coefficients between nodes are obtained through the use of compressed sensing theory, then the solving process is optimized by hyperbolic tangent function and revised Newton method, so as to realize the effective construction of the network topology. Experimental results show that, complete network construction requires less amount of time series information and the construction result has a certain rationality.

Keywords: isaster, Time series, Compressive sensing, Regional association, Complex network.

1 Introduction

Based on the time series information [1-2] of regional disaster, how to dig out the potential disaster relationship [3-5] between each area and then to construct a corresponding complex network, which is the problem to be solved in this paper. In fact, network construction is becoming a hot issue in the study of complex network and there has been proposed some construction methods such as reverse engineering method [6], perturbation response method [7], convergence mirroring system method [8], and noise-based correlation method [9] and so on, the basic idea of these methods is to derive the network topology by the behavior of individual units, but there are more or less the problem of excessive operational costs.

To solve this problem, [10-12] proposed a complex network construction method based on compressive sensing [13], which can realize network construction with a small amount of data relative to the size of the network topology, and have obtained good results in the construction of non-linear systems. However, the traditional compressive sensing algorithms still have some problems such as lower operational

---

* Corresponding author.

D.-S. Huang et al. (Eds.): ICIC 2014, LNAI 8589, pp. 771–778, 2014.

efficiency so that need for further improvements. To achieve a better network construction, this paper based on compressive sensing theory and made some improvements for its shortcomings.

2   Compressed Sensing

The core idea of compressive sensing theory [14] is that by using a small amount of measured values S ($S \in R^M$) to restore the sparse vector a ($a \in R^N$), and meet $S = D \cdot a$, where D is an $M * N$ dimensional full rank matrix, and the signal reconstruction process is to solve this following optimization problem, as shown in formula (1).

$$\min \|a\|_0 \quad \text{subject to} \quad D \cdot a = S \tag{1}$$

Where $\|a\|_0$ is the L_0 norm of vector a. The advantage of compressive sensing is that when a is a sparse vector and D satisfy the constraints, M can be much less than N, and the number of non-zero elements in a is less than M. Due to the sparsity exhibited by complex network itself, which also ensured the sparsity of vector a, so long as the construction problem of complex network can be changed into the form that compressive sensing theory can handle with, then this method can be used to accurately reconstruct network topology based on a small amount of time series information.

To solve the formula (1) directly is very difficult, so a common solving idea is to replace the formula (1) with the convex optimization model based on minimization L_1 norm, but this method has high computational complexity; another idea is to use the greedy algorithm to obtain an approximate solution for formula (1), the accuracy of these algorithms is lower than the former algorithms. In this paper, a smooth function is used to achieve the approximation to L_0 norm in order to improve the solving accuracy of compressive sensing algorithm, and revised Newton method is used to further improve the convergence speed.

3   Construction Algorithm

Set the number of node in the complex network of disaster regional association is n, the time series information of node i at a given time (e.g., time t) is divided into two parts, the first part is set as s_i, represents the affected results of region node i observed at the time t, the second part is set as x_i, which is an observation parameter vector that contains some associated disaster-environmental parameters at time t, set the vector dimension is m. Then the complex system disaster dynamics equation of node i can be represented by the formula (2).

$$s_i = f_i(x_i) + \sum_{j=1, j \neq i}^{n} G_{ij}(x_j - x_i) \tag{2}$$

Where $f_i(x_i)$ is disaster dynamics equation of the node i itself, and the second half of the formula (2) is relationship dynamics equation between node i and other nodes, the two dynamics equation together constitute the system dynamic equations of node i; G_{ij} is the association matrix of observation parameter vector between node i and node j, as shown in formula (3) below.

$$
G_{ij} = \begin{pmatrix} g_{ij}^{1,1} & g_{ij}^{1,2} & \cdots & g_{ij}^{1,m} \\ g_{ij}^{2,1} & g_{ij}^{2,2} & \cdots & g_{ij}^{2,m} \\ \cdots & \cdots & g_{ij}^{u,w} & \cdots \\ g_{ij}^{m,1} & g_{ij}^{m,2} & \cdots & g_{ij}^{m,m} \end{pmatrix}
$$
(3)

Where $g_{ij}^{u,w}$ represents the correlation between the u-th observation vector component of node i and the w-th observation vector component of node j. If there is at least an element is not 0 in this matrix, then it can consider of these two nodes interrelated, namely in complex network these two nodes exist an edge

For the formula (2), we can also represent in such a way as formula (4).

$$
s_i = \left[f_i(x_i) - \sum_{j=1, j \neq i}^{n} G_{ij}(x_i) \right] + \sum_{j=1, j \neq i}^{n} G_{ij} x_j
$$
(4)

Then the first part of the formula (4) becomes an equation only about x_i, and the second half of the equation is about the observation vector of other nodes. Then we use $Q_i(x_i)$ to represent the first part of formula (4), and use a power series of order N to expand $Q_i(x_i)$, as shown in formula (5) below.

$$
\begin{aligned}
Q_i(x_i) &= f_i(x_i) - \sum_{j=1, j \neq i}^{n} G_{ij}(x_i) \\
&= \sum_{l_1=0}^{N} \sum_{l_2=0}^{N} \cdots \sum_{lm=0}^{N} \left[(a_i)_k \right]_{l_1, l_2 \ldots lm} \\
&\quad \cdot \left[(x_i)_1 \right]^{l_1} \left[(x_i)_2 \right]^{l_2} \cdots \left[(x_i)_m \right]^{l_m}
\end{aligned}
$$
(5)

Where $(x_i)_1$ represents the first component of observation vector of the node i, $(a_i)_k$ represents the k-th scalar coefficient of the expansion term. Note that the formula (5) covered all possible expansion terms in the power series of order N, and the number of expansion terms is $(1+N)^m$, which means that many scalar coefficients of the expansion terms may be 0.

And now the formula (4) can also be represented as shown in formula (6) below.

$$
s_i = Q_i(x_i) + G_{i1} x_1 + G_{i2} x_2 + \ldots + G_{in} x_n
$$
(6)

From formula (6) it can be seen, in the case of the node time series information is known, the unknown things are the expansion scalar coefficients in $Q_i(x_i)$ and the association matrix G_{ij}. we take $G_{ij} x_j$ in formula (6) with the same power series expansion as in formula (5), and then the formula (6) can be expressed as formula (7) below.

$$S_i = Q_1(x_1) + Q_2(x_2) + Q_3(x_3) + \ldots + Q_n(x_n) \tag{7}$$

For each $Q(x)$ function in formula (7), for example to solve the $Q_i(x_i)$ in formula (5), set the coefficients of expansion terms be represented by the vector a_i, because $Q_i(x_i)$ is in accordance with the power series expansion, so the expanded form is fixed, as in the series $N = 3$, $m = 3$ case, the expansion of $Q_i(x_i)$ can be shown as formula (8).

$$Q_i(x_i) = (a_i)_{000} \, b_i^0 y_i^0 z_i^0 + (a_i)_{001} \, b_i^0 y_i^0 z_i^1 +$$
$$\ldots + (a_i)_{uvw} \, b_i^u y_i^v z_i^w + \ldots + (a_i)_{333} \, b_i^3 y_i^3 z_i^3 \tag{8}$$

Where b, y, z are the components of the vector x_i. The coefficient vector of formula (8) is a_i, and $a_i = [(a_i)_{000}, (a_i)_{001}, \ldots, (a_i)_{333}]^T$, then for observation vector $x_i(t)$ at time t, we get formula (9):

$$d_i(t) = \begin{bmatrix} b_i(t)^0 y_i(t)^0 z_i(t)^0, b_i(t)^0 y_i(t)^0 z_i(t)^1 \\ , \ldots, b_i(t)^3 y_i(t)^3 z_i(t)^3 \end{bmatrix} \tag{9}$$

Now $Q[x_i(t)] = d_i(t) \cdot a_i$, if we recorded the time series information of node i with L time, and set the observation vector as S, $S = [s_i(t_1), s_i(t_2), \ldots, s_i(t_L)]^T$, then we can get the equation $S = D \cdot a$ which compliance with compressive sensing theory, as shown in formula (10).

$$\begin{pmatrix} s_i(t_1) \\ s_i(t_2) \\ \ldots \\ s_i(t_L) \end{pmatrix} = \begin{pmatrix} d_1(t_1) & d_2(t_1) & \ldots & d_n(t_1) \\ d_1(t_2) & d_2(t_2) & \ldots & d_n(t_2) \\ \ldots & \ldots & \ldots & \ldots \\ d_1(t_L) & d_2(t_L) & \ldots & d_n(t_L) \end{pmatrix} \times \begin{pmatrix} a_1 \\ a_2 \\ .. \\ a_n \end{pmatrix} \tag{10}$$

In the formula (10), S and $d_i(t)$ can be obtained by the existing time series information, and then we can construct the network topology. However, because the traditional L_0 norm approximation methods have deficiencies in solving accuracy, this paper selected hyperbolic tangent function to improve the approximation performance, as shown in formula (11) below.

$$f_\sigma(a_x) = \frac{e^{\frac{a_x^2}{2\sigma^2}} - e^{\frac{a_x^2}{2\sigma^2}}}{e^{\frac{a_x^2}{2\sigma^2}} + e^{\frac{a_x^2}{2\sigma^2}}} \tag{11}$$

Where a_x represents the component of the vector $\boldsymbol{a}_i$, σ is an adjustment parameter, since in the case of σ close to 0, when a_x is equal to 0, the limit value of $f_\sigma(a_x)$ is equal to 0; when a_x not equal to 0, then the limit value of $f_\sigma(a_x)$ is equal to 1. If set $F_\sigma(a_i) = \sum_{x=1}^{n} f_\sigma(a_x)$, the function value will approximate to the number of nonzero component in vector $\boldsymbol{a}_i$, namely the approximation of L_0 norm.

Because the Hesse matrix in the Newton direction $\boldsymbol{d}$ in $F_\sigma(a_i)$ is not positive definite matrix which as shown formula (12), this will greatly affect the decrease effect of $\boldsymbol{d}$.

$$\nabla^2 \left(F_\sigma \left(\mathbf{a}_i \right) \right) = \begin{pmatrix} \dfrac{\partial^2 f_\sigma(a_1)}{\partial a_1^2} & 0 & \cdots & 0 \\ 0 & \dfrac{\partial^2 f_\sigma(a_2)}{\partial a_2^2} & \cdots & 0 \\ \cdots & \cdots & \cdots & \cdots \\ 0 & 0 & \cdots & \dfrac{\partial^2 f_\sigma(a_n)}{\partial a_n^2} \end{pmatrix} \tag{12}$$

To solve this problem, this paper constructs a correction matrix U to ensure its positive definiteness, as shown in formula (13) below.

$$\mathbf{U} = \nabla^2 \left(F_\sigma \left(\mathbf{a}_i \right) \right) + \varepsilon_x \mathbf{I} \tag{13}$$

Where I is the identity matrix, ε_x is the correction factor which is calculated as shown in formula (14).

$$\varepsilon_x = \frac{\dfrac{16 a_x^2}{\sigma^4} e^{\frac{a_x^2}{2\sigma^2}}}{e^{\frac{a_x^2}{2\sigma^2}} + e^{\frac{a_x^2}{2\sigma^2}}} \tag{14}$$

Then we get the Newton direction $\boldsymbol{d}$ modified with positive definiteness, as shown in formula (15) below.

$$\mathbf{d} = -\mathbf{U}^{-1} \nabla \left(F_\sigma \left(\mathbf{a}_i \right) \right) \tag{15}$$

4   Case Study

We choose earthquake disaster which with the characteristic of zonal distribution to evaluate the performance of the proposed method. There are five major earthquake regions in China. This paper selected totaling 60 cities as network nodes from these five regions, the time series data are take from the historical disaster databases which contain affected situation by earthquake and related disaster pregnant environmental parameters of each cities in a total of 40 years (from 1965 to 2004), namely $n = 60$, $L = 40$. Experimental platform is Matlab 7.0. On the parameter setting, the power series of order $N = 3$, and according to the experts recommend selected six parameters as earthquake-related parameters, namely the observation vector dimension $m = 6$.

Figure 1 showed the result of network topology obtained by the proposed algorithm and drawn by the software Pajek.

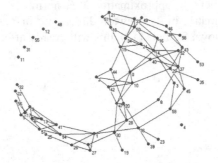

Fig. 1. Network topology of disaster regional association

Table 1 showed the connection rate statistical results of mapping the node-to-node links to the corresponding seismic zones based on the topological relationship between nodes. If the algorithm is valid, then firstly the nodes in same seismic zone should have a higher connection rate.

Table 1. Node connection rate of each seismic zone

Node number in seismic zone	number of seismic zone	Average connection rate
5	1	60%
4	4	75%
3	8	72%
2	6	83%
1	3	0%

Can be seen from Table 1, the node connection rate in seismic zones is in line with the expectation, which shows the network topology constructed by the algorithm had a certain scientific.

Figure 2 and Figure 3 showed the link matching degree between the obtained networks and the network in Figure 1 under different amount of time series data. we used two parameters to conduct this evaluation, respectively are non-empty link and empty link. As can be seen from these two figures, the proposed algorithm can obtain a consistent network topology in Figure 1 only need to use about 60% volume of the original time series data, namely can achieve an accurate construction of network with a small amount of data.

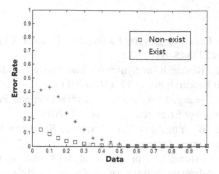

Fig. 2. Error link rate under different amount of data

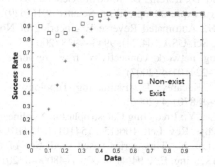

Fig. 3. Success link rate under different amounts of data

5   Conclusion

Based on compressive sensing theory, this paper tried to use a small amount of time series data to construct the complex network of disaster regional association, and used the hyperbolic tangent function and revised Newton method to improve the solving performance of compressive sensing algorithm. The direction of future work is to further expand the algorithm application fields in order to be able to effectively construct other social networks.

Acknowledgment.   This paper is sponsored by the National Natural Science Foundation of China (NSFC, Grant U1204703, U1304614), the Key Scientific and Technological Project of Henan Province (122102310004), the Innovation Scientists and Technicians Troop Construction Projects of Zhengzhou City (10LJRC190, 131PCXTD597), the Key Scientific and Technological Project of The Education Department of Henan Province (13A413355,13A790352, ITE12001) and 2012 year university subject of Zhengzhou Normal University (2012074).

References

1. Men, K.: Time-space Order of Severe Earthquake Disaster Chain and Its Prediction. Progress in Geophysics 22, 645–651 (2008)
2. Yan, J., Bai, J., Su, K.: Research on Symmetry and Tendency of Several Major Natural disasters. Geographical Research 30, 1159–1168 (2011)
3. Qiu, J., Xie, J., Li, W., Wang, J.: Research on Correlation and Periodicity of Moderate-strong Earthquake. Computer Engineering 37, 16–18 (2011)
4. Wu, S., Wu, G., Wang, W., Yu, Z.: A Time-Sequence Similarity Matching Algorithm for Seismological Relevant Zones. Journal of Software 17, 185–192 (2006)
5. Jia, Z., Hu, M., Dong, X., Hong, L.: Application of Ant-Clustering in Classification of Historical Disaster. AISS: Advances in Information Sciences and Service Sciences 4, 135–143 (2012)
6. Gardner, T.S., Bernardo, D., Lorenz, D.: Inferring Genetic Networks and identifying compound mode of action via expression profiling. Science 301(5629), 102–105 (2003)
7. Bongard, J., Lipson, H.: Automated Reverse Engineering of Nonlinear Dynamical systems. Proc. Natl. Acad. Sci. USA 104(24), 9943–9948 (2007)
8. Timme, M.: Revealing network connectivity from response dynamics. Phys. Rev. Lett. 98(22), 224101-1–224101-4 (2007)
9. Yu, D., Righero, M., Kocarev, L.: Estimating Topology of Networks. Phys. Rev. Lett. 97(18), 188701-1–188701-4 (2006)
10. Wang, W., Yang, R., Lai, Y.: Predicting Catastrophes in Nonlinear Dynamical Systems by compressive sensing. Phys. Rev. Lett. 106(15), 154101-1–154101-4 (2011)
11. Wang, W., Yang, R., Lai, Y.: Time-series based Prediction of Complex Oscillator networks via compressive sensing. EPL 94(4), 48006-1–48006-6 (2011)
12. Wang, W., Lai, Y., Grebogi, C.: Network Reconstruction Based on Evolutionary-game data via compressive sensing. Phys. Rev. X 1(2), 021021-1–021021-7 (2011)
13. Candès, E., Romberg, J., Tao, T.: Stable Signal Recovery From Incomplete and inaccurate measurements. Comm. Pure. Appl. Math. 59(8), 1207–1223 (2006)
14. Candès, E.: The Restricted Isometry Property and Its Implications for Compressed Sensing. C.R. Math. Acad. Sci. Paris 346(9-10), 589–592 (2008)

Cascading Failures in Power Grid under Three Node Attack Strategies*

Sui-Min Jia, Yun-Ye Wang, Cong Feng, Zhi-Juan Jia, and Ming-Sheng Hu

College of Information Science & Technology,
Zhengzhou Normal University, Zhengzhou 450044, China
fc_shallot@163.com

Abstract. This paper studies cascading failures of power grid under three node attack strategies based on the local preferential redistribution rule of the broken node's load. The initial load of a node with degree k is k^β, and β is a tunable parameter. We investigated the cascading propagation of US power grid under three node attack strategies and analyzed their attack effects. The three node attack strategies are HL (attack the node with the highest load), HPC (attack the node with the highest proportion between the capacity of the attacked node and the total capacities of the neighboring nodes), LL (attack the node with the lowest load). Study shows that the attack effect of HPC is the best one of the three in most cases. The attack effects of the three node attack strategies are compared with those of the three edge attack strategies. It is found that in a big range of β, the effect based on a node attack strategy is better than that based on the corresponding edge attack strategy. So in most cases, attacking the important nodes is more harmful than attacking the important edges in power grid.

Keywords: cascading failures, node attack, proportion of capacity, edge attack.

1 Introduction

In recent years, the cascading failure of power grid network has been attract high attention of researchers and becomes a hot topic in the research field. Many kinds of node attack strategies and edge attack strategies have been proposed, and a lot of research work has been done on US and European network [1-4]. In the reference[1,2], as the betweenness for the node's load, the author studied the spread of cascading failure of US power grid and the influence to the power grid transmission efficiency under different node attack strategies. Reference [3], using the mean-field theory,

---

* Project supported by the National Natural Science Foundation of China (Grant Nos. U1204703 and U1304614), the Key Scientific and Technological Project of Henan Province (122102310004), the Innovation Scientists and Technicians Troop Construction Projects of Zhengzhou City (10LJRC190 and 131PCXTD597), the Key Scientific and Technological Project of The Education Department of Henan Province (13A413355, 13A790352 and ITE12001) and the project of Zhengzhou Normal University (2012070).

D.-S. Huang et al. (Eds.): ICIC 2014, LNAI 8589, pp. 779–786, 2014.

studied the robustness of the European power grid under a deliberate attack. Reference [4], adopting the load model which is based on node degree, studied the cascading failure of US power grid under attack strategies of the minimum load and the maximum load. In this paper, we adopted the load model which based on node's degree and the local preferential redistribution rule to research the cascading failure of power grid [5]. According to this method, the load of failure nodes would be assigned to the neighbor nodes according to the distribution principle of prior probability. That is to say, the greater the degree of neighbor node, the more load to be assigned. This method has been applied to the study of European power grid cascading failures [6]. The local preferential redistribution rule also is appropriate for the cascading failure that is based on a node attack strategy [7].

2   The Model

In the study of power grid, we put the power plant and transformer substation as nodes, the transmission lines between power plants and transformer substations as inter-node even edge, thus an undirected network model was constructed. For large-scale blackout accidents, initially, only a single or a few nodes fail, then the failure is quickly spread in the power grid, causing larger failure. In the power grid, the spread of the failure is mainly caused by the redistribution of load which caused the power station successive overload. If an initial disturbance or event causes a node to go down, its power is automatically shifted to the neighboring nodes, which in most cases are able to handle the extra load. Sometimes, however, these nodes are also overloaded and must redistribute their increased load to their neighbors. This eventually leads to a cascade of failures: a large number of transmission lines are overloaded and malfunctioning at the same time. In normal conditions, the more transmission lines the power plant has, the higher the bearing load. Therefore, when the load model is set up, the initial load of a node is closely related to its degree. Here we define the initial load of each node i is [6]

$$F_i = k_i^{\beta}, \quad i = 1, 2, \cdots, N \tag{1}$$

where N is the initial number of nodes and k_i is the degree of node i, $\beta > 0$ is a tunable parameter, which controls the strength of the initial load of the node i. The capacity of a node is the maximum load that the node can handle, which is severely limited by cost. Thus, it is natural to assume that the capacity C_i of node i is proportional to its initial load F_i [8,9],

$$C_i = (1 + \alpha)F_i, \quad i = 1, 2, \cdots, N \tag{2}$$

where the constant $\alpha > 0$ is the capacity parameter. In this paper, we assume that the cascading failure is initially by a single node failure, and the load redistributes to its neighboring nodes caused the overload phenomenon. According to the distribution principle of prior probability [4-7,9], after a node failure, the increased load of the neighbor node j is ΔF_j,

$$\Delta F_j = F_i k_j^\beta / \sum_{m \in \Gamma_i} k_m^\beta \qquad (3)$$

where Γ_i represents the set of all neighboring nodes of node i. Since every node has a limited capacity to handle the load, so for the node j if

$$F_j + \Delta F_j > C_j \qquad (4)$$

then the node j will be broken and induce further the redistribution of the load $F_j + \Delta F_j$ and potentially cause other nodes breaking.

3   Simulation of Three Node Attack Strategies

In this section, we took US power grid as an example, researched the attack effect under three node attack strategies. The highest load node attack (HL) and the lowest load node attack (LL) are commonly attack strategies used on the study of power grid cascading failure [4,9]. Inspired by the reference [7] that based on the edge of the lowest proportion of capacity attack (LPC) strategy, we consider that when a node goes down, its load will be redistributed to its neighbor nodes, so whether it will lead to cascading failure not only associated with the size of the node load but also with the capacity size of the neighbor node. Therefore this paper proposed the highest pro-portion of capacity (HPC), calculating the ratio to the capacity of each node and the total capacity of its neighbor nodes. Because the capacity of node is proportional to its initial load, the ratio to the capacity of node can also be represented as the ratio to the load of node $F_i / \sum_{j \in \Gamma_i} F_j$, HPC attack strategy selects the largest ratio of capacity to attack.

In order to quantify denote the scale of cascading failures, we make the parame-ter $G = \sum_{i \in A} G_i / N_A N$, where A represents the collection of nodes which were at-tacked, N_A is the number of attack nodes, G_i is the number of failed nodes after the cascade process that node i was attacked complete, N is the total number of network nodes[7]. In order to make the experimental results more suitable for general, for the HL, we will first rank the load of all nodes from high to low, select the top 10 nodes in proper order to attack, finally, take the average of the results. The other two attack experiment methods are similar to HL, also take the average of 10 attacks, so $N_A = 10$ .We simulated change of US power grid cascading failure scale G with the capacity parameter α under three kinds of attack strategies, the result is shown in figure 1. In the figure, α_c is the critical threshold of the capacity parameter α .

In order to make the experimental results more applicable to general situation, α_c is the value α when the failure nodes number is one percent of the total number of network nodes. In some network attacks, the larger α_c, the better effect of the attack strategy.

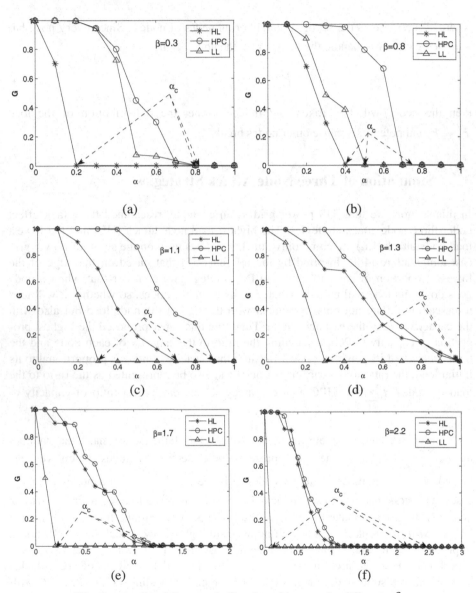

Fig. 1. cascading failures scale G varies with α under different β

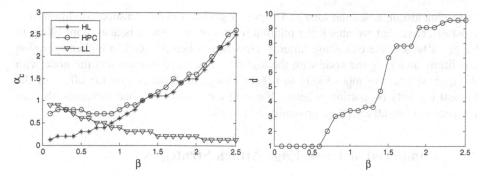

Fig. 2. The relation between α_c and β **Fig. 3.** The degree of HPC under different β

Figure 1 shows, when β is a fixed constant value, no matter what kind of attack strategies, the cascading failure scale of network G decreases with increasing of α, but under different attack strategies, the reduce speed are significantly different. For each attack strategy, the rule that G varies with α differs with different load parameters β. Through a large number of simulation experiments, we concluded the relationship between α_c and β under three attack strategies. As shown in Figure 2, for the attack strategy of HL, α_c is increased with the increase of β, that is to say, the attack effect of HL becomes more enhanced with the network load increased. Fig.1 (c) shows when $\beta = 1.3$, then $\alpha_c = 1$, when the node capacity is double to the initial load, cascading failure of the network still occurs. For the HPC attack strategy, when $\beta < 1$, α_c does not monotonically change with the increasing of β, when $\beta > 1$, on the whole, α_c is increased with the increasing of β. For the HPC attack strategy, α_c is decreased with the increasing of β. Comparing the three attack strategies, as in this paper we researched the parameter β in large range, studies have shown that the attack effect of HPC always be the best. When β is smaller, the attack effect of HPC is closed to LL, with the increase of β, the attack effect of HPC began to approach that of HL. Why is there such a phenomenon? We studied change of average degree of 10 attacked nodes with β under the HPC attack strategy, and the result is shown in Figure 3. We found that when β is smaller, the average degree of the selected attack nodes by HPC is smaller, with the increase of β, the average degree also followed increasing. According to definition of the node load in this paper, when β is smaller, the load of nodes selected by HPC is low, when β is larger, the load of nodes which selected by HPC is high. That is why when β is smaller, the attack effect of HPC is closed to LL, and when β is larger, the attack effect of HPC began to approach HL.

In conclusion, when the load of US power grid in lower operation condition, it is most effective that we attack the minimum load nodes, which is also more likely to trigger a large-scale cascading failures. However, when the load in higher operation condition, attacking the node with the highest capacity proportion and the node with the highest load are more likely to trigger cascading failures, and the effect of the highest capacity proportion is better, but with the power grid load increases, the gap between the two attack effect gradually decreases.

4   Compared of Three Edge Attack Strategies

In the power grid, like the substations and power plants, transmission lines also are indispensable to ensure the normal operation. In order to observe the effects which are caused by failure of the transmission line on grid, we define the edge load of network is $F_{ij} = \left(k_i k_j \right)^{\beta}$, the reference [7] studied the cascading failures under three edge attack strategies which are HL, LL and SPC. The SPC refers to that ratio of the total capacity of neighbor edges and the capacity of attacked edges is the smallest. But in this paper, the HPC refers to that ratio of the capacity of attacked nodes and the total capacity of neighbor nodes is the largest. So the two attack strategies are essentially the same. Here HL is the strategy of attacking the highest load edges, and LL is the strategy of attacking the lowest load edges. Through simulation experiments, the relationship between α_c and β is shown in Figure 4, we can see that the attack effect of HPC is decreased along with the increasing of the load parameter β, and some fluctuations appear in the process of decline. The attack effect of LL is decreased along with the increasing of β, but the HL is on the contrary. Compare of the three attack strategies, when $\beta \in [0.1, 1.4)$, relation of the attack effects is $HPC > LL > HL$, when $\beta \in [1.4, 2.2)$, $HPC > HL > LL$, and when $\beta \in [2.2, 2.5]$, $HL > HPC > LL$. So it can be concluded that when the power grid operation in low load condition, attacking the transmission lines which have the highest capacity ratio is most effective, when the power grid operation in high load condition, attacking the transmission lines which have the highest load most effective. From figure 4 and its subfigure which is the relation between α_c and β under three node attack strategies, we found that for the HL, no matter attacking nodes or attacking edges, the attack effects both increasing along with the load parameter β, and the corresponding curve α_c of the node attack strategy rises faster. For the same β, the α_c based on the node attack is larger than that on the edge attack.

Compare the curves of HL in figure 4 and its subfigure, for the node attack strategy, when $\beta = 1.3$, then $\alpha_c = 1$. It showed that if the load parameter in the power gird is 1.3, attacking the node with the highest load, the cascading failure still cannot be prevented even if the capacity of nodes rise to two times of the initial load. While for the edge attack strategy, in the study range of $\beta \in [0.1, 2.5]$, all α_c are less than 1, so

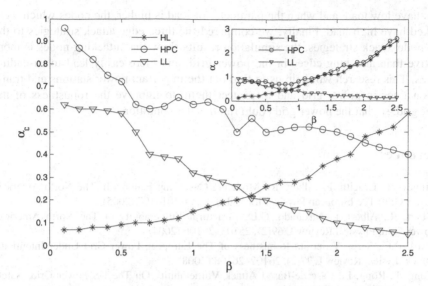

Fig. 4. The relation between α_c and β under three edge attack strategies
(The subfigure is the relation under three node attack strategies)

both the node and the edge strategies choose the highest load, the effect of node attack strategy is more harmful than edge attack strategy. For HPC, α_c is increased with the load parameter β under the node attack strategy, while α_c is decreased with β under the edge attack strategy. When $\beta \in [0.1, 0.3]$, α_c of the edge attack is larger than the node attack, when $\beta \in (0.3, 2.5]$, α_c of the node attack is larger than the edge attack. So when the load of the network is lower, the effect of edge attacking is more harmful than that of node attacking. The result is just the opposite when the load of the network is higher, and the gap becomes wider with the increase of load. For LL, the attack effects of node and edge strategies are weakened with the increase of network load. In this process, however, α_c of the node attack is always larger than the edge attack. So both the node and the edge strategies choose the lowest load, the effect of node attack strategy is more harmful than edge attack strategy. By comparing, we found that in most cases, attacking nodes is more effective than attacking edges in the power grid, and more easily lead to cascading failures.

5 Conclusion

In this paper, we took the US power grid as example, used the local preferential redistribution rule, and studied the cascading failure of power grid under three node attack strategies of HL, LL and HPC, of which HPC is proposed for the first time. This paper focuses on analyzing the effect of HPC change with the parameters of load. The result shows that, for HPC, when the parameter of load is lower, the attacked nodes have low load, and when the parameter of load is higher, the nodes which were attacked have high load. Finally, we compared the three edge attack strategies to the three node attack strategies, the simulation results show that attacking nodes is more effective than attacking edges in the power grid, and more easily lead to cascading failures. This research will help us to find out the important power stations and transmission lines in power grid and to protect them, to improve the robustness of the power grid, so that the power grid could run in safer condition.

References

1. Kinney, R., Crucitti, P., Albert, R.: Modeling Cascading Failures In The North American Power Grid. The European Physical Journal B 46(1), 101–107 (2005)
2. Albert, R., Albert, I., Nakarado, G.L.: Structural Vulnerability of The North American Power Grid. Physical Review E 69(2), 25103–25106 (2004)
3. Marti, R.C., Sergi, V., et al.: Robustness of The European Power Grid Under Intentional Attack. Pysical Review E 77(2), 26102–26108 (2008)
4. Wang, J., Rong, L.: Cascade-Based Attack Vulnerability On The US Power Grid. Safety Science 47(2), 1332–1336 (2009)
5. Wang, W., Chen, G.: Universal Robustness Characteristic of Weighted Networks Against Cascading Failure. Physical Review E 77(2), 26101–26105 (2008)
6. Wei, D., Luo, X.: Analysis of Cascading Failure In Complex Power Networks Under The Load Local Preferential Redistribution Rule. Physic A 391(9), 2771–2777 (2012)
7. Wang, J., Rong, L.: Robustness of The Western United States Power Grid Under Edge Attack Strategies Due To Cascading Failures. Safety Science 49(2), 807–812 (2011)
8. Motter, A., Lai, Y.: Cascade-Based Attacks On Complex Networks. Physical Review E 66(6), 65102–65105 (2002)
9. Lan, Q., Zou, Y., Feng, C.: Cascading Failure of Power Grids Under Three Attack Strategies. Chinese Journal of Computational Physics 29(6), 943–948 (2012)

Dynamical Distribution of Capacities Strategy for Suppressing Cascading Failure in Power Grid

Zhi-Juan Jia, Yu Zhang, Cong Feng[*], and Ming-Sheng Hu

College of Information Science & Technology,
Zhengzhou Normal University, Zhengzhou 450044, China
fc_shallot@163.com

Abstract. This paper studies the suppressing effect of Dynamical distribution of capacities (DDC) strategy in cascading failure of power grid. This strategy is proposed based on load characteristics and transmission method of power grid. Through research of suppressing effects under 3 different mode attack modes, it is found that this strategy achieves good effect under different attacking modes and in grid with different load, but suppressing characteristics are not the same. For the strategy of attacking the node with the highest load (HL), the suppressing effect is more obvious in high load power grid while for the strategy of attacking the node with the lowest load (LL), the suppressing effect is more obvious in low load power grid. For the strategy of attacking the node with the highest proportion between the capacity of the attacked node and the total capacities of the neighboring nodes (HPC), the suppressing effect is rather obvious in any kind of load conditions in power grid.

Keywords: power grid, cascading failure, node attack, suppressing of cascading failure.

1   Introduction

Over the past decade, a great deal of research has been done about robustness and stability of complicated network. Cascading failures, which a big threat to network security, receive more attention or the researchers and becomes core in the field of safety science. Many kinds of infrastructure networks are all at the risk of cascading failures damage, with power grid included. Vertex of power plant is the place where electric power is produced or transferred to other vertex. If one the vertex fails to handle overloaded electric power and distribute it to the other vertex, which can trigger breakdowns of newly overloaded vertices. This process will go on until all the loads of the remaining vertices are below their capacities. For large-scale blackout accidents, initially, it is only a single or a few vertices overloaded beyond their given capacity, leading to serious consequence. It once happened in the history in Ohio US, minor disturbance triggered black out affecting millions of people for 15 hours [1]. How to prevent and suppress cascading failures has been a greatly concerned and

---

[*] Corresponding author.

D.-S. Huang et al. (Eds.): ICIC 2014, LNAI 8589, pp. 787–794, 2014.
© Springer International Publishing Switzerland 2014

urgently need resolved problem. In the past study, many suppressing strategies were proposed, strategy of adjusting the network connection weights by Wang W X and Chen G R [2]; adopting optimum weight by Yang R et al [3]; the strategy proposed by Adilson E Motter that removing those easily overloaded nods and their neighbors to reduce accident risk achieves good effect in cutting down accident scale when load of the network not even [4]. Local protection strategy by Wang J W suppresses break-down significantly [5], and he also puts forward that to maximize network robustness by distributing load reasonably when setting out network [6]. Form researches mainly contain 2 points [7,8], one is changing network structure to stop malfunction from transmission, and another is in view of the network itself to set up reasonable and highly robust network. The two kinds of strategies were firstly applied in communications network and then adopted in power grid after corresponding change, though effectively worked under certain conditions, there are still some limitations. Based on load characteristics and transmission mode of power grid, this paper will introduce a new stressing strategy, dynamical distribution of capacities strategy, DDC.

2   Suppressing Strategy

Given each power station certain capacity $C_i = \lambda L_i$, normally the capacity is bigger than the load ($\lambda \geq 1$); and the beyond part is called excess capacity [9, 10]. When power consumption in certain part of the power grid suddenly increases greatly, the load of the power station which provides proper for this area would also increase substantially, likely beyond its given maximum capacity. At this time, if the neighbor power station automatically transmits power to this area, sharing part of load, the risk of power station overload would be effectively reduced. When the load of a node increasing in the network, the linked nodes automatically distribute certain capacity to it, so that to keep its capacity larger than its load and avoid overload accident, this is called Dynamical Distribution of Capacity, DDC. As it is internal dynamical distribution of the network, the total capacity is not increased. There is an interval between the node overloads and the malfunction occurs, during which the borrowed capacity for neighbor nodes will help it back to normal work from overloaded status [11,12]. Extra power gained by overloaded node [5]:

$$\Delta C_i = \sum_{m \in \Gamma_i} p(C_m - L_m) \tag{1}$$

Where Γ_i represents the set of all neighboring nodes of node i, while C_m and L_m are capacity and initial load of neighbor node m and $p \in [0,1]$ is intensity applied in controlling capacity distribution. When $p = 0$, neighbor node does not transmit redundant capacity to overload node, while when $p = 1$, neighbor node transmit all of its redundant capacity to the overload one. Base on DDC strategy, total capacity of an overloaded node after gain redundant capacity is

$$C_{i,t} = C_{i,t-1} + \sum_{m \in \Gamma_i} p(C_{m,t-1} - L_m) \tag{2}$$

Remaining capacity of neighbor node is

$$C_{m \in \Gamma_i, t} = C_{m \in \Gamma_i, t-1} - p(C_{m \in \Gamma_i, t-1} - L_{m \in \Gamma_i}) \tag{3}$$

This distribution strategy not only ensures the neighbor node has enough capacity to handle its own load, but also allow it transmit capacity to overload node, whether it can enhance robustness of network against cascading failure? Let us take the US power grid as example to simulate, to represent suppressing effect of cascading failures as parameter $G = \sum_{i \in A} N_i' / N_A N$, in which A is the set of attacked nodes, N_A is number of attacked nodes, N_i' is the number of remaining nodes after node i is attacked and cascading failures process completed, N is total number of nodes in the grid. To make sure the experiment result without loss of generality, we take the average result of the many experiments. For example, for the HL attack strategy, firstly, we take all nodes in the network according to the load from high to low ranking, and then select the 10 nodes in the front row, and in turn to attack, finally the experimental results are averaged, other attacks are similar, so N_A=10 .

3   Simulation and Discussion

In this section we will analyze the suppressing effect of DDC under three kinds of attack strategies on power grid cascading failure. First of all, we analyze the suppressing cascading failure for HPC attack strategy under different loads of power grid, the experimental results were shown in Figure 1. The curves of HLC represent the use of DDC, and the curve of HL does not. We can see that the strategy of DDC reduced the scale of cascading failure under different loads, especially in the high load condition is more obviously. As shown in Figure 1, when the parameter of capacity $\beta = 2$, the value of G in HL is less than 0.2, while the value of G in HLC reached 0.8, the number of surviving nodes increased 4 times. Thus, the DDC can effectively suppress the cascading failures caused by HL attack strategy in power grid.

Next, we analyze the suppressing cascading failure by using of DDC, for the HPC attack strategy under different loads of power grid , the experimental results were shown in Figure 2. Because the LL attack strategy is more likely to trigger a large-scale cascading failures in the low load condition, it focuses on analyzing the difference of attack effects between using DDC and not using DDC on low load state. The curves of LLC represent the use of DDC, and the curve of LL does not. We can see that LLC and LL have very significant differences in the low load state, as the load increases, the LLC curve and LL curve gradually approaches, the disparity of fault scale decreased gradually. This phenomenon is mainly due to the size of cascading failure by LLC is almost the same, while the size of cascading failure by LL decreases

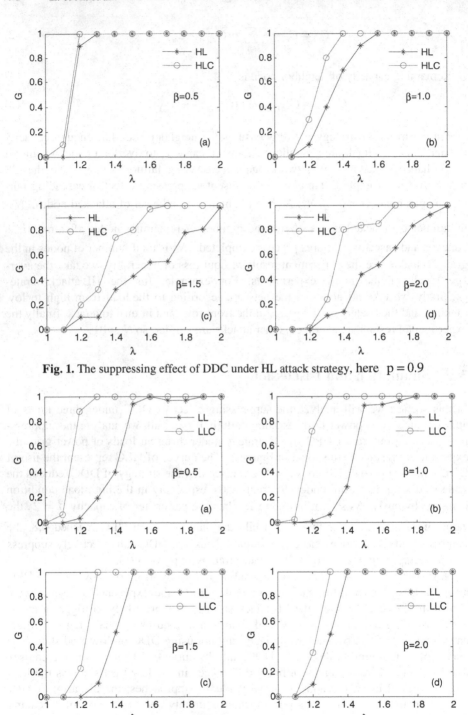

Fig. 1. The suppressing effect of DDC under HL attack strategy, here p = 0.9

Fig. 2. The suppressing effect of DDC under LL attack strategy, here p = 0.9

along with the increasing of the load, gradually to the LLC approximation, and finally close to equal. In the high load condition, cascading failure caused by LL is small. Therefore, in reality, if considering cost, we would better not to use DDC for this attack strategy.

Finally, we still use DDC to analyze the suppressing cascading failure, for the HPC attack strategy under different loads of power grid, the experimental results were shown in Figure 3, the curves of HPCC represent the use of DDC, but the curve of HPC does not. Since the HPC attack strategy in any load condition can cause large-scale cascading failures in power grid, which bring great harm to the power grid. Therefore, suppressing cascading failure in this attack strategy is extremely important. As can be seen from the figure that DDC showed the effects of suppressing cascading failure are very good in different load conditions of power grid. In figure 3(a), (b) and (c), when $\lambda = 1.3$, power grid nodes which without using the DDC are almost all failure, and those using the DDC are almost all maintaining normal working state. Thus, DDC strategy can effectively suppress the cascading failures caused by HPC.

The results of simulation experiment show that when the power grid was attacked by attack strategies of HL, LL and HPC, DDC can always better suppress cascading failures. But in the implementation of DDC strategy, we find that the proportion of capacity allocation can be controlled by adjusting the strength of

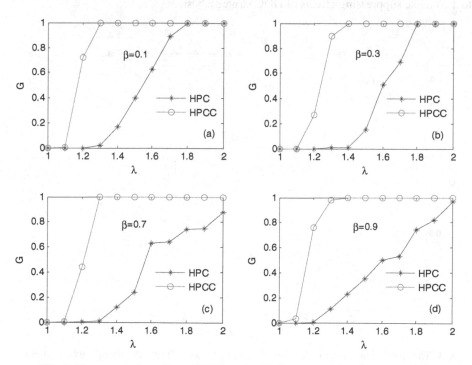

Fig. 3. The suppressing effect of DDC under HPC attack strategy, here $p = 0.9$

parameters p, and different values of p have different effects on suppressing cascading failure. Then we will analyze and compare suppressing effects for different loads of power grid in different value of p, here DDC takes three kinds of attack strategies that HL, LL and HPC respectively. The experimental results show that different values of p have different effects. As figure 4 shows, when $\beta = 1.5$, the experimental results of DDC under different values of p for HL attack strategy, we can find that the suppressing strategy of DDC is better when p=0.8 and p=0.9, but worst when p=1. When the value of p is small, the overloaded nodes only obtain few additional capacity, which may not be able to make the whole system back to normal state, that is to say it is more likely to occur cascading failures, therefore, increasing the value of p can reduce the possibility of cascading failure. With the value of p increases, although the overloaded node can receive a lot of additional capacity, the capacity of its neighbor nodes will reduce quickly, which leads to cascading failure. When p=1, the neighbor node puts all redundant capacity to the failure nodes, leading to its very prone to failure, which is why when p=1 the DDC has the worst effect. From the above, we can conclude that it is not the larger to p, the better to the suppressing effect, and we can find an appropriate value of p in the range of 0 to 1 to make suppressing effects of DDC strategy best.

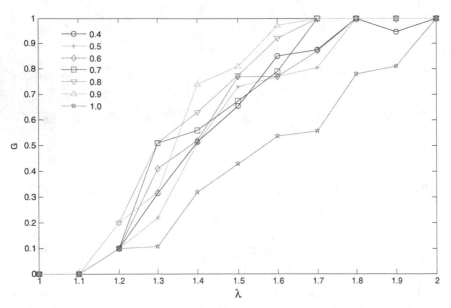

Fig. 4. The suppressing effect of DDC in different p under HL attack strategy, when $\beta = 1.5$

4   Conclusion

This paper mainly studied the suppressing effect of DDC strategy in power grid cascading failures. By adjusting the capacity allocation, protecting those overload nodes without changing the total capacity in the power grid, and carried out a large number of simulation experiments which take the US power grid for example. From the experimental results we can find that DDC is a potent strategy on suppressing cascading failure, the suppressing effects are reflected very well in the power grid both in different loads and different attacking strategies. It was also found that by adjusting the intensity of capacity allocation can help achieve the best effect of suppressing cascading failure. The findings above will help us to protect the power grid preferably, and avoid large scale failure.

Acknowledgment. This paper is sponsored by the National Natural Science Foundation of China (NSFC, Grant U1204703, U1304614), the Key Scientific and Technological Project of Henan Province (122102310004), the Innovation Scientists and Technicians Troop Construction Projects of Zhengzhou City (10LJRC190, 131PCXTD597) , the Key Scientific and Technological Project of The Education Department of Henan Province (13A413355,13A790352, ITE12001) and the project of Zhengzhou Normal University(2012070).

References

1. U.S.-CANADA POWER SYSTEM OUTAGE TASK FORCE, Final Report on the August 14th Blackout in the United States and Canada: Causes and Recommendations Canada (April 2004)
2. Wang, W.X., Chen, G.R.: Universal Robustness Characteristic of Weighted Networks Against Cascading Failure. Physical Review E 77, 26101, 1–5 (2008)
3. Yang, R., Wang, W.X., et al.: Optimal Weighting Scheme for Suppressing Cas-cades and Traffic Congestion in Complex Networks. Physical Review E 79, 26112, 1–6 (2009)
4. Adilson, E.M.: Cascade Control and Defense in Complex Networks. Physical Review Letters 93(9), 98701, 1–4 (2004)
5. Wang, J.W.: Robustness of Complex Networks with the Local Protection Strategy Against Cascading Failures. Safety Science 53, 219–225 (2013)
6. Wang, J.W.: Modeling Cascading Failures in Complex Networks Based on Radiate Circle. Physical A 391, 4004–4011 (2012)
7. Hu, K., Hu, T., Tang, Y.: Model for Cascading Failures with Adaptive Defense in Complex Networks. China. Phys. B 19, 80206, 1–7 (2010)
8. Hu, K., Hu, T., Tang, Y.: Cascade Defense via Control of The Fluxes in Complex Networks. Journal of Statistical Physics 141, 555–565 (2010)
9. Asztalos, A., Sreenivasan, S., Szymanski, B.K., Korniss, G.: Distributed Flow Optimization and Cascading Effects in Weighted Complex Networks. European Physical Journal B 85, 288, 1–10 (2012)

10. Wang, J.W., Rong, L.L.: Edge-based-attack Induced Cascading Failures on Scale-free Networks. Physica A 388, 1731–1737 (2009)
11. Huang, X.Q., Gao, J.X.: Robustness of Interdependent Networks Under Tar-geted Attack. Physical Review E 83, 65101, 1–4 (2011)
12. Jörg, L., Jakob, B.: Stochastic Load-redistribution Model for Cascading Failure Propagation. Physical Review E 81, 31129, 1-5 (2010)

Pinning Control of Asymmetrically Coupled Complex Dynamical Network with Heterogeneous Delays[*]

Fengli Ren and Hongyong Zhao

Department of Mathematics,
Nanjing University of Aeronautics and Astronautics,
Nanjing 210016, China
flren@nuaa.edu.cn, hongyongz@126.com

Abstract. In this paper, pinning control of a class of asymmetrically coupled complex dynamical network with heterogeneous delays is studied. Two kinds of controllers, adaptive controllers and linear feedback controllers, are presented and the Jordan canonical transformation method is used instead of the matrix diagonalization method. Therefore, it is not necessary for some relevant matrices to be diagonalizable. Moreover, a simply approximate formula is provided to estimate how many and which nodes should a network with fixed network structure and coupling strength be pinned to reach synchronization. Here, the inner-coupling matrix is not necessarily symmetric. One example is given to show the effectiveness of the proposed synchronization criteria.

Keywords: Complex dynamical networks, Heterogeneous delays, Pinning control.

1 Introduction

A complex network is a large set of interconnected nodes, in which a node is a fundamental unit with specific activity. The nature of complex networks is their inherent complexity which lead to tremendous difficulties in understanding various aspects of such complex networks, including the structural complexity, network evolution, node diversity, dynamical complexity and connection diversity, etc. Over the past one decade, analysis and control of complex behaviors in complex dynamical networks have received great attention of researchers in various fields.

Synchronization is one of the most important controlling activity to excite the collective behavior of complex dynamical networks [1]-[4]. Moreover, the synchronization of all dynamical nodes of a coupled complex dynamical network can well explain many natural phenomena observed. Recently, the controlled synchronization of complex dynamical networks is believed to be a rather significant topic in both theoretical research and practical applications. It is well known that there are usually a large number of nodes in a complex network in our real life. Generally, it is difficult to

---

[*] This work was supported by the National Natural Science Foundation of China under Grant No. 61104031.

D.-S. Huang et al. (Eds.): ICIC 2014, LNAI 8589, pp. 795–807, 2014.

control such complex networks by adding the controllers to all nodes. Therefore, there are attempts to control the dynamics of a complex network by pinning part of nodes. In [5], Wang and Chen further investigated the stabilization problem for a scale-free dynamical network via local feedback pinning on a small fraction of the network nodes and showed that the most effective arrangement of placing the controllers on the network nodes was to choose those "big" nodes with highest degrees. Based on master stability function, some stability criteria are derived for complex networks with asymmetric and heterogeneous connections by pinning a small fraction of nodes in [6]. Liu, Chen and Yuan proposed some effective criteria to pin a weighted complex dynamical network with time delay to a homogeneous stationary state [7]. Zhou, Lu and Lü investigated the pinning adaptive synchronization for a complex network and provide a approximate formula for estimating the detailed number of pinning nodes [8]. Some stability criteria were derived for complex dynamical networks with heterogeneous delays in both continuous- and discrete-time domains by pinning part of nodes in [9]. However, there are no coupling delays in complex network models in the most of the papers mentioned above. The references [7] and [9] required the coupling-configuration matrix to be symmetric despite time delays being considered.

Time delays commonly exist in complex network due to long-distance communication and traffic congestions, etc. The current work often focus on homogeneous time-delay (i.e., all the delays are same) in networks; obviously, heterogeneous (i.e., unequal and non-commensurate) time-delay complex networks are of practical importance but have some difficulties technically [9][10]. The coupling between nodes of many real-world networks is asymmetrical, such as the food web, metabolic network, World-Wide-Web, epidemic networks, document citation networks, etc [11]. Motivated by the above discussions, this paper will study the pinning control for a complex network with asymmetric coupling and heterogeneous delays. Several synchronization criteria for the complex network will be derived by using Lyapunov stability theory. We will provide a simply approximate formula for telling how many and which nodes should be pinned.

The rest of this paper is organized as follows. In Section 2, model description and mathematic preliminaries are given. Several pinning synchronization criteria are derived based on Lyapunov stability theory for the complex network in Section 3. In Section 4, an example is given to show the effectiveness of the proposed results. Finally, conclusions are drawn in Section 5.

Notations: $A \in M_n(\mathbb{C})(M_n(\mathbb{R}))$ denotes that A is an $n \times n$ complex (real) matrix; we use A^* to denote the transpose and complex conjugate of A. $\lambda_{\max}(A)$ and $\lambda_{\min}(A)$ are the maximum and minimum eigenvalues of A, respectively, if the spectrum of A is included in $\mathbb{R}$. I_n denotes the $n \times n$ identity matrix.

2   Model Description and Preliminaries

This section proposes a complex dynamical network model with asymmetrical coupling and heterogeneous delays and gives several necessary mathematic preliminaries.

2.1   A Generally Complex Dynamical Network Model

Consider a heterogeneous time-delay complex dynamical network consisting of N diffusively coupled identical nodes, where each node is an n-dimensional dynamical oscillator which can be stable, periodic, and even chaotic. The state equations of the whole network are described by the following differential equations

$$\dot{x}_i = f(x_i) + \sum_{j=1, j \neq i}^{N} c_{ij} A \overline{(x_j(t-\tau) - x_i(t-\tau))}, i = 1, 2, ..., N \tag{1}$$

where $x_i = [x_{i1}, x_{i2}, ..., x_{in}]^T \in \mathbb{R}^n$ is the ith node; $f(\cdot) \in \mathbb{R}^n$ is assumed sufficiently smooth nonlinear vector fields; node dynamics is $\dot{x} = f(x)$, $\overline{x_j(t-\tau)} = [\ x_{j1}(t-\tau_1), x_{j2}(t-\tau_2), ..., x_{jn}(t-\tau_n)]^T$, $\tau_i \geq 0$ $(i = 1, 2, ..., N)$ are the time delays. Here, all the nodes have the same time-delay vectors; however, different delay entries of node i may have different values. $A \in M_n(\mathbb{R})$ is the inner-coupling matrix and $C = (c_{ij})_{N \times N} \in M_n(\mathbb{R})$ is the coupling configuration matrix. If there is a link from node i to node $j (j \neq i)$, then $c_{ij} > 0$; otherwise, $c_{ij} = 0$. Assume that C is a diffusive matrix satisfying

$$c_{ii} = - \sum_{j=1, j \neq i}^{N} c_{ij} \tag{2}$$

In this model, we suppose that network (1) is strong connected in the sense of having no isolated clusters, which means that the coupling matrix C is irreducible. Here, C does not need to be symmetric.

2.2   Mathematic Preliminaries

To begin with, the definition of the ``matrix measure"[12] is introduced below.

Definition 1. Suppose that $\|\cdot\|_i$ is an induced matrix norm in $M_n(\mathbb{C})$, then the corresponding matrix measure of a given matrix A is a function $\mu_i : M_n(\mathbb{C}) \to \mathbb{R}$, defined by

$$\mu_i(A) = \lim_{\delta \to 0^+} \frac{\| I_n + \delta A \|_i - 1}{\delta} \tag{3}$$

in which I_n is the identity matrix. In general, the matrix measure is difficult to calculate. But for some special norms, it is easy to find the associate matrix measures.

For instance, $\mu_2(A)$ is half of the maximum eigenvalue of $A^* + A$. Here $\| A \|_2$ is the spectral norm of A, and A^* is the transpose and complex conjugate of A [10].

Lemma 1.[13] Let $G \in M_N(\mathbb{C})$ be a given complex matrix. There is a nonsingular matrix $\Phi \in M_N(\mathbb{C})$ such that

$$G = \Phi \begin{bmatrix} J_{n_1}(\lambda_1) & & & 0 \\ & J_{n_2}(\lambda_2) & & \\ & & \ddots & \\ 0 & & & J_{n_k}(\lambda_k) \end{bmatrix} \Phi^{-1} = \Phi J \Phi^{-1} \tag{4}$$

and $n_1 + n_2 + \cdots + n_k = N$. The Jordan matrix J of G is unique up to permutation of the diagonal Jordan blocks. The eigenvalues $\lambda_i, i = 1, 2, \ldots, k$, are not necessarily distinct. If G is a real matrix with only real eigenvalues, then the similarity matrix Φ can be taken to be real. $J_{n_i}(\lambda_i)$ is the Jordan blocks.

$$J_{n_i}(\lambda_i) = \begin{bmatrix} \lambda_i & 1 & & 0 \\ & \lambda_i & \ddots & \\ & & \ddots & 1 \\ 0 & & & \lambda_i \end{bmatrix}, i = 1, 2, \ldots, k. \tag{5}$$

Lemma 2.[14] Let $\tau \geq 0$ and $\alpha > \beta > 0$ be constants, $U(t)$ be a non-negative continuous function for $[-\tau, \infty)$ which satisfies

$$D^+ U(t) \leq -\alpha U(t) + \beta \sup_{-\tau \leq \theta \leq 0} U(t + \theta), \ t \geq t_0 \tag{6}$$

then

$$U(t) \leq \sup_{-\tau \leq \theta \leq 0} U(t_0 + \theta) e^{-\gamma(t - t_0)}, \quad t \geq t_0$$

where $\gamma > 0$ is a positive root of the equation $\gamma = \alpha - \beta e^{\gamma \tau}$.

Let $x = s(t; t_0, x_0) \in \mathbb{R}^n$ with $x_0 \in \mathbb{R}^n$, denoted as $s(t)$, be a solution of the node system $\dot{x} = f(x)$. The control objective is to make the states of the network (1) synchronize to a manifold defined by: $x_1(t) = x_2(t) = \cdots = x_N(t) = s(t)$, as $t \to \infty$ From (2), one has

$$\sum_{j=1}^{N} c_{ij} A \overline{s(t-\tau)} = 0$$

where $\overline{s(t-\tau)} = [s_1(t-\tau_1), s_2(t-\tau_2), \ldots, s_n(t-\tau_n)]^T$ the $\mathbf{s}(t) = [s(t), s(t), \ldots, s(t)]^T \in \mathbb{R}^{n \times N}$, is a synchronous solution of the complex dynamical network (1). Note that $s(t)$ can be an equilibrium point, a periodic orbit, an aperiodic orbit, even a chaotic orbit in the phase space.

3   Pinning Control of Complex Dynamical Networks

3.1   Pinning Control with Adaptive Controllers

To realize the goal (6), we firstly apply the pinning adaptive control strategy on a small fraction $\delta(0 < \delta \ll 1)$ of the nodes in network (1). Suppose that nodes $i_1, i_2, \ldots, i_l$ are selected and pinned with theadaptive controllers, and $i_{l+1}, i_{l+2}, \ldots, i_N$ are the unselected ones, where $l = [\delta N]$ stands for the smaller but nearest integer to the real number δN. Certainly, $l \geq 1$.Then the controlled network can be described as

$$\begin{cases} \dot{x}_{i_k} = f(x_{i_k}) + \sum_{j=1, j \neq i_k}^{N} c_{i_k j} A(\overline{x_j(t-\tau)} - \overline{x_{i_k}(t-\tau)}) + u_{i_k}, k = 1, 2, \ldots, l, \\ \dot{x}_{i_k} = f(x_{i_k}) + \sum_{j=1, j \neq i_k}^{N} c_{i_k j} A(\overline{x_j(t-\tau)} - \overline{x_{i_k}(t-\tau)}), k = l+1, l+2, \ldots, N. \end{cases}$$

(7)

Define error vectors as

$$e_i(t) = x_i(t) - s(t), i = 1, 2, \ldots, N. \tag{8}$$

The adaptive controllers are described by

$$\begin{cases} u_{i_k} = -p_{i_k} e_{i_k}, \dot{p}_{i_k} = \theta_{i_k} e_{i_k}^T R e_{i_k}, k = 1, 2, \ldots, l \\ u_{i_k} = 0, k = l+1, l+2, \ldots, N. \end{cases}$$

(9)

where $\theta_{i_k} (k = 1, 2, \ldots, l)$ are any positive constants, R is a positive definite matrix to be determined. Letting $p_i = 0$ if $i \in \{i_{l+1}, i_{l+2}, \ldots, i_N\}$, $p_i = 1$, if $i \in \{i_1, i_2, \ldots, i_l\}$ and combing (7) and (9), one obtains

$$\dot{x}_i = f(x_i) + \sum_{j=1, j\neq i}^{N} c_{ij} A(\overline{x_j(t-\tau) - x_i(t-\tau)}) - p_i e_i, i = 1, 2, \ldots, N. \quad (10)$$

Substituting (8) into (10), one has

$$\dot{e}_i = f(x_i) - f(s(t)) + \sum_{j=1, j\neq i}^{N} c_{ij} A(\overline{e_j(t-\tau) - e_i(t-\tau)}) - p_i e_i, \quad (11)$$

where $\overline{e_j(t-\tau)} = [e_{j1}(t-\tau_1), e_{j2}(t-\tau_2), \ldots, e_{jn}(t-\tau_n)]^T \in \mathbb{R}^n$ Its linearized

system reads as $\dot{e}_i = Df(s(t))e_i + \sum_{j=1, j\neq i}^{N} c_{ij} A(\overline{e_j(t-\tau) - e_i(t-\tau)}) - p_i e_i. \quad (12)$

Theorem 1. If there exist a natural number $1 \leq l < N$, a positive definite matrix R, a positive definite diagonal matrix Q and a constant a satisfying

$$\mu_2[RDf(s(t)) - c_{i_k i_k} R^T Q^{-1} R - (c_{i_k} + c_{i_k i_k}) A^T QA / 2] \leq a < 0, \ l+1 \leq k \leq N, \quad (13)$$

then the controlled network (7) is asymptotically stable about (6) under the pinning adaptive controllers

$$\begin{cases} u_{i_k} = -p_{i_k} e_{i_k}, \dot{p}_{i_k} = \theta_{i_k} e_{i_k}^T(t) Re_{i_k}(t), 1 \leq k \leq l, \\ u_{i_k} = 0, l+1 \leq k \leq N, \end{cases} \quad (14)$$

where $\theta_{i_k} (k = 1, 2, \ldots, l)$ are any positive constants; $Q = \text{diag}\{q_1, q_2, \ldots, q_n\} > 0$

$c_i = -\sum_{j=1, j\neq i}^{N} c_{ji} < 0 (i = 1, 2, \ldots, N)$.

Proof. Construct a Lyapunov-Krasovskii functional candidate as follows:

$$V(t) = \sum_{i=1}^{N} e_i^T(t) Re_i(t) - \sum_{i=1}^{N} (c_i + c_{ii}) \left(\sum_{s=1}^{n} \int_{t-\tau_s} q_s \eta_{is}^2(\mu) d\mu \right) + \sum_{k=1}^{l} \frac{(p_{i_k} - p)^2}{\theta_{i_k}},$$

where $\eta_i(t) = [\eta_{i1}(t), \eta_{i2}(t), \ldots, \eta_{in}(t)]^T = Ae_i(t)$, p is positive constant s atisfying

$$\mu_2[RDf(s(t)) - c_{i_k i_k} R^T Q^{-1} R - (c_{i_k} + c_{i_k i_k}) A^T QA / 2 - pR] \leq a < 0, 1 \leq k \leq l. \quad (15)$$

The time derivative of $V(t)$ along the solution of error system (12) is

$$\dot{V}(t) = \sum_{i=1}^{N}[\dot{e}_i^T(t)Re_i(t) + e_i^T(t)R\dot{e}_i(t)]$$

$$-\sum_{i=1}^{N}(c_i + c_{ii})\left(\sum_{s=1}^{n}q_s(\eta_{is}^2(t) - \eta_{is}^2(t-\tau_s))\right) + 2\sum_{k=1}^{l}\frac{(p_{i_k} - p)\dot{p}_{i_k}}{\theta_{i_k}}$$

$$= \sum_{i=1}^{N}e_i^T(t)[Df^T(s(t))R + RDf(s(t))]e_i(t) - 2\sum_{k=1}^{l}pe_{i_k}^T(t)Re_{i_k}(t)$$

$$+2\sum_{i=1}^{N}\sum_{j=1,j\neq i}^{N}c_{ij}e_i^T(t)R(\overline{\eta_j(t-\tau)} - \overline{\eta_i(t-\tau)}) - \overline{\eta_i(t-\tau)}^T Q\overline{\eta_i(t-\tau)})$$

$$-\sum_{i=1}^{N}(c_i + c_{ii})e_i^T(t)A^T QAe_i(t)$$

$$= \sum_{i=1}^{N}e_i^T(t)[Df^T(s(t))R + RDf(s(t))]e_i(t) + 2\sum_{i=1}^{N}\sum_{j=1,j\neq i}^{N}c_{ij}\xi_i^T(t)\overline{\eta_j(t-\tau)}$$

$$-2\sum_{i=1}^{N}\sum_{j=1,j\neq i}^{N}c_{ij}\xi_i^T(t)\overline{\eta_i(t-\tau)} - \sum_{i=1}^{N}(c_i + c_{ii})e_i^T(t)A^T QAe_i(t)$$

$$+\sum_{i=1}^{N}\left(-\sum_{j=1,j\neq i}^{N}c_{ji} - \sum_{j=1,j\neq i}^{N}c_{ij}\right)\overline{\eta_i(t-\tau)}^T Q\overline{\eta_i(t-\tau)}$$

$$-2\sum_{k=1}^{l}pe_{i_k}^T(t)Re_{i_k}(t) - 2\sum_{i=1}^{N}c_{ii}\xi_i^T(t)Q^{-1}\xi_i(t) - 2\sum_{i=1}^{N}\sum_{j=1,j\neq i}^{N}c_{ij}\xi_i^T(t)Q^{-1}\xi_i(t),$$

where $\overline{\eta_j(t-\tau)} = \overline{Ae_j(t-\tau)}$, $\xi_i(t) = Re_i(t)$. On the other hand, we have

$$\sum_{i=1}^{N}\sum_{j=1,j\neq i}^{N}c_{ji}\overline{\eta_i(t-\tau)}^T Q\overline{\eta_i(t-\tau)} = \sum_{i=1}^{N}\sum_{j=1,j\neq i}^{N}c_{ij}\overline{\eta_j(t-\tau)}^T Q\overline{\eta_j(t-\tau)}. \quad (16)$$

and

$$2\sum_{i=1}^{N}\sum_{j=1,j\neq i}^{N}c_{ij}\xi_i^T(t)\overline{\eta_j(t-\tau)} - \sum_{i=1}^{N}\sum_{j=1,j\neq i}^{N}c_{ij}\overline{\eta_j(t-\tau)}^T Q\overline{\eta_j(t-\tau)}$$

$$-\sum_{i=1}^{N}\sum_{j=1,j\neq i}^{N}c_{ij}\xi_i^T(t)Q^{-1}\xi_i(t) \leq 0, -2\sum_{i=1}^{N}\sum_{j=1,j\neq i}^{N}c_{ij}\xi_i^T(t)\overline{\eta_i(t-\tau)}$$

$$-\sum_{i=1}^{N}\sum_{j=1,j\neq i}^{N}c_{ij}\overline{\eta_i(t-\tau)}^T Q\overline{\eta_i(t-\tau)} - \sum_{i=1}^{N}\sum_{j=1,j\neq i}^{N}c_{ij}\xi_i^T(t)Q^{-1}\xi_i(t) \leq 0,$$

then one gets

$$\dot{V}(t) \le \sum_{i=1}^{N} e_i^T(t)[Df^T(s(t))R + RDf(s(t)) - 2c_{ii}R^TQ^{-1}R$$

$$-(c_i + c_{ii})A^TQA]e_i(t) - 2\sum_{k=1}^{l} pe_{i_k}^T(t)Re_{i_k}(t)$$

$$= \sum_{k=1}^{l} e_{i_k}^T(t)[Df^T(s(t))R + RDf(s(t)) - 2c_{i_k i_k}R^TQ^{-1}R$$

$$-(c_{i_k} + c_{i_k i_k})A^TQA - 2pR]e_{i_k}(t) + \sum_{k=l+1}^{N} e_{i_k}^T(t)[Df^T(s(t))R + RDf(s(t))$$

$$-2c_{i_k i_k}R^TQ^{-1}R - (c_{i_k} + c_{i_k i_k})A^TQA]e_{i_k}(t)$$

$$= \sum_{k=1}^{l} e_{i_k}^T(t)[(\Psi - pR)^T + (\Psi - pR)]e_{i_k}(t) + \sum_{k=l+1}^{N} e_{i_k}^T(t)(\Psi^T + \Psi)e_{i_k}(t)$$

where $\Psi = RDf(s(t)) - c_{i_k i_k}R^TQ^{-1}R - (c_{i_k} + c_{i_k i_k})A^TQA/2$. According to (13) and (15), it follows that

$$\dot{V}(t) \le \sum_{k=1}^{l} e_{i_k}^T(t)[(\Psi - pR)^T + (\Psi - pR)]e_{i_k}(t) + \sum_{k=l+1}^{N} e_{i_k}^T(t)(\Psi^T + \Psi)e_{i_k}(t)$$

$$\le 2\sum_{k=1}^{l} e_{i_k}^T(t)\mu_2(\Psi - pR)e_{i_k}(t) + 2\sum_{k=l+1}^{N} e_{i_k}^T(t)\mu_2(\Psi)e_{i_k}(t)$$

$$\le 2a\sum_{i=1}^{N} e_i^T(t)e_i(t) = 2a\sum_{i=1}^{N} \|e_i\|_2^2.$$

Since $a<0$, the error vector $e_i(i=1,2,\dots,N) \to 0$ as $t \to +\infty$. That is, the controlled network (7) is asymptotically stable about (6) under the pinning adaptive controllers.

Remark 1. Suppose that $C = c\bar{C}$, where $\bar{C}$ is a diffusive coupling matrix with $\bar{c}_{ij} = 0$ or $1(j \neq i)$. Then the condition (13) can be changed as

$$\mu_2[RDf(s(t)) + cd_{i_k}R^TQ^{-1}R - c(\bar{c}_{i_k} - d_{i_k})A^TQA/2] \le a < 0, l+1 \le k \le N$$
(17)

where $d_{i_k} = \sum_{j=1, j \neq i_k}^{N} \bar{c}_{i_k j}$ and $\bar{c}_i = -\sum_{j=1, j \neq i}^{N} c_{ji} < 0(i=1,2,\dots,N)$.

It should be noted that the coupling configuration matrix C does not need to be symmetric matrix in theorem 1. However, for some real networks with undirected graph topology structure, C is a symmetric matrix, and thus

$$c_{ii} = -\sum_{j=1, j\neq i}^{N} c_{ij} = -\sum_{j=1, j\neq i}^{N} c_{ji} = c_i. \tag{18}$$

From theorem 1, one can derive the condition for the system with C is a symmetric matrix. Here, we omitted it.

Remark 2. In [8], the authors investigated the problem of pinning adaptive synchronization for a general complex dynamical network. However, the coupling delays did not be considered in their paper.

Remark 3. In [7] and [9], the coupling configuration matrix is required to be symmetric and no coupling delays in [6].

3.2   Pinning Control with Linear Feedback Controllers

Let

$$\begin{cases} u_{i_k} = -p_{i_k} A \overline{e_{i_k}(t-\tau)}, k = 1, 2, \ldots, l \\ u_{i_k} = 0, k = l+1, l+2, \ldots, N. \end{cases} \tag{19}$$

then the controlled network (7) can be written as

$$\dot{e}_i = Df(s(t))e_i(t) + \sum_{j=1}^{N} c_{ij} A \overline{e_j(t-\tau)} - p_i A \overline{e_i(t-\tau)}, i = 1, 2, \ldots, N. \tag{20}$$

where $p_i = 0$ if $i \in \{i_{l+1}, i_{l+2}, \ldots, i_N\}$ and $p_i = 1$ if $i \in \{i_1, i_2, \ldots, i_l\}$. By using Kronecker product, one can rewritten (20) as

$$\dot{e}(t) = (I_N \otimes Df(s(t))e(t) + ((C-P) \otimes A)\overline{e(t-\tau)} \tag{21}$$

where $e(t) = [e_1^T(t), e_2^T(t), \ldots, e_N^T(t)]^T \in \mathbb{R}^{nN}, \overline{e(t-\tau)} = [\overline{e_1(t-\tau)}^T, \overline{e_2(t-\tau)}^T,$ $\ldots, \overline{e_N(t-\tau)}^T]^T \in \mathbb{R}^{nN}, P = \text{diag}\{p_1, p_2, \ldots, p_n\}$. From Lemma 1, there is a nonsingular matrix $\Phi \in M_N(\mathbb{C})$, such that $C - P = \Phi J \Phi^{-1}$. Let $e(t) = (\Phi \otimes I_n)\eta(t)$, then one has

$$\dot{\eta}(t) = (I_N \otimes Df(s(t))\eta(t) + (J \otimes A)\overline{\eta(t-\tau)}. \tag{22}$$

Considering the special form of J given in (4) and (5), system (22) can be easily transformed to the following equations

$$\dot{\zeta}(t) = (I_{n_i} \otimes Df(s(t)))\zeta(t) + (J_{n_i}(\lambda_i) \otimes A)\overline{\zeta(t-\tau)}, \tag{23}$$

where $i = 1, 2, \ldots, k (1 \le k \le N)$ and $n_1 + n_2 + \ldots + n_k = N$. Suppose $n_i = m$, then the i th m-node system in (23) can be described by the following m equations with $\zeta = (\zeta_1^T, \zeta_2^T, \ldots, \zeta_m^T)^T$,

$$\dot{\zeta}_1(t) = Df(s(t))\zeta_1(t) + \lambda_i A\overline{\zeta_1(t-\tau)} + A\overline{\zeta_2(t-\tau)} \tag{24}$$

$$\vdots$$

$$\dot{\zeta}_{m-1}(t) = Df(s(t))\zeta_{m-1}(t) + \lambda_i A\overline{\zeta_{m-1}(t-\tau)} + A\overline{\zeta_m(t-\tau)}, \tag{25}$$

$$\dot{\zeta}_m(t) = Df(s(t))\zeta_m(t) + \lambda_i A\overline{\zeta_m(t-\tau)}, \tag{26}$$

It is obviously that, if the zero solution of system (26) is exponentially stable, then the zero solution of the i thone-node system in

$$\dot{\xi}(t) = Df(s(t))\xi(t) + \lambda_i A\overline{\xi(t-\tau)}, i = 1, 2, \ldots, k, \tag{27}$$

is exponentially stable. On the other hand, if the zero solution of the i th equation in (27) is exponentially stable, then the zero solution of (26) is exponentially stable. It is easy to conclude that the zero solutions of the $m-1$ equations (24)-(25) are exponentially stable. That is, the exponential stability of the zero solution of the i th n_i-node system (23) is equivalent to that of the corresponding i th one-node system in (27). In the following, we only need to derive the criteria for the exponential stability of zero solutions of the k one-node systems in (27).

Theorem 2. If there exist k positive definite matrices $P_i \in M_n(\mathbb{C})$, $Q_i \in M_n(\mathbb{C})$ and k positive constants α_i such that

$$\frac{\tau^2 \lambda_{max}(P_i)}{\lambda_{min}(P_i)} \lambda_{max}(Q_i^{-1})(\lambda_{max}(D^T f(s(t))Df(s(t))) + |\lambda_i|^2 \lambda_{max}(A^T A))$$

$$+ \frac{|\lambda_i|^2 \lambda_{max}(P_i A Q_i(P_i A)^*)}{2} < \alpha_i, \text{ and} \tag{28}$$

$$\mu_2(P_i(Df(s(t)) + \lambda_i A)) \le -\alpha_i, \tag{29}$$

then, the zero solutions of the k individual systems in (27) are exponentially stable. That is to say the controlled network (7) is exponentially stable about (6) under the pinning controllers, where $\tau = \max_{1 \leq i \leq n} \tau_i, i = 1, 2, \ldots, k (1 \leq k \leq N)$.

Proof. Consider the linear time-varying system

$$\dot{\xi}(t) = Df(s(t))\xi(t) + \lambda_i A\overline{\xi(t-\tau)} \tag{30}$$

where $\xi(t) = [\xi_1(t), \xi_2(t), \ldots, \xi_n(t)]^T$. Define a Lyapunov function as follows

$$V_i(t) = \xi(t)^* P_i \xi(t).$$

Rewrite (30) as

$$\dot{\xi}(t) = [Df(s(t)) + \lambda_i A]\xi(t) - \lambda_i A \int_{t-\tau}^t \dot{\xi}(\theta)d\theta, \tag{31}$$

where $\overline{\int_{t-\tau}^t \dot{\xi}(\theta)d\theta} = [\int_{t-\tau_1}^t \dot{\xi}_1(\theta)d\theta, \int_{t-\tau_2}^t \dot{\xi}_2(\theta)d\theta, \ldots, \int_{t-\tau_n}^t \dot{\xi}_n(\theta)d\theta]^T$. Thus, the derivative of $V(t)$ along the trajectory of (31) is

$$\dot{V}_i(t) = \dot{\xi}^*(t)P_i\xi(t) + \xi(t)^* P_i\dot{\xi}(t)$$

$$= \xi(t)^*[(Df(s(t)) + \lambda_i A)^* P_i + P_i(Df(s(t)) + \lambda_i A)]\xi(t)$$

$$- (\lambda_i A \int_{t-\tau}^t \dot{\xi}(\theta)d\theta)^* P_i\xi(t) - \xi(t)^* P_i\lambda_i A \int_{t-\tau}^t \dot{\xi}(\theta)d\theta,$$

since

$$-(\lambda_i A \overline{\int_{t-\tau}^t \dot{\xi}(\theta)d\theta})^* P_i\xi(t) - \xi(t)^* P_i\lambda_i A \overline{\int_{t-\tau}^t \dot{\xi}(\theta)d\theta}$$

$$\leq \xi(t)^* \lambda_i P_i A Q_i (P_i\lambda_i A)^* \xi(t) + \left(\overline{\int_{t-\tau}^t \dot{\xi}(\theta)d\theta}\right)^* Q_i^{-1} \overline{\int_{t-\tau}^t \dot{\xi}(\theta)d\theta}, \tag{32}$$

then

$$\dot{V}_i(t) \leq \xi(t)^*[(Df(s(t)) + \lambda_i A)^* P_i + P_i(Df(s(t)) + \lambda_i A)]\xi(t)$$

$$+ \xi(t)^* \lambda_i P_i A Q_i (P_i\lambda_i A)^* \xi(t) + (\overline{\int_{t-\tau}^t \dot{\xi}(\theta)d\theta})^* Q_i^{-1} \overline{\int_{t-\tau}^t \dot{\xi}(\theta)d\theta}$$

$$\leq \xi(t)^*[(Df(s(t))+\lambda_i A)^* P_i + P_i(Df(s(t))+\lambda_i A)]\xi(t)$$

$$+\xi(t)^* \mid \lambda_i \mid^2 P_i A Q_i (P_i A)^* \xi(t) + \lambda_{\max}(Q_i^{-1})\tau^2 \sup_{t-\tau \leq \theta \leq t} ((\dot{\xi}(\theta))^* \dot{\xi}(\theta))$$

$$\leq \xi(t)^*[(Df(s(t))+\lambda_i A)^* P_i + P_i(Df(s(t))+\lambda_i A)]\xi(t)$$

$$+\xi(t)^* \mid \lambda_i \mid^2 P_i A Q(P_i A)^* \xi(t) + \lambda_{\max}(Q_i^{-1})\tau^2 \sup_{t-\tau \leq \theta \leq t} ((Df(s(\theta))\xi(\theta)$$

$$+\lambda_i A \overline{\xi(\theta-\tau)})^* (Df(s(\theta))\xi(\theta)+\lambda_i A \overline{\xi(\theta-\tau)}))$$

$$\leq \xi(t)^*[(Df(s(t))+\lambda_i A)^* P_i + P_i(Df(s(t))+\lambda_i A)]\xi(t)$$

$$+\xi(t)^* \mid \lambda_i \mid^2 P_i A Q_i (P_i A)^* \xi(t) + \lambda_{\max}(Q_i^{-1})\tau^2 (2\lambda_{\max}(D^T f(s(t))Df(s(t)))$$

$$+2 \mid \lambda_i \mid^2 \lambda_{\max}(A^T A)) \sup_{t-2\tau \leq \theta \leq t} (\xi^*(\theta)\xi(\theta))$$

$$\leq \frac{1}{\lambda_{\max}(P_i)}(-2\alpha_i + \mid \lambda_i \mid^2 \lambda_{\max}(P_i A Q_i A^* P_i^*))V(t) + \frac{2\tau^2}{\lambda_{\min}(P_i)}\lambda_{\max}(Q_i^{-1})$$

$$\times(\lambda_{\max}(D^T f(s(t))Df(s(t)))+ \mid \lambda_i \mid^2 \lambda_{\max}(A^T A)) \sup_{t-2\tau \leq \theta \leq t} (V(\theta)).$$

From (28) and Lemma 2, $V_i(t)(i=1,2,\ldots,k)$ are exponentially approach to zero. Therefore, the zero solutions of the k individual systems in (27) are exponentially stable.

4   Illustrative Example

In this section, we consider a complex dynamical network with heterogeneous delays composing of 50 is described by the following equations [9]:

$$\begin{pmatrix} \dot{x}_{i1} \\ \dot{x}_{i2} \\ \dot{x}_{i3} \end{pmatrix} = \begin{pmatrix} -x_{i1}+x_{i1}x_{i2} \\ -2x_{i2} \\ -3x_{i3} \end{pmatrix} + \sum_{j=1}^{N} c_{ij} A \begin{pmatrix} x_{j1}(t-\tau_1) \\ x_{j2}(t-\tau_2) \\ x_{j3}(t-\tau_3) \end{pmatrix} \tag{33}$$

The system without coupling term has one equilibrium point $[0,0,0]^T$. Take $s = [0\ 0,0]^T$, where $1 \leq i \leq 50$, $A = \text{diag}\{1,1,1\}$, $\tau_i = i(i=1,2,3)$, The coupling matrix C is a asymmetrical matrix with $c_{11}=-4.9$ and $c_{ii}=-0.3$ $(2 \leq i \leq 50)$. By taking $R = Q = I_n$ In Theorem 1, we only need to pin the first node of 50 nodes for realizing network synchronization. In the simulations, the parameters are taken as $\theta_1 = 0.01$ and $p_1(0)=1$.

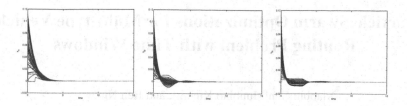

Fig. 1. The state response of the controlled complex network

5   Conclusions

In this paper, two synchronization criteria are derived for the heterogeneous time-delay complex network with asymmetric coupling. It is less conservative than the references. In [6][7] and [9], B or G-D are required to be diagonalizable, where G and B are coupling matrices and D is feedback gain matrix. In this paper, weshow that the diagonalization is not necessarily possible by using the Jordan canonical form. At the end of the paper, an example is given to show the effectiveness of the criteria.

References

1. Song, Q., Cao, J., Yu, W.: Second-order leader-following consensus of nonlinear multi-agents via pinning control. Syst. & Cont. Lett. 59, 553–562 (2010)
2. Yu, W., Chen, G., Lü, J.: On pinning synchronization of complex dynamical networks. Automatica 45(2), 429–435 (2009)
3. Wu, J., Jiao, L.: Synchronization in complex delayed dynamical networks with nonsymmetric coupling. Physica A 386, 513–530 (2007)
4. Wu, J., Jiao, L.: Synchronization in complex dynamical networks with nonsymmetric coupling. Physica D 237(19), 2487–2498 (2008)
5. Wang, X., Chen, G.: Pinning control of scale-free dynamical networks. Physica A 310(3-4), 521–531 (2002)
6. Xiang, L., Liu, Z., Chen, Z., Chen, F., Yuan, Z.: Pinning control of complex dynamical networks with general topology. Physica A 379, 298–306 (2007)
7. Liu, Z.X., Chen, Z.Q., Yuan, Z.Z.: Pinning control of weighted general complex dynamicalnetworks with time delay. Physica A 375, 345–354 (2007)
8. Zhou, J., Lu, J., Lü, J.: Pinning adaptive synchronization of a general complex dynamical network. Automatica 44, 996–1003 (2008)
9. Xiang, L., Chen, Z., Liu, Z., Chen, F., Yuan, Z.: Pinning control of complex dynamical networks with heterogeneous delays. Computers and Mathematics with Applications 56, 1423–1433 (2008)
10. Li, C.P., Sun, W.G., Kurths, J.: Synchronization of complex dynamical networks with time delays. Physica A 361, 24–34 (2006)
11. Williams, R.J., Martinez, N.D., Berlow, E.L., Dunne, J.A., Barabasi, A.-L.: Two degrees of separation in complex food webs. Proc. Natl. Acad. Sci. 99, 12913–12916 (2002)
12. Vidyasagar, F.: Nonlinear Systems Analysis. Prentice-Hall, Englewood Cliffs (1978)
13. Horn, R.A., Johnson, C.R.: Topics in Matrix Analysis. Post & Telecom Press, China (2005)
14. Gopalsamy, K.: Stability and Oscillations in Delay Differential Equations of Population Dynamics. Kluwer Academic, Dordrecht (1992)

Particle Swarm Optimizations for Multi-type Vehicle Routing Problem with Time Windows

Xiaobin Gan[1], Junbiao Kuang[1], and Ben Niu[1,2,3,*]

[1] College of Management, Shenzhen University, Shenzhen 518060, China
[2] Department of Industrial and System Engineering,
Hong Kong Polytechnic University, Hong Kong
[3] Hefei Institute of Intelligent Machine, Chinese Academy of Sciences, Hefei 230031, China
drniuben@gmail.com

Abstract. This paper presents a variant of vehicle routing problem with time windows (VRPTW) named multi-type vehicle routing problem with time windows (MT-VRPTW), which considers both multiple types of the vehicle and the uncertain number of vehicles of various types. As a consequence, the different combinations of multi-type vehicle will lead to diverse results, which should be evaluated by its own fitness function. In order to solve the proposed MT-VRPTW problem, six variants of particle swarm optimization (PSO) are used. The 2N dimensions encoding method is adopted to express the particle (N represents the number of demand point). In the simulation studies, the performances of the six PSO variants are compared and the obtained results are analyzed.

Keywords: Particle swarm optimization (PSO), Vehicle routing problem (VRP), Time windows.

1   Introduction

The vehicle routing problem (VRP) has been proposed and discussed for several decades. Considering different factors, researchers study this problem with various perspectives. In 1984, Solomon [1] designed a sequential insertion heuristics algorithm for vehicle routing and scheduling problems with time window constraints. In [2], Desrochers and Solomon introduced a branch-and-bound algorithm to get an excellent lower bound for the integer set partitioning formulation.

Compared with traditional optimization technique for the VRP, evolutionary computation has aroused increasing concerns in recent years. For instance, Prins [3] developed an effective evolutionary algorithm (EA) to tackle the VRP by spacing a smaller population of distinct solution. From another perspective, Ombuki and Hanshar [4] represented the VRP as a multiple objective problem and used Pareto ranking technique in genetic algorithm to obtain a set of equally valid solutions.

As one of effective swarm intelligence, particle swarm optimization (PSO) achieves a excellent effect when employed to solve some NP-hard problems such as

---

* Corresponding author.

D.-S. Huang et al. (Eds.): ICIC 2014, LNAI 8589, pp. 808–815, 2014.

traveling salesman problem (TSP) and knapsack problem. Similarly, many researchers implement PSO to solve the VRP. Chen and Yang [5] combined discrete PSO with simulated annealing (SA) algorithm to address capacitated vehicle routing problem. Considering multi-type vehicle and backhauls, Belmecheri et al. [6] introduced PSO with local search to solve vehicle routing problem with time windows (VRPTW).

Since the depot usually assigns a truck to meet a plurality of customer demands, single type vehicle cannot respond to the requirements flexibly. Therefore, we discuss a case of multi-type vehicle routing problem with time windows (MT-VRPTW) in this paper, which also considers a factor about the uncertain number of multi-type vehicles. Moreover, six PSO variants are employed to optimize this problem on account of the following aspects. First of all, we select global PSO (GPSO) with $w = 0.4$ [7] and GPSO with linearly decreasing w from 0.9 to 0.4 [7] as the fundamental variants altering the parameter. Secondly, three representative variants constructing neighborhood structure are expected to retain diversity of the individuals as well in this case, including unified particle swarm optimizer (UPSO) [8], fully informed particle swarm (FIPS) [9], and comprehensive learning particle swarm optimizer (CLPSO) [10]. Finally, multi-swarm cooperative particle swarm optimizer (MCPSO) [11, 12] utilizing the multiple swarms approach to search for global optimum is used.

The rest of this paper is organized as follows. Section 2 emphasizes to extend a mathematical model. Section 3 introduces various PSO variants. And Section 4 presents experiment study and the results. Conclusions are given in Section 5.

2   Description of MT-VRPTW

Based on the basic model for VRPTW [13], two extra key factors are involved. On one hand, we take multiple types of the vehicle into account. On the other hand, the number of vehicles of different types is uncertain. For simplifying the problem in the model, we only suppose and list two types of the vehicle. The number of vehicles of the first type is $K1$ and another type is $K2$. The extended mathematical formulation based on Kallehauge's basic VRPTW model [13] is presented as follows.

Define variable:

$$x_{ijk} = \begin{cases} 1 & \text{if vehicle k traverses route (i,j)} \\ 0 & \text{else} \end{cases} \tag{1}$$

Objective function

$$\text{Minimize } Z = \sum_{i=0}^{N} \sum_{j=0}^{N} \sum_{k=1}^{K} c_k * d_{ij} * x_{ijk} + pe * \sum_{i=1}^{N} \max(ET_i - t_i, 0) + pl * \sum_{i=1}^{N} \max(t_i - LT_i, 0) \tag{2}$$

where

$$t_j = \sum x_{ijk}(t_i + s_i + d_{ij} / v) \qquad \text{for } t_0 = 0, s_0 = 0 \tag{3}$$

Subject to

$$\sum_{j=1}^{N} x_{ijk} = \sum_{j=1}^{N} x_{jik} \leq 1 \qquad \text{for } i=0,\ 1 \leq k \leq K \tag{4}$$

$$\sum_{j=0}^{N} \sum_{k=1}^{K} x_{ijk} = 1 \qquad \text{for } 0 \leq i \leq N \tag{5}$$

$$\sum_{i=0}^{N} \sum_{k=1}^{K} x_{ijk} = 1 \qquad \text{for } 0 \leq j \leq N \tag{6}$$

$$\sum_{i=0}^{N} \sum_{j=0}^{N} x_{ijk} * q_i \leq Q_k \qquad \text{for } k=1,2,\ldots,K1 \text{ and } i \neq j \tag{7}$$

$$\sum_{i=0}^{N} \sum_{j=0}^{N} x_{ijk} * q_i \leq Q_k \qquad \text{for } k=1,2,\ldots,K2 \text{ and } i \neq j \tag{8}$$

$$\sum_{j=1}^{N} \sum_{k=1}^{K} x_{jik} = \sum_{j=1}^{N} \sum_{k=1}^{K} x_{ijk} \leq K \qquad \text{for } i=0 \tag{9}$$

$$0 \leq \sum_{i=0}^{N} \sum_{j=0}^{N} x_{ijk} \leq N \qquad \text{for } 1 \leq k \leq K, i \neq j \tag{10}$$

where let t_i denote the time when vehicle k arrives at the ith customer. And the demands of the ith customer is q_i and the corresponding unloading time is defined as s_i ($s_0=0$). The distance between the ith and jth customer is d_{ij} and the time window is described as $[ET_i, LT_i]$. There will be a cost pe incurred by waiting while a cost pl would be incurred for lateness. Besides, all vehicles have same velocity which is denoted as v. And we use c_1 and c_2 to represent unit freight of vehicles of different types ($c_k = c_1$ when $k=1, 2, \ldots, K1$; $c_k = c_2$ when $k=1, 2, \ldots, K2$).

Equation (2) is the goal of this model that minimizes the total cost. Equation (3) is used to calculate the time point. Constraint (4) means that each vehicle departs from the depot and comes back the same depot. Constraints (5) and (6) point out that each customer must be visited once and only once by exactly one vehicle. Then, constraints (7) and (8) restrict the vehicle load to the specified scope. And constraint (9) explains the number of delivery vehicles within limited scope. Finally, constraint (10) indicates that the customers served by the kth vehicle must not exceed total clients N.

3   PSO Variants for MT-VRPTW

In this section, we introduce six algorithms in detail and the key properties of those optimization algorithms are described specifically. In addition, the particle encoding scheme and design of fitness function are presented.

3.1   Description of the PSO Variants

By imitating the swarm behavior of birds flocking, Kennedy and Eberhart [14] proposed PSO in 1995. To achieve a balance between the global and local search, inertia weight w [7] was introduced. Subsequently, Shi and Eberhart [7] designed a linearly decreasing w to enhance prophase global search and anaphase local search.

In the past decade, various variants of PSO have sprung up via exploring a variety of types of topology. For instance, UPSO [8] obtained a good performance after combining the global PSO and local PSO together. Moreover, Mendes and Kennedy [9] introduced a new variant of PSO that is called the FIPS algorithm, which enables the particle to get information from all its neighbors to update the velocity. FIPS algorithm with U-ring topology will be adopted in the experiment. Firstly proposed by Liang et al. in 2006, CLPSO [10] is a good choice for tackling multimodal problems by harnessing all other particle's historical best information to update the particle velocity. Furthermore, inspired by the symbiotic interrelationships in nature, Niu et al. [11] developed MCPSO to emphasize the information exchange among multiple swarms. They took advantage of the relationship between a master swarm and several slave swarms to maintain the balance of global exploitation and local exploration.

The six PSO variants can improve the performance of algorithm in different manner, not only on the benchmark functions but in the realistic problems. With respect to the proposed MT-VRPTW, it is advisable to use these algorithms mentioned above for this complicated optimization problem and make a comparison between each other to find a better one.

3.2   Particle Encoding Scheme

Assumed that there are N customers in this distribution, 2N dimensions encoding mode [12] is adopted in this paper, that is, one of demand point corresponds with two dimensions: the first dimension denotes the marking number of vehicle and the second one is the sequence number of customers served by this vehicle. For example, we can define a particle as (N_v, N_c). Obviously, one can learn that N_v means the sequence number of vehicles and N_c represents the order of customers served by the corresponding vehicle in the N_v.

3.3   Design of Fitness Function

The penalty function is used to develop the fitness function as follows:

$$fitness = \min z + M * \max(G_k - Q_k, 0) \tag{11}$$

where G_k represents the actual vehicle loading. And Q_k is the given capacity of vehicle k. Most of all, M is set to an infinite number to restrict the vehicle load.

4   Experiments and Results

In this section, we consider a scenario of one depot to distribute goods for 14 customers. There are 8 vehicles in this depot and each vehicle is with the velocity of 60 units. The earliest departure time of each vehicle is half to seven. And the demands of customers, unloading time and time windows are shown in Table 1, respectively.

Besides, we assume that there are two types of the vehicle denoted as type A and type B. The specified vehicle load of type A and type B is 500 units and 400 units, respectively. The unit freight of type A is 3 and type B is 2.5. In addition, the waiting cost pe is 10 units per hour and the cost pl for being late is 100 units per hour. The distance between depot 0 and customer i (i=1, 2, ..., 14) is shown in Table 2.

Table 1. The demands and time windows of customers

Customer	Demands	Unloading time (minutes)	Time windows	
			ET	LT
1	120	60	8:00	17:00
2	200	90	8:00	10:30
3	120	60	8:00	17:00
4	130	25	8:00	16:00
5	150	80	8:00	12:00
6	140	70	8:00	8:30
7	60	40	8:00	17:00
8	110	60	8:00	17:00
9	100	120	8:00	18:00
10	180	90	8:00	10:45
11	90	50	8:00	8:30
12	160	90	8:00	13:00
13	140	60	8:00	12:30
14	150	70	8:00	17:00

Table 2. The distance between the depot and customers

d_{ij}	0	1	2	3	4	5	6	7	8	9	10	11	12	13	14
0	0	40	106	46	36	22	41	81	52	41	48	26	95	118	87
1		0	79	25	6	23	62	78	66	39	82	48	61	113	103
2			0	61	77	85	94	60	84	68	123	124	27	68	104
3				0	21	24	48	53	46	18	73	64	50	88	82
4					0	17	56	73	59	33	76	48	61	108	97
5						0	40	67	47	25	59	41	73	104	84
6							0	48	14	30	29	66	94	84	46
7								0	34	42	74	106	73	38	44
8									0	28	40	78	87	69	38
9										0	57	65	64	79	65
10											0	68	122	106	55
11												0	109	143	112
12													0	91	115
13														0	60
14															0

Based on MATLAB 7.14, every experiment conducted in this paper was run 30 times. For fair comparison, the population size is set to 100 and the maximum iteration is set to 500 in all algorithms. Furthermore, two acceleration coefficients are set to 2.0. Owing to the uncertain number of multi-type vehicles, there are seven kinds of combinations like ($K1, K2$) in total for this optimization problem. That is, ($K1, K2$) can be one of (1, 7), (2, 6), (3, 5), (4, 4), (5, 3), (6, 2), and (7, 1). The mean values and variance of each algorithm for different combinations of $K1$ and $K2$ is presented in Table 3. The computing results show that MCPSO got the best mean values for each combination of $K1$ and $K2$ when compared with five other PSO variants in this case. Especially, the minimum of total cost 2413.00 is obtained by MCPSO algorithm as the combination of ($K1, K2$) is equal to (5, 3).

Table 3. The mean and variance results for different situations

(K1,K2)		CLPSO	UPSO	FIPS	GPSO w: 0.9-0.4	GPSO w=0.4	MCPSO
(1,7)	Mean	2567.66	2599.01	2612.07	2534.46	2562.02	2464.50
	Variance	68.23	47.73	39.59	72.34	97.42	50.07
(2,6)	Mean	2581.97	2568.10	2568.90	2519.06	2605.32	2447.40
	Variance	51.34	66.37	66.69	94.05	83.64	56.31
(3,5)	Mean	2579.84	2577.15	2579.41	2558.57	2529.36	2438.60
	Variance	56.92	78.95	65.25	76.90	95.01	55.17
(4,4)	Mean	2555.53	2534.93	2586.38	2546.94	2525.05	2447.60
	Variance	75.56	62.41	78.44	97.34	99.71	65.01
(5,3)	Mean	2556.43	2531.82	2604.13	2538.79	2540.76	2413.00
	Variance	70.96	68.48	64.24	112.46	136.77	44.56
(6,2)	Mean	2552.99	2519.27	2551.39	2536.02	2567.01	2433.70
	Variance	65.98	88.85	98.14	94.60	107.06	78.79
(7,1)	Mean	2534.39	2546.57	2535.71	2540.51	2568.08	2434.70
	Variance	76.45	91.13	72.29	124.35	102.68	63.03

Table 4. The vehicle route and fee

Vehicle		Route of the vehicle	Fee
Type A	1	0→6→14→7→9→0	
	2	0→12→2→3→0	
Type B	1	0→10→5→0	2377.6
	2	0→1→11→0	
	3	0→13→8→4→0	

Table 5. The time when the corresponding vehicle arrives at customers

Customer	1	2	3	4	5	6	7	8	9	10	11	12	13	14
Time	7.2	10.0	12.3	11.9	9.2	7.2	11.0	10.3	12.4	7.3	8.6	8.1	8.5	9.1

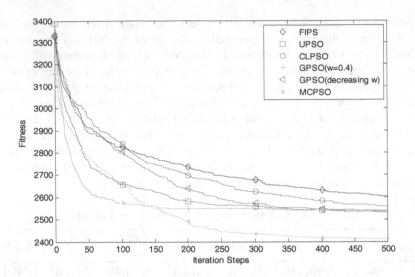

Fig. 1. The convergence graph of the six PSO variants

In particular, we discuss this situation where $K1$ is equal to 5 and $K2$ is equal to 3. And Fig. 1 presents the convergence graph for the six PSO variants. The optimal vehicle route and fee discovered by MCPSO are presented in Table 4. And the corresponding time point at which the vehicle reaches customers is showed in Table 5. From the Table 4, we can learn that there are only five vehicles assigned to serve customers actually. But, an excellent solution yielding the best fitness value can be accepted in the case by this time. This situation demonstrates a necessary optimization in the realistic vehicle allocation. In fact, there are more than vehicles needed to serve a group of customers in the depot. So it is highly significant that the depot could assign the transportation vehicle reasonably.

5   Conclusions

This paper develops an extension of VRPTW called MT-VRPTW and proposes an improved model. In the model, we consider two key realistic factors, namely multiple types of the vehicle and the uncertain number of vehicles of different types. By integrating those two aspects together, we can optimize the use of resources more efficiently in real world, especially in the arrangement of vehicles. Then, we select six representative PSO variants and apply those optimization algorithms to address a case of MT-VRPTW. According to the experiment results, we can observe that MCPSO performs better for this proposed problem with comparison to five other PSO variants.

In the future, we need to implement some other swarm intelligence algorithms [15, 16, 17] to solve this proposed problem. On the other hand, we can take more than one goal into account. And future work should focus on the multi-objective optimization of MT-VRPTW and the larger size simultaneously.

Acknowledgements. This work is partially supported by The National Natural Science Foundation of China (Grants nos. 71001072, 71271140), The Hong Kong Scholars Program 2012 (Grant no. G-YZ24), China Postdoctoral Science Foundation (Grant nos. 20100480705), Special Financial Grant from the China Postdoctoral Science Foundation (Grant nos. 2012T50584) and the Natural Science Foundation of Guangdong Province (Grant nos. S2012010008668, 9451806001002294).

References

1. Solomon, M.M.: Algorithms for The Vehicle Routing And Scheduling Problems With Time Window Constraints. Operations Research 35(2), 254–265 (1987)
2. Desrochers, M., Desrosiers, J., Solomon, M.: A New Optimization Algorithm for The Vehicle Routing Problem With Time Windows. Operations Research 40(2), 342–354 (1992)
3. Prins, C.: A Simple And Effective Evolutionary Algorithm for The Vehicle Routing Problem. Computers & Operations Research 31(12), 1985–2002 (2004)
4. Ombuki, B., Ross, B., Hanshar, F.: Multi-Objective Genetic Algorithms for Vehicle Routing Problem With Time Windows. Applied Intelligence 24(1), 17–30 (2006)
5. Chen, A.L., Yang, G.K., Wu, Z.M.: Hybrid Discrete Particle Swarm Optimization Algorithm for Capacitated Vehicle Routing Problem. Journal of Zhejiang University Science A 7(4), 607–614 (2006)
6. Belmecheri, F., Prins, C., Yalaoui, F., Amodeo, L.: Particle Swarm Optimization Algorithm for A Vehicle Routing Problem With Heterogeneous Fleet, Mixed Backhauls, And Time Windows. Journal of Intelligent Manufacturing 24(4), 775–789 (2013)
7. Shi, Y., Eberhart, R.: A Modified Particle Swarm Optimizer. In: Proceedings of Congress on Evolutionary Computation, pp. 73–79 (1998)
8. Parsopoulos, K., Vrahatis, M.: UPSO: A Unified Particle Swarm Optimization Scheme. Lecture Series on Computer and Computational Sciences 1, 868–873 (2004)
9. Mendes, R., Kennedy, J., Neves, J.: The Fully Informed Particle Swarm: Simpler, Maybe Better. IEEE Transactions on Evolutionary Computation 8(3), 204–210 (2004)
10. Liang, J.J., Qin, A., Suganthan, P.N., Baskar, S.: Comprehensive Learning Particle Swarm Optimizer for Global Optimization of Multimodal Functions. IEEE Transactions on Evolutionary Computation 10(3), 281–295 (2006)
11. Niu, B., Zhu, Y.L., He, X.X., Wu, H.: MCPSO: A Multi-Swarm Cooperative Particle Swarm Optimizer. Applied Mathematics and Computation 185(2), 1050–1062 (2007)
12. Gan, X.B., Wang, Y., Li, S.H., Niu, B.: Vehicle Routing Problem With Time Windows And Simultaneous Delivery And Pick-Up Service Based On MCPSO. Mathematical Problems in Engineering, 1–11 (2012)
13. Kallehauge, B., Larsen, J., Madsen, O., Solomon, M.M.: Vehicle Routing Problem With Time Windows. Column Generation, 67–98 (2005)
14. Eberhart, R., Kennedy, J.: A New Optimizer Using Particle Swarm Theory. In: International Symposium on Micro Machine and Human Scienc, pp. 39–43 (1995)
15. Niu, B., Wang, H., Chai, Y.J.: Bacterial Colony Optimization. Bacterial Colony Optimization, 1–28 (2012)
16. Niu, B., Wang, H., Wang, J.W., Tan, L.J.: Multi-Objective Bacterial Foraging Optimization. Neurocomputing 116, 336–345 (2012)
17. Niu, B., Wang, H., Duan, Q.Q., Li, L.: Biomimicry of Quorum Sensing Using Bacterial Lifecycle Model. BMC Bioinformatics 14, 1–13 (2013)

Prediction of Physical Time Series Using Spiking Neural Networks

David Reid[1], Abir Jaafar Hussain[2], Hissam Tawfik[1], and Rozaida Ghazali[3]

[1] Mathematics and Computer Science Department, Liverpool Hope University, UK
[2] Liverpool John Moores University, Byroom Street, Liverpool, L3 3AF, UK
[3] Universiti Tun Hussein Onn Malaysia,
Faculty of Computer Science and Information Technology, Malaysia
{reidd,tawfikh}@hope.ac.uk, a.hussain@ljmu.ac.uk,
rozaida@uthm.edu.my

Abstract. Forecasting the behavior of naturally occurring phenomena by the analysis of time series based data is the basis of scientific experimental design. In this paper, we consider a novel application of a Polychronous Spiking Network for the prediction of sunspot and auroral electrojet index by exploiting the inherent temporal capabilities of this spiking neural model. The performance of this network is benchmarked against two "traditional", rate-encoded, neural networks; a Multi-Layer Perceptron network and a Functional Link Neural Network. The results indicate that the inherent temporal characteristics of the Polychronous Spiking Network make it extremely well suited to the processing of time series based data.

Keywords: spiking neural network, polychronous spiking network, and physical time series prediction.

1   Introduction

A time series is a sequence of data values occurring at successive time intervals. A time series can be utilized in two ways:

- Looking backward – the use of historical data is used to analyze the previous behaviour of a system. Applications include diagnosis or recognition of machine fault or human disease.
- Looking forward – the use of time series data is used to predict or forecast the future behaviour of a system. Applications include stock or price prediction, market demand forecast, and prediction of natural phenomina,

Normally backward looking analysis is used to test an overarching theory about how the data is temporally correlated whereas forward looking analysis is primarily used to make forecasts. Time series forecasting assumes that future occurrences are based on present or past events, in which some aspects of the past patterns will continue into the future. Usually the goal of time series forecasting is to observe or model the existing data series to enable future unknown data values to be forecasted.

D.-S. Huang et al. (Eds.): ICIC 2014, LNAI 8589, pp. 816–824, 2014.

Time series forecasting takes an existing series of data $X_{t-n}, \ldots, X_{t-2}, X_{t-1}, X_t$ and forecasts $X_{t+1}, X_{t+2}, \ldots$ data values. Time is usually viewed in terms of discrete time steps. The size of the time interval depends on the problem at hand, and can be anything from milliseconds, hours to days, or even years. If the time series contain only one component, it is called a univariate time series; otherwise it is a multivariate time series. In a univariate series, the input variables are restricted to the signal being predicted, while in multivariate series, the raw data come from a variety of sources which will modify the input variables.

Traditional methods for time series forecasting are statistics-based, including moving average (MA), autoregressive (AR), autoregressive moving average (ARMA) models, linear regression and exponential smoothing [1]. These approaches do not produce fully satisfactory results, due to the nonlinear behaviour of most of the natural occurring time series. Recently more advanced techniques such as neural networks [2], [3], fuzzy logic [4] and fractals [5] have been successfully used in time series prediction.

Using neural networks for time series prediction has produced superior performance than statistical methods mainly due to their nonlinear nature and training capability; neural networks are universal approximators and have the ability to produce complex nonlinear mappings [6].

2   Spiking Neural Networks

2.1   Background

In recent years overwhelming experimental evidence indicates that many biological neural systems use the timing of single action potentials (or "spikes") to encode information [7], [8]. The so called "third generation" of neural networks use spike times to mimic these biological processes [9]. It is argued that these new generation of neural networks are potentially much more powerful, and, in reality, a superset, or the "traditional" or rate encoded neural networks hither to used [10]. By extension of this argument it is also suggested in this paper that employing spiking neurons as computational units, and by explicitly using temporal data in order to encode information, that Spiking Neural Networks are naturally suited to produce good time series prediction.

This new type of neural network can be defined as follows. Let N be a set of spiking neurons. Let P be a set of pre-synaptic neurons such that $P \subseteq N$, where the axon of each $p_j \in P$ makes contact with the dendrites or soma of each $n_i \in N$ to form a synapse. A weight $w_{pj,ni} > 0$ is given for a synapse, a response function $\varepsilon_{pi,ni} : R^+ \rightarrow R$ for each synapse and a threshold function $\Theta_{ni} : R^+ \rightarrow R$ for each neuron.

Thus, there exists set $s = \bigcup_{k=1}^{K} \{s_k\}$ of K synapses, with each synapse s_k forming a connection between each p_j to each n_i i.e. $s_k \subseteq PxN$. Each n_i receives a spike at a specific time t from each of the pre-synaptic neurons p_j .

The synapse s_k receiving this spike will have an inhibitory or excitatory response. This called the Post Synaptic Potential (PSP).

If the neuron has a threshold set to some specific value T_i then it will fire if at time t $\sum_{n=0}^{k} s_n \geq T_i$ or vice versa when this value is less. When firing does take place, there follows a relative refractory period when the possibility of firing again is diminished and also an absolute refractory period where there is no possibility of firing.

This model is commonly referred to as Leaky Integrate and Fire (or LIF).

2.2   Izhikevich Model

The Izhikevich model uses two variables, a variable representing voltage potential (v) and another representing membrane recovery (activation of potassium currents and inactivation of sodium currents) (u). Voltage is computed by integrating the following two differential equations (usually using Euler's method):

$$\frac{dv}{dt} = 0.04v^2 + 5v + 140 - u + W \tag{1}$$

$$\frac{du}{dt} = a(bv - u)$$

Where W is the weighted inputs (thalamic input) and a and b are control parameters for the model.

Like LIF when the voltage v exceeds a threshold value (usually 30) both v and u are reset as follows:

$$v \rightarrow c \tag{2}$$
$$u \rightarrow u + d$$

So in summary a and b are parameters effecting the recovery variable. c is the value for v which occurs after a spike and d is a constant value added to u after a spike.

2.3   Using Spike Time Dependant Plasticity for Learning

The learning rule most commonly used for spiking neural networks is derived from traditional Hebbian learning. This is Spike Time Dependant Plasticity (STDP) [12], [13]. STDP adjusts the connection strengths based on the relative timing of a particular neuron's output and input spikes. Synapses increase their efficacy if a pre-synaptic spike arrives just before the postsynaptic neuron is activated but are weakened if they occur afterwards. Repeated pre-synaptic spike arrival a few milliseconds before postsynaptic action potential leads to Long-Term Potentiation (LTP) of the synapses, whereas repeated spike arrival after postsynaptic spikes leads to Long-Term Depression (LTD). The STDP ability to rapidly react to the relative timing of spikes suggests the possibility of temporal coding schemes on a millisecond time scale. This effect has significant biological justification [14], [15].

The weight change Δw_j of a synapse from a pre-synaptic neuron j depends on the relative timing between pre-synaptic spike arrivals and postsynaptic spikes from neuron i. If we label the pre-synaptic spike arrival times at j as t_j^f where f is pre-synaptic spike count at time t, and similarly t_i^n is the postsynaptic spike count at time t then the total weight change Δw_j can be calculated as:

$$\Delta w_j = \sum_{j=1}^{N} \sum_{n=1}^{N} W(t_i^j - t_j^f)$$

(3)

where $W(....)$ denotes a specific STDP function. This function is often:

$$W(x) = A_+ \exp\left(\frac{-x}{\tau_+}\right) for\ x > 0$$

$$W(x) = A_- \exp\left(\frac{x}{\tau_-}\right) for\ x < 0$$

(4)

where parameters A_+ and A_- are parameters dependant on the value of the current synaptic weight and τ_+ and τ_- are time constants/boundaries normally in the order of 10 ms. However it should be noted that other STDP calculations can be used; there is nothing sacrosanct about this particular function [13].

2.4   Polychronous Delays

Izhikevich argues [11] that a potentially major contributing factor to learning is often neglected in spiking neural network research; that of axonal delay. He points out that there is often an implicit assumption that the axonal conduction delays between neurons are negligible or equal. In biological systems however axonal conduction delays can be as small as 0.1 ms and as large as 44 ms and yet the reaction delay between any individual pair of linked neurons often has sub millisecond precision. Therefore some timing must be taking place. When both the order and precise timing of firing is taken account an enormous number of spike patterns/trains can be encoded onto even a fairly modest number of neurons. Moreover this 'chain reaction' or cascade of spiking activity means that neurons in such a system are not synchronous but are time locked to each other. While most other spiking neural network research considers whole synchronization of the network to an input pattern as desirable, this type of network sees this modality as an undesirable property. It is the selection of sub groups of related neurons in a network that has significance, not the networks behavior on mass. This is the great strength of this type of network. Izhikevich calls such a network a polychronous neural network (or PSN). He goes on to prove that neuronal chains of firing patterns spontaneously emerge in such a network. This is referred to as a gamma rhythm [16], [17]. Moreover groups of neurons contributing to the gamma rhythm are both very precise in their firing pattern activity while also being remarkably tolerant to noise.

This gives the PSN the capability to very rapidly recognize complex patterns, often with a single neuron's spiking output being a flagged as a result of a pattern of other afferent neuron's reaction to stimulation.

This idea is called the theory of neuronal group selection (TNGS) or neural Darwinism [18]. It is this method that the authors of this paper utilize in order to try to predict values in an evolving and non-stationary physical time series environment.

It is argued in this paper that this type of polychromic delay calculation linked with STDP learning are very well suited mechanisms with which to manage time series based data.

3   Experiments and Results

3.1   SNN Used in the Experiments-Polychromous Spiking Network (PSN)

The data is presented to the networks directly without significant transformation. To do this the data is scaled between the upper and lower bounds of the transfer function in order to map their values to the available number of neurons. Data scaling is used to reduce the range difference in the data and to process outliers and.to accommodate the limits of the network's transfer function. Manipulation of the data using this process produces a new bounded dataset.

Several possible encoding methods were considered, especially inter spike interval representing data values at particular points in time; encoding time series data as a neuronal gray value bit pattern, however it was decided to directly map the time series values onto distinct neurons. The reasons for adopting this methodology are:

1   Simplicity, it is relatively easy to scale and then to map the data onto a set of neurons.
2   Easy to decode, the nature of the polychronous spiking neural network means that at any time interval many different patterns of neuronal activation can exist in the network representing possible "candidate" solutions. If a complex encoding scheme is used this may hide, or even destroy, the causal chain of neuronal activity.
3   Easy to encode; temporal information is encoded directly into the network without manipulation.
4   Easy to interpret; causal neuronal chains in the network correspond to different candidate solutions; these in turn can be mapped back to real data.

Training of the network consisted of:

1   Scaling and rounding the raw data to the nearest integer so that it would map onto the available neurons (100 in our case). These neurons represent the real values of the data

2   Presenting the scaled values to the network via "thalmic input". This is represented by the value W in equation (1) and amounts to the total influx on spiking information at a particular time step after synaptic delays have been accounted for. It is this variable that we use to fire the relevant neuron represented the scaled data value.

During experimentation three different training architectures were applied:

1. Randomly connected neurons with random delays, these along with their afferent neurons were updated after each training cycle.
2. Bands of connected neurons (with a 10 neuron neighbourhood), with each band and their afferent neurons being updated at each training cycle.
3. A focused single neuron and its afferent neurons being updated with each training cycle.

It was found that this last method was much quicker to train (by a factor of 4-5) and produced comparable results to the other 2 methods. It was noted that during training the system periodically entered bursts of activity that indicated that afferent neurons were being activated in a very focused way (in a manner suggesting a gamma cycle).

Training consisted of presenting the time series data values 20 times to the network (experimentation with lower numbers of training session (down to 5) produced comparable results).

The values to the midpoint of the data (200) were then taken as the "anchor" neurons. These are neurons that have been influenced, or will influence, other significant neurons. The network then looked for all possible pre-synaptic firing patterns in the previous 200 afferent neuron values that where similar to the previous 200 real data values (a tolerance of ±5ms was given). These were labelled as "candidate" paths.

After running the network the candidate path that most closely resembled the real data (by Euclidean distance) was chosen and a prediction made based on the continuation of firing of the candidate path to the efferent neuron most likely to fire in the 201st, 202nd 203rd, up to 400th time series spike time.

The forecast was made by considering different combinations of possible neuronal paths. Each path was ranked according to when neurons fired and which afferent neurons influenced them. The path though the network that most closely resembles the real time series data was deemed to be the best approximate forecast of the current and, via subsequent firings, future conditions.

3.2   Time Series Used in the Experiments

Two time series have been used for our experiments, namely the mean value of the auroral electrojet index (AE) index, and the, sunspot number time series.

The AE index is determined from various weather stations located in high latitude regions. At these stations, the north-south magnetic perturbations are determined as a function of time and the superposition of the measured data determines two components, the maximum negative and the maximum positive excursion in the north-south magnetic perturbations. The difference between the two components is called the AE index [19].

3.3   Experimental Testing

The performance of the spiking neural network was benchmarked with the performance of the multilayer perceptron (MLP) and the functional link (FLNN) neural networks (3rd order links where used as these are widely adopted in previous studies) [21, 22, 23]. The prediction performance of the networks was evaluated using the normalised mean square of the error (NMSE) and the signal to noise ratio (SNR) matrices.

The data sets used in this work were segregated in time order. In other words earlier period of data are used for training, and the data of the later period are used for testing. The main purpose of sorting them into this order is to discover the underlying structure or trend of the mechanism generating the data, that is to understand the relationship exist between the past, present and future data. For the MLP and FLNN each signal was divided into three data sets which are the training, validation and the out-of-sample which represent 25%, 25%, and 50% of the entire data, respectively (only the last data set is used for the PSN).

3.4   Simulation Results

The PSN was benchmarked with the MLP and the FLNN (trained with the incremental backpropagation learning algorithm) as well as with a Linear Predictor Coefficient (LPC) model. Early stopping with maximum number of 3000 epochs was used.

An average performance of 5 trials was used. The learning rate was selected between 0.1 and 0.5 and the momentum term was experimentally selected between 0.4 and 0.9. Two sets of random weights initializations were employed (in the range of [-0.5, 0.5] and [-1, 1]). Our primary interest is to assess the predictive ability of the PSN models against the other neural networks and against a standard linear model.

Table 1. The average simulation results over 5 trials using the benchmarked Neural Networks Structures and a Linear Predictor Coefficient (LPC) model

Mean Value of the AE index	MLP (Hidden nodes 3)	FLNN (Order 3)	LPC	PSN
NMSE	0.175805	0.133661	0.0615	0.0029796
SNR (dB)	28.3	29.45	31.3648	39.9127

Sunspot number	MLP (Hidden nodes 3)	FLNN (Order 3)	LPC	PSN
NMSE	0.1319	0.1366	0.1241	0.005011
SNR (dB)	25.16	25.01	22.6233	32.4729

The results (refer to Table 1) show that the Polychronous Spiking Network exhibits better performance both in terms the predication error minimisation signified by a relatively lower NMSE value and signal to noise ratio, when compared to MLP, FLNN and LPC predictors.

4   Conclusion

We have applied a specific type of spiking neural networks (Polychronous Spiking Network) to solve physical time series data prediction problems, in order to exploit the temporal characteristics of the spiking neural model.

Our spiking neural network model adopted the Izhikevich neural architecture using axonal delays, directly encoding the time series data into the model in such a way as to try to preserve valuable temporal predictive indicators.

Experiments using our SNN showed that it outperforms traditional, rate-encoded, neural networks, namely Multi-Layer Perceptrons and a higher order Functional Link Neural Network and a classic Linear Predictor Coefficient model. This was validated using two time series datasets: Mean Value of the Auroral Electrojet (AE) index, and a Sunspots dataset. The PSN favourable performance was evidenced by two key performance measures of Normalised Mean Squared Error and the Signal to Noise Ratio. Future work includes exploring the applicability of spiking neural network for long term prediction, exploring the effects of parameter tuning of the PSN on its performance, the potential incorporation of higher order terms in the spiking neural network paradigm and comparing performance with FLNNs with orders greater than 3.

References

1. Makhoul, J.: Linear prediction: A tutorial review. Proceedings of the IEEE 63(4), 561–580 (1975)
2. Conner, J., Atlas, L.: Recurrent neural networks and time series prediction. In: IEEE International Joint Conference on Neural Networks, New York, USA, pp. I 301- I 306 (1991)
3. Rape, R., Fefer, D., Drnovsek, J.: Time series prediction with neural networks: a case of two examples. In: IEEE Instrumentation and Measurement Technology Conference, Hammamatsu, Shizuoka, Japan, May 10-12, pp. 145–148 (1994)
4. Singh, S.: Fuzzy Nearest Neighbour Method for Time-Series Forecasting. In: Proc. 6th European Congress on Intelligent Techniques and Soft Computing (EUFIT 1998), pp. 1901–1905 (1998)
5. Tokinga, S., Moriyasu, H., Miyazaki, A., Shimazu, N.,, N.: A forecasting method for time series with fractal geometry and its application. Electronic and Communications in Japan, part 3, 82(3), 31–39 (1999)
6. Draye, J.S., Pavisic, D.A., Cheron, G.A., Libert, G.A.: Dynamic recurrent neural networks: a dynamic analysis. IEEE Transactions on Systems, Man, and Cybernetics-Part B 26(5), 692–706 (1996)

7. Hodgkin, A., Huxley, A.: A quantitative description of membrane current and its application to conduction and excitation in nerve. J. Physiol. 117, 500–544 (1952)
8. Abeles, M.: Corticonics: Neural circuits of the cerebral cortex. Cambridge University Press, New-York (1991)
9. Maass, W.: Networks of Spiking Neurons: The Third Generation of Neural Network Models. Neural Networks 10(9), 1659–1671 (1997)
10. Maass, W., Bishop, C.M.: Pulsed Neural Networks. MIT press (1998) ISBN 0-262-13350-4
11. Izhikevich, E.M.: Simple model of spiking neurons. IEEE Transactions on Neural Networks 14(6), 1569–1572 (2003)
12. Legenstein, R., Naeger, C., Maass, W.: What can a Neuron Learn with Spike-Timing-Dependent-Plasticty. Journal of Neural Computation 17(11), 2337–2382 (2005)
13. Nessler, B., Pfeiffer, M., Maass, W.: STDP enables spiking neurons to detect hidden causes of their inputs. In: Proc. of NIPS 2009: Advances in Neural Information Processing Systems, vol. 22, pp. 1357–1365. MIT Press (2010)
14. Levy, W.B., Steward, O.: Temporal contiguity requirements for long-term associative potentiation/depression in the hippocampus. Neuroscience 8(4), 791–797 (1983)
15. Markram, H., Lübke, J., Frotscher, M., Sakmann, B.: Regulation of synaptic efficacy by coincidence of postsynaptic APs and EPSPs. Science 275(5297), 5–213 (1997)
16. Izhikevich, E.M.: Polychronization: Computation with Spikes. Journal of Neural Computation 18(2), 245–282 (2006)
17. Arthur, J.V., Boahen, K.A.: Synchrony in Silicon: The Gamma Rhythm. IEEE Transactions on Neural Networks 18(6) (2007)
18. Edelman, G.M.: Neural Darwinism: Theory if Neuronal Group Selection. Basic Books, New York (1987)
19. Huang, C., Loh, C.: Nonlinear Identification of Dynamic Systems Using Neural Networks. Computer-Aided Civil and Infrastructure Engineering 16, 28–41 (2001)

The Application of Artificial Immune Systems for the Prediction of Premature Delivery

Rentian Huang[1], Hissam Tawfik[1], Abir Jaafar Hussain[2], and Haya Al-Askar[2]

[1] Mathematics and Computer Science Department, Liverpool Hope University,
Taggart Avenue, L16 9JD, UK
[2] School of Computing and Mathematical Sciences, Liverpool John Moores University,
Byroom Street, Liverpool, L3 3AF, UK
{huangr,tawfikh}@hope.ac.uk, a.hussain@ljmu.c.uk,
h.alaskar@2011.ljmu.ac.uk

Abstract. One of the most challenging tasks currently facing the healthcare community is the identification of premature labour. Premature birth occurs when the baby is born before completion of the 37-week gestation period. The incomplete understanding of the physiology of the uterus and parturition means that premature labour prediction is a difficult task. The reason for this may be that the initial symptoms of preterm labour occur commonly in normal pregnancies. There is some misclassification in regard to recognizing full-term and preterm labour; approximately 20% of women who are identified as reaching full-term labour actually deliver prematurely. This paper explores the applicability of Artificial Immune System (AIS) technique as a new methodology to classify term and preterm records. Our AIS approach shows better results when compared with Neural Network, Decision Tree, and Support Vector Machines, achieving more than 92% accuracy overall.

Keywords: Term delivery, preterm delivery, machine learning, classification, artificial immune system, Electrohysterography.

1 Introduction

Preterm birth has great impact on the new babies live, including health problems or increased risk of death. One million of preterm babies die each year. The earlier delivery has significant negative impacts on babies' later lives. Preterm infants are usually born at low weights of less than 2500 grams compared with full-term babies [1]. In their future lives, they might suffer from more neurological, mental and behavioural problems compared with full-term infants [2]. In other cases, preterm birth leads to increased probability of asthma, hearing and vision problems; some preterm infants may have difficulty with fine-motor and hand-eye coordination [3]. An early delivery impacts the development of Kidney and its function in the infant's future life [4]. Furthermore, 40% of survivors of extreme preterm infants can bed affected with chronic lung disease [5].

On the other hand, *preterm* births have a detrimental effect on the economy, families and community [6]. More than three quarters of premature babies can be saved with

D.-S. Huang et al. (Eds.): ICIC 2014, LNAI 8589, pp. 825–833, 2014.
© Springer International Publishing Switzerland 2014

feasible and cost-effective care, which result in more infant hospitalizations and a lot of healthcare expenditure. On the other hand, the reduced gestation duration increases the numbers of days spend in hospital. As result, preterm births have a negative economic effect. However, attempting better understanding of preterm deliveries can help to make right decision and prevention strategies to reduce the negatives effects of preterm deliveries on babies, families, societies and healthcare service [6].

Diagnosing preterm labour and predicting preterm birth have significant consequences, for both health and the economy [6]. Hence, this paper suggested the use of the artificial immune system for the prediction of premature delivery.

2   Electrohysterogram (EHG)

Electrohysterography (EHG) is a technique for measuring electrical activity of the uterus muscle during pregnancy, through uterine contractions [7]. EHG is one form of electromyography (*EMG*), the measurement of activity in muscular tissue. From medical point of view, the strengthening and increasing of uterine contractions over time is a sign of imminent labour and birth [9], and shoots up especially in the last three to four days, before delivery [10]. During parturition, the increasing of the contraction will help the body to prepare for the final stage of labour and parturition [6]. They will help shortening the cervix and force the foetus to descend into the birth canal. Therefore, the main function of uterine contractions is to produce the force and synchronicity required for true labour.

3   Term and Preterm Classification Methods

Computer algorithms, and visualization techniques, are fundamental in supporting the analysis of datasets. More recently, the medical domain has been using such techniques, extensively. Artificial Neural Networks (ANNs) have been used in a large number of studies to classify term and preterm deliveries [6].

A study by [8] used the Term-Preterm ElectroHysteroGram (TPEHG) database to evaluate classification accuracy. This occurred via sample entropy, against thirty cepstral coefficients and then against three. One of the feature sets, used for classification, consisted of calculating thirty cepstral coefficients, from each signal recording. The second feature set contained three cepstral coefficients. The selection of these values occurred by sequential forward selection and Fisher's discriminant. A multi-layer perceptron neural network classified the records, into term and preterm records. The results indicate that the reduced feature set, of three cepstral coefficients, gave the best classification accuracy of 72.73% (±13.5). This was in contrast to the entire thirty coefficients, whose accuracy was 53.11% (±10.5), and the sample entropy of 51.67% (±14.6).

4   Artificial Immune Systems (AISs)

The immune system is another biologically pattern recognition and classification system. By observing natural immune systems, researchers have identified many interesting processes and functions within them, which can provide useful metaphors for computation. AIS can learn to distinguish the 'self' from 'non self' and solve relevant classification problems. Additionally, the immune system performs important maintenance and repair functions. Artificial immune systems are one of the most rapidly emerging biologically motivated computing paradigms. There is significant growth in the applications of immune system models in various domains and across many fields. Examples include computer security, function optimization, control engineering, data mining, pattern recognition, image interpretation, anomaly detection, sensor fusion, and process monitoring.

In the last two decades, research has used classical immunology concepts to develop a set of algorithms, which including negative selection [12], immune networks [13], clonal selection [14], dendritic cells algorithm [15]. In this paper, we will investigate the applicability of the Artificial Immune Recognition System (AIRS) for the term and preterm classification under consideration. AIRS (Artificial Immune Recognition System) is an immune inspired supervised learning algorithm [16]. This algorithm uses immune mechanisms of resources competition, clone selection, maturation, mutation and memory cells generation. The training and test data items are viewed as 'antigens' in the system. These antigens induce the B-cells in the system to produce artificial recognition balls (ARBs) which compete with each other for the given resource number. The ARBs with higher resources will get more chances to produce the mutated offspring to improve the system. The memory cells generated after all training antigens have been introduced are used to classify the test data items. The AIRS algorithm's procedure is typically composed of five steps. These stages are initialization, memory cell identification and ARB generation, competition for resources and development of a candidate memory cell, memory cell introduction, and classification (Figure. 1) which can be described in details as follow.

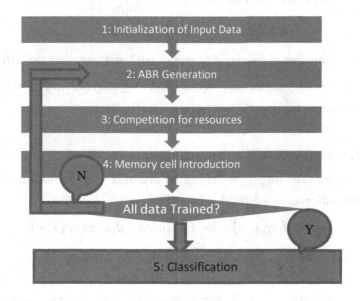

Fig. 1. Flowchart of AIRS

Step 1

Initialization is the first step of the algorithm. Which can also called data pre-processing. In this step, all data are normalized in order to comply with the Euclidean distance. The distance between any two data in the dataset is between 0 and 1. In this study, we use Eq. 1 to deal with normalization, and the calculation of the Euclidean distance is as Eq. 2.

$$x_i = \frac{\left(\dfrac{a_i}{max_i - min_i}\right)}{\sqrt{m}} \tag{1}$$

where a_i is the original datum of ith attribute, max_i and min_i are the maximum and minimum of ith attribute of all data, respectively, and m is the number of attributes in data.

$$Euclidean\_dis\tan ce = \sqrt{\sum_{i=1}^{m}(x_i - y_i)^2} \tag{2}$$

where x and y represent feature vectors.

After normalization, the affinity function, which used to calculation fitness is calculated as Eq. 3

$$affinity\_threshold = \frac{\sum_{i=1}^{n}\sum_{j=i+1}^{n} affinity(ag_i, ag_j)}{\dfrac{n(n-1)}{2}} \tag{3}$$

where n is the number of training data items, ag_i and ag_j are the ith and jth training antigen in the training vector, and $affinity(ag_i, ag_j)$ represents the Euclidean distance between the two antigens' feature vector.

Step 2

In this step, the first thing is to find out the memory cell, mc_{match} given a specific training antigen. The mc_{match} is defined as $\arg\max_{mc \in MC} stimulation(ag, mc)$. Where stimulation(x, y) is defined as Eq. 4.

$$simulation(x, y) = 1 - Euclidea\_dis\tan ce(x, y) \tag{4}$$

Given this definition, it can be assumed that the antibody mc_{match} is the most stimulated memory cell by the given antigen in the set of memory cells of the same category. When mc_{match} is identified, this cell is used to create new ARBs to be introduced to the population of pre-existing ARBs. The number of new ARBs depends on the stimulation value between the memory cell and the antigen.

Step 3
In this step, all ARBs presently existing in the system are awarded the resource numbers according to their affinity values. The ARBs with higher affinity values will get more resources than those with lower affinity values. If the sum of the number of resources of all ARBs exceeds the allowed number, the system will remove the ARBs and their awarded resources beginning with the lowest number of resources until the sum of the number of resources of all ARBs is lower than the allowed number in the system. After this process, the stimulation values of all remaining ARBs are calculated and maximal and minimal stimulation values are determined as $\max Stim$ and $\min Stim$, respectively. The stimulation level of each ARB is recalculated as Eq. 5.

$$arb.Stim = \begin{cases} \dfrac{arb.stim - \min-Stim}{\max Stim - \min Stim} & \text{if class of arb= antigen of antigen, otherwise} \\ 1 - \dfrac{arb.stim - \min-Stim}{\max Stim - \min Stim} \end{cases}$$

$$(5)$$

Then, we use Eq. 6 to calculate the average value of these levels for each class and verify if any of these average values is lower than a given stimulation threshold or not. If any of the average values is lower, the ARBs belonging to that class are mutated and the generated clones are added to ARB pool. This process is continuously done until the average stimulation levels of all classes are larger than the stimulation threshold.

$$s_i = \frac{\sum_{j=1}^{|ARB_i|} arb_j.stim}{|ARB_i|}, arb_j \in ARB_i \qquad (6)$$

where $i = 1,2,...,nc$, $s = \{s_1, s_2,..., s_{nc}\}$, $|ARB_i|$ is the number of ARBs belonging to ith class and $arb_j.stim$ is the stimulation level of jth ARB of ith class.

Step 4
After achieving the criterion described above, the ARB with the highest stimulation value in the same class with the presented training antigen is taken as a candidate memory cell $mc_{candidate}$. If the stimulation value of the $mc_{candidate}$ motivated by the training antigen is higher than the stimulation value of the mc_{match} , the candidate

memory cell is added to the set of memory cells. If this test is passed, a calculation of the affinity between $mc_{candidate}$ and mc_{match} must be obtained. If the affinity between this two memory cells is lower than the product of the affinity threshold and the affinity threshold scalar, $mc_{candidate}$ is replaced with mc_{match} in the set of memory cells.

Step 5
After repeating step 2 to step 4 to each training antigen, the developed memory cells are ready for exploitation and for classification. The classification is executed in a k-nearest neighbour approach. That is, the classification of a datum in the system is determined by the ballot of the results of the k most stimulated memory cells.

5   Methodology

The data used in this research were recorded at the Department of Obstetrics and Gynaecology, Medical Centre, Ljubljana between 1997 and 2006 [11]. In the Term-Preterm ElectroHysteroGram dataset database, there are 300 records of patients. Each record was collected by regular examinations at the 22nd week of gestation or around the 32nd week of gestation. The records were 30 minutes long, had a sampling frequency of 20Hz, and had a 16-bit resolution over a range of ±2.5 millivolts. Four electrodes placed on the abdominal surface. Each record is obtained from three channels, Channel 1, Channel 2 and Channel 3. The signals were filtered using 3 different 4-pole digital Butterworth filters. Signals have been filtered twice in both backwards and forwards directions in order to overcome the phase-shifting that can occur when utilizing these filters.

In the TPEHG dataset, records were generated using four features – *root mean squares*, *peak frequency*, *median frequency*, and *sample entropy*. *Mean frequency* and *sample entropy* have the most potential to discriminate between *term* and *preterm* records, while *root mean squares* and *peak frequencies* have had conflicting results, in the classification of *term* versus *preterm* records, but, nonetheless, have shown potential. In this paper, all these features are considered.

6   Evaluation

In this paper, the proposed AIRS was benchmarked against the multilayer MLP neural network, K-Nearest Neighbour Classifier (KNN), Decision tree classifier (TREEC) and Support Vector Classifier (SVC). Their performances have been evaluated using the sensitivity, specificity, positive, and negative predicted values that each classification algorithm produced to separate term and preterm signals. The data have been split up as follows – 60% of the data is randomly selected for training data, 25% for validation and 15% for testing. The experiments have been run five times to generate an average of the results obtained. The formulas used to measure sensitivity; specificity and accuracy are defined as follows:

$$Accuracy = ((TP + TN)/(TP + TN + FP + FN)) \times 100 \qquad (7)$$

$$Sensitivity = TP/(TP + FN) \qquad (8)$$

$$Specificity = TN/(FP + TN) \qquad (9)$$

6.1   Classifier Performance

The first evaluation uses the original TPEHG dataset (38 *preterm* and 262 *term*) – the preterm are oversampled using min and max to produce 262 *preterm* records). The simulation results indicated that the sensitivity for preterm records obtained with the AIRS is slightly better than the MLP; but the specificity of MLP is higher than other method. When compare with the true negative and true positive, AIRS outperform others.

Each TGEHG signal record contains clinical information relating to each of the patients that includes pregnancy duration, gestation duration at the time of recording; maternal age, number of previous deliveries (parity); previous abortions, weight at the time of recording, hypertension, diabetes, placental position, bleeding first trimester, bleeding second trimester, funneling, and whether they are a smoker. These eleven items of clinical information are added to the original features. Some information was missing for some of the patients, which led to unknown features on some recorders. Hence, the records with unknown information are removed for the next experiment, reducing the number of samples in the dataset to 19 *preterm* data samples and 108 *term* data samples. The re-sampling method was applied to generate 150 *preterm* data items. The results in Table 1 further show that the highest values obtained are by the AIRS in terms of mean error, standard deviation and classification accuracy followed by the KNNC network.

Table 1. Classifier Performance Results for the 0.3-3Hz Filter

	Sensitivity	Specificity	True Negative	True Positive	Mean Error	Std Error	Accuracy
MLP	0.8684	0.8255	0.8483	0.8526	0.1681	0.0491	84.87%
KNNC	0.8076	0.9047	0.7916	0.9130	0.2260	0.0505	85%
TREEC	0.8846	0.7619	0.8421	0.8214	0.2433	0.569	82%
SVC	0.8076	0.8571	0.7826	0.8750	0.1761	0.0549	83%
AIRS	0.9020	0.9235	0.9042	0.9401	0.0781	0.001	92.5%

7   Conclusions and Future Works

In this paper, we described an artificial immune inspired algorithm (AIRS) for term and pre-term classification for pregnant women. Most of the uterine EHG signal studies have concentrated on predicting true labour, which is based on the last stage of the pregnancy duration. This paper has evaluated the use of a machine learning approach, using records from earlier stages of gestation, to predict term and preterm deliveries. The method of classification used in this paper compares several existing classification algorithms and our AIRS. The evaluation of classifier performance has been measured using sensitivity and specificity, which are suitable evaluation measures for binary outputs (term/preterm). The proposed AIRS model outperformed several algorithms in the classification of uterine signal. Based on the experimental results, AIRS demonstrated higher detection rate, lower error rate and handled high dimensional data set better when compared with the results achieved by MLP, KNNC, TREEC, and SVM. Future extensions to this work include the optimisation of the parameters setting of the AIRS and investigating the applicability of other form of artificial immune systems such as the Dendritic Cell Algorithm for term and pre-term classification.

References

[1] Garfield, R.E., Maner, W.L.: Physiology and electrical activity of uterine contractions. Semin. Cell Dev. Biol. 18(3), 289–295 (2007)
[2] William, R.E.G., Maner, L.: Identification of human term and preterm labor using artificial neural networks on uterine electromyography data. Ann. Biomed. Eng. 35(3), 465–473 (2007)
[3] WebMd, Premature Labor (2012), http://www.webmd.com/baby/premature-labor
[4] Keijzer-Veen, M.G., van der Heijden, A.J.: The Effect of Preterm Birth on Kidney Development and Kidney Function over Time. In: Morrison, J.C. (ed.) Preterm Birth - Mother and Child (2012)
[5] Greenough, A.: Long Term Respiratory Outcomes of very Premature Birth (<32 weeks). Semin. Fetal Neonatal Med. 17(2), 73–76 (2012)
[6] Fergus, P., Cheung, P., Hussain, A., Al-Jumeily, D., Dobbins, C., Iram, S.: Prediction of Preterm Deliveries from EHG Signals Using Machine Learning. PLoS One 8(10), e77154 (2013)
[7] Garfield, R.E., Maner, W.L., MacKay, L.B., Schlembach, D., Saade, G.R.: Comparing uterine electromyography activity of antepartum patients versus term labor patients. Am. J. Obstet. Gynecol. 193(1), 23–29 (2005)
[8] Rabotti, C.: Characterization of uterine activity by electrohysterography. Technische Universteit Eindhoven (2010)
[9] WelcomeBabyHome, The Birth Process (2006), http://www.welcomebabyhome.com/pregnancy/pregnancy_birth_process.htm (2013)
[10] Lucovnik, M., Maner, W.L., Chambliss, L.R., Blumrick, R., Balducci, J., Novak-Antolic, Z., Garfield, R.E.: Noninvasive uterine electromyography for prediction of preterm delivery. Am. J. Obstet. Gynecol. 204(3), 228, e1–10 (2011)
[11] Rabotti, C., Mischi, M., Oei, S.G., Bergmans, J.W.M.: Noninvasive estimation of the electrohysterographic action-potential conduction velocity. IEEE Trans. Biomed. Eng. 57(9), 2178–2187 (2010)
[12] Forrest, S., Perelson, A., Allen, L., Cherukuri, R.: Self-Nonself Discrimination in a Computer. In: Proc. of IEEE Symposium on Research in Security and Privacy, Oakland, USA, pp. 202–212 (1994)

[13] Jerne, N.: Towards a network theory of the immune system. Annals of Immunology 125, 373–389 (1974)

[14] de Castro, L.N., Von Zuben, F.J.: The clonal selection algorithm with engineering applications. In: Proceedings of GECCO 2000, Workshop on Artificial Immune Systems and Their Applications, pp. 36–37 (2000)

[15] Greensmith, J., Twycross, J., Aickelin, U.: Dendritic cells for anomaly detection. In: Proc. of the Congress on Evolutionary Computation (CEC), pp. 664–671 (2006)

[16] Brownlee, J.: Artificial immune recognition system (AIRS): A review and analysis. CISCP, ICT. Swinburne University of Technology, Tech. Rep. TR-1-02 (2005)

Dynamic Hand Gesture Recognition Framework

Prashan Premaratne[1], Shuai Yang[1], ZhengMao Zhou[2], and Nalin Bandara[3]

[1] School of Electrical, Computer and Telecommunications Engineering,
University of Wollongong, North Wollongong, NSW Australia 2522
[2] Dalian Scientific and Technological Research Institute of Mining Safety,
No.20 Binhai West Road, Xigang District, Dalian, Liaoning, China
[3] Department Electrical, Electronic and Telecommunication Engineering,
Faculty of Engineering, General Sir John Kotelawala Defence University,
Sri Lanka

Abstract. Sign languages originated long before any speech-based languages evolved in the world. They contain subtleties that rival any speech-based languages conveying a rich source of information much faster than any speech-based languages. Similar to the diversity of speech-based languages, sign languages vary from region to region. However, unlike the speech counterpart, sign languages from diverse regions from the world have much common traits that originate from human evolution. Researchers have been intrigued by these common traits and have always wondered whether sign language-type communication is possible for instructing the computers opposed to the mundane keyboard and mouse. This trend is popularly known as Human Computer Interaction (HCI) and has used a subset of common sign language hand gestures to interact with machines through computer vision.Since the sign languages comprise of thousands of subtle gestures, a new sophisticated approach has to be initiated for eventual recognition of vast number of gestures. Hand gestures comprise of both static postures and dynamic gestures and can carry significantly rich vocabulary describing words in the thousands. In this article, we present our latest research that describes a mechanism to accurately interpret dynamic hand gestures using a concept known as 'gesture primitives' where each dynamic gesture is described as a collection of many primitives over time that can drive a classification strategy based on Hidden Markov Model to reliably predict the gesture using statistical knowledge of such gestures. We believe that even though our work is in its infancy, this strategy can be extended to thousands of dynamic gestures used in sign language to be interpreted by machines.

Keywords: Dynamic hand gestures, hidden Markov model, gesture primitives, sign language, hand postures.

1 Introduction

Almost all sign languages in the world can be considered as a superset of hand gestures used for human computer interaction. There has been significant success in controlling many electronic apparatus using hand gestures in the past [1]. HCI has featured prominently in most consumer-electronics control systems over the past decade [2].

D.-S. Huang et al. (Eds.): ICIC 2014, LNAI 8589, pp. 834–845, 2014.

However, these successes can be attributed to the limited use of hand gestures used in HCI for effective communication with very high accuracy [1, 3]. Hand-gesture-based sign languages can communicate information much faster than speech. If an accurate sign language communication with machines can be established, the outcome will allow multiple sign languages in the world to be translated automatically. Military will also be able to make use of silent sign language effectively for secure communication. In 2007, Premaratne *et al.* at the University of Wollongong demonstrated a real-time working prototype that could interact with a number of electronic devices via hand gestures [1, 3, 4-12]. There are significant obstacles that has stagnated the accuracy of hand gesture recognition which has confined the number of gestures suitable for communication to very few (less than 30) due to the difficulty in accurately recognizing hand gestures by different users. There are few obvious reasons behind the failure of many vision based classification systems to interpret sign languages over the years [2]. Some of the more prominent ones can be listed as follows:

1. Signs vary in time and space. Even if a sign is repeated by the same user, slight changes of speed and position of hands will occur.
2. Lack of depth information to the gesture(s) from camera complicates the problem computer vision problem.
3. Some fingers are occluded behind hand or arm.

The enormity of the number of gestures available and their similarity to many other gestures and the user's inability to sign them as in (1) above make the machines unreliable with the present feature extraction and classification approaches.

Sign languages have been in existence since the start of the humanity and would have been the first means of communication among the primitive humans. Before people communicated with a vocabulary and using sounds, it is conceivable that they communicated with various gestures using hand, face, mouth and body movements. However, today, the sign language is predominantly associated with disabilities from congenital to accidents. Most of the users are either hearing impaired or mute. There is also a subgroup of whom are children of such hearing impaired people whose senses are not affected but do use sign language because of the community needs in which they live.

Since the world population has surpassed 7 billion and the widespread use of technology, more and more work is done to advance the age-old sign languages which have developed from multitude of cultural backgrounds. Due to the advancement of societies and its welfare systems, more and more resources are allocated for the benefit of the disabled and their educational needs. Sign languages have seen the formal recognition of their use and are now making inroads to the new fields of usage such as in medicine so that disabled people will also benefit from the modern developments in the society.

Today the focus has shifted again from the mundane use of sign language to communicate with each other to the more advanced human machine interaction. This would in effect advance the interactions that disabled people would have with technology as well as make sign languages easily understandable by ordinary users. The technology can also pave way for automatic translation to other languages in other parts of the world making a silent communication revolution for the disability. Yet, the challenges are enormous and the different approaches taken by researchers around the world have shed light on difficulties ahead as well as the progress made so far. Sign languages, like the spoken languages, emerge and evolve naturally within hearing-impaired and mute communities. Within each country or region, wherever such communities exist, sign languages develop, independently from the spoken language of the region. Each sign language has its own grammar and rules, with a common property that they are all visually perceived.

2   Previous Work

One of the first attempts to interpret sign languages using a machine is reported by Kawai and Tamura [13] who developed a system which can recognize 20 Japanese hand gestures. This was the first instance of using computer vision and image processing techniques to recognize a sign language. They detected the motion of the hands in realtime by comparing a grey scale intensity of two consecutive image frames. They used temporal intensity changes, hand location tracking and classification of each sign using stop positions, simple shapes of trajectory and hand shape at each stop position. Few years prior to this development in 1984, they demonstrated a system that interpreted speech into computer generated hand gestures for machine interaction with deaf and mute [14, 15].

Heap and Samaria used deformable active shape models called 'smart snakes' developed by [16] for hand tracking [17]. Their research clearly shows how their tracking algorithm work but their disclosure on gesture recognition was very much limited. Fig. 1 shows 'smart snakes' approach for hand gesture recognition.

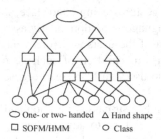

○ One- or two- handed   △ Hand shape
□ SOFM/HMM   ○ Class

Fig. 1. The hierarchical decision tree for sign language recognition using SOFM and HMM [27]

In 1995, Starner and Pentland published their research on dynamic gesture recognition and classification based on colored gloves and Hidden Makov Model classifier [18]. Their research highlighted that the finer details (pixel clarity) of gestures were not required and simple coarse outline along with their trajectory obtained using camera tracking color glove was sufficient to interpret gestures. They used 395 simple sentences as training set and separate 99 simple sentences for testing in which recorded a recognition accuracy of 92% with a limited 40 hand postures.

Grobel and Assan reported that vision based system using Hidden Markov Model (HMM) was able to recognize 262 isolated hand signs (non sentences) from Netherland sign language with an accuracy of 94% [19, 20]. They used both hands with cotton gloves. One hand which was the dominant hand making the signs used five colored fingers, separate colors for palm and another for back of the dominant hand whereas the non-dominant hand had a different color glove.

Vogler and Metasas of University of Pennsylvania devised both static gesture and dynamic gesture recognition system for ASL based on HMM and 3D motion analysis [21, 22]. They used a 3D image-based shape and motion tracking of a human's arm using deformable models. The output of these models consists of the three dimensional motion parameters of the subject's arms which can be used for recognizing American Sign Language (ASL) gestures.

Imagawa *et al.* argued that for accurate sign language recognition, both global and local information should be recognized [23]. Since the hand or arm movement conveys global information such as the hand and arm location and the path, static information such as the configuration of the hand and fingers should be combined with the information conveyed by the hand and arm movement. Their system first interprets the appropriate word from a dictionary using global features and then narrows word down to one by using detected local features. They recorded accuracy around 94% which was a significant achievement given that they relied on very low resolution images.

Even though, many hand gestures need dynamic tracking for meaningful interpretation, posture estimation results in very discrete recognition and is incomplete. Isaacs and Foo developed a similar hand pose recognition system that relied on wavelet decomposition for feature extraction and neural networks for classification [24]. They developed the system to represent around 2000 sign language sentences and phrases using a set of eighty words. One of the main limiting factors of their system was to remove the gestures that required facial gestures, since such signs represented a more complex problem. Isaacs and Foo successfully decomposed all static ASL alphabet images into two set of feature vectors with different dimensions in order to develop more insight into the classification problem. The first approach produced feature vectors by normalizing the energy and entropy measures of the resultant approximation matrix after two-level Daubechies 4 decomposition. Another set of feature vectors were obtained using decomposing each image to its lowest decomposition level of the particular wavelet used. The second set of feature vectors had an added advantage of longer length which represented a higher dimensional space that allowed for more variability when a large ASL vocabulary was incorporated. The ASL recognition system developed in this project had the ability to distinguish the reduced vocabulary set feature vectors from each other and from non-words. For classification purposes, neural networks were used. The classification system that was used had a two layer feed-forward neural network that recognized the 24 static letters in the ASL alphabet using still input images. Both wavelet- based decomposition methods discussed above were used for classification. The first produced an 8-element real-valued feature vector and the second 18-element feature vector. Each set of feature vectors were used to train a feed-forward neural network using Levenberg-Marquardt training. The system was capable of recognizing instances of static ASL finger spelling with 99.9% accuracy with an SNR as low as 2.

In 2012, Cooper *et al.* presented one of the most sophisticated sign language recognition system based on Kinect [25]. It consisted of several approaches to sub-unit based Sign Language Recognition (SLR). Their approach deviated from traditional classification of gesture to recognizing large lexicons of signs such as citation-form and dictionary look-up. Sub-unit based SLR uses a two stage recognition system where sign linguistic sub-units are identified followed by combining of sub-units to create a sign level classifier. The Cooper *et al.*'s method can be described as the most comprehensive effort to date for gesture recognition based on vision which combined many aspects of successful approaches over two decades. It described a combined gesture classification system using facial and bi-hand gestures known as linguistic sub-units were attempted by the best available vision technology available with Microsoft Kinect. They presented three types of sub-units for consideration; those learnt from appearance data as well as those inferred from both 2D and 3D tracking data. These sub-units were then combined using a sign level classifier which used a Markov Models to encode the temporal changes between sub-units and a Sequential Pattern Boosting to apply discriminative feature selection at the same time as encoding temporal information. This resulted in a more robust to noise and performed well in signer independent setups.

Hand-gesture recognition systems rely on computer vision to recognise hand gestures in the presence of any background clutter. The hand gestures must then be uniquely identified to issue appropriate control signals to the computer or other entertainment unit. During the past two decades, many attempts have been made to recognise hand gestures, with mixed success. Two of the drawbacks of the original systems were their inability to run in real time, because of the modest computing power available, and low classification scores for gesture recognition. The techniques used relied mostly on template matching or shape descriptors and required more processing time than could be provided in real time. Furthermore, the users were restricted to wearing gloves or markers to increase reliability, and were required to be at a preset distance from the camera. These restrictions meant that a practical system that could operate in a lounge room was unachievable. In 2007, a truly practical system that was capable of running with modest computing power in real time was unveiled at University of Wollongong [1, 3, 4]. This system was limited to 10 hand gestures for accuracy. However, the system can be improved to an even higher level of accuracy with more gestures by using multiple visual cues such as a "flock of features" model to represent a hand with fingers and the CI's proposed multi-cue hand-tracking algorithm, which is based on velocity-weighted features and colour cues [26].

3   The Approach to Classify Dynamic Gestures Using 'Gesture Primitives'

Human Computer Interaction is geared towards seamless human machine interaction without the need for touch LCDs, Keyboards or Sensor Gloves. Systems have already been developed to react to limited hand gestures especially in gaming and in consumer electronics control. Yet, it is a monumental task in bridging the well-developed sign languages in different parts of the world with a machine to interpret the meaning. One reason is the sheer extent of the vocabulary used in sign language and the sequence of gestures needed to communicate different words and phrases. Auslan, the Australian Sign Language is comprised of numbers, finger spelling for words-used-in-common practice and a medical dictionary. There are 7415 words listed in Auslan website. An effective sign language recognition system will be able to recognize any sign language of the world using computer vision in realtime. This will allow any sign language to be translated to any other sign language facilitating such users of ability to communicate with other such groups in the world without currently existing barriers. Furthermore, this will allow normal users to communicate with machines at speeds exceeding that is available with mouse and keyboards as well as seamless integration of entertainment and gaming experience unlike anything that they have experienced before.

The work described here has stemmed from the principal author's latest book on 'Human Computer Interaction using Hand Gestures' highlights the current drawbacks associated with HCI and fundamental research that it envisions [2]. As was outlined in the Introduction and Previous Work, existing approaches for gesture recognition has stagnated the sign language recognition beyond a limited number of gestures with significant accuracy. Given the enormity of the number of gestures used by the sign language users and their subtlety, current approaches have no hope of advancing the field of HCI. A significant breakthrough in HCI, as proposed, will not only benefit the sign language users but the HCI that governs a massive industry based on entertainment, computer gaming and developments in virtual reality.

The research on hand gesture recognition has two main categories such as static gesture (hand posture) recognition and gesture recognition (dynamic hand movement). The posture recognition refers to an isolated single hand pose or hand configuration and the gesture recognition or dynamic gesture recognition uses a variety of static gestures and hand movements sometimes accompanied by body or facial movements. In a continuous gesture (dynamic), there are many hand poses that changes in time, communicating a gesture such as letter 'z' in American Sign Language (ASL). In hand gesture controlled environments, the problem can be considered as a gesture spotting problem, where the task is to differentiate the meaningful gestures of the user from the unrelated ones. In sign language recognition, the continuous recognition problem includes the co-articulation problem. The preceding sign affects the succeeding one, which complicates the recognition task as the transitions between the signs should be explicitly modelled and incorporated to the recognition system. Moreover, language models are used to be able to perform on large-vocabulary databases.

As highlighted earlier, one of the major difficulties in accurate recognition of hand gestures is due to the enormity of the sign language vocabulary. Many feature extraction based methods rely on searching for matches with these large vocabulary databases for query matches. Fang *et al.* developed a hierarchical decision tree based method that drastically reduced this searching time increasing the efficiency of the classification system [27]. This method simplifies the complexity of gestures using 'divide and conquer' approach. A dynamic gesture is a sequence of many postures that is issued in a temporal fashion. Once few initial postures are identified, many hand gestures can be ruled out as not probable as every gesture has basic starting postures. Hand shapes are one of the primitives of sign language and reflect the information of hand configuration. They are known to very stable and can be used to distinguish most signs. Figure 1 illustrates the diagram of the hierarchical decision tree for sign language recognition, where each non-leaf node denotes a classifier associated with the corresponding word candidates and each branch at a node represents one class of this classifier. There are common elements among adjacent branches under one node, and their intersection is learned by a large amount of training samples. The input data are first fed into one- or two-handed classifier and then into left hand shape classifier and right hand shape classifier (no left hand shape classifier for the one-handed sign branch) and at last into the Self-Organizing Feature Maps (SOFM) or HMM classifier only with few candidates in which the final recognition results are obtained. Their hierarchical classification was further helped by certain unique characteristics of Chinese sign language. In the Chinese sign language dictionary, one-handed signs are always performed by right hand except for one sign "luo ma ni ya" by left hand [26]. The difference between one-handed sign and two-handed sign is whether signer's left hand performs the action. In the one-handed sign performance, signer's left hand usually puts on the left knee and remains motionless. However, in the two-handed sign, left hand may either stay a fixed posture or perform a movement trajectory. The position and orientation information of left hand plays a dominant part in determining one-or two- handed signs. Fig. 2 shows how 'gesture primitives' can be captured and used to describe words which can be recognized by a Hidden Markov Model as will be described in the final section.

Fig. 2. (Me Hearing Can Sign (four words describing 'I can hear and sign' using sign Language). In this sign communication, the motion between the words are not shown for simplicity.

An optimal context-dependent combination of information will be the key to the long-term goal of robust hand tracking. To enable long-term robustness, it is essential to use multiple cues simultaneously. Most approaches based on multiple cues use only one cue at a time, thereby optimising performance rather than robustness. Using multiple cues simultaneously will enable access not only to complementary and redundant information at all times but also to the robust failure detection that enables recovery.

This research which is undertaken currently at the University of Wollongong in Australia is to develop a general system framework for the integration of multiple cues as shown in the flow chart of Fig. 3. The ultimate goal of the framework is to enable robust hand-gesture tracking in dynamically changing environments. The general framework will exploit two methodologies for adaptation to different conditions. First, the integration scheme can be changed, and second, the visual cues themselves can be adapted. Adapting the visual cues enables direct adaptations to a changing environment. Changing the integration scheme reflects the underlying assumption that different cues may best suit different conditions. Based on this general framework, two different approaches are introduced and analysed experimentally. This research is expected to initiate the development of a system that aims at robust tracking through self-adaptive, multi-cue integration. The "democratic integration" introduced by Triesch and von der Malsburg [28] may be seen as an example of such a system. Our research is currently following the democratic-integration approach, but with its shortcomings mitigated. In particular, the original approach is limited to the tracking of a single target, which will then lead to a proposal for a second system. In this second system, multi-hypothesis tracking and multi-cue integration will be realised by means of "condensation", a conditional-density propagation algorithm proposed by Blake and Isard [29]. Expectation maximisation is used to obtain more reliable probability densities for the single-cue observations.

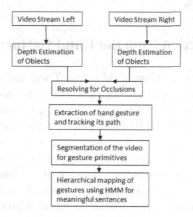

Fig. 3. Flow diagram of the proposed system that resolves the drawbacks of existing approaches

The system relies on identifying the words correctly in terms of their durations of signing from the measured signals. In this case, distinguishing words requires additional calculations due to noise and missed movements. A filter is used for noise reduction, and a velocity network is used to determine whether the sign is an ASL word or not. During this movement, the hand velocity can increase or decrease momentarily due to momentary starting or stopping of the hand. Hence, the use of a threshold value of velocity might not give a good solution for classification of hand movements. The ideal output for the velocity graph should be 1 from the time the sudden change in velocity is first seen until the time the velocity graph shows a series of low velocities.

Hence, the hand velocity is filtered and then input to the velocity neural network. The velocity neural network used in their approach was a three-stage network with two input neurons, ten hidden layer neurons, and two output neurons. The target vectors were created manually by observing the velocity graphs for different, randomly-selected signs. Twenty training samples of different signs were found to be Sufficient to achieve the desired accuracy. The network was trained using the Levenberg–Marquardt algorithm. A hyperbolic tangent sigmoid function was used in both layers. Their goal was to continuously recognize ASL signs using the glove and developed system in real time. They trained the ANN model for 50 ASL words with a different number of samples for every word and the classification results achieved 90% accuracy which demonstrated that their used successfully for isolated word recognition. They also concluded that some gestures in ASL required that both the right and left hands be manipulated simultaneously by applying the proposed model but with two data gloves and more motion trackers.

Today, ASL and many sign languages around the world see continuously being interpreted using either vision (with and without markers) or glove based systems as new technology develops. This trend will continue until it results in a highly reliable system in which, the mute and deaf would feel more natural to express their feelings like their able counterparts.

4   Feature Extraction from the Hand Gestures

Image classification is a very mature field today. There are many approaches to finding matches between images or image segments. From a basic correlation approach to scale-space techniques, they offer a variety of feature-extraction methods, with varying success. However, it is critical for hand gesture recognition that the feature extraction is fast and captures the essence of a gesture in a unique small data set. Neither the use of Fourier descriptors, which results in a large set of values for a given image, nor the scale-space approach succeed in this context. It is clear that effective real-time classification cannot be achieved using approaches such as template matching. Template matching is very prone to error whenever the user produces a hand gesture that is not exactly the same as one already stored in the library. It also fails because of scaling variability, where the varying distance to the camera may produce scaled versions of the gesture. Gesture variations caused by rotation, scaling, and translation can be circumvented by using sets of features that are invariant under these operations. Moment invariants encapsulate these properties.

Our research approach of using moment invariants stems from our success in developing the "Wave Controller". Gesture variations caused by rotation, scaling, and translation can be circumvented by using a set of features, such as moment invariants, that are invariant to these operations. In this project, we will be evaluating four moment invariants using the well-known Hu moments [32, 33]. In this project, moment invariants will be used as the major set of features, and the number of identifiable gestures will be increased by using dynamic gestures [32].

5   Classification of Hand Gestures

After features are extracted from valid gestures, they can be classified using a new approach using Hidden Markov Model approach. As shown in Fig. 4, a typical HMM approach for classifying gesture primitives will allow the statistically most probable gesture to be recognized as the one offered. As illustrated in the diagram, any gesture that deviates from a know gesture will be discarded at any stage of dynamic gesture recognition. This is a significant improvement over the existing static gesture recognition and other dynamic gesture recognition strategies. Even though, HMM is not new in hand gesture recognition, hand gesture primitives have not been widely used for effective classification. Yet, combining our research experience in such classifications, we are hopeful that the HMM will allows the ultimate classification strategy to incorporate large number of dynamic gestures into hand gesture recognition.

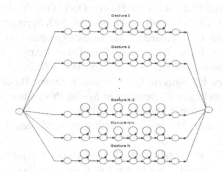

Fig. 4. A Typical HMM classification for dynamic hand gesture recognition

6   Experimental Results and Conclusion

Our current approach has indicated promise and the viability of using 'gesture primitives' for large number of gesture classification. Currently, we are using only 3 moment invariant values for each gesture primitive which is captured every 20 millisecond when dynamic gestures are performed very slowly. The gesture trajectory is tracked using Kalman filtering in a OpenCV environment and we are only attempting to develop the best strategy to capture gesture primitives in order to establish a framework. Currently, the HMM is used to predict a dynamic gesture set of 20 which has few stability issues. Yet, there is success in certain gestures around 70% when the signers perform the gesture with great care resembling any previous attempt. We are very hopeful that this framework will achieve a significant breakthrough for dynamic hand gestures as well as for sign language communication in foreseeable future.

References

1. Premaratne, P., Nguyen, Q.: Consumer Electronics Control System Based on Hand Gesture moment invariants. IET Computer Vision 1-1, 35–41 (2007)
2. Premaratne, P.: Human Computer Interaction using Hand Gestures, Springer Publication (2014), http://www.amazon.com/Computer-Interaction-Gestures-Cognitive-Technology/dp/9814585688, ISBN-10: 9814585688
3. Hutcheon, S.: Last Hurrah for Lost Remote, Sydney Morning Herald (July 18, 2007) http://www.smh.com.au/articles/2007/07/18/1184559833067.html
4. International Reporter (July16, 2007) http://www.internationalreporter.com/News-2402/Now,-seven-simple-hand-gestures-to-switch-your-TV-on.html
5. Premaratne, P., Yang, S., Zou, Z., Vial, P.: Australian Sign Language Recognition Using Moment Invariants. In: Huang, D.-S., Jo, K.-H., Zhou, Y.-Q., Han, K. (eds.) ICIC 2013. LNCS (LNAI), vol. 7996, pp. 509–514. Springer, Heidelberg (2013)
6. Premaratne, P., Ajaz, S., Premaratne, M.: Hand Gesture Tracking and Recognition System for Control of Consumer Electronics. In: Huang, D.-S., Gan, Y., Gupta, P., Gromiha, M.M. (eds.) ICIC 2011. LNCS (LNAI), vol. 6839, pp. 588–593. Springer, Heidelberg (2012)
7. Premaratne, P., Nguyen, Q., Premaratne, M.: Human computer interaction using hand Gestures. In: Advanced Intelligent Computing-Theories and Applications – Communications in Computer nd Information Science, vol. 93, pp. 381–386 (2010)
8. Premaratne, P., Safaei, F., Nguyen, Q.: Moment Invariant Based Control System Using Hand Gestures. In: Intelligent Computing in Signal Processing and Pattern Recognition, vol. 345, pp. 322–333. Springer, Heidelberg (2006)
9. Premaratne, P., Premaratne, M.: Image Matching using Moment Invariants. Neurocomputing 137, 65–70 (2014)
10. Premaratne, P., Ajaz, S., Premaratne, M.: 'Hand Gesture Tracking and Recognition System Using Lucas-Kanade Algorithm for Control of Consumer Electronics. For Neurocomputing Journal 116(20), 242–249 (2013)
11. Premaratne, P., Nguyen, Q.: Consumer Electronics Control System Based on Hand Gesture Moment Invariants. IET Computer Vision 1-1, 35–41 (2007)
12. Yang, S., Premaratne, P., Vial, P.: Hand Gesture Recognition: An Overview. In: 5th IEEE International Conference on Broadband Network and Multimedia Technology (2013)
13. Kawai, H., Tamura, S.: Recognition of Sign Language Motion Images. Pattern Recognition 21(4), 343–353 (1988)
14. Kawai, H., Tamura, S.: Deaf-And-Mute Sign Language Generation System. In: Proc. Medical Images and Icons SPIE 0515 (1984)
15. Kawai, H., Tamura, S.: Deaf-and-Mute Sign Language Generation System. Pattern Recognition 18(3/4), 199–205 (1985)
16. Cootes, T.F., Taylor, C.J.: Active shape models- 'Smart Snakes'. In: Proceedings of the British Machine Vision Conference, pp. 266–275 (1992)
17. Heap, T., Samaria, F.: Real-time Hand Tracking and Gesture Recognition Using Smart Snakes. In: Proceedings of Interface to Real and Virtual Worlds, pp. 1–13 (1995)
18. Starner, T., Pentland, A.: Real-Time American Sign Language Recognition From Video Using Hidden Markov Models, Technical Report 375. MIT Media Lab (1995)
19. Assan, M., Grobel, K.: Video-based Sign Language Recognition Using Hidden Markov models. In: Wachsmuth, I., Fröhlich, M. (eds.) GW 1997. LNCS (LNAI), vol. 1371, pp. 97–109. Springer, Heidelberg (1998)

20. Assan, M., Grobel, K.: Isolated Sign Language Recognition using Hidden Markov Models. In: IEEE International Conference on Computational Cybernetics and Simulation, pp. 162–167 (1997)
21. Vogler, C., Metaxas, D.: Handshapes and movements: Multiple-channel american sign language recognition. In: Camurri, A., Volpe, G. (eds.) GW 2003. LNCS (LNAI), vol. 2915, pp. 247–258. Springer, Heidelberg (2004)
22. Vogler, C., Metaxas, D.: ASL Recognition based on a Coupling between HMMs and 3D Motion Analysis. Technical Reports (CIS). Department of Computer and Information Science, University of Pennsylvania (1998)
23. Imagawa, K., Matsuo, H., Taniguchi, R., Arita, D.: Recognition of Local Features for Camera-based Sign Language Recognition System. In: Proceedings of 15th International Conference on Pattern Recognition, pp. 849–853 (2000)
24. Isaacs, J., Foo, S.: Hand pose estimation for American sign Language Recognition. In: Proceedings of the Thirty-Sixth Southeastern Symposium on System Theory, pp. 132–136 (2004)
25. Cooper, H., Ong, E., Pugeau, N., Bowden, R.: Sign Language Recognition Using Sub-Units. Journal of Machine Learning Research 13, 2205–2231 (2012)
26. Premaratne, P., Ajaz, S., Premaratne, M.: Hand Gesture Tracking and Recognition System Using Lucas-Kanade Algorithm for Control of Consumer Electronics. Neurocomputing Journal 116(20), 242–249 (2013)
27. Fang, G., Gao, W., Zhao, D.: Large Vocabulary Sign Language Recognition Based on Hierarchical decision trees. In: Proceedings of the 5th International Conference on Multimodal Interfaces, pp. 125–131 (2003)
28. Triesch, J., von der Malsburg, C.: Self-organized Integration of Adaptive Visual Cues for Face Tracking. In: Proc. of the 4th IEEE Int'l Conf. on Automatic Face and Gesture Recognition (2000)
29. Blake, A., Isard, M.: The Application of Techniques from Graphics, Vision, Control Theory and statistics to visual tracking of shapes in Motion. Springer (1998)
30. Oz, C., Leu, M.C.: American Sign Language word Recognition with a Sensory Glove Using Artificial neural networks. Engineering Applications of Artificial Intelligence 24(7), 1204–1213 (2011)
31. Pan, Z., Li, Y., Zhang, M., Sun, C., Guo, K., Tang, X., Zhou, S.: A real-time Multi-cue Hand tracking algorithm based on computer vision. IEEE Virtual Reality, 219–222 (2000)
32. Premaratne, P., Nguyen, Q., Premaratne, M.: Human Computer Interaction Using Hand Gestures. In: Huang, D.-S., McGinnity, M., Heutte, L., Zhang, X.-P. (eds.) ICIC 2010. CCIS, vol. 93, pp. 381–386. Springer, Heidelberg (2010)
33. Zhongliang, Q., Wenjun, W.: Automatic Ship Classification by Superstructure Moment Invariants and Two-stage Classifier. In: ICCS/ISITA 1992 Communications on the Move, pp. 544–547 (1992)

No-Reference Fingerprint Image Quality Assessment

Kamlesh Tiwari and Phalguni Gupta

Department of Computer Science and Engineering,
Indian Institute of Technology Kanpur, Kanpur 208016, India
{ktiwari,pg}@cse.iitk.ac.in

Abstract. Quality of a fingerprint image is assessed to control the registration of poor quality images in the database so that a good accuracy of fingerprint recognition system can be achieved. This paper proposes a quality assessment scheme for digital fingerprint image. It makes use of complete ridge line of a thinned fingerprint image for quality assessment. It introduces three robust measures (1) ridge-line smoothness (2) inter ridge-line distance and (3) minutiae extractability for the quality assessment. Experiments are performed on a database comprising of 1000 fingerprint images of 500 subjects of various age group lying between 18 and 75. It has been found that the performance of the fingerprint recognition system is improved from *CRR* of 74.6% to 100% and *EER* of 13.08% to 0.01% by controlling the registration of inferior quality fingerprint in the database. Quality of a fingerprint images is an important indicator of its performance in automatic fingerprint based recognition system.

Keywords: Biometrics, Fingerprint, Quality, Minutiae, Matching, ROC Curve.

1   Introduction

Fingerprints have been used as an authentic evidence of person's identity in the court of law from very old time because of its properties like persistence and uniqueness. Fingerprint is the print pattern developed on a surface while touched by the tip of a finger. The pattern consists of black lines called *ridges* and in between while space called *valley*. Fingerprint can be categorized in two ways, 1) exemplar fingerprints where the person cooperatively provides samples by rolling his finger on a paper or placing it on a fingerprint scanner and 2) latent fingerprints which are covertly collected from a already touched surface like glass, table top, handle or a weapon etc. specially obtained from a crime scenes.

Fingerprint quality is usually defined as a measure of the clarity of the ridge and valley structures, as well as the extractability of features [6]. Quality of a fingerprint image expresses the usability of its features in a recognition system. It has very little to do with the perceptual visual appearance of the fingerprint which is typically dependent upon sharpness, contrast, blur, *etc*. Some of the important factors affecting the quality of fingerprint acquired are type of sensor used, behavior of the user, skin conditions, aging, dryness, noise etc. A high quality fingerprint contains low inter-class but high intra-class

D.-S. Huang et al. (Eds.): ICIC 2014, LNAI 8589, pp. 846–854, 2014.

similarity. Quality can serve as a direct indicator of matching performance. Error rates of the system can significantly be improved by preventing the use of poor quality fingerprints. Awareness of the fingerprint quality can also be used to device a quality based fusion strategy in a multi-algorithm recognition system.

Various quality measures which are based on local features, global features or their combination are available in literature. Sine-wave model of ridge valley pattern is used in [9] to determine fingerprint quality. A Gabor-feature based method for determining the quality of the fingerprint images has been proposed in [17]. It uses standard deviation of block wise Gabor features to determine the quality. Orientation certainty level (OCL), proposed in [12], uses localized texture pattern and ridge to valley structure for quality. Two new quality measures, ridge and valley clarity (GCS) and global orientation flow (GOQS), are introduced in [5] to determine overall image quality. A hybrid scheme that takes local texture features, global foreground area, central position of foreground, the number of minutiae and the existence of singular points into account to discriminate the quality has been presented in [15]. Parameters developed in [6] measure the energy concentration in the frequency domain and the spatial coherence in local regions. In [7] the use of power spectrum in the spatial frequency domain for quality estimation has been proposed. Model in [11] assumes sinusoidal-wave structure of ridges and valleys in a fingerprint and similarity between the sinusoidal-waves and the input fingerprint image is measured to determine its quality. In [8] orientation tensor of fingerprint images is used to quantify the signal impairments and quality. A hierarchical scheme for estimating fingerprint quality has been presented in [16] to evaluate the overall quality of fingerprint image by computing effective area, Fourier spectrum and contrast entropy. In [26] effective area, energy concentration, spatial consistency and directional contrast in a back propagation neural network are used to estimate the quality of fingerprint. Penalty function to combine the orientation coherence and core position for deciding fingerprint quality has been used in [25]. A fuzzy inference system to combine image contrast, curvature and ridge flow is proposed in [27]. It has shown that by removing 15% of poorer images, there is an improvement of equal error rate from 1.69% to 1.04% for FVC 2006 database. Comparative study of fingerprint image quality estimating techniques has been provided in [10], [2] and [27]. National Institute of Standards and Technology (NIST) [1] has provided an open source biometric image software distribution that contains a neural network based overall fingerprint image quality estimation algorithm *nifq*. The algorithm is an implementation of [19] that assigns quality values in the range of 1 to 5 where 1 signifies the highest quality and 5 is the lowest quality. One of the major factors which cause the failure of many algorithms is that the fingerprint image quality does not have a linear relationship with the local features [26]. Use of fingerprint for human recognition has a deep impact on the development of biometrics research. There exist several biometric based personal recognition systems in literature which are based on single or a combination of traits such as face [13], fingerprint [18,24,23], palmprint [4,22,21], iris [14], knuckleprint [3] *etc.*

This paper presents an approach to quantitatively estimate the quality of a fingerprint image which can be used for its automatic classification of fingerprint. The approach applies a preprocessing technique to thin the fingerprint image and subsequently all ridges in the fingerprint are examined one by one. It proposes three measures for goodness of ridge lines and their estimation technique. Results have shown a strong correlation between the computed quality values and the performance of fingerprint recognition system. The paper is organized as follow. Next section explains the proposed fingerprint quality estimation scheme. Experimental setup and results obtained are provided in Section 3. Finally, conclusions are presented in last section.

2   Proposed Scheme

This section proposes an efficient scheme for fingerprint image quality assessment. The scheme comprises of preprocessing to get a thinned image and finding smoothness of ridge line, determining distance between two consecutive ridge lines and estimating extractability of minutiae. Preprocessing is done to get a thinned image. A ridge in fingerprint is called ridge-line in thinned image. Thinning is performed such that every pixel belonging to ridge-line except end points can have only two neighbors in its eight neighborhoods. To select a single ridge-line of the fingerprint, we also consider junction points where two or more ridge-line meet as the end point of the line. Therefore, in our consideration a ridge-line starts from one junction or end point and terminates at another junction or end point.

2.1   Ridge-Line Smoothness

Every ridge-line of the thinned image is analyzed for its smoothness. It is based on the assumption that the orientation of fingerprint ridge changes gradually. To test the smoothness of a point on ridge line, ten adjacent backward pixels are considered to estimate the location of the next three pixels. Euclidean distance between estimated pixel locations and the actual pixel location is summed for all three estimated pixels to get forward pixel estimation error. Similar procedure is followed for backward pixels as well. Both these errors, forward pixel estimation error and backward pixel estimation error are added. If this sum is greater than a empirical threshold then the current pixel is considered to be rough. Every rough pixel is invalidated.

2.2   Inter Ridge-Line Distance

Inherent structure of a fingerprint image is such that one ridge uniformly covers other. Inter ridge-line distance between two ridge-lines is homogeneous and lies on small variation range. An appropriate ridge lies smoothly between two other ridge-lines. Its distance with other two ridge-lines is uniform and does not varies much. To estimate the

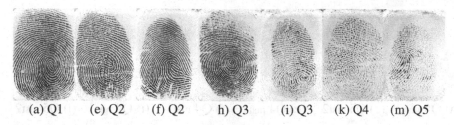

(a) Q1   (e) Q2   (f) Q2   h) Q3   (i) Q3   (k) Q4   (m) Q5

Fig. 1. Fingerprint Images Samples of Quality Q1, Q2, Q3, Q4 and Q5 by Proposed Scheme

distance of ridge-line with its neighboring ridge, perpendicular lines are drawn on every point of the ridge-line. The distance from the ridge-line to the first nearest black pixel is called as inter ridge-line distance. This step produces a series of ridge-line distances corresponding to every point of the ridge-line in both of its sides. A ridge-line originating due to noise produces a mangled ridge-line distance pattern which can be identified and corresponding ridge-line portion is marked invalid by using the following heuristics.

1. Range heuristic: if the inter ridge-line distance on a particular point of ridge-line is found to be larger or smaller than an empirically determined range, the corresponding point on ridge-line is treated as invalid.
2. Distance consistency: if difference of ridge-line distance for two consecutive points differ more than an empirical threshold, then both the points are considered invalid.
3. Invalid propagation: if a point on the ridge-line becomes invalid due to range heuristics or distance consistency, it also invalidates points of its local neighborhood.
4. Length invalidation: if the length of a ridge-line is less than an empirical threshold then the complete ridge-line is invalidated.
5. Line invalidation ratio: if more than half of a ridge-line becomes invalid due to above factors then the complete ridge-line is made invalid.

2.3 Minutiae Extractability

Every ridge-line can produce two minutiae points which are at its end points. But not all ridge-lines contribute to minutiae. This is because of the fact that the available noise and other effects may contribute to some spurious ridges in the thinned fingerprint image. These spurious lines can be identified by the properties of ridge line smoothness and inter ridge line distance. A line which contains majority of invalid pixels is not considered for minutiae extraction. End points of valid ridge-lines are considered for the candidacy of minutiae. An end point of the ridge-line can either be a termination or a bifurcation point.

If the end point of the ridge-line is not a junction point then following is done. All black pixels of ridge-line are inverted to white and a circle of radius one pixel is drawn at the end point of ridge-line under consideration. This circle may pass through a black pixel. If it is not, the radius is incremented iteratively by one pixel until it

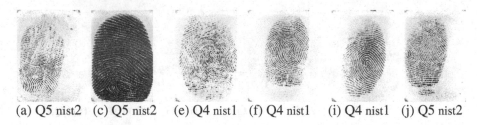

(a) Q5 nist2   (c) Q5 nist2   (e) Q4 nist1   (f) Q4 nist1   (i) Q4 nist1   (j) Q5 nist2

Fig. 2. Example of some Fingerprints having Disagreement on Quality Labels between Proposed Scheme and *nifq*

passes through a black pixel. This radius is analyzed. If the radius lies in some empirically selected range then the ridge-line is said to have a valid minutiae at end point location, otherwise it is not. In second case when the end point of the ridge-line is a junction point then we proceed as follows. A circle is similarly grown step by step at the point until there is exactly three black pixels on its circumference. When the condition gets violated the circle growing is stopped. The radius of the circle is again tested for lying in an empirical range. In case the radius passes the test, the pixel location is called to have a valid bifurcation minutia at the end point of the ridge-line, otherwise it is not.

Fraction of the fingerprint image pixels occupied by valid minutiae circles and perpendicular lines on valid point of the ridge-line is the quality score of the fingerprint image.

3   Experimental Results

To test the proposed scheme, we have used a in-house database, IITK-sel500FP. It contains fingerprint images of 500 persons of age group from 18 to 75 year. Every person has provided two fingerprint samples with a gap of two month. First sample is acquired in phase-1 and the second sample is acquired in phase-2. This database has a special importance because it has been collected from the persons who are not very careful about their skin condition and are involved in laborious tasks. Persons are divided into five age groups viz. 18-24, 25-34, 35-44, 45-59 and 60+. According to NIST quality assessment, this database contains 34.7%, 19%, 23.6%, 11.2% and 11.5% images with quality 1, 2, 3, 4 and 5 respectively where 1 stands for the highest quality and 5 for the poorest quality.

Experimental results   are analyzed using standard performance metrics   like false accept rate, false rejection rate, equal error rate, correct recognition rate and decidability index. False Accept Rate (FAR) refers to the rate at which unauthorized individual (imposers) are accepted by the biometric system as a valid user. False Reject Rate (FRR) refers to the probability that a biometric system fails to identify a genuine subject. Equal Error Rate (EER) is the rate at where FAR and FRR are equal. Lower the value of EER, better is the system. Receiver operating characteristic (ROC) curve

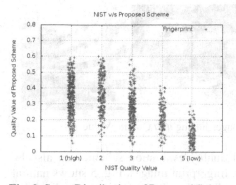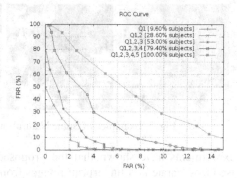

Fig. 3. Score Distribution of Proposed Scheme with Respect to NIST

Fig. 4. ROC Plots while Removing Poor Quality Images

is a plot between *FAR* and *FRR*. Lower the area under the curve better is the system. Correct Recognition Rate (*CRR*) represents rank-1 accuracy and signifies the percentage of the best match which pertains to the same subject. Decidability Index (*DI*) signifies how well the distribution of genuine and imposter matching scores are separated.

3.1 Fingerprint Quality Results

Distribution of the proposed quality scores on IITK-sel500FP database is shown in Fig. 3 with reference to the quality values assigned by *nifq*[1] of NIST [1]. It can be seen that the quality value of an image obtained through *nfiq* may not confirm with that of the proposed scheme. A *k*-means clustering is applied on the observed fingerprint quality scores to get the thresholds for dividing the database into the five classes. Experimentally we have obtained four thresholds which are 0.1380, 0.2399, 0.3360 and 0.4417. A fingerprint image is labeled to have quality Q5 if its score lies below 0.1380. If the score is greater than 0.1380 but is less than 0.2399, it is labeled to have quality Q4. In case the score lies between 0.2399 and 0.3360, it is labeled to have quality Q3. Similarly for quality Q2, images have the score in range 0.3360 to 0.4417. All images having the score greater than 0.4417 are labeled as quality Q1. Fig. 1 shows some of the images of different quality as classified by the proposed quality accessing scheme.

There are 345 images where the proposed scheme and NIST have agreed on same classification label. At 422 places, the classification difference is found to be 1. It shows that the consistency of the proposed scheme with *nifq* at many places. The classification difference of 2 is found at 161 images while difference of 3 is at 60 fingerprint images. Also there are 12 fingerprints with classification difference of 4. Fig. 2 shows some of the fingerprint images where a large classification difference is observed. It is evident from the Fig. 2 that the proposed scheme has a capacity to separates those bad quality images which get high quality values from *nfiq*.

---

[1] A negligible amount of random noise is added to *nifq* values in order to get a visualization effects in the Fig. 3.

Fig. 5. Examples of Some Natural Images having *nifq* Quality Value 1

Experiments have shown that the proposed quality estimation scheme can also be used to separate non fingerprint images from fingerprint images. Fig. 5 shows natural images (non-fingerprint) which are labeled quality 1 by *nfiq* but assigned quality label Q5 by the proposed scheme.

To test the performance of a fingerprint recognition system, an open source fingerprint matcher *bozorth3* of NIST has been used. This matcher provides the similarity score between two fingerprint impressions. It uses minutiae features of the fingerprint for matching. Minutiae features are also extracted by using *mindtct* of NIST [1]. Effect of restricting the registration of poor quality fingerprint images according to the proposed quality estimation scheme has been analyzed critically. Initially it has considered all fingerprints in the database for matching and has evaluated its performance. It has observed that the system has *EER* of 13.1% and *CRR* of 75% with *DI* of 1.8. In the next iteration, all fingerprints having quality Q5 are removed from the database and the experiments have been performed on the reduced database. It has achieved *CRR* of 86% and *EER* of 8.3% with *DI* of 1.8. This process is repeated by discarding the fingerprints of quality Q4 from the database. Now the system performance has been improved. It has achieved *EER* of 4.5%, *CRR* of 94% with *DI* of 1.9. In the next stage the experiment has been performed on the image having quality Q1 and quality Q2. It has achieved *EER* of 2.7% and CRR of 97% with *DI* of 2.1. Finally, only images of quality Q1 are considered to test the performance of the proposed scheme. It has been found that *EER* and *CRR* of the system are 0.02% and 100% respectively with *DI* of 3.0. Fig. 4 shows the receiver operating characteristic (*ROC*) curve for the five experiments.

In Table 1 the performance parameters correct recognition rate (*CRR*), equal error rate (*EER*) and decidability index (*DI*) are reported. The fingerprint images that are marked as poor quality by the proposed algorithm are removed from the database and hence the percentage of subjects decreases as the quality threshold increases. It can be seen that the values of *CRR* and *EER* increases and the value of *DI* decreases monotonically. This indicates that the fingerprint images which are removed in out experiment are actually of poor quality and it confirms the quality awareness of the proposed scheme. It should be noted that the increase in system performance cannot be gained only by reducing the size of database. If good quality images are removed and bad quality images are kept in the database then the performance of the system would decrease. Table 1 has also shown that a significant gain in the performance can be achieved by using the proposed quality estimation scheme for restricting the inclusion of poor quality fingerprint images.

Table 1. Percentage of subjects, CRR and ERR against threshold by the proposed scheme

Threshold	Subjects %	CRR %	EER %	DI
0.00	100.00	74.60	13.08	1.78
0.10	86.60	82.22	9.70	1.81
0.20	63.80	91.85	5.50	1.79
0.30	37.40	96.26	4.55	1.99
0.40	15.80	100.00	0.01	2.71

4   Conclusion

An efficient fingerprint quality estimation scheme has been proposed in this paper. It critically analyses the smoothness, sprawl and feature potential of a thinned fingerprint image to estimate its quality. Experimental results have proved the quality awareness of the scheme. It have also shown the capability to successfully distinguish a fingerprint image from non-fingerprint image. Quality informed removal of bad quality fingerprint images have shown that the performance of an automatic fingerprint based recognition system improves *EER* from 13.08% to 0.01% and CRR from 74.60% to 100%.

Acknowledgements. Authors like to acknowledge the support provided by the Department of Information Technology, Government of India to carry out this work.

References

1. NIST biometric image software,
   http://www.nist.gov/itl/iad/ig/nbis.cfm
2. Alonso, F.F., Fierrez, J., Ortega, G.J., Gonzalez, R.J., Fronthaler, H., Kollreider, K., Bigun, J.: A Comparative study of fingerprint image quality estimation methods. Information Forensics and Security 2(4), 734–743 (2007)
3. Badrinath, G.S., Nigam, A., Gupta, P.: An efficient finger-knuckle-print based recognition system fusing sift and surf matching scores. In: Qing, S., Susilo, W., Wang, G., Liu, D. (eds.) ICICS 2011. LNCS, vol. 7043, pp. 374–387. Springer, Heidelberg (2011)
4. Badrinath, G.S., Tiwari, K., Gupta, P.: An efficient palmprint based recognition system using 1D-DCT features. In: Huang, D.-S., Jiang, C., Bevilacqua, V., Figueroa, J.C. (eds.) ICIC 2012. LNCS, vol. 7389, pp. 594–601. Springer, Heidelberg (2012)
5. Chen, T.P., Jiang, X., Yau, W.Y.: Fingerprint image quality analysis. In: Intl. Conf. on Image Processing, vol. 2, pp. 1253–1256. IEEE (2004)
6. Chen, Y., Dass, S.C., Jain, A.K.: Fingerprint quality indices for predicting authentication performance. In: Kanade, T., Jain, A., Ratha, N.K. (eds.) AVBPA 2005. LNCS, vol. 3546, pp. 160–170. Springer, Heidelberg (2005)
7. Fiérrez-Aguilar, J., Chen, Y., Ortega-Garcia, J., Jain, A.K.: Incorporating image quality in multi-algorithm fingerprint verification. In: Zhang, D., Jain, A.K. (eds.) ICB 2005. LNCS, vol. 3832, pp. 213–220. Springer, Heidelberg (2005)

8. Fronthaler, H., Kollreider, K., Bigun, J., Fierrez, J., Alonso, F.F., Ortega, G.J., Gonzalez, R.J.: Fingerprint image-quality estimation and its application to multialgorithm verification. Transactions on Information Forensics and Security 3(2), 331–338 (2008)

9. Hong, L., Wan, Y., Jain, A.: Fingerprint image enhancement: algorithm and performance evaluation. Pattern Analysis and Machine Intelligence 20(8), 777–789 (1998)

10. Jin, C., Kim, H., Cui, X., Park, E., Kim, J., Hwang, J., Elliott, S.: Comparative assessment of fingerprint sample quality measures based on minutiae-based matching performance. In: Intl. Symposium on Electronic Commerce and Security, vol. 1, pp. 309–313. IEEE (2009)

11. Lee, S., Lee, C., Kim, J.: Model-based quality estimation of fingerprint images. In: Zhang, D., Jain, A.K. (eds.) ICB 2005. LNCS, vol. 3832, pp. 229–235. Springer, Heidelberg (2005)

12. Lim, E., Jiang, X., Yau, W.: Fingerprint quality and validity analysis. In: Intl. Conf. on Image Processin, pp. 465–469. IEEE (2002)

13. Nigam, A., Gupta, P.: Comparing human faces using edge weighted dissimilarity measure. In: Intl. Conf. on Control, Automation, Robotics and Vision (ICARCV), pp. 1831–1836 (2010)

14. Nigam, A., Gupta, P.: Iris recognition using consistent corner optical flow. In: Lee, K.M., Matsushita, Y., Rehg, J.M., Hu, Z. (eds.) ACCV 2012, Part I. LNCS, vol. 7724, pp. 358–369. Springer, Heidelberg (2013)

15. Qi, J., Abdurrachim, D., Li, D., Kunieda, H.: A hybrid method for fingerprint image quality calculation. In: Automatic Identification Advanced Technologies, pp. 124–129 (2005)

16. Saquib, Z., Soni, S., Vig, R.: Hierarchical fingerprint quality estimation scheme. In: Intl. Conf. on Computer Design and Applications, vol. 1, pp. 493–500. IEEE (2010)

17. Shen, L., Kot, A., Koo, W.: Quality measures of fingerprint images. In: Bigun, J., Smeraldi, F. (eds.) AVBPA 2001. LNCS, vol. 2091, pp. 266–271. Springer, Heidelberg (2001)

18. Singh, N., Tiwari, K., Nigam, A., Gupta, P.: Fusion of 4-slap fingerprint images with their qualities for human recognition. In: World Congress on Information and Communication Technologies, pp. 925–930 (2012)

19. Tabassi, E., Wilson, C., Watson, C.: NIST fingerprint image quality. NIST Res. Rep. NISTIR7151 (2004)

20. Tada, C.H.M., Zaghetto, A., Macchiavello, B.: Fingerprint image quality estimation using a fuzzy inference system. In: Intl. Conf. on Forensic Computer Science, pp. 41–45 (2012)

21. Tiwari, K., Arya, D.K., Badrinath, G., Gupta, P.: Designing palmprint based recognition system using local structure tensor and force field transformation for human identification. Neurocomputing 116, 222–230 (2013)

22. Tiwari, K., Arya, D.K., Gupta, P.: Palmprint based recognition system using local structure tensor and force field transformation. In: Huang, D.-S., Gan, Y., Gupta, P., Gromiha, M.M. (eds.) ICIC 2011. LNCS, vol. 6839, pp. 602–607. Springer, Heidelberg (2012)

23. Tiwari, K., Mandal, J., Gupta, P.: Segmentation of slap fingerprint images. Emerging Intelligent Computing Technology and Applications, 182–187 (2013)

24. Tiwari, K., Mandi, S., Gupta, P.: A heuristic technique for performance improvement of fingerprint based integrated biometric system. In: Huang, D.-S., Bevilacqua, V., Figueroa, J.C., Premaratne, P. (eds.) ICIC 2013. LNCS, vol. 7995, pp. 584–592. Springer, Heidelberg (2013)

25. Wu, M., Yong, A., Zhao, T., Guo, T.: A systematic algorithm for fingerprint image quality assessment. In: Huang, D.-S., Gan, Y., Gupta, P., Gromiha, M.M. (eds.) ICIC 2011. LNCS (LNAI), vol. 6839, pp. 412–420. Springer, Heidelberg (2012)

26. Yang, X.K., Luo, Y.: A classification method of fingerprint quality based on neural network. In: Intl. Conf. on Multimedia Technology, pp. 20–23. IEEE (2011)

27. Zhao, Q., Liu, F., Zhang, D.: A comparative study on quality assessment of high resolution fingerprint images. In: Intl. Conf. on Image Processing, pp. 3089–3092. IEEE (2010)

Author Index

Printed in the United States
By Bookmasters